Point-Set Topology with Topics

Basic General Topology for Graduate Studies

Point-Set Topology with Topics

Basic General Topology for Graduate Studies

Robert André
University of Waterloo, Canada

World Scientific

NEW JERSEY • LONDON • SINGAPORE • BEIJING • SHANGHAI • HONG KONG • TAIPEI • CHENNAI • TOKYO

Published by

World Scientific Publishing Co. Pte. Ltd.
5 Toh Tuck Link, Singapore 596224
USA office: 27 Warren Street, Suite 401-402, Hackensack, NJ 07601
UK office: 57 Shelton Street, Covent Garden, London WC2H 9HE

Library of Congress Cataloging-in-Publication Data
Names: André, Robert (Mathematician), author.
Title: Point-set topology with topics : basic general topology for graduate studies /
Robert André, University of Waterloo, Canada.
Description: New Jersey : World Scientific, [2024] | Includes bibliographical references and index.
Identifiers: LCCN 2023021913 | ISBN 9789811277337 (hardcover) |
ISBN 9789811277344 (ebook for institutions) | ISBN 9789811277351 (ebook for individuals)
Subjects: LCSH: Topology--Textbooks. | Point set theory--Textbooks.
Classification: LCC QA611 .A537 2024 | DDC 514/.322--dc23/eng20230819
LC record available at https://lccn.loc.gov/2023021913

British Library Cataloguing-in-Publication Data
A catalogue record for this book is available from the British Library.

For any available supplementary material, please visit
https://www.worldscientific.com/worldscibooks/10.1142/13439#t=suppl

Desk Editors: Logeshwaran Arumugam/Rosie Williamson

Typeset by Stallion Press
Email: enquiries@stallionpress.com

To

R.C.

Je me plaisois surtout aux Mathématiques à cause de la certitude et de l'évidence de leurs raisons. Mais je ne remarquois point encore leur vray usage.

René Descartes
Discours de la Méthode (1637)

For though we love both the truth and our friends, piety requires us to honor the truth first.

Aristotle
Nicomachean Ethics (350 BC)

$$\sim \int tion \sim$$

I like it when the longing becomes belonging, dance of polarities ultimately becomes union, feeling the other like the lost part of oneself.
So... union becomes "$\int$ *tion.*"

Roberta Colzani
(2021)

Preface

This text was prepared to serve as an introduction to the study of general topology. Most students in mathematics are required, at some point in their study, to have knowledge of the fundamentals of general topology and master topological techniques that may be useful in their area of specialization. "Trust us, not only will you be glad to be skilled at using these tools, but a lot of concepts you study in your field of interest will eventually be seen as specific cases investigated in a more general context in general topology" they are told. It sometimes occurs that some students develop a particular fascination for general topology in spite of it occasionally being described as being "no longer in fashion anymore". Recall that mathematicians such as George Cantor and Felix Hausdorff, were also told that some of the mathematics they spent time investigating was "not in fashion". Reasons for the continued study of this topic often go beyond the simple perception that it is a practical or useful tool. The intricate beauty of the mathematical structures that are derived in this field become its main attraction. This is, of course, how this writer perceives the subject and served as the main motivation to prepare a textbook that will help the reader enjoy its study. Of course, one would naturally hope that an author would write a book only about something he or she feels passionate about. I often heard some students describe general topology as being "hard". Well, some of it is. But I have often thought that maybe this perception was developed because they did not approach it quite in the right way. In this text, we try to improve on the ways that students are introduced to it.

But first, I should at least write a few words about the mathematical content of this textbook. The choice of content as well as the order and pace of the presentation of the concepts found in the text were developed with senior math undergraduate or math graduate students in mind. The targeted reader will have been exposed to some mathematical rigor to a level normally found in an introduction to mathematical analysis texts or as presented in an introduction to linear algebra or abstract algebra texts. The first two sections of Part I consist mostly of a review in the form of a summarized presentation of very basic ideas on normed vector spaces and metric spaces. These are meant to ease the reader into the main subject matter of general topology (in Chapters 3–20 of Parts II–VI). Parts II–VI normally form the core material contained in most, one or two semester, Basic General Topology course. Once we have worked through the most fundamental concepts of topology in Chapters 1–20, the reader will be exposed to brief introductions to more specialized or advanced topics. These are presented in Part VII in the form of a sequence of chapters many of which can be read or studied, independently, or in short sequences of two or three chapters, provided the student has mastered Chapters 3–20.

Chapters related to the more basic ideas of general topology are followed by a list of *Concept review* type questions. These questions highlight for students the main ideas presented in that section and will help test their understanding of these concepts. The answers to all *Concept review* questions are in the main body of the text. Attempting to answer these questions will help the student discover essential notions which are often overlooked when first exposed to these ideas. Reading a section provides a certain level of understanding, but answering questions, even simple ones, related to its content requires a much deeper understanding. The efforts required in answering correctly such questions leads the student to the ability to solve more complex problems in the Exercise sections. If the student desires a more in-depth study of a topic in Part VII, there are many excellent topology books that can satisfy this need.

Textbook examples will serve as solution models to most of the exercise questions at the end of each section.

In certain sections, we make use of elementary set theory. A student who feels a bit rusty when facing the occasional references to set theory notions may want to review some of these. For convenience, a summary of the main set theory concepts appear at the end of the text in the form of an appendix to the book. A more extensive coverage of naive set theory is offered in the book *Axioms and Set Theory* by this author. It is highly recommended and will serve as an excellent companion to this book.

As we all know, any textbook, when initially published, will contain some errors, some typographical, others in spelling or in formatting and, what is even more worrisome, some mathematical. Critical or alert readers of the text can help weed out the most glaring mistakes by communicating suggestions and comments directly to the author. This will be much appreciated by this writer as well as by future readers.

Robert André
University of Waterloo, Ontario

About the Author

Robert André is a Ph.D. graduate from the University of Manitoba with a specialization in Point-Set topology under the supervision of Dr. R. Grant Woods. He has lectured at the University of Manitoba and the University of Regina and 17 years as a faculty member of the Department of Pure Mathematics at the University of Waterloo in Ontario, Canada. He also lectured mathematics for 13 years at the high school level in Manitoba, Saskatchewan, and New Brunswick.

Contents

Part I
Norms and Metrics

Chapter 1

Norms on Vector Spaces

Abstract
In this section, we review a few basic notions about inner products on vector spaces and how they are used as a mechanism to construct a distance measuring tool called "norm" We then define "norm on an abstract vector space" with no reference to an inner product. We show how to distinguish between norms that are induced by an inner product and those that are not. We then provide a few examples of norms on vectors spaces of continuous functions, $C[a, b]$. We end this section with a formal definition of the compact property in normed vector spaces.

1.1 Review of Inner Product Spaces

We begin by reviewing a few basic notions about those vector spaces we refer to as *inner product spaces.* Recall that the set $\mathbb{R}^n = \{(x_1, x_2, x_3, \ldots, x_n) : x_i \in \mathbb{R}\}$ equipped with two operations, addition and scalar multiplication, is known to be a *vector space.* We can also equip a vector space with a third operation called "dot product" which maps pairs of vectors in $\mathbb{R}^n$ to some real number. The dot-product is a specific real-valued operation on $\mathbb{R}^n$ which belongs to a larger family of vector space operations called *inner products.* We briefly remind ourselves of a few facts about inner products on abstract vector spaces.

Definition 1.1 Let V be a vector space over the reals. An *inner product* is an operation which maps pairs of vectors in V to a real number. We denote an inner product on V by $\langle \vec{v}, \vec{w} \rangle$. A real-valued function on $V \times V$ is referred to as an *inner product* if and only if it satisfies the following four axioms:

IP1: The number $\langle \vec{v}, \vec{v} \rangle$ is greater than or equal to 0 for all $\vec{v}$ in V. Equality holds if and only if $\vec{v} = 0$. (Hence, if $\vec{v}$ is not 0, $\langle \vec{v}, \vec{v} \rangle$ is *strictly* larger than 0.)
IP2: For all $\vec{v}, \vec{w} \in V$, $\langle \vec{u}, \vec{v} \rangle = \langle \vec{v}, \vec{u} \rangle$ (Commutativity)
IP3: For all $\vec{u}, \vec{v}, \vec{w} \in V$, $\langle \vec{u} + \vec{w}, \vec{v} \rangle = \langle \vec{u}, \vec{v} \rangle + \langle \vec{w}, \vec{v} \rangle$
IP4: For all $\vec{u}, \vec{v} \in V$, and $\alpha \in \mathbb{R}$, $\langle \alpha\vec{u}, \vec{v} \rangle = \alpha\langle \vec{u}, \vec{v} \rangle$

A vector space V is called an *inner product space* if it is equipped with some specified inner product.

Definition 1.2 Let V be an inner product space. If $\vec{v}$ is a vector in V we define

$$\|\vec{v}\| = \sqrt{\langle \vec{v}, \vec{v} \rangle}$$

The expression $\|\vec{v}\|$ is called the *norm* (or length) of the vector $\vec{v}$ induced by the inner product $\langle \vec{u}, \vec{v} \rangle$ on V.

Example 1. It is easily verified that, for $\vec{x} = (x_1, x_2, x_3, \ldots, x_n)$ and $\vec{y} = (y_1, y_2, y_3, \ldots, y_n)$ in $\mathbb{R}^n$, its well-known dot-product

$$\langle \vec{x}, \vec{y} \rangle = \vec{x} \cdot \vec{y} = x_1y_1 + x_2y_2 + \cdots + x_ny_n$$

satisfies the four axioms above and so is an inner product on the vector space $\mathbb{R}^n$. This is often referred to as the *Euclidean inner product* or the *standard inner product* on $\mathbb{R}^n$. The norm on $\mathbb{R}^n$ induced by the dot product is

$$\begin{aligned}\|\vec{x}\| &= \|(x_1, x_2, x_3, \ldots, x_n)\| \\ &= \sqrt{(x_1, x_2, x_3, \ldots, x_n) \cdot (x_1, x_2, x_3, \ldots, x_n)} \\ &= \sqrt{\sum_{i=1}^{n} x_i^2}\end{aligned}$$

This particular norm is referred to as the *Euclidean norm* on $\mathbb{R}^n$. It is also referred to as the L_2-norm on $\mathbb{R}^n$, in which case, it will be represented as $\|\vec{x}\|_2$. We will use this particular norm to measure distances between vectors in $\mathbb{R}^n$. That is, the distance between $\vec{x} = (x_1, x_2, x_3, \ldots, x_n)$ and $\vec{y} = (y_1, y_2, y_3, \ldots, y_n)$ is defined to be

$$\|\vec{x} - \vec{y}\| = \sqrt{\sum_{i=1}^{n} (x_i - y_i)^2}$$

In the case where $n = 2$ or 3, this represents the usual distance formula between points in 2-space and 3-space, respectively. In the case where $n = 1$, it represents the absolute value of the difference of two numbers.

Example 2. Consider the vector space, $V = C[a, b]$, the set of all continuous real-valued functions on the closed interval $[a, b]$ equipped with the usual addition and scalar multiplication of functions. We define the following inner product on $C[a, b]$ as:

$$\langle f, g \rangle = \int_a^b f(x)g(x)\,dx$$

Showing that this operation satisfies the inner product axioms IP1 to IP4 is left as an exercise. In this case, the norm of f, induced by this inner product, is seen to be

$$\|f\| = \sqrt{\int_a^b f(x)^2\,dx}$$

It is also referred to as the L_2-norm on $C[a, b]$ and we represent it as $\|f\|_2$.

The following theorem called the *Cauchy–Schwarz inequality* offers an important problem solving tool when working with inner product spaces. The norm which appears in the inequality is the one induced by the given inner product. This inequality holds true for any well-defined inner product on a vector space.

Theorem 1.3 (Cauchy–Schwarz inequality). *Let V be a vector space equipped with an inner product and its induced norm. Then, for vectors $\vec{x}$ and $\vec{y}$ in V,*

$$|\langle \vec{x}, \vec{y} \rangle| \leq \|\vec{x}\| \|\vec{y}\|$$

Equality holds true if and only if $\vec{x}$ and $\vec{y}$ are collinear (i.e., $\vec{x} = \alpha\vec{y}$ or $\vec{y} = \alpha\vec{x}$).

Proof. The statement clearly holds true if $\vec{y} = 0$.

Let $\vec{x}, \vec{y}$ be two (not necessarily distinct) vectors in V where $\vec{y} \neq 0$. For any real number t,

$$\begin{aligned} 0 &\leq \langle \vec{x} - t\vec{y}, \vec{x} - t\vec{y} \rangle \\ &= \|\vec{x}\|^2 - 2t\langle \vec{x}, \vec{y} \rangle + t^2\|\vec{y}\|^2 \end{aligned}$$

Choosing

$$t = \frac{\langle \vec{x}, \vec{y} \rangle}{\|\vec{y}\|^2}$$

in the above equation we obtain

$$0 \leq \|\vec{x}\|^2 - \frac{\langle \vec{x}, \vec{y} \rangle^2}{\|\vec{y}\|^2}$$

The inequality, $|\langle \vec{x}, \vec{y} \rangle| \leq \|\vec{x}\| \|\vec{y}\|$, follows.

We now prove the second part of the statement. If $\vec{x}$ and $\vec{y}$ are collinear, say $\vec{x} = \alpha\vec{y}$, then

$$\begin{aligned}
|\langle \vec{x}, \vec{y}\rangle| &= |\langle \alpha\vec{y}, \vec{y}\rangle| \\
&= |\alpha|\langle \vec{y}, \vec{y}\rangle \\
&= |\alpha|\|\vec{y}\|\|\vec{y}\| \\
&= \|\alpha\vec{y}\|\|\vec{y}\| \\
&= \|\vec{x}\|\|\vec{y}\|
\end{aligned}$$

Conversely, suppose $|\langle \vec{x}, \vec{y}\rangle| = \|\vec{x}\|\|\vec{y}\|$. If $\vec{y}$ is zero then $\vec{0} = t\vec{y}$ and so $\vec{x}$ and $\vec{y}$ are collinear. Suppose $\vec{y} \neq \vec{0}$. Consider $t = \dfrac{\langle \vec{x}, \vec{y}\rangle^2}{\|\vec{y}\|^2}$.

$$\begin{aligned}
\langle \vec{x} - t\vec{y}, \vec{x} - t\vec{y}\rangle &= \|\vec{x}\|^2 - \frac{\langle \vec{x}, \vec{y}\rangle^2}{\|\vec{y}\|^2} \text{(as described above)} \\
&= 0
\end{aligned}$$

So, by IP1, $\langle \vec{x} - t\vec{y}, \vec{x} - t\vec{y}\rangle = 0$ implies $\vec{x} - t\vec{y} = 0$ so $\vec{x}$ and $\vec{y}$ are collinear. We have shown that equality holds if and only if $\vec{x}$ and $\vec{y}$ are collinear.

We now verify that a norm, $\| \ \|$, which is induced by an inner product $\langle \ , \ \rangle$ on the vector space V will always satisfy the following three fundamental properties:

(1) For all $\vec{x} \in V$, $\|\vec{x}\| \geq 0$, equality holds if and only if $\vec{x} = 0$.
(2) For all $\vec{x} \in V$ and scalar $\alpha \in \mathbb{R}$, $\|\alpha\vec{x}\| = |\alpha|\|\vec{x}\|$.
(3) For all $\vec{x}, \vec{y} \in V$, $\|\vec{x} + \vec{y}\| \leq \|\vec{x}\| + \|\vec{y}\|$.

The first property follows from the fact that the norm is defined as being the square root of a number. The second property follows from the straightforward argument:

$$\begin{aligned}
\|\alpha\vec{x}\| &= \sqrt{\langle \alpha\vec{x}, \alpha\vec{x}\rangle} \\
&= \sqrt{\alpha^2}\sqrt{\langle \vec{x}, \vec{x}\rangle} \\
&= |\alpha|\|\vec{x}\|
\end{aligned}$$

The third property is referred to as the *triangle inequality.* It's non-trivial proof invokes the Cauchy–Schwarz inequality. We prove it below.

Corollary 1.4 The triangle inequality for norms induced by inner products. *For any pair of vectors $\vec{x}$ and $\vec{y}$ in an inner product space, $\|\vec{x}+\vec{y}\| \leq \|\vec{x}\| + \|\vec{y}\|$. If equality holds then the two vectors $\vec{x}$ and $\vec{y}$ are collinear.*

Proof.

$$\begin{aligned}
\|\vec{x}+\vec{y}\|^2 &= \langle \vec{x}+\vec{y}, \vec{x}+\vec{y}\rangle \\
&= \|\vec{x}\|^2 + 2\langle \vec{x}, \vec{y}\rangle + \|\vec{y}\|^2 \\
&\leq \|\vec{x}\|^2 + 2|\langle \vec{x}, \vec{y}\rangle| + \|\vec{y}\|^2 \\
&\leq \|\vec{x}\|^2 + 2\|\vec{x}\|\|\vec{y}\| + \|\vec{y}\|^2 \quad \text{(Cauchy–Schwarz)} \\
&= (\|\vec{x}\| + \|\vec{y}\|)^2
\end{aligned}$$

Thus $\|\vec{x}+\vec{y}\| \leq \|\vec{x}\| + \|\vec{y}\|$, as required.

In the case where we have equality: Suppose we have $\|\vec{x}+\vec{y}\| = \|\vec{x}\| + \|\vec{y}\|$. Then $\|\vec{x}+\vec{y}\|^2 = (\|\vec{x}\| + \|\vec{y}\|)^2$. From the development above, the inequalities must be equalities, so we must have $\langle \vec{x}, \vec{y}\rangle = \|\vec{x}\|\|\vec{y}\|$. This means $|\langle \vec{x}, \vec{y}\rangle| = \|\vec{x}\|\|\vec{y}\|$. By the Cauchy–Schwarz theorem, $\vec{x}$ and $\vec{y}$ are collinear.

Example 3. Use the the dot product properties and the law of cosines, $c^2 = a^2 + b^2 - 2ab\cos\theta$ (where a, b, c represent the lengths of the sides of a triangle $\triangle ABC$ and θ is the angle between to the sides AB and AC) to prove the Cauchy–Schwarz inequality in $\mathbb{R}^2$.

Solution: Consider the two vectors $\vec{a}$ and $\vec{b}$ in $\mathbb{R}^2$ where one is not a scalar multiple of the other. Then the extremities of the two vectors form a triangle with angle θ between $\vec{a}$ and $\vec{b}$. Then the norms $\|\vec{a}\|$, $\|\vec{b}\|$ and $\|\vec{b}-\vec{a}\|$ are numbers which can represent the length of the sides of the triangle. Let θ represent the angle between $\vec{a}$ and $\vec{b}$.

We apply the cosine law to obtain

$$\|\vec{b}-\vec{a}\|^2 = \|\vec{a}\|^2 + \|\vec{b}\|^2 - 2\|\vec{a}\|\,\|\vec{b}\|\cos\theta$$

If $\langle \vec{a}, \vec{b}\rangle$ represents the dot-product, we obtain, by applying the dot-product properties and the property $\|\vec{a}\|^2 = \langle \vec{a}, \vec{a}\rangle$

$$\begin{aligned}
\|\vec{a}\|^2 - 2\langle \vec{b}, \vec{a}\rangle + \|\vec{b}\|^2 &= \|\vec{a}\|^2 + \|\vec{b}\|^2 - 2\|\vec{a}\|\,\|\vec{b}\|\cos\theta \\
-2\langle \vec{b}, \vec{a}\rangle &= -2\|\vec{a}\|\,\|\vec{b}\|\cos\theta \\
\langle \vec{b}, \vec{a}\rangle &= \|\vec{a}\|\,\|\vec{b}\|\cos\theta \\
\frac{\langle \vec{b}, \vec{a}\rangle}{\|\vec{a}\|\,\|\vec{b}\|} &= \cos\theta
\end{aligned}$$

Since $\|\cos\theta\| \le 1$,

$$\frac{|\langle \vec{b}, \vec{a}\rangle\,|}{\|\vec{a}\|\,\|\vec{b}\|} \le 1$$

Example 4. Suppose $\vec{a}$ and $\vec{b} = (1,0)$ are two vectors in $\mathbb{R}^2$ where $\|\vec{a}\| = 1$ with angle between $\vec{a}$ and $\vec{b} = \theta$. Show that $\langle \vec{a}, \vec{b}\rangle = \cos\theta$, as is represented in Figure 1.1.

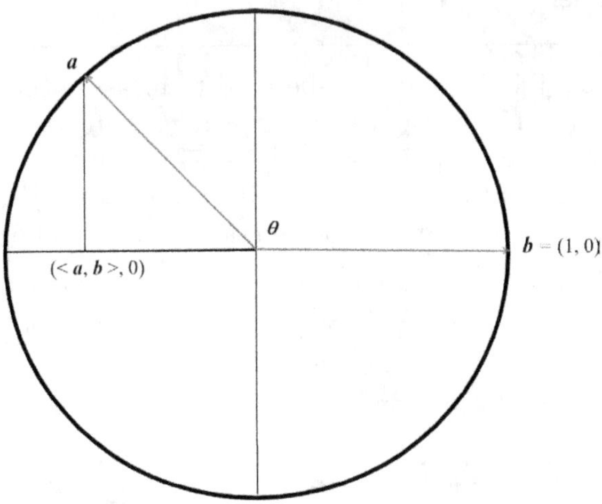

Figure 1.1. Dot-product in the plane.

Solution: This follows from the above example and

$$\cos\theta = \frac{\langle \vec{a}, \vec{b} \rangle}{\|\vec{a}\| \, \|\vec{b}\|} = \frac{\langle \vec{a}, \vec{b} \rangle}{1 \times 1}$$

1.2 Norms on Arbitrary Vector Spaces

We now show that norms on a vector space can exist independently from inner products.

Definition 1.5 Let V a vector space over the reals. A norm on V is a function, $\| \ \| : V \to \mathbb{R}$, which satisfies the three following norm axioms:

N1: For all $\vec{v} \in V$, $\|\vec{v}\| \geq 0$. The equality $\|\vec{v}\| = 0$ holds true if and only if $\vec{v} = 0$. Hence, if $\vec{v}$ is not $\vec{0}$, $\|\vec{v}\| > 0$.
N2: For all $\vec{v} \in V$, $\alpha \in \mathbb{R}$, $\|\alpha\vec{v}\| = |\alpha|\|\vec{v}\|$.
N3: For all $\vec{v}, \vec{u} \in V$, $\|\vec{u} + \vec{v}\| \leq \|\vec{u}\| + \|\vec{v}\|$ (Triangle inequality).

A vector space V is called a *normed vector space* if it is equipped with some specified norm satisfying these three norm axioms. It is denoted by $(V, \| \ \|)$. Note that the vector space V need not be an inner product space, nor need there be any relationship between $\| \ \|$ and some inner product.

This function, $\| \ \| : V \to \mathbb{R}$, will be used to measure the "distance", $\|\vec{x} - \vec{y}\|$, between any two vectors $\vec{x}$ and $\vec{y}$. Note that $\|\vec{x} - \vec{y}\| = \|(-1)(\vec{y} - \vec{x})\| = |-1|\|\vec{y} - \vec{x}\| = \|\vec{y} - \vec{x}\|$. That is, "the distance between $\vec{x}$ and $\vec{y}$ is the same as the distance between $\vec{y}$ and $\vec{x}$".

We provide a few examples of functions known to be norms.

Example 5. The norm $\| \ \|$ induced by the inner product of an inner product space $(V, \langle \ , \ \rangle)$ has already been shown to satisfy the axioms N1, N2, and N3.

Example 6. The 1-norm on $\mathbb{R}^n$ is defined as

$$\|\vec{x}\|_1 = \|(x_1, x_2, x_3, \ldots, x_n)\|_1 = \sum_{i=1}^{n} |x_i|$$

It can be shown to satisfy the three norm axioms. (Left as an exercise.)

Example 7. The ∞-norm on $\mathbb{R}^n$ is defined as

$$\|\vec{x}\|_\infty = \|(x_1, x_2, x_3, \ldots, x_n)\|_\infty = \max\{|x_1|, |x_2|, \ldots, |x_n|\}$$

It can be shown to satisfy the three norm axioms. (Proving that N1 and N2 are satisfied is left as an exercise.)

We prove that the ∞-norm on $\mathbb{R}^n$ satisfies the triangle inequality N3.

Proof of $\|\vec{x} + \vec{y}\|_\infty \leq \|\vec{x}\|_\infty + \|\vec{y}\|_\infty$. Note that $|x_i| \leq \max_{i=1\ldots n}\{|x_i|\}$ for each i, and $|y_i| \leq \max_{i=1\ldots n}\{|y_i|\}$ for each i. Then for each $i = 1$ to n,

$$\begin{aligned} |x_i + y_i| &\leq |x_i| + |y_i| \\ &\leq \max_{i=1\ldots n}\{|x_i|\} + \max_{i=1\ldots n}\{|y_i|\} \\ &= \|\vec{x}\|_\infty + \|\vec{y}\|_\infty \end{aligned}$$

Then $\max_{1\ldots n}\{|x_i - y_i|\} \leq \|\vec{x}\|_\infty + \|\vec{y}\|_\infty$ and so $\|\vec{x}+\vec{y}\|_\infty \leq \|\vec{x}\|_\infty + \|\vec{y}\|_\infty$, as required.

Example 8. Let $p \geq 1$. The p-norm on $\mathbb{R}^n$ is defined as

$$\|\vec{x}\|_p = \|(x_1, x_2, x_3, \ldots, x_n)\|_p = \left(\sum_{i=1}^{n} |x_i|^p\right)^{1/p}$$

It can be shown that this function satisfies the three norm axioms. We will not prove this for p in general.

Note that when $p = 2$ the p-norm is simply the Euclidean norm on $\mathbb{R}^n$. Since the Euclidean norm is induced by an inner product, it automatically satisfies the three norm axioms. Proving that, for any $p \geq 1$, the p-norm satisfies N1 and N2 is straightforward. But proving that the triangle inequality holds true for all $p \geq 1$ is not easy. The interested readers can look the proof up in most real analysis texts or find it online.

A natural question comes to mind. Are all norms on a vector space V induced by some inner product? The answer is no!

The following theorem tells us how to recognize those norms which are induced by some inner product.

Theorem 1.6 *Suppose* $\|\ \|$ *is a norm on a vector space* V. *There exists an inner product* $\langle\ ,\ \rangle$ *such that* $\|\vec{x}\| = \sqrt{\langle \vec{x}, \vec{x}\rangle}$, *for all* $\vec{x} \in V$, *if and only if, for all* $\vec{x}, \vec{y} \in V$, *the norm* $\|\ \|$ *satisfies the parallelogram identity*

$$2(\|\vec{x}\|^2 + \|\vec{y}\|^2) = \|\vec{x}+\vec{y}\|^2 + \|\vec{x}-\vec{y}\|^2$$

Proof. The proof involves showing that, if $\|\ \|$ satisfies the parallelogram identity then the identity

$$\langle \vec{x}, \vec{y}\rangle = \frac{1}{4}\left(\|\vec{x}+\vec{y}\|^2 - \|\vec{x}-\vec{y}\|^2\right)$$

is a valid inner product on V which induces $\|\ \|$. The proof is routine and so is not presented here.

Examples of norms on $C[a, b]$

Example 9. The L_p-norm. Let $V = C[a, b]$ denote the family of all real-valued continuous functions on the closed interval $[a, b]$ equipped with the usual $+$ and scalar multiplication. If $p \geq 1$, then we define the L_p-norm on $C[a, b]$ as follows:

$$\|f\|_p = \left(\int_a^b |f(x)|^p \, dx\right)^{1/p}$$

In the case where $p = 1$, we have

$$\|f\|_1 = \int_a^b |f(x)| \, dx$$

which is easily shown to satisfy the three norm axioms. In the case where $p = 2$, then this norm is one which is induced by the inner product

$$\langle f, g\rangle = \int_a^b f(x)g(x) \, dx$$

on $C[a, b]$.

It is straightforward to show that, for all $p \geq 1$, $\|f\|_p$ satisfies N1 and N2. But proving that $\|f+g\|_p \leq \|f\|_p + \|g\|_p$, for all $p \geq 1$, is difficult. This inequality is referred to as the *Minkowski inequality.* The interested readers will find a proof in most real analysis texts or online.

Example 10. The sup-norm: We define another norm on the vector space, $C^*(S)$, of all bounded continuous real-valued functions on the space S. Recall that the *supremum* of a subset A of an ordered set S, written as "sup A" is the *least upper bound* of A which is contained in S. Note that sup A may or may not belong to A. We define the sup-norm on $C^*(S)$ as

$$\|f\|_\infty = \sup\{|f(x)| : x \in S\}$$

The sup-norm is also referred to as the "infinity-norm" or "uniform norm". Showing that this is a valid norm is left as an exercise.

1.3 Convergence and Completeness in a Normed Vector Space

Convergence of sequences forms and important part of analysis. In what follows, norms on $\mathbb{R}^n$ are always assumed to be the Euclidean norm, unless otherwise stated. An infinite sequence in a normed vector space, V, is a function which maps $\mathbb{N}$ (not necessarily one-to-one) onto a subset, S, of V. This means that a sequence is the indexation of the elements of a countable subset, S, by using the natural numbers. For example, $S = \{\vec{x}_0, \vec{x}_1, \vec{x}_2, \ldots, \vec{x}_n, \ldots\}$. Note that some of these may be repeated (since the function mapping $\mathbb{N}$ into S need not be one-to-one). We generalize the notion of a "convergent sequence and its limit" in $\mathbb{R}$ to a "convergent sequence and its limit" in a normed vector space $(V, \|\ \|)$.

Definition 1.7 Let $(V, +, \alpha, \|\ \|)$ be a vector space equipped with a norm $\|\ \|$. We say a sequence of vectors $\{\vec{x}_n\}$ in V converges to a vector $\vec{a}$ in V with respect to the norm $\|\ \|$ if

> For any $\varepsilon > 0$, there exists an integer $N > 0$ such that $\|\vec{x}_n - \vec{a}\| < \varepsilon$ whenever $n > N$.

If the sequence of vectors $\{\vec{x}_n\}$ converges to the vector $\vec{a}$, then we write

$$\lim_{n\to\infty} \vec{x}_n = \vec{a}$$

The following proposition lists the standard properties respected by limits in normed vector spaces.

Proposition 1.8 *Let $\{\vec{x}_n\}$ and $\{\vec{y}_n\}$ be two sequences in the normed vector space V. Suppose $\{\vec{x}_n\}$ converges to $\vec{a}$. Let α and β be scalars.*

(a) $\lim_{n\to\infty} (\alpha\vec{x}_n \pm \beta\vec{y}_n) = \alpha \lim_{n\to\infty} \vec{x}_n \pm \beta \lim_{n\to\infty} \vec{y}_n$.
(b) $(\lim_{n\to\infty} \vec{x}_n = \vec{a}) \Leftrightarrow (\lim_{n\to\infty} \|\vec{x}_n - \vec{a}\| = 0)$.
(c) *If $\vec{x} = (x_1, x_2, x_3, \ldots, x_n) \in \mathbb{R}^n$ then $|x_i| \leq \|\vec{x}\|_2$, for $i = 1$ to n.*
(d) *A sequence of vectors $\{\vec{x}_k : k = 1, 2, 3, \ldots\}$ in $\mathbb{R}^n$ converges to the vector $\vec{a} = (a_1, a_2, a_3, \ldots, a_n)$ (with respect to the Euclidean norm) if and only if the components of the vectors in $\{\vec{x}_k\}$ converge to the corresponding components of $\vec{a}$ (with respect to the absolute value).*

Proof. The proof is omitted.

If a sequence, $T = \{f_n : n \in \mathbb{N}\}$, in $C^*(S)$ converges to f with respect to the sup-norm, $\| \ \|_\infty$, then we will say that

"T converges uniformly to f"

and that f is the "uniform limit" of the sequence T.

Suppose $C^*(S)$ is given for some space S and $\{T_x : x \in S\}$ is a family of sequences in $\mathbb{R}$ where,

$$T_x = \{f_n(x) : n \in \mathbb{N}\}$$

each of which converges (pointwise) to some number u_x. Suppose that, for a given $\varepsilon > 0$, there exists an $N > 0$ such that $n > N$ implies $|u_x - f_n(x)| < \varepsilon$, independent of the value of x. Define the function, $f : S \to \mathbb{R}$ as $f(x) = u_x$. Then, when $N > 0$,

$$\sup\{|f_n(x) - f(x)| : x \in S\} = \|f_n - f\|_\infty \leq \varepsilon$$

so $\{f_n : n \in \mathbb{N}\}$ converges uniformly to f. This provides another way of recognizing the uniform limit of a sequence of functions. That is, we show that, the existence of an N depends only on the value of ε and not the particular value of x.

Example 11. Show that the uniform limit, f, of a sequence, $\{f_n\}$, in $C^*(S)$, is continuous on S and so belongs to $C^*(S)$.

Solution: Let $T = \{f_n : n \in \mathbb{N}\}$ be a sequence in $C^*(S)$ which converges uniformly to f. Let $a \in S$ and $\varepsilon > 0$. It suffices to show that there is a $\delta > 0$ such that $\|x - a\| < \delta$ implies that $|f(x) - f(a)| < \varepsilon$. See that, there is $N > 0$ such that $n > N$ implies

$$\begin{aligned}|f(x) - f(a)| &\leq |f(x) - f_n(x)| + |f_n(x) - f_n(a)| + |f_n(a) - f(a)| \\ &< \|f_n - f\|_\infty + |f_n(x) - f_n(a)| + \|f_n - f\|_\infty \\ &< \varepsilon/3 + |f_n(x) - f_n(a)| + \varepsilon/3\end{aligned}$$

For $n > N$, f_n is continuous so there exists $\delta > 0$ such that $|x-a| < \delta$ implies $|f_n(x) - f_n(a)| < \varepsilon/3$.

Then $|x - a| < \delta$ implies $|f(x) - f(a)| < \varepsilon$. So f is continuous on S.

We now define those sequences in a normed vector space called *Cauchy sequences.* As we shall soon see all convergent sequences of $(V, \| \ \|)$ must be Cauchy sequences, but some "subsets" S of V have Cauchy sequences which do not converge in S.

Definition 1.9 A sequence of vectors $\{\vec{x}_n\}$ in a normed vector space V is said to be a *Cauchy sequence* in V, or simply said to be *Cauchy*, if for any $\varepsilon > 0$, there exists an integer $N > 0$ such that $\|\vec{x}_m - \vec{x}_n\| < \varepsilon$ whenever $m, n > N$.

At first glance, the reader may feel that a Cauchy sequence is just another way of referring to a "convergent sequence". In many situations, this is indeed the case. But there are sets, S, containing Cauchy sequences which do not converge to a point inside S. For example, we may refer to the sequence $\{1, 1/2, 1/3, \ldots, 1/n, \ldots\}$ as a sequence in the vector space $\mathbb{R}$, or, as a sequence in the subset $S = \{x : 0 < x \leq 1\}$ where S inherits the absolute value norm from $\mathbb{R}$. The given sequence certainly converges to 0 in $\mathbb{R}$ and so can easily be proven to be Cauchy in $\mathbb{R}$. Using the same norm on S we must conclude that it is also Cauchy in S. However, there is one property of S which clearly distinguishes it from $\mathbb{R}$: It is that $\mathbb{R}$ contains the limit of this Cauchy sequence while S does *not* contain the limit of this same sequence. So we will use Cauchy sequences to help us categorize subsets of a vector space based on whether they contain the limits of all their Cauchy sequences, or not.

This motivates the following definition of *complete subsets* of a normed vector space.

Definition 1.10 A subset S of a normed vector space V is a *complete subset of* V if every Cauchy sequence in S converges to a vector in S.

Definition 1.10.1 A complete normed vector space is referred to as *Banach space.*

Definition 1.10.2 If V is a complete normed vector space whose norm is induced by an inner product then V is referred to as a *Hilbert space.* That is, if $(V, +\alpha, \langle\rangle, \|\ \|)$ is complete then V is a Hilbert space.

Even if all Hilbert spaces are Banach spaces not all Banach spaces are Hilbert spaces.

Theorem 1.11 *Completeness of* $(\mathbb{R}, |\ |)$. *The space* $(\mathbb{R}, |\ |)$ *is complete with respect to* $|\ |$.

Proof. Let $T = \{x_n : n \in \mathbb{N}\}$ be a Cauchy sequence in $\mathbb{R}$.

We claim that T is bounded: For $\varepsilon = 1$, there exists N such that $n, m > N$ implies $|x_n - x_m| < 1$. If $k > N$ then $|a_k| = |a_k - a_N + a_N| < |a_k - a_N| + |a_N| < 1 + |a_N|$. Then T is bounded by $\max\{|a_0|, |a_1|, |a_2|, \ldots, |a_N| + 1\}$. So T is bounded as claimed.

We are required to show that T converges to a point in $\mathbb{R}$. By the Bolzano–Weierstrass theorem[1] T has a subsequence $M = \{x_{n_k} : k \in \mathbb{N}\}$ which converges to, say, $L \in \mathbb{R}$.

We claim that T must also converge to L: There exists N such that $n, m > N$ implies $|x_n - x_m| < \varepsilon/2$. We can pick $K > 0$ such that $|x_{n_k} - L| < \varepsilon/2$ whenever $k > K$. We can choose $M = \max\{K, N\}$. Since $n_k \geq n$ (by definition of subsequence), if $n > M$,

$$|x_n - L| \leq |a_n - a_{n_k}| + |a_{n_k} - L| < \varepsilon/2 + \varepsilon/2 = \varepsilon$$

So T converges to L, as required.

Theorem 1.12 (Completeness of $(C_b(S), \|\ \|_\infty)$). *Let S be a space and $C_b(S)$ be the set of all bounded real-valued functions on S equipped with the sup-norm, $\|\ \|_\infty$. Then $C_b(S)$ is a complete space with respect to $\|\ \|_\infty$.*

Proof. Let $\{f_n : n \in \mathbb{N}\}$ be a Cauchy sequence in $C_b(S)$. We are required to show that $\{f_n : n \in \mathbb{N}\}$ converges to a function f in $C_b(S)$ with respect to $\|\ \|_\infty$.

Let $x \in S$. Then

$$|f_n(x) - f_m(x)| \leq \|f_n - f_m\|_\infty < \varepsilon \quad \text{for all } m, n > N$$

Then $U = \{f_n(x) : n \in \mathbb{N}\}$ is Cauchy in $\mathbb{R}$. Since $\mathbb{R}$ is complete (see Theorem 1.11), U converges to, say, u_x. For each $x \in S$ define the function, $f : S \to \mathbb{R}$, as $f(x) = u_x \in \mathbb{R}$. Then $\{f_n : n \in \mathbb{N}\}$ converges pointwise to $f(x)$. For $m \geq N$,

$$\begin{aligned} |f(x) - f_m(x)| &= |\lim_{n\to\infty} f(x) - f_m(x)| \\ &= \lim_{n\to\infty} |f(x) - f_m(x)| \\ &\leq \lim_{n\to\infty} \|f_n - f_m\|_\infty \\ &\leq \varepsilon \end{aligned}$$

[1]Bolzano–Weierstrass: Every bounded sequence in $\mathbb{R}$ has a convergent subsequence.

N independent of the value of x in S. Then $\|f - f_n\|_\infty \leq \varepsilon$. Therefore, f is the uniform limit of the Cauchy sequence, $\{f_n : n \in \mathbb{N}\}$. Then $f \in C_b(S)$. So $C_b(S)$ is complete.

We will not dwell on the notions of convergence and completeness of subsets at this time. We will return to discuss these concepts in the more general contexts of metric spaces and topological spaces.

1.4 The Compact Property on Normed Vector Spaces

We will present immediately the formal definition of the *compact property* in normed vector spaces. The notion of *compactness* will also be discussed in a the more general context of metric spaces and topological spaces. This definition will be followed by an important characterization of the compact property valid for normed vector spaces with a finite basis. The proof is non-trivial and so is omitted here. The complete proof is usually presented in texts which serve as an introduction to real analysis.

One of the main reasons for an early introduction of compact subsets in such a course is to access the important "Extreme value theorem" which states that "A continuous function on a compact subset, T, of a normed vector space, V, attains its maximum value at a point inside T". The *Heine–Borel theorem* allows us to refer to *compact subsets* as being those subsets that are "closed and bounded". In abstract topological spaces, we will not have access to this tool, since the Heine–Borel theorem applies only to those normed vector spaces with a finite basis.

Definition 1.13 A subset, T, of a normed vector space, V, is said to be *compact*[2] if and only if every sequence, $\{x_i : i \in \mathbb{N}\}$, in T has a subsequence, $\{x_{f(i)} : i \in \mathbb{N}\}$, which converges to some point, p, which belongs to T.

[2]When we revisit the property of compactness in our study of topological spaces, we will replace the word "compact" with the words "sequentially compact".

Theorem 1.14 (The generalized Heine–Borel theorem). *A subset of normed vector space with a finite base is compact if and only if it is both closed and bounded.*

Proof. The proof is omitted.

Exercises

1. Prove that, if $\vec{x}$ and $\vec{y}$ are vectors in $\mathbb{R}^n$, then $|\, \|\vec{x}\| - \|\vec{x} - \vec{y}\| \,| \leq \|\vec{y}\|$.
2. Exercise questions on norms.

   (a) Consider the vector space, $C[a, b]$, of all continuous functions on the interval $[a, b]$. Recall that $\| \ \|_1 : C[a, b] \to \mathbb{R}$ is defined as

   $$\|f\|_1 = \int_a^b |f(x)| dx$$

   Show that $\| \ \|_1$ is a valid norm on $C[a, b]$.

   (b) Let $\{f_n : n \in \mathbb{N}, n \neq 0\}$ be a sequence of constant functions defined as $f_n(x) = 2 + \frac{1}{n}$ for each n. That is, $f_1(x) = 3$, $f_2(x) = 5/2$, $f_3(x) = 7/3$, and so on. If $\{f_n\}$ is viewed as a subset of $C[-1, 2]$ equipped with the norm $\| \ \|_1$ determine,

   $$\lim_{n\to\infty} \int_{-1}^{2} |f_n(x)|\, dx$$

   Justify all your steps carefully.

3. Recall that $C[a, b]$ is the vector space of all continuous functions on the interval $[a, b]$.

   (a) Show that $\langle f, g\rangle = \int_a^b f(x)g(x)\, dx$ is a valid inner product.

   (b) Let $\| \ \|$ be the norm on $C[0, 1]$ which is induced by the inner product given in part (a) of this question. Compute $\|e^x\|$.

4. Recall the definition of "p-norm": $\|\vec{x}\|_p = \left(\sum_{i=1}^n |x_i|^p\right)^{1/p}$, on $\mathbb{R}^n$.

   (a) Compute $\|(-1, 0, 1)\|_3$.

(b) *Hölder's inequality* on $\mathbb{R}^n$ says that "if $p, q \geq 1$ and $1/p + 1/q = 1$, then $\|(x_1y_1, x_2y_2, \ldots, x_ny_n)\|_1 = \sum_{i=1}^n |x_iy_i| \leq \|\vec{x}\|_p\|\vec{y}\|_q$". Note that p and q are not assumed to be integers; Hölder's inequality holds true as long as the condition "$1/p + 1/q = 1$" is satisfied. Although it is difficult to see why this inequality is of any interest or is of any value at this point, we will assume it holds true and use it to practice using simpler concepts. Show that the *Cauchy–Schwarz inequality* on $\mathbb{R}^n$, $|\langle\vec{x}, \vec{y}\rangle| \leq \|\vec{x}\|_2\|\vec{y}\|_2$, (where $\langle\vec{x}, \vec{y}\rangle$ is the Euclidean inner product) follows from the *Hölder's inequality* on $\mathbb{R}^n$.

(c) *Hölder's inequality* on $C[a, b]$ says that "if $p, q \geq 1$ and $1/p + 1/q = 1$, then $\|fg\|_1 \leq \|f\|_p\|g\|_q$". Show that the *Cauchy–Schwarz inequality* on $C[a, b]$,

$$|\langle f, g\rangle| \leq \|f\|_2\|g\|_2$$

follows from the *Hölder's inequality* on $C[a, b]$.

5. Invoke the Cauchy–Schwarz inequality to show that

$$-\left[\frac{\sqrt{2}-1}{2}\right]^{1/2} \leq \int_0^{\pi/4} \sqrt{\cos x}\sqrt{\sin x}\, dx \leq \left[\frac{\sqrt{2}-1}{2}\right]^{1/2}$$

6. Let f be a function in $C[a, b]$. Show that $\lim_{p\to\infty} \|f\|_p \leq \|f\|_\infty$.
7. A subset S of a normed vector space is bounded if there exists a positive number M such that $\|\vec{x}\| < M$ for all $\vec{x}$ in S. Show that any Cauchy sequence $\{\vec{x}_n\}$ in a normed vector space $(V, \|\ \|)$ when viewed as a set is bounded.
8. Suppose $(V, \|\ \|)$ is a normed vector space and $\{\vec{x}_n\}$ is a sequence in V which does not converge to $\vec{0}$. If $\delta \in \mathbb{R}$, let $B_\delta(\vec{0}) = \{\vec{x} \in V : \|\vec{x}\| < \delta\}$. Show that there exists some $\delta > 0$ and a subsequence $\{x_{n_i}\}_{i=1}^\infty$ of $\{\vec{x}_n\}$ such that $\{\vec{x}_{n_i}\} \cap B_\delta(\vec{0}) = \varnothing$.
9. Convergence of vectors in a normed vector space.

(a) Review the definition of "convergence of a sequence of vectors in a normed vector space". Let $(V, \|\ \|)$ be an abstract normed vector. Suppose V contains a sequence of vectors,

$\{\vec{u}_n : n \in \mathbb{N}\}$, which converges to some vector in V. Show that

$$\lim_{n\to\infty} \|\vec{u}_n\| = \|\lim_{n\to\infty} \vec{u}_n\|$$

Hence, whenever the sequence, $\{\vec{u}_n : n \in \mathbb{N}\}$, converges in V then so does the sequence, $\{\|\vec{u}_n\| : n \in \mathbb{N}\}$, of real numbers.

(b) Let $\vec{v} \in \mathbb{R}^2$. Construct a sequence $\{\vec{u}_n\}$ in $\mathbb{R}^2$ such that $\lim_{n\to\infty} \|\vec{u}_n\| = \|\vec{v}\|$ where $\lim_{n\to\infty} \vec{u}_n \neq \vec{v}$.

(c) Suppose $\{\vec{x}_n\}$ and $\{\vec{u}_n\}$ are two sequences in $\mathbb{R}^n$ which both converge to the same point $\vec{y}$. Show that the sequence $\{\|\vec{x}_n - \vec{u}_n\|\}$ must converge to 0.

10. Let $\{\vec{x}_n : n \in \mathbb{N}\}$ and $\{\vec{y}_n : n \in \mathbb{N}\}$ be sequences in a normed vector space $(V, +, \alpha, \|\ \|)$. Let $\{\alpha_n : n \in \mathbb{N}\}$ be a sequence of real numbers.

    (a) Show that, if $\{\vec{x}_n : n \in \mathbb{N}\}$ and $\{\vec{y}_n : n \in \mathbb{N}\}$ converge to $\vec{x}$ and $\vec{y}$, respectively, then the sequence $\{\vec{x}_n + \vec{y}_n : n \in \mathbb{N}\}$ converges to $\vec{x} + \vec{y}$.
    (b) Show that, if $\{\vec{x}_n : n \in \mathbb{N}\}$ converges to the vector $\vec{x}$ and $\{\alpha_n : n \in \mathbb{N}\}$ converges to the real number α then the sequence of vectors, $\{\alpha_n\vec{x}_n : n \in \mathbb{N}\}$, converges to $\alpha\vec{x}$.

11. Suppose $(V, \|\ \|)$ is a normed vector space and $\{\vec{x}_n\}$ is a Cauchy sequence in V. Show that if $\{\vec{x}_n\}$ has a convergent subsequence, say $\{\vec{x}_{n_i}\}_{i=1}^{\infty}$, then $\{\vec{x}_n\}$ converges in V.
12. Let $\delta > 0$. Suppose $(V, \|\ \|)$ is a normed vector space and $\{\vec{x}_n\}$ is a sequence in V such that $\|\vec{x}_n\| > \delta$ for all n. Show that if $\{\vec{x}_n\}$ is Cauchy then so is the sequence $\left\{\frac{x_n}{\|x_n\|}\right\}$ on $B = \{\vec{x} \in V : \|\vec{x}\| = 1\}$.
13. Suppose $(V, \|\ \|)$ is a normed vector space and $B = \{\vec{x} \in V : \|\vec{x}\| = 1\}$.

    (a) Show that if V is complete with respect to the norm $\|\ \|$ then B is complete with respect to the same norm inherited from V.
    (b) Show that if B is known to be complete with respect to the norm $\|\ \|$, then V must also be complete with respect to $\|\ \|$.

14. Let M be an n-dimensional subspace of the inner product space $(V, \langle\ ,\ \rangle, \|\ \|)$. Prove that M is complete with respect to the norm $\|\ \|$. Essentially this says that all subspaces of finite dimensional inner product spaces are complete subsets.
15. Suppose $(V, \|\ \|)$ is normed vector space whose norm satisfies the parallelogram identity

$$2(\|\vec{x}\|^2 + \|\vec{y}\|^2) = \|\vec{x} + \vec{y}\|^2 + \|\vec{x} - \vec{y}\|^2$$

   (a) Show that $\langle \vec{x}, \vec{y} \rangle = \frac{1}{4}(\|\vec{x} + \vec{y}\|^2 - \|\vec{x} - \vec{y}\|^2)$ (referred to as the *polarizing identity*) satisfies the three inner product axioms IP1, IP2, and IP4.
   (b) Show that the polarizing identity in part (a) satisfies the inner product property $\langle \vec{x} + \vec{y}, \vec{z} \rangle = \langle \vec{x}, \vec{z} \rangle + \langle \vec{y}, \vec{z} \rangle$ hence, the polarizing identity is a valid inner product.
   (c) Show that this inner product induces the norm of V.
16. It is known that for any $\vec{x} \in \mathbb{R}^n$, $\|\vec{x}\|_\infty = \lim_{p\to\infty} \|\vec{x}\|_p$. Prove this for the case where $n = 2$.

Chapter 2

Metrics on Sets

Abstract

In this section, we define the concept of a "metric" on a set, giving rise to "metric spaces". The metric is defined as a tool for measuring distances between points in the given set. Many metrics can be defined on a set as long as they each satisfy three metric axioms. Metrics are then used to determine whether a given sequence converges to a point or not. The subsets of a set called "open ball" and "open sets" are then defined. Functions are then introduced to map points from one metric space to another metric space. We are particularly interested in those functions which are continuous on their respective domains. The topological version of the notion of continuity is presented in terms of "open sets".

2.1 Measuring Distance in Arbitrary Sets

We now generalize a few notions of distances between vectors in a normed vector space to distances between points in an arbitrary set. Arbitrary sets, in their most rudimentary form, do not usually appear with some algebraic structure defined on them (such as vector spaces, for example, on which addition and scalar multiplication are defined). When a set is equipped with addition and scalar multiplication at the onset, subtraction $\vec{a} - \vec{b}$ of two points $\vec{a}$ and $\vec{b}$ is easily given meaning. With a previously defined norm, we measured the length, $\|\vec{a} - \vec{b}\|$, of the difference $\vec{a} - \vec{b}$ to obtain the number which represents a notion of distance between $\vec{a}$ and $\vec{b}$. This method is inspired by the way we normally determine the distance between two real numbers, say -7 and 2.5, in the vector space, $\mathbb{R}$, for example. In $\mathbb{R}$, distance between points is expressed by referring to the absolute value. To measure distances between points in arbitrary sets we will proceed differently. Given a set S, rather than defining a "norm" function $\| \ \| : S \to \mathbb{R}$ on S, we will define a function, $\rho : S \times S \to \mathbb{R}$, which maps pairs of points x and y in S to a number which will represent the distance between these two points in a that particular set. This function is what we will call a "metric" on S. There will be certain restrictions on the properties possessed by ρ. We would, of course, not want ρ to give us a distance $x \to y$ which is different from the distance $y \to x$. Also, we definitely do not want ρ to give a "negative" distance between two points. With different metrics, ρ_1 and ρ_2, the set, (S, ρ_1), may be different in nature from the set, (S, ρ_2).

Definition 2.1 Let S be a non-empty set. A metric, ρ on S, is a function, $\rho : S \times S \to \mathbb{R}$, which satisfies three *metric axioms*:

M1: For every x, $y \in S$, $\rho(x, y) \geq 0$. Also,
- *Identity of indiscernibles*: $\rho(x, y) = 0$ implies $x = y$.
- *Indiscernibility of identicals*: $x = y$ implies $\rho(x, y) = 0$.

M2: *Symmetry*: For every x, $y \in S$, $\rho(x, y) = \rho(y, x)$.

M3: *Triangle inequality*: For every x, y, $z \in S$, $\rho(x, y) \leq \rho(x, z) + \rho(z, y)$.

A set S equipped with a metric, ρ, is called a *metric space*. It is expressed as (S, ρ).

Many metrics can be defined on a given set S. For a given set, S, if ρ_1 and ρ_2 are different metrics then (S, ρ_1) and (S, ρ_2) are considered to be different metric spaces. We provide a few examples.

Example 1. Let $(V, \| \ \|)$ be a normed vector space. Define $\rho : V \times V \to \mathbb{R}$ as $\rho(\vec{x}, \vec{y}) = \|\vec{x} - \vec{y}\|$. The function, ρ, is easily seen to satisfy the two first metric axioms, M1 and M2. The triangle inequality for norms guarantees that M3 also holds true. Then ρ, thus defined on V, is a metric induced by a norm and so (V, ρ) is a metric space (it can simultaneously be viewed as a normed vector space depending on the context) when considering the normed vector space, $(\mathbb{R}^2, \| \ \|_2)$, the metric induced by the norm $\|\ \|_2$ is the distance formula: For $\vec{x} = (x_1, x_2)$ and $\vec{y} = (y_1, y_2)$,

$$\rho(\vec{x}, \vec{y}) = \|\vec{x} - \vec{y}\|_2 = \sqrt{(x_1 - y_1)^2 + (x_2 - y_2)^2}$$

This particular example shows that, given any normed vector space $(V, \|\ \|)$, we can always express it as a metric space, (V, ρ), by defining $\rho(\vec{x}, \vec{y}) = \|\vec{x} - \vec{y}\|$. However, one should remember that a metric space (S, ρ) need not be a normed vector space, since S itself need not be a vector space equipped with addition and scalar multiplication, both of which are necessary to define a norm on S.

Example 2. Let $\rho : \mathbb{R}^2 \times \mathbb{R}^2 \to \mathbb{R}$ be defined as:

$$\rho(\vec{x}, \vec{y}) = \rho((x_1, x_2), (y_1, y_2)) = \sup\{|x_1 - y_1|, |x_2 - y_2|\}$$

It is left to the reader to verify that this function satisfies all three metric axioms, and so is a valid metric on $\mathbb{R}^2$.

Example 3. Let S be any non-empty set. The function $\rho : S \times S \to \mathbb{R}$ defined as

$$\rho(x, y) = \begin{cases} 0 & \text{if } x = y \\ 1 & \text{if } x \neq y \end{cases}$$

can be verified to satisfy all three metric axioms and so is a valid metric on S. Any pair of distinct points are at a distance of one from each other while a point is at a distance zero from itself. This metric is referred to as the *discrete metric.*

Example 4. Consider the set of all integers $\mathbb{Z}$. Let $k : \mathbb{Z} \times \mathbb{Z} \to \mathbb{N}$ be defined as

$$k(x, y) = \max\{2^n : 2^n \text{ divides } x - y\}$$

We define $\rho : \mathbb{Z} \times \mathbb{Z} \to \mathbb{R}$ as follows:

$$\rho(n, m) = \begin{cases} 0 & \text{if } n = m \\ \dfrac{1}{k(m, n)} & \text{if } m \neq n \end{cases}$$

We verify that ρ is indeed a metric on $\mathbb{Z}$.

— It follows directly from the definition that $\rho(n,m)$ is non-negative and $\rho(n,m) = 0$ if and only if $m = n$. So ρ satisfies property M1.
— We see that 2^n divides $x - y$ if and only if 2^n divides $y - x$. Then $k(x,y) = k(y,x)$. Then, if $m \neq n$, $\rho(m,n) = \frac{1}{k(m,n)} = \frac{1}{k(n,m)} = \rho(n,m)$. So ρ satisfies M2.
— Suppose $\rho(m,n) = 1/2^d$ and $\rho(n,t) = 1/2^e$. Then $2^d|(m-n)$ and $2^e|(n-t)$. We will suppose, without loss of generality, that $e \leq d$. It is easily verified that $2^e|(m-t)$. It follows that $k(m,t) \geq 2^e$, hence,

$$\begin{aligned}\rho(m,t) &= \frac{1}{k(m,t)} \\ &\leq \frac{1}{2^e} \\ &= \rho(n,t) \\ &\leq \rho(m,n) + \rho(n,t)\end{aligned}$$

So $\rho(m,t) \leq \rho(m,n) + \rho(n,t)$. We have shown that ρ satisfies M3.

2.2 Metric Subspaces

We know that certain subsets of a vector space are referred to as "subspaces" provided they satisfy specific conditions. There are no required conditions on a subset, T, of (S,ρ) to be called a "metric subspace" provided we know what its metric will be.

Definition 2.1.1 Suppose (S,ρ) is a metric space and $T \subseteq S$. Then T can inherit the metric ρ of its superset, S, and declare itself to be a metric space $(T, \rho|_T)$, simply by restricting the function $\rho : S \times S \to \mathbb{R}$ to $\rho|_T : T \times T \to \mathbb{R}$. In this case, (T, ρ_T) is referred to as a *metric subspace* of (S,ρ). The metric, ρ_T, is referred to as the *subspace metric.*

For example, suppose $(\mathbb{R}, \| \ \|)$ is a normed vector space. Suppose ρ is the metric on $\mathbb{R}$ induced by its norm, $\| \ \|$, and $T = (-3, 5] \subset \mathbb{R}$. Then (T, ρ_T) is a metric subspace of $\mathbb{R}$.[1]

[1]Of course, T is not a vector subspace of the vector space $\mathbb{R}$.

Of course, metric subspaces of a metric space S are themselves metric spaces.

2.3 Convergence and Completeness in a Metric Space

We now direct our attention to those subsets of the set S which are sequences. The definition of a *sequence*, $\{x_n\}$, does not involve the notions of "norm" or "metric" and so is precisely as we previously defined it in the section on normed vector spaces. However, *convergence* or *divergence* of a sequence depends very much on the tool we use to measure distances in the set. The following definitions of *limits* and convergence in a metric space are in many ways identical to those involving norms.

Definition 2.2 Let (S, ρ) be a metric space. We say that a sequence, $\{x_n\}$, of points in S converges to the point, a, in S with respect to the metric ρ if and only if

> for any $\varepsilon > 0$, there exists an integer $N > 0$ such that $\rho(x_n, a) < \varepsilon$ whenever $n > N$.

If the sequence of points $\{x_n\}$ converges to the point a (with respect to the metric ρ), then we write

$$\lim_{n\to\infty} x_n = a$$

Just as for limits in normed vector spaces, we can say that $\{x_n\}$ converges to the point a with respect to a metric ρ in different ways:

$$\lim_{n\to\infty} x_n = a \qquad \Leftrightarrow \qquad \lim_{n\to\infty} \rho(x_n, a) = 0$$

Definition 2.2.1 Suppose (S, ρ) is a metric space, $T \subseteq S$ and $\{x_n\}$ is a sequence in S which converges to the point a with respect to ρ. Then we will say that the point a is a *limit point of* T.

Note that, if $\{x_n\} \subseteq T \subseteq (S, \rho)$ and a is the limit of this sequence with respect to ρ, this does not guarantee that a also belongs to T.

Consider, for example, the metric space $(\mathbb{R}, \rho)$ where $\rho(x, y) = |x - y|$ and $T = \{\pi + 1/n : n = 1, 2, 3, \ldots\}$. Then (T, ρ) is also a metric space. We see that the point π is clearly a limit point of T but $\pi \notin T$.

Just as in normed vector spaces, the notion of "Cauchy sequence" can also be defined in terms of metrics.

Definition 2.3 A sequence of points $\{x_n\}$ in a metric space (S, ρ) is said to be a *Cauchy sequence* in S, with respect to the metric ρ (or simply said to be *Cauchy*) if and only if, for any $\varepsilon > 0$, there exists an integer, $N > 0$, such that $\rho(x_m, x_n) < \varepsilon$, whenever $m, n > N$.

Example 5. *Every convergent sequence is a Cauchy sequence.* The following brief argument confirms what we intuitively feel must be true: If $\{x_i\}$ converges to the point, a, with respect to the metric, ρ, then $\{x_i\}$ is a Cauchy sequence. To prove this, we let $\varepsilon > 0$. By hypothesis, there exists $N > 0$ such that $n > N \Rightarrow \rho(x_n, a) < \varepsilon/2$. If $m > N$, then, invoking the metric axiom M3,

$$\rho(x_m, x_n) \leq \rho(x_m, a) + \rho(x_n, a) < \varepsilon/2 + \varepsilon/2 = \varepsilon$$

So $\{x_i\}$ is Cauchy with respect to the metric ρ.

Definition 2.4 Let (S, ρ) be a metric space. The set S is *complete with respect to the metric* ρ if every Cauchy sequence in S converges to some point in S.

Example 6. Consider the set $S = (0, \infty)$. Let ρ_1 denote the discrete metric on S and ρ_2 be defined as $\rho_2(x, y) = |x - y|$. We compare the two metric spaces (S, ρ_1) and (S, ρ_2).

Claim 1: The metric space, (S, ρ_2), is *not* complete. See that the sequence $\{1/n : n = 1, 2, 3, \ldots\}$ is a sequence in $(\mathbb{R}, \rho_2)$ which converges to 0, with respect to ρ_2 and so is a Cauchy sequence in $(\mathbb{R}, \rho_2)$. It is then a Cauchy sequence inside S. Since the Cauchy sequence $\{1/n\}$ does not converge in S, then S is not a complete subset of S.

Claim 2: The metric space, (S, ρ_1), is a complete metric space. Let $\{x_i\}$ be a Cauchy sequence in S with respect to the discrete metric ρ_1. Let $\varepsilon = 1/2$. Then there exists $N > 0$ such that $m, n > N$ implies $\rho_1(x_m, x_n) < \varepsilon = 1/2$. Then if $m, n > N$, $\rho_1(x_m, x_n) = 0$. This means $x_n = x_{N+1}$, for all $n > N$. Then for any ε, $n > N \Rightarrow \rho_1(x_{N+1}, x_n) = 0 < \varepsilon$ so $\{\vec{x}_i\}$ converges to $x_{N+1} \in S$. We conclude that (S, ρ_1) is a complete metric space.

Note that a metric space (S, ρ) may be a complete metric space and still have proper subsets which are not complete metric subspaces.

2.4 Open and Closed Subsets of a Metric Space

We now define a few fundamental subsets of metric spaces.

Definition 2.5 Let (S, ρ) be a metric space.

(a) If $y \in S$, and $\varepsilon > 0$ we define an *open ball of radius* ε *of center* y as being the set

$$B_\varepsilon(y) = \{x \in S : \rho(y, x) < \varepsilon\}$$

(b) Let $U \subseteq S$. We say that U *is an open subset in* S if U is the union of open balls each of which is entirely contained in U. That is, for each $x \in U$, there exists a real number number $\varepsilon_x > 0$ such that

$$U = \cup\{B_{\varepsilon_x}(x) : x \in U\}$$

Instead of saying "U is an open subset of S" we often simply say "U is open in S".

(c) Let $F \subset S$. We say that F *is a closed subset of* S if and only if every limit point of F belongs to F.[2] That is,

$$F = \{x \in S : x = \lim_{n\to\infty} x_n \text{ for some sequence } \{x_n\} \subseteq F\}$$

Instead of saying "F is a closed subset of S" we often simply say "F is closed in S".

Example 7. We define the metric ρ on $\mathbb{R}$ as $\rho(x, y) = |x - y|$. If U is the the interval $(-5, 3) = \{x \in \mathbb{R} : -5 < x < 3\}$, then U is an open subset in $\mathbb{R}$ since, for every point $x \in U$, we can find $\varepsilon > 0$ such that $(x - \varepsilon, x + \varepsilon) \subseteq (-5, 3)$. However, $V = (-4, 3]$ is not an open subset in $\mathbb{R}$ since V contains a point 3 such that the open ball $(3 - \varepsilon, 3 + \varepsilon) \nsubseteq V$ no matter how small we choose ε to be.

Example 8. We define the metric ρ on the set $U = (-5, 3]$ as $\rho(x, y) = |x - y|$. Consider the subset $S = (0, 3]$. The subset S is open in the metric space U since, for each $y \in S$ there is ε_y such that, $S = \cup\{B_{\varepsilon_y}(y) : y \in S\}$. Some readers might object, stating that there can be no ε such that $B_\varepsilon(3) \subseteq S$. But we must be careful and see that we are viewing S as a subset of the metric space $U = (-5, 3]$. See that, if we choose $\varepsilon = 1/2$, $B_\varepsilon(3) = \{x \in U : (3 - \varepsilon, 3 + \varepsilon)\} = (3 - \varepsilon, 3] \subseteq (2, 3]$ is indeed entirely contained in S.

Example 9. We define the metric ρ on $\mathbb{R}^2$ as $\rho(\vec{x}, \vec{y}) = \|\vec{x} - \vec{y}\|_2$. The x-axis, $U = \{(x, 0) : x \in \mathbb{R}\}$ is closed in $\mathbb{R}^2$ since if $\vec{a} = (a, b)$ is a limit point of U then there must exist a sequence $\{(x_n, 0)\}$ such that $(a, b) = \lim_{n\to\infty}(x_n, 0) = (\lim_{n\to\infty} x_n, \lim_{n\to\infty} 0)$; this implies $b = 0$, hence, $(a, b) \in U$. It then follows that U is closed in $\mathbb{R}^2$.

Example 10. Let S be an arbitrary non-empty set equipped with the discrete metric ρ. For any $x \in S$, $B_{1/2}(x) = \{x\} \subseteq \{x\}$ hence, $\{x\}$ is an open subset of S. Let x be a limit point of $\{x\}$. The only sequence in $\{x\}$ is the constant sequence $\{x, x, x, \ldots\}$ which converges to $x \in \{x\}$. Hence, $\{x\}$ is closed in S.

---

[2]Recall that x is a limit point of a set F if there exists a sequence inside F which converges to x.

Remark. We can similarly define the notion of an "open subset S" of a normed vector space $(V, \| \ \|)$ by defining $\rho(\vec{x}, \vec{y}) = \|\vec{x} - \vec{y}\|$, in which case U is open in $(V, \| \ \|)$ if and only if U is open in (V, ρ). In $(V, \| \ \|))$, the open ball, $B_\varepsilon(\vec{y})$, centered at $\vec{y}$ is defined as $B_\varepsilon(\vec{y}) = \{\vec{x} \in V : \|\vec{y} - \vec{x}\| = \rho(\vec{y}, \vec{x}) < \varepsilon\}$. Similarly, if $F \subseteq (V, \| \ \|)$, F is said to be closed in V if and only if F contains all its limit points with respect to the norm $\| \ \|$.

Theorem 2.6 *Let (S, ρ) be a metric space.*

(a) *Both the empty set, $\varnothing$, and S are open subsets of S.*
(b) *Both the empty set, $\varnothing$, and S are closed subsets of S.*
(c) *Finite intersections of open subsets of S are open subsets of S.*
(d) *Arbitrarily large unions*[3] *of open subsets of S are open in S.*

Proof. The proofs of all four parts are left as an exercise.

Suppose (S, ρ) is a metric space and $T \subseteq S$. Let $\rho_T : T \times T \to \mathbb{R}$ be the restriction of the function ρ to $T \times T$. Then (T, ρ_T) is a metric space. Or we simply say that T is a subspace of (S, ρ). One may wonder what relationship exists between the open sets of the subspace T and the open sets of the space S. The following theorem answers this question.

Theorem 2.7 *Suppose (S, ρ) is a metric space and $T \subseteq S$ equipped with the metric ρ_T inherited from the set S. Also suppose $U \subseteq T \subseteq S$. Then U is open in T (with respect to the metric ρ_T) if and only if there exists an open subset U^* in S (with respect to the metric ρ) such that $U = U^* \cap T$.*

Proof. We are given that (S, ρ) is a metric space and $T \subseteq S$ equipped with the metric ρ_T inherited from the set S. Suppose $U \subseteq T \subseteq S$.

[3]The words "arbitrarily large unions" include the notion of the union of a countably or uncountably infinite number of sets. For example, $\cup\{A_n : n \in \mathbb{N}\}$.

($\Leftarrow$) Suppose U^* is open in S and $U = U^* \cap T$. If $x \in U$, then there exists $\varepsilon > 0$ such that $B_\varepsilon(x) \subseteq S$. Now

$$B_\varepsilon(x) \cap T = \{y \in T : \rho(x,y) < \varepsilon)\} = \{y \in T : \rho_T(x,y) < \varepsilon)\} = B'_\varepsilon(x)$$

by definition, an open ball radius ε center x, in T. Since $U = \cup\{B'_\varepsilon(x) : x \in U\}$, then U is open in T.

($\Rightarrow$) Suppose U is an open subset of $T \subseteq S$. If $x \in U$ then there exists $\varepsilon > 0$ such that $B_\varepsilon(x) \subseteq U$. Let $B'_\varepsilon(x) = \{y \in S : \rho(x,y) < \varepsilon\}$. Then $B'_\varepsilon(x)$ is an open subset of S such that $B'_\varepsilon(x) \cap T = \{y \in T : \rho_T(x,y) < \varepsilon\} = B_\varepsilon(x)$. Then

$$\begin{aligned}[\cup\{B'_\varepsilon(x) : x \in T\}] \cap T &= \cup\{B'_\varepsilon(x) \cap T : x \in T\}\\ &= \cup\{B_\varepsilon(x) : x \in T\}\\ &= U\end{aligned}$$

If $U^* = \cup\{B'_\varepsilon(x) : x \in T\}$, then U^* is an open subset of S such that $U = U^* \cap T$.

Theorem 2.8 *A subset F is a closed subset of a metric space (S, ρ) if and only if its complement, $S \backslash F$, is an open subset of S.*

Proof. Recall that closed subsets of S are those subsets F which contain all their limit points. So the statement says that F is closed in S if and only if $S \backslash F$ contains no limit point of F. The proof is left as an exercise.

2.5 Characterizations of Continuous Functions on a Metric Space

In what follows we study *continuous functions* mapping one metric space, (S_a, ρ_a) to another metric space (S_b, ρ_b).

Definition 2.9 Let (S_a, ρ_a) and (S_b, ρ_b) be two metric spaces.

(a) We say that a function $f : S_a \to S_b$ is *continuous at the point* $u \in S_a$ if and only if it satisfies the following condition:

> For every $\varepsilon > 0$, there exist $\delta > 0$ such that $\rho_a(u, x) < \delta \Rightarrow \rho_b(f(u), f(x)) < \varepsilon$.

(b) If $A \subseteq S_a$, we say that a function $f : A \to S_b$ is *continuous on the set* A if and only if it is continuous at each point in A.

There are various ways of recognizing continuous functions on a subset T of a metric space (S, ρ). The following theorem illustrates the most important ones.

Theorem 2.10 *Let (S_a, ρ_a) and (S_b, ρ_b) be two metric spaces and $f : S_a \to S_b$ be a function mapping the S_a into the set S_b. Then the following are equivalent:*

(1) *The function f is continuous on S_a.*
(2) *Whenever a sequence $\{x_n\}$ in S_a converges to a point $u \in S_a$, then the sequence $\{f(x_n)\}$ in S_b converges to $f(u)$ in S_b.*
(3) *Suppose $f[S_a]$ denotes the range of f on the domain S_a. For any open subset U of $f[S_a] \subseteq S_b$, the subset $f^{\leftarrow}[U] = \{x \in S_a : f(x) \in U\}$ is open in S_a.*

Proof. We are given two metric spaces, (S_a, ρ_a) and (S_b, ρ_b) and a function and $f : S_a \to S_b$.

$(1 \Rightarrow 2)$ Suppose $f : S_a \to S_b$ satisfies the formal definition of continuity on S_a. Let $u \in S_a$ and $\varepsilon > 0$. There exists δ such that $\rho_a(u, y) < \delta \Rightarrow \rho_b(f(u), f(y)) < \varepsilon$. Suppose $\{x_n\}$ is a sequence in S_a such that $\lim_{n\to\infty} x_n = u$. Then the sequence $\{f(x_n)\}$ and the point $f(u)$ are defined in S_b. We claim that $\{f(x_n)\}$ converges to $f(u)$. Since $\{x_n\}$ converges to u, there exist an $N > 0$ such that $n > N$ implies $\rho_a(u, x_n) < \delta$. Then, for $n > N$, $\rho_b(f(u), f(x_n)) < \varepsilon$. It then follows that $\lim_{n\to\infty} f(x_n) = f(u)$, as required.

(2 ⇒ 1) Suppose the function $f : S_a \to S_b$ is such that, for $u \in S_a$ and $\{x_n\} \subseteq S_a$,

$$\lim_{n\to\infty} x_n = u \qquad \Rightarrow \qquad \lim_{n\to\infty} f(x_n) = f(u)$$

Let $\varepsilon > 0$. Suppose f is not continuous at u. Then there exists $\varepsilon > 0$ such that for any δ there exists $x \in S_a$ such that $\rho_a(u, x) < \delta$ and $\rho_b(f(u), f(x)) > \varepsilon$. Then, for any $\delta_n = \frac{1}{n}$, there exists $x_n \in S_a$ such that $\rho_a(u, x_n) < \delta_n = \frac{1}{n}$ and $\rho_b(f(u), f(x_n)) > \varepsilon$. We can then construct a sequence $\{x_n\}$ in S_a such that $\lim_{n\to\infty}(x_n - u) = 0$ and $\lim_{n\to\infty} f(x_n) \neq f(u)$, contradicting our hypothesis.[4] The source of our contradiction is our supposition that f is not continuous at u. So f must be continuous at u, as required.

(1 ⇒ 3) Suppose $f : S_a \to S_b$ satisfies the formal definition of continuity on S_a. Let $u \in S_a$ and $\varepsilon > 0$. Let U be an open subset of $f[S_a] \subseteq S_b$. Let $u \in f^{\leftarrow}[U]$. Then $f(u) \in U$. Since U is open in $f[S_a]$, there exists $\varepsilon > 0$ such that $B_\varepsilon(f(u)) \subseteq U$ where $B_\varepsilon(f(u)) = \{x \in f[S_a] : \rho_b(f(u), x) < \varepsilon\}$. Since f is continuous at u, there exists $\delta > 0$ such that $\rho_a(u, x) < \delta \Rightarrow \rho_b(f(u), f(x)) < \varepsilon$. This means that, if $x \in B_\delta(u)$, then $f(x) \in B_\varepsilon(f(u))$. Then

$$u \in B_\delta(u) \subseteq f^{\leftarrow}[B_\varepsilon(f(u))] \subseteq f^{\leftarrow}[U]$$

So $f^{\leftarrow}[U]$ is open in S_a, as required.

(3 ⇒ 1) Suppose $f : S_a \to S_b$ is a function on S_a such that whenever U is open in $f[S_a]$, $f^{\leftarrow}[U]$ is open in S_a. Let $u \in S_a$ and $\varepsilon > 0$. By hypothesis, $f^{\leftarrow}[B_\varepsilon(f(u))]$ is open in S_a and contains u. Then there exists $\delta > 0$ such that $u \in B_\delta(u) \subseteq f^{\leftarrow}[B_\varepsilon(f(u))]$. This means that if $\rho_a(u, x) < \delta$ then $\rho_b(f(u), f(x)) < \varepsilon$, as required.

Note that the third characterization of a continuous function on a set, "U is open $\Rightarrow f^{\leftarrow}[U]$ is open", is normally not used to determine continuity of f at a *point* in the domain. This characterization "U is open $\Rightarrow f^{\leftarrow}[U]$ is open" is referred to as the *topological definition of continuity on a set* while the second characterization "$\{x_n\} \to a \Rightarrow \{f(x_n)\} \to f(a)$" is referred to as the *sequential definition of continuity at a point.*

---

[4] Note that the axiom of choice is invoked here.

Exercises

1. Prove parts (a)–(d) of Theorem 2.6.
2. (a) Show that the function which appears in Example 2 on page 25 is a valid metric.
   (b) Show that the function which appears in Example 3 on page 25 is a valid metric.
3. Define $\rho : \mathbb{R} \times \mathbb{R} \to \mathbb{R}$ as $\rho(x, y) = |x - y|$. Prove that, thus defined, ρ is a valid metric on $\mathbb{R}$.
4. Let ρ_1 and ρ_2 be two metrics on the set S.
   (a) For a given $k > 0$ we define the function $k\rho : S \times S \to \mathbb{R}$ as $(k\rho)(x, y) = k \times \rho(x, y)$. Show that $k\rho$ is valid metric on S.
   (b) If $\rho_1 + \rho_2 : S \times S \to \mathbb{R}$ is defined as $(\rho_1 + \rho_2)(x, y) = \rho_1(x, y) + \rho_2(x, y)$ show that $\rho_1 + \rho_2$ is a valid metric on S.
   (c) If $\rho : S \times S \to \mathbb{R}$ is defined as $\rho(x, y) = \min\{1, \rho_1(x, y)\}$ show that ρ is a valid metric on S.
5. Suppose $\{x_n\}$ is a sequence in a metric space (S, ρ). Show that if $\lim_{n\to\infty} x_n = a$ and $\lim_{n\to\infty} = b$ where both a and b belong to S then $a = b$.
6. We define two functions $\rho_1 : \mathbb{R}^2 \times \mathbb{R}^2 \to \mathbb{R}$ and $\rho_2 : \mathbb{R}^2 \times \mathbb{R}^2 \to \mathbb{R}$ as follows:

$$\rho_1((a_1, a_2), (b_1, b_2)) = |a_1 - b_1| + |a_2 - b_2|$$

$$\rho_2((a_1, a_2), (b_1, b_2)) = \max\{|a_1 - b_1|, |a_2 - b_2|\}$$

   (a) Show that ρ_1 thus defined is a valid metric on $\mathbb{R}^2$.
   (b) Show that ρ_2 thus defined is a valid metric on $\mathbb{R}^2$.
   (c) Show that $\lim_{n\to\infty}\{\vec{x}_n\} = \vec{x} \in \mathbb{R}^2$ with respect to ρ_1 if and only if $\lim_{n\to\infty}\{\vec{x}_n\} = \vec{x} \in \mathbb{R}^2$ with respect to ρ_2.
7. Let a, b be distinct points in the metric space (S, ρ). Find disjoint open sets A and B such that $a \in A$, $b \in B$ and $A \cap B = \varnothing$.
8. Suppose $g : S \to \mathbb{R}$ and $h : S \to \mathbb{R}$ are two continuous functions on the metric space (S, ρ). Show that the set $U = \{x \in S : h(x) < g(x)\}$ is an open subset of M.
9. Prove that a finite subset of a metric space (S, ρ) is always closed in S.
10. Prove Theorem 2.8: A subset F is a closed subset of a metric space (S, ρ) if and only if its complement, $S\backslash F$, is an open subset of S.

Part II

Topological Spaces: Fundamental Concepts

Chapter 3

A Topology on a Set

Abstract
In this section, we define the notion of a topological space. We will begin by describing those families of subsets of a set which form a topology, τ, on a set, showing along the way how to recognize"open subsets" and "closed subsets". There can be many topologies on a set. Given a pair of topologies τ_1 and τ_2 on S one can sometimes be seen as being "weaker" or "stronger" than the other. We have provided many examples both in the main body of the text as well as in the given exercises.

3.1 Introduction

In our review of metric spaces we realized, in Theorem 2.10, that a function, $f : S \to T$, from a metric space S to another metric space T is *continuous* whenever the following condition is satisfied: "The set, $f^{\leftarrow}[U]$, is open in S whenever the set, U, is open in T". We learned that, some sets are described as being "open", others as being "closed", some are both open and closed, while some are neither. We then learned how to distinguish one from the other. For example, we recognized closed sets as being those whose complement is open. We can recognize a closed set, F, as being one that "contains the limit point of every convergent sequence in F". It seems that knowing various characterizations for each type of set is not only useful, but important. How did this notion of an "open set" originate? Recall that we constructed open sets in a set S with the help of a norm or metric previously defined on S. These two distance measuring tools defined on S allowed us to define the notion of an "open ball" in S which, in turn, allowed us to recognize those subsets, T, of S which are open.

We would now like to define "open set" in a much more abstract context, in a way that is independent of any distance measuring tool such as a norm or metric. We will gradually add more structure to what we will call, *topological spaces.* We will then proceed to classify the family of all topological spaces according to some of their intrinsic properties. Thus gathering together those spaces with similar properties. It may occur that a particular class of topological spaces, is actually a subclass of some other class.

Definition 3.1 Let S be a non-empty set.

(a) A *topology* on S is a collection, τ, of subsets of S which possesses the following properties:

O1: Both the empty set, $\varnothing$, and S belong to τ.
O2: If $\mathscr{C} \subseteq \tau$, then $\cup\{C \in \mathscr{C}\}$ is a set which also belongs to τ.
O3: If $\mathscr{F}$ is a finite subset of τ, then $\cap\{C \in \mathscr{F}\}$ is also a set which belongs to τ.

We will refer to O1, O2, and O3 as the *open set axioms.*[1]

(b) If τ is a topology on a set S (as defined above) then each member, $U \in \tau$, is called an *open subset* of S. We often simply say that "U is *open* in S".

(c) Suppose we have defined a topology, τ, on some set S. Then this set, S, when considered together with this topology τ, is called a *topological space* and can be represented as (S, τ). The condition, O2, can also be expressed by the phrase "τ is closed under arbitrary unions". The condition, O3, can also be expressed by the phrase "τ is closed under finite intersections".

---

[1]Some readers may notice that, if $\mathscr{C} = \varnothing \subseteq \tau$, then, by O2, $\cup\{C \in \varnothing\} = \varnothing \in \tau$. Also, if the finite subset, $\mathscr{F}$, is empty then, by O3, $\cap\{C \in \varnothing\} = S \in \tau$. So, theoretically, it would be sufficient to axiomatize "open sets" with O2 and O3 where O1 would logically flow from these two. The axiom O1 is normally included for convenience to make it easier to identify a topology.

It is important to remember that the elements of a topology τ on S are all subsets of S. To say that S is an open subset of S it is to assume that a topology τ has been previously defined; that is, that τ satisfies the three properties stated above. There can be many topologies defined on a given set. If τ_1 and τ_2 are two different topologies on a set S, then (S, τ_1) and (S, τ_2) are considered to be distinct topological spaces even though they contain the same points. When we view (S, τ_1) and (S, τ_2) simply as *sets* they are of course equal. We speak of U as being an open subset of S when it is clearly understood from the context that U is a member of a predefined topology τ on S.

Example 1. Suppose (S, ρ) is a metric space and $B_\varepsilon(y) = \{x \in S : \rho(x, y) < \varepsilon\}$ represents an open ball of radius ε and center y in S. Let

$$\tau_\rho = \{\varnothing\} \cup \{U \subseteq S : U \text{ is the union of open balls in } S\}$$

Then τ_ρ is a topology on S. (Showing that τ_ρ satisfies the three open set axioms is left as an exercise.) We say that τ_ρ is the *topology induced by the metric* ρ on S.

If a metric, ρ, is defined on S, a set A belongs to τ_ρ if and only if A is open in the metric space (S, ρ). In this sense, every metric space is a topological space.

Note: There exist topological spaces whose topology is not derived from a metric ρ. Those topological spaces whose open sets can be derived some metric have a special name.

Definition 3.2 Let (S, τ) be a topological space. We will say that (S, τ) is *metrizable* if there exists a metric, ρ, on S such that $\tau = \tau_\rho$ (where τ_ρ is induced on S by ρ).

From this definition, we can state that the class of all topological spaces can be subdivided into two subfamilies: One consists of all metrizable spaces the other of non-metrizable ones.

For example, the usual metric $\rho(x, y) = |x - y|$ on $\mathbb{R}$ induces a topology, τ_ρ, on $\mathbb{R}$. We normally refer to this topology as being the *usual topology on* $\mathbb{R}$. In many textbooks it is also referred to as the

Euclidean topology on $\mathbb{R}$. In this case, the open subsets of $\mathbb{R}$ are the sets which are unions of open intervals $B_\varepsilon(y) = (y - \varepsilon, y + \varepsilon)$. For example, the sets $U_1 = \cup\{(n, n+1) : n \in \mathbb{Z}\}$ and $U_2 = (-4, 9)$ can be shown to belong to τ_ρ, but the subset $U_3 = [-7, -3) \notin \tau_\rho$. (It is left to the reader to verify this.)

Example 2. (A non-metrizable space). Consider the set $S = \mathbb{R}^2$ equipped with the topology

$$\tau = \{B \subseteq \mathbb{R}^2 : \mathbb{R}^2 \backslash B \text{ is countable}\}^2 \cup \{\varnothing, \mathbb{R}^2\}$$

(a) Verify that the family, τ, of subsets of $\mathbb{R}^2$ is indeed a topology on $\mathbb{R}^2$.
(b) Verify that the topological space, $\mathbb{R}^2$, equipped with the topology, τ, described above is *not* metrizable.

Solution: Given: $\tau = \{B \subseteq \mathbb{R}^2 : \mathbb{R}^2 \backslash B \text{ is countable}\} \cup \{\varnothing, \mathbb{R}^2\}$.

(a) Verification that the sets in τ satisfy the three open sets axioms O1, O2, and O3 is left as an exercise.
(b) Suppose ρ is a metric on $\mathbb{R}^2$ such that $\tau = \tau_\rho$. We will show that, given $\tau = \tau_\rho$, then $\rho(\vec{x}, \vec{y}) > \rho(\vec{x}, \vec{z}) + \rho(\vec{y}, \vec{z})$, contradicting the fact that ρ is a valid metric on $\mathbb{R}^2$.

Suppose $\vec{x}$ and $\vec{y}$ are distinct points in $\mathbb{R}^2$ and $\rho(\vec{x}, \vec{y}) = \alpha \neq 0$. Consider the open balls $B_{\alpha/2}(\vec{x})$ and $B_{\alpha/2}(\vec{y})$. Since both balls belong to τ, each ball has a countable complement; so each ball is an uncountable subset. Hence, $B_{\alpha/2}(\vec{y})$ cannot be entirely contained in $\mathbb{R}^2 \backslash B_{\alpha/2}(\vec{x})$. That means that $B_{\alpha/2}(\vec{y}) \cap B_{\alpha/2}(\vec{x}) \neq \varnothing$. Let $\vec{z} \in B_{\alpha/2}(\vec{y}) \cap B_{\alpha/2}(\vec{x})$. Then

$$\begin{aligned} \rho(\vec{x}, \vec{z}) + \rho(\vec{y}, \vec{z}) &< \alpha/2 + \alpha/2 \\ &= \alpha \\ &= \rho(\vec{x}, \vec{y}) \end{aligned}$$

We see that $\rho(\vec{x}, \vec{y}) > \rho(\vec{x}, \vec{z}) + \rho(\vec{y}, \vec{z})$ and so ρ does not satisfy the triangle inequality, contradicting the fact that ρ is a metric on $\mathbb{R}^2$.

[2] If A is a subset of a space S, then $S \backslash A$ is the complement of A. It denotes the set of all points in S which don't belong to A.

So there can be no metric ρ on $\mathbb{R}^2$ such that $\tau = \tau_\rho$. We conclude that the topological space, $(\mathbb{R}^2, \tau)$, is not a metrizable space.[3]

We introduce the following supplementary definition.

Supplementary definition: If (S, τ) is a topological space and $x \in U$ where U is an open subset of S, then we say that U is an

open neighborhood of x.

Furthermore, if U is open and T is a subset of S such that $x \in U \subset T$, then we say that T is a *neighborhood of x.*

Note that a point, x, in a topological space, (S, τ), always has at least one open neighborhood, namely S. If $T = (-2, 4) \cup (4, 7]$ in $\mathbb{R}$, equipped with the usual topology, we see that T is a neighborhood of 3 but not a neighborhood of 4 nor of 7 (since there is no open set U such that $7 \in U \subseteq T$).

Definition 3.3 Suppose τ_1 and τ_2 are two topologies on a given set S. If $\tau_1 \subseteq \tau_2$, then we say that τ_1 is a *weaker topology than* τ_2 on S or that τ_2 is a *stronger topology than* τ_1 on S. We can also say that τ_1 is a *coarser topology than* τ_2 on S, or that τ_2 is a *finer topology than* τ_1 on S. We will say that the two topological spaces (S, τ_1) and (S, τ_2) are *equivalent* if and only if $\tau_1 = \tau_2$.

Example 3. Suppose S is a non-empty set and $\mathscr{P}(S)$ denotes the power set of S (i.e., $\mathscr{P}(S)$ denotes the collection of all subsets of S). Let $\tau_d = \mathscr{P}(S)$. Then τ_d is a topology on S. It is left to the reader to verify this. In this case, for every single point $x \in S$, $\{x\}$ is open in S.

---

[3]Later, once we covered the concept of "Hausdorff" this problem will be more easily solved by stating that "this space is not metrizable because it not Hausdorff".

Supplementary definition: For a given space S, if

$$\tau_d = \mathscr{P}(S)$$

the topology, τ_d, is referred to as the *discrete topology on S.*

If S is equipped with the discrete topology then every non-empty subset of S is an open neighborhood of the elements it contains.

Example 4. For a non-empty set S, if $\tau_i = \{\varnothing, S\}$, then the two-element set, τ_i, is a topology on S.

Supplementary definition: If $\tau_i = \{\varnothing, S\}$, the topology, τ_i, is normally referred to as the *indiscrete topology on S.* In this case, for any $x \in S$, S is its only neighborhood.

Since all topologies on S must at least contain $\varnothing$ and S, τ_i is the weakest (coarsest) of all topologies on S. On the other hand, $\tau_d = \mathscr{P}(S)$ is the strongest (finest) of all topologies on S. For any topology τ on S, we then have

$$\tau_i \subseteq \tau \subseteq \tau_d$$

It is left for the reader to verify the following important fact.

Fact: Given a family $\{\tau_k : k \in I\}$ of topologies on a set S, it is an easy exercise to show that the family $\cap_{k\in I}\tau_k$ is also a topology on S.[4] This fact can also be expressed as *The set of all topologies is closed under arbitrary intersections.*

But $\cup\{\tau_k : k \in I\}$ need not necessarily be a topology on S. For example, verify that $\tau_1 = \{\varnothing, S, \{b\}, \{a, b\}\}$ and $\tau_2 = \{\varnothing, S, \{c\}, \{a, c\}\}$ are two topologies on S; but their union, $\tau_1 \cup \tau_2$, is not a topology on S. (Since $\{a, b, c\} \notin \tau_1 \cup \tau_2$. Verify this.)

[4]Caution: However, arbitrary unions of topologies may not form a topology on a set.

3.2 Closed Subsets of a Topological Space

In our brief overview of metric spaces, (S, ρ), we defined a closed subset, F, of S as one which contains all its limit points. We then saw that, in a metric space S, a subset, F, of S is closed if and only if its complement, $S\backslash F$, is open.[5] Metric spaces were equipped with a "metric" so that we could discuss the much needed notions of "convergence of a sequence" and "closed set" in a set S. We now formally define the analogous notions of "closed subsets in a topological space". We will not discuss the concepts of "convergence" and "limit points" in a topological space immediately. We will be doing an in-depth study of this topic in a section later in this book.

Definition 3.4 Let F be a subset of a topological space (S, τ). If the complement, $S\backslash F$, of F is an open subset in S then we say that F is a *closed subset in S*.

The definition of "closed" states that "($S\backslash F$ is open) $\Rightarrow$ (F is closed"). Conversely,

$$\begin{aligned} A \text{ is closed} \quad &\Rightarrow \quad A = S\setminus[S\backslash A] \text{ is closed} \\ &\Rightarrow \quad [S\backslash A] \text{ is open} \end{aligned}$$

We can then actually write

$$(F \text{ is closed}) \quad \Leftrightarrow \quad (S\backslash F \text{ is open})$$

Suppose (S, τ) is a topological space and $\mathscr{F} = \{A \subset S : A \text{ is closed in } S\}$. Then we can define the topology in terms of $\mathscr{F}$,

$$\tau = \{A : S\backslash A \in \mathscr{F}\} \quad \text{and} \quad \mathscr{F} = \{A : S\backslash A \in \tau\}$$

Theorem 3.5 *Let (S, τ) be a topological space and I be any indexing set. Then,*

(a) *Both $\varnothing$ and S are closed in S.*

[5] $S\backslash F = \{x \in S : x \notin F\}$.

(b) *If* $\{F_i : i \in I\}$ *is a family of closed subsets of* S *then* $\cap\{F_i : i \in I\}$ *is a closed subset of* S.
(c) *If* $\{F_i : i = 1, 2, 3, \ldots, k\}$ *is a finite family of closed subsets of* S *then* $\cup\{F_i : i = 1, 2, 3, \ldots, k\}$ *is a closed subset of* S.

Proof.

(a) Since $\varnothing$ is open, then $S = S\backslash\varnothing$ is closed. Since S is open, $\varnothing = S\backslash S$ is closed.
(b) Let $\{F_i : i \in I\}$ be a family of closed subsets of S. Then, for each $i \in I$, $S\backslash F_i$ is open. Since (by De Morgan's law) $S\backslash\cap\{F_i : i \in I\} = \cup\{S\backslash F_i : i \in I\}$ is open (being the union of open sets), then $\cap\{F_i : i \in I\}$ is a closed subset of S.
(c) This part is left as an exercise.

Supplementary proposition: *Defining a topology on S in terms of "closed sets".* Suppose we are given a set S and family, $\mathscr{F} = \{F : F \subseteq S\}$ of elements from $\mathscr{P}(S)$ which satisfies the following three conditions:

F1: The sets $\varnothing$, S both belong to $\mathscr{F}$.
F2: If $\{F_i : i \in I\} \subseteq \mathscr{F}$ then $\cap\{F_i : i \in I\} \in \mathscr{F}$.
F3: If $\{F_i : i = 1, 2, 3, \ldots, k\}$ is a finite subset of $\mathscr{F}$ then $\cup\{F_i : i = 1, 2, 3, \ldots k\} \in \mathscr{F}$.

Then the family, $\tau = \{S\backslash F : F \in \mathscr{F}\}$, forms a topology on S.

Proof. The proof showing that τ is a topology is left to the reader. It easily follows from an application of De Morgan's law.[6]

Supplementary definition: The conditions F1, F2, and F3 described above are referred to as the

closed set axioms.

---

[6]That is, $S \setminus [\cup_{i\in I} F_i] = \cap_{i\in I}[S\backslash F_i]$ and $S \setminus [\cap_{i\in I} F_i] = \cup_{i\in I}[S\backslash F_i]$.

Example 5. Suppose S is a non-empty set and

$$\mathscr{F} = \{F : F \text{ is a finite subset of } S\} \cup \{\varnothing, S\}$$

Then the set $\mathscr{F}$ satisfies the three closed sets axioms F1, F2, and F3. (Verify this!) Then $\tau = \{S \backslash F : F \in \mathscr{F}\}$ forms a topology on S. The elements of τ are

$$\tau = \{\varnothing, S\} \cup \{U \subset S : S \backslash U \text{ is finite}\}$$

Given a set S, the family of subsets,

$$\tau = \{A : A \subseteq S \text{ and } S \backslash A \text{ is finite}\} \cup \{S, \varnothing\}^7$$

forms a topology of S called the *cofinite topology on* S or the *Zariski topology on S.*

3.3 Subspace Topology on a Subset

We previously defined the notion of a "metric subspace", T, as being a subset of a metric space, (S, ρ), equipped with the subspace metric, ρ_T.

In the more general case of a topological space, (S, τ), any subset T can be declared to be a "topological subspace" provided the reader understands what topology is defined on T. Suppose H is a non-empty subset of a topological space (S, τ). Then H can inherit its topology from τ, in a natural way, as shown the following theorem.

Theorem 3.6 *Let (S, τ) be topological space and $H \subseteq S$.*

(a) *Let*

$$\tau_H = \{U \subset H : U = K \cap H, \text{ for some } K \in \tau\}$$

Then τ_H is a topology on H.

---

[7]If $S \backslash U$ is finite, we say U is cofinite.

(b) *Let*

$$\mathscr{F}_H = \{F \subset H : F = M \cap H \text{ where } M = S \setminus K, K \in \tau\}$$

Then $\mathscr{F}_H$ *represents* all *closed subsets of the topological space* (H, τ_H).

Proof. The proof is left as an exercise.

Definition 3.7 If (S, τ) is topological space and $H \subseteq S$ and

$$\tau_H = \{U \subset H : U = K \cap H,\ K \in \tau\}$$

then τ_H is called the *subspace topology* or *relative topology* on H induced by S, or inherited from S. In such a case, we will say that (H, τ_H) is a *subspace of* S.

Some subsets of a topological space (S, τ) can be both open and closed. For example, if $(\mathbb{R}, \tau_i)$ is equipped with the indiscrete topology, $\tau_i = \{\varnothing, \mathbb{R}\}$, both $\varnothing$ and $\mathbb{R}$ are the only open subsets of $\mathbb{R}$ and so both $\varnothing$ and $\mathbb{R}$ are the only closed subsets of $\mathbb{R}$. That is, $\varnothing$ and $\mathbb{R}$ are simultaneously open and closed in $\mathbb{R}$. We consider a less trivial example.

Example 6. Let $H = [3, 5)$. Consider the subset

$$T = H \cup \{9\} = [3, 5) \cup \{9\}$$

of $(\mathbb{R}, \tau)$ where $\mathbb{R}$ is equipped with the usual topology, τ. Suppose the subspace (T, τ_T) is equipped with the subspace topology inherited from τ. Now H is a subset of both T and $\mathbb{R}$. Since $H = T \cap [2, 6]$, then H is closed in T with respect to the subspace topology τ_T. Since $H = T \cap (2, 5)$, then H is open in T with respect to the subspace topology τ_T. So the subset, H, is both open and closed in the subspace, T. Note, however, that H is neither open nor closed in S.

There is a special adjective used to refer to those subsets which are both open and closed. Since its use is fairly common, we formally define it below.

Definition 3.8 If T is a subset of a topological space (S, τ), and T is both open and closed with respect to τ, we say that...

T is clopen in S.

Suppose (S, τ) is a topological space which contains the non-empty subset (F, τ_F) equipped with the subspace topology, τ_F, inherited from S. If one speaks of an open subset, U, of F, it may sometimes not be obvious whether $U \in \tau$ or $U \in \tau_F$. To be more specific, we can write "U is an S-open subset in F" to mean $U \in \tau$ and "U is an F-open subset of F" to mean $U \in \tau_F$.[8]

3.4 Other Examples

We provide a few more examples of topological spaces.

Example 7. Let S be a set and $B \subseteq S$. Let $\tau_B = \{A \in \mathscr{P}(S) : B \subseteq A\} \cup \{\varnothing\}$.

(a) Verify that τ_B is indeed a topology.
(b) Describe the closed subsets of (S, τ_B).

Solution: We are given that $B \subseteq S$ and $\tau_B = \{A \in \mathscr{P}(S) : B \subseteq A\} \cup \{\varnothing\}$.

(a) We verify that τ_B is a topology on S by confirming that it satisfies the open set axioms O1, O2, and O3.
 — By definition, $\varnothing \in \tau_B$; also, since $B \subseteq S$ then $S \in \tau_B$.
 — Suppose $\mathscr{U}$ is a non-empty subset of τ. Then $B \subseteq U$ for each $U \in \mathscr{U}$. So $B \subseteq \cup\{U : U \in \mathscr{U}\}$. Hence, $\cup\{U : U \in \mathscr{U}\} \in \tau_B$.
 — Suppose $\mathscr{F}$ is a finite subset of τ_B. Then $B \subseteq F$ for each $F \in \mathscr{F}$. So $B \subseteq \cap\{F : F \in \mathscr{F}\}$. Hence, $\cap\{F : F \in \mathscr{F}\} \in \tau_B$.

(b) We now describe the closed subsets of S. We consider the sets, A, such that $A \cap B = \varnothing$, and those satisfying $A \cap B \neq \varnothing$.
 — If $A \subseteq S$ such that $A \cap B = \varnothing$, then $B \subseteq S \backslash A$; this means $S \backslash A$ is open, hence, A is a closed subset of S.

---

[8]Note that if F is S-open and U is an F-open subset of F then U is also S-open.

— Suppose, on the other hand, that for $A \subseteq S$ and $A \cap B \neq \varnothing$. We claim that S is the only closed subset which contains A. To see this, note that if F is closed then $S \backslash F$ is open and so either contains B or is $\varnothing$. If $A \cap B \neq \varnothing$ and $A \subseteq F$ then $S \backslash F$ cannot contain B so $S \backslash F = \varnothing$, which means that $F = S$.

So the closed subsets of S are the family

$$\mathscr{F} = \{F \subset S : F \cap B = \varnothing\} \cup \{S\}$$

Example 8. Let (S, τ) be a topological space and suppose B is a fixed subset of S. Let

$$\tau_B = \{A \in \mathscr{P}(S) : A = C \cup (D \cap B) \text{ where } C, D \in \tau\}$$

(a) Verify that τ_B is another topology on S.
(b) Is one of the two topologies, τ_B, τ, stronger than the other? Are these two topologies equivalent topologies?

Solution: We are given that (S, τ) is a topological space and B is a fixed subset of S.

(a) We begin by showing that τ_B satisfies the three open set axioms O1, O2, and O3.

— We know $\varnothing, S \in \tau$. So we have $\varnothing = \varnothing \cup (\varnothing \cap B) \in \tau_B$ and $S = S \cup (S \cap B) \in \tau_B$.

— Suppose $\mathscr{U} = \{C_i \cup (D_i \cap B) : i \in I\} \subseteq \tau_B$, where $C_i, D_i \in \tau$ for all $i \in I$. Then $\cup_{i \in I} C_i$, and $\cup_{i \in I} D_i$ both belong to τ. Then

$$\bigcup_{i \in I} [C_i \cup (D_i \cap B)] = [\cup_{i \in I} C_i] \ \cup \ ([\cup_{i \in I} D_i] \cap B) \in \tau_B$$

Thus, τ_B is closed under arbitrary unions.

— Suppose $\mathscr{F} = \{C_i \cup (D_i \cap B) : i = 1, 2, \ldots, n\} \subseteq \tau_B$. Then

$$\begin{aligned}
\bigcap_{i=1,\ldots,n} [C_i \cup (D_i \cap B)] &= \bigcap_{i=1,\ldots,n} [(C_i \cup D_i) \cap (C_i \cup B)] \\
&= [\cap_{i=1,\ldots,n} (C_i \cup D_i)] \ \cap \ [(\cap_{i=1,\ldots,n} C_i) \cup B] \\
&= [\cap_{i=1,\ldots,n} (C_i \cup D_i)] \cap [(\cap_{i=1,\ldots,n} C_i)] \\
&\quad \cup [\cap_{i=1,\ldots,n} (C_i \cup D_i) \cap B] \\
&\in \tau_B
\end{aligned}$$

So τ_B satisfies the three open set axioms O1, O2, and O3.

(b) We claim that $\tau \subseteq \tau_B$: See that, if $C \in \tau$, then $C = C \cup (\varnothing \cap B) \in \tau_B$, so $\tau \subseteq \tau_B$. So τ_B is a topology on S which is *finer* (*stronger*) than τ.

We claim that these two topologies are not equivalent: Suppose $B \notin \tau$, $D \in \tau$ and $B \subset D$. Then $\varnothing \cup (D \cap B) = B \notin \tau$. Since $B \in \tau_B$, then τ_B contains elements which are not in τ. So $\tau \subset \tau_B$.

In the above example, we say that "the topology τ_B *extends* τ *over* B".

3.5 Free Union of Topological Spaces

Suppose we are given a family of topological spaces. There is a way to unite them into one single new larger topological space without altering their individual topology. This is referred to a being the "free union" of these topological spaces. We define this concept.

Definition 3.9 Let $\{S_i : i \in I\}$ be a family of topological spaces. For each space, S_i, we associate a space, $S_i^* = \{i\} \times S_i$ in such a way that S_i^* and S_i are identical except for the fact that $\{i\} \times S_i$ has a label i attached to S_i. This is to guarantee that if $i \neq j$ then S_i^* and S_j^* are entirely different sets and so have empty intersection. This allows us to view the family, $\{S_i^* : i \in I\}$, as being pairwise disjoint, in the sense that no two spaces have elements in common. We define the *free union of the family* $\{S_i : i \in I\}$,[9] denoted as, $\sum_{i \in I} S_i^*$, as being the topological space

$$\sum_{i \in I} S_i^* = \cup\{S_i^* : i \in I\}$$

in which U is open in $\sum_{i \in I} S_i^*$ if and only if $U \cap S_i^*$ is open for each $i \in I$. This topology, thus defined, is referred to as the

disjoint union topology.

---

[9]Some texts may refer to this set as *direct sum* or *free sum* or *topological direct sum.*

Example 9. For each $n \in \mathbb{N}\backslash\{0\}$, let L_n denote the set,

$$L_n = \left\{(x, y_n) : y_n = g_n(x) = \left(\frac{\sin\left(\frac{\pi}{4n}\right)}{\cos\left(\frac{\pi}{4n}\right)}\right)x,\ x \neq 0\right\}$$

Let $L_0 = \{(x, 0) : x \in \mathbb{R}\}$ and

$$T = \cup\{L_n : n \in \mathbb{N}\backslash\{0\}\} \cup \{L_0\}$$

It is possible to view T as a single subset of $\mathbb{R}^2$ and equip it with the subspace topology, in which case open neighborhoods of points on L_0 would intersect points on other lines. For a different topology, we could view T as a free union of disjoint subspaces of $\mathbb{R}^2$, where each line is equipped with the subspace topology. With the free union topology, an open neighborhood of a point on a line is restricted to the line itself. Each line would be clopen in T. For example, $\{(x, 0) : 1 < x < 7\}$ would be an open neighborhood of $(5, 0)$ in T.

3.6 Topics: *G*-Delta and *F*-Sigma Sets

Besides the fundamental *open* and *closed* subsets of a topological space introduced earlier, there are other subsets defined in terms of open and closed sets with special properties. Two of these are called G-delta's, F-sigma's. Developing some familiarity with these now will be good practice. As well, it will allow us to freely refer to them in various examples, theorems and exercise questions to come.

The G-delta and F-sigma sets in a topological space. We have seen that arbitrary unions of open subsets of a topological space are open. However, only the intersection of (at most) finitely many open sets are guaranteed to be open. Similarly, arbitrary intersections of closed sets are closed, but the union of (at most) finitely many closed sets are guaranteed to be closed. The intersection of countably many open sets and the union of countably many closed sets may be relevant in the study of certain types of spaces.

Before we continue, we remind the reader that a non-empty set S is said to be "countable" if it is finite or, in the case where it is infinite, the elements of S can be indexed by the natural numbers. That is, $S = \{x_i : i = 0, 1, 2, 3, \ldots\}$. We can also say that S is countable if there exists a function $f : \mathbb{N} \to S$ mapping the natural numbers *onto* S. We now formally define those special subsets of a topological space we call *G-delta*'s and *F-sigma*'s.

Definition 3.10 The sets in a topological space, (S, τ), which are the intersection of at most countably many open sets are called G_δ-sets (or simply G_δ). Those sets in S which are the union of at most countably many closed sets are called F_σ-sets (or simply F_σ). Neither of these special sets need be open or closed.

Trivially, if F is closed in (S, τ) then F is an F_σ and if U is open then U is a G_δ.

Example 10. If $\mathbb{R}$ is equipped with the usual topology, the set $T = [2, 7]$ is obviously a F_σ. It is also a G_δ since

$$[2, 7] = \cap\{(2 - 1/n, 7 + 1/n) : n = 1, 2, 3, \ldots\}$$

So some sets can be both a G_δ and an F_σ with respect to the same topology τ.

Example 11. On the other hand, suppose (S, τ_i) is equipped with the indiscrete topology, $\tau_i = \{\varnothing, S\}$. If T is a proper non-empty subset of S, we see that $T \not\subseteq \varnothing$ and $T \neq S$; so T is neither a G_δ nor an F_σ with respect to τ_i.

Example 12. We consider the set of all rationals, $\mathbb{Q}$, as a subset of $\mathbb{R}$ equipped with the usual topology. It is known that $\mathbb{Q}$ is countably infinite and so can be expressed in the form $\mathbb{Q} = \{x_i : i = 1, 2, 3, \ldots\}$. Then $\mathbb{Q} = \cup\{\{x_i\} : i = 1, 2, 3, \ldots\}$ where each $\{x_i\}$ is a closed subset of $\mathbb{R}$. So $\mathbb{Q}$ is an F_σ.[10]

The following theorem exhibits properties respected by each F_σ and G_δ and the families of all G_δ's and F_σ's of a topological space.

Theorem 3.11 *Suppose F is an F_σ and G is a G_δ in S.*

(a) *The complement of F in S is a G_δ and the complement of G in S is an F_σ.*

---

[10] Once we have the necessary tools we will prove that $\mathbb{Q}$ is not a G_δ.

(b) *There exists a sequence,* $\{F_i : i = 1, 2, 3, \ldots\}$*, of closed subsets of* S *such that*

$$F_i \subseteq F_{i+1} \text{ for all } i = 1, 2, 3, \ldots, \text{ and } F = \cup\{F_i : i = 1, 2, 3, \ldots\}$$

(c) *There exists a non-increasing sequence,* $\{G_i : i = 1, 2, 3, \ldots\}$*, of open subsets of* S *such that*

$$G_{i+1} \subseteq G_i \text{ for all } i = 1, 2, 3, \ldots, \text{ and } G = \cap\{G_i : i = 1, 2, 3, \ldots\}$$

(d) *Suppose* $\mathscr{F}$ *denotes the family of all* F_σ*'s in* S *and* $\{F_i : i = 1, 2, 3, \ldots\}$ *represents at most countably many elements in* $\mathscr{F}$*. Then*

$$\cup\{F_i : i = 1, 2, 3, \ldots\} \in \mathscr{F} \text{ and } \cap\{F_i : i = 1, 2, 3, \ldots, k\} \in \mathscr{F}$$
$$\text{for any } k$$

(e) *Suppose* $\mathscr{G}$ *denotes the family of all* G_δ*'s in* S *and* $\{G_i : i = 1, 2, 3, \ldots\}$ *represents at most countably many elements in* $\mathscr{G}$*. Then*

$$\cap\{G_i : i = 1, 2, 3, \ldots\} \in \mathscr{G} \text{ and } \cup\{G_i : i = 1, 2, 3, \cdots, k\} \in \mathscr{G}$$
$$\text{for any } k$$

Proof. Given: (S, τ) be topological space; F is an F_σ and G is a G_δ in S.

(a) Suppose
$F = \cup\{K_i : i = 1, 2, 3, \ldots, \text{ where each } K_i \text{ is closed in } S\}$

$$\begin{aligned} S\backslash F &= S\backslash(\cup\{K_i\}) \\ &= \cap\{S\backslash K_i\} \quad \text{(By De Morgan's rule)} \\ &= \text{ a } G_\delta\text{-set} \end{aligned}$$

The proof of the second part of (a) follows by a similar application of De Morgan's rule.

(b) The proof is left as an exercise for the reader.

(c) The proof is left as an exercise for the reader.

(d) For countable unions of F_σ's:

$$\bigcup_{j=1}^{\infty}[\cup\{F_{(i,j)} : i = 1, 2, 3, \ldots\}] = \cup\{F_{(i,j)} : (i,j) \in \mathbb{N} \times \mathbb{N}\}$$

where $\mathbb{N} \times \mathbb{N}$ is known to be countable.
For finite intersections of F_σ's:

$$\bigcap_{i=1}^{k}[\cup_{j=1}^{\infty}\{F_{(i,j)}\}] = \bigcup_{(j_1,\ldots,j_k)\in\mathbb{N}\times\mathbb{N}\times\cdots\times\mathbb{N}} \{F_{(1,j_1)} \cap \cdots \cap F_{(k,j_k)}\}$$

where $\mathbb{N} \times \mathbb{N} \times \cdots \times \mathbb{N}$ is known to be countable.

(e) This part is proved similarly to part (d).

We summarize two of the results in the above theorem:

> The family, $\mathscr{F}$, of all F_σ's of a topological space is closed under countable unions and closed under *finite* intersections.
>
> The family, $\mathscr{G}$, of all G_δ's of a topological space is closed under countable intersections and closed under *finite* unions.

We will see a bit later that, in certain specific classes of topological spaces, F_σ's are closed and G_δ's are open. We have to do some groundwork before we can discuss these.

3.7 Topics: The Family of Borel Sets[11]

Topological spaces each contain a particular family of subsets of $\mathscr{P}(S)$ which plays a role in certain fields of study where topology is applied. Particularly in analysis. Before we formally define it, we begin by defining a special type of subset of $\mathscr{P}(S)$ called a "σ-ring".

> A subset, $\mathscr{K}$, of $\mathscr{P}(S)$ is called a *σ-ring* if:
>
> (1) For any $A \in \mathscr{K}$, $S\backslash A \in \mathscr{K}$.
> (2) Whenever $\{A_i : i \in \mathbb{N}\} \subseteq \mathscr{K}$ then $\cup\{A_i : i \in \mathbb{N}\} \in \mathscr{K}$.

[11]This is a more specialized topic. It can be omitted without loss of continuity.

To summarize, a σ-ring is simply a family of sets which is "closed under complements" and "countable unions of its sets". See that $\mathscr{P}(S)$, itself a σ-ring, may contain many σ-rings. Given a topological space, (S, τ), we will consider all those σ-rings in $\mathscr{P}(S)$, which contain τ. To obtain the *unique smallest σ-ring*, $\mathscr{B}$, in $\mathscr{P}(S)$ that contains τ, we then take the intersection of all σ-rings in $\mathscr{P}(S)$ which contain τ,

$$\mathscr{B} = \cap\{\mathscr{K} \subseteq \mathscr{P}(S) : \mathscr{K} \text{ is a } \sigma\text{-ring}, \tau \subseteq \mathscr{K}\}$$

The reader should verify that this "intersection of all σ-rings in $\mathscr{P}(S)$ which contain τ" is itself a σ-ring containing τ; we emphasize, that this intersection is the *unique* and *smallest* such σ-ring. We have a name for this particular set.

Definition 3.12 Given a topological space (S, τ), we call the smallest σ-ring, $\mathscr{B}$, in $\mathscr{P}(S)$ which contains τ,...

the family of Borel sets in S.

Each member, $A \in \mathscr{B}$, is referred to as a *Borel set*. That is,

$$\{A = \text{"a } \textit{Borel set} \text{ in } S\text{"}\} \Leftrightarrow \{A \in \mathscr{B}\}$$

Every topological space, S, has its unique family, $\mathscr{B}$, of Borel sets. We identify a Borel set by confirming that it belongs to $\mathscr{B}$. To help us identify Borel sets we list a few properties of $\mathscr{B}$. The reader is left to verify that $\mathscr{B}$...

— is closed under complements,
— is closed under countable unions and countable intersections,
— contains all G_δ's and all F_σ's.

The definition of a "Borel set" in S makes it difficult to recognize such a subset of S. But it also makes it easy to recognize a large subfamily of $\mathscr{B}$: "All open subsets (including $\varnothing$ and S itself), all G_δ's and all F_σ's are Borel sets". But there may be others.

Theorem 3.13 *Let (S, τ) be a topological space. The family, $\mathscr{B}$, of Borel sets is the* unique smallest subfamily of $\mathscr{P}(S)$, *that*

(a) *contains* τ,
(b) *is closed under complements,*
(c) *is closed under countable unions.*

Furthermore, $\mathscr{B}$ *satisfies the following three properties:*

(1) $\mathscr{B}$ *contains all* F_σ*'s of* S,
(2) $\mathscr{B}$ *is closed under countable intersections,*
(3) $\mathscr{B}$ *contains all* G_δ*'s of* S.

Proof. Given: (S, τ) is a topological space.

Suppose $\mathscr{B}$ is the family of Borel sets in $\mathscr{P}(S)$. By definition, $\mathscr{B}$ is the intersection of all σ-rings that

(a) contain τ,
(b) that are closed under complements,
(c) closed under countable unions.

So $\mathscr{B}$ is itself the *unique smallest* σ-ring of subsets of S which satisfies these three properties.

Since $\mathscr{B}$ contains τ, it contains all open subsets of S and since it is closed under complements, it contains all closed subsets of S. Since it is closed under countable unions, then it must contain all F_σ's. This establishes property 1.

We now verify that $\mathscr{B}$ is closed under countable intersections: Let $\{A_i : i = 1, 2, 3, \ldots\}$ be a countable family of subsets in $\mathscr{B}$. Then

$$\begin{aligned}\cap\{A_i : i = 1, 2, 3, \ldots\} &= S\backslash S\backslash(\cap\{A_i : i = 1, 2, 3, \ldots\}) \\ &= S\backslash\cup[S\backslash\{A_i : i = 1, 2, 3, \ldots\}] \\ &\in \mathscr{B}\end{aligned}$$

This establishes property 2.

It then follows that, since $\mathscr{B}$ contains all open sets, it follows from property two that it must also contain all G_δ's of S. This establishes property 3.

The above theorem guarantees that every open set, closed set, G_δ and F_σ in a topological space S can be referred to as a Borel set in S. It is sometimes difficult to identify subsets of a topological space (S, τ) which are not Borel sets (with respect to τ). Consider for example, the topological space (S, τ_i) equipped with the indiscrete topology. If A is a non-empty proper subset of S then A is not a Borel set since $\{\varnothing, S\} = \tau_i$ is the smallest σ-ring which contains τ_i and does not contain the element A. On page 76 of this text we provide another example.

Concept Review

1. Given a set S, what does a topology τ on S represent? How does one verify whether a family of subsets is a topology?
2. Given an open subset, U, of a topological space, (S, τ), what is the relationship between U and τ?
3. What are the three open set axioms of a topological space (S, τ)?
4. Given a metric space, (S, ρ), describe a topology on S which is induced by ρ.
5. Describe the *usual topology* on $\mathbb{R}$.
6. Given a point, x, in a topological space, (S, τ), what is a *neighborhood* of x?
7. Given two topologies, τ_1 and τ_2, what does it mean to say that τ_1 is weaker than τ_2?
8. Given two topologies, τ_1 and τ_2, what does it mean to say that τ_1 is finer than τ_2?
9. Given a non-empty set, S, describe the discrete topology on S.
10. Given a non-empty set, S, describe the indiscrete topology on S.
11. Suppose F is a closed subset of the topological space (S, τ). What is the relation between F and τ?
12. What are the three closed set axioms, F1, F2, and F3, of a topological space (S, τ).
13. If S is a non-empty set what do we mean by the *cofinite* or *Zariski* topology on S?
14. If T is a subset of the topological space (S, τ) what is the subspace topology on T?
15. Suppose B is a subset of the topological space (S, τ) such that $B \notin \tau$. Describe the topology τ_B which extends τ over B?
16. What does it mean to say that a set is metrizable?
17. What is a G_δ of a topological space? What is an F_σ of a topological space?
18. Describe the family of Borel sets in a topological space (S, τ).

19. Provide a few examples of Borel sets in $\mathbb{R}$ equipped with the usual topology. Is $\mathbb{Q}$ a Borel set? Why?
20. Define the topological space called the "free union" of the spaces, $\{S_i : i \in I\}$.

Exercises

1. Prove the statement in Theorem 3.5.
2. Consider the open interval $S = (-3, 7)$ in $\mathbb{R}$.
   (a) Construct a topology τ on S which contains five elements.
   (b) Consider the subset $T = (-2, 4] \subset S$. For the topology τ constructed in part (a) what is the subspace topology τ_T on T inherited from S.
   (c) Are the open subsets of T necessarily open subsets of S?
3. Consider $\mathbb{R}$ equipped with the usual topology τ (induced by the Euclidian metric). Let $(\mathbb{Q}, \tau_{\mathbb{Q}})$ be the set of all rational numbers equipped with the subspace topology inherited from $\mathbb{R}$. Consider the subset $T = [-\pi, \pi) \cap \mathbb{Q}$. Determine whether T is open in $\mathbb{Q}$, closed in $\mathbb{Q}$, both open and closed in $\mathbb{Q}$, or none of these.
4. Construct a topology other than the *discrete* or *indiscrete topology* on the set $S = \{\triangle, \lozenge, \square\}$.
5. Let $\mathscr{F} = \{A \subseteq \mathbb{R} : A \text{ is countable}\} \cup \{\varnothing, \mathbb{R}\}$. Show that $\mathscr{F}$ satisfies the three conditions F1, F2, and F3 described on page 46. Then use this to construct a topology on $\mathbb{R}$. (This is referred to as being the *cocountable topology.*
6. Suppose τ_A and τ_B are two topologies on a set S. Determine whether $\tau_A \cap \tau_B$ is a topology on S.
7. If $\mathbb{R}$ is equipped with the usual topology and $\mathbb{Z}$ represents the set of all integers determine whether $\mathbb{Z}$ is open or closed (or both or neither) in $\mathbb{R}$.
8. Suppose $\mathbb{R}$ is equipped with the usual topology and $T = [1, 4] \cup (6, 10) \subset \mathbb{R}$ where T is equipped with the subspace topology. Determine whether [1, 4] is open in T, closed in T or both open and closed in T. Determine whether (6, 10) is open in T, closed in T or both open and closed in T.

Chapter 4

Set Closures, Interiors and Boundaries

Abstract
In this section, we introduce the notions of closure and interior of subsets of a topological space. The concept of the boundary of a set is then defined in terms of its interior and closure. Based on their properties, we derive the "closure axioms" and "interior axioms". We then begin viewing closure and interior of sets as being operators on $\mathscr{P}(S)$. From this perspective, we better see how closure and interior operators on $\mathscr{P}(S)$ can be used to topologize a set, providing examples on how this can be done.

4.1 The Closure of a Set

If $T = (2,7]$ is viewed as a subset of the topological space $\mathbb{R}$ (equipped with the usual topology), we easily see that it is not closed since its complement, $\mathbb{R}\backslash T = (-\infty, 2] \cup (7, \infty)$, is not open in $\mathbb{R}$. And yet we feel that it wouldn't take very much for us to "make it closed": We need only add the element, 2, to T to obtain the closed subset $T^* = [2,7]$. Adding the *fewest number* of points possible to a set T to obtain a closed set is what we will refer to as obtaining the *closure of* T. The key words here are "fewest number" of points, and no more. In this case, we would say that the "closure of $T = (2,7]$ is the set $T^* = [2,7]$, the smallest closed subset of $\mathbb{R}$ which contains all the elements of T. With this example in mind, we will now formally define a concept called "closure of a subset".

Definition 4.1 Let S be a topological space and $T \subseteq S$. We define the *closure of T in S*, denoted by, $\mathrm{cl}_S T$ (or by $\mathrm{cl}_S(T)$), as

$$\mathrm{cl}_S T = \cap\{F : F \text{ is closed in } S \text{ and } T \subseteq F\}$$

The reader should first be aware of the following verifiable *facts* for any subset, T, of the topological space S.

(1) *The closure of T, $\mathrm{cl}_S T$, is closed in S*: This follows from the fact that arbitrary intersections of closed sets are closed.
(2) *The set $T \subseteq \mathrm{cl}_S T$*: This follows from the definition of closure of T.
(3) *The set, $\mathrm{cl}_S T$, is the smallest closed set which contains T*: Suppose A is a closed set containing T. Then $A \in \{F : F$ is closed in S and $T \subseteq F\}$. Hence, $\mathrm{cl}_S T \subseteq A$.
(4) *If T is closed then $T = \mathrm{cl}_S T$*: This is true since T is the smallest closed set containing T.

Supplementary definition. Let A be a subset of a topological space S.

If x is a point in S such that, for every S-open neighborhood U of x, $U \cap A$ contains some point *other than* x, then we say that...

x is a *cluster point of A*.

The set of all cluster points of A is called the *derived set of A*.

This definition provides us with another way of describing a closed set:

> *The set B is closed if and only if it contains all its cluster points.*

Verification of this fact is left to the reader.

For example, if $B = (1, 3) \cup (3, 5] \cup \{6\}$ in $\mathbb{R}$, the derived set (i.e., the set of all cluster points of B) is $[1, 5]$. The element, 6, is not a cluster point of B since there is an open neighborhood, $(5.5, 7)$ of $\{6\}$

which does not meet other elements of B. The set B is not closed since it doesn't contain the cluster points 1 and 3.

Example 1. Let T be the open interval, $(0, 1)$, viewed as a subset of $\mathbb{R}$ equipped with the usual topology. Then $\text{cl}_{\mathbb{R}}T = [0, 1]$. To prove this we must show that:

(1) $[0, 1]$ is closed by showing that $\mathbb{R}\backslash[0, 1]$ is open.
(2) $\{0, 1\} \subseteq A$ for any closed set A containing the interval $(0, 1)$.

This is left as an exercise.

Example 2. If $\mathbb{Q}$ is the set of all rational numbers then $\text{cl}_{\mathbb{R}}(\mathbb{Q}) = \mathbb{R}$.

Proof. To prove this we will show that, if F is a closed subset of $\mathbb{R}$ such that $\mathbb{Q} \subseteq F$, then $F = \mathbb{R}$.

Suppose F is a closed subset of $\mathbb{R}$ such that $\mathbb{Q} \subseteq F$. Then $\mathbb{R}\backslash F$ is open. Suppose $\mathbb{R}\backslash F \neq \varnothing$. Then $\mathbb{R}\backslash F$ is the union of open intervals each of which must contain a rational number. Since $\mathbb{Q} \cap (\mathbb{R}\backslash F) \neq \varnothing$, this contradicts $\mathbb{Q} \subseteq F$. Then $\mathbb{R}\backslash F = \varnothing$. This means that the only closed set containing $\mathbb{Q}$ is $\mathbb{R}$. So $\text{cl}_{\mathbb{R}}(\mathbb{Q}) = \mathbb{R}$.

Example 3. Suppose S is a topological space induced by the metric ρ (i.e., the elements of τ are unions of open balls of the form $B_\varepsilon(x) = \{y : \rho(x, y) < \varepsilon\}$). Suppose F is a non-empty subset of S. We define

$$\rho(x, F) = \inf \{\rho(x, u) : u \in F\}$$

Show that $\text{cl}_S F = \{x : \rho(x, F) = 0\}$.

Solution: To do this, we must show

(1) $F \subseteq \{x : \rho(x, F) = 0\}$,
(2) $S \setminus \{x : \rho(x, F) = 0\}$ is open in S,
(3) If $F \subseteq A$ where A is a closed subset of S, then $\{x : \rho(x, F) = 0\} \subseteq A$.

The details are left as an exercise.

We now list and prove a few of the most fundamental closure properties.

Theorem 4.2 *Let A and B be two subsets of a topological space (S, τ). Then,*

(1) $cl_S(\varnothing) = \varnothing$.
(2) *If $A \subseteq B$ then $cl_S(A) \subseteq cl_S(B)$.*
(3) $cl_S(A \cup B) = \mathrm{cl}_S(A) \cup \mathrm{cl}_S(B)$ *(Closure "distributes" over finite unions).*
(4) $cl_S(\mathrm{cl}_S(A)) = \mathrm{cl}_S(A)$.

Proof.

(1) Since $\varnothing$ is closed, $\mathrm{cl}_S(\varnothing) \subseteq \varnothing$. Since $\varnothing \subseteq \mathrm{cl}_S\varnothing$, then $\mathrm{cl}_S(\varnothing) = \varnothing$.
(2) We are given that $A \subseteq B$. If F is closed in S and $B \subseteq F$ then $A \subseteq B \subseteq F$. Then

$$A \subseteq \cap\{F : F \text{ is closed in } S \text{ and } B \subseteq F\} = \mathrm{cl}_S(B)$$

By (1) of the *facts* above, $\mathrm{cl}_S(B)$ is closed in S and so

$$\mathrm{cl}_S(A) = \cap\{F : F \text{ is closed in } S \text{ and } A \subseteq F\} \subseteq \mathrm{cl}_S(B)$$

We have shown that $\mathrm{cl}_S(A) \subseteq \mathrm{cl}_S(B)$.
(3) Since $A \subseteq A \cup B$ and $B \subseteq A \cup B$, then $\mathrm{cl}_S(A) \subseteq cl_S(A \cup B)$ and $\mathrm{cl}_S(B) \subseteq cl_S(A \cup B)$ (by parts (1) and (2)). So,

$$\mathrm{cl}_S(A) \cup cl_S[B) \subseteq \mathrm{cl}_S(A \cup B)$$

Since $A \subset \mathrm{cl}_S(A)$ and $B \subset \mathrm{cl}_S(B)$, $A \cup B \subseteq \mathrm{cl}_S(A) \cup \mathrm{cl}_S(B)$, a closed subset in S. Since $\mathrm{cl}_S(A \cup B)$ is the smallest closed set containing $A \cup B$, then

$$\mathrm{cl}_S(A \cup B) \subseteq \mathrm{cl}_S(A) \cup \mathrm{cl}_S[B)$$

We conclude that $\mathrm{cl}_S(A \cup B) = \mathrm{cl}_S(A) \cup \mathrm{cl}_S(B)$.
(4) By part (1) $A \subseteq \mathrm{cl}_S(A) \subseteq \mathrm{cl}_S(\mathrm{cl}_S(A))$. Since $\mathrm{cl}_S(A)$ is closed (see Fact (1)), $\mathrm{cl}_S(\mathrm{cl}_S(A)) \subseteq \mathrm{cl}_S(A)$. It then follows that $\mathrm{cl}_S(\mathrm{cl}_S(A)) = \mathrm{cl}_S(A)$.

Example 4. Closure does not distribute over intersections. Suppose $A = (2, 5)$ and $B = (5, 7)$. Then $\mathrm{cl}_{\mathbb{R}}(A \cap B) = \mathrm{cl}_{\mathbb{R}}(\varnothing) = \varnothing$. On the other hand, $\mathrm{cl}_{\mathbb{R}}(A) = [2, 5]$ and $\mathrm{cl}_{\mathbb{R}}(B) = [5, 7]$ hence,

$\mathrm{cl}_{\mathbb{R}}(A) \cap \mathrm{cl}_{\mathbb{R}}(B) = \{5\}$. This shows that $\mathrm{cl}_S(A) \cap \mathrm{cl}_S(B) \neq \mathrm{cl}_S(A \cap B)$ may sometimes occur.

It is however possible to prove that

$$\mathrm{cl}_S(A \cap B) \subseteq \mathrm{cl}_S(A) \cap \mathrm{cl}_S(B)$$

Proving this is left as an exercise.

Remark on closures of arbitrary unions. We have seen in Theorem 4.2 that $\mathrm{cl}_S(A \cup B) = \mathrm{cl}_S A \cup \mathrm{cl}_S B$, so the "closure distributes over finite unions". This does not hold true for arbitrary unions. Consider the sets of the form

$$A_i = \left(\frac{1}{i}, 3\right] \text{ where } \mathrm{cl}_{\mathbb{R}} A_i = \left[\frac{1}{i}, 3\right]$$

for $i = 1, 2, 3, \ldots$ Verify that $\cup\{\mathrm{cl}_{\mathbb{R}}(A_i) : i = 1, 2, 3, \ldots\} = (0, 3]$ (left as an exercise).

Since $\mathrm{cl}_{\mathbb{R}}[\cup\{A_i : i = 1, 2, 3, \ldots\}] = [0, 3]$ (left as an exercise), then

$$\cup\{\mathrm{cl}_{\mathbb{R}}(A_i) : i = 1, 2, 3, \ldots\} \neq \mathrm{cl}_{\mathbb{R}}[\cup\{A_i : i = 1, 2, 3, \ldots\}]$$

4.2 Closure Viewed as an Operator on $\mathscr{P}(S)$

The closure of a set can be viewed as an action

$$\mathrm{cl}_S : \mathscr{P}(S) \to \mathscr{P}(S)$$

performed on a set. It takes an arbitrary set, A, and associates to it another set, $\mathrm{cl}_S(A)$, obtained by adding sufficiently many points (but no more) so as to produce a "closed set". It can then be viewed as a function. With this in mind, we define the *Kuratowski closure operator.*

Definition (Kuratowski closure operator). Suppose S is a non-empty set and $K : \mathscr{P}(S) \to \mathscr{P}(S)$ is a function which satisfies the four conditions:

K1: $K(\varnothing) = \varnothing$ and $A \subseteq K(A)$
K2: If $A \subseteq B$ then $K(A) \subseteq K(B)$

K3: $K(A \cup B) = K(A) \cup K(B)$
K4: $K(K(A)) = K(A)$

A function, $K : \mathscr{P}(S) \to \mathscr{P}(S)$, satisfying these four properties is referred to as a *Kuratowski closure operator* where K1–K4 are the *Kuratowski closure operator axioms.*

Topologizing a set S by using a closure operator: The reader should notice that, in the above definition, the set S is not described as being a "topological space" since no topology is defined on it. It is just a set. The following theorem shows that, if we are given a Kuratowski operator, $K : \mathscr{P}(S) \to \mathscr{P}(S)$, on $\mathscr{P}(S)$ then we can use K to generate a topology, τ_K, on S such that

$$\mathrm{cl}_S A = K(A)$$

for all $A \in \mathscr{P}(S)$.

Theorem 4.3 *Let S be a set and suppose $K : \mathscr{P}(S) \to \mathscr{P}(S)$ satisfies the four* Kuratowski closure operator axioms. *Define $\mathscr{F} \subseteq \mathscr{P}(S)$ as*

$$\mathscr{F} = \{A \subseteq S : K(A) = A\}$$

(a) *Then $\mathscr{F}$, is the set of all closed subsets of some topology, τ_K, on S. That is,*

$$\tau_K = \{S \backslash A : \text{ where } A \in \mathscr{F}\}$$

(b) *Furthermore, in (S, τ_K), $cl_S(A) = K(A)$, for any $A \subseteq S$.*

Proof. Given: The operator $K : \mathscr{P}(S) \to \mathscr{P}(S)$. Also, $\mathscr{F} = \{A \subseteq S : K(A) = A\}$.

(a) To prove the statement (a), it will suffice to show that $\mathscr{F}$ satisfies the three "closed sets" conditions F1, F2, and F3 on page 46. If so, then we can define τ_K by invoking the statement of the proposition on page 46.

— Note that $\{\varnothing, S\} \subseteq \mathscr{F}$. To see this note that, by K1, $S \subseteq K(S) \subseteq S \Rightarrow K(S) = S$. Both $\varnothing$ and S belong to $\mathscr{F}$. The set $\mathscr{F}$ satisfies condition F1.

— Suppose $\mathscr{U} = \{F_i : i \in I\}$ is an arbitrarily large family of sets in $\mathscr{F}$. By K1, $\cap\mathscr{U} \subseteq K(\cap\mathscr{U})$. Also, by K2, $K(\cap\mathscr{U}) \subseteq K(F_i) = F_i$ for each $i \in I$. So $K(\cap\mathscr{U}) \subseteq \cap\{F_i : i \in I\} = \cap\mathscr{U}$. Hence, $K(\cap\mathscr{U}) \subseteq \cap\mathscr{U}$. So $K(\cap\mathscr{U}) = \cap\mathscr{U}$. We then have $K(\cap\mathscr{U}) \in \mathscr{F}$. The set $\mathscr{F}$ is then closed under arbitrary intersections. The set $\mathscr{F}$ satisfies condition F2.

— We now show that $\mathscr{F}$ is closed under finite unions. We must show that if A and $B \in \mathscr{F}$ then $K(A \cup B) = A \cup B$. Consider $A, B \in \mathscr{F}$. By K2,

$$A \cup B \subseteq K(A \cup B)$$

$$\begin{aligned} A \cup B \subseteq K(A) \cup K(B) &\Rightarrow K(A \cup B) \subseteq K(K(A) \cup K(B)) \text{ (By K2)} \\ &\Rightarrow K(A \cup B) \subseteq K(K(A)) \cup K(K(B)) \text{ (By K3.)} \\ &\Rightarrow K(A \cup B) \subseteq A \cup B \text{ (By K4.)} \end{aligned}$$

We conclude that $K(A \cup B) = K(A) \cup K(B)$. The set $\mathscr{F}$ satisfies condition F3.

So $\mathscr{F}$ is the set of all closed sets in S. This means the topology τ_K on S is

$$\tau_K = \{S \backslash A : \text{ where } A \subseteq \mathscr{F}\}$$

Hence, the set, $\mathscr{F} = \{A \subseteq S : K(A) = A\}$, represents all closed subsets of S (with respect to τ_K).

(b) We now prove the second statement, $\text{cl}_S(A) = K(A)$.
Let $A \subseteq (S, \tau_K)$. Then $\text{cl}_S(A) \in \mathscr{F}$. So $K(\text{cl}_S(A)) = \text{cl}_S(A)$. We claim that, from this we can obtain $\text{cl}_S(A) = K(A)$.

Proof of claim:

Let $A \subseteq S$.

$$\begin{aligned} K(K(A)) &= K(A) \quad \text{(By K4)} \\ &\Rightarrow K(A) \in \mathscr{F} \\ &\Rightarrow S \backslash K(A) \in \tau_K \\ &\Rightarrow K(A) \text{ is closed with respect to } \tau_K \\ &\Rightarrow \text{cl}_S(A) \subseteq K(A) \quad \text{(By K2, } A \subseteq K(A).) \end{aligned}$$

$$\begin{aligned} A \subseteq \mathrm{cl}_S(A) &\Rightarrow K(A) \subseteq K(\mathrm{cl}_S(A)) \quad \text{(By K2)} \\ &\Rightarrow K(A) \subseteq \mathrm{cl}_S(A) \quad (\text{Since } K(\mathrm{cl}_S(A)) = \mathrm{cl}_S(A)) \end{aligned}$$

We conclude that $\mathrm{cl}_S(A) = K(A)$, as claimed.

We have shown that any Kuratowski closure operator $K : \mathscr{P}(S) \to \mathscr{P}(S)$ can be used to construct a topology, τ_K, on S in such a way that, for any $A \subseteq S$, $K(A) = \mathrm{cl}_S(A)$. We illustrate this in the following example.

Example 5. We consider the set $\mathbb{R}^2$, a set with uncountably many elements. We define a function $K : \mathscr{P}(\mathbb{R}^2) \to \mathscr{P}(\mathbb{R}^2)$ as follows:

$$K(A) = A \text{ if } A \text{ is countable}$$

$$K(A) = \mathbb{R}^2 \text{ if } A \text{ is uncountable}$$

$$K(\varnothing) = \varnothing$$

Show that K is a Kuratowski operator. Then find the topology on $\mathbb{R}^2$ induced by the operator K.

Solution: Proving that K satisfies the properties K1–K4 is routine and so is left as an exercise. Then, thus defined, K is a Kuratowski closure operator. By Theorem 4.3 the set

$$\mathscr{F} = \{A \subseteq \mathbb{R}^2 : K(A) = A\} = \{A \subseteq \mathbb{R}^2 : A \text{ is countable}\} \cup \{\varnothing, \mathbb{R}^2\}$$

represents the set of all closed subsets of the topological space $(\mathbb{R}^2, \tau_K)$. We deduce that

$$\tau_K = \{B \subseteq \mathbb{R}^2 : \mathbb{R}^2 \backslash B \text{ is countable}\} \cup \{\varnothing, \mathbb{R}^2\}$$

Definition 4.4 We will refer to τ_K in this example as the *cocountable topology* on $\mathbb{R}^2$.

4.3 The Interior of a Set

Given a non-empty set A the "closure of A" has been defined as being the intersection of all closed subsets of S which contain A. We now wish to consider the union of all open subsets of S which are entirely contained in A.

Definition 4.5 Let A be a non-empty subset of the topological space (S, τ). We say that a point x is an *interior point* of A if there exist an open subset, U, of S such that $x \in U \subseteq A$. We define the *interior of A*, denoted $\text{int}_S A$ (or as $\text{int}_S(A)$) as follows:

$$\text{int}_S A = \{x \in S : x \in U \subseteq A \text{ for some open } U \text{ in } S.\}$$

If A contains no interior points then we will say that the interior, $\text{int}_S A$, of A is empty.

Clearly $\text{int}_S A \subseteq A$. For example, if $A = (2, 4] \cup \{5\}$, $\text{int}_{\mathbb{R}} A = (2, 3)$. The element 5 does not belong to $\text{int}_{\mathbb{R}} A$ since there is no open interval U in $\mathbb{R}$ such that $5 \in U \subseteq A$. The fact that $\{5\}$ is open in A is irrelevant.

The reader is left to verify that $\text{int}_S A$ can equivalently be described as being

the largest open subset of S which is entirely contained in A.

The following theorem shows a relationship between the interior and closure of a set. It also proposes a method to determine the interior of a set, A, by considering the closure of its complement, $S \backslash A$.

Theorem 4.6 *Let (S, τ) be a topological space and A be a subset of S. Then,*

$$S \setminus \text{int}_S(A) = \text{cl}_S(S \backslash A)$$

Proof. Given: (S, τ) is a topological space and A is a subset of S.

Since $\text{int}_S(A) \subseteq A$, then $S \backslash A \subseteq S \backslash \text{int}_S(A)$. But $S \setminus \text{int}_S(A)$ is closed in S, hence,

$$\text{cl}_S(S \backslash A) \subseteq S \setminus \text{int}_S(A)$$

Also,

$$\begin{aligned} S \backslash A \subseteq \text{cl}_S(S \backslash A) \quad &\Rightarrow \quad S \backslash (\text{cl}_S(S \backslash A)) \subseteq A \\ &\Rightarrow \quad S \backslash (\text{cl}_S(S \backslash A)) \subseteq \text{int}_S(A) \\ &\Rightarrow \quad S \backslash \text{int}_S(A) \subseteq (\text{cl}_S(S \backslash A)) \end{aligned}$$

We thus obtain $S \setminus \text{int}_S(A) = \text{cl}_S(S \backslash A)$

Using this theorem, we let the reader verify that the following three statements are equivalent:

(a) $\text{int}_S(A) = S \backslash \text{cl}_S(S \backslash A)$,

(b) $\text{int}_S(S \backslash A) = S \backslash \text{cl}_S(A)$,

(c) $\text{cl}_S(A) = S \backslash (\text{int}_S(S \backslash A))$.

Just as for the closure of a set, we have four basic similar properties for the interior of sets.

Theorem 4.7 *Let (S, τ) be a topological space and A and B be subsets of S.*

(a) *The set,* $\text{int}_S(A)$, *is open in S. Also,* $\text{int}_S(A)$ *is the largest open subset of S which is entirely contained in A.*
(b) *If $B \subseteq A$ then* $\text{int}_S(B) \subseteq \text{int}_S(A)$.
(c) *The set* $\text{int}_S(A \cap B) = \text{int}_S(A) \cap \text{int}_S(B)$. *(Int$_S$ "distributes" over finite intersections.)*
(d) *The set* $\text{int}_S(\text{int}_S(A)) = \text{int}_S(A)$.

Proof. The proofs of statements (a), (b), and (d) are left as an exercise.

Proof of $\text{int}_S(A \cap B) = \text{int}_S(A) \cap \text{int}_S(B)$:

$$
\begin{aligned}
S \backslash \text{int}_S(A \cap B) &= \text{cl}_S(S \backslash (A \cap B)) \\
&= \text{cl}_S[(S \backslash A) \cup (S \backslash B)] \\
&= \text{cl}_S(S \backslash A) \cup \ \text{cl}_S(S \backslash B) \\
&= [S \backslash \text{int}_S(A)] \cup [S \backslash \text{int}_S(B)] \\
&= S \backslash [\text{int}_S(A) \cap \text{int}_S(B)] \\
&\Rightarrow \\
\text{int}_S(A \cap B) &= \text{int}_S(A) \cap \text{int}_S(B)
\end{aligned}
$$

Example 6. Given that $\mathbb{R}$ is equipped with the usual topology what is $\text{int}_{\mathbb{R}}(\mathbb{Q})$?

Solution: We consider the subset, $\mathbb{Q}$, of all rationals in $\mathbb{R}$. By Theorem 4.7 part (a), $\text{int}_{\mathbb{R}}(\mathbb{Q}) \subseteq \mathbb{Q}$. If $\text{int}_{\mathbb{R}}(\mathbb{Q})$ is non-empty, it should be a union of non-empty open intervals. But every open interval in $\mathbb{R}$ contains an irrational; then $\text{int}_{\mathbb{R}}(\mathbb{Q}) = \varnothing$.

Example 7. If $\mathbb{R}$ is equipped with the usual topology, then $\text{int}_{\mathbb{R}}([0, 1]) = (0, 1)$.

Example 8. The set $\text{int}_{\mathbb{R}}(A \cup B)$ need *not* be equal to $\text{int}_{\mathbb{R}}(A) \cup \text{int}_{\mathbb{R}}(B)$: If $\mathbb{R}$ is equipped with the usual topology, then

$$\text{int}_{\mathbb{R}} \left([0, 1] \cup [1, 2] \right) = (0, 2)$$

while

$$\text{int}_{\mathbb{R}}[0, 1] \ \cup \text{int}_{\mathbb{R}}[1, 2] = (0, 1) \cup (1, 2)$$

4.4 The Interior Viewed as an Operator $\text{int}_S : \mathscr{P}(S) \to \mathscr{P}(S)$

Just as for closures of sets we can view "int_S" as a function, $I : \mathscr{P}(S) \to \mathscr{P}(S)$.

Definition 4.8 Let S be a non-empty set and $I : \mathscr{P}(S) \to \mathscr{P}(S)$ be a function satisfying the properties:

I1: $I(S) = S$
I2: $I(A) \subseteq A$, for all $A \subset S$
I3: $I(I(A)) = I(A)$, for all $A \subset S$
I4: $I(A \cap B) = I(A) \cap I(B)$, for all $A, B \in \mathscr{P}(S)$ (I distributes over finite intersections).

The function $I : \mathscr{P}(S) \to \mathscr{P}(S)$ satisfying the listed properties is called an *interior operator.* We refer to I1–I4 as being the *interior operator axioms.*

Topologizing a set S by using an interior operator. Again, just as for the closure operator, the definition of the function, I : $\mathscr{P}(S) \to \mathscr{P}(S)$, doesn't refer to any topology on S. But we will show that the function, I, can be used to define a topology on S by choosing appropriate sets in its range. We desire a topology, τ, such that A is open in S if and only if $A = \text{int}_S A = I(A)$.

Theorem 4.9 *Let S be a set and suppose $I : \mathscr{P}(S) \to \mathscr{P}(S)$ satisfies the four* interior operator axioms. *Define $\mathscr{U} \subseteq \mathscr{P}(S)$ as*

$$\mathscr{U} = \{A \subseteq S : I(A) = A\}$$

Then,

(a) *The set $\mathscr{U}$ forms a topology on S.*
(b) *Furthermore, if S is equipped with topology $\mathscr{U}$,* $\text{int}_S(A) = I(A)$, *for any $A \subseteq S$.*

Proof. Let S be a set and $I : \mathscr{P}(S) \to \mathscr{P}(S)$ be a function satisfying the four *interior operator axioms* I1–I4 listed above. Let $\mathscr{U} = \{A \in \mathscr{P}(S) : I(A) = A\}$.

(a) We are required to prove that $\mathscr{U}$ forms a topology on S.
We see that:

— By I1, $I(S) = S$, so $S \in \mathscr{U}$. By I2, $I(\varnothing) \subseteq \varnothing$. Since $\varnothing \subseteq I(\varnothing)$, then $I(\varnothing) = \varnothing$ and so $\varnothing \in \mathscr{U}$.

— Suppose A and B belong to $\mathscr{U}$. By property I4, $I(A \cap B) = I(A) \cap I(B) = A \cap B$. So $\mathscr{U}$ is closed under finite intersections.
— To show that $\mathscr{U}$ is closed under arbitrary unions we first verify that

$$(A \subseteq B) \Rightarrow (I(A) \subseteq I(B)) \quad (*)$$

$$\begin{aligned} A \subseteq B &\Rightarrow A = B \cap A \\ &\Rightarrow I(A) = I(B \cap A) = I(B) \cap I(A) \\ &\Rightarrow I(A) \subseteq I(B) \end{aligned}$$

Let $\{A_i\}_{i \in M}$ be a collection of sets in $\mathscr{U}$. It suffices to show $I(\cup\{A_i\}_{i \in M}) = \cup\{A_i\}_{i \in M}$.

$$\begin{aligned} I(A_i) \subseteq \cup\{I(A_i)\}_{i \in M} &\Rightarrow I(I(A_i)) \subseteq I(\cup\{I(A_i)_{i \in M}) \quad \text{(By *)} \\ &\Rightarrow I(A_i) \subseteq I(\cup\{I(A_i)\}_{i \in M}) \\ &\qquad \text{(since } A_i \in \mathscr{U} \text{ for all } i \in M) \\ &\Rightarrow \cup\{I(A_i)\}_{i \in M} \subseteq I(\cup\{I(A_i)\}_{i \in M}) \end{aligned}$$

By I2, $I(\cup\{I(A_i)\}_{i \in M}) \subseteq \cup\{I(A_i)\}_{i \in M}$. Then $I(\cup\{I(A_i)\}_{i \in M}) = \cup\{I(A_i)\}_{i \in M}$ so $\cup\{I(A_i)\}_{i \in M} \in \mathscr{U}$.

Then set $\mathscr{U}$ satisfies the three open set axioms O1, O2, and O3. So $\mathscr{U}$ is a topology on S, as required.

We will denote the topology $\mathscr{U}$ on S induced by the operator I, by τ_I.

(b) We are now required to show that $I(A) = \text{int}_S(A)$ with respect to τ_I.

Suppose $A \subseteq S$.

By I3, $I(I(A)) = I(A)$, so $I(A) \in \mathscr{U} = \tau_I$; so $I(A)$ is open. Since $\text{int}_S(A)$ is the largest open subset of S contained in A, and $I(A) \subseteq A$ (by I2),

$$I(A) \subseteq \text{int}_S(A)$$

Also see that, since $\text{int}_S(A) \in \tau_I = \mathscr{U}$ and $\text{int}_S(A) \subseteq A$,

$$\begin{aligned} \text{int}_S(A) &= I(\text{int}_S(A)) \\ &\subseteq I(A) \quad \text{(By * above } A \subseteq B \Rightarrow I(A) \subseteq I(B)) \end{aligned}$$

then $\text{int}_S(A) \subseteq I(A)$.

We conclude that $I(A) = \text{int}_S(A)$.

We provide a few examples.

Example 9. Let $I : \mathscr{P}(\mathbb{R}) \to \mathscr{P}(\mathbb{R})$ be a function defined as

$$I(\mathbb{R}) = \mathbb{R}$$
$$I(A) = A\backslash\mathbb{Q} \quad \text{otherwise}$$

(a) Show that $I : \mathscr{P}(\mathbb{R}) \to \mathscr{P}(\mathbb{R})$, thus defined, is an interior operator on $\mathscr{P}(\mathbb{R})$.
(b) Use the interior operator described in part (a) to define a topology, τ_I, on $\mathbb{R}$.
(c) For the topology, τ_I, on $\mathbb{R}$ shown in part (b), describe the open subsets, the closed subsets of $\mathbb{R}$, the closure of sets and the interior of sets.
(d) A *Borel sets* example: If $\mathscr{F}$ represents the set of all closed subsets with respect to the topology, τ_I, on $\mathbb{R}$, show that

$$\mathscr{B} = \tau_I \cup \mathscr{F}$$

is the smallest σ-ring containing τ_I and so is a family of Borel sets.

Solution:

(a) We show I is an interior operator.

1. By definition, $I(\mathbb{R}) = \mathbb{R}$ so I1 is satisfied.
2. Also, if $A \neq \mathbb{R}$, $I(A) = A\backslash\mathbb{Q} \subseteq A$. So I2 is satisfied.
3. If $A \neq \mathbb{R}$

$$\begin{aligned} I(I(A)) &= I(A)\backslash\mathbb{Q} \\ &= (A\backslash\mathbb{Q})\backslash\mathbb{Q} \\ &= A\backslash\mathbb{Q} \\ &= I(A) \end{aligned}$$

So, $I(I(A)) = A$. So I3 is satisfied.

4. If neither A nor B is $\mathbb{R}$,

$$\begin{aligned} I(A \cap B) &= (A \cap B) \backslash \mathbb{Q} \\ &= (A \backslash \mathbb{Q}) \cap (B \backslash \mathbb{Q}) \\ &= I(A) \cap I(B) \end{aligned}$$

So, $I(A \cap B) = I(A) \cap I(B)$. Then I4 is satisfied.

This means that $I : \mathscr{P}(\mathbb{R}) \to \mathscr{P}(\mathbb{R})$ is an interior operator.

(b) We now use the interior operator described in part (a) to topologize $\mathbb{R}$.
Since $I : \mathscr{P}(\mathbb{R}) \to \mathscr{P}(\mathbb{R})$ has been shown to be an interior operator, then

$$\begin{aligned} \tau_I &= \{A : I(A) = A\} \\ &= \{A : A \backslash \mathbb{Q} = A\} \cup \{\mathbb{R}\} \\ &= \{A : A \text{ does not contain any points of } \mathbb{Q}\} \cup \{\mathbb{R}\} \end{aligned}$$

is a topology on $\mathbb{R}$.

(c) For the topology, τ_I, on $\mathbb{R}$ we now describe the open subsets, the closed subsets of $\mathbb{R}$, the closure of sets and the interior of sets.

Open sets in $\mathbb{R}$. Open sets in $\mathbb{R}$, are $\mathbb{R}$ itself and all sets which do not contain any rationals, including $\varnothing$.

For example, if $\mathbb{J}$ is the set of irrationals and $r \in \mathbb{J}$, then, since $\{r\}$ contains no rationals, $\{r\}$ is an open singleton set. Also, if $q \in \mathbb{Q}$, $\mathbb{R}$ is the only open set containing q. So, $\{q\}$ is *not* an open singleton set.

Closed sets in $\mathbb{R}$. Suppose B is not $\mathbb{R}$. We claim that B is closed in $\mathbb{R}$ with respect to τ_I if and only if $\mathbb{Q} \subseteq B$:

$$\begin{aligned} \mathbb{Q} \subseteq B &\Leftrightarrow \mathbb{R} \backslash B = (\mathbb{R} \backslash B) \backslash \mathbb{Q} \\ &\Leftrightarrow \mathbb{R} \backslash B = I(\mathbb{R} \backslash B) \\ &\Leftrightarrow \mathbb{R} \backslash B \text{ is open (with respect to } \tau_I) \\ &\Leftrightarrow B \text{ is closed (with respect to } \tau_I) \end{aligned}$$

For example, if $r \in \mathbb{J}$, since $\mathbb{Q} \not\subseteq \{r\}$, the singleton set, $\{r\}$, is *not* a closed set.

Closure of a set. Then taking the closure of a subset A of $\mathbb{R}$ comes down to adding all of $\mathbb{Q}$ to A. That is, if $A \neq \mathbb{R}$, $\text{cl}_{\mathbb{R}} A = A \cup \mathbb{Q}$. For example, if $q \in \mathbb{Q}$, since $\text{cl}_{\mathbb{R}}\{q\} = \mathbb{Q}$, $\{q\}$ is *not* closed.

Interior of a set in $\mathbb{R}$. $\text{Int}_{\mathbb{R}} A = A \cap \mathbb{J}$. Finding the interior of A comes down to removing any trace of $\mathbb{Q}$ in A.

(d) We are required to show: The set, $\mathscr{B} = \tau_I \cup \mathscr{F}$, is the smallest σ-ring containing τ_I and so is a family of Borel sets of $\mathbb{R}$.

Given: $\mathscr{F}$ is the set of all closed subsets in $\mathbb{R}$ with respect to τ_I. We claim that $\mathscr{B}$ is a σ-ring.

Closure of $\mathscr{B}$ under countable unions. If $U \in \tau_I$ and $V \in \mathscr{F}$ then $U \cup V \in \mathscr{F}$ (since an open subset union a closed subset containing $\mathbb{Q}$ contains $\mathbb{Q}$. So $U \cup V$ is a closed subset.)

The set $\mathscr{F}$ is closed under countable unions (since closed subsets are those subsets of $\mathbb{R}$ which contain all of $\mathbb{Q}$, arbitrary unions of elements of $\mathscr{F}$ are closed with respect to τ_I). This actually show that $\mathscr{F}$ contains all its F_σ's. Since all F_σ's are closed, then all G_δ's are open and so belong to τ_I.

Trivially, τ_I is closed under arbitrary unions and so is closed under countable unions.

Closure of $\mathscr{B}$ under "complements". Let $U \in \mathscr{B}$. If $U \in \tau_I$ then $\mathbb{R}\backslash U \in \mathscr{F}$. If $U \in \mathscr{F}$ then $\mathbb{R}\backslash U \in \tau_I$.

So, $\mathscr{B}$ is a σ-ring, as claimed. Since it must contains τ_I and all complements, it is the smallest σ-ring containing τ_I. By definition, it is a family of Borel sets of $\mathbb{R}$ with respect to τ_I.[1]

4.5 The Boundary of the Subset of a Space

We have seen that, for any subset, A, of a topological space (S, τ),

$$\text{int}_S(A) \subseteq A \subseteq \text{cl}_S(A)$$

[1]Note that, if T is the closed interval [1, 3] in $\mathbb{R}$ then $T \notin \tau_I \cup \mathscr{F}$ (since $[1, 3]$ contains some elements of $\mathbb{Q}$, but does not contain all of $\mathbb{Q}$) and so is not an element of the unique family, $\mathscr{B}$, of all Borel sets, with respect to τ_I. We can then refer to it as a *non-Borel set.*

Often, $\text{cl}_S(A) \setminus \text{int}_S(A) \neq \varnothing$. We will now briefly discuss those sets whose points belong to $\text{cl}_S(A)\setminus\text{int}_S(A)$.

Definition 4.10 Let A be a subset of a topological space (S, τ). We define the *boundary of* A, denoted as $\text{bd}_S(A)$, as

$$bd_S(A) = \text{cl}_S(A) \setminus \text{int}_S(A)^2$$

The expressions, $\text{Fr}_S(A)$, $\partial_S(A)$, $\text{Bd}_S(A)$ are also commonly used to represent the "boundary of A". It is easily verified that

$$\text{bd}_S(A) = \text{cl}_S(A) \cap \text{cl}_S(S\setminus\text{int}_S A)$$

Since the finite intersection of closed sets is closed, we see that the boundary, $\text{bd}_S(A)$ of a set A, is always closed. Furthermore, for any set A in S, both A and $S \setminus A$ share the same boundary (like adjacent neighbors who share the same fence). It is always the case that $\text{int}_S(A) \cap \text{bd}_S(A) = \varnothing$ and that $\text{int}_S(A)$, $\text{bd}_S(A)$ and $\text{int}_S(\text{cl}_S(S\setminus A))$ are pairwise disjoint sets whose union is S. The reader is left to verify this.

Example 10. If $\mathbb{Q}$ is viewed as a subset of $\mathbb{R}$ equipped with the usual topology then

$$\begin{aligned}\text{bd}_{\mathbb{R}}(\mathbb{Q}) &= \text{cl}_{\mathbb{R}}(\mathbb{Q})\setminus \text{int}_{\mathbb{R}}(\mathbb{Q})\\ &= \mathbb{R}\setminus\varnothing\\ &= \mathbb{R}\end{aligned}$$

Example 11. If $B = [0, 1]$ is a closed interval viewed as a subset of $\mathbb{R}$ equipped with the usual topology then $\text{bd}_{\mathbb{R}}(B) = \{0, 1\}$. It is left to the reader to verify this.

Example 12. Let $B = \{(x, y) \in \mathbb{R}^2 : x^2 + y^2 < 1\}$be a subset of $\mathbb{R}^2$ equipped with the topology induced by the Euclidean metric. Then

$$\text{bd}_{\mathbb{R}^2}(B) = \text{cl}_{\mathbb{R}^2}(B)\setminus\text{int}_{\mathbb{R}^2}(B) = \{(x, y) \in \mathbb{R}^2 : x^2 + y^2 = 1\}$$

The reader is left to verify the details.

---

[2]The word "Frontier" is also sometimes used instead of "boundary". When the term "Frontier of A" is used, it is denoted by $\text{Fr}_S A$.

Example 13. Suppose the set of natural numbers $\mathbb{N}$ is equipped with the cofinite topology.[3] Let E denote the set of all even natural numbers. Then

$$\begin{aligned}\mathrm{bd}_{\mathbb{N}}(E) &= \mathrm{cl}_{\mathbb{N}}(E)\backslash\mathrm{int}_{\mathbb{N}}(E)\\ &= \mathbb{N}\backslash\varnothing\\ &= \mathbb{N}\end{aligned}$$

Example 14. Suppose $B = (0,1) \cup (1,2]$ viewed as a subset of $\mathbb{R}$ equipped with the usual topology. We compute the boundary to be,

$$\begin{aligned}\mathrm{bd}_{\mathbb{R}}(B) &= \mathrm{cl}_{\mathbb{R}}(B)\backslash\mathrm{int}_{\mathbb{R}}(B)\\ &= [0,2] \setminus (0,1) \cup (1,2)\\ &= \{0,2\}\end{aligned}$$

It is interesting to note that

$$\begin{aligned}\mathrm{int}_{\mathbb{R}}(\mathrm{cl}_{\mathbb{R}}(B)) &= \mathrm{int}_{\mathbb{R}}([0,2])\\ &= (0,2) \neq B\end{aligned}$$

4.6 On Dense Subsets of a Topological Space

Suppose B and A are subsets of (S,τ) such that $A \subseteq B$ and $B\backslash A$ contains only boundary points of A. When two sets relate to each other in this way we say that "A is dense in B". We define this formally.

Definition 4.11 Suppose A and B are subsets in a topological space (S,τ). If

$$A \subseteq B \text{ and } B \subseteq \mathrm{cl}_S(A)$$

[3]The open subsets are those whose complement is finite.

then we will say that

A is a dense subset of B.

In the case where $B = S$, then A is dense in S if and only if $\text{cl}_S A = S$.

If $A \subseteq S$ is such that $\text{int}_S \text{cl}_S A = \varnothing$ then we say that A is *nowhere dense in S.*

Example 15. Suppose

$$A = \{(x, y) : x^2 + y^2 < 1\} \text{ and } B = A \cup \{(0,1), (1,0), (0,-1), (-1,0)\}$$

Since

$$A \subseteq B \text{ and } B \subset \text{cl}_{\mathbb{R}^2} A = \{(x, y) : x^2 + y^2 \leq 1\}$$

then A is dense in B.

Example 16. *A nowhere dense subset.* Let $A = \{6\}$ be a subset of $\mathbb{R}$. Then $\text{cl}_{\mathbb{R}} A = \{6\}$ and $\text{int}_S \text{cl}_S A = \varnothing$. So A is nowhere dense in the space $\mathbb{R}$.

Another example: The set of all integers, $\mathbb{Z}$, is nowhere dense in $\mathbb{R}$, since $\text{int}_{\mathbb{R}} \text{cl}_{\mathbb{R}} \mathbb{Z} = \varnothing$.

Example 17. Verify that the set $C = \{(x, y) : x^2 + y^2 = 1\}$ is nowhere dense in $\mathbb{R}^2$.

On countable dense subsets of a space. The property in the following definition is one which refers to an upper bound for the cardinality of a dense subset of a topological space. Hence, in a way it expresses a restriction on its size.

Definition 4.12 We will say that a topological space (S, τ) is *separable* if and only if S contains a countable dense subset.

For example, the topological space, $(\mathbb{R}, \tau)$, equipped with the usual topology is *a separable space* since, $\text{cl}_{\mathbb{R}} \mathbb{Q} = \mathbb{R}$, and the subset of all rational numbers is a countable subset of $\mathbb{R}$. Note how the border of $\mathbb{Q}$ is much larger than $\mathbb{Q}$ itself.

We will subdivide the class of all topological spaces into two subclasses: Those that are separable and those that are not.

4.7 Topic: Regular Open Sets and Regular Closed Sets

Suppose A is a non-empty open subset of a topological space (S, τ). We know that $A \subseteq \text{cl}_S A$ is always true. Also, it is always true that, if A is open, $A = \text{int}_S A \subseteq \text{int}_S\text{cl}_S A$ (by part b) of Theorem 4.7). So it is always true that

$$A \in \tau \Rightarrow A \subseteq \text{int}_S\text{cl}_S A \quad (\dagger)$$

One may wonder whether, when A is open, must we have equality between A and $\text{int}_{\mathbb{R}}\text{cl}_{\mathbb{R}} A$. The following simple example will help answer this question.

Let $A = (1,3) \cup (3,7)$, an open subset of $\mathbb{R}$. Then by applying the above reasoning we have,

$$A = (1,3) \cup (3,7) \subseteq \text{int}_{\mathbb{R}}\text{cl}_{\mathbb{R}} A = \text{int}_{\mathbb{R}}[1,7] = (1,7)$$

So in the case where $A = (1,7)$, we do have equality; but if $A = (1,3) \cup (3,7)$ we have an open subset, A, such that $A \neq \text{int}_{\mathbb{R}}\text{cl}_{\mathbb{R}} A$. Then the best we can then do is affirm that, when A is open, A is a subset of $\text{int}_S\text{cl}_S A$. We have a special name for those open subsets where equality holds.

Definition 4.13 An open subset, A, of a topological space (S, τ) is called a *regular open subset of* S if and only if $A = \text{int}_S\text{cl}_S A$. In this book, we denote the set of all regular open subsets of S by $\mathscr{R}o(S)$.

We know that for an open set A, it is always true that, $A \subseteq \text{int}_S\text{cl}_S A$. Then

$$\begin{aligned} S\backslash(\text{int}_S\text{cl}_S A) \subseteq S\backslash A &\Rightarrow \text{cl}_S[S\backslash(\text{cl}_S A)] \subseteq S\backslash A \\ &\Rightarrow \text{cl}_S\text{int}_S(S\backslash A) \subseteq S\backslash A \end{aligned}$$

So it is always true that,

$$F \text{ closed } \Rightarrow \text{cl}_S\text{int}_S(F) \subseteq F$$

Then, if A is regular open, $A = \text{int}_S\text{cl}_S A$, so for the complement, $F = S\backslash A$, we can say that

$$F = \text{cl}_S\text{int}_S(F)$$

So, $F = \text{cl}_S\text{int}_S(F)$ if and only if the complement of F is regular open.

We have a name for the complements of regular open subsets.

Definition 4.14 A closed subset, F, of a topological space (S, τ) is called a *regular closed subset of* S if and only if

$$F = \text{cl}_S\,\text{int}_S F$$

Hence, the regular closed subsets of S are precisely the complements of the regular open subsets. We denote the set of all regular closed subsets of S by $\mathscr{Rc}(S)$.

Example 18. Let S be a topological space and $\mathscr{Ro}(S)$ be the set of all regular open sets in S.

(a) Verify that $\mathscr{Ro}(S)$ is closed under finite intersections but is not closed under finite unions.
(b) We know that, if $A = \text{int}_S\text{cl}_S A$, then A belongs to $\mathscr{Ro}(S)$. Suppose A is some non-empty subset of S which does not belong to $\mathscr{Ro}(S)$. Verify that $\text{int}_S\text{cl}_S A$ belongs to $\mathscr{Ro}(S)$.
(c) If $\mathscr{Rc}(S)$ denotes the set of all regular closed sets in S, show that $\mathscr{Rc}(S)$ is closed under finite unions.

Solution: Given: $\mathscr{Ro}(S)$ is the set of all regular open sets in S.

(a) Suppose A and B are open subsets. Then $A \cap B$ is open. Suppose $A = \text{int}_S\text{cl}_S A$ and $B = \text{int}_S\text{cl}_S B$. Then

$$\begin{aligned}\text{int}_S\text{cl}_S(A \cap B) &\subseteq \text{int}_S(\text{cl}_S A \cap \text{cl}_S B) \quad \text{By Theorem 4.7}\\ &= \text{int}_S\text{cl}_S A \cap \text{int}_S\text{cl}_S B\\ &= A \cap B\end{aligned}$$

Since $A \cap B \subseteq \text{int}_S\text{cl}_S(A \cap B)$ and $\text{int}_S\text{cl}_S(A \cap B) \subseteq A \cap B$, then $A \cap B$ is regular open.

So, the family of all regular open sets is closed under finite intersections.

On the other hand, $(1, 5)$ and $(5, 9)$ are both regular open but $(0, 5) \cup (5, 9)$ is not. So, the family of all regular open sets is not closed under finite unions.

(b) We are given $A \subseteq S$. We are required to show that $\text{int}_S\text{cl}_S A$ is regular open. See that

$$\begin{aligned}\text{int}_S\text{cl}_S(\text{int}_S\text{cl}_S A) &\subseteq \text{int}_S\text{cl}_S(\text{cl}_S A) \quad (\text{Since } \text{int}_S\text{cl}_S A \subseteq \text{cl}_S A)\\ &= \text{int}_S\text{cl}_S A\end{aligned}$$

Since $\text{int}_S\text{cl}_S A \subseteq \text{int}_S\text{cl}_S(\text{int}_S\text{cl}_S A)$, then $\text{int}_S\text{cl}_S A$ is regular open.

(c) This part is left an exercise.

Concept Review

1. Given a topological space (S, τ) and $T \subseteq S$, define the closure, $\text{cl}_S(T)$, in S, with respect to the topology τ.
2. Does cl_S "distribute" over finite unions? How about finite intersections?
3. List the four Kuratowski closure operator axioms, K1–K4.
4. If $K : \mathscr{P}(S) \to \mathscr{P}(S)$ satisfies the four Kuratowski closure operator axioms describe a topology on S which can be constructed from K.
5. Define the cocountable topology on $\mathbb{R}^2$. Describe the Kuratowski operator used to develop this topology.
6. Given a topological space (S, τ) and $A \subseteq S$, define "interior point" of A with respect to τ. Define the interior, $\text{int}_S(A)$, of A with respect to τ.
7. Does int_S distribute over finite unions? How about finite intersections?
8. State the four interior operator axioms I1–I4 for $I : \mathscr{P}(S) \to \mathscr{P}(S)$.
9. Given a set S, describe a topology that can be constructed on S by using the operator $I : \mathscr{P}(S) \to \mathscr{P}(S)$.
10. Describe the topology induced on $\mathbb{R}$ by the operator $I(A) = A\backslash\mathbb{Q}$.

11. Define the boundary of a set with respect to a topology τ on S.
12. Define what we mean when we say that "A is dense in B".
13. Define what we mean when we say that "A is nowhere dense in B".
14. Define a regular open subset and a regular closed subset of a space.
15. Show that for any subset U of S, $\text{int}_S\text{cl}_S U$ is regular open in S.

Exercises

1. Let $T = (0,1)$ be viewed as a subset of $\mathbb{R}$ equipped with the usual topology. Show that $\text{cl}_{\mathbb{R}}(T) = [0,1]$. This is Example 1 on page 63.
2. Suppose S is a topological space induced by the metric ρ (i.e., the elements of τ are unions of open balls of the form $B_\varepsilon(x) = \{y : \rho(x,y) < \varepsilon\}$). Suppose F is a non-empty subset of S. We define $\rho(x,F) = \inf\{\rho(x,u) : u \in F\}$. Show that $\text{cl}_S(F) = \{x : \rho(x,F) = 0\}$. This is Example 3 on page 63.
3. Consider $\mathbb{R}$ equipped with the usual topology τ (induced by the Euclidian metric). Let $(\mathbb{Q}, \tau_{\mathbb{Q}})$ be the set of all rational numbers equipped with the subspace topology inherited from $\mathbb{R}$. Consider the subset $T = [-\pi, \pi) \cap \mathbb{Q}$. Determine whether T is open in $\mathbb{Q}$, closed in $\mathbb{Q}$, both open and closed in $\mathbb{Q}$, or none of these.
4. In Example 4 on page 65, it is show that $\text{cl}_S(A) \cap \text{cl}_S(B) \neq \text{cl}_S(A \cap B)$ sometimes occurs. Show that $\text{cl}_S(A \cap B) \subseteq \text{cl}_S(A) \cap \text{cl}_S(B)$ is always true.
5. Let (S,τ) be a topological space and A and B be subsets of S. Show that:

   (a) $\text{int}_S(A)$ is open in S. Also, $\text{int}_S(A)$ is the largest open subset of S which is entirely contained in A.
   (b) If $B \subseteq A$ then $\text{int}_S(B) \subseteq \text{int}_S(A)$.
   (c) $\text{int}_S(A \cap B) = \text{int}_S(A) \cap \text{int}_S(B)$.
   (d) $\text{int}_S(\text{int}_S(A)) = \text{int}_S(A)$.

   (This is Theorem 4.7, part (c) is already proven.)
6. Let S be a set and $I : \mathscr{P}(S) \to \mathscr{P}(S)$ be an interior operator satisfying the four conditions I1–I4 listed on page 72. Show that,

if $\tau = \{A \in \mathscr{P}(S) : I(A) = A\}$, (S, τ) is a topological space such that $I(A) = \text{int}_S(A)$ for all $A \in \mathscr{P}(S)$.

7. If $B = [0, 1]$ is a closed interval viewed as a subset of $\mathbb{R}$ equipped with the usual topology show that $\text{bd}_{\mathbb{R}}(B) = \{0, 1\}$.
8. Let $B = \{(x, y) \in \mathbb{R}^2 : x^2 + y^2 \langle 1, x \rangle 0\} \cup \{(x, y) \in \mathbb{R}^2 : x^2 + y^2 \leq 1, x \leq 0\}$ be a subset of $\mathbb{R}^2$ equipped with the topology induced by the Euclidean metric. Show that

$$\text{bd}_{\mathbb{R}^2}(B) = \text{cl}_{\mathbb{R}^2}(B) \backslash \text{int}_{\mathbb{R}^2}(B) = \{(x, y) \in \mathbb{R}^2 : x^2 + y^2 = 1\}$$

9. Show that A is both open and closed in the topological space (S, τ) if and only if $\text{bd}_S(A)$ is empty.
10. Let A and B be two subsets of the topological space (S, τ). Show that if $\text{bd}_S(A) \cap \text{bd}_S(B) = \varnothing$ then $\text{int}_S(A \cup B) = \text{int}_S(A) \cup \text{int}_S(B)$.
11. Let $B = [0, 7) \cup \{9\}$ where B is equipped with the subspace topology inherited from $\mathbb{R}$ itself equipped with the usual topology. If $A = \{0\} \cup \{3\} \cup (5, 7)$, find (i) $\text{cl}_B(A)$, (ii) $\text{bd}_B(A)$, and (iii) $\text{int}_B(A)$.

Chapter 5

Bases of Topological Spaces

Abstract
In this section, we define a neighborhood system of $x \in S$ with respect to a given topology on S. Given a topology, τ, on S, we define a "base for the topology τ". We deduce a set-theoretic property called the "base property" possessed by any base. Those subsets of $\mathscr{P}(S)$ which satisfy the described property will be shown to be a base for some topology on S. We introduce the notion of a "subbase for the topology τ" by describing its properties. We then show how to topologize a set both from a collection of sets which possesses the "base property" and also from an arbitrary collection of subsets.

5.1 Neighborhoods of Points

It is not always easy to confirm that a given subset, $\mathscr{S}$, of $\mathscr{P}(S)$ is a well-defined topology, τ, on a set S. Ultimately, we would prefer a topology on S to be described in a way that provides some intuitive idea about what its open subsets are like. In this section, we describe a subfamily, $\mathscr{B}$, of open sets from which every open set in τ is constructed. With this objective in mind, we begin by introducing the concept of *neighborhoods* of a point in (S, τ).

Definition 5.1 Let (S, τ) be a topological space and $x \in A \subseteq S$. We will say that A is a *neighborhood of x with respect to τ* if $x \in \text{int}_S(A)$.

For a given $x \in S$, the set

$$\mathscr{U}_x = \{A \in \mathscr{P}(S) : A \text{ is a neighborhood of } x\}$$

is called a *neighborhood system of x with respect to τ.*

A subfamily, $\mathscr{B}_x$, of a neighborhood system, $\mathscr{U}_x$, such that, for any open set, A, containing x there exists $U_x \in \mathscr{B}_x$ such that $x \in U_x \subseteq A$, is called a *neighborhood base at x*. If every element of $\mathscr{B}_x$ is open then we can say it is an *open neighborhood base at x.*

Note that, since the definition of a neighborhood refers to the "interior" of sets, a *neighborhood system* is always expressed with respect to some topology, τ, defined on the set S. Distinguish the two concepts defined in the above statement. We see that a neighborhood base, $\mathscr{B}_x$, at x is a subset of a neighborhood system at x. The set, $\mathscr{B}_x$, must be such that, any open neighborhood of x, contains an element of $\mathscr{B}_x$. While there can be only one neighborhood system at x we will see that there can be more than one neighborhood base at this point.

For example, $B = [-1, 5) \cup [6, \infty)$ is a neighborhood of 0 with respect to the usual topology of $\mathbb{R}$. The set

$$\mathscr{U}_3 = \{U \in \mathscr{P}(\mathbb{R}) : 3 \in \text{int}_{\mathbb{R}}(U)\}$$

is a neighborhood system of 3. If A belongs to $\mathscr{U}_3$, the element "3" must belong to the interior of A in the sense that 3 cannot be sitting on A's boundary. For example, $[0, 4] \in \mathscr{U}_3$, but $[1, 3] \notin \mathscr{U}_3$. Note that, for a given $x \in S$, its neighborhood system, $\mathscr{U}_x$, with respect to τ is, by definition, unique.

Based on the definition, we can make the following comments about neighborhoods of a point in a topological space, (S, τ).

— The empty set, $\varnothing$, is not the neighborhood of any point, so $\varnothing \notin \mathscr{U}_x$ for any $x \in S$.
— In general, a neighborhood, U, of a point x is not necessarily open, but it must contain x in its interior, $\text{int}_S(U)$.
— If $x \in S$, then $\mathscr{U}_x$ is not empty since S is a neighborhood of x. If τ is the *indiscrete topology*, $\{\varnothing, S\}$, then S is the only neighborhood of every point $x \in S$.

The "neighborhood of a point" concept allows us to come up with another characterization of open sets provided we have predefined "interior of a set". We propose:

"[A is open in S] $\Leftrightarrow$ [$x \in A \Rightarrow \exists$ neighborhood, U_x,

where $x \in U_x \subseteq A$]"

This will work since this would imply that

$$A = \cup_{x\in A}\{\text{int}_S(U_x) : U_x \in \mathscr{U}_x \text{ and } U_x \subseteq A\}$$

Or, we could simply say that "A is open in S provided A contains an x-neighborhood, U_x, for each point x in A". This definition of "open set" (in terms of neighborhoods) is equivalent to its formal definition of "open set".

5.2 A Base for a Topology

In our study of normed vector spaces (as well as of metric spaces) we have seen that, introducing the notion of "open ball, $B_\varepsilon(x)$, center x with radius ϵ", led to a convenient way of constructing a subfamily, $\mathscr{B}_x = \{B_\varepsilon(x) : \varepsilon > 0\}$, of open neighborhoods of a point x. Every open subset of S can then simply be described as being the union of open neighborhoods of the form $B_\varepsilon(x)$ contained in the set. The collection of open sets,

$$\mathscr{B} = \cup\{\mathscr{B}_x : x \in S\}$$

is sufficient to generate all open subsets of S.

Definition 5.2 Let (S, τ) be a topological space. Suppose $\mathscr{B}$ is a subset of τ satisfying the property:

For any $A \in \tau$, A is the union of elements of a subset, $\mathscr{C}$, of $\mathscr{B}$.

We call the subset, $\mathscr{B}$, a *base for open sets* or an *open base for the topology* τ (often abbreviated by simply saying *a base for* S). The word *basis* is sometimes used instead of "base".[1]

The elements of a base are referred to as *basic open sets.*[2]

If $\mathscr{F}$ is a family of closed subsets of S satisfying the property:

> For any closed subset B in S, B is the intersection of elements of a subset, $\mathscr{C}$, of $\mathscr{F}$.

We call the subset, $\mathscr{F}$, a *base for closed sets* of S.

A base for open sets is generally not unique. Given the topological space $(\mathbb{R}, \tau)$, where τ is the usual topology, both τ and $\mathscr{B} = \{(a, b) : a < b\}$ are bases for $\mathbb{R}$.

It is easily verified that if $\mathscr{F}$ is a base for closed subsets of S then the family, $\mathscr{B}$, of all complements of the elements of $\mathscr{F}$ will form a base for open sets.

How does one go about constructing a useful base for a topology? A good way to start is to determine some properties possessed by a known useful open base. The following theorem characterizes a base for a topology on a set S.

Theorem 5.3 *Let (S, τ) be a topological space and $\mathscr{B} \subset \tau$. Then the following are equivalent:*

(1) *The family $\mathscr{B}$ is a base for τ.*
(2) *For any $x \in S$, $\mathscr{B}$ contains an open neighborhood base, $\mathscr{B}_x$, of sets at x.*
(3) *If $U \in \tau$ and $x \in U$, there exists $B \in \mathscr{B}$ such that $x \in B \subseteq U$.*

---

[1]Both words "base" and "basis" are commonly used; thus the reader can assume these have the same meaning.

[2]Note that, if $A = \varnothing \in \tau$, then A is the union of all elements from $\mathscr{C} = \varnothing = \{\,\} \subseteq \mathscr{B}$. So $\mathscr{B}$ also generates the empty set.

Proof. We are given that (S, τ) is a topological space and $\mathscr{B} \subset \tau$.

($1 \Rightarrow 2$) Suppose the set $\mathscr{B}$ is a base for a topology τ on S and $x \in S$. Let

$$\mathscr{B}_x = \{B \in \mathscr{B} : x \in B\}$$

Suppose A is an open subset of S which contains x. It suffices to show that $\mathscr{B}_x$ contains an open neighborhood U of x such that $x \in U \subseteq A$. By hypothesis and Definition 5.2, $A = \cup\{B : B \in \mathscr{B}\}$. Then, there exists some $B_x \in \mathscr{B}$ such that $x \in B_x \subseteq A$. By definition, $B_x \in \mathscr{B}_x$. So, $\mathscr{B}_x$ is an open neighborhood base of x contained in $\mathscr{B}$.

($2 \Rightarrow 3$) This follows immediately from the definition of "neighborhood base".

($3 \Rightarrow 1$) Suppose $A \in \tau$. If A is empty then A is the union of all open sets in $\varnothing = \{\ \} \in \mathscr{B}$. Suppose $x \in A$. By hypothesis, there exists $B \in \mathscr{B}$ such that $x \in B \subseteq A$. Then A is the union of sets from $\mathscr{B}$. So $\mathscr{B}$ is a base for open subsets of S.

5.3 The "Base Property"

We have seen how an arbitrary set can be topologized by two different techniques. One by using a *Kuratowski closure operator*, the other by using an *interior operator*. These operators must satisfy certain axioms. If they do, then we can define a particular topology corresponding to the operator. In this section, we will propose two other methods for topologizing a set. We will first discuss another characterization of a base for a topology on a set S.

Theorem 5.4 *Let S be a non-empty set and $\mathscr{B} \subseteq \mathscr{P}(S)$. The set $\mathscr{B}$ is a base for a topology τ on S if and only if $S = \cup\{B : B \in \mathscr{B}\}$ and, if $x \in A \cap B$ for some $A, B \in \mathscr{B}$, then there exists $C \in \mathscr{B}$ such that $x \in C \subseteq A \cap B$.*

Proof. We are given that S is a non-empty set and $\mathscr{B} \subseteq \mathscr{P}(S)$.

($\Rightarrow$) Suppose the set $\mathscr{B}$ is a base for a topology τ on S.

Since $S \in \tau$, then $S = \cup\{B : B \in \mathscr{C}\}$ where $\mathscr{C} \subseteq \mathscr{B}$. Since $\cup\{B : B \in \mathscr{B}\} \subseteq S$, then $S = \cup\{B : B \in \mathscr{B}\}$.

Suppose $x \in A \cap B$ for some $A, B \in \mathscr{B}$. Then $A, B \in \tau$ and so $A \cap B \in \tau$. Since $A \cap B$ is open and (by Theorem 5.3) $\mathscr{B}$ contains an open neighborhood base, $\mathscr{B}_x$, there exists $B_x \in \mathscr{B}_x \subseteq \mathscr{B}$ such that $x \in B_x \subseteq A \cap B$. So B_x plays the role of C is the statement, as required.

($\Leftarrow$) We are given that $\mathscr{B} \subseteq \mathscr{P}(S)$ and satisfies the two property: $S = \cup\{B : B \in \mathscr{B}\}$ and if $x \in A \cap B$ for some $A, B \in \mathscr{B}$ then there exists $C \in \mathscr{B}$ such that $x \in C \subseteq A \cap B$.

Let $\mathscr{T} = \{A \subseteq S : A = \cup\{C : C \in \mathscr{C}\}$ for some subset $\mathscr{C}$ of $\mathscr{B}\}$.

We are required to show that $\mathscr{T}$ is a topology on S and that $\mathscr{B}$ is base for $\mathscr{T}$.

If we show that $\mathscr{T}$ is a topology on S then, by definition, $\mathscr{B}$ is a base for $\mathscr{T}$.

O1: The set S belongs to $\mathscr{T}$: Since $S = \cup\{B : B \in \mathscr{B}\}$, then $S \in \mathscr{T}$. We claim the set $\varnothing$ belongs to $\mathscr{T}$: The union of all elements in $\varnothing \subset \mathscr{B}$ is $\varnothing$. So $\varnothing \in \mathscr{T}$.

O2: *Claim*: The set $\mathscr{T}$ is closed under unions. Let $\{A_\alpha : \alpha \in \Gamma\} \subseteq \mathscr{T}$. For $\alpha \in \Gamma$, $A_\alpha = \cup\{B : B \in \mathscr{B}_\alpha \subseteq \mathscr{B}\}$. Then $\cup_{\alpha\in\Gamma}\{A_\alpha\} = \cup_{\alpha\in\Gamma}\{\cup_{B\in\mathscr{B}_\alpha}\{B\}\}$, a union of elements in $\mathscr{B}$. So $\cup_{\alpha\in\Gamma}\{A_\alpha\} \in \mathscr{T}$.

O3: *Claim*: The set $\mathscr{T}$ is closed under finite intersections. Let $A, C \in \mathscr{T}$. It suffices to show that $A \cap C \in \mathscr{T}$. Let $x \in A \cap C$. See that

$$A = \cup\{B : B \in \mathscr{B}_A \subseteq \mathscr{B}\} \text{ and } C = \cup\{B : B \in \mathscr{B}_C \subseteq \mathscr{B}\}$$

There exist $B_A \in \mathscr{B}_A$ and $B_C \in \mathscr{B}_C$ such that $x \in B_A \cap B_C$. By hypothesis, there exists $B_x \in \mathscr{B}$ such that $x \in B_x \subseteq B_A \cap B_C \subseteq A \cap C$. Then for every $x \in A \cap C$ there exists $B_x \in \mathscr{B}$ such that $x \in B_x \subseteq A \cap B$. Then $A \cap B = \cup_{x\in A\cap C}\{B_x\}$; so $A \cap C \in \mathscr{T}$, as required.

So, $\mathscr{T}$ is a topology on S. By definition of $\mathscr{T}$, every element of $\mathscr{T}$ is a union of elements of $\mathscr{B}$ so, by definition of "open base", $\mathscr{B}$ is base of $\mathscr{T}$.

In this text, we will call the special property which is satisfied by $\mathscr{B} \subseteq \mathscr{P}(S)$ in the theorem statement, the "base property". We formally define this below.

Definition 5.4.1. Let S be a non-empty set and let $\mathscr{B}$ be a subset of $\mathscr{P}(S)$.

Base property: The set $\mathscr{B}$ satisfies the *base property*, if $S = \cup\{B : B \in \mathscr{B}\}$ and, if $x \in A \cap B$ for some $A, B \in \mathscr{B}$, then there exists $C \in \mathscr{B}$ such that $x \in C \subseteq A \cap B$.

Theorem 5.4 guarantees that, given any non-empty set S, if a subset, $\mathscr{B}$, of $\mathscr{P}(S)$ satisfies the *base property* then $\mathscr{B}$ forms a base for some topology, τ, on S. The family, τ, is made precisely of the arbitrary unions of the elements of subsets of $\mathscr{B}$. In this case, we say that $\mathscr{B}$ *generates the topology* τ. The above statement is a powerful tool for topologizing sets.

Throughout our study of general topology, it will be very useful to have a variety of examples of topological spaces at hand. The reader is encouraged to take note of these, or at least bookmark them, for future reference in this book. For this purpose, most are given a particular name so that they can be more easily be referred to in the index of this book. The following examples, illustrate how the above result is used to topologize a subset of $\mathbb{R}^2$.

Example 1. Recall that a G_δ-set is a subset of a space, (S, τ) which is a countable intersection of elements of τ. Let

$$\mathscr{G} = \{G \in \mathscr{P}(S) : G \text{ is a } G_\delta\text{-set}\} \subseteq \mathscr{P}(S)$$

(the set of all G_δ-sets in a topological space, (S, τ)). Show that $\mathscr{G}$ is an open base for some topology, $\tau_{\mathscr{G}}$, on S.

Solution: Since every element of τ is an element of $\mathscr{G}$, then $S = \cup\{G : G \in \mathscr{G}\}$. Suppose $A, B \in \mathscr{G}$ and $x \in A \cap B$. Then there exists open subsets, $\{U_i : i \in \mathbb{N}\}$ and $\{V_i : i \in \mathbb{N}\}$, such that $A = \cap\{U_i : i \in \mathbb{N}\}$ and $B = \cap\{V_i : i \in \mathbb{N}\}$. Then

$$\begin{aligned} x &\in (\cap\{U_i : i \in \mathbb{N}\}) \cap (\cap\{V_i : i \in \mathbb{N}\}) \\ &= \cap\{U_i \cap V_i : i \in \mathbb{N}\} \quad \text{(A } G_\delta.\text{)} \\ &\subseteq A \cap B \end{aligned}$$

Since $\cap\{U_i \cap V_i : i \in \mathbb{N}\}$ is a G_δ, then $\mathscr{G}$ satisfies the base property and so generates a topology, $\tau_{\mathscr{G}}$.

We will refer to $\tau_{\mathscr{G}}$ as the

G_δ-topology generated by τ.

Example 2. The *Moore plane.* Also called, *Niemytzki's topology* (Figure 5.1).

Let $S = \{(x, y) \in \mathbb{R}^2 : y \geq 0\}$. The set $B_\varepsilon(a, b)$ represents the usual open ball of center, (a, b), and radius ε. Let

$$\mathscr{A} = \{B_\varepsilon(x_0, y_0) : x_0 \in \mathbb{R}, y_0 > 0, \text{ and } \varepsilon < y_0\}$$

For each $x \in \mathbb{R}$ and $\varepsilon > 0$, let

$$D_\varepsilon(x, 0) = \{(x, 0)\} \cup B_\varepsilon(x, \varepsilon)$$

That is, $D_\varepsilon(x, 0)$ is an open ball, $B_\varepsilon(x, \varepsilon)$, of radius ε tangent to the horizontal axis at $(x, 0)$ with the point $(x, 0)$ attached to it. Let

$$\mathscr{D} = \{D_\varepsilon(x, 0) : x \in \mathbb{R}, \varepsilon > 0\}$$

Let $\mathscr{B} = \mathscr{A} \cup \mathscr{D}$. Show that $\mathscr{B}$ is the base for some topology on S.

Solution: If we show that $\mathscr{B}$ satisfies the "base property" then $\mathscr{B}$ is a base which generates a topology, τ, on the half-plane S. It is first easily seen that $S = \cup\{B : B \in \mathscr{B}\}$.

We consider case 1: Suppose $A, B \in \mathscr{B}$ and $(x, y) \in A \cap B$ where $y > 0$. Then the situation is analogous to what occurs in $\mathbb{R}^2$ with the usual topology, and so

$$(x, y) \in B_\varepsilon(x, y) \subseteq A \cap B$$

for some $\varepsilon > 0$.

We consider case 2: Suppose $A, B \in \mathscr{B}$ and $(x, 0) \in A \cap B$. Then $A = \{(x, 0)\} \cup B_{\varepsilon_1}(x, \varepsilon_1)$ and $B = \{(x, 0)\} \cup B_{\varepsilon_2}(x, \varepsilon_2)$. Let $\varepsilon = \frac{\min\{\varepsilon_1, \varepsilon_2\}}{2}$ and $C = \{(x, 0)\} \cup B_\varepsilon(x, \varepsilon)$. Then

$$(x, 0) \in C \subseteq A \cap B$$

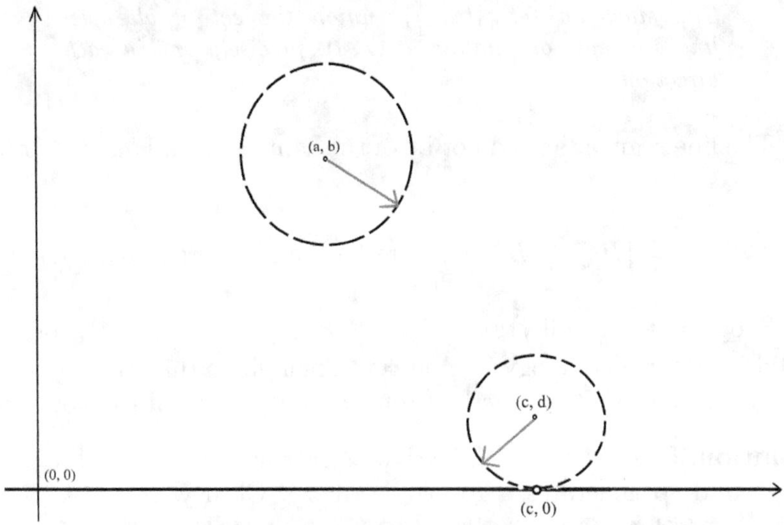

Figure 5.1. Moore plane.

By invoking Theorem 5.4, we conclude that $\mathscr{B}$ is a base for a topology, τ_M, on S.

This well-known topological space, (S, τ_M), is referred to as the *Moore plane*. Some authors also refer to this topology, τ_M, on S, as the *Niemytzki's tangent disc topology.*

The topology τ_M is strictly stronger than τ, the usual topology. To see this, let τ represent the subspace topology on the set S, inherited from $\mathbb{R}^2$ equipped with the usual topology. We verify that τ_M is strictly stronger than τ. Let $B(0,0)$ denote a basic open neighborhood of the point $(0,0)$ in τ_M. Then $B(0,0) \cap \{(0,0)\} = \{(0,0)\}$. The set $B_\varepsilon(0,0) \cap S$ is a half-open disc which contains the interval, $(-\varepsilon, \varepsilon)$, and so cannot be contained in $B(0,0)$. Then $B(0,0) \notin \tau$ so $T_M \not\subseteq \tau$. On the other hand, given $(u,0) \in B_\varepsilon(0,0) \cap S$ one can construct a neighborhood base element, $B(u,0)$, of radius small enough so that it is contained in $B_\varepsilon(0,0) \cap S$. So $\tau \subseteq \tau_M$.

So the Moore plane topology is strictly stronger than the usual topology on S.

Example 3. *The radial plane.* Consider the set $\mathbb{R}^2$. We will equip $\mathbb{R}^2$ with what is called the *radial plane topology.* If $(x,y) \in \mathbb{R}^2$, we define a "radial ball centered at (x,y)", $B_{(x,y)}$, in $\mathbb{R}^2$, as follows:

The singleton set, $\{(x, y)\}$, union the set of all open line segments originating at (x, y), precisely one in each direction.

The line segments need not be of the same length. For each (x, y), let

$$\mathscr{B}_{(x,y)} = \{B_{(x,y)} : B_{(x,y)} \text{ is a radial ball centered at } (x, y)\}$$

Show that the collection, $\mathscr{B} = \cup\{\mathscr{B}_{(x,y)} : (x, y) \in \mathbb{R}^2\}$, forms a valid base for a topology, τ^*, on $\mathbb{R}^2$. Then show that the topology, τ^*, generated by $\mathscr{B}$ is strictly stronger than the usual topology, τ.

Solution: First part: We *claim* that $\mathscr{B}$ satisfies the "open base property" and so generates a topology on $\mathbb{R}^2$. Clearly, $\mathbb{R}^2 = \cup\{B_{(x,y)} : (x, y) \in \mathbb{R}^2\}$. Suppose $(x, y) \in U \cap V$ where U, $V \in \mathscr{B}_{(x,y)}$. Let K_U be a line segment in a particular direction originating at (x, y) such that $K_U \subset U$ and K_V be a line segment originating at (x, y) pointing in the same direction as K_U but $K_V \subset V$. Let $K_{U\cap V} = K_U \cap K_V$. Then $K_{U\cap V}$ is an open line segment originating at (x, y) which is contained in $U \cap V$. This can be repeated for all open lines in all directions originating at (x, y) to obtain a radial ball, $M_{(x,y)}$. So there exists $M_{(x,y)} \in \mathscr{B}_{(x,y)}$ such that

$$(x, y) \in M_{(x,y)} \subseteq U \cap V$$

So, $\mathscr{B}$ satisfies the open base property.

Hence, by Theorem 5.4, $\mathscr{B}$ is an open base which generates a topology on $\mathbb{R}^2$, say τ^*. We are done with the first part of the question.

Second part: We *claim* that the usual topology, τ, is contained in τ^*. Suppose $(x, y) \in U$ where U is an open base element of τ centered at (x, y). Then $U = B_\varepsilon(x, y)$ for some ε. Let $(a, b) \in B_\varepsilon(x, y)$. Then there exists $\delta > 0$ such that $B_\delta(a, b) \subseteq B_\varepsilon(x, y)$. Let V be a set such that all open line segments originating at (a, b) are of length less than $\delta/2$. Then $V \in \mathscr{B}$ and $V \subseteq B_\delta(a, b) \subseteq B_\varepsilon(x, y)$. So $U \in \tau^*$. Then $\tau \subseteq \tau^*$, as claimed.

We *claim* that τ^* is strictly stronger than τ. To show this it suffices to show that τ^* contains an element which does not belong to τ.

Consider the sets $U = B_1(0,1)$ and $V = B_1(0,-1)$.[3] Then, if $L = \{(x,0) : x \in \mathbb{R}, |x| \leq k\}$ the set,

$$W = U \cup L \cup V$$

representing the two balls U and V with a straight horizontal line through $(0,0)$ added to them does not belong to τ. However, the set $M(0,0)$ which has rays emanating from $(0,0)$, in all directions, each of which is contained in W is an open neighborhood of $(0,0)$ in τ^*. So τ is a proper subset of τ^*, as claimed.

5.4 The Subbase of a Topology

We have seen that, when given an arbitrary set S, a subfamily of $\mathscr{P}(S)$ which satisfies the "base property" can be used to generate a topology on S. We shall soon see that any subfamily family of $\mathscr{P}(S)$ can be used to generate a base which, in turn, will generate a topology, τ, on S. We will refer to such a family as a "subbase" for τ. Even if this may, at first, appear to be a very convenient way to generate topologies on a set, it does make it more difficult to predict, from a subbase, what the topology generated by such a collection of sets will be like. So, for this convenience, there is a price to pay. We know that we can obtain a topology from a collection of sets which satisfies the "base property" in a single step: base $\to \tau$. But to obtain a topology, τ, from a subbase requires two steps:

$$\text{subbase} \to \text{base} \to \tau$$

But the notion of a subbase is nevertheless interesting to investigate since, if a space is already equipped with a topology, τ, it can be useful to understand what kind of subbase generates τ. It may allow us to dig deeper into the background of τ to better see why τ satisfies certain properties.

Definition 5.5 Let (S,τ) be a topological space. A *subbase for the topology* τ is a non-empty subfamily, $\mathscr{S}$, of τ such that the set, $\mathscr{B}$,

[3] The set $B_1(0,1)$ refers to the open ball center $(0,1)$ and radius 1 while $B_1(0,-1)$ refers to the open ball center $(0,-1)$ with radius 1.

defined as

$$\mathscr{B} = \{B : B = \cap\{U : U \in \mathscr{F}\} \text{ where } \mathscr{F} \text{ is a finite subset of } \mathscr{S}\}$$

forms a base for τ.

The definition states that, if a base of τ can be obtained from the set of all finite intersections of the elements of a collection, $\mathscr{S}$, then $\mathscr{S}$, is called a subbase of τ. This of course means that every element of a subbase of τ belongs to τ and so is open.

In the following two examples, we are given a particular topology on a set. For each topological space we identify a subbase for the given topology.

Example 4. *A subbase for the usual topology on* $\mathbb{R}$. Consider the sets

$$(a, \infty) = \{x \in \mathbb{R} : a < x\}$$
$$(-\infty, b) = \{x \in \mathbb{R} : x < b\}$$

and the family

$$\mathscr{S} = \{(a, \infty) : a \in \mathbb{R}\} \cup \{(-\infty, b) : b \in \mathbb{R}\}$$

We see that $\mathscr{S}$ forms a subbase for the usual topology τ on $\mathbb{R}$ since the set of all finite intersections of elements of $\mathscr{S}$ form the set, $\mathscr{B} = \{(a, b) : a < b\}$, known to be a base of τ.

Example 5. Consider $(\mathbb{N}, \tau_d)$ where τ_d is the *discrete topology* (see the definition on page 44). Then $\tau_d = \mathscr{P}(\mathbb{N})$.

The set of all singleton sets, $\mathscr{B} = \{\{n\} : n \in \mathbb{N}\}$, forms a base for τ_d since every subset of $\mathbb{N}$ is the union of elements from some subset $\mathscr{C} \subseteq \mathscr{B}$. If

$$N_a = \{n \in \mathbb{N} : n \leq a\}$$
$$M_b = \{n \in \mathbb{N} : n \geq b\}$$

and

$$\mathscr{S} = \{N_a : a \in \mathbb{N}\} \cup \{M_b : b \in \mathbb{N}\}$$

then $\mathscr{B}$ is a subset of the family of finite intersections of elements of $\mathscr{S}$. So, $\mathscr{S}$ is a subbase for the topology τ_d on $\mathbb{N}$.

5.5 A Subbase as a Generator of a Topology

We now show how *any* non-empty subset, $\mathscr{S}$, of $\mathscr{P}(S)$ will generate a topology, $\tau_{\mathscr{S}}$ on S. This topology, will have $\mathscr{S}$ as "subbase".

Suppose $\{\tau_j : j \in I\}$ represents the class of all topologies on S. Suppose S is a set and $\mathscr{S}$ is a non-empty subset of $\mathscr{P}(S)$. Let

$$\mathscr{I} = \{\tau_j : j \in I, \mathscr{S} \subseteq \tau_j\}$$

denote *all* topologies on S which contain $\mathscr{S}$. The set $\mathscr{I}$ is non-empty since it at least contains the discrete topology, $\tau_d = \mathscr{P}(S)$, which, by definition, contains $\mathscr{S}$. Since the family of all topologies on a set is closed under intersections (see page 44), (but not necessarily under unions) then the family

$$\tau_{\mathscr{S}} = \cap\{\tau_j : j \in I,\ \mathscr{S} \subseteq \tau_j\}$$

is also a topology on S which contains all elements of $\mathscr{S}$. In fact, it is easily verified that $\tau_{\mathscr{S}}$ *is the smallest possible topology on* S *which contains* $\mathscr{S}$.

In the following theorem, we show that $\mathscr{S}$ is a subbase for the constructed topology, $\tau_{\mathscr{S}}$, on S.

Theorem 5.6 *Let* S *be a non-empty set and* $\mathscr{S} \subseteq \mathscr{P}(S)$. *Suppose*

$$\tau_{\mathscr{S}} = \cap\{\tau_j : \tau_j \text{ is a topology on } S,\ \mathscr{S} \subseteq \tau_j\}$$

Then $\mathscr{S}$ *is a subbase for* $\tau_{\mathscr{S}}$ *on* S *which contains* $\mathscr{S}$.

Proof. Given: $\mathscr{S} \subseteq \mathscr{P}(S)$ and $\tau_{\mathscr{S}} = \cap\{\tau_j : j \in I, \tau_j$ is a topology on S, $\mathscr{S} \subseteq \tau_j\}$.

We have already seen that $\tau_{\mathscr{S}}$ is the smallest topology on S with contains $\mathscr{S}$.

Let

$$\mathscr{B} = \{B : B = \cap_{M \in \mathscr{F}} M \text{ where } \mathscr{F} \text{ is a finite subset of } \mathscr{S}\}$$

Claim: We claim that $\mathscr{B}$ is a basis for a topology τ on S.

Note that, the empty set, $\varnothing$, is a finite subset of $\mathscr{S}$. Then

$$\cap_{M \in \varnothing} M = \{x \in S : x \in M \text{ for every } M \in \varnothing\} = S$$

So, $S \in \mathscr{B}$.

Let $A, D \in \mathscr{B}$ and $\mathscr{A}$ and $\mathscr{D}$ be finite subfamilies of $\mathscr{S}$ such that $A = \cap_{M \in \mathscr{A}} M$ and $D = \cap_{M \in \mathscr{D}} M$. Then $\mathscr{A} \cup \mathscr{D}$ is a finite subfamily of $\mathscr{S}$. Suppose $x \in A \cap D$. Then

$$x = (\cap_{M \in \mathscr{A}} M) \cap (\cap_{M \in \mathscr{D}} M) = \cap_{M \in \mathscr{A} \cup \mathscr{D}} M = A \cap D$$

If $E = \cap_{M \in \mathscr{A} \cup \mathscr{D}} M$, $x \in E \subseteq A \cap D$.

Then $\mathscr{B}$ is a base for *some* topology τ on S, as claimed. Also $\mathscr{S}$ is a subbase for τ.

Since $\tau_{\mathscr{S}}$ is the smallest topology which contains $\mathscr{S}$, then $\tau_{\mathscr{S}} \subseteq \tau$. On the other hand, if $U \in \tau$ then U is the union of elements of $\mathscr{B} \subseteq \tau_{\mathscr{S}}$. So $U \in \tau_{\mathscr{S}}$. We conclude that $\tau \subseteq \tau_{\mathscr{S}}$.

Then $\tau_{\mathscr{S}} = \tau$ and so $\mathscr{S}$ is a subbase of $\tau_{\mathscr{S}}$.

The above theorem guarantees that any non-empty subset of $\mathscr{P}(S)$ is a subbase for some topology on S. We provide the following examples where this principle is applied.

Example 6. Consider the sets $X_a = \{x \in \mathbb{R} : a < x\}$ and $Y_b = \{x \in \mathbb{R} : x \leq b\}$ and the family $\mathscr{S} = \{X_a : a \in \mathbb{R}\} \cup \{Y_b : b \in \mathbb{R}\}$. We see that the intersections of finite subsets of $\mathscr{S}$ are either $\varnothing$ or of the form $(a, b]$. So,

$$\mathscr{B} = \{(a, b] : a < b\} \cup \{\varnothing\}$$

forms a base for a topology, $\tau_{\mathscr{S}}$, on $\mathbb{R}$ generated by the subbase $\mathscr{S}$. This topology is referred to as the

upper limit topology on $\mathbb{R}$

or the *Sorgenfrey line.*[4]

Example 7. Consider the set, $\mathscr{S} = \{[a, b] : a < b\}$, of all non-empty closed and bounded intervals in $\mathscr{P}(\mathbb{R})$.

We see that $\mathscr{S}$ does not satisfy the "base property" (since there does not exist $[x, y]$ in $\mathscr{S}$ such that $[x, y] \subseteq [a, b] \cap [b, c]$) and so cannot form a base for a topology on $\mathbb{R}$.

---

[4]If $\mathscr{B} = \{[a, b) : a < b\} \cup \{\varnothing\}$, this topology is referred to as the *lower limit topology on* $\mathbb{R}$.

However, the theorem guarantees that $\mathscr{S}$ is a subbase for some topology on $\mathbb{R}$. Describe this topology.

Solution: We see that non-empty finite intersections of elements of $\mathscr{S}$ are of the form $[c, d]$ where $c \leq d$. In particular, $[u, x] \cap [x, v] = \{x\}$ is an open base element for all $x \in \mathbb{R}$. So, the subbase $\mathscr{S}$ will generate an open base, $\mathscr{B} = \{\{x\} : x \in \mathbb{R}\}$, for the discrete topology, τ_d.

We see that the subbase, $\mathscr{S}$, holds more sets then is really required to generate τ_d. We can reduce its size to

$$\mathscr{S}^* = \{[x-1, x] : x \in \mathbb{R}\} \cup \{[x, x+1] : x \in \mathbb{R}\}$$

Since, for each $x \in \mathbb{R}$, $[x-1, x] \cap [x, x+1] = \{x\}$, $\mathscr{S}^*$ still generates τ_d.

Example 8. Consider two topological spaces (S, τ_S) and (T, τ_T). Using the sets S and T we can construct a new set, the Cartesian product $S \times T$, defined as $S \times T = \{(x, y) : x \in S, y \in T\}$. We can topologize the set $S \times T$ by defining a suitable subbase.

We will proceed as follows. Define the two projection functions $\pi_S : S \times T \to S$ and $\pi_T : S \times T \to T$ as $\pi_S(x, y) = x$ and $\pi_T(x, y) = y$, respectively. We will define as subbase for $S \times T$

$$\mathscr{S} = \{\pi_S^{-1}(U) : U \in \tau_S\} \cup \{\pi_T^{-1}(V) : V \in \tau_T\}$$

where $\pi_S^{-1}(U) = U \times T$ and $\pi_T^{-1}(V) = S \times V$. By referring to the principle $(A \times B) \cap (C \times D) = (A \cap C) \times (B \cap D)$ (verify this!), the basis $\mathscr{B}$ induced by this subbase is of the form

$$\mathscr{B} = \{U \times V : U \in \tau_S, V \in \tau_T\}$$

The topology on $S \times T$ whose base is $\mathscr{B}$ is called the *product topology on $S \times T$*. Such topological spaces created by combining known topological spaces are important and so will be discussed in depth in the next few chapters.

Supplementary definition: Two bases $\mathscr{B}$ and $\mathscr{B}^*$ are said to be *equivalent bases* for a particular set S if they generate the same topology.

Note that the usual basis for $\mathbb{R}$ and the basis for the *upper limit topology* described in the example above are not equivalent bases since $(a, b]$ does not belong to the usual topology on $\mathbb{R}$. However, since

$$(a, b) = \cup\{(a, b - 1/n] : n = 1, 2, 3, \ldots\} \in \tau_{\mathscr{S}}$$

then every base element for the usual topology belongs to $\tau_{\mathscr{S}}$ and so... the usual topology is weaker than $\tau_{\mathscr{S}}$.

In the following example, we illustrate a different topology, τ_{scat} on $\mathbb{R}$. The topological space, $(\mathbb{R}, \tau_{\text{scat}})$, is known as the *scattered line* or the *discrete irrational extension of* $\mathbb{R}$.

Example 9. Consider the real line, $\mathbb{R}$. Let τ denote the usual topology on $\mathbb{R}$ and $\mathbb{J}$ denote the subset of all irrationals. Let

$$\tau_{\text{scat}} = \{U \cup V : U \in \tau \text{ and } V \subseteq \mathbb{J}\}$$

Verify that τ_{scat} is indeed a topology on $\mathbb{R}$. Find a base for this topology.

Solution: O1: Since $\varnothing \in \tau$ and $\varnothing \subseteq \mathbb{J}$, $\varnothing = \varnothing \cup \varnothing \in \tau_{\text{scat}}$. So $\varnothing \in \tau_{\text{scat}}$. Since $\mathbb{R} = \mathbb{R} \cup \mathbb{J}$, then $\mathbb{R} \in \tau_{\text{scat}}$.

O2: Let $\mathscr{A} = \{U_i \cup V_i : i \in I\}$ be a subfamily of τ_{scat}. Then

$$\begin{aligned} \cup\mathscr{A} &= \cup\{U_i \cup V_i : i \in I\} \\ &= \bigcup_{i\in I} U_i \ \cup \ \bigcup_{i\in I} V_i \end{aligned}$$

where $\bigcup_{i\in I} U_i \in \tau$ and $\bigcup_{i\in I} V_i \subseteq \mathbb{J}$. So $\cup\mathscr{A} \in \tau_{\text{scat}}$.

O3: Let $\mathscr{D} = \{U_i \cup V_i : i = 1 \text{ to } n\}$ be a finite subfamily of τ_{scat}. Then

$$(U_1 \cup V_1) \cap (U_2 \cup V_2) = (U_1 \cap U_2) \ \cup \ [(U_1 \cap V_2) \cup (V_1 \cap (U_2 \cup V_2))]$$

The right-hand side is an element of τ_{scat} (where $U_1 \cap U_2 \in \tau$ and $(U_1 \cap V_2) \cup (V_1 \cap (U_2 \cup V_2)) \subseteq \mathbb{J}$). Proceeding by finite induction on n, we conclude O3 is satisfied.

So τ_{scat} is indeed a topology on $\mathbb{R}$.

Suppose $A \in \tau_{\text{scat}}$ contains the rational number q. Then $A = U \cup V$ for some $U \in \tau$ and $V \subseteq \mathbb{J}$. Since q is not irrational, $q \in U$. Then there exists $\varepsilon > 0$ such that $q \in (q-\varepsilon, q+\varepsilon) \subseteq U \subseteq A$. So, the family $\{(q-\varepsilon, q+\varepsilon) : \varepsilon > 0\}$ forms a neighborhood base for q in τ_{scat}.

On the other hand, if $r \in \mathbb{J}$, and $A \in \tau_{\text{scat}}$ contains r, then $\{r\} = \varnothing \cup \{r\} \in \tau_{\text{scat}}$. So, $\{\{r\} : r \in \mathbb{J}\}$ forms neighborhood base of r.

Then

$$\{\{r\} : r \in \mathbb{J}\} \cup \{(q-\varepsilon, q+\varepsilon) : \varepsilon > 0, q \in \mathbb{Q}\}$$

forms a base for the scattered line.

5.6 Topic: Spaces with Countable Bases

Some properties of a topological space, S, may depend on the cardinality of an open base on S. We will see that some spaces have a countable open base while others simply have a countable neighborhood base *at each point.* It will be useful to discuss these notions as soon as possible in the text.

Definition 5.7 Let (S, τ) be a topological space. The topological space, S, is said to be

first countable

if and only if every point, $x \in S$, has a countable neighborhood base. The topological space, S, is said to be

second countable

if and only if the smallest of its open bases is a countable base.

Countability properties. We have discussed a countability property which is indirectly related to the topology of a space before. Recall (from Definition 4.10) that a space is

separable

if it has at least one countable dense subspace. The three properties, *separable*, *first countable* and *second countable* are often referred to as "the countability properties" of a topological space. When introduced to an unfamiliar topological space we often try to determine which property (or properties) apply to the space in question. We begin by discussing in which ways these properties relate to each other.

Theorem 5.8 *A second countable topological space is a first countable space.*

Proof. Suppose (S, τ) is a second countable space. Then S has a countable base, $\mathscr{B}$. Then, for each $x \in S$, $\mathscr{B}_x = \{B \in \mathscr{B} : x \in B\}$ is a neighborhood base at x. Since $\mathscr{B}$ is countable and $\mathscr{B}_x \subseteq \mathscr{B}$, $\mathscr{B}_x$ cannot be uncountable. So (S, τ) is first countable.

The class of all first countable spaces is a subclass of all second countable spaces. However, there are some uncountably large topological spaces which are first countable but not second countable. The following example illustrates such a space.

Example 10. Suppose τ_S represents the *upper limit topology* on $\mathbb{R}$ (the Sorgenfrey line) (see Example 4 on page 98). Recall that, when $\mathbb{R}$ is equipped with this topology, $\mathscr{B} = \{(x, y] : x, y \in \mathbb{R}\}$ is an open base for $\mathbb{R}$. Show that $(\mathbb{R}, \tau_S)$ is first countable but not second countable.

Solution: The space, $(\mathbb{R}, \tau_S)$ is first countable: For each $x \in \mathbb{R}$,

$$\mathscr{B}_x = \{(x - 1/n, x] : n \in \mathbb{N}\}$$

is a countable neighborhood base at x. So, $(\mathbb{R}, \tau_S)$ is first countable.

The space, $(\mathbb{R}, \tau_S)$ is not second countable: Let the function $\phi : \mathbb{R} \to \mathscr{B}$ be defined as

$$\phi(x) = (x - 1, x]$$

So, $\phi[\mathbb{R}] \subseteq \mathscr{B}$. If $x \neq y$, then

$$\phi(x) = (x - 1, x] \neq (y - 1, y] = \phi(y)$$

Since $\mathbb{R}$ is uncountable and ϕ is one-to-one on its domain, $\mathbb{R}$, then $\phi[\mathbb{R}]$ is an uncountable subset of $\mathscr{B}$. Also, $\{(\infty, b] : a \in \mathbb{R}\} \cup \{(c, \infty) : c \in \mathbb{R}\}$ forms a subbase for τ_S generating a base which contains every element of $\phi[\mathbb{R}]$. So $\mathscr{B}$ is uncountable.

So, $(\mathbb{R}, \tau_S)$ is not second countable. As required, metrizable spaces turn out to be first countable spaces. That is, every point of a metrizable space has a countable neighborhood base. We already have the tools needed to prove this immediately.

Theorem 5.9 *Any metrizable space is first countable.*

Proof. Suppose (S, τ) is a metrizable space whose open sets are generated by the metric ρ. Then the set

$$\mathscr{B}_x = \{B_{1/n}(x) : n \in \mathbb{N}, n > 0\}$$

(an open ball center x with radius $1/n$) forms a countable neighborhood base at x.

Hence, S is a first countable space.

Example 11. Since $\mathbb{R}$ (equipped with usual topology) is metrizable, then it is first countable.

Example 12. Verify that the space, $\mathbb{R}$, when equipped with the usual topology, is second countable.

Solution: We are required to show $\mathbb{R}$ has a countable open base.

Suppose U is an open subset of $\mathbb{R}$ and $x \in U$. Then there exists an open interval, $B_\varepsilon(x) = (x - \varepsilon, x + \varepsilon)$, such that $x \in B_\varepsilon(x) \subseteq U$.

We consider two cases:

— If $x \in \mathbb{Q}$, then there exists integer m such that $x \in B_{\frac{1}{m}}(x) \subseteq B_\varepsilon(x) \subseteq U$.

— If x is an irrational, we know that there is a sequence of rationals which converges to x, so we can find a rational y in $B_{\frac{1}{4m}}(x)$, such that

$$x \in B_{1/4m}(y) \subseteq B_\varepsilon(x)$$

We have just shown that $\mathscr{B} = \{B_m(y) : y, m \in \mathbb{Q}\}$ forms a base for the open sets in $\mathbb{R}$. Since $\mathbb{Q}$ is countable, then so is $\mathbb{Q} \times \mathbb{Q}$[2] so

$$|\mathscr{B}| = |\{B_m(y) : (y, m) \in \mathbb{Q} \times \mathbb{Q}\}| = |\mathbb{Q} \times \mathbb{Q}| = \aleph_0$$

We have shown that, when equipped with the usual topology, $\mathbb{R}$ has a countable base for open sets.

[2]See appendix on cardinalities.

Example 13. Recall that the radial plane, $(\mathbb{R}^2, \tau_r)$, is equipped with what is called the *radial plane topology*, τ_r (see the example on page 93). The open sets in the radial plane are defined as follows: The subset U is an open neighborhood of q if and only if $q \in U$, and U is the union of a set of open line segments, precisely one in each direction, each one originating at q. It is shown on page 93, that this is a valid topology and that it is strictly stronger than the usual topology on $\mathbb{R}^2$. Verify that the radial plane is *not* first countable.

Solution: Let $p \in \mathbb{R}^2$. Suppose $(\mathbb{R}^2, \tau_r)$ is first countable. Then p has a countable open neighborhood base,

$$\mathscr{B}_p = \{B_i : i \in \mathbb{N}\backslash\{0\}\}$$

We will construct an element, V, of τ_r which contains p but does not contain any element of $\mathscr{B}_p$.

For each radial set, $B_i \in \mathscr{B}_p$, let r_i be the length of the longest ray originating at p which appears in B_i and $|r_i|$ denote its length. We then inductively construct a sequence of rays, $R = \{r_i^* : i \in \mathbb{N}\backslash\{0\}\}$ each originating at p in the direction of r_i and of length

$$|r_i^*| = |r_i/2^i|$$

Since we have only countably many rays, the set

$$\{r_i^* : i \in \mathbb{N}\backslash\{0\}\} \cup \{p\}$$

is not yet a complete radial set centered at p. We will add uncountably many more rays of arbitrary length to the ones we have to obtain a complete radial open set, V, centered at p. No open neighborhood B_i can be contained in this V (since the rays become arbitrarily small, shrinking down to p). So, $\mathscr{B}_p$ cannot be an open neighborhood base for p. So, p must have an uncountable open neighborhood base.

5.7 Topic: Relating the Countable Base Property with the Separable Property

Recall that in Definition 4.10, we defined a *separable* topological space as being a space which has a countable dense subset. We saw, for example, that, since $\mathbb{Q}$ is a countable dense subset of the reals, $\mathbb{R}$ is separable.[5]

In the following theorem, we see that all second countable spaces are guaranteed to have a countable dense subset. The proof of this statement follows from a more general result (in part (a)) which states that, given any space S, given any open base, $\mathscr{B}$, S has a dense subset D whose cardinality, $|D|$, is less than or equal to the cardinality of $\mathscr{B}$.

Theorem 5.10 ("Second countable ⇒ separable" theorem). *Suppose $\mathscr{B}$ is an open base of the space (S, τ_s) such that, amongst all bases of S, it is one of least cardinality.*

(a) *There exists in S a dense subset, D such that $|D| \leq |\mathscr{B}|$.*
(b) *Any second countable topological space is a separable space.*

Proof. We are given that $\mathscr{B} = \{B_i : i \in I\}$ is a base such that if any other base has cardinality J, $J \geq |I|$.

(a) For each $i \in I$, choose $x_i \in B_i$. (Choice!)[6] Let $D = \{x_i : i \in I\}$. Then $|D| \leq |\mathscr{B}|$.

We *claim* that D is dense in S: Let U be a non-empty open subset of S.

Then, for $x \in U$ there exists $B_i \in \mathscr{B}$ such that $x \in B_i \subseteq U$.

Then $x_i \in B_i \subseteq U$. Then $U \cap D \neq \varnothing$. So every open set in S intersects D. This means that D is dense in S, as *claimed.*

Then D is a dense subset of S such that $|D| \leq |I|$. Since I is the smallest indexing set for any open base, then D has a cardinality which is less than or equal to the cardinality of any base.

(b) It follows immediately from part (a) that "second countable property" implies the "separable property". For, if $|\mathscr{B}| = \aleph_0$, then D is a countable dense subset of S. So, S is separable.

We have just shown that those spaces that have a countable base of open sets must have a countable dense subset. In general, the

[5]Where $\mathbb{R}$ is equipped with the usual topology.

[6]Existence theorems often (but not always) suggest an application of the Axiom of choice in the proof. Keep an eye open for it.

converse does not hold true. That is, there are spaces that have a countable dense subset which do not have a countable base of open sets. But if S is a metrizable space then the converse holds true, as we shall now see.

Theorem 5.11 *Separable metrizable spaces are second countable.*

Proof. We are given that S is a metrizable separable topological space. Then there is a metric, ρ, on S such that the space, S, is equivalent to (S, ρ).

Since S is separable, then S has a countable dense subset, $D = \{x_i : i \in \mathbb{N}\backslash\{0\}\}$.

For each i and n in $\mathbb{N}\backslash\{0\}$ let $B_{(i,n)} = B_{1/n}(x_i)$, be an open ball center x_i and radius $1/n$. Consider

$$\mathscr{B} = \{B_{(i,n)} : i, n \in \mathbb{N}\backslash\{0\}\}$$

For $x \in S$, let U_x be an open neighborhood of x. Then there exists, $j \in \mathbb{N}\backslash\{0\}$ such that $B_{1/j}(x) \subseteq U_x$. Since D is dense in S, $B_{1/j}(x) \cap D$ is non-empty. Say,

$$x_k \in B_{\frac{1}{2j}}(x) \cap D$$

Then

$$x \in B_{\frac{1}{2j}}(x_k) \subseteq B_{\frac{1}{2j}}(x) \subseteq U_x$$

Then $\mathscr{B}$ is a countable base for open sets in S. So, the separable metric space, S, is second countable.

The following diagram summarizes the above results.

$$\begin{array}{ccc} \text{metrizable} + \text{separable} & & \text{metrizable} \\ \Downarrow & & \Downarrow \\ \text{second countable} & \Longrightarrow & \text{first countable} \\ \Downarrow & & \\ \text{separable} & & \end{array}$$

Example 14. From the above statements, we have another proof that the set of all reals equipped with the usual topology is both separable and second countable.

5.8 Topic: Hereditary Topological Properties

Some properties on spaces are carried over from the whole space to their subspaces, while others are not. Those properties that do are called "hereditary properties".

Definition 5.12 A topological property, say P, of a space (S, τ_S) is said to be a *hereditary topological property* provided every subspace, (T, τ_T), of S also has P.

Example 15. Show that metrizability is a hereditary property.

Solution: Suppose (S, τ) is metrizable. Then there exists a metric ρ such that (S, τ) and (S, ρ) have the same open sets. Suppose $T \subseteq S$ has the subspace topology and $\rho_t : T \times T \to \mathbb{R}$ is the subspace metric on T. Then (T, τ_t) and (T, ρ_t) have the same open sets and so T is metrizable. So, subspaces of metrizable spaces are metrizable.

"First countable" is another example of a hereditary topological property. However, "separable" is not a hereditary topological property. Witness the separable space, $\mathbb{R}$ with the usual topology; its subspace of all irrationals, $\mathbb{J}$, has no countable dense subset.

However, the reader may want to verify that, if S is separable and V is an *open* subspace of S, then V is separable.

We now show that the "second countable" property is hereditary.

Theorem 5.13 *Suppose (S, τ_S) is a second countable topological space. Then any non-empty subspace of S is also second countable. So "second countable" is a hereditary property.*

Proof. Suppose (S, τ_S) has a countable base $\mathscr{B} = \{B_i : i \in \mathbb{N}\}$. Suppose (T, τ) is a non-empty subspace of S. Let U be an open subset of T. Then there exists an open subset U^* of S such that $U = U^* \cap T$. Then there exists $N \subseteq \mathbb{N}$ such that $U^* = \cup\{B_i : i \in N\}$. Then $U = \cup\{B_i \cap T : i \in N\}$. So $\mathscr{B}_T = \{B_i \cap T : i \in \mathbb{N}\}$ is a countable basis of T. Hence, T inherits the second countable property from its superset S.

5.9 Topic: Ordinal Space

Ordinal numbers play a useful role in general topology. This "linearly well-ordered" set often serves as a counterexample to certain conjectures. Recall that a linearly ordered set is said to be "well-ordered" if,

every non-empty subset contains its least element.

We state some of the most fundamental facts about ordinal numbers.

Definition 5.14 An *ordinal number* is a set whose elements are themselves sets.[7] We say that the set, α, is an ordinal number if it satisfies the two properties:

— If u, v are elements of α, they are themselves "sets of sets" such that either $u \in v$, $v \in u$, or $u = v$.
— If $u \in v$, and $v \in \alpha$, then $u \in \alpha$.

The elements of an ordinal number, α, are linearly ordered by "$\in$ or $=$". The first countable ordinal numbers are in fact the natural numbers. Each ordinal is the set whose elements are all of its predecessors.

$$\begin{array}{lcl}
\varnothing & = & 0 \\
\{\varnothing\} & = & \{0\} = 1 \\
\{\varnothing, \{\varnothing\}\} & = & \{0,1\} = 2 \\
\{\varnothing, \{\varnothing\}, \{\varnothing, \{\varnothing\}\}\} & = & \{0,1,2\} = 3 \\
\{\varnothing, \{\varnothing\}, \{\varnothing, \{\varnothing\}\}, \{\varnothing, \{\varnothing\}, \{\varnothing, \{\varnothing\}\}\}\} & = & \{0,1,2,3\} = 4 \\
\vdots & & \vdots \\
n \cup \{n\} & = & \{0,1,2,\ldots,n\} = n+1 \\
\vdots & & \vdots \\
\{0,1,2,3,\ldots\} & = & \omega \\
\omega \cup \{\omega\} & = & \omega + 1 \\
\vdots & & \vdots
\end{array}$$

See that $n+1 = \{0,1,2,3,\ldots,n\}$ is both a subset of $n+2$ and an ordinal contained in the next ordinal, $n+2 = \{0,1,2,3,\ldots,n,n+1\}$.

---

[7]The family of *all* ordinal numbers is too large to be called a "set". So, we refer to it as the *class* of all ordinal numbers.

Also see that any $n + 1$ is both a subset of ω and an element of ω (where ω is the first countably infinite ordinal).

All elements of an ordinal are themselves ordinals (a statement which requires proof). Those ordinals that do not contain a maximal ordinal (such as ω, for example) are referred to as *limit ordinals.* These can also be recognized by the fact that limit ordinals don't have an immediate predecessor. The *limit* ordinal, ω, can be expressed as a half-open interval,

$$\omega = [0, \omega) = \{\text{ordinal } \alpha : \alpha < \omega\}$$

where "$\alpha < \omega$" is interpreted as $\alpha \in \omega$. A non-limit ordinal, $\beta + 1 = \{0, 1, 2, \ldots, \beta\}$ can be expressed as an interval

$$\beta + 1 = [0, \beta] = [0, \beta + 1) = \{\alpha : \alpha \leq \beta\}$$

where "$\alpha \leq \beta$" is interpreted as "$\alpha \in \omega$ or $\alpha = \beta$".

In the case of non-limit ordinals, $\beta + 1 = [0, \beta]$,

$$\cup (\beta + 1) = \cup [0, \beta] = \cup \{\alpha : \alpha \text{ ``}{\in_{\text{or}}}{=}\text{''}\ \beta\} = \beta$$

$$\sup (\beta + 1) = \sup [0, \beta] = \sup\{\alpha : \alpha \text{ ``}{\in_{\text{or}}}{=}\text{''}\ \beta\} = \beta$$

Limit ordinals are quite different in nature. We can recognize non-zero limit ordinals by the following characterization. The following three statements are equivalent:

— The ordinal, $\gamma = [0, \gamma) = \{\alpha : \alpha \in \gamma\}$, is a limit ordinal.
— For the ordinal γ, $\cup\gamma = \cup[0, \gamma) = \gamma$.
— For the ordinal γ, $\sup \gamma = \sup[0, \gamma) = \gamma$.

From this characterization, we can eventually argue (this is not immediate) that the union of all countable ordinals must be an uncountable ordinal, ω_1, which we refer to as the *first uncountable ordinal.* The ordinal $\omega_1 \cup \{\omega_1\} = \omega_1 + 1$ is its successor.

5.10 Topic: Topologizing an Ordinal Set

We will topologize an initial segment of ordinals, $S = [0, \omega_\alpha]$, (either with an open or closed right end) by defining an appropriate subbase, $\mathscr{S}$, which will generate a collection, $\mathscr{B}$, satisfying the "base property". The collection, $\mathscr{B}$, will, in turn, generate a topology on S.

Definition 5.15 Let ω_γ be an ordinal and $S = [0, \omega_\gamma] = \{\text{ordinals } \alpha : \alpha \leq \omega_\gamma\}$.

Suppose β and μ are both ordinals which belong to S. Let

$$S_\mu = (\mu, \omega_\gamma] = \{\alpha \in S : \alpha > \mu\}$$
$$S_\beta = [0, \beta) = \{\alpha \in S : \alpha < \beta\}$$

The standard subbase of S is defined as

$$\mathscr{S} = \{S_\mu : \mu \in S\} \cup \{S_\beta : \beta \in S\} = \{(\mu, \omega_\gamma] : \mu \in S\} \cup \{[0, \beta) : \beta \in S\}$$

This subbase will generate a base, $\mathscr{B}$, which in turn will generate the topology, τ_ω, of S.

When the space S is equipped with the topology τ_ω, (S, τ_ω), is referred to as an

ordinal space.

The topology that is generated by this subbase is called the

interval topology on the set of ordinals.

A few facts about an ordinal space. When we say "ordinal space" we mean a set of ordinals with the topology generated by the described subbase $\mathscr{S}$. But the best way to memorize the topology of the ordinal space is to remember what the elements of its base for open sets look like. Remember that there are two types of ordinals: limit ordinals and successor (non-limit) ordinals. Every ordinal number, α, without exception, has an immediate successor, $\alpha + 1$, by definition. Some ordinals, γ, have an immediate predecessor, say β, provided

$$\gamma = \beta + 1$$

In this case, $\sup [0, \gamma) = \sup [0, \beta + 1) = \sup [0, \beta] = \beta$.

While some ordinals, μ don't have an immediate predecessor (limit ordinal). In this case,

$$\sup \{\delta : \delta < \mu\} = \sup [0, \mu) = \mu$$

So, when we consider the intersection of two elements, $(\mu, \omega_\gamma]$ and $[0, \beta)$, of the subbase, $\mathscr{S}$, we get the open-ended interval (μ, β). At this point, we see only two possible cases for β:

— **Case 1:** Suppose β has an immediate predecessor, say γ (because $\beta = \gamma + 1$). In this case, we can express (μ, β) as the half-open interval, $(\mu, \gamma]$.
— **Case 2:** Suppose β doesn't have an immediate predecessor. Then $\beta = \{\delta : \delta < \beta\}$. In this case, (μ, β) can be expressed as the half-open interval, $(\mu, \beta]$.

In *both* cases, we have an open base element which is a half-open interval. So, we conclude that a base, $\mathscr{B}$, for open sets in the ordinal space, $S = [0, \omega_\gamma]$, is the set

$$\mathscr{B} = \{(\alpha, \beta] : \alpha, \beta \in S,\ \alpha < \beta\}$$

This is easier to remember.[8]

5.11 Topic: On Bases Generated by Regular Open Sets

Suppose we are given a topological space (S, τ). Recall that, in 4.12, we defined:

> *An open subset, U, of S is called a* regular open *subset if it satisfies the property,*
>
> $$U = \text{int}_S\text{cl}_S U$$

In the expression, $\text{int}_S\text{cl}_S U$, the interior and closure are with respect to the topology τ on S. The symbol

$$\mathscr{Ro}(S) = \{U \in \tau : U \text{ is regular open}\}$$

represents the set of all regular open subsets of S. So $\mathscr{Ro}(S) \subseteq \tau$. But since $\mathscr{Ro}(S)$ is not closed under unions (see the example on page 81) then it is not, by itself, a topology on S.

Clearly, $\varnothing$ and S belong to $\mathscr{Ro}(S)$. On page 81, we showed that $\mathscr{Ro}(S)$ is closed under finite intersections. Then, if $x \in S$ and $x \in A \cap B$ where $\{A, B\} \subseteq \mathscr{Ro}(S)$, since $A \cap B \in \mathscr{Ro}(S)$, then $\mathscr{Ro}(S)$

[8] A more detailed study of the ordinals in the context of set theory is found in Axioms and set theory, by Robert André (can be found online).

satisfies the "base property". Then, by Theorem 5.4,

$$\mathscr{Ro}(S) \text{ is an open base for some topology, say } \tau_s, \text{ on } S$$

The elements of $\mathscr{Ro}(S)$ are all open with respect to τ but not all elements in τ_s are necessarily regular open in S. Furthermore, there may be some open sets in τ which are not unions of elements in $\mathscr{Ro}(S)$ and so these are not in τ_s. So we have possibly distinct topologies, τ_s and τ, where

$$\mathscr{Ro}(S) \subseteq \tau_s \subseteq \tau$$

Note that the topology generated by $\mathscr{Ro}(S)$ is weaker than τ, but under certain conditions, τ and τ_s may be equivalent topologies.

Definition 5.16 Let (S, τ) be a topological space and τ_s denote the topology whose base is $\mathscr{Ro}(S)$. We have seen that τ_s and τ may be distinct topologies for the same underlying set, S.

If τ_s is a proper subset of τ, the topological space, (S, τ_s), is called the *semiregularization of S with respect to τ*.

If $\tau = \tau_s$, then we will call (S, τ) a *semiregular* topological space.[9] That is, a semiregular topological space is a space such that for all $U \in \tau$, U is the union of regular open sets.[10]

The semiregularization of, (S, τ), is akin to running τ through a strainer where only the elements of τ_s successfully make it through. The reader should verify that the set of real numbers, $\mathbb{R}$, equipped with the usual topology is an example of a semiregular topological space. Hence, it is its own semiregularization. We will refer to various properties of semiregular spaces further on in the text.

[9]We will specify later in this text that, in the literature, semiregular spaces are assumed to be Hausdorff. But for now this is irrelevant.

[10]The choice of the name "semiregular space" suggests that we will be introduced to a "regular space" at some point. This is the case. We will show in Chapter 9 that "regular spaces" (to be defined in that chapter) are semiregular spaces, but there exists semiregular spaces which are not regular spaces.

5.12 Topic: Spaces with a Base of Clopen Sets

We now consider the set, $\mathscr{B}(S) = \{U \in \mathscr{P}(S) : U \text{ is clopen}\}$, of all clopen sets in the topological space, (S, τ). The set $\mathscr{B}(S)$ is never empty since $\varnothing$ and S are elements of that set. Furthermore, if U and V belong to $\mathscr{B}(S)$ and $x \in U \cap V$, given that $U \cap V$ also belongs to $\mathscr{B}(S)$ then $\mathscr{B}(S)$ satisfies the "base property".

This means that $\mathscr{B}(S)$ forms a base for some topology, τ_b, on S. Since $\mathscr{B}(S) \subseteq \tau$, then the topology, τ_b, is weaker than τ, but may, under certain conditions, be equivalent to it.

Definition 5.17 Let (S, τ) be a topological space and τ_b denote the topology whose base is

$$\mathscr{B}(S) = \{U \in \mathscr{P}(S) : U \text{ is clopen}\}$$

If $\tau = \tau_b$, then we say that (S, τ) is a *zero-dimensional* topological space.

Example 17. Consider the subspace, $(\mathbb{Q}, \tau)$, of rational numbers with the subspace topology of $\mathbb{R}$ itself equipped with the usual topology. Verify that $\mathbb{Q}$ is a zero-dimensional topological space.

Solution: Consider $\mathscr{U} = \{(a, b) : a < b, a \text{ and } b \text{ irrationals}\} \subseteq \mathbb{R}$. This forms a base for open sets in $\mathbb{R}$. Then $\mathscr{U}_Q = \{U \cap \mathbb{Q} : U \in \mathscr{U}\}$ forms a base for open sets for $\mathbb{Q}$. Each element of $\mathscr{U}_Q$ is clopen in $\mathbb{Q}$ so $\mathbb{Q}$ is zero-dimensional.

This example confirms that there are non-discrete zero-dimensional spaces.

Concept Review

1. Define a neighborhood system of x with respect to τ.
2. Define a neighborhood base of x with respect to the topology τ.
3. Is $\varnothing$ a neighborhood of a point x? Is a neighborhood of x necessarily open?
4. Define a base for a topology τ.

5. Find a base for the usual topology τ on $\mathbb{R}$.
6. Give a characterization of a base of τ in terms of "neighborhoods".
7. What does it mean to say that a subset $\mathscr{P}(S)$ satisfies the "base property"?
8. Describe the Moore plane and the base for its topology.
9. Given an arbitrary subset $\mathscr{S}$ of $\mathscr{P}(S)$ explain how it can be used to construct a topology on S.
10. Define a *subbase for a topology.*
11. Describe a subbase for the usual topology on $\mathbb{R}$.
12. Describe a topology generated by a subbase $\mathscr{S}$ in terms of other topologies on S.
13. Given two topological spaces S and T and a Cartesian product $S \times T$. Find a useful subbase involving projection maps that can be used to generate a topology on $S \times T$.
14. Describe the *upper limit topology* (or Sorgenfrey topology) on $\mathbb{R}$ in terms of its base and subbase.
15. Is $\mathbb{R}$ with the upper limit topology first countable? Is it second countable.
16. What does it mean to say that two subsets of $\mathscr{P}(S)$ are equivalent topologies for the set S.
17. What can we say about the size of the neighborhood bases at the points of $\mathbb{R}$ with respect to the usual topology?
18. hat can we say about the size of the base of $\mathbb{R}$ with respect to the usual topology?
19. Define *first countable topological space.*
20. Define *second countable topological space.*
21. Is $\mathbb{R}$ equipped with the usual topology first countable? What about second countable?
22. Describe a topological space which is first countable but not second countable.
23. State a relationship between "second countable" and "separable".
24. Describe the topology on an ordinal space.
25. Metrizable space are necessarily first countable. Describe a neighborhood base.
26. Describe how the set of regular open subsets can be used to generate a topology on a space.
27. What does it mean to say that a property is hereditary?

28. Define a semiregular space.
29. Given a topological space, (S, τ), what is the semiregularization of (S, τ)?
30. Define a zero-dimensional space. As an example provide a subspace of $\mathbb{R}$ which is zero-dimensional.

Exercises

1. Show that, if F is a closed subset of the topological space (S, τ) and $x \notin F$, then x has a neighborhood which does not intersect F.
2. Let $\vec{x} = (a, b) \in \mathbb{R}^2$ equipped with the usual (Euclidean) topology. For each $q \in \mathbb{R}$, let $B_{\vec{x}}(q) = \{(x, y) : \max\{|a - x|, |b - y|\} < q\}$. Show that $\mathscr{B}_{\vec{x}} = \{B_{\vec{x}}(q) : q \in \mathbb{R}, q \geq 0\}$ forms a neighborhood base at $\vec{x}$.
3. Suppose τ_1 and τ_2 are two topologies on the set S with respective bases $\mathscr{B}_1$ and $\mathscr{B}_2$. Show that $\tau_1 \subseteq \tau_2$ if and only if whenever $x \in B_1$ there exists $B_2 \in \mathscr{B}_2$ such that $x \in B_2 \subseteq B_1$.
4. Let A and B be two infinite sets each equipped with the cofinite topology. Describe a basis of $A \times B$ equipped with the product topology.
5. Let (S, τ) be a topological space with base $\mathscr{B}$. If $A \subseteq S$, show that $\mathscr{B}_A = \{B \cap A : B \in \mathscr{B}\}$ is a base for the open sets of A.
6. Let (S, τ_S) and (T, τ_T) be two first countable topological spaces. Show that $S \times T$ equipped with the product topology is first countable.
7. Prove that, if (A, τ_1) is a subspace of (B, τ_2) and (B, τ_2) is a subspace of (C, τ_3) then (A, τ_1) is a subspace of (C, τ_3).

Chapter 6

Continuity on Topological Spaces

Abstract
In this section, we formally define the notion of a continuous function mapping one topological space into another. We discuss various characterizations of these. An important subclass of continuous functions is the one called "homeomorphisms". We will see why these are fundamental in the study of topological spaces.

6.1 Basic Notions and Notation Associated to Functions Mapping Sets to Sets

We begin by establishing the notation and terminology we will use in our discussion of functions. Suppose $f : S \to T$ is a well-defined function mapping a topological space, S, into another topological space, T. For $f : S \to T$ the expressions, "f is one-to-one and onto" and "f is a bijection between S and T" are simply different ways of conveying the same idea.

The function, $f : S \to T$, induces another function, $f : \mathscr{P}(S) \to \mathscr{P}(T)$, where

$$f[A] = \{y \in T : y = f(x) \text{ for some } x \in A \subseteq S\}$$

Essentially, $f[A]$ is the image of the set A under the function $f : S \to T$. The function $f : S \to T$ also induces the function

$f^{\leftarrow} : \mathscr{P}(T) \to \mathscr{P}(S)$ where

$$f^{\leftarrow}[B] = \{x \in S : \text{ where } f(x) \in B\}$$

We can also say that, if $f^{\leftarrow}[B] = D$, then D is the "inverse image" or "pre-image" of B under f, or that the function $f^{\leftarrow}$ "pulls back" the set B onto the set D inside the domain S of f. In the case where $f^{\leftarrow}[\{y\}] = \{x\}$, if there is no risk of confusion, we will simply write $f^{\leftarrow}(y) = x$.[1] In the case where $f : S \to T$ is one-to-one and onto T then $f^{\leftarrow}$ can itself be seen as a well-defined function, $f^{\leftarrow} : T \to S$, and so we can write "$f^{\leftarrow}(x) = y$ if and only if $f(y) = x$" (without using the square brackets).

In this book, if $f : S \to \mathbb{R}$ is a function and $0 \notin f[S]$, f^{-1} will be interpreted as follows:

$$f^{-1}(x) = \frac{1}{f(x)}$$

In the following theorem statement, we review some basic principles on how functions act on sets. In particular, we are reminded of the following three principles:

(1) A function "respects arbitrary unions of sets".
(2) If a function is *one-to-one* it will "respect arbitrary intersections". Otherwise, a function "does *not* always respect intersections of sets".
(3) An inverse function, $f^{\leftarrow}$, "always respects unions, intersections and complements of sets".

Theorem 6.1 *Let* $f : A \to B$ *be a* function *mapping the set* A *to the set* B. *Let* $\mathscr{A}$ *be* a set *of subsets of* A *and* $\mathscr{B}$ *be* a set *of subsets of* B. *Let* $D \subseteq A$ *and* $E \subseteq B$. *Then:*

(a) $f\left[\bigcup_{S\in\mathscr{A}} S\right] = \bigcup_{S\in\mathscr{A}} f[S]$.
(b) $f\left[\bigcap_{S\in\mathscr{A}} S\right] \subseteq \bigcap_{S\in\mathscr{A}} f[S]$ *where equality holds true only if* f *is one-to-one.*

[1]Note that f need not be one-to-one on all of the domain in order for us to speak of $f^{\leftarrow}$ in this way.

(c) $f[A \backslash D] \subseteq B \backslash f[D]$. *Equality holds true only if f is one-to-one and onto B.*
(d) $f^{\leftarrow}\left[\bigcup_{S \in \mathscr{B}} S\right] = \bigcup_{S \in \mathscr{B}} f^{\leftarrow}[S]$.
(e) $f^{\leftarrow}\left[\bigcap_{S \in \mathscr{B}} S\right] = \bigcap_{S \in \mathscr{B}} f^{\leftarrow}[S]$.
(f) $f^{\leftarrow}[B \backslash E] = A \backslash f^{\leftarrow}[E]$.

Proof.

(a)
$$\begin{aligned} x \in f\left[\bigcup_{S \in \mathscr{A}} S\right] &\Leftrightarrow \quad x = f(y) \text{ for some } y \in \bigcup_{S \in \mathscr{A}} S \\ &\Leftrightarrow \quad x = f(y) \text{ for some } y \text{ in some } S \in \mathscr{A} \\ &\Leftrightarrow \quad x = f(y) \in f[S] \text{ for some } S \in \mathscr{A} \\ &\Leftrightarrow \quad x \in \bigcup_{S \in \mathscr{A}} f[S] \end{aligned}$$

(b) It will be helpful to first prove this statement for the intersection of only two sets U and V. The use of a Venn diagram will also help visualize what is happening.

So, we first prove the statement: $f[U \cap V] \subseteq f[U] \cap f[V]$ with equality only if f is one-to-one on $U \cup V$.

Case 1: We consider the case where $U \cap V = \varnothing$.
Then $f[U \cap V] = \varnothing \subseteq f[U] \cap f[V]$. So, the statement holds true.

Case 2: We now consider the case where $U \cap V \neq \varnothing$.

$$\begin{aligned} x \in f[U \cap V] &\Leftrightarrow x = f(y) \text{ for some } y \in U \cap V \\ &\Leftrightarrow x = f(y) \text{ for some } y \text{ contained in both } U \text{ and } V \\ &\Rightarrow x = f(y) \in f[U] \text{ and } f[V] \\ &\Leftrightarrow x \in f[U] \cap f[V] \end{aligned}$$

We now show that if f is one-to-one on $U \cup V$, then $f[U] \cap f[V] \subseteq f[U \cap V]$ and so equality holds true.

— Suppose $x = f(y) \in f[U] \cap f[V]$. Then there exists $u \in U$ and $v \in V$ such that $f(u) = f(v) = f(y)$. Since f is one-to-one, $u = v = y$. This implies $y \in U \cap V$. Hence, $f[U \cap V] = f[U] \cap f[V]$.

The proof of the general statement is left as an exercise.

(c) Proof is left as an exercise.

(d) $$x \in f^{\leftarrow}\left[\bigcup_{S\in\mathscr{B}} S\right] \Leftrightarrow x = f(y) \text{ for some } y \in \bigcup_{S\in\mathscr{B}} S$$

$$\text{(By definition of } f^{\leftarrow}.)$$

$$\Leftrightarrow x = f(y) \text{ for some } y \text{ in some } S \in \mathscr{B}$$

$$\Leftrightarrow x \in f^{\leftarrow}[\{y\}] \subseteq f^{\leftarrow}[S] \text{ for some } S \in \mathscr{B}$$

$$\Leftrightarrow x \in \bigcup_{S\in\mathscr{B}} f^{\leftarrow}[S]$$

Thus, $f^{\leftarrow}\left(\bigcup_{S\in\mathscr{B}} S\right) = \bigcup_{S\in\mathscr{B}} f^{\leftarrow}(S)$.

(e) Proof is left as an exercise.

(f) Proof is left as an exercise.

6.2 Continuous Functions on Topological Spaces

Given two topological spaces, (S, τ_S) and (T, τ_T), we will discuss various types of functions, $f : S \to T$, which map S into T. The reader is already familiar with those functions called "continuous functions" mapping $\mathbb{R}$ to $\mathbb{R}$. We will generalize this notion of continuity to topological spaces.

Our formal definition of a continuous function mapping a topological space into another is presented below. Those readers who are familiar with the "epsilon-delta" definition of a continuous function (normally presented in any *Introduction to analysis* course) will notice the analytical approach cannot be used in topology since topological spaces, in their most rudimentary form are not equipped with distance functions such as absolute values, norms or metrics.

Definition 6.2 Let $f : S \to T$ be a function mapping (S, τ_S) into (T, τ_T). We say that

$f : S \to T$ is continuous on S if, for any open subset $U \in T$, $f^{\leftarrow}[U]$ is open in S.

If $x \in S$, we will say that "f *is continuous at the point* x" if, for any neighborhood, U, of $f(x)$ (inside T) there exists a neighborhood V of x such that $f(x) \in f[V] \subseteq U$.

So, a function $f : S \to T$ is continuous if f pulls back open sets in T to open sets in S".

The above definition has two parts to it. The first describes continuity of f on a set while the second describes continuity of f at a point x. Clearly, we cannot apply the first definition to determine continuity of f at a point. But a set A is simply a collection of points. If a function f can be shown to be continuous at every point x in a set A we would hope that, we would obtain continuity on the set as defined in part (a). The next theorem confirms that this is the case.

Theorem 6.3 *Let (S, τ_S) and (T, τ_T) be two topological spaces and $f : S \to T$ be a function. Then f is continuous on S if and only if f is continuous at every point of S.*

Proof. ($\Rightarrow$) Suppose f is continuous on S, $x \in S$ and $y = f(x) \in f[S]$. We are required to show that f is continuous at the point x.

Suppose U is a neighborhood of y. By definition of continuity on a set, $f^{\leftarrow}[\text{int}_T U]$ is an open neighborhood of $x \in f^{\leftarrow}[\{y\}] \subseteq f^{\leftarrow}[\text{int}_T U]$ in S. By definition of neighborhood, there exists an open set, $V \subseteq f^{\leftarrow}[\text{int}_T U]$, such that $x \in V$. Then $y = f(x) \in f[V] \subseteq U$. So, we have found the required neighborhood, V, of x. So f is continuous at the point x.

($\Leftarrow$) Suppose that f is continuous at every point of S. Let U be a non-empty open subset of $f[S] \subseteq T$. We are required to show that $f^{\leftarrow}[U]$ is open in S.

Let $x \in f^{\leftarrow}[U]$. Then $f(x) \in U$. Since f is continuous at x, then there exists a neighborhood V of x such that $f[V] \subseteq U$. Now $x \in \text{int}_T V \subseteq f^{\leftarrow}[f[V]] \subseteq f^{\leftarrow}[U]$. Then $f^{\leftarrow}[U]$ is an open subset of S. So f is continuous on S.

There are other ways of recognizing those functions which are continuous on a set. For example, if $f : S \to T$ satisfies the property,

"$f^{\leftarrow}[F]$ is closed in S whenever F is closed in the codomain T"

then, when U is open in T, $S \setminus f^{\leftarrow}[U] = f^{\leftarrow}[T \setminus U]$ is closed in S. Then $f^{\leftarrow}[U]$ is open in S and so f is continuous on S. The reader is left to verify that the converse also holds true.

Other useful characterizations of continuity on a topological space are given as follows.

Theorem 6.4 *Let (S, τ_S) and (T, τ_T) be two topological spaces and $f : S \to T$ be a function.*

(a) *The function f is continuous on S if and only if f pulls back subbase elements of the topological space T to open sets in S.*[2]
(b) *The function f is continuous on S if and only if f pulls back open base elements of the topological space T to open sets in S.*[3]
(c) *The function f is continuous on S if and only if, for any subset U of S,*

$$f[cl_S U] \subseteq cl_T f[U]$$

Proof. The proofs of parts (a) and (b) are left as an exercise.

(c) ($\Rightarrow$) Suppose f is continuous on S. To show that $f[\text{cl}_S U] \subseteq \text{cl}_T f[U]$ it will suffice to show that $x \notin \text{cl}_T f[U]$ implies $x \notin f[\text{cl}_S U]$.

Suppose $x \notin \text{cl}_T f[U]$. Then there musts exist an open V in T such that $x \in V \subseteq T \setminus \text{cl}_T f[U]$. Then $f^{\leftarrow}(x) \subseteq f^{\leftarrow}[V] \subseteq S \setminus U$. By continuity of f, $f^{\leftarrow}[V]$ is open in S and so $\text{cl}_S U \cap f^{\leftarrow}[V] = \varnothing$. Then $f[f^{\leftarrow}(x)] = x \notin f[\text{cl}_S U]$. So $f[\text{cl}_S U] \subseteq \text{cl}_T f[U]$.

The proof of ($\Leftarrow$) is left as an exercise.

Suppose A is a subspace of the topological space, (S, τ), equipped with the subspace topology, τ_A. Suppose $f : S \to T$ is known to be continuous on its domain. Then, when f is restricted to $f|_A$ on A, $f|_A$ preserves the continuity property on A. This is confirmed by the following theorem statement.

Theorem 6.5 *Let (S, τ_S), (T, τ_T), and (Z, τ_Z) be topological spaces.*

[2]That is, $f^{\leftarrow}[B]$ is open in S whenever B is a subbase element of T.
[3]That is, $f^{\leftarrow}[B]$ is open in S whenever B is a base element of T.

(a) *If* $f : S \to T$ *and* $g : T \to Z$ *are both continuous on their domains then* $g \circ f : S \to Z$ *is continuous on* S. *(That is, the composition of continuous functions is continuous.)*

(b) *Suppose* $f : S \to T$ *is a continuous function on* S *and* $A \subseteq S$. *Let* $f|_A : A \to T$ *denote the restriction of* f *to the subset* A. *Then* $f|_A$ *is continuous on* A.

Proof. (a) The proof of part (a) is left as an exercise.

(b) Let U be an open subset of $f[A]$ with respect to the subspace topology $\tau_{f[A]}$ in T. Then there exists $U^* \in \tau_T$ such that $U = U^* \cap f[A]$. Now

$$\begin{aligned} x \in f|_A^{\leftarrow}[U] &\Rightarrow \{x \in A : f(x) \in U\} \\ &\Rightarrow x \in \{x \in S : f(x) \in U^*\} \cap A \\ &\Rightarrow x \in f^{\leftarrow}[U^*] \cap A \end{aligned}$$

Since f is continuous on S, $f^{\leftarrow}[U^*] \cap A$ is an open subset of A with respect to the subspace topology. So $f|_A^{\leftarrow}[U]$ is open in A.

We have seen that continuous functions are those functions which "pull back" open sets to open sets, or, equivalently, "pull back" closed sets to closed sets. We will encounter at least two other similar types of functions, which are not necessarily continuous.

Definition 6.6 Let $f : S \to T$ be a function mapping the topological space, (S, τ_S), into the space, (T, τ_T). We say that $f : S \to T$ is *an open function on* S, if for any open subset U of S, $f[U]$ is open in the space, (T, τ_T). We say that $f : S \to T$ is *a closed function on* S, if for any closed subset F of S, $f[F]$ is closed in the space, (T, τ_T).[4]

The reader should be alerted to the fact that an open function need *not* be a continuous function; similarly, a closed function need *not* be

[4]Whether the function $f : S \to T$ is open or closed depends on the topology defined on S and T.

continuous. Also, we caution the reader by pointing out that, even if $f : S \to T$ is an open function, it does *not* necessarily follow that $f^{\leftarrow} : T \to S$ is a continuous function on T. (See one of the examples that will follow.) However, for one-to-one functions f, "f is closed if and only if f is open" is true, as we shall now prove.

Supplementary theorem. Suppose $f : S \to T$ is *one-to-one and onto* T. Then f is an open function if and only if f is a closed function.

Proof. ($\Rightarrow$) Suppose $f : S \to T$ is a one-to-one and onto open function. We are required to show that f is also a closed function. Let F be a closed subset of S. Then $U = S \backslash F$ is open in S and so $f[U]$ is open in T. Then

$$\begin{aligned} f[U] &= f[S\backslash F] \\ &= f[S]\backslash f[F] \quad \text{(Since } f \text{ is one-to-one)} \\ &= T\backslash f[F] \end{aligned}$$

Since $T\backslash f[F]$ is open in T, then $f[F]$ is closed in T. We conclude that f is a closed function, as required.

Proof of the converse ($\Leftarrow$) is left to the reader.

Example 1. Suppose $(\mathbb{R}, \tau)$ denotes the real line with the usual topology, τ, and $(\mathbb{R}, \tau_s)$ denotes the real line with the *upper limit topology* (Sorgenfrey line), τ_S.

Let $i : (\mathbb{R}, \tau) \to (\mathbb{R}, \tau_s)$ denote the identity map, $i(x) = x$. Verify that this identity map is open on $(\mathbb{R}, \tau)$ but is not continuous on its domain.

Solution: Since the function $i(x) = x$ maps the open base element, (a, b), of τ to the open set $(a, b) \in \tau_s$ (as we have seen earlier, $\tau \subset \tau_s$), then i maps open sets to open sets hence, i is an open function. But $(a, b] \notin \tau$. So $i^{\leftarrow}[(a, b]] = (a, b] \notin \tau$. So, the open identity map i is *not* continuous on $\mathbb{R}$ with respect to τ.

This example illustrates that, if the codomain has more open sets then the domain, then the identity function will not pass the test of continuity.

Example 2. Let $(\mathbb{R}^2, \tau)$ be equipped with the usual topology. Suppose the open ball center (0, 0) of radius 1,

$$B = \{(x, y) \in \mathbb{R}^2 : x^2 + y^2 < 1\}$$

is equipped with the subspace topology, τ_B. We define the function $f : B \to \mathbb{R}^2$ as follows:

$$f(x, y) = (x, y)$$

If U is an open subset of B then $f[U] = U \cap B$ an open subset of $\mathbb{R}^2$. So f is an open map. But f is not a closed map. To see this note that B is a closed subset of itself with respect to τ_B. On the other hand, $f[B] = B$ is not closed in the codomain $\mathbb{R}^2$.

If f is one-to-one and onto, we arrive at a different conclusion, as we shall now see.

Theorem 6.7 *Let (S, τ_S) and (T, τ_T) be topological spaces. For a one-to-one and onto continuous function $f : S \to T$,*

$$f^{\leftarrow} : T \to S \text{ is continuous} \quad \Leftrightarrow \quad f : S \to T \text{ is open}$$

Proof. The proof is left as an exercise.

6.3 Homeomorphic Topological Spaces

For a continuous function, $f : S \to T$, its inverse function, $f^{\leftarrow} : T \to S$, may, *or may not*, be continuous, even if f is one-to-one. Those one-to-one continuous functions $f : S \to T$ where $f^{\leftarrow}$ is continuous on $f[S] \subseteq T$ are fundamental in the study of topology. They have a special name.

Definition 6.8 Let (S, τ_S) and (T, τ_T) be topological spaces and $f : S \to T$ be a function. If f simultaneously satisfies all three of the following conditions,

(1) f is one-to-one on S and onto T
(2) f is continuous on S
(3) $f^{\leftarrow}$ is continuous on T

then the function, f, is called a *homeomorphism from* S *onto* T. If $f : S \to T$ is a homeomorphism then S and T are said to be *homeomorphic topological spaces*, or f is said to *map* S *homeomorphically onto* T.

Example 3. Let the open interval, $S = (-\pi/2, \pi/2)$, be equipped with the usual subspace topology. The one-to-one and onto function, $\tan : S \to \mathbb{R}$, is well-known to be continuous on its domain, S. Similarly its inverse (arctan), $\tan^{\leftarrow} : \mathbb{R} \to S$, is continuous on its domain, $\mathbb{R}$. By definition, tan is a homeomorphism. We can then say that $S = (-\pi/2, \pi/2)$ and $\mathbb{R}$ are homeomorphic topological spaces. In fact, as we shall see, any non-empty open interval of $\mathbb{R}$ is homeomorphic to $\mathbb{R}$.

Theorem 6.9 *Let* (S, τ_S) *and* (T, τ_T) *be topological spaces. Suppose* $f : S \to T$ *is* one-to-one and onto S. *The following are equivalent:*

(1) *The function* f *is a homeomorphism.*
(2) *The function* f *is both continuous and open.*
(3) *The function* f *is both continuous and closed.*

Proof. The proof follows from the statement in Theorem 6.7. It is left as an exercise.

When the function $f : S \to T$ is one-to-one and open but is *not* onto T, the pair S and T cannot be said to be homeomorphic spaces. In this case, S is homeomorphic to the proper subspace, $f[S]$, of T.

We often express this situation by saying that "f embeds S in T". We define this formally.

Supplementary definition. Let (S, τ_S) and (T, τ_T) be topological spaces and $f : S \to T$ be a function mapping S into T (which may or may not be "onto"). If f is a homeomorphism then we say that

f embeds S into T

or, simply that "f is an *embedding*". When the specific homeomorphism in question is not explicitly described, we might simply say that

T contains a homeomorphic copy of S.

This homeomorphic copy is, of course, the image of the hypothesized homeomorphism.

Also note that, if $f : S \to T$ embeds S into T, then $f^{\leftarrow} : f[S] \to S$ is also a homeomorphism.

We have seen that, if $f : S \to T$ is a homeomorphism between the two topological spaces (S, τ_S) and (T, τ_T), the one-to-one function, f, maps each open base element in τ_S to a unique set in τ_T, and vice-versa. Since f is one-to-one, it respects both arbitrary unions and arbitrary intersections; then every element of τ_S will be paired, under f, to exactly one element in τ_T. This suggests that some properties in S which involve open sets will be "mirrored" inside T. We will refer to such properties as "topological properties". We formally define this notion.

Definition 6.10 Suppose P is property which is satisfied on a topological space S. If P is satisfied in any homeomorphic copy of S, then P is called a *topological property* or a *topological invariant.*

If $f : S \to T$ is known to be a homeomorphism and P is known to be a topological property, then, by definition,

P is satisfied in $f[S]$ if and only if P is satisfied in S.

Example 4. Recall that in a metric space, (S, ρ), a sequence, $\{x_n : n \in \mathbb{N}\}$, is referred to as being a Cauchy sequence if and only if, for every $\varepsilon > 0$, there exists N such that whenever $n, m > N$ then $\rho(x_n, x_m) < \varepsilon$. Being a "Cauchy sequence" in a metric space property which does not translate easily to a topological space. To see this consider the continuous function $g(x) = 1/x$ on $\mathbb{R}\backslash\{0\}$. The function g maps $\mathbb{R} \setminus \{0\}$ homeomorphically onto itself. The sequence $T = \{1,\ 1/2,\ 1/3, \ldots\}$ is a Cauchy sequence in the domain of g. But $g[T] = \{1, 2, 3, \ldots\}$ is not Cauchy in the image of g in $\mathbb{R}\backslash\{0\}$.[5]

Example 5. Consider the following property, P, on a space S:

> Every real-valued continuous function on S assumes its maximum value on S.

Verify that P is a topological property.

Solution: The negation, $\neg P$, of the property P is: "There is a continuous real-valued function on S which does not attain its maximum value in S". We will show that $\neg P$ is a topological property. The desired result will follow.

Suppose S is a space which satisfies $\neg P$. Then there exists a continuous function,

$$f : S \to \mathbb{R}$$

which does not attain its maximum value in S.

Let $h : S \to T$ be a homeomorphism mapping S onto T. We claim that T also satisfies $\neg P$.

Proof of claim: Consider the continuous function, $f{\circ}h^{\leftarrow} : T \to \mathbb{R}$. If $b \in T$, then $h^{\leftarrow}(b) \in S$. Then, by hypothesis, there must exist $a \in S$ such that $f(a) > (f{\circ}h^{\leftarrow})(b)$. Note that $(f{\circ}h^{\leftarrow})(h(a)) = f(a) > (f{\circ}h^{\leftarrow})(b)$. Thus $f{\circ}h^{\leftarrow}$ does not assume a maximum value on T. Then T also satisfies $\neg P$, as claimed.

We conclude that S satisfies $\neg P$ if and only if T satisfies $\neg P$. Then S satisfies P if and only if T satisfies P. So, P is a topological property.

It is interesting to note that the closed interval $[0, 1]$ cannot be homeomorphic to $\mathbb{R}$ since elementary calculus shows that $[0, 1]$ satisfies property P and $\mathbb{R}$ does not.

---

[5]We will eventually see that the notion of uniform continuity of a function on a subset of a metric space also does not translate automatically to a topological space.

6.4 Continuity and Countability Properties

We have previously defined the three countability properties,

separable
first countable
second countable.

We would like to verify whether or not a continuous function will always carry over each of these properties from its domain into its codomain.

Recall that a topological space, (S, τ), is *separable* if and only if S contains a countable dense subset. We expect that "separable" is a topological property. This fact is confirmed by the following result which shows that separability is carried over by continuous functions. Hence, if S is separable, every topological space which is homeomorphic to S is also separable.

Theorem 6.11 *Suppose (S, τ_S) is a separable topological space. Then any continuous image of S is also separable.*

Proof. Suppose (S, τ_S) is separable and $f : S \to T$ is a continuous function mapping S onto the topological space (T, τ_T).

We claim that the image, T, of S under f is also separable. Given that S is separable, then it contains a countable dense subset, say D. Since the cardinality of the image of a function is less than or equal to the cardinality of its domain, then $f[D]$ is a countable subset of T. It now suffices to show that $f[D]$ is dense in T. Suppose U is a non-empty open subset of T. Continuity of f guarantees that $f^{\leftarrow}[U]$ is open and so there must exist $x \in f^{\leftarrow}[U] \cap D$. Then

$$\begin{aligned} f(x) &\in f[f^{\leftarrow}[U] \cap D] \\ &\subseteq f[f^{\leftarrow}[U]] \cap f[D] \\ &= U \cap f[D] \end{aligned}$$

Hence, every open subset U of T intersects $f[D]$ in a non-empty set. We conclude that $f[D]$ is dense in T. So, T is separable.

A simple continuous function is (by itself) not quite strong enough to carry over the second countable property or the first countable property from its domain to its codomain. We suspect that homeomorphisms can certainly do so. But, as we shall see, we don't need the full power of a homeomorphism for this. The following theorem shows that a continuous *open* function will suffice. If so, since homeomorphisms are both continuous and open, then "second countable" and "first countable" are topological properties.

Theorem 6.12 *Let $f : S \to T$ be a continuous* open *function mapping S onto a space T. If the space S is second countable then T is second countable. If the space S is first countable then T is first countable.*

Proof. Let (S, τ_S) and (T, τ_T) be topological spaces and $f : S \to T$ be a *continuous open function* mapping S onto T.

If S is second countable then S has a countable open base, $\mathscr{B}$. Consider the family, $\mathscr{B}_T = \{f[B] : B \in \mathscr{B}\}$. By hypothesis, $\mathscr{B}_T$ is a set of countably many open subsets of T. We claim that the countable set, $\mathscr{B}_T$, is a base for open sets in T.

Proof of claim: Let U be a non-empty open subset of T and let $y \in U$. Let $x \in f^{\leftarrow}(y) \in f^{\leftarrow}[U]$. Since f is continuous, $f^{\leftarrow}[U]$ is open in S, so there exist an open $V \in \mathscr{B}$ such that $x \in V \subseteq f^{\leftarrow}[U]$. Since f is declared to be open, $f(x) = y \in f[V] \in \mathscr{B}_T$, an open subset of U. We can then conclude that U is the union of elements from $\mathscr{B}_T$. So, $\mathscr{B}_T$ forms a base for T, so T is second countable.

The proof that the "first countable" property is carried over by continuous open functions is left as an exercise.

6.5 The Weak Topology Induced by a Family of Functions

Suppose (T, τ_T) is a space and we are given a function, $f : S \to T$, mapping the set S onto the topological space, T (where S is not yet topologized). With this premise alone, we would normally not discuss the continuity of f since "continuity" is defined in terms of a topology on both the function's domain and codomain. But we could first hypothesize "continuity of f" and then force on S enough open sets so that this family of open sets would support the continuity of f.

Is this hard to do? It is quite easy. We need only equip the domain, S of f, with the discrete topology, τ_d. Then, no matter how f is defined, since every subset of S is open then f is, by definition, continuous on S. But this is not entirely satisfactory — nor an interesting problem — since the discrete topology on S doesn't depend on the function, f, at all. We can tighten up our question a bit so that the topology on S will relate to the specific function f.

Let $f : S \to T$ be some function mapping the set S into the topological space T. We will topologize S so that, not only is the function f guaranteed to be continuous on S, but that the chosen topology is the smallest such topology. To do this, we will define

$$\mathscr{S}_f = \{f^{\leftarrow}[U] : U \in \tau_T\}$$

a subset of $\mathscr{P}(S)$, declaring it to be a subbase for a topology. By taking all finite intersections of elements from $\mathscr{S}_f$ we will generate the subset, $\mathscr{B}_f$, of $\mathscr{P}(S)$ which satisfies the "base property". By taking all unions of elements from $\mathscr{B}_f$ we will generate the smallest topology, τ_f, with subbase $\mathscr{S}_f$. The family τ_f is the weakest topology possible that will guarantee the continuity property on f. We will refer to τ_f as being the *weak topology on S induced by f*. Eliminating just one element from this topology would be sufficient to prevent f from being continuous on S. Also the topology we obtain is directly related to the function f. If we consider a different function we may obtain a different topology.

If we want the two functions $f : S \to T$ and $g : S \to T$ to be continuous on S, we would require more open sets on S. That is, the required subbase would have to be,

$$\mathscr{S}_{\{f,g\}} = \{f^{\leftarrow}[U] : U \in \tau_T\} \cup \{g^{\leftarrow}[U] : U \in \tau_T\}$$

In this case, the topology generated by the subbase $\mathscr{S}_{\{f,g\}}$ would be called the *weak topology induced by $\{f, g\}$*.

We can generalize this even more. Rather than restrict ourselves to only two functions each with the same domain S, we will consider the more general *weak topology on S induced by a family of functions* $\{f_\alpha : S \to T_\alpha\}_{\alpha \in I}$, formally defined as follows.

Definition 6.13 Let S be a non-empty set and $\{(T_\alpha, \tau_\alpha) : \alpha \in \Gamma\}$ denote a family of topological spaces. For each $\alpha \in \Gamma$, suppose

$$f_\alpha : S \to T_\alpha$$

is a function mapping S onto T_α. Let

$$\mathscr{S} = \{f_\alpha^{\leftarrow}[U_\alpha] : U_\alpha \in \tau_\alpha\}_{\alpha \in \Gamma}$$

Then $\mathscr{S}$ can be declared to be the subbase of a topology, $\tau_{\mathscr{S}}$, on S. The topology $\tau_{\mathscr{S}}$ is called the

weak topology induced by the family of functions $\{f_\alpha : \alpha \in \Gamma\}$ on S.

The reader is encouraged to keep this definition in mind. We will refer to the weak topology induced by a set of functions when we will discuss the question of a suitable topology on the Cartesian product of a family of topological spaces.

6.6 Topic: Continuous Functions on a Dense Subset

We examine particular properties of a continuous function on a dense subset of (S, τ).

Note that, if A is dense in B and $A \subseteq D \subseteq B$, then D must be dense in B. The reader is left to verify this. It is also easy to verify that, if D is dense in S, then $S \backslash D$ cannot contain any non-empty open subsets of S.

Before we state the next theorem, we provide the following definition. If $f : S \to T$ and $g : S \to T$ are two functions such that $f(x) = g(x)$ for all $x \in A \subseteq S$ then we will say that

f and g agree on A.

We present the following theorem concerning continuous functions which agree on a dense subset of topological space.

Theorem 6.14 *Suppose (S, τ) is a topological space and (T, τ_ρ) is a metrizable topological space induced by the metric ρ. Suppose $f : S \to T$ and $g : S \to T$ are two continuous functions which agree on some dense subset D of (S, τ). Then f and g must agree on all of S.*[6]

Proof. Let (T, ρ) be the metric space which is equivalent to (T, τ_ρ). Let

$$D = \{x \in S : f(x) = g(x)\}$$

By hypothesis, f and g agree on D. We are required to show that $D = S$.

Suppose $a \in S \setminus D$; then $f(a) \neq g(a)$. Then there exists in T, *disjoint* basic open balls, $B_\varepsilon(f(a))$ and $B_\varepsilon(g(a))$, of S of radius

$$\varepsilon = \frac{\rho(\,f(a),\, g(a)\,)}{3}$$

with center $f(a)$ and $g(a)$, respectively. Since both f and g are continuous on S, both $f^{\leftarrow}[B_\varepsilon(f(a))]$ and $g^{\leftarrow}[B_\varepsilon(g(a))]$ are open in S each containing at least the point a. So the open subset

$$f^{\leftarrow}[B_\varepsilon(f(a))] \cap g^{\leftarrow}[B_\varepsilon(g(a))] \neq \varnothing$$

in S. We claim that $f^{\leftarrow}[B_\varepsilon(f(a))] \cap g^{\leftarrow}[B_\varepsilon(g(a))] \cap D = \varnothing$. For if,

$$x \in f^{\leftarrow}[B_\varepsilon(f(a))] \cap g^{\leftarrow}[B_\varepsilon(g(a))]$$

$f(x) \in B_\varepsilon(f(a))$ and $g(x) \in B_\varepsilon(g(a))$ so $f(x) \neq g(x)$ (since $B_\varepsilon(f(a))$ and $B_\varepsilon(g(a))$ are disjoint). So $a \notin S \backslash D$. So $S \backslash D$ is empty. So, $D = S$, as required.

---

[6]Once we have introduced the concept of "Hausdorff", this statement generalizes from "metrizable topological spaces" to "Hausdorff topological spaces".

6.7 Topic: On Algebras of Continuous Real-Valued Functions

We now take a quick glance at the algebra,[7] $(C(S), +, \cdot, \text{scalar mult})$, of those continuous functions on a topological space, S, with range in $\mathbb{R}$. If S is a topological space, we denote the set of all continuous functions, $f : S \to \mathbb{R}$, mapping S into $\mathbb{R}$ by $C(S)$. The set $C(S)$ is normally considered equipped with the algebraic operations $+$, $\cdot$ and scalar multiplication defined as

$$(f+g)(x) = f(x) + g(x)$$
$$(f \cdot g)(x) = f(x)g(x)$$
$$(\alpha f)(x) = \alpha f(x)$$

We will assume that the reader is able to write proofs showing that $C(S)$ is closed under sums, multiplication and real scalar multiplication and that $|f|$ defined as

$$|f|(x) = |f(x)|$$

also belongs to $C(S)$ whenever f is in $C(S)$. We will also refer to $(C(S), +, \cdot)$ as a *ring of continuous functions*. The elements of $C(S)$ can be *partially ordered* with "$\leq$"[8] where

$$f \leq g \text{ if and only if } f(x) \leq g(x)$$

for all $x \in S$. We also define the operations $\vee$ and $\wedge$ on $C(S)$ as

$$(f \vee g)(x) = \max\{f(x), g(x)\} \text{ for } x \in S$$
$$(f \wedge g)(x) = \min\{f(x), g(x)\} \text{ for } x \in S$$

Verification of the following formulas is left to the reader.

$$f \vee g = \frac{f + g + |f - g|}{2}$$

$$f \wedge g = \frac{f + g - |f - g|}{2}$$

[7]An *algebra* is a set with $+$ and $\cdot$ and scalar multiplication by elements of a field.
[8]A partially ordered set is a set P on which is defined a binary relation "$\leq$" which is reflexive, antisymmetric and transitive.

From these, we can conclude that both $f \vee g$ and $f \wedge g$ are continuous whenever f and g are continuous.

6.8 Topic: The Pasting Lemma and a Generalization

The continuity of a function, f, on a given space can sometimes be more easily confirmed by determining the continuity of f on some of its subspaces. The "pasting lemma" is a statement which says that two continuous functions can be "pasted together" to create some other continuous function. The lemma is implicit in the use of piecewise functions.

Lemma 6.15 *Let A and B be closed subsets of a topological space S such that $S = A \cup B$. Then $f : S \to T$ is continuous if and only if $f|_A$ and $f|_B$ are both continuous.*

Proof. Let $f : S \to T$ be a function and A and B are both closed in S.

($\Rightarrow$) Suppose f is continuous on S. Then, by Theorem 6.5 both $f|_A$ and $f|_B$ are continuous on A and B, respectively.

($\Leftarrow$) Suppose both $f|_A$ and $f|_B$ are continuous on the closed subsets A and B, respectively. Let F be a closed subset of T. Then both $f|_A^{\leftarrow}[F]$ and $f|_B^{\leftarrow}[F]$ are closed since each is the pre-image of f when restricted to A and B, respectively. Then their union, is also closed, being a finite union of closed sets. So, $f^{\leftarrow}[F] = f|_A^{\leftarrow}[F] \cup f|_B^{\leftarrow}[F]$, a closed subset of S. So, f is continuous on S.

We generalize the pasting lemma to the case where f acts on an infinite collection of closed sets. The statement in the following theorem involves a family (possibly infinite) of subsets referred to as being a "locally finite collection".

Definition 6.16 A *locally finite* collection, $\mathscr{C}$, of subsets of a topological space, S, is one for which every point in S has at least one neighborhood which meets only finitely many elements of $\mathscr{C}$.

Such families of subsets satisfy an interesting property, presented in the following lemma.

Lemma 6.17 *Suppose (S,τ) is a topological space. Let $\mathscr{U} = \{U_i : i \in I\}$ be a locally finite collection of sets in S. Then $\mathscr{U}^* = \{cl_S U_i : i \in I\}$ is also locally finite.*

Furthermore,

$$cl_S[\cup\{U_i : i \in I\}] = \cup\{cl_S U_i : i \in I\}$$

Proof. Let $\mathscr{U} = \{U_i : i \in I\}$ be a locally finite collection of sets in S.

Then, for $x \in S$, there is an open neighborhood B such that $B \cap U \neq \varnothing$ for finitely many members of $\mathscr{U}$ and $B \cap U = \varnothing$ for all others. For these "other". U's, since B is open in S, $B \cap \mathrm{cl}_S U = \varnothing$. In the case where $B \cap U \neq \varnothing$, then $B \cap \mathrm{cl}_S U \neq \varnothing$ for finitely many U's. The existence of this open neighborhood B of x is all that is required to prove the local finiteness property of $\{\mathrm{cl}_S U_i : i \in I\}$.

We now prove the second part. Let

$$M = \cup\{U_i : i \in I\}$$

We are required to prove $\mathrm{cl}_S M = \cup\{\mathrm{cl}_S U_i : i \in I\}$.

Clearly, $\cup\{\mathrm{cl}_S U_i : i \in I\} \subseteq \mathrm{cl}_S M$. So, we need only prove $\mathrm{cl}_S M \subseteq \cup\{\mathrm{cl}_S U_i : i \in I\}$.

Let $p \in \mathrm{cl}_S M$. We claim that $p \in \cup\{\mathrm{cl}_S U_i : i \in I\}$. There exists an open neighborhood B of p such that $B \cap \mathrm{cl}_S U_i \neq \varnothing$ for finitely many elements, $\mathscr{F}$, in $\mathscr{U}^*$.

See that, for $p \in \mathrm{cl}_S M$,

$$\begin{aligned} p \in \mathrm{cl}_S M &= \mathrm{cl}_S \left[\cup\{U_i \in \mathscr{U} : \mathrm{cl}_S U_i \notin \mathscr{F}\} \bigcup \cup\{U_i \in \mathscr{F}\}\right] \\ &= \mathrm{cl}_S \left[\cup\{U_i \in \mathscr{U}^* : \mathrm{cl}_S U_i \notin \mathscr{F}\}\right] \bigcup \mathrm{cl}_S \left[\cup\{U_i \in \mathscr{F}\}\right] \\ &= \mathrm{cl}_S \left[\cup\{U_i \in \mathscr{U}^* : \mathrm{cl}_S U_i \notin \mathscr{F}\}\right] \bigcup \cup\{\mathrm{cl}_S U_i \in \mathscr{F}\} \end{aligned}$$

($\mathrm{cl}_S U_i$ is closed and $\mathscr{F}$ is finite)

Now p cannot belong to $\text{cl}_S\,[\cup\{U_i \in \mathscr{U}^* : \text{cl}_S U_i \notin \mathscr{F}\}]$ since $B \cap \text{cl}_S U_i = \varnothing$ for all $\text{cl}_S U_i \notin \mathscr{F}$. Then $p \in \cup\{\text{cl}_S U_i \in \mathscr{F}\} \subseteq \cup\{\text{cl}_S U_i : i \in I\}$, as claimed.

So, $\text{cl}_S M = \cup\{\text{cl}_S U_i : i \in I\}$, as required.

It follows from the lemma that, if $\mathscr{F} = \{F : F \in \mathscr{F}\}$ is any locally finite collection of closed sets, then

$$\text{cl}_S[\cup\{F : F \in \mathscr{F}\} = \cup\{F : F \in \mathscr{F}\}$$

Theorem 6.18 *Suppose (S, τ) is a topological space. Let $\mathscr{F}$ be a locally finite family of closed subsets which covers all of S (i.e., $S = \cup\mathscr{F}$). Let $f : S \to T$ be a function mapping S into a space T. Then f is continuous on S if and only if $f|_F : F \to T$ is continuous on each $F \in \mathscr{F}$.*

Proof. By Theorem 6.5, part b), we need only prove ($\Leftarrow$).

($\Leftarrow$) Suppose $\mathscr{F}$ is a locally finite collection of closed subsets of S which covers all of S. Also suppose $f : S \to T$ is a function mapping S into a space T such that $f|_F : F \to T$ is continuous on each $F \in \mathscr{F}$. We are required to show that f is continuous on all of S. To prove this, it suffices to show that, if A is closed in T, then $f^{\leftarrow}[A]$ is closed in S.

Let A be a closed subset of T. Then,

$$\begin{aligned} f^{\leftarrow}[A] &= \cup\{F \cap f^{\leftarrow}[A] : F \in \mathscr{F}\} \\ &= \cup\{f|_F^{\leftarrow}[A] : F \in \mathscr{F}\} \end{aligned}$$

Since $f|_F$ is continuous for each $F \in \mathscr{F}$, then $f|_F^{\leftarrow}[A]$ is a closed subset of F. Then there exists a closed subset G of S such that $f|_F^{\leftarrow}[A] = G \cap F$, a closed subset of S.

Consider the collection

$$\mathscr{C} = \{f|_F^{\leftarrow}[A] : F \in \mathscr{F}\}$$

of closed subsets of S. Note that $f_F^{\leftarrow}[A] = \cup\mathscr{C}$. We will show that this set is closed in S.

Claim. We claim that $\mathscr{C}$ is locally finite.

Proof. Let $p \in S$. Then there exists an open neighborhood, V, of p and a finite subcollection, $\mathscr{G}$, of $\mathscr{F}$ such that $V \cap F \neq \varnothing$ if $F \in \mathscr{G}$ and $V \cap F = \varnothing$ for $F \in \mathscr{F} \backslash \mathscr{G}$.

Note that for $\mathscr{F} \backslash \mathscr{G}$,

$$\begin{aligned} V \cap f|_F^{\leftarrow}[A] &= V \cap (F \cap f^{\leftarrow}[A]) \\ &= (V \cap F) \cap f^{\leftarrow}[A] \\ &= \varnothing \cap f^{\leftarrow}[A] \\ &= \varnothing \end{aligned}$$

Then at most finitely many elements of $\mathscr{C}$ intersect the neighborhood V of p. So, $\mathscr{C}$ is locally finite collection of closed subsets, as claimed.

Then $f^{\leftarrow}[A] = \cup\mathscr{C}$ is closed (see paragraph preceding the theorem).

Then $f : S \to T$ is continuous on S.

In the above theorem, if the members of the collection $\mathscr{F}$ are all open (rather than all closed as hypothesized in the theorem) the family $\mathscr{F}$ need not be locally finite and so the statement may not hold true.

Corollary 6.19 *Suppose (S, τ) is a topological space. Let $\mathscr{F} = \{F_i : i \in F\}$ be a* finite *family of closed subsets which covers all of S (i.e., $S = \cup\mathscr{U}$). Let $f : S \to T$ be a function mapping S into a space T. Then f is continuous on S if and only if $f|_{F_i} : F_i \to T$ is continuous on each $F_i \in \mathscr{F}$.*

Proof. See that, if $\mathscr{F} = \{F_i : i \in F\}$ be a *finite* family of closed subsets which covers all of S, then $\mathscr{F}$ is a σ-locally finite cover of closed sets on S. The result follows immediately from the theorem.

Concepts Review

1. Given the function $f : S \to T$ and $A \subseteq S$, define $f[A]$.
2. Given the function $f : S \to T$ and $B \subseteq S$, define $f^{\leftarrow}[B]$.

3. State the formal topological definition of "f is continuous function on the set A".
4. State the formal topological definition of "f is continuous function at the point x".
5. Give a formal theorem statement which link the above two definitions of continuity.
6. A continuous function $f : S \to T$ pulls back open subbase elements in $\mathscr{P}(T)$ to what kind of set in S?
7. A continuous function $f : S \to T$ pulls back open base elements in $\mathscr{P}(T)$ to what kind of set in S?
8. If $f : S \Rightarrow T$ is continuous on S and $A \subseteq S$, show that $f|_A$ is continuous on A.
9. Is it correct to say "$f : S \to T$ is continuous if and only if f pulls back closed sets to closed sets"?
10. What does it mean to say $f : S \to T$ is an open functions?
11. What does it mean to say $f : S \to T$ is a closed functions?
12. Is it okay to say that "open functions are always closed"? What about "continuous"?
13. What does it mean to say that $f : S \to T$ is a homeomorphism?
14. What does it mean to say that S and T are homeomorphic spaces?
15. Give two characterizations of homeomorphic functions.
16. What is a topological property?
17. Is "first countable" a topological property? What about "second countable"?
18. Do continuous functions necessarily carry over the second countable property to its codomain? What about the first countable property?
19. What does it mean to say that A is a dense subset of B?
20. What can we say about two continuous functions which agree on a dense subset of a metrizable topological space?
21. What does it mean to say that a topological space is separable?
22. What can we say about continuous images of separable spaces?
23. What can we say about second countable spaces in reference to the "separable property"?
24. Provide examples of hereditary and non-hereditary topological properties.
25. Define the weak topology induced by a family of functions $\{f_\alpha : S \to T_\alpha\}_{\alpha \in I}$.

26. Provide two characterizations of "$f : S \to T$ is continuous on S" involving families of subsets which cover S.

Exercises

1. Suppose $S = \{a, b\}$ is a set with topology $\tau_S = \{\varnothing, \{a\}, S\}$. Let $T = \{a, b\}$ where T is equipped with the discrete topology τ_d. Let $i : S \to T$ denote the identity map. What can we say about the continuity or non-continuity of the functions $i : S \to T$ and $i^{\leftarrow} : T \to S$?
2. Let $f : S \to T$ be a continuous function mapping (S, τ_S) onto (T, τ_T). If U is a G_δ in T, is $f^{\leftarrow}[U]$ necessarily a G_δ in S? Show that f pulls back F_σ's in T to F_σ's in S.
3. Let $f : (S, \tau_S) \to (T, \tau_T)$ be a continuous function mapping S onto T. Show that f is continuous on S if and only if $f[\text{cl}_S[U]] \subseteq \text{cl}_T f[U]$ for any $U \in \mathscr{P}(S)$.
4. Recall that *infinite countable sets* are those sets which can be mapped one-to-one and onto the natural numbers $\mathbb{N}$. Suppose X and Y are both countable dense subsets of $(\mathbb{R}, \tau)$ where τ is the usual topology. Show that X and Y must be homeomorphic subspaces of $\mathbb{R}$.
5. Consider the topological spaces $S = (\mathbb{R}^2, \tau_1)$ and $T = (\mathbb{R}, \tau)$ where τ_1 and τ represent the usual topology. (The open base of τ_1 are the open balls, $B_\varepsilon(x, y)$, with center (x, y) and radius ε).
   Consider the function $f : S \to T$ defined as $f(x, y) = x$.
   (a) Show that f is an open function.
   (b) Show that f is not a closed function. (Hint: See that $F = \{(x, y) : xy = 1\}$ is a closed subset of S. Consider the image of F under f.)
6. Prove: The function $f : S \to T$ is a closed function if and only if whenever F is a closed subset of S then $\{t \in T : f^{\leftarrow}[\{t\}] \cap F \text{ is non-empty}\}$ is a closed subset of T.
7. Suppose $f : S \to T$ is a closed function. Show that whenever V is an open subset of S and $f^{\leftarrow}[\{x\}]$ is a subset of V then $x \in \text{int}_T(f[\text{cl}_S V])$.

8. Suppose $f : S \to T$ is a closed function and F is closed subset of S. Show that the restriction, $f|_F : F \to f[F]$, is also a closed function.
9. Suppose U and V are subsets of S such that $U \cup V = S$ and $x \in U \cap V$. Suppose $f : S \to T$ is a function such that both $f|_U$ and $f|_V$ are continuous on U and V, respectively. Show that f is continuous at x.
10. Suppose that $f : S \to T$ is a one-to-one and onto function. Show that f is a homeomorphism if and only if, for any $U \in \mathscr{P}(S)$, $f[\mathrm{cl}_S U] = \mathrm{cl}_T f[U]$.

Chapter 7

Product Spaces

Abstract
In this section, we will review some fundamental facts about those sets that are "Cartesian products of sets". We will then consider two topologies on a Cartesian product of topological spaces and study some of their most fundamental properties. In the last half of the chapter, we will look at some applications where product spaces play an important role. In particular, we prove the existence of a continuous function which maps the closed interval $[0, 1]$ onto the product space, $[0, 1] \times [0, 1] \times [0, 1]$, a cube in a three-dimensional space.

7.1 Fundamentals of Cartesian Products

In topology, Cartesian products are a rich and important source of examples of various known topological properties. Given a family of known topological spaces, we can construct a new larger topological space with it's own particular properties. For this reason, it is crucial to understand them well. The best way to do this is to practice using them in various contexts.

Most students are exposed to the notion of a Cartesian product with two or three factors, in some form or other, at the high school level. For example,

$$A \times B \times C = \{(a, b, c) : a \in A, b \in B, c \in C\}$$

When discussing *infinite* Cartesian products of sets we must proceed cautiously while deciding which topology we will adopt to best suit our purposes. We thought it would be best if we begin by presenting a formal definition of products of sets. Those readers already well familiar with these concepts can skim through this section and go directly to Section 7.2.

Definition 7.1 Let $\{S_\alpha : \alpha \in I\}$ be an indexed family of sets. The Cartesian product of these sets, denoted by $\prod_\alpha S_\alpha$, or in more detail as

$$\prod_{\alpha \in I} S_\alpha = \{\ f \mid f : I \to \cup_{\alpha \in I} S_\alpha\}$$

is the set of all functions f mapping the index set, I, into the union, $\cup_{\alpha \in I} S_\alpha$, such that, for $\beta \in I$, $f(\beta) \in S_\beta$.[1] So, if u is an element of $\prod_\alpha S_\alpha$ and $y_\alpha = f(\alpha)$, then we can express it in the form

$$u = \{f(\alpha) : \alpha \in I\} = \{y_\alpha : \alpha \in I\} = \langle y_\alpha \rangle_{\alpha \in I}$$

or we can write it more simply as $\{y_\alpha\} = \langle y_\alpha \rangle$. If $\beta \in I$, the set,

S_β, is called the βth *factor*

of the Cartesian product, $\prod_\alpha S_\alpha$, while

y_β is called the βth *coordinate* of the element, $\langle y_\alpha \rangle_{\alpha \in I}$

For example, if $I = \{1, 2, 3\}$,

$$\prod_{\alpha \in I} S_\alpha = \{(a, b, c) : a \in S_1, b \in S_2, c \in S_3\} = S_1 \times S_2 \times S_3$$

The size of $\prod_{\alpha \in I} S_\alpha$ depends on the size of the respective sets, S_α, and the size of the index set.

If we are given an indexed family of sets, say $\{S_\alpha : \alpha \in J\}$, where all factors, S_α, represent the same set, S, the reader should be familiar with the following notation:

$$S^J = \prod_{\alpha \in J} S$$

---

[1]We will assume that a verification involving a combination of the *Axiom of union* and the *Axiom of power set* guarantees that $\prod_\alpha S_\alpha$ is indeed a "set".

The expression, S^J, is normally interpreted as meaning "the set of all functions mapping J into S". For example, if $J = \mathbb{N}$ and $S_\alpha = \mathbb{R}$ for all $\alpha \in J$, then

$$\mathbb{R}^{\mathbb{N}} = \prod_{\alpha \in \mathbb{N}} \mathbb{R}$$

It thus represents all countably infinite sequences of real numbers, $\{a_0, a_1, a_2, a_3, \ldots\}$, or equivalently, the set of all functions mapping $\mathbb{N}$ into $\mathbb{R}$. Another example is, $\mathbb{R}^{\mathbb{R}}$, which represents the set of all functions mapping $\mathbb{R}$ into $\mathbb{R}$. Since $\mathbb{R}$ is not normally viewed as an index set, the set, $\mathbb{R}^{\mathbb{R}}$, of all functions mapping $\mathbb{R}$ into $\mathbb{R}$ is not often expressed as a Cartesian product.[2]

7.1.1 *A few words on the definition of Cartesian product*

In our definition of Cartesian product we refer to an indexed family, $\{S_\alpha : \alpha \in I\}$, of sets. We didn't state explicitly that each one of these is non-empty. Should we include this requirement in the definition? What happens if, say $S_\beta = \varnothing$, for some $\beta \in I$? Since there is nothing in S_β, then there cannot exist a function which will map β to some element in S_β and so the cautious reader will conclude that $\prod_\alpha S_\alpha$ must be empty. This is not catastrophic since to say the product is non-empty is to assume that each factor is non-empty. So, we can leave it as is. On the other hand, if no S_α is empty, are we guaranteed that there exists at least one function, $f : I \to \cup_{\alpha \in I} S_\alpha$, such that f assigns β in I to a particular element in S_β? If so, is there a way to decide which element should be selected by f? The assumption that at least one such function exists invokes the statement in the Axiom of choice:

> Given any set $\mathscr{A}$ of non-empty sets, there is a rule f which associates to each set A in $\mathscr{A}$ some element $a \in A$.

So, the Axiom of choice grants us permission to assume that at least one function f will select a point $f(\beta) = m_\beta$ in S_β for us. However, other than this guarantee that at least one f exists, we have no way

[2]If we write the product, $\prod_{\alpha \in \mathbb{R}} S_\alpha$, then the reader can assume that the Well-ordering theorem (equivalent to the Axiom of Choice) has been invoked to linearly order $\mathbb{R}$.

of ever determining what that function f is. We are assuming the existence of a mathematical entity we will never ever see. Invoking the Axiom of choice is not ideal, but it is the best we can do. For most people, the fact that the Cartesian product of non-empty sets is non-empty is obvious and so they don't lose any sleep over it. Throughout this section, we will, as a rule, assume the Axiom of choice holds true, and not point out it's application at each place it is involved, unless it is of particular interest to do so.

In this book, we declare that the "Cartesian product of topological spaces, $S = \prod_{i\in I} S_i$", we will always mean "the *non-empty* Cartesian product, S, of spaces".

7.1.2 *A few set-theoretic properties of Cartesian products*

We have yet to topologize Cartesian products of spaces. But before we do so, we present a few of the Cartesian product's most fundamental properties.

Definition 7.2 Let $S = \prod_{i\in I} S_i$ be a Cartesian product of sets. The family of functions $\{\pi_i : i \in I\}$ where $\pi_i : \prod_{i\in I} S_i \to S_i$, is defined as

$$\pi_j(\langle x_i\rangle_{i\in I}) = x_j$$

We will refer to π_j as being the jth *projection* which maps $\prod_{i\in I} S_i$ *onto* S_j.

Theorem 7.3 *Let $\{S_i : i \in I\}$ and $\{T_i : i \in I\}$ be two families of sets and $S = \prod_{i\in I} S_i$ and $T = \prod_{i\in I} T_i$ be their corresponding Cartesian products.*

(a) *If $\prod_{i\in I} S_i \subseteq \prod_{i\in I} T_i$, then $S_i \subseteq T_i$ for each $i \in I$.*
(b) *For U_i, V_i, both subsets of S_i,*

(i) $\prod_{i\in I} U_i \cap \prod_{i\in I} V_i = \prod_{i\in I}(U_i \cap V_i)$
(ii) $\prod_{i\in I} U_i \cup \prod_{i\in I} V_i = \prod_{i\in I}(U_i \cup V_i)$

Proof. (a) Let $\beta \in I$. Since the βth projection map, π_j, is onto the (S_j)th factor,

$$S_j = \pi_j\left(\prod_{i \in I} S_i\right) \subseteq \pi_j\left(\prod_{i \in I} T_i\right) = T_j$$

The proofs of part (b) are left for the reader.

7.2 Topologizing the Cartesian Product of Topological Spaces

Taking Cartesian products of large numbers of topological spaces is a powerful way to construct new topological spaces from "old" ones. There is more than one topology for us to choose from, some of which will eventually be more useful than others. For finite products, there is one which most users would consider as being the most intuitive and as being the "standard" one. For infinite products the choice of topology may depend on the context.

The third example on page 99, illustrates how one might proceed to topologize the Cartesian product of two spaces (S, τ_S) and (T, τ_T). In that example, the topology, τ, on $S \times T$ was generated by choosing, as subbase for τ, the family of sets

$$\mathscr{S} = \{\pi_S^{\leftarrow}[U] : U \in \tau_S\} \cup \{\pi_T^{\leftarrow}[V] : V \in \tau_T\}$$

where $\pi_S : S \times T \to S$ and $\pi_T : S \times T \to T$ are projection maps. Recall that $\pi_S(a, b) = a$ for all $b \in T$ and $\pi_T(a, b) = b$ for all $a \in S$ and so $\pi_S^{\leftarrow}(a) = \{a\} \times T$ and $\pi_T^{\leftarrow}(b) = S \times \{b\}$. So, the set of all finite intersections of the elements in $\mathscr{S}$ forms a base, $\mathscr{B}$, for this topology. Thus, the base elements of τ are of the form

$$\begin{aligned}\pi_S^{\leftarrow}[U] \cap \pi_T^{\leftarrow}[V] &= (U \times T) \cap (S \times V)\\ &= (U \cap S) \times (V \cap T)\\ &= U \times V\end{aligned}$$

where $U \in \tau_S$ and $V \in \tau_T$. That is, $\mathscr{B} = \{U \times V : U \in \tau_S, V \in \tau_T\}$.

Note that, in this particular example, the topology constructed for $S \times T$ is the

weak topology induced by the projection functions $\{\pi_S, \pi_T\}$

This "weak topology" on $S \times T$ comes with the guarantee of continuity for each projection map, π_α, on $S \times T$. This motivates the choice of the "standard topology" for arbitrary products described in the following definition.

Definition 7.4 Let $\{(S_\alpha, \tau_\alpha) : \alpha \in I\}$ be an indexed family of non-empty topological spaces and

$$S = \prod_{\alpha \in I} S_\alpha$$

be the Cartesian product of these spaces. Let $\{\pi_\alpha : \alpha \in I\}$ be the family of the associated projection maps

$$\pi_\beta : \prod_{\alpha \in I} S_\alpha \to S_\beta$$

We define the *product topology* or the *standard topology* as being the weak topology on $\prod_{\alpha \in I} S_\alpha$ induced by the family of functions $\{\pi_\alpha : \alpha \in I\}$. The Cartesian product of topological spaces, when equipped with this weak topology, is referred to as a *product space.* The family of all sets of the form,

$$\{\pi_\alpha^{\leftarrow}[U] : U \in \tau_\alpha\}$$

is the subbase, $\mathscr{S}$, for the product space. While each element of the base of open sets, $\mathscr{B}$, is of the form

$$\cap_{\alpha \in F}\{\pi_\alpha^{\leftarrow}[U_\alpha] : U_\alpha \in \tau_\alpha\}$$

where F is a finite subset of I.

Since $\pi_\alpha^{\leftarrow}[U_\alpha] \cap \pi_\alpha^{\leftarrow}[V_\alpha] = \pi_\alpha^{\leftarrow}[U_\alpha \cap V_\alpha]$, we can assume that all the α's in the finite set, F, are distinct. This assumption does not alter the definition of *product topology.* Also, $\pi_\beta^{\leftarrow}[U_\beta]$ is a subset of $\prod_{\alpha \in I} S_\alpha$ where, if $\alpha \neq \beta$, the αth factor is, S_α, itself, and only the βth factor is U_β. Hence, for any open base element, every factor is S_α itself except for finitely many factors, U_α, as proper subsets of S_α.

It is also worth noting that the product topology is the absolute smallest topology on $S = \prod_{\alpha \in I} S_\alpha$ which guarantees that each and every projection map in

$$\{\pi_\alpha : \prod_{\alpha \in I} S_\alpha \to S_\alpha\}$$

is continuous on its domain, S.

Finally, note that the product topology depends on the topology of each of the factors. It does not depend on some topology defined on the index set, I.

7.2.1 *Projection maps are open*

In the following proposition, we will show that each map, π_α, is also open. This is a useful fact to remember.

Proposition 7.5 *Given a product space, $S = \prod_{\alpha \in I} S_\alpha$, the projection map,*

$$\pi_\beta : S \to S_\beta$$

is an open map.

Proof. Given: A product space, $S = \prod_{\alpha \in I} S_\alpha$, and a projection map, $\pi_\beta : S \to S_\beta$.

Since functions respect arbitrary unions, it suffices to show that the projection map sends basic open sets to open sets.

Let

$$\cap\{\pi_{\alpha_i}^{\leftarrow}[U_{\alpha_i}] : i = 1, 2, \ldots, k\}$$

be a basic open set in S.

$$\text{If } \beta = \alpha_j \Rightarrow \pi_\beta[\cap\{\pi_{\alpha_i}^{\leftarrow}[U_{\alpha_i}] : i = 1, 2, \ldots, k\}] = U_{\alpha_j}$$

$$\text{If, for } i = 1, 2, \ldots, k,\ \beta \neq \alpha_i \Rightarrow \pi_\beta[\cap\{\pi_{\alpha_i}^{\leftarrow}[U_{\alpha_i}] : i = 1, 2, \ldots, k\}] = S_\beta$$

So, π_β is an open map, as required.

Theorem 7.6 *Let $\{(S_\alpha, \tau_\alpha) : \alpha \in I\}$ be an indexed family of non-empty topological spaces and $\prod_{\alpha \in I} S_\alpha$ be the Cartesian product space* *equipped with the product topology, τ. For each $\alpha \in I$, S_α has an open base represented by, $\mathscr{B}_\alpha$. Then the set*

$$\mathscr{B}^* = \{ \ \cap_{\alpha \in F}\{\pi_\alpha^{\leftarrow}[B_\alpha] : B_\alpha \in \mathscr{B}_\alpha\} \ \}_{\alpha \in I}$$

where F is finite, forms a base for τ.

Proof. Let $V = \cap_{\alpha \in F}\{\pi_\alpha^{\leftarrow}[U_\alpha] : U_\alpha \in \tau_\alpha\}$ (with F finite) be an open base element of the product topology, τ. Suppose $\langle x_\alpha \rangle_{\alpha \in I} \in V$. Then, for each $\alpha \in F$, there exists an open base element, $B_\alpha \in \mathscr{B}_\alpha$, such that $x_\alpha \in B_\alpha \subseteq U_\alpha$. Then

$$\langle x_\alpha \rangle_{\alpha \in I} \in \cap_{\alpha \in F}\{\pi_\alpha^{\leftarrow}[B_\alpha] : B_\alpha \in \mathscr{B}_\alpha\} \subseteq V$$

Then every open base element of τ is the union of elements from $\mathscr{B}^*$. So $\mathscr{B}^*$ forms a base for the product topology.

7.2.2 *The "box topology" on a Cartesian product*

The product topology is not the only topology we can define on (infinite) products of topological spaces. Some readers may have noticed that, for an infinite product, $S = \prod_{\alpha \in I} S_\alpha$, the collection

$$\mathscr{B}^* = \cap_{\alpha \in I}\{\pi_\alpha^{\leftarrow}[U_\alpha] : U_\alpha \in \tau_\alpha\}$$

(note the intersection of *infinitely* many sets) satisfies the "base property". Hence, this set will be a base for a topology τ^* which is different from the product topology, τ, on S. In the literature, it is referred to as the *box topology.* Note that every open base element, $B \in \mathscr{B}$, for the product topology is an open base element in $\mathscr{B}^*$. Hence, the "box topology" is stronger (finer) than the "product topology". In fact, if every factor of $\prod_{\alpha \in I} U_\alpha$ is a proper open subset of S_α and $\langle x_\alpha \rangle_{\alpha \in I} \in \prod_{\alpha \in I} U_\alpha \in \tau^*$, then there does not exist a $V \in \mathscr{B}$ such that $\langle x_\alpha \rangle_{\alpha \in I} \in V \subseteq \prod_{\alpha \in I} U_\alpha$. Can you see why? So $\mathscr{B}^* \not\subseteq \mathscr{B}$.

Note that, for the Cartesian product of finitely many spaces, the product topology and the box topology are equivalent topologies.

Example 1. Show that the product space $\mathbb{R}^3 = \mathbb{R} \times \mathbb{R} \times \mathbb{R}$ is metrizable.

Solution: This can be seen since the basic open sets in $(\mathbb{R}^3, \tau)$ equipped with the product topology are three-dimensional open rectangular boxes. While the basic open sets of $(\mathbb{R}^3, \rho)$, where $\rho(\vec{x}, \vec{y})$ is the distance between the two given points, are open balls. Since the rectangular boxes can be filled with open balls and vice versa, the metric space $(\mathbb{R}^3, \rho)$ and the product space $\mathbb{R}^3 = \mathbb{R} \times \mathbb{R} \times \mathbb{R}$ (equipped with the product topology) are equivalent topological spaces. So $(\mathbb{R}^3, \tau)$ is metrizable.

Example 2. The product space $[0,1]^3$ is metrizable. Since $[0,1]^3$ is a subspace of the metrizable, $\mathbb{R}^3$, then by the example on page 107, $[0,1]^3$ is metrizable.

The notion of a *uniform metric* on a product $S = \prod_{j \in J}[0,1]$ is discussed in the following example.

Example 3. Let $S = \prod_{j \in J}[0,1]$. We define the following metric on S,

$$\rho\left(\langle x_j \rangle_{j \in J}, \langle y_j \rangle_{j \in J}\right) = \sup\left\{|x_j - y_j| : j \in J\right\}$$

The verification that this is a valid metric on S is left to the reader. Clearly, the induced metric topology on S is not equivalent to the product topology (since, for small ε, no j-factor of the open base element, $B_\varepsilon\left(\langle x_j \rangle_{j \in J}\right)$, is equal to $[0,1]$). The metric, ρ, is referred to at the

uniform metric

on a product on $S = \prod_{j \in J}[0,1]$. The topology induced by the uniform metric is referred to at the

$$\textit{uniform topology} \text{ on } \prod_{j \in J}[0,1]$$

A word of caution: Theorems in which the a Cartesian product equipped with the product topology is involved may not hold true if that product is equipped with the uniform topology.

7.3 On Interiors, Closures and Boundaries of Product Spaces

It is sometimes useful to know how to handle the interior and the closure of a product. The interior of a product is well-behaved in the case of a finite product space, while the closure of a product distributes nicely over its factors even for infinite products.

Theorem 7.7 *Let $\{(S_\alpha, \tau_\alpha) : \alpha \in I\}$ be an indexed family of non-empty topological spaces and $S = \prod_{\alpha\in I} S_\alpha$ be a Cartesian product equipped with the product topology. Let $T = \prod_{\alpha\in I} A_\alpha$ where each A_α is a subspace of S_α. Then,*

$$\mathrm{cl}_S T = \prod_{\alpha\in I} \mathrm{cl}_{S_\alpha} A_\alpha$$

Proof. The Cartesian product space $S = \prod_{\alpha\in I} S_\alpha$ and $A_\alpha \subseteq S_\alpha$ are given.

Let $T = \prod_{\alpha\in I} A_\alpha$.

We claim that $\prod_{\alpha\in I} \mathrm{cl}_{S_\alpha} A_\alpha \subseteq \mathrm{cl}_S T$.

Let $\langle x_\alpha\rangle \in \prod_{\alpha\in I} \mathrm{cl}_{S_\alpha} A_\alpha$. To show that $\langle x_\alpha\rangle \in \mathrm{cl}_S T$ we must show that any open neighborhood of $\langle x_\alpha\rangle$ in S meets $\prod_{\alpha\in I} A_\alpha$.

Let $U = \prod_{\alpha\in I} U_\alpha$ be a basic open neighborhood of $\langle x_\alpha\rangle$ in $\prod_{\alpha\in I} S_\alpha$.

Then, for each α, $x_\alpha \in \mathrm{cl}_{S_\alpha} A_\alpha \cap U_\alpha$. Then $U_\alpha \cap A_\alpha \neq \varnothing$.

For each α, there exists $y_\alpha \in U_\alpha \cap A_\alpha$. Then

$$\langle y_\alpha\rangle \in \prod_{\alpha\in I} A_\alpha \cap \prod_{\alpha\in I} U_\alpha = \prod_{\alpha\in I} A_\alpha \cap U$$

So,

$$\langle x_\alpha\rangle \in \mathrm{cl}_S \prod_{\alpha\in I} A_\alpha$$

Then $\prod_{\alpha\in I} \mathrm{cl}_{S_\alpha} A_\alpha \subseteq \mathrm{cl}_S T$ as claimed.

We claim that $\mathrm{cl}_S T \subseteq \prod_{\alpha\in I} \mathrm{cl}_{S_\alpha} A_\alpha$.

Let $\langle x_\alpha\rangle \in \mathrm{cl}_S \prod_{\alpha\in I} A_\alpha$ and U_β be an open neighborhood of x_β in S_β. Then $\pi_\beta^{\leftarrow}[U_\beta]$ is open in $\prod_{\alpha\in I} S_\alpha$. The set $\pi_\beta^{\leftarrow}[U_\beta]$ contains some point $\langle y_\alpha\rangle \in \prod_{\alpha\in I} A_\alpha$. Then $y_\beta \in U_\beta \cap A_\beta$. So, $x_\beta \in \mathrm{cl}_{S_\alpha} A_\beta$.

Therefore, $\langle x_\alpha\rangle \in \prod_{\alpha\in I} \mathrm{cl}_{S_\alpha} A_\alpha$.

We conclude that $\mathrm{cl}_S T \subseteq \prod_{\alpha\in I} \mathrm{cl}_{S_\alpha} A_\alpha$ as claimed.

We then have

$$\mathrm{cl}_S \prod_{\alpha\in I} A_\alpha = \prod_{\alpha\in I} \mathrm{cl}_{S_\alpha} A_\alpha$$

as required.

Note that the above statement holds true also if the product is equipped with the box topology.

Theorem 7.8 *Let S and T be topological spaces which contain subsets A and B, respectively. Then,*

$$\text{int}_{S\times T}(A \times B) = \text{int}_S A \times \text{int}_T B$$

Proof. We are given the Cartesian product $M = S\times T$ and $A \subseteq S$, $B \subseteq T$.

Since $\text{int}_S A$ and $\text{int}_T B$ are both open in S and T, respectively, then $\text{int}_S A \times \text{int}_T B$ is open in M. Since $\text{int}_S A \times \text{int}_T B \subseteq A \times B$, then

$$\text{int}_S A \times \text{int}_T B \subseteq \text{int}_M(A \times B)$$

Conversely, let $(a, b) \in \text{int}_M(A \times B)$. Then there is an M-open neighborhood, U, of (a, b) such that

$$(a, b) \in U \subseteq \text{int}_M(A \times B) \subseteq A \times B$$

Then $U = \cup_{i\in I}(V_i \times W_i)$, where V_i is open in S and W_i is open in T. Let $j \in I$ such that $(a, b) \in V_j \times W_j \subseteq U \subseteq A \times B$. Then $a \in \text{int}_S A$ and $b \in \text{int}_T B$. Then $(a, b) \in \text{int}_S A \times \text{int}_T B$.

We conclude that $\text{int}_M(A \times B) \subseteq \text{int}_S A \times \text{int}_T B$.

So, $\text{int}_M(A \times B) = \text{int}_S A \times \text{int}_T B$.

We should be careful not to generalize the statement in the theorem immediately above to arbitrary products.

Consider, $M = \prod_{n\in\mathbb{N}} \mathbb{R}$ equipped with the product topology, and the subset, $\prod_{n\in\mathbb{N}}(0, 1)$, of M. Then $\prod_{n\in\mathbb{N}}\text{int}_{\mathbb{R}}(0, 1) = \prod_{n\in\mathbb{N}}(0, 1)$.

On the other hand consider,

$$W = \text{int}_M \left[\prod_{n\in\mathbb{N}}(0, 1)\right]$$

Let $U = \prod_{n\in\mathbb{N}} Y_n$ be a basic open subset (with respect to the product topology) of M, where $Y_n = \mathbb{R}$ for all but finitely many n's. Since $\prod_{n\in\mathbb{N}}(0, 1)$ cannot contain even a single basic M-open set, $\prod_{n\in\mathbb{N}}(0, 1)$ can have no interior.

We conclude that $\text{int}_M\left[\prod_{n\in\mathbb{N}}(0,1)\right] = \varnothing \neq \prod_{n\in\mathbb{N}}\text{int}_{\mathbb{R}}(0,1)$.

Example 4. Let S and T be topological spaces which contain subsets A and B, respectively. Verify that

$$\text{bd}_{S\times T}(A\times B) = [\text{cl}_S A \times \ \text{bd}_T B] \cup [\ \text{bd}_S A \times \text{cl}_T B]$$

Solution: See that

$$\begin{aligned}\text{bd}_{S\times T}(A\times B) &= \text{cl}_{S\times T}(A\times B)\backslash\text{int}_{S\times T}(A\times B)\\ &= (\text{cl}_S A\times \text{cl}_T B)\backslash(\text{int}_S A\times \text{int}_T B)\\ &= [\text{cl}_S A\backslash\text{int}_S A\times \text{cl}_T B)]\cup[\text{cl}_S A\times \text{cl}_T B\backslash\text{int}_T B]\\ &= [\text{bd}_S A\times \text{cl}_T B]\cup[\text{cl}_S A\times \ \text{bd}_T B]\end{aligned}$$

7.4 Countability Properties Involving Product Spaces

In this text, when we say "product space", we will always mean the Cartesian product equipped with the product topology. The product topology is to be considered as the default topology unless stated otherwise.

In the following two theorems, we investigate whether the countably properties carry over from factors to product and vice versa. Recall that a topological space, (S,τ_s), is *second countable* if it has at least one countable base for open sets. It is *first countable* if each point, $x\in S$, has a countable open neighborhood base at x.

Theorem 7.9 *Let* $\{(S_\alpha,\tau_\alpha):\alpha\in\mathbb{N}\}$ *be an indexed* countable *family of non-empty topological spaces and* $S=\prod_{\alpha\in\mathbb{N}}S_\alpha$ *be the corresponding product space.*

(a) *Then the product space, S, is second countable if and only if each S_α is second countable.*
(b) *Then the product space, S, is first countable if and only if each S_α is first countable.*

Proof. We are given that $S=\prod_{\alpha\in\mathbb{N}}S_\alpha$ is a product space of countably many factors.

We will prove part (a).

($\Rightarrow$) Suppose S is second countable. We claim that each S_α is second countable.

Recall (from Proposition 7.5) that the projection map π_α is continuous is and open for each $\alpha \in \mathbb{N}$. Hence, by Theorem 6.12, given that π_α is continuous each $S_\alpha = \pi_\alpha[S]$ is second countable. We are done with this direction. (Note that the countability of the index set plays no role in this direction.)

($\Leftarrow$) We are given that, for each $\alpha \in \mathbb{N}$, (S_α, τ_α) is a second countable topological space, and $\prod_{\alpha \in \mathbb{N}} S_\alpha$ is a countable product equipped with the product topology, τ.

By hypothesis, for each $\alpha \in \mathbb{N}$, S_α has a countable base, $\mathscr{B}_\alpha$, of open sets. Then, by Theorem 7.6, the elements of τ of the form

$$\cap_{\alpha \in F}\{\pi_\alpha^{\leftarrow}[B_\alpha] : B_\alpha \in \mathscr{B}_\alpha\}$$

when gathered together form a base, $\mathscr{B}$, for open sets in τ. Since each $\mathscr{B}_\alpha$ is countable, then so is the subfamily

$$\{\pi_\alpha^{\leftarrow}[B_\alpha] : B_\alpha \in \mathscr{B}_\alpha, \alpha \in \mathbb{N}\}$$

of the base, $\mathscr{B}$, of S. Let $\mathscr{F} = \{F : F \subset \mathbb{N},\ F \text{ is finite}\}$. Note that the set, $\mathscr{F}$, has countably many elements (see footnote[3]). (This is where the countable the number of factors is countable is taken into consideration.)

Since the cardinality of $\mathscr{F} \times \mathscr{B}_\alpha \times \mathbb{N}$ is $\aleph_0$, then the base

$$\mathscr{B} = \{\ \cap_{\alpha \in F}\{\pi_\alpha^{\leftarrow}[B_\alpha] : B_\alpha \in \mathscr{B}_\alpha\}\ \}_{F \text{ finite},\ \alpha \in \mathbb{N}}$$

is countable. This means $\prod_{\alpha \in \mathbb{N}} S_\alpha$ is second countable.

The proof of part (b) is similar and so is left as an exercise.

In the following analogous theorem, the "separable property" will carry over from factors to the product and vice-versa, provided the

[3]The number of finite subsets of $\mathbb{N}$ is countable: For each $n \in \mathbb{N}$, let $\mathscr{U}_n = \{A \in \mathscr{P}(\mathbb{N}) : \max A \leq n\}$, the set of all subsets of $\{0, 1, 2, \ldots, n\}$. Then, for each n, $|\mathscr{U}_n| = 2^{n+1}$. The countable union of countable sets is countable (see 19.3 of Appendix A.2), so $\cup_{n \in \mathbb{N}}\{\mathscr{U}_n\}$ is countable. If F is a finite subset of $\mathbb{N}$, $F \in \mathscr{U}_m$ for some m. Then $\cup_{n \in \mathbb{N}}\{\mathscr{U}_n\}$ contains all finite subsets of $\mathbb{N}$. Then the number of finite subsets of $\mathbb{N}$ is countable.

number of factors in the product is restricted to no more than $c = |\mathbb{R}| = 2^{\aleph_0}$. The provided proof is notationally involved, but exhibits an interesting technique.

Theorem 7.10 *Let $\{(S_\alpha, \tau_\alpha) : \alpha \in I\}$ be a family of sets indexed by the set I, where $|I| \leq c = 2^{\aleph_0}$. Also suppose that S_α is a non-singleton set, for all $\alpha \in I$. Let $S = \prod_{\alpha \in I} S_\alpha$ be the corresponding product space. Then S is separable if and only if each S_α is separable.*

Proof. We are given that $S = \prod_{\alpha \in I} S_\alpha$ is a product space of non-empty non-singleton sets where $|\, I\,| \leq c = 2^{\aleph_0}$.

($\Rightarrow$) Suppose S is separable. Recall from the proof that appears on page 149, the projection map π_α is continuous for each $\alpha \in \mathbb{N}$. Hence, by Theorem 6.11, the continuous image of a separable space is separable, so each $S_\alpha = \pi_\alpha[S]$ is separable. We are done with this direction. The cardinality of I is irrelevant for this direction.

($\Leftarrow$) Suppose each S_α is separable and $|I| \leq c$. Then $|I| \leq |\,[0, 1]\,|$. So, there is a one-to-one function $f : I \to [0, 1]$ which maps I *into* $[0, 1]$. We can then index the product space S with $f[I]$. That is, $S = \prod_{\alpha \in I} S_{f(\alpha)}$. So, if $x \in S$ then x is of the form $\langle x_{f(\alpha)} \rangle_{\alpha \in I}$.

Since each factor of S is separable, each $S_{f(\alpha)}$ contains a countable dense subset

$$D_{f(\alpha)} = \{d_{f(\alpha)_n} : n = 1, 2, 3, \ldots\}$$

We will now construct a countable subset, D^*, of S which is dense in S.

For each $k \in \mathbb{N}\backslash\{0\}$ we can construct in $[0, 1]$ a finite sequence of pairwise disjoint closed intervals

$$\mathscr{A}_k = \{[a_j, b_j]_k : j \in \{1, 2, \ldots, k\}\ \}$$

such that $a_j, b_j \in \mathbb{Q} \cap [0, 1]$ and $a_{j+1} > b_j \leq 1$.

For each k, if $\alpha \in I$, $f(\alpha) \in [a_j, b_j]_k$ for, *at most*, one $j \in \{1, 2, \ldots, k\}$.

See that $\mathscr{W} = \cup\{\mathscr{A}_k : k = 1, 2, 3, \ldots\}$ is a countable set of intervals. We can then index the elements of $\mathscr{W}$ as

$$\mathscr{W} = \{[a_j, b_j]_{k,n} : j \in \{1, 2, \ldots, k\},\ k \in \{1, 2, 3, \ldots\},$$
$$n \in \{1, 2, 3, \ldots\}\}$$

where, for each n, $[a_j, b_j]_{k,n} = [a_j, b_j]_k \in \mathscr{A}_k$.

We define a function $h : \mathscr{W} \to \prod_{\alpha \in I} D_{f(\alpha)} \subseteq \prod_{\alpha \in I} S_{f(\alpha)}$ as follows:

$$h([a_j, b_j]_{k,n}) = \langle y_{f(\alpha)} \rangle_{\alpha \in I} \quad \text{where} \begin{cases} y_{f(\alpha)} = d_{f(\alpha)_n} & \text{if } f(\alpha) \in [a_j, b_j]_{k,n} \\ y_{f(\alpha)} = d_{f(\alpha)_1} & \text{otherwise} \end{cases}$$

Is $h : \mathscr{W} \to S$ a well-defined function? Yes it is, since (as mentioned earlier) for each k, if $\alpha \in I$, $f(\alpha) \in [a_j, b_j]_k$ for, *at most*, one $j \in \{1, 2, \ldots, k\}$.

Also note that every component of $h([a_j, b_j]_{k,n}) = \langle y_{f(\alpha)} \rangle_{\alpha \in I}$ belongs to some $D_{f(\alpha)}$ (which is dense in $U_{f(\alpha)}$).

Let $D^* = h[\mathscr{W}]$.

Claim 1: That D^* is countable in S.

Since the domain of h, $\mathscr{W}$, is countable then so is its range, $D^* = h[\mathscr{W}]$, in $\prod_{\alpha \in I} S_{f(\alpha)}$, as claimed.

Claim 2: That D^* is dense in S.

Let $B = \pi_{\alpha_1}^{\leftarrow}[U_{f(\alpha_1)}] \cap \cdots \cap \pi_{f(\alpha_k)}^{\leftarrow}[U_{f(\alpha_k)}]$ be a basic open neighborhood of a point in $S = \prod_{\alpha \in I} S_{f(\alpha)}$.

We need to verify that $B \cap D^* \neq \varnothing$.

Since $D_{f(\alpha)} = \{d_{f(\alpha)_n} : n = 1, 2, 3, \ldots\}$ is dense in $S_{f(\alpha)}$, then for each $f(\alpha_i)$ there exists $d_{f(\alpha_i)_{n_i}} \in U_{f(\alpha_i)}$, for $i = 1 \ldots, k$.

If $d_{f(\alpha_i)_m} \in S_{f(\alpha_i)}$, then

$$\pi_{f(\alpha_i)}{}^{\leftarrow}(d_{f(\alpha_i)_m}) = \{\langle y_{f(\alpha)} \rangle : f(\alpha_i) \in [a_j, b_j]_{k,m}\} = h([a_j, b_j]_{k,m})] \in h[\mathscr{W}]$$
$$\pi_{f(\alpha_i)}{}^{\leftarrow}(d_{f(\alpha_i)_1}) = \{\langle y_{f(\alpha)} \rangle : f(\alpha_i) \notin [a_j, b_j]_{k,m}\} = h([a_j, b_j]_{k,m}) \in h[\mathscr{W}]$$

Then, $h[\mathscr{W}] \cap \pi_{\alpha_1}^{\leftarrow}[U_{f(\alpha_1)}] \cap \cdots \cap \pi_{f(\alpha_k)}^{\leftarrow}[U_{f(\alpha_k)}] \neq \varnothing$. Then $B \cap D^* \neq \varnothing$, so D^* is dense in S, as claimed.

So, $h[\mathscr{W}]$ is a countable dense subset of $\prod_{\alpha\in I} S_{f(\alpha)}$. Then $\prod_{\alpha\in I} S_{f(\alpha)}$ is separable, as required.

One should exercise caution when transferring a topological property possessed by all factors of a product to the product itself. The statements of the above two theorems were proven with particular restrictions on the number of factors.

7.5 On Continuous Functions Involving Product Spaces

The choice of the "product topology" on a Cartesian product, $\prod_{\alpha\in I} S_\alpha$, is due mostly to the fact that it guarantees the continuity on a product space of all its projection maps, $\{\pi_\alpha : \alpha \in I\}$. It is the smallest topology that will guarantee this.

Lemma 7.11 *Let $S = \prod_{\alpha\in I} S_\alpha$ be a product space and (T, τ) be a space. Suppose $g : T \to S$ is a function mapping the space T to the product space, S. For each $\alpha \in I$, let*

$$f_\alpha = \pi_\alpha \circ g$$

Then

$$[g \text{ is continuous on } T] \Leftrightarrow [f_\alpha : T \to S_\alpha \text{ is continuous for all } \alpha]$$

Proof. Recall that, by definition of "product space", each π_α is guaranteed to be continuous on S.

($\Rightarrow$) If $g : T \to S$ is continuous then, for all $\alpha \in I$, so is $(\pi_\alpha \circ g) : T \to S_\alpha$ (since, by hypothesis, π_α is continuous, and the composition of continuous functions was shown to be continuous).

($\Leftarrow$) We fix $\beta \in I$. We are given that $(\pi_\beta \circ g) : T \to S_\beta$ is continuous. We are required to show that g is continuous. If V_β is open in S_β, since $(\pi_\beta \circ g)$ is continuous, then

$$g^{\leftarrow}\left[\pi_\beta^{\leftarrow}[V_\beta]\right] = (\pi_\beta \circ g)^{\leftarrow}[V_\beta]$$

is open in $S = \prod_{\alpha \in I} S_\alpha$. Since $\prod_{\alpha \in I} S_\alpha$ is equipped with the product topology, $\pi_\beta^{\leftarrow}[V_\beta]$ is a subbase element for this topology. So, g pulls back an open subbase element, $\pi_\beta^{\leftarrow}[V_\beta]$, to an open subset of $S = \prod_{\alpha \in I} S_\alpha$. By Theorem 6.4, g is continuous on T, as required.

Remark. In the above theorem, the fact that the Cartesian product, $S = \prod_{\alpha \in I} S_\alpha$, is equipped with the product topology is what guarantees that each π_α is continuous. If S is equipped with the box topology the theorem may not hold true.

The following example illustrates how the above theorem may break down if $\prod_{\alpha \in I} S_\alpha$ is equipped with the box topology.

Example 5. Consider the Cartesian product $\prod_{n \in \mathbb{N}} \mathbb{R}$ and suppose $\mathbb{R}$ is equipped with the usual topology. Let $i : \mathbb{R} \to \mathbb{R}$ denote the identity function $i(x) = x$ (a continuous function on $\mathbb{R}$). Let $g : \mathbb{R} \to \prod_{n \in \mathbb{N}} \mathbb{R}$ be a function where, for each $n \in \mathbb{N}$,

$$(\pi_n \circ g)(x) = i(x)$$

(where π_n is the nth projection map). If $\prod_{n \in \mathbb{N}} \mathbb{R}$ is equipped with the product topology we know, by the theorem above, that g must be continuous on $\mathbb{R}$.

Suppose that $\prod_{n \in \mathbb{N}} \mathbb{R}$ is equipped with the box topology. Show that g need not be continuous.

Solution: If $\prod_{n \in \mathbb{N}} \mathbb{R}$ is equipped with the box topology, consider the open subset

$$B = (-1, 1) \times (-1/2, 1/2) \times (-1/3, 1/3) \times (-1/4, 1/4) \cdots$$

with respect to the box topology. Suppose $g^{\leftarrow}[B]$ was open in $\mathbb{R}$. Then there would exist ε, such that $g[(-\varepsilon, \varepsilon)] \subseteq B$. Then $(\pi_n \circ g)[(-\varepsilon, \varepsilon)] = i[(-\varepsilon, \varepsilon)] = (-\varepsilon, \varepsilon) \subseteq (-1/n, 1/n)$ for all n. This can't possibly be. So $g^{\leftarrow}[B]$ is not open. So g is not continuous on $\mathbb{R}$.

Theorem 7.12 *Let $\{S_\alpha\}_{\alpha \in I}$ and $\{T_\alpha\}_{\alpha \in I}$ be two sets of topological spaces and let $\{f_\alpha\}_{\alpha \in I}$ be a family of functions, $f_\alpha : S_\alpha \to T_\alpha$. Then the function, $g : \prod_{\alpha \in I} S_\alpha \to \prod_{\alpha \in I} T_\alpha$, defined as*

$$g(\langle x_\alpha \rangle_{\alpha \in I}) = \langle f_\alpha(x_\alpha) \rangle_{\alpha \in I}$$

is continuous if and only if f_α is continuous on S_α for each $\alpha \in I$.

Proof. Suppose $S = \prod_{\alpha \in I} S_\alpha$, $T = \prod_{\alpha \in I} T_\alpha$ and $\{f_\alpha\}_{\alpha \in I}$ is a family of functions, $f_\alpha : S_\alpha \to T_\alpha$. Let $\pi_{\beta_S} : S \to S_\beta$ and $\pi_{\beta_T} : T \to T_\beta$ be βth projection maps. Let $g : \prod_{\alpha \in I} S_\alpha \to \prod_{\alpha \in I} T_\alpha$ be defined as $g(\langle x_\alpha \rangle_{\alpha \in I}) = \langle f_\alpha(x_\alpha) \rangle_{\alpha \in I}$.

($\Leftarrow$) For this direction, we are given that each function f_α is continuous on S_α. We want to show that the continuity of g follows.

To show continuity of g it will suffice to show that g pulls back subbase elements of $\prod_{\alpha \in I} T_\alpha$ to open sets in $\prod_{\alpha \in I} S_\alpha$. Let U_β be an open subset of T_β. Then $\pi^{\leftarrow}_{\beta_T}[U_\beta]$ is a subbase element for open sets in $T = \prod_{\alpha \in I} T_\alpha$.

By Lemma 7.11, if $\pi_\alpha \circ g$ is continuous for all α, then g is continuous. It then suffices to show that $g^{\leftarrow}[\pi^{\leftarrow}_{\beta_T}[U_\beta]]$ is open in S.

See that,

$$\begin{aligned}
\pi^{\leftarrow}_{\beta_S}[f^{\leftarrow}_\beta[U_\beta]] &= \pi^{\leftarrow}_{\beta_S}[\{x \in S_\beta : f_\beta(x) \in U_\beta\}] \\
&= \{\langle x_\alpha \rangle_{\alpha \in I} \in S : f_\beta(x_\beta) \in U_\beta\} \\
&= \{\langle x_\alpha \rangle_{\alpha \in I} \in S : \{f_\alpha(x_\alpha)\} \in \pi^{\leftarrow}_{\beta_T}[U_\beta]\} \\
&= \{\langle x_\alpha \rangle_{\alpha \in I} \in S : g(\langle x_\alpha \rangle_{\alpha \in I}) \in \pi^{\leftarrow}_{\beta_T}[U_\beta]\} \\
&= g^{\leftarrow}\left[\langle x_\alpha \rangle_{\alpha \in I} \in S : \{x_\alpha\} \in \pi^{\leftarrow}_{\beta_T}[U_\beta]\}\right] \\
&= g^{\leftarrow}[\pi^{\leftarrow}_{\beta_T}[U_\beta]]
\end{aligned}$$

so,

$$g^{\leftarrow}[\pi^{\leftarrow}_{\beta_T}[U_\beta]] = \pi^{\leftarrow}_{\beta_S}[f^{\leftarrow}_\beta[U_\beta]]$$

Since both π_{β_S} and f_β are continuous, then the right-hand side is open; so $g^{\leftarrow}[\pi^{\leftarrow}_{\beta_T}[U_\beta]]$ is open. So, g is continuous, as required for $\Leftarrow$.

($\Rightarrow$) Suppose $g : \prod_{\alpha \in I} S_\alpha \to \prod_{\alpha \in I} T_\alpha$ is continuous, where

$$g(\langle x_\alpha \rangle_{\alpha \in I}) = \langle f_\alpha(x_\alpha) \rangle_{\alpha \in I}$$

We are required to show that each $f_\alpha : S_\alpha \to T_\alpha$ is continuous.

Then for each β, $(\pi_\beta \circ g)(\langle x_\alpha \rangle_{\alpha \in I}) = \pi_\beta(\langle f_\alpha(x_\alpha) \rangle_{\alpha \in I}) = f_\beta(x_\beta) \in T_\beta$. Since both π_β and g are continuous on their domain, then so is f_β on S_β, as required. This proves $\Rightarrow$.

We expect that, *if two product spaces, S and T, have corresponding factors which are homeomorphic then S and T are homeomorphic.* In the proof of the following theorem, we confirm this by explicitly exhibiting the required homeomorphism.

Theorem 7.13 *Let $S = \prod_{\alpha\in I} S_\alpha$ and $T = \prod_{\alpha\in I} T_\alpha$ be two product spaces. Suppose that, for each $\alpha \in I$, S_α and T_α are homeomorphic. Then the spaces S and T are homeomorphic.*

Proof. For each $\alpha \in I$, let $f_\alpha : S_\alpha \to T_\alpha$ be a homeomorphism mapping S_α onto T_α.

Let $g : \prod_{\alpha\in I} S_\alpha \to \prod_{\alpha\in I} T_\alpha$ be defined as

$$g(\langle x_\alpha\rangle_{\alpha\in I}) = \langle f_\alpha(x_\alpha)\rangle_{\alpha\in I} \in \prod_{\alpha\in I} T_\alpha$$

Since each f_α is continuous and one-to-one, then so is g (by Theorem 7.12).

We claim that g is open.

By hypothesis, each f_α is an open map. Let $\pi^{\leftarrow}_{\alpha_1}[U_{\alpha_1}] \cap \pi^{\leftarrow}_{\alpha_2}[U_{\alpha_2}] \cdots \cap \pi^{\leftarrow}_{\alpha_k}[U_{\alpha_k}]$ be an open base element in $\prod_{\alpha\in I} S_\alpha$. Since g is one-to-one, see that

$$\begin{aligned}
&g\left[\pi^{\leftarrow}_{\alpha_1}[U_{\alpha_1}] \cap \pi^{\leftarrow}_{\alpha_2}[U_{\alpha_2}] \cdots \cap \pi^{\leftarrow}_{\alpha_k}[U_{\alpha_k}]\right]\\
&\quad = g\left[\pi^{\leftarrow}_{\alpha_1}[U_{\alpha_1}]\right] \cap g\left[[\pi^{\leftarrow}_{\alpha_2}[U_{\alpha_2}]\right] \cdots \cap g\left[[\pi^{\leftarrow}_{\alpha_k}[U_{\alpha_k}]\right]\\
&\quad = f_{\alpha_1}\left[\pi^{\leftarrow}_{\alpha_1}[U_{\alpha_1}]\right] \times \cdots \times f_{\alpha_k}\left[\pi^{\leftarrow}_{\alpha_k}[U_{\alpha_k}]\right]
\end{aligned}$$

with all other factors equal to T_γ.

Since each f_{α_i} is open, RHS is open in $\prod_{\alpha\in I} T_\alpha$.

Since the right-hand side is open, then g is open. So g is a homeomorphism.

We have shown that $g(\langle x_\alpha\rangle_{\alpha\in I}) = \langle f_\alpha(x_\alpha)\rangle_{\alpha\in I}$ is the required homeomorphism mapping S onto T.

We now show that every product space contains a homeomorphic copy of each of its factors.

Theorem 7.14 *Let $S = \prod_{\alpha \in I} S_\alpha$ be a product space. Then, for each $\alpha \in I$, S contains a subspace which is a homeomorphic copy of S_α.*

Proof. Let $\beta \in I$. For $\alpha \neq \beta$, we choose and fix $k_\alpha \in S_\alpha$. Let

$$T = \prod_{\alpha \in I} T_\alpha = \{\langle x_\alpha \rangle_{i \in I} \ : \ T_\beta = S_\beta \text{ and } [\alpha \neq \beta] \Rightarrow [T_\alpha = \{k_\alpha\}]\}$$

So, every T is a subspace of S. We claim that T is homeomorphic to S_β.

We define $g : S_\beta \to T$ as $g(x) = \langle x_\alpha \rangle_{\alpha \in I}$, where $x_\beta = x$, and if $\alpha \neq \beta$, $x_\alpha = k_\alpha$. Then g maps S_β one-to-one and onto $T \subseteq S$. Then the function, $(\pi_\beta \circ g) : S_\beta \to S_\beta$ is the continuous identity map onto S_β. Invoking Theorem 7.11, we conclude that g is continuous on S_β. So, g embeds S_β in the proper subset T of $\prod_{\alpha \in I} S_\alpha$ and so embeds S_β in $\prod_{\alpha \in I} S_\alpha$, as required.

Example 6. Let $\{(S_n, \rho_n) : n \in \mathbb{N}\}$ be a countable family of metric spaces. Suppose that, for each $n \in \mathbb{N}$, $\sup\{\rho_n(x, y) : x, y \in S_n\} \leq 1$. Let

$$S = \prod_{n \in \mathbb{N}} S_n$$

the set of all sequences with the nth entry in S_n. Let $\rho : S \times S \to \mathbb{R}$ be defined as

$$\rho(f, g) = \sum_{n \in \mathbb{N}} \frac{\rho_n\,(\,f(n),\, g(n)\,)}{2^n}$$

Verify that $\rho : S \times S \to \mathbb{R}$ is a metric on S. Furthermore, verify that the topology on $S = \prod_{n \in \mathbb{N}} S_n$ derived by the metric ρ is the product topology.

Solution: We are given that $S = \prod_{n \in \mathbb{N}} S_n$ where each S_n is a metric space and, for each n,

$$\sup\{\rho_n(x, y) : x, y \in S_n\} \leq 1$$

That is, the distance between any two elements of S_n does not exceed 1.

We claim that $\rho : S \times S \to \mathbb{R}$ thus defined, is a valid metric on S. For $f, g \in S$ let $f = \langle f(n)\rangle_{n\in\mathbb{N}}$ and $g = \langle g(n)\rangle_{n\in\mathbb{N}}$.

For M1, $\rho(f,g) = \sum_{n\in\mathbb{N}} \frac{\rho_n(\,f(n), g(n)\,)}{2^n} \leq \sum_{n\in\mathbb{N}} \frac{1}{2^n} = 2 \geq 0$.

If $\rho(f,g) = \sum_{n\in\mathbb{N}} \frac{\rho_n(\,f(n), g(n)\,)}{2^n} = 0$, then $\rho_n(\,f(n),\, g(n)\,) = 0$ for all n. Since $f = \langle f(n)\rangle_{n\in\mathbb{N}}$ and $g = \langle g(n)\rangle_{n\in\mathbb{N}}$, then $f = g$.

For M2, $\rho(f,g) = \sum_{n\in\mathbb{N}} \frac{\rho_n(\,f(n), g(n)\,)}{2^n} = \sum_{n\in\mathbb{N}} \frac{\rho_n(\,g(n), f(n)\,)}{2^n} = \rho(g,f)$.

For M3, we compare $\rho(f,h)$ and $\rho(f,g)+\rho(g,h)$. Since, for each n, $\rho_n(\,f(n),\, h(n)\,) \leq \rho_n(\,f(n),\, g(n)\,) + \rho_n(\,g(n),\, h(n)\,)$, then

$$
\begin{aligned}
\rho(f,h) &= \sum_{n\in\mathbb{N}} \frac{\rho_n(f(n),h(n))}{2^n} \\
&\leq \sum_{n\in\mathbb{N}} \frac{\rho_n(f(n),g(n)) + \rho_n(g(n),h(n))}{2^n} \\
&= \sum_{n\in\mathbb{N}} \left[\frac{\rho_n(f(n),g(n))}{2_n} + \frac{\rho_n(g(n),h(n))}{2^n}\right] \\
&= \sum_{n\in\mathbb{N}} \frac{\rho_n(f(n),g(n))}{2_n} + \sum_{n\in\mathbb{N}} \frac{\rho_n(g(n),h(n))}{2^n} \\
&\qquad \text{(all terms are positive)} \\
&= \rho(f,g) + \rho(g,h)
\end{aligned}
$$

So, ρ is a valid metric on $S = \prod_{n\in\mathbb{N}} S_n$, as claimed.

We will now compare the the metric topology on (S,ρ) to the product topology on S.

Let $f = \langle f(n)\rangle \in S$. Let $V = B_{1/2^p}(f)$ be an open ball in S with respect to ρ. We claim that V is open in S with respect to the product topology.

Proof of claim: Let

$$U = \{g \in S : g = \langle g(n)\rangle,\ \rho(f(n), g(n)) < 2^{-p-n-2} \text{ for } n \geq p+2\}$$

Note that U is a union of open base elements and so is an open subset of S with respect to the product topology. It suffices to show that

$U \subset V$. If $y \in U$, then

$$\begin{aligned}
\rho(f,g) &= \sum_{n\in\mathbb{N}} \frac{\rho_n(f(n),g(n))}{2^n} \\
&= \left(\frac{\rho_0(f(0),g(0))}{2^0} + \frac{\rho_1(f(1),g(1))}{2^1} + \cdots \right.\\
&\quad \left. + \frac{\rho_{p+2}(f(p+2),g(p+2))}{2^{p+2}}\right) \\
&\quad + \frac{\rho_{p+3}(\,f(p+3),\ g(p+3)\,)}{2^{p+3}} + \cdots \\
&< \left(\frac{2^{-p-0-2}}{2^0} + \frac{2^{-p-1-2}}{2^1} + \frac{2^{-p-2-2}}{2^2} + \cdots + \frac{2^{-p-(p+2)-2}}{2^{p+2}}\right) \\
&\quad + \frac{1}{2^{p+3}} + \frac{1}{2^{p+4}} + \cdots + \\
&= \left(\frac{1}{2^{p+2}} + \frac{1}{2^{p+4}} + \frac{1}{2^{p+6}} + \cdots + \frac{1}{2^{3p+6}}\right) \\
&\quad + \left(\frac{1}{2^{p+3}} + \frac{1}{2^{p+4}} + \frac{1}{2^{p+5}} + \cdots + \right) \\
&< \frac{1}{2^{p+1}} + \frac{1}{2^{p+1}} \\
&= \frac{1}{2^p}
\end{aligned}$$

So, $g \in V = B_{1/2^p}(f)$.

So, V is open in S with respect to the product topology.

Suppose now that U is an open subbase element in S with respect to the product topology. Suppose U contains the point $a = \langle a(n)\rangle_{n\in\mathbb{N}}$. It will suffice to find an ε such that $B_\varepsilon(a)$ (with respect to ρ) is contained in U.

Now, U is of the form $\{x : f(n) \in W_n\}$ where $W_n = S_n$ for all n except, possibly, for some $k \in \mathbb{N}$ where W_k is open in S_k. Then there is an ε_1 such that

$$B_a = \{g \in S : \rho_k(a(k), g(k)) < \varepsilon_1\}$$

is a subset of U containing a. Then

$$\frac{\rho_k(a(k), g(k))}{2^k} < \frac{\varepsilon_1}{2^k}$$

Choose $\varepsilon = \varepsilon_1/2^k$. We then claim that $B_\varepsilon(a) \subseteq B_a \subseteq U$ (with respect to the metric, ρ).

Proof of claim: Let $g \in B_\varepsilon(a)$. Then

$$\rho(a, g) = \sum_{n \in \mathbb{N}} \frac{\rho_n(a(n), g(n))}{2^n} < \varepsilon = \frac{\varepsilon_1}{2^k}$$

Since

$$\frac{\rho_k(a(k), g(k))}{2^k} \leq \sum_{n \in \mathbb{N}} \frac{\rho_n(a(n), g(n))}{2_n} = \rho(a, g) < \varepsilon = \varepsilon_1/2^k$$

then this $g \in B_a$. So, $B_\varepsilon(a) \subseteq B_a \subseteq U$, with respect to ρ. We conclude that U is open in (S, ρ).

Then, the metric space, (S, ρ), and the product space, (S, τ), have the same topology.

7.6 Topic: Product Spaces are "Commutative"

What effect does it have on the topology of a product space if we alter the order of its factors? We investigate this question by first considering a particular case in the following example.

Example 7. Show that for topological spaces S_1, S_2, and S_3,

$$S = S_1 \times S_2 \times S_3 \text{ and } T = S_3 \times S_1 \times S_2$$

are homeomorphic.

Solution: Let $q : \{1, 2, 3\} \to \{3, 1, 2\}$ be the one-to-one function where $q(1) = 3$, $q(2) = 1$, $q(3) = 2$. Let $q^* : S_1 \times S_2 \times S_3 \to S_3 \times S_1 \times S_2$ be defined as

$$q^*\langle x_1, x_2, x_3\rangle = \langle x_3, x_1, x_2\rangle$$

We claim that $\pi_{q(3)}\circ q^*$ is continuous. If so we can apply Theorem 7.11, to conclude q^* is continuous. Let U be an open subset of $S_{q(3)}$. Then $U \times S_1 \times S_2$ is a basic open set in T. Then

$$\begin{aligned}(\pi_{q(3)}q^*)^{\leftarrow}[U] &= q^{*\leftarrow}[\pi_{q(3)}{}^{\leftarrow}[U]] \\ &= q^{*\leftarrow}[U \times S_1 \times S_2] \\ &= S_1 \times S_2 \times U \text{ (Since } q^*[S_1 \times S_2 \times U] = U \times S_1 \times S_2\text{)} \\ &\quad \text{(a basic open subset of } S\text{).}\end{aligned}$$

So, $\pi_{q(3)}\circ q^*$ is continuous, as claimed. Similarly, $\pi_{q(2)}\circ q^*$ and $\pi_{q(1)}\circ q^*$ are continuous. By Theorem 7.11, $q^* : S \to T$ is continuous. Similarly $q^{*\leftarrow} : T \to S$ is continuous. So, $q^* : S \to T$ is a homeomorphism, as required.

In the next theorem, we generalize the proof in the example to one involving larger product spaces.

Theorem 7.15 *Let $S = \prod_{\alpha\in I} S_\alpha$ and let $q : I \to J$ be a one-to-one function mapping I onto an indexing set, J. Then there is a homeomorphism mapping S onto $T = \prod_{\alpha\in I} S_{q(\alpha)}$.*

Proof. Let $S = \prod_{\alpha\in I} S_\alpha$ and let $q : I \to J$ be a one-to-one function mapping the indexing set, I, onto the indexing set, J. We are required to show that $T = \prod_{\alpha\in I} S_{q(\alpha)}$ and $S = \prod_{\alpha\in I} S_\alpha$ are homeomorphic.

Let $q^* : S \to T$ be defined as $q^*(\langle x_\alpha\rangle_{\alpha\in I}) = \langle x_{q(\alpha)}\rangle_{\alpha\in I}$.

Since q is one-to-one and onto J, then both $q^* : S \to T$ and $q^{*\leftarrow} : T \to S$ are one-to-one and onto. Then $\pi_{q(\alpha)}\circ q^* : S \to S_{q(\alpha)}$, where

$$(\pi_{q(\alpha)}\circ q^*)(\langle x_\alpha\rangle_{\alpha\in I}) = \pi_{q(\alpha)}(\langle x_{q(\alpha)}\rangle_{\alpha\in I}) = x_{q(\alpha)}$$

Let $U_{q(\beta)}$ be an open neighborhood of $x_{q(\beta)}$ in $S_{q(\beta)}$.

Then

$$\begin{aligned}(\pi_{q(\beta)}\circ q^*)^{\leftarrow}[U_{q(\beta)}] &= q^{*\leftarrow}[\pi_{q(\beta)}{}^{\leftarrow}[U_{q(\beta)}]] \\ &= q^{*\leftarrow}[\cdots \times S_{q(i)} \times \cdots \times U_{q(\beta)} \times \cdots \times S_{q(i)} \times \cdots] \\ &= \cdots \times S_i \times \cdots \times U_\beta \times \cdots \times S_i \times \cdots \\ &\quad \text{(a basic open subset of } S\text{).}\end{aligned}$$

So, $\pi_{q(3)} \circ q^*$ is continuous, as claimed. By Theorem 7.11, $q^* : S \to T$ is continuous.

Similarly $q^{*\leftarrow} : T \to S$ is continuous. So, $q^* : S \to T$ is a homeomorphism, as required.

The above theorem confirms that "altering the order of the factors of a product space produces another product space which is homeomorphic to the original one". We can summarize the statement by saying that...

product spaces are commutative.

7.7 Example: The *Tychonoff Plank*

We now briefly discuss the product of two ordinal spaces, for future reference. The ordinal space was introduced in Definition 5.14.[4] In the following example, ω_1 represents the first uncountable ordinal, while, ω_0 represents the first countable infinite ordinal. Let W represent the ordinal space, $[0, \omega_1]$, and T represent the ordinal space, $[0, \omega_0]$. Recall (from the discussion which appears after Definition 5.14) that the elements of the open base for an ordinal space are of the form $(\alpha, \beta]$.

Let

$$S = W \times T = [0, \omega_1] \times [0, \omega_0]$$

be the product space of the two given ordinal spaces. Then the elements of S can be viewed as ordered pairs $(\alpha, \beta) \in W \times T$. Since both sets are linearly ordered, it doesn't hurt to visualize the product space, S, as a Cartesian plane of numbers where W represents the horizontal axis and T represents the vertical axis. We would then have $(0, 0)$ in the lower left corner and (ω_1, ω_0) in the top right corner. The topological space $S = [0, \omega_1] \times [0, \omega_0]$ equipped with this topology is commonly referred to by topologists as the

Tychonoff plank.

---

[4]A more detailed study of the ordinals in the context of set theory is found in Axioms and set theory, by Robert André (can be found online).

The subspace S^* of S defined as

$$S^* = S\backslash\{(\omega_1, \omega_0)\}$$

simply obtained by deleting the top right corner from the Tychonoff plank is appropriately referred to as the

Deleted Tychonoff plank.

As an open neighborhood base of the point, $(\beta, \mu) \in S$, we can use elements of the form

$$\mathscr{B}_{(\beta,\mu)} = \{(\alpha, \beta] \times (\gamma, \mu] : \alpha < \beta \text{ and } \gamma < \mu\}$$

The Tychonoff plank may appear to be a topological space that is, in many ways, similar to the the product space $\mathbb{R} \times \mathbb{N}$ but, as we will eventually see, has quite different properties. Both the Tychonoff plank and the Deleted Tychonoff plank are useful topological spaces to remember.

7.8 Topic: Embedding a Topological Space in a Product Space

In this section, we will show how a particular family, $\mathscr{F}$, of continuous functions on a topological space, (S, τ), can be used to embed S in a product space whose factors are the range of the functions in $\mathscr{F}$. This method is exhibited in a theorem titled "The Embedding Theorem I". The *Embedding theorem I* applies only to topological spaces, S, in which singleton sets are closed. In order to understand this theorem one must be familiar with two very important notions in topology:

— Family of functions which *separates points and closed sets* of S.
— *Evaluation map* with respect to a family, $\mathscr{F}$, of functions.

Definition 7.16 Let (S, τ_s) be a topological space and $\{(S_\alpha, \tau_\alpha) : \alpha \in \Gamma\}$ be an indexed family of non-empty topological spaces. Let $\mathscr{F} = \{f_\alpha : \alpha \in \Gamma\}$ be a set of *continuous functions*, $f_\alpha : S \to S_\alpha$, each one mapping S onto its range $f_\alpha[S] \subseteq S_\alpha$.

(a) We say that $\mathscr{F}$ *separates points and closed sets* if, whenever F is a closed subset of S and $x \notin F$, then there exists at least one function $f_\beta \in \mathscr{F}$ such that $f_\beta(x) \notin \mathrm{cl}_{S_\beta} f_\beta[F]$.[5]

(b) We define a function, $e : S \to \prod_{\alpha\in\Gamma} S_\alpha$, as follows:

$$e(x) = \langle f_\alpha(x)\rangle_{\alpha\in\Gamma} \in \prod_{\alpha\in\Gamma} f_\alpha[S] \subseteq \prod_{\alpha\in\Gamma} S_\alpha$$

We refer to the function $e : S \to \prod_{\alpha\in\Gamma} S_\alpha$ as the

evaluation map of S into $\prod_{\alpha\in\Gamma} S_\alpha$ with respect to $\mathscr{F}$

Theorem 7.17 The embedding theorem I. *Let (S, τ_s) be a topological space in which every singleton set in S is a closed subset of S. Given an indexed family of non-empty topological spaces, $\{(S_\alpha, \tau_\alpha) : \alpha \in \Gamma\}$, let $\mathscr{F} = \{f_\alpha : \alpha \in \Gamma\}$ be a set of* continuous functions *where each f_α maps S onto its range, $f_\alpha[S]$, inside S_α.*

(a) *If $\mathscr{F}$* separates points and closed sets *of its domain, S, and $\prod_{\alpha\in\Gamma} S_\alpha$ is equipped with the product topology, then the evaluation map,*

$$e(x) = \langle f_\alpha(x)\rangle_{\alpha\in\Gamma} \in \prod_{\alpha\in\Gamma} f_\alpha[S] \subseteq \prod_{\alpha\in\Gamma} S_\alpha$$

with respect to $\mathscr{F}$, is both continuous and one-to-one on S.

(b) *Furthermore, the function, $e : S \to \prod_{\alpha\in\Gamma} S_\alpha$, maps S homeomorphically onto*

$$e[S] \subseteq \prod_{\alpha\in\Gamma} f_\alpha[S] \subseteq \prod_{\alpha\in\Gamma} S_\alpha$$

Hence, this evaluation map embeds a homeomorphic copy of S into the product space, $\prod_{\alpha\in\Gamma} S_\alpha$.

---

[5] Similar expressions such as "the subsets A and B are completely separated" and "the real-valued function, $f : S \to \mathbb{R}$ separates A and B if $A \subseteq Z(f)$ and $B \subseteq Z(f-1)$ for some $f \in C(X)$" will be defined when discussing normal spaces in Definition 10.2.

Proof. We are given that (S, τ_s) is a topological space in which all singleton sets, $\{x\}$, are closed in S and a family of topological spaces, $\{S_\alpha : \alpha \in \Gamma\}$. For the set,

$\mathscr{F} = \{f_\alpha : S \to S_\alpha\}_{\alpha \in \Gamma}$, of continuous functions on S, we define

$$e : S \to \prod_{\alpha\in\Gamma} f_\alpha[S] \subseteq \prod_{\alpha\in\Gamma} S_\alpha$$

as an evaluation map with respect to $\mathscr{F}$.

(a) Note that, for each $\alpha \in \Gamma$ and $x \in S$, $(\pi_{\alpha}\circ e)(x)) = \pi_\alpha(\langle f_\alpha(x)\rangle_{\alpha\in\Gamma}) = f_\alpha(x)$. Since, for each $\alpha \in \Gamma$, f_α is continuous, then so is $\pi_\alpha \circ e : S \to f_\alpha[S]$. By Lemma 7.11, $e : S \to \prod_{\alpha\in\Gamma} f_\alpha[S]$ is continuous.

We *claim* that $e : S \to \prod_{\alpha\in\Gamma} f_\alpha[S]$ is one-to-one on S. Suppose a and b are distinct points in S. Then, since the single set $\{b\}$ is closed and $\mathscr{F}$ separates points and closed sets, there exists $\beta \in \Gamma$ such that $f_\beta(a) \notin \text{cl}_{S_\beta} f_\beta[\{b\}]$. Then the βth component of $e(a) = \langle f_\alpha(a)\rangle_{\alpha\in\Gamma}$ and $e(b) = \langle f_\alpha(b)\rangle_{\alpha\in\Gamma}$ are distinct and so $e(a) \neq e(b)$. We conclude that the evaluation map $e : S \to \prod_{\alpha\in\Gamma} f_\alpha[S]$ is one-to-one on S, as *claimed.*

(b) To prove that the evaluation map $e : S \to \prod_{\alpha\in\Gamma} f_\alpha[S]$ embeds S into $\prod_{\alpha\in\Gamma} S_\alpha$, it will suffice to show that it is an open function and then invoke Theorem 6.9.

Let U be a non-empty open subset of S with the point $u \in U$. Then $F = S\setminus U$ is closed in S. Since $\mathscr{F}$ separates points and closed sets, there exists $\beta \in \Gamma$ such that $f_\beta(u) \notin \text{cl}_{S_\beta} f_\beta[F]$. That means, $f_\beta(u) \in S_\beta\setminus[\text{cl}_{S_\beta} f_\beta[F]]$.

We now show that $e[U]$ is open in $\prod_{\alpha\in\Gamma} S_\alpha$.

Note that,

$$\begin{aligned}(\pi_{\beta}\circ e)(u) &= \pi_\beta(e(u))\\ &= \pi_\beta(\langle f_\alpha(u)\rangle_{\alpha\in\Gamma})\\ &= f_\beta(u)\\ &\in S_\beta\setminus\left[\text{cl}_{S_\beta} f_\beta[F]\right]\end{aligned}$$

Since $e(u) \in \pi_\beta^{\leftarrow}\left[S_\beta\setminus\left[\text{cl}_{S_\beta} f_\beta[F]\right]\right]$ and since π_β is continuous, then $\pi_\beta^{\leftarrow}\left[S_\beta\setminus\left[\text{cl}_{S_\beta} f_\beta[F]\right]\right]$ is an open neighborhood of $e(u)$ in $\prod_{\alpha\in\Gamma} S_\alpha$. It now suffices to show that

$$\pi_\beta^{\leftarrow}\left[S_\beta \backslash \left[\text{cl}_{S_\beta} f_\beta[F]\right]\right] \subseteq e[U]$$

Suppose $e(a) \in \pi_\beta^{\leftarrow}\left[S_\beta \backslash \left[\text{cl}_{S_\beta} f_\beta[F]\right]\right]$.

$$\begin{aligned} e(a) \in \pi_\beta^{\leftarrow}\left[S_\beta \backslash \left[\text{cl}_{S_\beta} f_\beta[F]\right]\right] &\Rightarrow (\pi_\beta \circ e)(a) \in \pi_\beta\left[\pi_\beta^{\leftarrow}\left[S_\beta \backslash \text{cl}_{S_\beta} f_\beta[F]\right]\right] \\ &\Rightarrow f_\beta(a) \in \left[S_\beta \backslash \text{cl}_{S_\beta} f_\beta[F]\right] \\ &\Rightarrow f_\beta(a) \notin \text{cl}_{S_\beta} f_\beta[F] \\ &\Rightarrow a \in S\backslash F = S\backslash(S\backslash U) = U \\ &\Rightarrow e(a) \in e[U] \end{aligned}$$

So, $\pi_\beta^{\leftarrow}\left[S_\beta \backslash \left[\text{cl}_{S_\beta} f_\beta[F]\right]\right]$ is an open neighborhood of $e(u)$ which is entirely contained in $e[U]$. We conclude $e[U]$ is open in $\prod_{\alpha\in\Gamma} S_\alpha$ and so $e : S \to \prod_{\alpha\in\Gamma} S_\alpha$ is a homeomorphism.

We have thus shown that $e : S \to \prod_{\alpha\in\Gamma} S_\alpha$ embeds S into the product space, $\prod_{\alpha\in\Gamma} S_\alpha$.

7.9 Topic: The Cantor Set: Viewed a the Continuous Image of a Product Space

The Cantor set is a notion which is part of general mathematical culture. We often refer to it because of its interesting set-theoretic and topological properties. It can be described in various ways. We will formally provide a set-theoretic definition of the Cantor set and then study it from a topological point of view.

7.9.1 *A prologue to a definition of the Cantor set*

Recall that $\mathbb{Z}^+ = \{1, 2, 3, \ldots\}$. Let $D = \{0, 1, 2, \ldots, 8, 9\}$. So, $\prod_{n\in\mathbb{Z}^+} D = D^{\mathbb{Z}^+}$ represents the set of all functions mapping $\mathbb{Z}^+$ into D. The set, $\prod_{n\in\mathbb{Z}^+} D$ can be described as the set of all countably infinite ordered strings, of the form,

$$\langle m_n \rangle_{n\in\mathbb{Z}^+}, \quad \text{where } m_n \in \{0,1,2,\ldots,9\}$$

Suppose we define the function, $\varphi : \prod_{n\in\mathbb{Z}^+} D \to [0,1]$ (the closed interval in $\mathbb{R}$ from 0 to 1) as follows:

$$\varphi(\langle m_1, m_2, m_3, \ldots, m_n, \ldots, \rangle) = \sum_{n=1}^{\infty} \frac{m_n}{10^n}$$

Note that every number $x \in [0,1]$ (represented in its infinite decimal expansion $0.m_1m_2m_3\cdots$) can be expressed in the form,

$$x = \sum_{n=1}^{\infty} \frac{m_n}{10^n} = 0.m_1m_2m_3\cdots$$

For example,

$$\varphi(\langle 0,0,0,,\ldots\rangle) = \sum_{n=1}^{\infty} \frac{0}{10^n} = 0$$

$$\varphi(\langle 9,9,9,\ldots\rangle) = \sum_{n=1}^{\infty} \frac{9}{10^n} = 0.9999\ldots = 1$$

$$\varphi(\langle 2,9,9,\ldots\rangle) = 0.2999\ldots = 0.3000\ldots = \varphi(\langle 3,0,0,\ldots\rangle)$$

and so φ maps $\prod_{n\in\mathbb{Z}^+} D$ *onto* [0,1] but is not necessarily one-to-one (since an endless string of 9's and an endless string of 0's may be mapped to the same element in [0, 1]). But the entire set [0,1] is, indeed, the image of $\prod_{n\in\mathbb{Z}^+} D$ under φ.

However, if we reduce the size of D to say, $D^* = \{0,1,2,5,7,9\}$, then the function, $\varphi : \prod_{n\in\mathbb{Z}^+} D^* \to [0,1]$, defined similarly, would produce a range which is a *proper* subset of $[0,1]$ which would contain multiple gaps in it (since the digits 3, 6, and 8 are lacking in the ordered strings of the domain).

If we set $D = \{0,1,2\}$ and every number x in $[0,1]$ is expressed in its triadic expansion form[6] then the set

$$\varphi\left[\prod_{n\in\mathbb{Z}^+}\{0,1,2\}\right] = [0.000\ldots_3,\ 0.2222\ldots_3] = [0,1]$$

[6] …, 0, 1, 2, 10, 11, 12, 100, 101, 102, 110,…

would look like,

$$\varphi(\langle m_n\rangle_{n\in\mathbb{Z}^+}) = \sum_{n=1}^{\infty}\frac{m_n}{3^n} = 0.m_1m_2m_3\cdots_3 \in [0.000\ldots_3,\ 0.2222\ldots_3]$$
$$= [0,1]$$

For example,

$$\varphi(\langle 0,0,0,\ldots\rangle) = 0.0000\ldots_3 = \sum_{n=1}^{\infty}\frac{0}{3^n}$$

$$\varphi(\langle 2,2,2,\ldots\rangle) = 0.2222\ldots_3 = \sum_{n=1}^{\infty}\frac{2}{3^n} = 1$$

$$\varphi(\langle 1,1,1,\ldots\rangle) = 0.1111\ldots_3 = \sum_{n=1}^{\infty}\frac{1}{3^n} = \frac{1}{2} \quad \text{(Verify!)}$$

$$\varphi(0,1,2,2,\ldots) = 0.012222_3\ldots = 0.0200000000\ldots_3$$
$$= \varphi(0,2,0,0,0,\ldots)$$

Note that, φ *is not one-to-one onto* $\prod_{n\in\mathbb{Z}^+}\{0,1,2\}$.

If we shrink the set $\{0,1,2\}$ to $D=\{0,2\}$, with $m_n\in\{0,2\}$ and $\varphi:\prod_{n\in\mathbb{Z}^+}\{0,2\}\to[0,1]$ is defined as

$$\varphi(\langle m_n\rangle_{n\in\mathbb{Z}^+}) = \sum_{n=1}^{\infty}\frac{m_n}{3^n} = 0.m_1m_2m_3\ldots_3 \in [0.000\ldots_3,\ 0.2222\ldots_3]$$
$$= [0,1]$$

where $0.m_1m_2m_3\ldots_3$ contains only 0's and 2's. The set, $\varphi\left[\prod_{n\in\mathbb{Z}^+}\{0,2\}\right]$ is now a *proper* subset of $[0.000\ldots_3,\ 0.2222\ldots_3] = [0,1]$.

In this case, φ *is indeed onto* $[0,1]$ *its domain.* For example, even if $0.02000\ldots_3 = 0.01222\ldots_3$

$$\varphi^{\leftarrow}(0.012222\ldots_3) = \varphi^{\leftarrow}(0.02000\ldots_3) = \{0,2,0,0,0,\ldots\}$$

So, every point in $[0,1]$ can be expressed with simply 0's and 2's in ternary expansion form.

The "Cantor set" involves the particular function

$$\varphi : \prod_{n\in\mathbb{Z}^+} \{0,2\} \to [0,1]$$

who's range, $\varphi\left[\prod_{n\in\mathbb{Z}^+}\{0,2\}\right]$, is a proper subset of [0, 1].

Definition 7.18 The Cantor set: A definition. Suppose $\mathbb{Z}^+ = \mathbb{N}\backslash\{0\}$ and $\prod_{n\in\mathbb{Z}^+}\{0,2\} = \{0,2\}^{\mathbb{Z}^+}$, the set of all countably infinite sequences of 0's and 2's. Let the function

$$\varphi : \prod_{n\in\mathbb{Z}^+} \{0,2\} \to [0,1]$$

be defined as

$$\varphi(\langle m_n\rangle_{n\in\mathbb{Z}^+}) = \sum_{n=1}^{\infty} \frac{m_n}{3^n}$$

where $\varphi(\langle m_n\rangle_{n\in\mathbb{Z}^+}) = 0.m_1m_2m_3\ldots_3$ is an element of $[0,1]$ expressed in its triadic expansion form. Since the digit 1 is lacking in the ordered strings, φ is one-to-one on its domain and the range, $\varphi[\prod_{n\in\mathbb{Z}^+}\{0,2\}]$, is a *proper* subset of the interval, $[0.000\ldots_3, 0.2222\ldots_3] = [0,1]$, with multiple gaps in it. The one-to-one image,

$$C = \varphi[\prod_{n\in\mathbb{Z}^+}\{0,2\}]$$

of $\prod_{n\in\mathbb{Z}^+}\{0,2\}$ is a proper subset of $[0.000\ldots_3, 0.2222\ldots_3] = [0,1]$. The set, C, is referred to as the *Cantor set.*

Then the Cantor set can be seen as

$$C = \left\{\sum_{n=1}^{\infty} \frac{m_n}{3^n} : m_n \in \{0,2\}\right\}$$

all possible values of the infinite sum $\sum_{n=1}^{\infty} \frac{m_n}{3^n}$ where $m_n = 0$ or 2.

7.9.2 *What does the Cantor set look like?*

The definition of the "Cantor Set", C, states that C is a particular subset of the closed unit interval, $[0,1]$. The Cantor set is the range, $\varphi[\prod_{n\in\mathbb{Z}^+}\{0,2\}]$, under the a well-defined function, φ, with domain, $\prod_{n\in\mathbb{Z}^+}\{0,2\}$. This definition, by itself, is not very useful when trying to visualize what kind of subset of $[0,1]$ it represents. If the reader is willing to go through the trouble of verifying this, C does not contain any of the points in the open intervals, $(1/3,2/3)$, $(1/3^2,2/3^2)$, $(7/3^2,8/3^2)$. Since φ maps $\prod_{n\in\mathbb{Z}^+}\{0,1,2\}$ onto $[0,1]$, then φ will map $\prod_{n\in\mathbb{Z}^+}\{0,2\}$ only onto a proper subset of [0, 1]. Such gaps will occur. It doesn't take long before one sees a pattern emerge. We normally view the Cantor set as a set which is inductively constructed in stages by successively defining a nested sequence, $C_0, C_1, C_2, \ldots$, of subsets of [0,1]. Ultimately we define the Cantor set as being the intersection, $C = \cap_{n\in\mathbb{N}} C_n$, of all of these (after convincing ourselves that C would not be empty). We can visualize this procedure as follows.

$$C_0 = [0,1]$$

$$C_1 = C_0 \backslash (1,3)$$

$$C_2 = C_1 \backslash \left(\frac{1}{3^2}\frac{2}{3^2}\right) \cup \left(\frac{7}{3^2}\frac{8}{3^2}\right)$$

$$C_3 = C_2 \backslash \ldots \text{open middle thirds in } C_2$$

$$\vdots$$

$$C_{n+1} = C_n \backslash \ldots \text{open middle thirds of the remaining intervals in } C_n$$

$$\vdots$$

The construction of C_n is normally described by saying "...to obtain C_n, subtract open middle thirds from C_{n-1}" After inductively obtaining an infinite sequence of nested sets, $\{C_n\}_{n\in\mathbb{N}}$, in this way we define the Cantor set as being the infinite intersection

$$C = \cap_{n=0}^{\infty} C_n$$

We visualize the Cantor set geometrically (up to the construction of C_5) as follows.

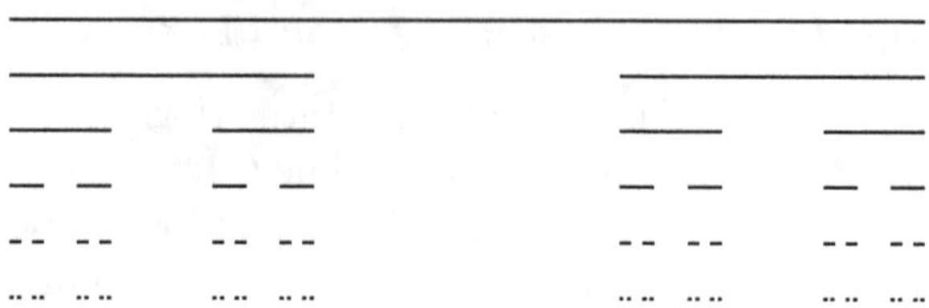

In Definition 7.18 of the Cantor set, $C = \varphi[\prod_{n\in\mathbb{Z}^+}\{0,2\}] \to [0,1]$ where

$$\varphi(\langle m_n\rangle_{n\in\mathbb{Z}^+}) = \sum_{n=1}^{\infty}\frac{m_n}{3^n}$$

the topology of the sets involved was not discussed. So, we didn't speak of the "continuity" of φ on the product $\prod_{n\in\mathbb{Z}^+}\{0,2\}$. We will define topologies of all sets involved in the most natural way. We will equip the set $\{0,2\}$ with the discrete topology, the set $\prod_{n\in\mathbb{Z}^+}\{0,2\}$ with the product topology, and finally, the Cantor set C, with the subspace topology inherited from $\mathbb{R}$. We will now show that, the Cantor set, C, is a homeomorphic copy of $\prod_{n\in\mathbb{Z}^+}\{0,2\}$.

Theorem 7.19 *The one-to-one function,* $\varphi : \prod_{n\in\mathbb{Z}^+}\{0,2\} \to [0,1]$, *defined as*

$$\varphi(\langle m_n\rangle_{n\in\mathbb{Z}^+}) = \sum_{n=1}^{\infty}\frac{m_n}{3^n}$$

which maps the product space $\prod_{n\in\mathbb{Z}^+}\{0,2\}$ *homeomorphically onto the Cantor set,*

$$C = \varphi\left[\prod_{n\in\mathbb{Z}^+}\{0,2\}\right]$$

a subset of $[0,1]$. *So, the Cantor set is a topological copy of the product space,* $\prod_{n\in\mathbb{Z}^+}\{0,2\}$, *contained in* $[0,1]$.

Proof. Given the one-to-one function $\varphi : \prod_{n\in\mathbb{Z}^+}\{0,2\} \to [0,1]$ where

$$\varphi(\langle m_n\rangle_{n\in\mathbb{Z}^+}) = \sum_{n=1}^{\infty}\frac{m_n}{3^n}$$

Let $\varepsilon > 0$. First note that, for all $\langle m_n \rangle_{n\in\mathbb{Z}^+} \in \prod_{n\in\mathbb{Z}^+}\{0,2\}$,

$$|\varphi(\langle m_n \rangle_{n\in\mathbb{Z}^+})| \leq \sum_{n=1}^{\infty} \frac{2}{3^n} = 1$$

Then the sequence, $\{\sum_{n=k}^{\infty} \frac{2}{3^k}\}_{n\in\mathbb{Z}^+}$, converges to zero. So, there exists N such that

$$\sum_{n=N+1}^{\infty} \frac{2}{3^n} < \varepsilon \quad (*)$$

Claim 1: The function φ is continuous. If $\langle k_n \rangle_{n\in\mathbb{Z}^+} \in \prod_{n\in\mathbb{Z}^+}\{0,2\}$ then

$$\varphi(\langle k_n \rangle_{n\in\mathbb{Z}^+}) = \sum_{n=1}^{\infty} \frac{k_n}{3^n} = x \in C$$

Let $B_\varepsilon(x)$ be and open interval in $\mathbb{R}$ with center x and radius ε. Then $B_\varepsilon(x) \cap C$ is an open neighborhood of x in C. To show continuity of φ at $\langle k_n \rangle_{n\in\mathbb{Z}^+}$, it will suffice to find an open neighborhood, U, of $\langle k_n \rangle_{n\in\mathbb{Z}^+}$ such that $\varphi[U] \subseteq B_\varepsilon(x) \cap C$.

Since the series, $\sum_{n=1}^{\infty} \frac{k_n}{3^n}$, converges to x, then the sequence $\{x - \sum_{n=1}^{m} \frac{k_n}{3^n}\} = \{\sum_{n=m+1}^{\infty} \frac{k_n}{3^n}\}$ converges to zero as m tends to infinity. Then,

$$\left|\sum_{n=N+1}^{\infty} \frac{k_n}{3^n}\right| \leq \sum_{n=N+1}^{\infty} \frac{2}{3^n} < \varepsilon \quad \text{(By *)}$$

Let $U = \pi_1^{\leftarrow}(k_1) \cap \cdots \cap \pi_N^{\leftarrow}(k_N)$ be an open base neighborhood of $\{k_n\}$ in $\prod_{n\in\mathbb{Z}^+}\{0,2\}$ and suppose $\langle b_n \rangle_{n\in\mathbb{Z}^+}$ is some element in U. This implies $b_1 = k_1, b_2 = k_2, \ldots, b_N = k_N$. So,

$$\begin{aligned}
|x - \varphi(\langle b_n \rangle_{n\in\mathbb{Z}^+})| &= \left|x - \sum_{n=1}^{\infty} \frac{b_n}{3^n}\right| \\
&= \left|\sum_{n=1}^{\infty} \frac{k_n}{3^n} - \sum_{n=1}^{\infty} \frac{b_n}{3^n}\right|
\end{aligned}$$

$$\leq 0 + \sum_{n=N+1}^{\infty} \frac{|k_n - b_n|}{3^n}$$

$$\leq \sum_{n=N+1}^{\infty} \frac{2}{3^n}$$

$$< \varepsilon$$

Then $\varphi[U] \subseteq B_\varepsilon(x) \cap C$. Then φ is continuous at $\langle k_n \rangle_{n\in\mathbb{Z}^+}$, and so at all points of $\prod_{n\in\mathbb{Z}^+}\{0,2\}$. Then φ is continuous on $\prod_{n\in\mathbb{Z}^+}\{0,2\}$, as claimed.

Claim 2: The function φ is homeomorphic. To prove that φ is a homeomorphism it will suffice to show it is an open function.

For $y = \langle k_n \rangle_{n\in\mathbb{Z}^+} \in \prod_{n\in\mathbb{Z}^+}\{0,2\}$, $\sum_{n=1}^{\infty} \dfrac{k_n}{3^n}$ converges to $\varphi(y)$.

Let $U = \pi_{j_1}^{\leftarrow}(k_{j_1}) \cap \cdots \cap \pi_{j_m}^{\leftarrow}(k_{j_m})$ be an arbitrary open neighborhood base element of $\langle k_n \rangle_{n\in\mathbb{Z}^+}$.

To show that φ is open, it will suffice to find some ε such that $B_\varepsilon(\varphi(y)) \cap C \subseteq \varphi[U]$.

Let $N = \max\{j_1, j_2, \ldots, j_m\}$ and let $\varepsilon = \frac{1}{3^{N+1}}$. (**)

We claim that, if $|\varphi(\langle k_n \rangle_{n\in\mathbb{Z}^+}) - \varphi(\langle z_n \rangle_{n\in\mathbb{Z}^+})| < \varepsilon$, then $\varphi(\langle z_n \rangle_{n\in\mathbb{Z}^+}) \in \varphi[U]$.

$$|\varphi(\langle k_n \rangle_{n\in\mathbb{Z}^+}) - \varphi(\langle z_n \rangle_{n\in\mathbb{Z}^+})| = \left|\sum_{n=1}^{\infty} \frac{k_n}{3^n} - \sum_{n=1}^{\infty} \frac{z_n}{3^n}\right|$$

$$= \left|\sum_{n=1}^{\infty} \frac{k_n - z_n}{3^n}\right|$$

$$= \left|\sum_{n=1}^{N} \frac{k_n - z_n}{3^n} + \sum_{n=N+1}^{\infty} \frac{k_n - z_n}{3^n}\right| < \varepsilon$$

Then

$$-\varepsilon = \frac{-1}{3^{N+1}} < \sum_{n=1}^{N} \frac{k_n - z_n}{3^n} + \sum_{n=N+1}^{\infty} \frac{k_n - z_n}{3^n} < \frac{1}{3^{N+1}} = \varepsilon$$

$$\frac{-1}{3^{N+1}} - \sum_{n=N+1}^{\infty} \frac{k_n - z_n}{3^n} < \sum_{n=1}^{N} \frac{k_n - z_n}{3^n} < \frac{1}{3^{N+1}} - \sum_{n=N+1}^{\infty} \frac{k_n - z_n}{3^n}$$

After some computation we obtain,

$$\frac{-4}{3^{N+1}} < \sum_{n=1}^{N} \frac{k_n - z_n}{3^n} < \frac{4}{3^{N+1}} \quad \text{(Verify this!)}$$

To show that $\varphi(\langle z_n \rangle_{n \in \mathbb{Z}^+}) \in \varphi[U]$, it suffices to show that, $y_n - z_n = 0$ for $n = 1$ to N.

For $m = 1$ to N, let

$$S_m = \sum_{n=1}^{m} \frac{k_n - z_n}{3^n}$$

Suppose $k_1 - z_1 = 2$ or -2; then $S_1 = \pm 2/3 \notin (-\frac{4}{3^{N+1}}, \frac{4}{3^{N+1}})$. So $k_1 - z_1 = 0$.

Suppose $k_m - z_m = 0$ for $m < N$ and $k_{m+1} - z_{m+1} = 2$ or -2. Then $S_{m+1} = \pm 2/3 \notin (-\frac{4}{3^{N+1}}, \frac{4}{3^{N+1}})$. So $k_{m+1} - z_{m+1} = 0$.

We then have $y_n - z_n = 0$ for $n = 1$ to N.

This is precisely what was needed to show that $\varphi(\{z_n\}) \in \varphi[U]$. This means that, for our choice of $\varepsilon = \frac{1}{3^{N+1}}$ (at (**)), $B_\varepsilon(\varphi(\{k_n\}) \cap C \subseteq \varphi[U]$. So, φ is an open map on $\prod_{n \in \mathbb{Z}^+} \{0, 2\}$. So, φ is a homeomorphism, as claimed.

The theorem is proved, as required.

The Cantor set was defined as being the range $\varphi\left[\prod_{n \in \mathbb{Z}^+} \{0, 2\}\right]$ of $\prod_{n \in \mathbb{Z}^+} \{0, 2\}$. But the above statement shows it is much more. Since we have shown $\prod_{n \in \mathbb{Z}^+} \{0, 2\}$ is a homeomorphic copy of C then, topologically speaking, the subspace, C, of $[0, 1]$ we call the Cantor set "is" the product space, $\prod_{n \in \mathbb{Z}^+} \{0, 2\}$. The topological point of view certainly provides much more insight on the nature and various properties of C as well as those sets that are linked to it via continuous functions. The set-theoretic definition of C is blind to this aspect.

7.9.3 *About the cardinality of C*

Our geometric description of the construction of the Cantor set, C, showed that C was obtained by successively removing open middle-third interval from a previous set, leaving behind, at least, the endpoints of countably many closed intervals. The endpoints, all of them of the form, $\frac{m_n}{3_n}$, $(m_n \in \{0, 2\})$ are rationals and are never removed and so must belong to C. This may lead one to believe that C is countably infinite. But our formal definition of C (as well as the previous theorem) shows that this cannot be so. It states that C can be mapped homeomorphically onto the uncountable set $\prod_{n\in\mathbb{Z}^+}\{0,2\}$. Then, $|C| = 2^{|Z^+|} = 2^{\aleph_0} = c$. So C is uncountable. We leave the reader to reflect on the more challenging question: If x is a point in C which is not an endpoint of a middle third, what does it look like? How can it be that a "non-endpoint" is left behind?

7.9.4 *Is the Cantor topological space discrete?*

By this question, we are wondering whether every single point of C is both open and closed. We have shown that the Cantor set, C, is the homeomorphic image of the product space, $\prod_{n\in\mathbb{Z}^+}\{0,2\}$. If every point of C was open then every point of $\prod_{n\in\mathbb{Z}^+}\{0,2\}$ would be open. But this is easily verified not to be the case. Since no point of the product space can contain a basic open set of the product space, $\prod_{n\in\mathbb{Z}^+}\{0,2\}$.

So, C can be viewed as a sprinkling of an uncountable set of points in $[0,1]$. It interesting to note that C can continuously be mapped onto $[0,1]$ as the following theorem shows.[7]

Theorem 7.20 *There is a continuous function, $\delta : C \to [0,1]$, which maps the Cantor set, C, onto the closed interval $[0,1]$.*

Proof. Recall that the function, $\varphi : \prod_{n\in\mathbb{Z}^+}\{0,2\} \to C$, maps $\prod_{n\in\mathbb{Z}^+}\{0,2\}$ homeomorphically onto C. It is defined as

$$\varphi(\langle m_n\rangle_{n\in\mathbb{Z}^+}) = \sum_{n=1}^{\infty}\frac{m_n}{3^n}$$

We begin by defining a similar function, $\psi : \prod_{n\in\mathbb{Z}^+}\{0,2\} \to [0,1]$ which maps the same set, $\prod_{n\in\mathbb{Z}^+}\{0,2\}$, onto $[0,1]$. It is defined as

$$\psi(\langle m_n\rangle_{n\in\mathbb{Z}^+}) = \sum_{n=1}^{\infty}\frac{m_n}{2^{n+1}}$$

[7]The rest of this section can be omitted without loss of continuity.

We claim that ψ is continuous on its domain and onto the closed interval $[0,1]$. To see that ψ is onto $[0,1]$ simply note that the expression

$$\sum_{n=1}^{\infty} \frac{m_n}{2^{n+1}} = \sum_{n=1}^{\infty} \frac{m_n/2}{2^n} = \sum_{n=1}^{\infty} \frac{s_n}{2^n} = 0.s_1s_2s_3s_4\ldots_2 \in [0,1]$$

where each s_n is 0 or 1 and $0.s_1s_2s_3s_4\ldots_2$ is simply the dyadic expansion for an element in $[0,1]$.[8]

To prove continuity of ψ proceed precisely as for φ in the previous theorem.

We define $\delta : C \to [0,1]$ as

$$\delta = \psi \circ \varphi^{\leftarrow}$$

where $\varphi^{\leftarrow}$ continuously maps C one-to-one and onto $\prod_{n\in\mathbb{Z}^+}\{0,2\}$ and ψ continuously maps $\prod_{n\in\mathbb{Z}^+}\{0,2\}$ onto $[0,1]$.[9] So, δ continuously maps C onto $[0,1]$, as required.

7.10 Topic: Peano's Space Filling Curve[10]

As a final example of an application of product spaces we show the somewhat surprising result which states that a cube

$$[0,1]^3 = \{(x,y,z) : x,y,z \in [0,1]\}$$

equipped with the product topology, in $\mathbb{R}^3$ is the continuous image of the closed interval [0, 1] with the usual topology inherited from $\mathbb{R}$.

---

[8]For example, if $y = \langle m_n \rangle_{n\in\mathbb{Z}^+}$ is the string $\{2,0,2,0,2,0,\ldots\}$, then $\psi(y) = 0.10101010\ldots$ a point in [0, 1] in dyadic expansion form. The function ψ is easily seen to be onto $[0,1]$. For example, given $x = [0.001001001\ldots,]$, $\psi(\{0,0,2,0,0,2,\ldots\}) = x$.

[9]Note that ψ is not one-to-one since ψ maps the two distinct strings $\{0,0,2,2,2,\ldots\}$ and $\{0,1,0,0,0,\ldots\}$ to the same point $0.001111\ldots = 0.01000000\ldots$

[10]Pronounced: pay-an-o. This section can be omitted without loss of continuity.

In the first example, we show that there is a continuous function which maps the Cantor set onto the cube, $[0,1]^3$.

Example 8. Find a function, $t : C \to [0,1]^3$, which maps the Cantor set, C, continuously onto the cube $[0,1]^3$.

Solution: We first list a few relevant functions presented in this chapter up to now.

— $\varphi : \prod_{n\in\mathbb{Z}^+}\{0,2\} \to \varphi\left[\prod_{n\in\mathbb{Z}^+}\{0,2\}\right] = C$:
This function maps $\prod_{n\in\mathbb{Z}^+}\{0,2\}$ homeomorphically onto C. (Theorem 7.19).

— $\varphi^{\leftarrow} : C \to \prod_{n\in\mathbb{Z}^+}\{0,2\}$:
The function $\varphi^{\leftarrow}$ maps C homeomorphically onto $\prod_{n\in\mathbb{Z}^+}\{0,2\}$.

— $\delta : \prod_{n\in\mathbb{Z}^+}\{0,2\} \to [0,1]$:
There exists a continuous function δ mapping $\prod_{n\in\mathbb{Z}^+}\{0,2\}$ onto [0,1]. (By Theorem 7.20.)

— $q^* : \prod_{n\in\mathbb{Z}^+}\{0,2\} \to \prod_{n\in\mathbb{Z}_A^+\cup\mathbb{Z}_B^+\cup\mathbb{Z}_C^+}\{0,2\}$:
Since $\mathbb{Z}^+ = \mathbb{N}\backslash\{0\}$ is countably infinite, it has the same cardinality as $3\mathbb{Z}^+$. Then there is a one-to-one function mapping $\mathbb{Z}^+$ onto $3\mathbb{Z}^+ = \mathbb{Z}_A^+ \cup \mathbb{Z}_B^+ \cup \mathbb{Z}_C^+$ (the disjoint union of three copies of $\mathbb{Z}^+$). Then there is a homeomorphism, q^* mapping $\prod_{n\in\mathbb{Z}^+}\{0,2\}$ onto $\prod_{n\in\mathbb{Z}_A^+\cup\mathbb{Z}_B^+\cup\mathbb{Z}_C^+}\{0,2\}$. (By Theorem 7.15.)

— $f : \prod_{n\in\mathbb{Z}_A^+\cup\mathbb{Z}_B^+\cup\mathbb{Z}_C^+}\{0,2\} \to \prod_{n\in\mathbb{Z}_A^+}\{0,2\} \times \prod_{n\in\mathbb{Z}_B^+}\{0,2\} \times \prod_{n\in\mathbb{Z}_C^+}\{0,2\}$:
There
is an onto homeomorphism f with domain $\prod_{n\in\mathbb{Z}_A^+\cup\mathbb{Z}_B^+\cup\mathbb{Z}_C^+}\{0,2\}$ defined as

$$f(\langle x_\alpha\rangle_{\alpha\in 3\mathbb{Z}^+}) = \langle f_A(\langle x_\alpha\rangle_{\alpha\in 3\mathbb{Z}^+}),\, f_B(\langle x_\alpha\rangle_{\alpha\in 3\mathbb{Z}^+}),\, f_C(\langle x_\alpha\rangle_{\alpha\in 3\mathbb{Z}^+})\rangle$$

where f_A, f_B and f_C are defined in proposition 7.21 as

$$f_A(\langle x_\alpha\rangle_{\alpha\in 3\mathbb{Z}^+}) = \langle x_\alpha\rangle_{\alpha\in\mathbb{Z}_A^+}$$

$$f_B(\langle x_\alpha\rangle_{\alpha\in 3\mathbb{Z}^+}) = \langle x_\alpha\rangle_{\alpha\in\mathbb{Z}_B^+}$$

$$f_C(\langle x_\alpha\rangle_{\alpha\in 3\mathbb{Z}^+}) = \langle x_\alpha\rangle_{\alpha\in\mathbb{Z}_C^+}$$

— $h : \prod_{n\in\mathbb{Z}_A^+}\{0,2\} \times \prod_{n\in\mathbb{Z}_B^+}\{0,2\} \times \prod_{n\in\mathbb{Z}_C^+}\{0,2\} \to [0,1] \times [0,1] \times [0,1]$:

If $T_A = \prod_{n\in\mathbb{Z}_A^+}\{0,2\}$, and $h[T_A \times T_B \times T_C] = \delta[T_A] \times \delta[T_B] \times \delta[T_C]$ then h continuously maps $T_A \times T_B \times T_C$ onto $[0,1] \times [0,1] \times [0,1]$. (By 7.12 and 7.20.)

By appropriately composing these functions, we have,

$$
\begin{aligned}
C \to \text{1-1 onto } \varphi^{\leftarrow} \to & \prod_{n\in\mathbb{Z}^+}\{0,2\} \\
\to \text{1-1 onto } q^{*} \to & \prod_{n\in\mathbb{Z}_A^+\cup\mathbb{Z}_B^+\cup\mathbb{Z}_C^+}\{0,2\} \\
\to \text{1-1 onto } f \to & \prod_{n\in\mathbb{Z}_A^+}\{0,2\} \times \prod_{n\in\mathbb{Z}_B^+}\{0,2\} \times \prod_{n\in\mathbb{Z}_C^+}\{0,2\} \\
\to \text{onto } h \to & [0,1] \times [0,1] \times [0,1]
\end{aligned}
$$

Let

$$t = h\circ f\circ q^{*}\circ\varphi^{\leftarrow}$$

Then the function, $t : C \to [0,1]^3$, continuously maps C onto $[0,1]^3$, as required.

Example 9. Show that there exists a function which continuously maps $[0,1]$ onto the cube $[0,1]^3$.

Solution: We adopt the notation introduced in the previous example.

Each of the three functions

$$
\begin{aligned}
f_A &= \pi_A\circ t = \pi_A\circ h\circ f\circ q^{*}\circ\varphi^{\leftarrow} : C \to [0,1] \\
f_B &= \pi_B\circ t = \pi_B\circ h\circ f\circ q^{*}\circ\varphi^{\leftarrow} : C \to [0,1] \\
f_C &= \pi_C\circ t = \pi_C\circ h\circ f\circ q^{*}\circ\varphi^{\leftarrow} : C \to [0,1]
\end{aligned}
$$

continuously maps the points in the Cantor set, C, onto the points of $[0,1]$.

Recall that C is a subset of $[0,1]$ with multiple open intervals missing. Also recall that a function $g : [0,1] \to [0,1]$ is a continuous extension of f_A if g is continuous on $[0,1]$ and $g|_C = f_A$. We can then *extend* the continuous function $f_A : C \to [0,1]$ to a continuous function $F_A : [0,1] \to [0,1]$ so that, on each missing open interval (a, b) in C, F_A is defined as the being the function which is linear from $(a, F_A(a)) = (a, f_A(a))$ to $(b, F_A(b)) = (b, f_A(b))$. The continuous extension, F_A, of f_A then maps $[0, 1]$ continuously onto $[0, 1]$. The same holds true for the similarly defined functions, $F_B : [0, 1] \to [0, 1]$ and $F_C : [0, 1] \to [0, 1]$.

We then have the three continuous onto functions,

$$F_A : [0, 1] \to [0, 1]$$
$$F_B : [0, 1] \to [0, 1]$$
$$F_C : [0, 1] \to [0, 1]$$

We define the function $F : [0.1] \to [0, 1]^3$, as

$$F(x) = \langle\, F_A(x), F_B(x), F_C(x) \,\rangle \in [0, 1] \times [0, 1] \times [0, 1]$$

Since all three functions F_A, F_B, and F_C are continuous, then by Theorem 7.12, F is continuous on $[0, 1]$. So, F continuously maps the closed interval $[0, 1]$ onto $[0, 1]^3$.

We normally refer to the continuous image of the closed interval $[0, 1]$ as a

curve.

In this sense, saying that "there is a curve that can fill the cube $[0, 1]^3$" or "there is a curve that goes through every point of the cube $[0, 1]^3$" is another way of saying that "$[0, 1]$ can be mapped continuously onto the cube $[0, 1]^3$". The reader who follows through the proof carefully will notice that it can be generalized to the statement "For any integer n there is a curve which goes through each point of the cube $[0, 1]^n$". The set

$$[0, 1]^n$$

is referred to as a *cube*, for any natural number, $n \geq 1$. The curve is referred to as *Peano's space filling curve.*[11]

[11]Interested readers are referred to *Mathematical Intelligencer* 6(3) (1984), p. 78 for a graphic representation of a space filling curve.

7.11 Topic: Product Spaces Are "Associative"

We address the question: How can we simplify the product of a product spaces? We begin by illustrating what we mean by this question.

Let $I = \{1, 2, \ldots, 12\}$, $A = \{1, 2, 3, 4\}$, $B = \{5, 6\}$, and $C = \{7, 8, \ldots, 12\}$. The set I is the disjoint union the of sets A, B and C. Let $\{(S_\alpha, \tau_\alpha) : \alpha \in \{1, 2, \ldots, 12\}$ be a set of 12 topological spaces. Consider the two product spaces S and Y defined as follows:

$$S = \prod_{\alpha \in I} S_\alpha \quad \text{and} \quad Y = \prod_{\alpha \in A} S_\alpha \times \prod_{\alpha \in B} S_\alpha \times \prod_{\alpha \in C} S_\alpha$$

We wonder, "How do the two product spaces compare?". We see that S is a product space with twelve factors, while Y is a product space of only three factors. This observation is sufficient to conclude that $S \neq Y$. On the other hand, we see that each of the three factors of Y are themselves product spaces, all three with factors, S_α, identical to the ones found in S. While we have excluded, the possibility that S equals Y, it would be reasonable to suspect that these two spaces are homeomorphic copies of each other. We will present this conjecture as a proposition (in just a slightly more general form) followed by its proof.

Proposition 7.21 *Let $\{(S_\alpha, \tau_\alpha) : \alpha \in I\}$ be a set of non-empty topological spaces. Suppose I is the disjoint union of three non-empty subsets, A, B, and C where the elements in $A \cup B \cup C$ respect the order in which they appear in I. Then*

$$S = \prod_{\alpha \in I} S_\alpha \quad \textit{and} \quad Y = \prod_{\alpha \in A} S_\alpha \times \prod_{\alpha \in B} S_\alpha \times \prod_{\alpha \in C} S_\alpha$$

are homeomorphic product spaces.

Proof. We define three functions, $f_A : S \to \prod_{\alpha \in A} S_\alpha$, $f_B : S \to \prod_{\alpha \in B} S_\alpha$ and $f_C : S \to \prod_{\alpha \in C} S_\alpha$ as follows:

$$f_A(\langle x_\alpha \rangle_{\alpha \in I}) = \langle x_\alpha \rangle_{\alpha \in A}$$

$$f_B(\langle x_\alpha \rangle_{\alpha \in I}) = \langle x_\alpha \rangle_{\alpha \in B}$$

$$f_C(\langle x_\alpha \rangle_{\alpha \in I}) = \langle x_\alpha \rangle_{\alpha \in C}$$

All three functions are easily seen to be onto $\prod_{\alpha \in A} S_\alpha$, $\prod_{\alpha \in B} S_\alpha$ and $\prod_{\alpha \in C} S_\alpha$, respectively.

Note that,

$$\text{For } \beta \in A,\ (\pi_\beta \circ f_A)(\langle x_\alpha \rangle_{\alpha \in I}) = x_\beta = \pi_\beta(\langle x_\alpha \rangle_{\alpha \in I})$$
$$\text{For } \beta \in B,\ (\pi_\beta \circ f_B)(\langle x_\alpha \rangle_{\alpha \in I}) = x_\beta = \pi_\beta(\langle x_\alpha \rangle_{\alpha \in I})$$
$$\text{For } \beta \in C,\ (\pi_\beta \circ f_C)(\langle x_\alpha \rangle_{\alpha \in I}) = x_\beta = \pi_\beta(\langle x_\alpha \rangle_{\alpha \in I})$$

By Theorem 7.11, the functions f_A, f_B, and f_C are all continuous on S.

We now consider the function

$$f : \prod_{\alpha \in I} S_\alpha \to \prod_{\alpha \in A} S_\alpha \times \prod_{\alpha \in B} S_\alpha \times \prod_{\alpha \in C} S_\alpha$$

defined as

$$f(\langle x_\alpha \rangle_{\alpha \in I}) = \langle\ f_A(\langle x_\alpha \rangle_{\alpha \in I}),\ f_B(\langle x_\alpha \rangle_{\alpha \in I}),\ f_C(\langle x_\alpha \rangle_{\alpha \in I})\ \rangle$$

We claim f is one-to-one and continuous on $\prod_{\alpha \in I} S_\alpha$.

Since f_A, f_B, and f_C are all three continuous, then, by Theorem 7.12, $f : S \to Y$ is continuous on its domain S. Also, since I is the disjoint union of A, B, and C, then f maps S one-to-one and onto Y. (Verify this!). So, f is one-to-one and continuous as claimed.

To show that f is a homeomorphism it now suffices to show that it is open.

Let $V = \pi^{\leftarrow}_{\alpha_1}[U_{\alpha_1}] \cap \pi^{\leftarrow}_{\alpha_2}[U_{\alpha_2}] \cap \cdots \cap \pi^{\leftarrow}_{\alpha_k}[U_{\alpha_k}]$ be an open base element in $S = \prod_{\alpha \in I} S_\alpha$. Then

$$f[V] = \langle\, f_A[V],\ f_B[V],\ f_C[V]\,\rangle$$

For $f_A[V] = f_A[\pi^{\leftarrow}_{\alpha_1}[U_{\alpha_1}] \cap \pi^{\leftarrow}_{\alpha_2}[U_{\alpha_2}] \cap \cdots \cap \pi^{\leftarrow}_{\alpha_k}[U_{\alpha_k}]]$:

$$\begin{aligned} f_A[\pi^{\leftarrow}_{\alpha_i}[U_{\alpha_i}]] &= U_{\alpha_i} \quad \text{if } \alpha_i \in A \\ &= \prod_{\alpha \in A} S_\alpha \quad \text{if } \alpha_i \notin A \end{aligned}$$

The same holds true for f_B and f_C.

So $f[V] = \langle\, f_A[V],\ f_B[V],\ f_C[V]\,\rangle$ is open.

The given homeomorphism, f, confirms that the product spaces

$$\prod_{\alpha \in A \cup B \cup C} S_\alpha \quad \text{and} \quad \prod_{\alpha \in A} S_\alpha \times \prod_{\alpha \in B} S_\alpha \times \prod_{\alpha \in C} S_\alpha$$

are homeomorphic.

Concept Review

1. Give a general definition of a Cartesian product of sets.
2. Provide two definitions of the Cantor set.
3. Define the *product topology* on a Cartesian product.
4. Define the *box topology* on a Cartesian product.
5. Define what we mean by *product space.*
6. What do we mean by "partitioning" a product space and how does it change its topology.
7. What effect does changing the order of the factors have on the topology of a product space.
8. Describe the *Tychonoff plank* and its topology.
9. What does it mean to say that a set of functions $\mathscr{F}$ separates points and closed sets of a set S.
10. Define evaluation map with respect to a set of functions $\mathscr{F}$.
11. State the *Embedding theorem.*
12. The embedding theorem describes a homeomorphism between a topological space and a product space. What are these spaces? What is the homeomorphism?
13. The Cantor set equipped with the usual topology is shown to be homeomorphic to a product space. Which one? Describe the homeomorphism.
14. Describe a continuous function which maps the Cantor set, C, continuously onto the closed interval $[0, 1]$.
15. Is the Cantor set a countable subset of $\mathbb{R}$?
16. Is the Cantor set, equipped with the subspace topology, discrete? Is it second countable?
17. Does there exist a continuous function which maps C onto the cube, $[0, 1]^3$?
18. Does there exist a continuous function which maps $[0, 1]$ onto the cube, $[0, 1]^3$?

Exercises

1. Show that $\prod_{\alpha\in\Gamma} S_\alpha$ is dense in $\prod_{\alpha\in\Gamma} T_\alpha$ if and only if, for each $\alpha \in \Gamma$, S_α is dense in T_α.
2. Suppose that, for each $\alpha \in \Gamma$, $S_\alpha \subseteq T_\alpha$. Then $\prod_{\alpha\in\Gamma} S_\alpha$ is a subset of the product space, $\prod_{\alpha\in\Gamma} T_\alpha$. Show that the product topology

on $\prod_{\alpha\in\Gamma} S_\alpha$ is the same as the subspace topology $\prod_{\alpha\in\Gamma} S_\alpha$ inherits from $\prod_{\alpha\in\Gamma} T_\alpha$.

3. Given the product space $\prod_{\alpha\in\Gamma} S_\alpha$, show that the projection map $\pi_\alpha : \prod_{\alpha\in\Gamma} S_\alpha \to S_\alpha$ is an open function. Is it a closed function?
4. If in the product space, $S = \prod_{\alpha\in\Gamma} S_\alpha$, each S_α is discrete, describe the open subsets of S.
5. If in the product space, $S = \prod_{\alpha\in\Gamma} S_\alpha$, each S_α is indiscrete, describe the open subsets of S.
6. Let $X = \prod_{\alpha\in\Gamma} S_\alpha$ and $Y = \prod_{\gamma\in\Phi} T_\gamma$ be two product spaces where Γ and Φ have the same cardinality confirmed by the one-to-one function $q : \Gamma \to \Phi$. For each α, S_α and $T_{q(\alpha)}$ are homeomorphic topological spaces. Show that X and Y are homeomorphic.
7. Suppose we are given a topological space (S, τ_s) and a family of topological spaces, $\{(T_\alpha, \tau_\alpha) : \alpha \in \Gamma\}$. Suppose $\mathscr{F} = \{f_\alpha : \alpha \in \Gamma\}$ is a family of functions, $f_\alpha : S \to T_\alpha$, where each f_α maps its domain S into T_α. Let

$$\mathscr{B} = \{f_\alpha^{\leftarrow}[U] : (\alpha, U) \in \Gamma \times \tau_\alpha\}$$

   Show that $\mathscr{B}$ is a base for open sets of S if and only if $\mathscr{F}$ separates points and closed sets in S.
8. Suppose $U \subseteq S$ and $V \subseteq T$. Show that $\text{int}_{S\times T}(U \times V) = \text{int}_S U \times \text{int}_T V$.
9. Show that, if the product space, $\prod_{\alpha\in\Gamma} S_\alpha$, is first countable, then S_α is first countable for each $\alpha \in \Gamma$.

Chapter 8

The Quotient Topology

Abstract
In this section, we will present a method to topologize the range, T, of a function, f, whose domain, S, is a topological space. The topology on the range is referred to as the "quotient topology induced by f". When the function f is used for this purpose, it is referred to as the associated quotient map.

8.1 The Strong Topology Induced by a Function

We have previously discussed a function, f, with an untopologized domain, S, which is mapped to a topological space (T, τ_T). We topologized S in such a way that guarantees the continuity of the function $f : S \to T$ on S. To do this, we must be sure that S is provided with enough open sets so that $f : S \to T$ is continuous. The easiest way to do this is to assign to S the discrete topology, since, on such a space any function will be continuous on S. But, ideally, we preferred a topology which is custom-made for f and τ_T. We then opted for the *weak topology induced by* f and τ_T. Namely,

$$\tau_S = \{U \subseteq S : U = f^{\leftarrow}[V], \text{ for some open } V \text{ in } T\}$$

In this section, we will work the other way around. We are given a topological space, S, and a function $f : S \to T$. We wish to assign to T a topology, τ_T, that will guarantee that f is continuous on S.

Again, we could take the easy way out by assigning to T the indiscrete topology, $\{\varnothing, T\}$. It only has one non-empty open set, T, pulled back by f to the open set, $S \in \tau_S$. We again obtain continuity, but one which is independent of the function, f. We opt for choosing a topology on T which is custom made for the given function, f, and the topology on the domain. That is, we will choose the *strongest topology on* T that will guarantee continuity of f on the given topological space (S, τ_S). With this mind, we present the following formal definition.

Definition 8.1 Let $f : S \to T$ be a function mapping the topological space (S, τ_S) *onto* the set T. We assign to the set T the following topology

$$\tau_f = \{U \subseteq T : f^{\leftarrow}[U] \text{ is open in } S\}$$

The set, τ_f, is referred to as the *quotient topology induced by* f and τ_S. When T is equipped with the quotient topology, then (T, τ_f) is referred to as the *quotient space of* S *induced by* f and $f : S \to T$ is referred to as its associated *quotient map.*[1]

The set τ_f is referred to as being the *strongest (or finest) topology on* T that will guarantee continuity of $f : S \to T$ on the given topological space (S, τ_S).

Since $f^{\leftarrow}$ respects both infinite unions and intersections of sets then τ_f is a well-defined topology on T. Also since τ_f contains precisely those sets which are pulled back to some set in τ_S, and no other sets, then τ_f is indeed the largest topology that guarantees continuity for f on S. The quotient topology, τ_f, induced by f is then unique.

It is important for the reader to notice that, in the above definition, we declare the function $f : S \to T$ to be *onto* T. This fact may be relevant in some of the proofs that follow. If $f : S \to T$ was not declared to be *onto* T, we would at least have to modify our definition of quotient topology to

$$\tau_f = \{U \subseteq f[S] : f^{\leftarrow}[U] \in \tau_S\}$$

Suppose we are given two topological spaces (S, τ_S) and (T, τ_T) and a continuous function, f, mapping S onto T. If we are given no more

---

[1]Some authors use the terms "identification topology" instead of *quotient topology* and "identification map" instead of *quotient map.*

information about τ_T, it may or may not be the quotient topology induced by f. The following theorem shows that there are various ways of recognizing a quotient topology induced by a given function. By definition, quotient maps must at least be continuous.

Theorem 8.2 *Suppose (S, τ_S) and (T, τ_T) are topological spaces and $f : S \to T$ is a continuous function* onto T.

(a) *If $f : S \to T$ is an open map, then τ_T is the quotient topology on T induced by f. That is, $\tau_T = \tau_f$.*
(b) *If $f : S \to T$ is a closed map, then τ_T is the quotient topology on T induced by f. That is, $\tau_T = \tau_f$.*
(c) *If there is a continuous function $g : T \to S$ such that $(f \circ g)(x) = x$ on T, then $\tau_T = \tau_f$.*

Proof. We are given that $f : S \to T$ is a continuous function mapping S *onto* T with topology τ_T. To show that τ_T is the quotient topology induced by f it suffices to show that τ_T is the strongest topology, τ_f that will guarantee continuity to this function, f. Since $f : S \to T$ is continuous, and τ_f is the largest topology on T for which f is continuous, then $\tau_T \subseteq \tau_f$. For each of the three parts, it will then suffice to show that $\tau_f \subseteq \tau_T$.

(a) We are given that $f : S \to T$ is open.
We claim that $\tau_f \subseteq \tau_T$. Let $U \in \tau_f$. By definition of τ_f, $f^{\leftarrow}[U]$ is open in S. Since f is both open and onto, then $f[f^{\leftarrow}[U]] = U \in \tau_T$. So $\tau_f \subseteq \tau_T$. Hence, $\tau_T = \tau_f$.
(b) The proof of part (b) is similar to (a) and so is left to the reader.
(c) We are given that $g : T \to S$ is continuous such that $(f \circ g)(x) = x$ (where $f \circ g : T \to T$).
We claim that $\tau_f \subseteq \tau_T$. Suppose $U \in \tau_f$. Then, by definition of τ_f, $f^{\leftarrow}[U]$ is open in S. We have

$$\begin{aligned} U &= \{x \in T : (f \circ g)(x) = x \in U\} \\ &= (f \circ g)^{\leftarrow}[U] \\ &= g^{\leftarrow}[f^{\leftarrow}[U]] \\ \Rightarrow &\quad U \text{ is open in } T \text{ since } g \text{ and } f \text{ are continuous} \\ \Rightarrow &\quad U \in \tau_T \end{aligned}$$

Hence, $U \in \tau_T$. So $\tau_f \subseteq \tau_T$, as claimed.
Then $\tau_f = \tau_T$, as required.

Example 1. Consider the Cartesian product $\mathbb{R} \times \mathbb{R}$ equipped with the product topology. If π_1 and π_2 are the projection maps $\pi_1(a,b) = a$ and $\pi_2(a,b) = b$ mapping $\mathbb{R} \times \mathbb{R}$ onto $\mathbb{R}$ equipped with the usual topology, then the subsets of the form

$$\pi_1^{\leftarrow}[U] \cap \pi_2^{\leftarrow}[V]$$

(where U and V are open in $\mathbb{R}$) constitute the open base elements of $\mathbb{R} \times \mathbb{R}$. We have shown that the projection maps $\pi_i : \mathbb{R} \times \mathbb{R} \to \mathbb{R}$ are open functions. Then by part (a) of the above theorem, $\mathbb{R}$ equipped with the usual topology is the same as the quotient space induced by $\pi_1 : \mathbb{R} \times \mathbb{R} \to \mathbb{R}$. Since $\pi_1[\{a\} \times \mathbb{R}] = \{a\}$ we often say that

π_1 collapses the vertical lines in $\mathbb{R} \times \mathbb{R}$ to a point

Since there can be many different open maps from (S, τ_S) into (T, τ_T), it is possible to have distinct maps f and g associated to the same quotient topology, $\tau_T = \tau_f = \tau_g$.

8.2 An Equivalence Relation Induced by a Function f

Any function, f, mapping a set S onto a set T can be used to partition the domain into subsets we call "fibers in S induced by the function f." We will first formally define this set theoretic notion.

Definition 8.3 Let $f : S \to T$ be a function mapping a set S *onto* a set T. If $w \in T$, we will refer to $f^{\leftarrow}[\{w\}]$ as the *fiber of w under the map f*. Fibers in the domain are the preimages of singleton sets with respect to f.[2]

---

[2]The word "fiber" (or "fibre") usually interpreted in this way in the field of set theory. But it can have an entirely different meaning in other mathematical fields. Its interpretation is determined by the context.

Let $f : S \to T$ be a function which maps the topological space (S, τ_s) *onto* the quotient space, (T, τ_f). Using this function we will construct a new set by defining a relation R_f on S as follows:

$$[\, u \text{ is related to } v \,] \qquad \Leftrightarrow \qquad [\, f(u) = f(v) \text{ in } T \,]$$

The phrase, "u is related to v" in S, can be more succinctly expressed as

$$uR_f v \quad \text{or} \quad (u, v) \in R_f$$

This essentially means that u and v are related if and only if they both belong to the same fiber under the map $f : S \to T$.[3] The relation, R_f, on S is easily seen to be reflexive, symmetric and transitive, and so R_f is an *equivalence relation* on S. We will denote an equivalence class of x under R_f by

$$S_x = \{y \in S : xR_f y\}$$

and the set of all such equivalence classes in S by

$$S/R_f = \{\, S_x : x \in S \,\}$$

Each element, S_x, is a fiber under f, and the set, S/R_f, can be viewed as the set of all fibers in S under f. In set theory, S/R_f, is normally called the *quotient set of S induced by R_f*. Just like the set of all fibers of a function, $f : S \to T$, partitions the domain, we see that the set of all equivalence classes partitions the set S. By this we mean that the elements of S/R_f are pairwise disjoint subsets which cover all of S. One should remember that, when S_x is in S, it is a subset of S, but, whenever S_x is in S/R_f, it is viewed as one of its elements.

There is a "natural" function, $\theta : S \to S/R_f$, defined as

$$\theta(x) = S_x$$

mapping the points of the topological space (S, τ_S) onto the "elements" of the set S/R_f. One might more figuratively say that θ collapses each fiber, $f^{\leftarrow}(x)$, in S down to a unique element, S_x, in S/R_f, or, equivalently, maps fibers in S to singleton sets, $\{S_x\}$, in S/R_f.

[3]In this context, a fiber is a set theoretic concept and is independent of the topologies involved.

As long as no topology is declared on S/R_f our discussion remains within the bounds of set theory. We will topologize S/R_f as being the *quotient space induced by the function,* $\theta : S \to S/R_f$, a function we can now refer to as a *quotient map.* So, S/R_f will be equipped with the quotient topology, τ_θ. More specifically,

$$\tau_\theta = \{U : \theta^{\leftarrow}[U] \text{ is open in } S\}$$

The topology on S/R_f will be the strongest topology that guarantees the continuity of $\theta : S \to S/R_f$.

We now have two quotient spaces induced by two functions with domain (S, τ_S):

$$f : S \to (T, \tau_f)$$
$$\theta : S \to (S/R_f, \tau_\theta)$$

An insightful reader may already have a feeling that the two quotient spaces are topologically the same. To confirm this, we have to show that they are linked by some homeomorphism. We will connect S/R_f to T by defining a third function, $\phi_f : S/R_f \to T$ as

$$\phi_f(S_x) = f(x) \quad (\text{where } S_x = \theta(x).)$$

which maps points of S/R_f to points of T (remembering that f is *onto* T and so every element of T can be represented by $f(x)$ for some x in S). The following diagram illustrates the relationship between S, S/R_f, and T.

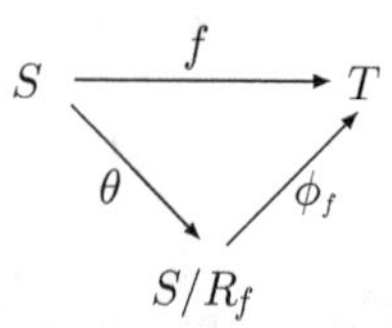

This allows us to express the function, f, as a composition of functions $\phi_f \circ \theta = f$. The following definition refers to expressions of the form $f^{\leftarrow}[f[U]]$. One should keep in mind that, even though it may occur that $U = f^{\leftarrow}[f[U]]$, it is always the case that $U \subseteq f^{\leftarrow}[f[U]]$.

In theorem below we will show that

$$\phi_f : S/R_f \to T \text{ maps } S/R_f \text{ homeomorphically onto } T$$

Definition 8.4 Let $g : S \to T$ be a function mapping the topological space (S, τ_S) *onto* a topological space (T, τ_T). A subset, $U \subset S$, is said to be *g-saturated* whenever U is the *complete* inverse image of some subset V in T. Equivalently, U is g-saturated if and only if $U = g^{\leftarrow}(g(U))$.

Theorem 8.5 *Suppose $f : S \to T$ is continuous on S and U is a non-empty open subset of T. Then $f^{\leftarrow}[U]$ is θ-saturated.*

Proof. It suffices to show that $f^{\leftarrow}[U] = \theta^{\leftarrow}[\theta\,[f^{\leftarrow}[U]]]$. Since $f^{\leftarrow}[U] \subseteq \theta^{\leftarrow}[\theta\,[f^{\leftarrow}[U]]]$, we need only show inclusion in the opposite direction.

$$\begin{aligned}
x \in \theta^{\leftarrow}[\theta\,[f^{\leftarrow}[U]]] \quad &\Rightarrow \quad \theta(x) \in \theta\,[f^{\leftarrow}[U]] \\
&\Rightarrow \quad \theta(x) \in \{\theta(y) : y \in f^{\leftarrow}[U]\} \\
&\Rightarrow \quad S_x \in \{S_y : f(y) \in U\} \\
&\Rightarrow \quad f(x) \in U \\
&\Rightarrow \quad x \in f^{\leftarrow}[U]
\end{aligned}$$

Then $\theta^{\leftarrow}[\theta\,[f^{\leftarrow}[U]]] \subseteq f^{\leftarrow}[U]$. Then $f^{\leftarrow}[U]$ is θ-saturated. That is,

$$\theta^{\leftarrow}[\theta\,[f^{\leftarrow}[U]]] = f^{\leftarrow}[U]$$

Theorem 8.6 *Suppose (S, τ_S) is a topological space. Let (T, τ_f) and $(S/R_f, \tau_\theta)$ be two quotient spaces induced by the quotient maps $f : S \to T$ and $\theta : S \to S/R_f$, where $\theta(x) = S_x$. Then the function, $\phi_f : S/R_f \to T$ defined as*

$$\phi_f(S_x) = f(x)$$

maps S/R_f homeomorphically onto T.

Proof. We are given that $f : S \to T$ and $\theta : S \to S/R_f$ and are quotient maps, hence are continuous on S. We are required to show that the function, $\phi_f : S/R_f \to T$, defined as, $\phi_f(S_x) = (\phi_f \circ \theta)(x) = f(x)$, is a homeomorphism.

We claim that ϕ_f is one-to-one and onto T: See that, since f is onto T, ϕ_f is onto T. Also, if $S_x \neq S_y$, then $f(x) \neq f(y)$, which implies $\phi_f(x) \neq \phi_f(y)$; so ϕ_f is one-to-one and onto.

We claim that ϕ_f is continuous on S/R_f: Let U be open in T. Since f is continuous $f^{\leftarrow}[U]$ is open in S.

$$\begin{aligned}\phi_f^{\leftarrow}[U] &= \{S_x : \phi_f(S_x) \in U\}\\ &= \{S_x : \phi_f(\theta(x)) \in U\}\\ &= \{\theta(x) : f(x) \in U\}\\ &= \{\theta(x) : x \in f^{\leftarrow}[U]\}\\ &= \theta[f^{\leftarrow}[U]]\end{aligned}$$

We have shown above that $f^{\leftarrow}[U]$ is θ-saturated hence $\theta^{\leftarrow}[\theta\,[f^{\leftarrow}[U]]] = f^{\leftarrow}[U]$. Since $\theta : S \to S/R_f$ is a quotient map and $f^{\leftarrow}[U]$ is open then $\theta\,[f^{\leftarrow}[U]]$ is open. So, $\phi_f^{\leftarrow}[U]$ is open. This establishes the claim that ϕ_f is continuous.

We claim that ϕ_f is open on S/R_f: From the diagram above we know that $f = \phi_f \circ \theta$. So, $\theta = \phi_f^{\leftarrow} \circ f$. Since θ is continuous then $(\phi_f^{\leftarrow} \circ f) : S \to S/R_f$ is continuous on S. Let U be an open subset of S/R_f.

$$\begin{aligned}\theta^{\leftarrow}[U] \quad &= \quad (\phi_f^{\leftarrow} \circ f)^{\leftarrow}[U] \in \tau_S\\ &\Rightarrow \quad f^{\leftarrow}[(\phi_f^{\leftarrow})^{\leftarrow}[U]] \in \tau_S\\ &\Rightarrow \quad f^{\leftarrow}[\phi_f[U]] \in \tau_S\end{aligned}$$

Since T has the quotient topology induced by f, $\phi_f[U]$ is open in T. We have shown that ϕ_f is an open map, as claimed.

We have shown that the one-to-one onto function, $\phi_f : S/R_f \to T$, is both continuous and open, hence it maps S/R_f homeomorphically onto T.

We summarize the main ideas behind the above result. If we are given a topological space, (S, τ_S), and a function f mapping S onto a set T we have the two main ingredients necessary to topologize T with the quotient topology, τ_f. We do so in a way that guarantees continuity of the function, $f : S \to T$. The fibers of the function, f, covers its domain. By collapsing, via the map θ, each fiber down to a point we create another set, S/R_f, and topologize it with the quotient topology, τ_θ. The expression "identifying the points of the fibers" is also of common usage. This new topological space, $(S/R_f, \tau_\theta)$ has been proven to be a homeomorphic copy of (T, τ_f).

8.3 Decomposition Spaces

We have seen that the quotient space induced by $\theta : S \to S/R_f$ involves the decomposition of the set S into non-overlapping sets, $\{S_x\}$. We will now slightly generalize this notion of "decomposition of a set." Suppose we are given a topological space (S, τ_s). Rather than partition S into equivalence classes, each of which contains all the elements that belong to a particular fiber, S_x, of a function f, we can also partition the set S in an arbitrary way. By this we mean that we express S as the union of non-intersecting subsets. We will denote this set of subsets by $\mathscr{D}_S$. Each element x of S belongs to some element, labeled D_x, of $\mathscr{D}_S$. Note that, if $y \in D_x$ then D_x and D_y are simply different labels for the same element of $\mathscr{D}_S$. The set

$$\mathscr{D}_S = \{D_x : x \in S\}$$

is reminiscent of a quotient set whose elements are equivalent classes of some relation R on S. We can similarly define a function $\theta : S \to \mathscr{D}_S$, on the topological space, S, where $\theta(u) = D_u$, the unique element of $\mathscr{D}_S$ which contains u. Then $\theta^{\leftarrow}(D_u) = \{x \in S : x \in D_u\} = D_u \subseteq S$. The set, $\mathscr{D}_S$, is not yet topologized, but we know of a procedure to topologize it, since it is, after all, the range of a function θ on S. We can equip $\mathscr{D}_S$ with the quotient topology,

$$\tau_\theta = \{U \subseteq \mathscr{D}_S : \theta^{\leftarrow}[U] \text{ is open in } S\}$$

induced by θ. We formally define the concepts we have just presented.

Definition 8.7 Let (S, τ_s) be a topological space. If $\mathscr{D}_S$ is collection of pairwise disjoint subsets of S such that every element of S belongs to some element of $\mathscr{D}_S$, then we say $\mathscr{D}_S$ is a

decomposition of S.

The function, $\theta : S \to \mathscr{D}_S$, defined as $\theta(u) = D_x$ if and only if $u \in D_x$, mapping every element of S onto $\mathscr{D}_S$ is called the *decomposition map of S onto $\mathscr{D}_S$*. The function, θ, is also referred to as the *identification map*. The function, θ, is said to *identify the elements of the subset* $D_x \subseteq S$. If τ_θ is the quotient topology induced on $\mathscr{D}_S$ by θ, then we say that $(\mathscr{D}_S, \tau_\theta)$ is a

decomposition space or *quotient space.*

Why does all of this sound familiar? Recall that, if $f : S \to T$, then the function, f, decomposes its domain, S, into fibers. In this case, $(\mathscr{D}_S, \tau_\theta)$ is a copy of $(S/R_f, \tau_\theta)$.

The following theorem will help recognize the open subsets of the decomposition space, $\mathscr{D}_S$. It states that open subsets $\mathscr{U} = \{D_x : D_x \in \mathscr{U}\}$ of $\mathscr{D}_S$ correspond to open θ-saturated subsets $\cup\{D_x : D_x \in \mathscr{U}\}$ of S.

Theorem 8.8 *Suppose $(\mathscr{D}_S, \tau_\theta)$ is a decomposition space of the topological space (S, τ_s) and let $\mathscr{U} = \{D_x : x \in I \subseteq S\}$ be a subset of $\mathscr{D}_S$. Then $\mathscr{U}$ is open in $\mathscr{D}_S$ if and only if*

$$\cup\{D_x : x \in I\}$$

is open in S.

Proof. We are given that $\mathscr{U} = \{D_x : x \in I \subseteq S\}$ be a subset of $\mathscr{D}_S$.

($\Rightarrow$) Suppose $\mathscr{U}$ is open in $\mathscr{D}_S$.

Then $\theta^{\leftarrow}[\mathscr{U}] = \{x \in S : \theta(x) = D_x \in \mathscr{U}\} = \{x \in S : x \in I\}$ is open in S. We are required to show that $\cup\{D_x : x \in I\}$ is open in S.

$$\begin{aligned} \theta^{\leftarrow}[\mathscr{U}] &= \theta^{\leftarrow}[\cup\{\{D_x\} : x \in I\}] \\ &= \cup\{\theta^{\leftarrow}[\{D_x\}] : x \in I\} \\ &= \cup\{D_x \subseteq S : x \in I\} \end{aligned}$$

is open since θ is continuous.

($\Leftarrow$) Suppose $\cup\{D_x \subseteq S : x \in I\}$ is open in S. We are required to show that $\mathscr{U} = \{D_x : x \in I \subseteq S\}$ is open in $\mathscr{D}_S$. To show that $\mathscr{U}$

is open in $\mathscr{D}_S$ it suffices to show that $\theta^{\leftarrow}[\mathscr{U}]$ is open in S (since $\mathscr{D}_S$ is equipped with the quotient topology induced by θ).

$$\begin{aligned}\cup\{D_x \subseteq S : x \in I\} &= \cup\{\theta^{\leftarrow}[\{D_x\}] : x \in I\}\\ &= \theta^{\leftarrow}[\cup\{D_x : x \in I\}]\\ &= \theta^{\leftarrow}[\mathscr{U}]\end{aligned}$$

Since $\cup\{D_x : x \in I\} = \theta^{\leftarrow}[\mathscr{U}]$ is open then $\mathscr{U}$ is open.

Example 2. Consider the Cartesian product $\mathbb{R} \times \mathbb{R}$ equipped with the product topology. We can partition this product into its vertical lines. The projection function $\pi_1 : \mathbb{R} \times \mathbb{R} \to \mathbb{R}$ can be seen as a decomposition map which collapses the vertical lines to a point by

$$\pi_1[\{a\} \times \mathbb{R}] = \{a\}$$

In this case (since π_1 is an open map), the quotient topology induced by π_1 on $\mathbb{R}$ is the usual topology. It is the finest (strongest) topology on $\mathbb{R}$ such that π_1 is continuous on $\mathbb{R} \times \mathbb{R}$.

On the other hand, if we are given that $\mathbb{R}$ is equipped with the usual topology the Cartesian product $\mathbb{R}\times\mathbb{R}$ can be equipped with the weakest (coarsest) topology induced by the function $\pi_1 : \mathbb{R}\times\mathbb{R} \to \mathbb{R}$. It has as open base

$$\mathscr{B} = \{\pi_1^{\leftarrow}[U] : U \text{ is open in } \mathbb{R}\}$$

Example 3. Let $\mathbb{J}$ denote the set of all irrational numbers in the closed interval $[0, 1]$ and let

$$T = \{1\} \cup \mathbb{J}$$

a proper subset of $[0, 1]$. We define a function $f : [0, 1] \to T$ as follows

$$f(x) = \begin{cases} x & \text{if } x \text{ is irrational in } [0,1] \\ 1 & \text{if } x \text{ is rational in } [0,1] \end{cases}$$

Let the domain, $[0, 1]$, be equipped with the usual topology and the range, $T = \{1\} \cup \mathbb{J}$, be equipped with the quotient topology induced by the function f.

Describe the open subsets of the quotient space, T, induced by the function f on $[0,1]$.

Solution: We are given that f fixes the irrationals in $[0,1]$ and collapses all rationals to 1. The set $T = \{1\} \cup \mathbb{J}$ is equipped with the quotient topology induced by f.

The open subsets of $\{1\} \cup \mathbb{J}$ are defined as

$$\tau_f = \{U \subseteq \{1\} \cup \mathbb{J} : f^{\leftarrow}[U] \text{ is open in } [0,\ 1]\}$$

We have $f^{\leftarrow}[\varnothing] = \varnothing$ and $f^{\leftarrow}[\{1\} \cup \mathbb{J}] = [0,1]$ so $\{\varnothing,\ \{1\} \cup \mathbb{J}\} \subseteq \tau_f$.

Consider, for example, the open interval $U = (1/5, \pi/4)$. Then $[U \cap \mathbb{J}] \cup \{1\} \subseteq \{1\} \cup \mathbb{J}$. So,

$$\begin{aligned} f^{\leftarrow}\left[\,[U \cap \mathbb{J}] \cup \{1\}\,\right] &= f^{\leftarrow}[U \cap \mathbb{J}] \cup f^{\leftarrow}[\{1\}] \\ &= [\,U \cap \mathbb{J}\,] \cup [\,[0,1] \cap \mathbb{Q}\,] \end{aligned}$$

Since $[\,U \cap \mathbb{J}\,] \cup [\,[0,1] \cap \mathbb{Q}\,]$ is not open in $[0,1]$, then $[\,U \cap \mathbb{J}\,] \cup \{1\}$ is not open in $\{1\} \cup \mathbb{J}$.

It seems that the complete, open, *preimage*, $f^{\leftarrow}[V]$, of any non-empty subset, V, of $\{1\} \cup \mathbb{J}$ should contain open intervals and, at least, both 0 and 1. Are there such proper subsets of $[0,1]$? We test the following subset:

$$f^{\leftarrow}[V] = [0, \pi/7) \cup (\pi/7,\ \pi/5) \cup (\pi/5,\ \pi/4) \cup (\pi/4, 1]$$

is open in $[0,1]$. We seek a set V in $\{1\} \cup \mathbb{J}$ that f pulls back to the subset, $f^{\leftarrow}[V]$. We test

$$f\,[f^{\leftarrow}[V]] = \{1\} \cup \mathbb{J} \setminus \{\pi/7,\ \pi/5,\ \pi/4\}$$

to find

$$\begin{aligned} &f^{\leftarrow}[\{1\} \cup \mathbb{J} \setminus \{\pi/7,\ \pi/5,\ \pi/4\}] \\ &\quad = [0, \pi/7) \cup (\pi/7,\ \pi/5) \cup (\pi/5,\ \pi/4) \cup (\pi/4, 1] \\ &\quad \text{An open subset of } [0,1]. \end{aligned}$$

So, $\{1\} \cup \mathbb{J} \setminus \{\pi/7,\ \pi/5,\ \pi/4\} \in \tau_f$.

Then

$$\tau_f = \{\varnothing\} \cup \{\{1\} \cup D : D \text{ is open and dense in } \mathbb{J}\}$$

Example 4. Suppose we are given the subset, $S = [0, 2\pi] \times [0, 2\pi]$ of $\mathbb{R}^2$. We will decompose S as follows: For each point $(x, 0)$ on the line $[0, 2\pi] \times \{0\}$ let

$$D_{(x,0)} = \{(x, 0), (2\pi - x, 2\pi)\}$$

So for each $x \in [0, 2\pi]$, $D_{(x,0)}$ and $D_{(2\pi - x, 2\pi)}$ represent the same element of the decomposition, $\mathscr{D}_S$. For example, $D_{(0,0)} = D_{(2\pi,2\pi)} = \{(0, 0), (2\pi, 2\pi)\}$.

For each $(x, y) \notin [0, 2\pi] \times \{0\} \cup [0, 2\pi] \times \{2\pi\}$ let

$$D_{(x,y)} = \{(x, y)\}$$

Each subset of the form $D_{(x,0)}$ is a subset of S which contains two points. All other subsets of the decomposition are singleton sets. The decomposition space, $\mathscr{D}_S$, of the subspace S obtained is referred to as the *Möbius strip.*[4] See Figure 8.1.

The two lines $L_1 = \{(x, 0) : 0 \leq x \leq 2\pi\}$ and $L_2 = \{(x, 2\pi) : 0 \leq x \leq 2\pi\}$ are collapsed together (after inverting one of the lines) to form the Möbius strip.

Without the inversion of one of the lines, the decomposition space becomes a topological representation of a cylindrical shell.

8.3.1 *Moore plane decomposition*

Example 5. Let (S, τ_S) denote the Moore plane. (Review the description of the Moore plane by looking over the example on page 92.) Let $F = \{(x, 0) | x \in \mathbb{R}\}$ and $W = S \setminus F$. We can then express S as the disjoint union of the sets F and W. Let $\mathscr{D} = \{\{x\} : x \in W\} \cup \{F\}$, be the decomposition space of S which results from collapsing the subset, F, down to a point and identifying all other points to themselves.

---

[4]In reference to August Ferdinand Möbius, 1790–1868, a German mathematician and astronomer.

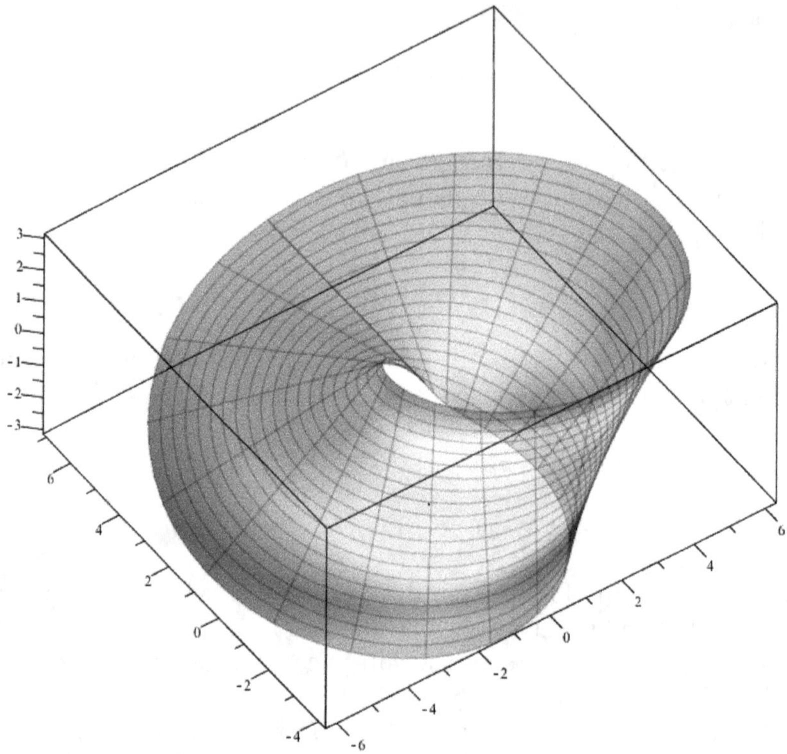

Figure 8.1. Topological representation of the Möbius strip.

Describe an open neighborhood base of each point in the decomposition space $\mathscr{D}$.

Solution: We are given the Moore plane, (S, τ_s), and the decomposition $\mathscr{D} = \{\{x\} : x \in W\} \cup \{F\}$. It easily verified that points in S are closed and so S is T_1.

If $\theta : S \to \mathscr{D}$ is the corresponding quotient map, then $\theta(a, b) = (a, b)$ if $(a, b) \in W$ and $\theta(u, v) = F$ when $(u, v) \in F$. The quotient topology on $\mathscr{D}$ is, by definition,

$$\tau_\theta = \{U \subseteq \mathscr{D} : \theta^{\leftarrow}[U] \text{ is open in } S\}$$

The open neighborhood base of points, $\{x\} \in \mathscr{D}$ for $x \in W$ is the same as for τ_W. If τ_W is the subspace topology on the open subspace, W, inherited from S, then $\tau_W \subseteq \tau_\theta$.

We now describe an open neighborhood base for the element, $F \in \mathscr{D}$.

Let $N = \mathbb{N}\backslash\{0\}$. Consider the set, $N^{\mathbb{R}}$, of all functions which map $\mathbb{R}$ into N.

We now construct a basic open neighborhood of the element F. For $f \in N^{\mathbb{R}}$, let

$$U_f = \cup\left\{B_{\frac{1}{f(x)}}\left(x, \frac{1}{f(x)}\right) : (x, 0) \in \mathbb{R} \times \{0\}\right\}$$

where $\frac{1}{f(x)}$ is the radius of each open ball of center $\left(x, \frac{1}{f(x)}\right)$ tangent to the x-axis. Since, for each f, $\theta^{\leftarrow}[U_f \cup \{F\}] = U_f \cup F$ is open in S, then

$$\mathscr{B}_F = \{U_f \cup \{F\} : f \in N^{\mathbb{R}}\}$$

is an uncountably large base of open neighborhoods for F, since for any open neighborhood V of the x-axis, there exists $f \in N^{\mathbb{R}}$ such that $\{(x, 0) : x \in \mathbb{R}\} \subseteq \mathscr{B}_F \subseteq V$.

8.3.2 *On subspaces of a decomposition space*

Example 6. Let (S, τ_s) be a topological space. Suppose $(\mathscr{D}, \tau_{\mathscr{D}})$ is a decomposition space induced by some decomposition of S. Let T be a non-empty subset of S. Then the decomposition $\mathscr{D}$ of S will induce a decomposition space $(\mathscr{D}_T, \tau_T)$,

$$\mathscr{D}_T = \{D \cap T : D \in \mathscr{D}\}$$

of T. We can identify another decomposition space $(\mathscr{D}|_T, \tau_{|_T})$ associated to T:

$$\mathscr{D}|_T = \{D \in \mathscr{D} : D \cap T \neq \varnothing\}$$

The space, $(\mathscr{D}|_T, \tau_{|_T})$ is equipped with the subspace topology, $\tau_{|_T}$, inherited from $(\mathscr{D}, \tau_{\mathscr{D}})$.

Provide an example that shows that $(\mathscr{D}|_T, \tau_{|_T})$ and $(\mathscr{D}_T, \tau_T)$ are generally not homeomorphic spaces.

Solution: Consider the decomposition, $\mathscr{D} = \{\{x\} \times \mathbb{R} : x \in \mathbb{R}\}$, of $\mathbb{R}^2$. Let $T = \{(x, 0) \in \mathbb{R}^2 : x < 0\} \cup \{(0, 1)\}$, a subset of $\mathbb{R}^2$, with the subspace topology.

Let $\mathscr{D}_T = \{D \cap T : D \in \mathscr{D}\}$. Then $\mathscr{D}_T = \{\{(x,0)\} : x < 0\} \cup \{(0,1)\}$.

Let $\theta_T : T \to \mathscr{D}_T$ be the quotient function which maps a point x in T to the unique element of $\mathscr{D}_T$ which contains it. Then θ_T induces a the quotient topology, $\tau_{\mathscr{D}}$ on $\mathscr{D}_T$. That is, U is open in $\mathscr{D}_T$ if and only if, $\theta_T^{\leftarrow}[U]$ is open in T. The function θ_T is clearly one-to-one on T. Also, since $\theta_T^{\leftarrow}[\{(0,1)\}] = \{(0,1)\}$ is open in T (with respect to the subspace topology), the point $\{(0,1)\}$ is a clopen element of $\mathscr{D}_T$. The point $\{(0,1)\}$ in $\mathscr{D}_T$ is referred to as an *isolated point.*

We now consider the decomposition,

$$\mathscr{D}|_T = \{D \in \mathscr{D} : D \cap T \neq \varnothing\}$$

We claim that $\mathscr{D}|_T$ does not contain an isolated point, and so cannot be homeomorphic to $\mathscr{D}_T$. Note that $\mathscr{D}|_T = \{\{x\} \times \mathbb{R} : x \leq 0\}$. Since every neighborhood of a vertical line in $\cup\mathscr{D}|_T$ intersects another vertical line then $\mathscr{D}|_T$ cannot contain an isolated point. So, $\mathscr{D}|_T$ and $\mathscr{D}_T$ cannot be homeomorphic.

Example 7. Let S^n denote the n-sphere in the space, $T = \mathbb{R}^{n+1}\backslash\{\vec{0}\}$ (equipped with the usual topology). That is, $S^n = \{\vec{x} \in \mathbb{R}^{n+1}\backslash\{\vec{0}\} : \|\vec{x}\| = 1\}$. For example, S^1 is the unit circle in $\mathbb{R}^2\backslash\{(0,0)\}$, S^2 is the sphere surface in $\mathbb{R}^3\backslash\{(0,0)\}$. Let

$$L_\lambda = \{\lambda\vec{x} : \vec{x} \in T\}$$

represent a line in T passing through the point, $\vec{0}$, and $\vec{x}$ with "slope" λ. Then the collection of lines, $\{L_\lambda : \lambda \in \mathbb{R}_{>0}\}$, constitutes a decomposition, $\mathscr{D}_T$, of the space T. The elements of $\mathscr{D}_T = \{D_\lambda : \lambda \in \mathbb{R}_{>0}\}$, are of the form $D_\lambda = L_\lambda$ for $\lambda \in \mathbb{R}_{>0}$.

Let $f : T \to S^n$ be defined as

$$f(\vec{x}) = \frac{\vec{x}}{\|\vec{x}\|}$$

a continuous function on T. Then

$$f[L_\lambda] = f[\{\lambda\vec{x} : \vec{x} \in T\}] = \lambda\vec{x}/\|\lambda\vec{x}\| \in S^n \cap L_\lambda$$

So, f is a function which collapses each set L_λ down to the singleton set $L_\lambda \cap S^n$. Then, f induces a map

$$f^* : \mathscr{D}_T \to S^n$$

which is one-to-one on $\mathscr{D}_T$ onto S^n. If U is open in S^n, $f^{*\leftarrow}[U]$ is open in $\mathscr{D}_T$.

If V is open in $\mathscr{D}_T$, then $f^*[V]$ is open in S^n. So f^* is a homeomorphism mapping $\mathscr{D}_T = T/R_f$ onto S^n. So, S^n is homeomorphic to $\mathbb{R}^{n+1}/R_f$.

Note: See that S^n cannot be homeomorphic to $\mathbb{R}^{n+1}$ nor to $\mathbb{R}^n$ since S_n is closed and bounded (compact) and neither $\mathbb{R}^{n+1}$ nor $\mathbb{R}^n$ are.

Concept Review

1. Define the quotient topology induced by a function.
2. If τ_T is a topology on T for which $f : S \to T$ is continuous, how does τ_T compare with the quotient topology, τ_f, induced by f?
3. If τ_T is a topology on T for which $f : S \to T$ is a continuous open map, how does τ_T compare with the quotient topology, τ_f, induced by f?
4. If τ_T is a topology on T for which $f : S \to T$ is a continuous closed map, how does τ_T compare with the quotient topology, τ_f, induced by f?
5. What is a *fiber* of a point under a map f?
6. Define the set S/R_f of all equivalence classes induced by a function $f : S \to T$ by referring to a function $\theta : S \to S/R_f$.
7. What topology is defined on S/R_f?
8. How is the function $\phi_f : S/R_f \to T$ defined?
9. If $g : S \to T$ is a function and $U \subseteq S$, what does it mean to say that U is g-saturated?
10. If $g : S \to T$ is a function and $U \subseteq S$ is g-saturated, is there a way to describe the subset U in terms of g?
11. Given $f : S \to T$ and $\theta : S \to S/R_f$ describe the homeomorphism which links S/R_f to T.
12. Describe how to construct a decomposition space, $\mathscr{D}_S$, from a topological space (S, τ_S).

Exercises

1. Let (S, τ_S) be a topological space. We define a relation, R, on S as follows: $(u, v) \in R$ if and only if $\text{cl}_S\{u\} = \text{cl}_S\{v\}$.

(a) Show that R is an equivalence relation on S.

(b) Let $(S/R, \tau_\theta)$ denote the quotient space induced by the natural map $\theta : S \to S/R$.

(i) If S_x is an element in S/R, is $\{S_x\}$ necessarily a closed subset of S/R?

(ii) If S_x and S_y are distinct elements of S/R, does there exist an open set in τ_θ which contains S_x but not S_y?

2. Describe the quotient space, $\mathbb{R}/R$, if $(u, v) \in R$ if and only if $x - y$ is an integer.
3. Suppose $\mathscr{D}_{\mathbb{R}^2}$ is a decomposition of $\mathbb{R}^2$ (with the usual topology) whose elements are circles with center at the origin. Show that the corresponding decomposition space, $(\mathscr{D}_{\mathbb{R}^2}, \tau_\theta)$, and the set of all non-negative real numbers, $\{x \in \mathbb{R} : x \geq 0\}$, equipped with the usual topology are homeomorphic topological spaces.

Part III

Topological Spaces: Separation Axioms

Chapter 9

Separation with Open Sets

Abstract
In this section we will present five classes of topological spaces which, together, form a hierarchy of separation axioms. They differ by the way their open sets can be used to separate their points and their closed sets. For each of these spaces, we present a few of their characterizations and basic properties. These five types of topological spaces are called, T_i-spaces, where $i = 0, 1, 2, 3, 4$.

9.1 The T_0 Separation Axiom

In this chapter, we will introduce five other classes of topological spaces. We will also discuss various characterizations and properties of these spaces. We distinguish these five classes by the separation axioms they satisfy. The first one is referred to as the T_0 axiom.

Definition 9.1 Let (S, τ_s) be a topological space. We say that S is a T_0-*space* if, for every pair of distinct points u and v, there exists at least one open set in τ_S, which contains one of these two points, but not the other.[1]

---

[1] A T_0-space is also known as a *Kolmogorov space* (named after the mathematician Andrey Kolmogorov).

Most of the topological spaces we have been exposed to are T_0-spaces, so finding an example of a T_0-space is easy, as long as we can justify why a particular space is declared to be T_0. The space, $\mathbb{R}$, with the usual topology, is T_0 since, for any $a, b \in \mathbb{R}$ where $a < b$, $(a-1, b)$ contains a but not b. We will soon see that the Euclidean space, $\mathbb{R}$, can separate points in much more intricate ways. For many, the topological spaces of interest are, at the very least, T_0.

Are there any non-T_0-spaces? The indiscrete space (S, τ_i) (where S is not a singleton) with topology,

$$\tau_i = \{\varnothing, S\}$$

is not T_0. For, if $a, b \in S$, since S is the only non-empty open set, it contains all points and so is unable to separate a point a from a point b in the way prescribed by the T_0-property.

We provide another example of a non-T_0 space. Consider the projection function $\pi_1 : \mathbb{R} \times \mathbb{R} \to \mathbb{R}$ on the Cartesian product $\mathbb{R} \times \mathbb{R}$. This function collapses each vertical line in the product to a point. We equip the product $\mathbb{R} \times \mathbb{R}$ with the weak topology induced by π_1. Then $\{\pi_1^{\leftarrow}[U] : U \text{ is open in } \mathbb{R}\}$ forms a base for open sets in $\mathbb{R} \times \mathbb{R}$. Consider the distinct points (a, b) and (a, c). Any basic open set of $\mathbb{R} \times \mathbb{R}$ which contains (a, b) contains $\{a\} \times \mathbb{R}$. So, no open set contains (a, b) but note (a, c) nor does it contains (a, c) but not (a, b). So, $\mathbb{R} \times \mathbb{R}$ equipped with the weak topology induced by π_1 is not T_0.

9.1.1 *Combining the T_0 separation property with other topological properties*

Even though T_0-spaces, by themselves, may be considered by some as being of little interest, it is good practice to see what happens when we combine the T_0-property with other topological properties. Consider the following example in which we show that infinite second countable T_0-spaces cannot be arbitrarily large.

Example 1. Suppose (S, τ_s) is an infinite second countable T_0-space. Show that the cardinality, $|S|$, of S is less than or equal to $2^{\aleph_0}$ (where, $\aleph_0 = |\mathbb{N}|$).

Solution: Since S is an infinite second countable space, then, by definition, it has an open base, $\mathscr{B}$, such that $|\mathscr{B}| \leq \aleph_0$. For each

$x \in S$, let

$$\mathscr{B}_x = \{B \in \mathscr{B} : x \in B\} \subseteq \mathscr{B}$$

If $y \neq x$ then $\mathscr{B}_x$ cannot equal $\mathscr{B}_y$, for, if it did, then $y \in B_x$ for all $B_x \in \mathscr{B}_x$, and there would be no $B_x(B_y)$ which can separate x from y, a contradiction of the T_0 property. So there is a one-to-one function, $f : S \to \mathscr{P}(\mathscr{B})$, defined as

$$f(x) = \mathscr{B}_x$$

Then $|S| = |\{\mathscr{B}_x : x \in S\}|$. Recall that the cardinality of the power set, $\mathscr{P}(T)$, of a set T is $2^{|T|}$. Since $|\mathscr{B}| \leq \aleph_0$, then $|\mathscr{P}(\mathscr{B})| = 2^{|\mathscr{B}|} \leq 2^{\aleph_0}$.

Since, for each $x \in S$, $\mathscr{B}_x \in \mathscr{P}(\mathscr{B})$, then $|\{\mathscr{B}_x : x \in S\}| \leq |\mathscr{P}(\mathscr{B})|$.

We conclude that

$$|S| = |\{\mathscr{B}_x : x \in S\}| \leq |\mathscr{P}(\mathscr{B})| \leq 2^{\aleph_0}$$

so the cardinality of a second countable T_0-space, S, is less than or equal to $2^{\aleph_0}$, as required.[2]

9.2 The T_1 Separation Axiom

In general, we prefer that topological a space (S, τ) has points which are closed subsets of S. The next separation axiom called, T_1, requires that τ has sufficiently many open sets to guarantee that its singleton sets, $\{x\}$, must be closed subsets.

Definition 9.2 Let (S, τ) be a topological space. We say that S is a T_1-*space* if, for every pair of distinct points u and v, there exists an open set, U, which contains u but not v, *and* an open set, V, which contains v but not u.[3]

---

[2]Note that $2^{\aleph_0}$ is the cardinality of $\mathbb{R}$.

[3]The T_1-spaces are also referred to as Fréchet spaces, named after the mathematician Maurice Fréchet.

Of course, every T_1-space is T_0. Also, $\mathbb{R}^n$, when equipped with the usual topology, is a T_1-space. Are there T_0-spaces which are not T_1? In the following example we exhibit a T_0-space which is not T_1.

Example 2. Let S be a set which contains more than a single point. Suppose $u \in S$ and $\mathscr{F} \subseteq \mathscr{P}(S)$ where

(1) For non-empty F, $[F \in \mathscr{F}] \Leftrightarrow [u \in F]$.
(2) $\varnothing \in \mathscr{F}$.

Verify that the two given conditions on $\mathscr{F}$ together satisfy the closed sets axioms and so define a topology, τ_s, on S. Then show that (S, τ_S) is T_0, but not T_1.

Solution: We claim that $\mathscr{F}$ describes all closed subsets with respect to the topology $\tau_s = \{U : u \notin U\} \cup \{S\}$. Clearly, $\{S, \varnothing\} \subseteq \mathscr{F}$. Finite unions of sets each of which contains the point u will contain the point u and arbitrary intersections of sets each of which contain u will also contain u. Then $\mathscr{F}$ satisfies the closed set axioms. Then the set, $\mathscr{F}$, represents all closed sets for the topology

$$\tau_s = \{U : u \notin U\} \cup \{S\}$$

We see that, if $x \neq u$, then $S \backslash \{u\}$ is an open neighborhood of x and does not contain u. So, S is T_0. But since the only open neighborhood of u is all of S, then S cannot be T_1. So, (S, τ_S) is T_0 not T_1.

9.2.1 *Characterizations of T_1-spaces*

The most useful characterization of T_1-spaces is given as follows. This characterization is such that some may use it as a definition of T_1.

Theorem 9.3 *Let (S, τ_s) be a topological space. The space S is T_1 if and only if every singleton set, $\{x\}$, of S is a closed subset of S.*

Proof. Let (S, τ_s) be a topological space. If S is a singleton set then, trivially, S is T_1 and its only singleton set is S, which is closed. Suppose S contains more than one point.

($\Rightarrow$) Suppose S is T_1. Let $u \in S$. We claim that $\{u\}$ is closed. For each, $x \neq u$, there exists open, U_x such that, $x \in U_x$ and $u \in S \backslash U_x$.

Then $\{u\} = \cap\{S\backslash U_x : x \neq u\}$, the intersection of closed subsets of S. So, $\{u\}$ is closed in S.

($\Leftarrow$) Suppose every point of S is closed. Then, then for any $u, y \in S$, if $y \neq u$, $S\backslash\{u\}$ is an open neighborhood of y which does not contain u and $S\backslash\{y\}$ is an open neighborhood of u which does not contain y. Hence, S is T_1.

Theorem 9.4 *Let (S, τ_S) and (T, τ_T) be topological spaces.*

(a) *If S is a T_1-space then so are all its subspaces. So T_1 is a hereditary property.*
(b) *Suppose S is a T_1 space and $f : S \to T$ is a closed function. Then $f[S]$ is a T_1 subspace of T.*
(c) *Suppose $S = \prod_{i\in I} S_i$ is a product space. The space S is a T_1 product space if and only if each of its S_i-factors is a T_1-space.*

Proof. Let (S, τ_S) and (T, τ_T) be topological spaces.

(a) Suppose S is T_1 and $u \in W \subseteq S$. Since $\{u\}$ is a closed subset of S, it is a closed subset of its subspace W (with respect to the subspace topology). So, W is T_1 (by Theorem 9.3).
(b) Suppose S is T_1 and $f : S \to f[S]$ is a closed map. Let $v \in f[S]$. Then there exists $x \in S$ such that $f[\{x\}] = \{v\}$. By hypothesis, $\{x\}$ is closed and f is closed so, $\{v\}$ is closed. Hence, $f[S]$ is T_1.
(c) ($\Rightarrow$) Suppose $S = \prod_{i\in I} S_i$ is a T_1 product space. Recall that each factor, S_i, is homeomorphic to a subspace, U_i, of S. By part (a), U_i is T_1 and so S_i is T_1.

    ($\Leftarrow$) Suppose each factor, S_i, of $S = \prod_{i\in I} S_i$ is T_1. Let $q = < a_i >_{i\in I}$ be a point in S. Then no S_i is empty since, for each $i \in I$, $a_i \in S_i$. Since each S_i is T_1, for each coordinate a_i, the singleton set, $\{a_i\}$, is a closed subset of S_i. For all $j \in I$, $\pi_j : S \to S_j$ is continuous; then $\pi_j^{\leftarrow}(\{a_j\})$ is closed in S. Then $\{q\} = \cap_{j\in I}\left\{\pi_j^{\leftarrow}(\{a_j\})\right\}$ is closed in S. So, S is T_1.

In the following example, we show that the decomposition space, $\mathscr{D}$, whose elements are the fibers of $f : S \to T$, is T_1 if and only if the fibers of f are closed subsets of S.

Example 3. Let $f : S \to T$ be a function. Suppose τ_f is the quotient topology on S induced by the quotient map f. Let $\mathscr{D}_S = \{D_x : x \in T\}$ be a decomposition space of S, with elements, $D_x = f^{\leftarrow}(x)$. Show that $\mathscr{D}_S$ is T_1 if and only if each of its elements, $D_x = f^{\leftarrow}(x)$, is a closed subset of S.

Solution: Let $f : S \to (T, \tau_T)$ be a function. $f^{\leftarrow}(x) = D_x$ for each $x \in T$ and $\mathscr{D}_S = \{D_x : x \in T\}$ be a decomposition space of S induced by f, equipped with the quotient topology τ_f. Let $q : S \to \mathscr{D}_S$ be the quotient map $q(x) = D_x$.

($\Leftarrow$) We are given that, for each $x \in T$, $D_x = f^{\leftarrow}(x)$ is a closed subset of S. We are required to show that $\mathscr{D}_S$ is T_1. Let $S_y \in \mathscr{D}_S$. It suffices to show that $\{S_y\}$ is closed in $\mathscr{D}_S$. Consider

$$q^{\leftarrow}[\mathscr{D}_S \backslash \{S_y\}] = S \backslash f^{\leftarrow}[\{y\}]$$

where by hypothesis, $S \backslash f^{\leftarrow}[\{y\}]$ is open in S. Since $\mathscr{D}_S$ is equipped with the quotient topology,

$$q^{\leftarrow}[\mathscr{D}_S \backslash \{S_y\}] \text{ open in } S \quad \Rightarrow \quad \mathscr{D}_S \backslash \{S_y\} \text{ open in } \mathscr{D}_S$$

So, $\{S_y\}$ is closed in $\mathscr{D}_S$. This implies $\mathscr{D}_S$ is T_1.

($\Rightarrow$) We are given that $\mathscr{D}_S$ is a T_1 quotient space. Then, for each $x \in T$, $\{D_x\}$ is a closed subset of $\mathscr{D}_S$. We are required to show that $D_x = f^{\leftarrow}(x)$ is a closed subset of S. Since $q : S \to \mathscr{D}_S$ is continuous, then $q^{\leftarrow}[\{D_x\}] = f^{\leftarrow}(x)$ is closed in S, as required.

9.3 The T_2 Separation Axiom or Hausdorff Topological Space

We will now define a T_2-space. The terminology "T_2-space" helps remind us that $T_2 \Rightarrow T_1 \Rightarrow T_0$. Although both T_2 and Hausdorff mean the same thing, the word "Hausdorff" is more commonly used.[4]

---

[4]Sadly, the talented mathematician, Felix Hausdorff (along with many other very talented European mathematicians) did not survive persecution by the Nazi regime in Germany.

Definition 9.5 Let (S, τ_s) be a topological space. We say that S is a T_2-*space* or *Hausdorff space* if, for every pair of distinct points u and v, there exists non-intersecting open neighborhoods, U and V such that $u \in U$ and $v \in V$.

Clearly, any Hausdorff space, (S, τ_s), satisfies the T_1 axiom. Hence,

> *if S is Hausdorff then, for each $x \in S$, $\{x\}$ is a closed subset of S.*

Example 4. An infinite set, S, with the cofinite topology (also referred to as the Zariski topology, see page 47) is T_1 but not Hausdorff since every pair of open neighborhoods will intersect with infinitely many points in their intersection. However, it is easily verified to be T_1.

Example 5. Any metrizable topological space, (S, τ_ρ), is Hausdorff since, given distinct points a and b, $B_\varepsilon(a) \cap B_\varepsilon(b) = \varnothing$ where $\varepsilon < \frac{\rho(a,b)}{3}$. A standard approach to verifying whether a topological space is metrizable is to first check if it is Hausdorff. If it is not Hausdorff then the question is settled.

9.3.1 *Hausdorff characterizations*

There are a few equivalent ways of expressing the Hausdorff separation property. We present a few of the more common characterizations.

Theorem 9.6 *Let (S, τ_S) be a topological space. The following statements are equivalent.*

(a) *The space, S, is Hausdorff.*
(b) *If u and v are distinct points in S then there exists an open neighborhood, U, of u such that $v \notin cl_S U$.*
(c) *If $u \in S$, then $\cap\{cl_S U : U$ is an open neighborhood of $u\} = \{u\}$*
(d) *The set $D = \{(u, u) : u \in S\}$ is closed in the product space, $S \times S$.*

Proof. Let (S, τ_s) be a topological space.

(a $\Rightarrow$ b) We are given that S is Hausdorff. Suppose u and v are distinct points in S. Then there exists disjoint open neighborhoods, U and V, that contain u and v, respectively. See that $U \subseteq S \backslash V$, a closed subset of S; hence, $\mathrm{cl}_S U \subseteq S \backslash V$. Then $v \notin U \subseteq \mathrm{cl}_S U$.

(b $\Rightarrow$ c) Suppose that, if $u \neq v$ in S, there exists an open neighborhood, U, of u such that $v \notin \mathrm{cl}_S U$. Then $v \in \cap\{\mathrm{cl}_S U : U \text{ is a neighborhood of } u\}$ is impossible. So, $\cap\{\mathrm{cl}_S U : U \text{ is a neighborhood of } u\} = \{u\}$.

(c $\Rightarrow$ d) Suppose that $\cap\{\mathrm{cl}_S U : U \text{ is a neighborhood of } u\} = \{u\}$, for all $u \in S$. Let $D = \{(a, a) \in S \times S : a \in S\}$. We are required to show that D is closed in $S \times S$. Let $(u, v) \in (S \times S)\backslash D$. Then u and v are distinct elements of S. By hypothesis, there exists an open neighborhood, U, of u such that $v \notin \mathrm{cl}_S U$. Then $U \times (S\backslash \mathrm{cl}_S U)$ is an open neighborhood of (u, v). If $a \in U$, then $a \notin S\backslash \mathrm{cl}_S U$ so $(a, a) \notin U \times (S\backslash \mathrm{cl}_S U)$. So, $U \times (S\backslash \mathrm{cl}_S U)$ does not intersect D. So, $(S \times S)\backslash D$ is open. Hence, D is closed.

(d $\Rightarrow$ a) Suppose D is closed in $S \times S$. Let u and v be distinct points in S. Then $(u, v) \in (S \times S)\backslash D$. Since D is closed there exists and open neighborhood, $U \times V$ of (u, v) which does not intersect D. Then U and V are disjoint open neighborhoods of u and v, respectively. So, S is Hausdorff.

In the following theorem, we show that a subspace will always inherit the Hausdorff property from its topological space. Also, the Hausdorff property is carried over from a set of factors, $\{S_i\}_{i\in I}$, to the Cartesian product space, $\prod_{i\in I} S_i$, they generate. The Hausdorff property is also carried over from the domain, S, of a one-to-one closed function, f, to its range, $f[S]$.

However, continuous images of Hausdorff spaces need not necessarily be Hausdorff, as the example following the theorem will show.

Theorem 9.7 *Let (S, τ_S) be a Hausdorff topological space and $\{S_i\}_{i\in I}$ be a family of Hausdorff topological spaces.*

(a) *If U is a subspace of S, then U is Hausdorff. That is, "Hausdorff" is a hereditary property.*

(b) *If $f : S \to T$ is a one-to-one closed function onto T, then T is Hausdorff.*
(c) *The product space, $\prod_{i\in I} S_i$, generated by the Hausdorff factors, $\{S_i\}_{i\in I}$, is Hausdorff.*

Proof. Let (S, τ_s) be a Hausdorff topological space.

(a) Suppose u and v are distinct points in the subspace, W, of S. Then there exists disjoint open neighborhoods, U and V, which respectively contain the points u and v. Then $U \cap W$ and $V \cap W$ are disjoint open neighborhoods in W, which respectively contain the points u and v.
(b) Suppose $f : S \to T$ is a one-to-one closed function mapping S onto T. Recall that a one-to-one closed function is also an open function. (See the corresponding theorem on page 124.) Then disjoint open neighborhoods U and V of u and v are mapped to disjoint open neighborhoods of $f[U]$ and $f[V]$ of $f(u)$ and $f(v)$.
(c) We are given that $\{S_i\}_{i\in I}$ is a family of Hausdorff topological spaces. Let $\{u_i\}_{i\in I}$ and $\{v_i\}_{i\in I}$ be distinct points of $\prod_{i\in I} S_i$. Then, for some $j \in I$, $u_j \neq v_j$. Let U_j and V_j be disjoint open neighborhoods of u_j and v_j in S_j. Then $\pi_j^{\leftarrow}[U_j]$ and $\pi_j^{\leftarrow}[V_j]$ are disjoint open base elements containing $\{u_i\}_{i\in I}$ and $\{v_i\}_{i\in I}$. So, $\prod_{i\in I} S_i$ is Hausdorff.

9.3.2 *Examples involving the Hausdorff property*

Example 6. Consider the space $S = \{0, 1\}$ equipped with the discrete topology,

$$\tau_d = \{\varnothing, \{0\}, \{1\}, \{0, 1\}\}$$

and $T = \{0, 1\}$ equipped with the topology,

$$\tau_T = \{\varnothing, \{0\}, \{0, 1\}\}$$

Let $f : S \to T$ be a one-to-one onto function defined as $f(x) = x$ on S. Since S is discrete, each point can serve as it own open neighborhood and so is Hausdorff. Also, every function is continuous on a discrete space so f is continuous on S. Use this example to show that the continuous image of a Hausdorff space need not be Hausdorff.

Solution: In τ_T, there is no open neighborhood that contains 1 but not 0. So, T is not Hausdorff. In fact, T is not even T_1. So continuous images of Hausdorff spaces need not be Hausdorff.

Example 7. Show that, if a continuous function, $f : U \to V$, maps U one-to-one into a Hausdorff space V then its domain, U, must also be Hausdorff.

Solution: Let $f : U \to V$ be a one-to-one continuous function mapping U into a Hausdorff space, V. Then, since Hausdorff is a hereditary property, $f[U]$ is a Hausdorff subspace of V. Also, since f is continuous, $f^{\leftarrow} : f[U] \to U$ is a one-to-one closed function onto U. By Theorem 9.7, part (b), the space U must be Hausdorff. This is what we were required to show.

In the following two examples we show how the Hausdorff characterization in Theorem 9.6, part (d)

$$[S \text{ is Hausdorff}] \Leftrightarrow [D = \{(u,u) : u \in S\} \text{ is a closed subset of } S \times S]$$

can serve as a useful tool for solving certain types of problems.

Example 8. Show that, if f and g are two continuous functions each mapping a topological space S into a Hausdorff space T, then the set, $U = \{u \in S : f(u) = g(u)\}$, is a closed subset of S. That is, the set on which f and g agree is closed in S.

Solution: Let $h : S \to T \times T$ be a function defined as $h(x) = (f(x), g(x))$, and let $U = \{u \in S : f(u) = g(u)\}$. By Theorem 9.6, the statement,

$$\text{“}T \text{ Hausdorff”} \Leftrightarrow \text{“}D = \{(u,u) : u \in T\} \text{ is a closed subset of } T \times T\text{”}$$

Since both f and g are continuous on S, then so is h (see Theorem 7.12). Hence, $h^{\leftarrow}[D]$ is closed in S. Since

$$\begin{aligned} h^{\leftarrow}[D] &= \{u \in S : h(u) = (f(u), g(u)) \in D\} \\ &= \{u \in S : f(u) = g(u)\} \\ &= U \end{aligned}$$

the set, U, is closed in S, as required.

Example 9. Show that, if $f : S \to T$ is continuous, where T is Hausdorff then the graph of f,

$$G = \{(u, f(u)) : u \in S\} \subseteq S \times T$$

is closed in $S \times T$.

Solution: Let $t : S \times T \to T \times T$ be defined as

$$t(u, v) = (f(u), v)$$

Then, since f and the identity map are continuous on S, t is continuous on $S \times T$. Hence, since T is Hausdorff, by Theorem 9.6, $D = \{(u, u) : u \in T\}$ is a closed subset of $T \times T$. Thus $t^{\leftarrow}[D]$ is closed in $S \times T$. But

$$\begin{aligned} t^{\leftarrow}[D] &= \{(u, v) \in S \times T : t(u, v) \in D\} \\ &= \{(u, v) \in S \times T : (f(u), v) \in D\} \\ &= \{(u, v) \in S \times T : f(u) = v\} \\ &= \{(u, f(u)) \in S \times T : u \in S\} \\ &= G \end{aligned}$$

So the graph G is closed in $S \times T$.

Theorem 9.8 *Let S and T be two spaces where T is Hausdorff. If D is a dense subset of S and $f : S \to T$ and $g : S \to T$ are continuous on S which agree on D, then $f = g$ on S.*

Proof. Let D be a dense subset of S such that $f(x) = g(x)$ for all $x \in D$.

Suppose $a \in S \backslash D$. Then $f(a) \neq g(a)$ in T. Since T is Hausdorff, then there exists *disjoint* basic open sets, $B_{f(a)}$ and $B_{g(a)}$ with empty intersection.

Both $f^{\leftarrow}[B_{f(a)}]$ and $g^{\leftarrow}[B_{g(a)}]$ are open in S and each contains the point a. But they should have empty intersection. Contradiction. So, $S \backslash D$ must be empty. So, $D = S$.

9.4 Completely Hausdorff Spaces

We will now slightly strength the *Hausdorff property* to one called the *completely Hausdorff property.* So that we understand why this stronger property is significant we will confirm that there are Hausdorff spaces which are not completely Hausdorff.

Definition 9.9 Let (S, τ_s) be a topological space. We say that a space is *completely Hausdorff* if and only if for any pair of points a and b in S, there exist open neighborhoods U and V of a and b, respectively, such that

$$\text{cl}_S U \cap \text{cl}_S V = \varnothing$$

The completely Hausdorff property is also represented by $T_{2\frac{1}{2}}$. It is also sometimes referred to as a *Urysohn space.*

Clearly every completely Hausdorff space is Hausdorff. But,

$$\text{Hausdorff} \not\Rightarrow \text{completely Hausdorff}$$

as the following example shows. Equivalently, $T_2 \not\Rightarrow T_{2\frac{1}{2}}$.

Example 10. Let

$$T = \text{int}_{\mathbb{R}^2}(\,[0,1] \times [0,1])$$

and

$$S = T \cup \{(0,0), (1,0)\}$$

The points in T have basic neighborhoods which are the usual Euclidean open balls. The basic open neighborhoods of the point $(0,0)$ are defined to be open rectangles of the form

$$\mathscr{B}_{(0,0)} = \{(0,1/2) \times (0,1/n) \cup \{(0,0)\} : n > 0\}$$

where $(0,1/2)$ and $(0,1/n)$ are open intervals. The basic open neighborhoods of the point $(1,0)$ are defined to be open rectangles of the form

$$\mathscr{B}_{(1,0)} = \{(1/2,1) \times (0,1/m) \cup \{(1,0)\} : m > 0\}$$

Verify that the basic open neighborhoods of (0, 0) and (1, 0) can have empty intersection but that their closures will always have a point of the form $(1/2, x)$ in common. So, the topological space, S, is Hausdorff but not completely Hausdorff.

9.5 T_3-Spaces and "Regular Spaces"

The separation axioms, T_0, T_1, T_2 and $T_{2\frac{1}{2}}$ illustrate four different ways that open sets can separate two sets of points. The completely Hausdorff ($T_{2\frac{1}{2}}$-space) space imposes the strongest conditions since $T_{2\frac{1}{2}} \Rightarrow T_2 \Rightarrow T_1 \Rightarrow T_0$.

We now want to use open sets to separate a non-empty closed set, F, from a point, x, such that $x \in S \backslash F$. To do this we will define a T_3-space by using the definition of a T_2-space as a model.

Definition 9.10 Let (S, τ_s) be a topological space.

(a) We say that S is a T_3-*space* if, for any non-empty closed subset, F and point $v \in S \backslash F$, there exists non-intersecting open subsets, U and V such that $F \subseteq U$ and $v \in V$.
(b) If (S, τ_s) is both T_1 and T_3 then we will say that S is a *regular space*.

Let's consider the most trivial of topological spaces, the indiscrete space, (S, τ_i), where $|S| > 1$. There does not exist a "closed F and $x \in S\backslash F$" in S and so S is (vacuously) a T_3-space. But distinct points u and v in S are not contained in disjoint open sets and so, S is not Hausdorff. So, we have a situation here which needs fixing: $T_3 \not\Rightarrow T_2$. We can avoid this situation by defining "regular space $= T_1 + T_3$". If S is both T_1 and T_3 and u, v are distinct points then, since S is T_1, they are both closed, and, since S is T_3, they are, respectively, contained in disjoint open sets and so S is T_2.

Then "$[T_3 + T_1] \Rightarrow T_2 \Rightarrow T_1 \Rightarrow T_0$". Equivalently,

$$\text{Regular} \quad \Rightarrow \quad \text{Hausdorff} \quad \Rightarrow \quad T_1 \Rightarrow T_0$$

In the following example, we produce a topological space which is Hausdorff but not regular.

Example 11. Let $(\mathbb{R}, \tau)$ be the space of real numbers equipped with the usual topology, τ. Then $\mathscr{B} = \{(a, b) \subset \mathbb{R} : a < b\}$ denotes the standard open base of $\mathbb{R}$. Let

$$\mathscr{S} = \mathscr{B} \cup \{\mathbb{Q}\}$$

the standard open base be increased by one element, $\{\mathbb{Q}\}$. That is, the set of all rationals is a subbase element and so is declared to be an open set. We declare $\mathscr{S}$ to be a subbase for some topology $\tau_{\mathscr{S}}$ on $\mathbb{R}$. Show that the topological space, $(\mathbb{R}, \tau_{\mathscr{S}})$, is Hausdorff but not regular.

Solution: Note that $\mathbb{Q}$ belongs to $\tau_{\mathscr{S}}$ but is not an element of the usual topology, τ, and so $\mathscr{S} = \mathscr{B} \cup \{\mathbb{Q}\} \not\subseteq \tau$. Since every element of $\mathscr{B}$ is a base element of τ, then $\tau \subset \tau_{\mathscr{S}}$. So $\tau_{\mathscr{S}}$ is strictly stronger then the usual topology on $\mathbb{R}$. Since $(\mathbb{R}, \tau)$ is Hausdorff, then so is $(\mathbb{R}, \tau_{\mathscr{S}})$.

We now claim that $(\mathbb{R}, \tau_{\mathscr{S}})$ is *not* regular. Since $\mathbb{R}$ is Hausdorff, $\mathbb{R}$ is T_1. So, we are required to show that the topological space generated by $\mathscr{S}$ is not T_3. Let $\mathbb{J} = \mathbb{R}\backslash\mathbb{Q}$. Since $\mathbb{Q} \in \tau_{\mathscr{S}}$, $\mathbb{Q}$ is an open subset of $\mathbb{R}$ with respect to $\tau_{\mathscr{S}}$. So, $\mathbb{J}$ is a closed subset of $\mathbb{R}$ with respect to $\tau_{\mathscr{S}}$.

Then 1 is a point which does not belong to the closed subset, $\mathbb{J}$, of $\mathbb{R}$. Suppose U is any open neighborhood of 1. Then there exists and open interval (a, b) such that $1 \in (a, b) \subseteq U$. Since $(a, b) \cap \mathbb{J} \neq \varnothing$, then there can be no pair of disjoint open sets which contain 1 and $\mathbb{J}$, respectively. So $(\mathbb{R}, \tau_{\mathscr{S}})$ is not regular.

We now present two useful characterizations of "regular".

Theorem 9.11 *Let (S, τ_s) be a T_1 topological space. The following statements about S are equivalent.*

(a) *The space S is regular.*
(b) *For every $x \in S$ and open neighborhood, U of x, there exists an open neighborhood, V of x, such that $x \in \mathrm{cl}_S V \subseteq U$.*
(c) *For every $x \in S$ and closed subset F disjoint from $\{x\}$, there exists an open neighborhood, V of x, such that $\mathrm{cl}_S V \cap F = \varnothing$.*

Proof. Let (S, τ_s) be a T_1 topological space.

(a $\Rightarrow$ b) We are given that S is regular and that U is an open neighborhood of $x \in S$. Then $S \backslash U$ and $\{x\}$ are disjoint closed subsets of S. Since S is regular, there exists disjoint open sets, V and W, containing x and $S \backslash U$, respectively. Then $S \backslash W$ is a closed set, entirely contains V and is entirely contained in U. So, $\mathrm{cl}_S V \subseteq S \backslash W \subseteq U$.

(b $\Rightarrow$ c) We are given that for every $x \in S$ and open neighborhood, U of x, there exists an open neighborhood, V of x, such that $x \in \mathrm{cl}_S V \subseteq U$. Let $x \in S$ and F be a closed subset disjoint from $\{x\}$. By hypothesis, there exists an open neighborhood, V, of x such that $\mathrm{cl}_S V \subseteq S \backslash F$.

(c $\Rightarrow$ a) We are given that, for every $x \in S$ and closed subset F disjoint from $\{x\}$, there exists an open neighborhood, V of x, such that $\mathrm{cl}_S V \cap F = \varnothing$. Let $x \in S$ and F be a closed subset disjoint from $\{x\}$. By hypothesis, there exists an open neighborhood, V, of x such that $\mathrm{cl}_S V \cap F = \varnothing$. Then V and $S \backslash (\mathrm{cl}_S V)$ are disjoint open sets of containing x and F respectively. So, S is T_3. Since S is Hausdorff, it is T_1, hence, S is regular.

Example 12. Show that any metrizable space is regular.

Solution: Suppose (S, τ_ρ) is a metrizable space whose open sets are generated by a metric, ρ. Metrizable spaces are known to be Hausdorff and so are T_1.

Let $y \in S$ and U be an open neighborhood of y. Then there exists $\varepsilon > 0$ such that the ball, $B_\varepsilon(y)$, (center y and radius ε) is entirely contained in U. Then, if $V = B_{\varepsilon/3}(y)$, $\text{cl}_S V = \{x : \rho(x, y) \leq \varepsilon/3\} \subseteq B_\varepsilon(x) \subseteq U$. By Theorem 9.11, S is regular.

9.5.1 *Basic properties of regular spaces*

In the following theorems, we confirm that "regular" is a hereditary property and that it carries over products, in both directions.

Theorem 9.12 *Subspaces of regular spaces are regular.*

Proof. Let (S, τ_S) be a regular topological space and (T, τ_T) be a subspace of S.

Any point, x, in T is a point in S and so is closed in S. Then $\{x\}$ is closed in T. We must conclude that T is T_1.

Let $\{x\}$ and F be disjoint closed subsets of T. Then there exists a closed subset, F^*, of S such that $F = F^* \cap T$. By hypothesis, there exists disjoint open subsets, U and V, of S which contain x and F^*, respectively. Then the disjoint open subsets, $U \cap T$ and $V \cap T$, of T contain x and F, respectively. We can then conclude that T is regular and so "regular" is a hereditary topological property.

9.5.2 *Semiregular spaces revisited*

Recall that an open set, U, is *regular open* in S, if

$$U = \text{int}_S \text{cl}_S U$$

Also, from Definition 5.15,

> a space (S, τ) is said to be *semiregular* if the set, $\mathscr{Ro}(S)$, of all regular open subsets of S forms a base for the open sets of S.

Although it is not explicitly stated in the original definition 5.15, semiregularity is traditionally defined on spaces which are assumed to be Hausdorff only. We will adopt this tradition here. We will show that the "regular" separation axiom on a space, S, forces semiregularity on S.

Theorem 9.13 *A regular space is semiregular.*

Proof. Let (S, τ) be a regular space. To prove that S is semiregular it will suffice, by definition, to show that $\mathscr{Ro}(S)$ is a base for the open sets of S.

Suppose $u \in S$ and B is an open neighborhood of u. By Theorem 9.11, there exists an open subset, U, of S such that $u \in \mathrm{cl}_S U \subseteq B$. Then $u \in \mathrm{int}_S \mathrm{cl}_S U \subseteq B$. Since $\mathrm{int}_S \mathrm{cl}_S U \in \mathscr{Ro}(S)$, then $\mathscr{Ro}(S)$ is a base for open sets of S, as required.

Theorem 9.14 *A regular space is completely Hausdorff.*

Proof. The proof is straightforward and so is left an exercise.

In Example 10 on page 220, we presented a topological space which is easily verified to be semiregular but is not completely Hausdorff (since the rectangular neighborhoods do not have closures which separate the points (0, 0) and (1, 0). From this example we can conclude that

"semiregular" does not imply "completely Hausdorff."

We then have the followings diagram of implications,

$$\begin{array}{c} \text{completely Hausdorff} \\ \not\Uparrow \\ \text{regular} \Rightarrow \text{semiregular} \Rightarrow \text{Hausdorff} \Rightarrow T_1 \Rightarrow T_0 \\ \Downarrow \\ \text{completely Hausdorff} \end{array}$$

Equivalently,

$$T_3 + T_1 \Rightarrow T_{2\frac{1}{2}} \Rightarrow T_2 \Rightarrow T_1 \Rightarrow T_0$$

We will now examine regular separation properties involving products.

Theorem 9.15 *Let $\{S_i\}_{i\in I}$ be a family of topological spaces and $S = \prod_{i\in I} S_i$ be a corresponding product space. Then S is regular if and only if each factor, S_i, is regular.*

Proof. In Theorem 9.4, it is shown that the T_1-property carries over both from products to factors and from factors to products. It will then suffice to show that this holds true for the T_3 property.

($\Rightarrow$) We are given that $\prod_{i\in I} S_i$ is a T_3-space. We have shown that $\prod_{i\in I} S_i$ contains a homeomorphic copy of each S_i. Since "regular" is a hereditary property, each S_i is T_3 and so is regular.

($\Leftarrow$) We are given that $S = \prod_{i\in I} S_i$ is a product space and each S_i is T_3.

Let $\langle u_i\rangle_{i\in I} \in S$ and W be any open neighborhood of $\langle u_i\rangle_{i\in I}$. Then there exists a base element, $\cap\{\pi_i^{\leftarrow}[U_i] : i \in F_{\text{finite}} \subseteq I\}$, of S such that

$$\langle u_i\rangle_{i\in I} \in \cap\{\pi_i^{\leftarrow}[U_i] : i \in F_{\text{finite}} \subseteq I\} \subseteq W$$

Since each S_i is regular, there exists open S_i-neighborhoods, $\{V_i\}_{i\in F\subseteq I}$, such that $u_i \in V_i \subseteq \mathrm{cl}_{S_i} V_i \subseteq U_i$ for $i \in F_{\text{finite}}$. Then

$$\begin{aligned}\langle u_i\rangle_{i\in I} \in \cap\{\pi_i^{\leftarrow}[V_i] : i \in F\} &\subseteq \mathrm{cl}_S\left[\cap\{\pi_i^{\leftarrow}[V_i] : i \in F\}\right]\\ &= \cap\{\mathrm{cl}_S \pi_i^{\leftarrow}[V_i] : i \in F\}\\ &= \cap\{\pi_i^{\leftarrow}[\mathrm{cl}_{S_i} V_i] : i \in F\}\\ &\subseteq \cap\{\pi_i^{\leftarrow}[U_i] : i \in F\}\\ &\subseteq W\end{aligned}$$

Since

$$\langle u_i\rangle_{i\in I} \in \cap\{\pi_i^{\leftarrow}[V_i] : i \in F\} \subseteq \mathrm{cl}_S\left[\cap\{\pi_i^{\leftarrow}[V_i] : i \in F\}\right] \subseteq W$$

then, by the characterization theorem above, S is T_3 and so is regular.

9.5.3 *A few more examples involving the regular separation axiom*

Example 13. Show that the *Moore plane* is regular. (For a description of the Moore plane, see the example on page 92.)

Solution: Let (S, τ_S) denote the Moore plane. The proof that S is T_1 is straightforward and so is left as an exercise. We now show that S is T_3.

In the Moore plane there are two types of points. Those points, (x, y), whose basic open neighborhoods are of the form $B_\varepsilon(x, y)$ and the points of the form, $(x, 0)$, which have basic open neighborhoods of the form

$$B_\varepsilon(x, y) \cup \{(x, 0)\} \quad \text{where } \varepsilon = y$$

So, we consider these two cases separately.

Let U be an open base neighborhood, $B_\varepsilon(x, y)$, of (x, y), $y \neq 0$. Then

$$(x, y) \in \text{cl}_S B_{\varepsilon/3}(x, y) \subset U$$

Let V be an open base neighborhood of $(x, 0)$ of the form, $B_\varepsilon(x, y) \cup \{(x, 0)\}$, where $\varepsilon = y$. Then

$$(x, 0) \in \text{cl}_S B_{\varepsilon/3}(x, y/3) \cup \{(x, 0)\} \subset V$$

We conclude that the Moore plane is T_3. So, S is regular.

Example 14. Recall from Definition 5.16 that a space is said to be zero-dimensional if it has a base of clopen sets. Show that zero-dimensional spaces which are T_1 are regular spaces.

Solution: Let F be a closed subset of a zero-dimensional T_1-space S. Then S has a base of clopen subsets. If $x \in S \backslash F$, then there is a subset, B, which is both open and closed such that $x \in B \subseteq S \backslash F$. Then B is an open neighborhood of x which is disjoint from the open neighborhood, $S \backslash B$ of F. So, S is regular.

9.6 T_4-Spaces and Normal Spaces

Our final axiom of separation "by open sets" involves, what we will call, "T_4-space". "Normal" spaces will be those T_4-spaces which are also T_1.

Definition 9.16 Let (S, τ_S) be a topological space. We say that S is a *T_4-space* if, for any pair of non-empty disjoint closed subsets, F and W, in S, there exists non-intersecting open subsets, U and V containing F and W, respectively.

Example 15. As our first example of a T_4 space we consider the topological space, $(\mathbb{R}, \tau_u)$, where $\tau_u = \{(x, \infty) : x \in \mathbb{R}\} \cup \{\varnothing, \mathbb{R}\}$. Since the closed subsets of $\mathbb{R}$ are of the form $(-\infty, y]$, disjoint pairs of non-empty closed sets do not exist in $\mathbb{R}$ and so $\mathbb{R}$ is vacuously T_4. Consider the subsets, $F = (-\infty, 5]$ and $\{6\}$. Any open subset U which contains F will contain the point 6. So, $(\mathbb{R}, \tau_u)$ is not T_3. Then, $T_4 \not\Rightarrow T_3$.

We would prefer that spaces satisfying our next level of closed set separation be, at least, regular spaces. We rectify this situation by defining *normal spaces* as ones respecting the rule,

$$\text{“normal”} \Leftrightarrow T_4 + T_1$$

Let us verify why this will work. If S is $T_1 + T_4$ and F is a non-empty closed subset of S, for $\{u\} \not\subset F$, we would be guaranteed the existence of disjoint open subsets, U an V, which, respectively, contain the closed subsets $\{u\}$ and F, hence, S would be both T_1 and T_3; by definition, S would be regular. So we would have the desired chain of implications,

$$\text{normal} \Rightarrow \text{regular} \Rightarrow \text{completely Hausdorff} \Rightarrow \text{Hausdorff} \Rightarrow T_1 \Rightarrow T_0$$

We will then proceed with this definition of a *normal topological space.*

Definition 9.17 If (S, τ_s) is both T_1 and T_4 then we will say that S is a *normal space.*[5]

So,

$$T_4 + T_1 \quad \Rightarrow \quad T_3 + T_1 \quad \Rightarrow \quad T_{2\frac{1}{2}} \quad \Rightarrow \quad T_2 \quad \Rightarrow \quad T_1 \quad \Rightarrow \quad T_0$$

[5]Readers may find that, in the literature, many authors invert the definitions of “T_4” and “normal” as well as for “T_3” and “regular”. So, when consulting other texts we caution the reader to verify carefully which version the particular author has a preference for. Some writers may feel strongly about their chosen version. But as long as one is aware of such possible discrepancies, most readers will easily adapt to a particular version, as long as the author is consistent with its use throughout the body of the text.

Before we proceed with various examples, we provide a few important characterizations of those spaces we refer to as being "normal".

Theorem 9.18 *Let (S, τ_s) be a topological space. The following statements about S are equivalent.*

(a) *The space S is normal.*
(b) *For every proper closed subset, F in S, and proper open set, U containing F, there exists an open set, V containing F, such that $F \in cl_S V \subseteq U$.*
(c) *For every pair of disjoint closed sets, F and W in S, there exists and open subset U such that $F \subseteq U$ and $cl_S U \cap W = \varnothing$.*
(d) *For every pair of disjoint closed sets, F and W in S, there exists disjoint open subsets U and V containing F and W, respectively, whose closures, $cl_S U$ and $cl_S V$, do not intersect.*

Proof. We are given that (S, τ_S) be a topological space.

($a \Rightarrow b$) Suppose S is normal. Let F be a proper closed subset of S and U be a proper open neighborhood of F. Then $S \backslash U$ is a closed subset of S which does not meet F. Since S is normal, there exists disjoint open subsets V and W containing F and $S \backslash U$, respectively. Then $\text{cl}_S V \subseteq S \backslash W$, where $\text{cl}_S V \cap S \backslash U = \varnothing$. Hence, $F \subseteq \text{cl}_S V \subseteq U$.

($b \Rightarrow c$) Suppose that, for every proper closed subset, F in S, and proper open set, U containing F, there exists an open set, V containing F, such that $F \in \text{cl}_S V \subseteq U$. Let F and W be a pair of disjoint closed sets in S. Then $S \backslash W$ is an open subset which contains F. By hypothesis, there exists an open set, V containing F, such that $F \in \text{cl}_S V \subseteq S \backslash W$. Then $\text{cl}_S V \cap W = \varnothing$.

($c \Rightarrow d$) Suppose that for every pair of disjoint closed sets, F and W in S, there exists and open subset U such that $F \subseteq U$ and $\text{cl}_S U \cap W = \varnothing$. Let F and W be a pair of disjoint closed sets in S. By hypothesis, there exists and open subset U such that $F \subseteq U$ and $\text{cl}_S U \cap W = \varnothing$. Again by hypothesis, there exists and open subset V such that $W \subseteq V$ and $\text{cl}_S V \cap \text{cl}_S U = \varnothing$.

($d \Rightarrow a$) Suppose that for every pair of disjoint closed sets, F and W in S, there exists disjoint open subsets U and V containing F and W, respectively, whose closures, $\text{cl}_S U$ and $\text{cl}_S V$, do not intersect. Then there exists disjoint open subsets U and V containing F and W, respectively, such that U and V, do not intersect. So, S is normal.

It is useful to know that the class of all normal spaces contains the class of all metrizable spaces.

Theorem 9.19 *Metrizable topological spaces are normal spaces.*

Proof. Metrizable spaces are Hausdorff, hence T_1. Let ρ be a metric on S.

We now show that S is T_4. Let F and W be disjoint closed subsets of S. If $x \in F$ and $y \in W$ there exists α_x and β_y such that $B_{\alpha_x}(x) \cap W = \varnothing$ and $B_{\beta_y}(y) \cap F = \varnothing$. Then

$$U = \cup\{B_{\alpha_x/3}(x) : x \in F\}$$
$$V = \cup\{B_{\beta_y/3}(y) : y \in W\}$$

are open sets containing F and W, respectively.

We claim that $U \cap V = \varnothing$. Suppose not. Suppose $q \in U \cap V$. Then there exists some $a \in F$ and $b \in W$ such that $q \in B_{\alpha_a/3}(a) \cap B_{\beta_b/3}(b)$. Suppose, WLOG, $\alpha_a \geq \beta_b$. Then

$$\begin{aligned} \rho(a,b) &\leq \rho(a,q) + \rho(b,q) \\ &< \alpha_a/3 + \beta_b/3 \\ &\leq 2\alpha_a/3 \\ &< \alpha_a \end{aligned}$$

But α_a was chosen so that $B_{\alpha_a}(a) \cap W = \varnothing$. Contradiction. So, $U \cap V = \varnothing$, as claimed. So, S is T_4. We conclude that S is normal, as required.

Since, for every $n \geq 1$, $\mathbb{R}^n$ equipped with the usual topology is metrizable, then...

for every natural number n, $\mathbb{R}^n$ is a normal topological space.

In Example 5 on page 162, we showed that the countably infinite product of a metric space is metrizable. Then $S = \prod_{n\in\mathbb{N}} \mathbb{R}$ is metrizable; so, S is normal.

Unlike regular spaces normal spaces are not hereditary. But they are closed-hereditary (that is, closed subsets inherit the normal property). Also, unlike regular spaces, the normal property does not generally carry over from factors to their product space, even when the product is finite. However, if the product space is normal then so will be each of its factors.

Theorem 9.20 *Let (S, τ_S) be a normal topological space.*

(a) *If the function $f : S \to f[S] = T$ is a closed continuous function onto the space (T, τ_T) then T is normal.*
(b) *If T is a closed subspace of S then T is normal.*
(c) *If $S = \prod_{i\in I} S_i$ is a normal product space of topological spaces $\{S_i : i \in I\}$, then each of its factors, S_i, is normal.*

Proof. Let (S, τ_S) be a normal topological space.

(a) Suppose $f : S \to f[S] = T$ is a closed continuous function. Since S is T_1 and f is closed, then T is T_1.

Let F and W be two disjoint closed subsets of T. Continuity of f guarantees that $f^{\leftarrow}[F]$ and $f^{\leftarrow}[W]$ are disjoint closed subsets of S. Then there exists disjoint open sets U and V containing F and W, respectively. Let $A = T\backslash f[S\backslash U]$ and $B = T\backslash f[S\backslash V]$. Since f is closed, A and B are disjoint open neighborhoods of F and W, respectively. It quickly follows that T is T_4. So, T is normal.
(b) Let T be a closed subspace of S. Then T is T_1. If F and W are disjoint closed subsets of the closed subspace T, then they are disjoint closed subsets of S. It quickly follows that T is T_4. So, T is normal.
(c) We are given that $S = \prod_{i\in I} S_i$ is normal. There exists a homeomorphism, $h : S_i \to S$, which embeds each S_i in S (Theorem 7.14). Then $S\backslash h[S_i]$ is easily verified to be open in S and so $h[S_i]$ is a closed subspace of S. Since closed subspaces of normal spaces are normal, then $h[S_i]$ is normal and so, S_i is normal.

9.6.1 *The ordinal spaces* $[0, \omega_\alpha]$ *and* $[0, \omega_\alpha)$ *are normal*

We now investigate whether ordinal spaces such as $[0, \omega_\alpha]$ or $[0, \omega_\alpha)$ satisfy the normal property. The ordinal space has a topology, τ_ω, which has an open base whose elements are of the form $(\alpha, \beta]$. Also, since the ordinal space is well-ordered, non-empty sets have an ordinal which is a supremum (infimum) of that set.

Example 16. Let ω_α be any ordinal number and let $S = [0, \omega_\alpha]$. Show that the ordinal space (S, τ_ω) is normal, independent of the value ω_α. The proof that shows $[0, \omega_\alpha)$ that is normal is practically the same. (See the reference to "ordinal space" at Definition 5.14 on page 110.)

Solution: Verifying that S is T_1 is left to the reader. Let F and W be disjoint non-empty closed subsets of S. If $\mu \in F$ and $\mu \neq 0$ let

$$S_\mu = \{x \in W : x < \mu\} = [0, \mu) \cap W$$

If S_μ is empty, let $x_\mu = 0$. Then $(x_\mu, \mu]$ is an open neighborhood of $\mu \in F$.

Suppose S_μ is not empty. In this case, let $x_\mu = \sup S_\mu$.[6] Since W is closed, then $x_\mu \in W$. Then the open set, $(x_\mu, \mu]$ is a basic open neighborhood of $\mu \in F$ which does not intersect W.

This is repeated for each element, μ, of F, to obtain, the open neighborhood

$$U_F = \cup\{(x_\mu, \mu] : \mu \in F\}$$

of F. The set U_F clearly contains no elements of W.

By applying precisely the same procedure we obtain an open neighborhood

$$U_W = \cup\{(x_\alpha, \alpha] : \alpha \in W\}$$

of W, which contains no elements of F.

---

[6]The existence of the supremum can be justified by the fact that the class of all ordinals is well-ordered. See the related statements from set theory in the appendix.

We need now only verify that $U_W \cap U_F = \varnothing$. If so, then S is T_4 and hence is normal. The standard approach is to suppose $U_F \cap U_W \neq \varnothing$.

Then there exists some $\mu \in F$ and some $\alpha \in W$, such that $(x_\mu, \mu] \cap (x_\alpha, \alpha]$ is non-empty. Without loss of generality, suppose $\alpha < \mu$. But, this can't be, since this would force $\alpha \in (x_\mu, \mu]$, when we know that $(x_\mu, \mu] \cap W = \varnothing$.

So, $U_W \cap U_F = \varnothing$. We conclude that S is T_4 hence is normal. Proceed similarly for $S = [0, \omega_\alpha)$.

Example 17. Show that the *deleted Tychonoff plank*

$$T = [0, \omega_1] \times [0, \omega_0] \setminus \{(\omega_1, \omega_0)\}$$

is not normal. (See page 167 for a description of this space.)

Solution: Recall that T is topologized as a product space. Consider $U = [0, \omega_1)$ viewed as a closed interval subset of $[0, \omega_1)$ itself and consider $V = \{\omega_0\}$, a singleton set. Then the subset $F = U \times V$ is a closed subset (the top edge) of T with respect to the subspace topology (since its complement is easily seen to be open). We will also consider the disjoint closed subset,

$$K = \{\omega_1\} \times [0, \omega_0) \subseteq T$$

If T is normal then we should be able to separate F from K with a pair of non-intersecting open sets.

Suppose M is an open set in T which contains K.

A basic neighborhood of u in K which would be entirely contained in M would be of the form

$$W_u = (\alpha_u, \gamma_u] \times (\beta_u, \mu_u]$$

If $W_u = (\alpha_u, \gamma_u] \times (\beta_u, \mu_u] \subseteq M$ for each $u \in K$

$$K \subseteq M = \cup\{W_u : u \in K\}$$

We claim that $M \cap F \neq \varnothing$. Since there can be at most countably many u's in K,

$$\sup\{\alpha_u : u \in K\} = \rho < \omega_1$$

since the sup of a countable set cannot reach the uncountable ordinal ω_1. So $(\rho, \omega_1) \cap M \neq \varnothing$. So, no open neighborhood M of K can miss F, as claimed.

So, T is not normal.

Example 18. Show that the subspace of a normal space need not be normal.

Solution: We showed in the examples above that the Tychonoff plank is normal but that its subspace, the Deleted Tychonoff plank, is not.

Concept Review

1. What is the formal definition of a T_0-space?
2. Is there a topological space which is not T_0? If so provide an example.
3. Show that an infinite second countable T_0-space cannot have a cardinality larger than $2^{\aleph_0}$.
4. What is the formal definition of a T_1-space, emphasizing in the process how it is different from a T_0-space.
5. Give a characterization of a T_1-space.
6. Does a subspace of a T_1-space necessarily inherit the T_1 property?
7. Does a closed function f carry the T_1 property of its domain, S, to its range $f[S]$.
8. Does the T_1 property of all its factors, S_i, carry over to the product space, $\prod_{i \in I} S_i$?
9. If $\prod_{i \in I} S_i$ possesses the T_1 property does each of the factors necessarily possess the T_1 property?
10. Define the Hausdorff property (T_2 property) on a topological space.
11. Provide three different characterizations of the Hausdorff topological property.
12. Is the Hausdorff property carried over from the domain of a continuous function, $f : S \to T$, to its range $f[S]$? What if the function f was just closed? How about if f was both one-to-one and closed?

13. If all factors of a product space are Hausdorff, is the product space in question necessarily Hausdorff?
14. If $f : S \to T$ is a function and $G = \{(u, f(u)) : u \in S\}$ is the graph of f in $S \times T$, give a condition on S or T that will guarantee that the graph, G, is closed in $S \times T$.
15. Define both a T_3-space and a regular space.
16. Does T_3 imply Hausdorff? Does "regular" imply Hausdorff?
17. Give two characterizations of "regular space".
18. Are metrizable spaces necessarily "regular"? Are regular spaces always metrizable?
19. Is "regular" a hereditary property?
20. Does the "regular" property always carry over from a product to its factors? What about, from the factors to the product?
21. Define both a T_4-space and a normal space.
22. Does T_4 imply "regular"? Does "normal" imply "regular"?
23. Give three characterizations of "normal space".
24. Are metrizable spaces necessarily "normal"?
25. What kind of function will always carry a normal property from its domain to its range?
26. Is the normal property hereditary? If not, what kind of subspace will inherit the normal property from its superset?
27. Is the product of normal spaces necessarily normal?
28. Are open ordinal spaces $[0, \omega_\alpha)$ normal spaces?
29. Is the deleted Tychonoff plank normal?

Exercises

1. Let (S, τ_S) and (T, τ_T) be two topological spaces and $f : S \to T$ and $g : T \to S$ be two continuous functions satisfying the property: $(g \circ f)(x) = x$ for all x in S.

   (a) Show that if T is Hausdorff, then S must also be Hausdorff.
   (b) Show that $f[S]$ is closed subset of T.

2. Suppose $S = \{x_0, x_1, \ldots, x_n\}$ is a finite set equipped with a topology, τ_S, which makes of (S, τ_S) a Hausdorff space. Describe all such topologies on S.

3. Let (S, τ_S) and (T, τ_T) be topological spaces where S is Hausdorff and contains a dense subset D. Suppose $f : S \to T$ is a continuous function such that $f|_S : D \to f[D]$ is a homeomorphism. Show that $f[S \backslash D] \subseteq T \backslash f[D]$.
4. Suppose (S, τ_S) is regular and (T, τ_T) is a topological space. Suppose the function $f : S \to T$ is a continuous function mapping S onto T. Show that if f is both an open and closed function as well, then T is Hausdorff.
5. Suppose (S, τ_S) is a topological space and K is a non-empty subset of S. Suppose $\tau_S = \{U \subseteq S : K \subseteq U\} \cup \{\varnothing, S\}$. Verify that this is indeed a valid topology on S. If so, are there any other properties that must be satisfied if we want S to be regular?
6. Suppose (S, τ_S) is a finite regular topological space. Is (S, τ_S) necessarily normal? Why?
7. Suppose (S, τ_S) is a regular topological space and K is an infinite subset of S. Show that there exists an infinite family of open subsets, $\{U_i \subseteq S : i \in \mathbb{N},\ U_i \cap K \neq \varnothing\}$, such that, if $i \neq j$, then $\mathrm{cl}_S U_i \cap \mathrm{cl}_S U_j = \varnothing$.
8. Suppose (S, τ_S) is a normal topological space and that R is an equivalence relation on S. Let (S, τ_θ) be the quotient space induced by the natural map $\theta : S \to S/R$. Show that, if θ is both open and closed, then (S, τ_θ) is a normal topological space.
9. Let $S = [0, 1]$ and τ denote the usual topology on S. If $T = \mathbb{Q} \cap [0, 1]$, let τ_s be the topology on S which is generated by the subbase, $\mathscr{S} = \tau \cup \mathscr{P}(T)$. Show that (S, τ_s) is a normal topological space.

Chapter 10

Separation with Continuous Functions

Abstract
In this section we will investigate another method for separating closed sets. A function, $f : S \to [0, 1]$, which continuously maps a topological space, S, onto the closed interval $[0, 1]$ in such a way that f has different constant values on a pair of disjoint closed subsets can be used to separate sets.

10.1 Normal Spaces Revisited

Suppose we are given a space (S, τ_S) and a continuous function $f : S \to [0, 1]$. If, for a pair of disjoint closed subsets A and B of S, $A \subseteq f^{\leftarrow}[\{0\}]$ and $B \subseteq f^{\leftarrow}[\{1\}]$ then we will say that

$$f \textit{ separates } A \textit{ and } B.$$

Suppose that, for a pair of disjoint closed subsets, A and B, there exists a *continuous* function, $f : S \to [0, 1]$ which separates A and B, then we can easily generate a pair of disjoint open sets, $U = f^{\leftarrow}[[0, 1/2)]$ and $V = f^{\leftarrow}[(1/2, 1]]$, which "separate" A and B. If this holds true for every pair of disjoint closed sets we could conclude that such a space, S, is a normal topological space. This provides us with another way of recognizing a normal space.

On the other hand, suppose a space, S, is known to be normal. If A and B are disjoint closed sets in S, are we guaranteed the existence of a "separating function", $f : S \to [0, 1]$, which will separate A and B? The answer to this question is not obvious. Proving the existence of a separating real-valued continuous function on S which serves this particular purpose is not a trivial task. In fact, it can be quite hard. The statement,

> For each pair of disjoint closed subsets, A and B, of S there exists a separating function for A and B if and only if S is normal.

is titled *Urysohn's lemma.*[1]

Its proof invokes the previously proven characterization of the normal property:

> The topological space S is normal if and only if, for every non-empty closed set F and open neighborhood U of F, there exists another open neighborhood V of F such that $\mathrm{cl}_S V \subseteq U$.

Constructing the required continuous function on a normal topological space just from this characterization of normality requires quite a bit of mathematical prowess. In this chapter, we will carefully work our way through this proof.

10.2 Urysohn's Lemma

The proof of Urysohn's lemma will be presented in two parts. We will first discuss a method for constructing a function on a topological space and then prove continuity of this function. The proof of Urysohn's lemma will follow.

We will begin by highlighting a few properties of a countable union of subsets, V_n, of $[0, 1]$. For each $n \in \mathbb{N}\backslash\{0\}$, we define a subset V_n as

$$V_n = \left\{\frac{m}{2^n} : m = 1, 2, 3, \ldots, 2^n - 1\right\}$$

[1]Named after the Russian mathematician Pavel Urysohn (1898–1924). Urysohn was a gifted mathematician destined to make great contributions to mathematics. He drowned while on holiday at the young age of 26 years old. The proof of Urysohn's lemma is an illustration of his deep insight in a mathematical problem.

For example,

$$V_1 = \{1/2^1\}$$

$$V_2 = \left\{\frac{1}{2^2}, \frac{2}{2^2}, \frac{3}{2^2}\right\}$$

$$V_3 = \left\{\frac{1}{2^3}, \frac{2}{2^3}, \frac{3}{2^3}, \frac{4}{2^3}, \frac{5}{2^3}, \frac{6}{2^3}, \frac{7}{2^3}\right\}$$

$$\vdots$$

$$V_{100} = \left\{\frac{1}{2^{100}}, \frac{2}{2^{100}}, \frac{3}{2^{100}}, \ldots, \frac{2^{100}-1}{2^{100}}\right\}$$

$$\vdots$$

Note that, for each n, $V_n \subset V_{n+1}$. So, every V_n contains the elements of its predecessors and some. Let

$$J = \cup\{V_n : n = 1, 2, 3, \ldots\}$$

The set, J, is referred to as the dyadic rationals in $[0,1]$. Then J is countably infinite, dense[2] and *linearly ordered* in $[0,1]$. That is, $[0,1]\backslash J$ contains no intervals and, even though J *contains neither* 0 *nor* 1, we have $0 = \inf(J)$ and $1 = \sup(J)$. So, J can be used as a countably infinite indexing set.

With this definition in mind, we are now set to formally state and prove *Urysohn's lemma.* Even though we presented it as a theorem it is generally referred to, in the literature, as a lemma.

Theorem 10.1 Urysohn's lemma.[3] *Let (S, τ_S) be a T_1 topological space. The topological space (S, τ_S) is* normal *if and only if given a pair of disjoint non-empty closed sets, F and W, in S there exists a continuous function $f : S \to [0,1]$ such that, $F \subseteq f^{\leftarrow}[\{0\}]$ and $W \subseteq f^{\leftarrow}[\{1\}]$.*

---

[2]To see "dense": For any interval (a, b), $\frac{1}{2^n} < b - a$, for some n. Then, $1 < b2^n - a2^n$ implies $a2^n + 1 < b2^n$, hence there must exist an integer, m, such that $a2^n < m < b2^n \Rightarrow a < m/2^n < b$.

[3]This should not be confused with Urysohn's Extension theorem.

Proof. Let (S, τ_s) be a T_1 topological space.

($\Leftarrow$) We are given a pair of disjoint closed sets, F and W. Also, we are given that there exists a continuous function $f : S \to [0, 1]$ such that $F \subseteq f^{\leftarrow}[\{0\}]$ and $W \subseteq f^{\leftarrow}[\{1\}]$. Then $f^{\leftarrow}[\,[0, 1/3)\,]$ and $f^{\leftarrow}[\,(2/3, 1]\,]$ are disjoint open sets containing F and W, respectively. It follows that S is normal.

($\Rightarrow$) Suppose S is normal and F and W are disjoint non-empty closed sets in S. Since $S\backslash W$ is an open neighborhood of F, there exists and open subset $U_{1/2}$ such that $F \subseteq U_{1/2} \subseteq \text{cl}_S U_{1/2} \subseteq S\backslash W$. Normality of S allows us to repeat this step with dyadic rationals, $\{\frac{1}{2^2}, \frac{2}{2^2}, \frac{3}{2^2}\}$ as index set, to obtain the chain

$$F \subseteq U_{1/4} \subseteq \text{cl}_S U_{1/4} \subseteq U_{2/4} \subseteq \text{cl}_S U_{2/4} \subseteq U_{3/4} \subseteq \text{cl}_S U_{3/4} \subseteq S\backslash W$$

More generally, for each n, let $V_n = \{U_{\frac{i}{2^n}} : i = 1 \text{ to } 2^n - 1\}$ where

$$F \subset U_{\frac{1}{2^n}} \subseteq \text{cl}_S U_{\frac{1}{2^n}} \subset \cdots \subset U_{\frac{i-1}{2^n}} \subseteq \text{cl}_S U_{\frac{i-1}{2^n}} \subset U_{\frac{i}{2^n}} \subset \cdots \subset U_{\frac{2^n-1}{2^n}} \subset S\backslash W$$

From V_n we inductively construct V_{n+1} with the same properties and where $V_n \subset V_{n+1}$. Let $V = \cup\{V_n : n = 1, 2, 3, \ldots\}$. We obtain the countably infinite set, $\{U_i : i \in J\}$ where, for all $j < i$ in J,

$$F \subset U_j \subseteq \text{cl}_S U_j \subseteq U_i \subseteq \text{cl}_S U_i \subset S\backslash W$$
$$W \subseteq S\backslash \text{cl}_S U_i \subseteq S\backslash U_i \subset S\backslash U_j \subseteq S\backslash \text{cl}_S U_j \subset S\backslash F$$

Also, for all i, $W \subset S\backslash \text{cl}_S U_i$.
Define

$$U_0 = \cap\{U_i : i \in J\} = \cap\{\text{cl}_S U_i : i \in J\}$$
$$U_1 = \cap\{S \setminus \text{cl}_S U_i : i \in J\} = \cap\{S \setminus U_i : i \in J\}$$

where neither U_0 nor U_1 is empty. Then

$$S\backslash U_1 = S\backslash\cap\{S \setminus \text{cl}_S U_i : i \in J\} = \cup\{\text{cl}_S U_i : i \in J\}$$

So, we have the strictly increasing chains of inclusions,

$$F \subseteq U_0 \subset \cdots \subset U_i \subseteq \text{cl}_S U_i \subset \cdots \subseteq \cup\{U_i\}_{i\in J} \subseteq \cup\{\text{cl}_S U_i\}_{i\in J}$$
$$= S\backslash U_1 \subset S\backslash W$$

$$W \subseteq U_1 = \cap\{S\backslash \text{cl}_S U_i\}_{i\in J} \subset \cdots \subset S\backslash \text{cl}_S U_i \subseteq S\backslash U_i \subset \cdots$$
$$\subseteq S\backslash U_0 \subseteq S\backslash F$$

Note that for a point x in S

$$x \in U_0 \quad \Leftrightarrow \quad x \in U_i \text{ for all } i \;\Leftrightarrow\; x \in \text{cl}_S U_i \text{ for all } i$$
$$x \in U_1 \quad \Leftrightarrow \quad x \in S\backslash \text{cl}_S U_i \text{ for all } i \in J \;\Leftrightarrow\; x \in S\backslash U_i \text{ for all } i$$

So,

$$\begin{aligned} x \notin U_0 \cup U_1 &\Leftrightarrow x \notin U_0 \text{ and } x \notin U_1 \\ &\Leftrightarrow x \notin \text{cl}_S U_t \text{ and } x \notin S\backslash U_k \text{ for some } k, t \in J \\ &\Leftrightarrow x \in S\backslash \text{cl}_S U_t \text{ and } x \in U_k, \text{ for some } k, t \in J \text{ (*)} \\ &\Leftrightarrow x \in S\backslash \text{cl}_S U_t \cap U_k, \text{ for some } k, t \in J \text{ (**)} \end{aligned}$$

So, if $x \notin U_0 \cup U_1$ $x \in S\backslash \text{cl}_S U_t \cap U_k$, an open neighborhood of x in S.

For any given $x \in S$, let

$$J_x = \{i \in J : x \in U_i\}$$
$$J_x^* = \{i \in J : x \in S\backslash U_i\}$$

a proper subset of [0, 1]. See that $\text{glb}J_x = \text{lub}J_x^*$.

We will define a function, $f : S \to [0, 1]$, as follows:

$$f(x) = \begin{cases} 0 & \text{If } x \in U_i \text{ for all } i \in J \\ \text{glb}J_x = \text{lub}J_x^* & \text{otherwise} \end{cases}$$

See that

$$\begin{aligned} x \in U_0 &\Leftrightarrow x \in \cap\{U_i : i \in J\} \\ &\Leftrightarrow x \in U_i \text{ for all } i \\ &\Leftrightarrow \text{glb}\{i : x \in U_i\} = \text{glb}(0, 1] = 0 = f(x) \end{aligned}$$

So, $f^{\leftarrow}(0) = U_0$

Also, see that

$$\begin{aligned} x \in U_1 \quad &\Leftrightarrow \quad x \in \cap\{S\backslash U_i : i \in J\} \\ &\Leftrightarrow \quad x \in S\backslash U_i \text{ for all } i \\ &\Leftrightarrow \quad \text{lub}\{i : x \in S\backslash U_i\} = \text{lub}[0,1) = 1 = \text{lub}J_x^* = f(x) \end{aligned}$$

So, $f^{\leftarrow}(1) = U_1$. We then have,

$$F \subseteq U_0 = f^{\leftarrow}[\{0\}]$$
$$W \subseteq U_1 = f^{\leftarrow}[\{1\}]$$

Claim 1: For each $k \in J$,

$$f[\text{cl}_S U_k] \subseteq [0, k]$$

Proof of claim: Let $x \in \text{cl}_S U_k$, then for any $i > k$, $x \in U_i$ (since $\text{cl}_S U_k \subset U_i$). So, J_x contains $(k, 1]$. So, $\text{glb}J_x \leq k$. That is, $f(x) \leq k$. So, $f[\text{cl}_S U_k] \subseteq [0, k]$, as claimed.

So, if $f(x) > k$ then $x \notin \text{cl}_S U_k$.

Claim 2: For each $k \in J$,

$$f[S\backslash U_k] \subseteq [k, 1]$$

Proof of claim: Let $x \in S\backslash U_k$, then for any $i < k$, $x \in S\backslash U_i$ (since $S\backslash U_i \subset S\backslash U_k$). So, J_x^* contains $[0, k)$. Then $\text{lub}J_x^* \geq k$. That is $f(x) \geq k$. So, $f[S\backslash U_k] \subseteq [k, 1]$, as claimed. So, if $x \notin U_k$, then $f(x) > k$.

We now prove that $f : S \to [0, 1]$ is a continuous function on S.

Let $q \in S$ and $a, b \in [0, 1]$ such that $f(q) \in (a, b)$. Then there exist $r, t \in J$ such that

$$a < r < f(q) < t < b$$

Let $W = U_t \backslash \text{cl}_S U_r$.

Claim 3: W is an open neighborhood of q.

Proof of claim: We are given $r < f(q) < t$. Since, by Claim 2, if $q \notin U_t$, then $f(q) \geq t$, we must then have

$$q \in U_t$$

Since, by Claim 1, if $f(q) \geq r$, then q cannot belong to $\text{cl}_S U_r$. Then $q \in U_t \backslash \text{cl}_S U_r$. So, W is an open neighborhood of q in S, as claimed.

Claim 4: That $f[W] \subseteq (a, b)$.

Proof of claim: Suppose $x \in U_t \backslash \text{cl}_S U_r$. Then $x \in \text{cl}_S U_t$. So, by Claim 1, $f(x) \in [0, t]$. Since $x \notin \text{cl}_S U_r$, by Claim 2, $f(x) \in [r, 1]$. So, $f(x) \in [r, t] \subseteq (a, b)$.

So, $f[W] \subset (a, b)$, as claimed.

So,

$$f(q) \in f[W] \subseteq [r, t] \subset (a, b)$$

We conclude f is continuous on S.

Since $f[F] \subseteq f[U_0] = \{0\}$ and $f[W] \subseteq f[U_1] = \{1\}$, we are done.

Urysohn's lemma provides another way of recognizing normal topological spaces.

Definition 10.2 Given a pair of disjoint closed sets, F and W, in a space S, we will refer to the continuous function, $f : S \to [0, 1]$, introduced in Urysohn's proof, as a

Urysohn separating function for F and W.

We also say that this function is one which *separates the disjoint closed sets F and W*. For future reference, we define these formally.

Definition 10.3 Suppose A and B are disjoint non-empty subsets of a space S. We will say that A and B are *completely separated in S* if and only if there is a continuous function $f : S \to [0, 1]$ such that $A \subseteq f^{\leftarrow}(0)$ and $B \subseteq f^{\leftarrow}(1)$. In such a case, we say the function, f, (completely) *separates A and B*.[4]

---

[4]Compare this definition with a similar one provided in Definition 7.16, in another context, one where we speak of a family of continuous functions, not necessarily real-valued. Here we speak of a single real-valued continuous function. The reader is however encouraged to verify that when the family is one of continuous real-valued functions then the two definitions are equivalent (since $\mathbb{R}$ is completely regular).

We now include the following useful fact.

Fact: Given a real-valued continuous function, $f : S \to \mathbb{R}$, we define a zero-set, $Z(f)$, of f as

$$Z(f) = \{x \in S : f(x) = 0\}$$

Two sets in S are completely separated if and only if they are contained in disjoint zero-sets in S.

Proof of this fact: For($\Leftarrow$), suppose $A \subseteq Z(f)$ and $B \subseteq Z(g)$ where $Z(f)$ and $Z(g)$ are disjoint zero-sets in S. Then

$$h(x) = \frac{|f(x)|}{|f(x)| + |g(x)|}$$

is a well-defined continuous function on S. See that $h[A] = \{0\}$ and $h[B] = \{1\}$. So, A and B are (completely) separated.

For ($\Rightarrow$), suppose A and B are *completely* separated. Then there exists a continuous real-valued function, $f : S \to [0, 1]$, such that $f[A] = \{0\}$ and $f[B] = \{1\}$. Then $A \subseteq Z(f)$ and $B \subseteq Z(f-1)$, $Z(f)$ and $Z(f-1)$ are disjoint zero-sets of the functions, f and $f-1$, respectively.

We will say more on zero-sets further down in this chapter.

10.3 Completely Regular Spaces

Note that, if there exists a Urysohn separating function for each pair of disjoint closed subsets in a normal space, then we can find a Urysohn function that can separate a point from a closed subset or separate any pair of points in any T_1 space. But there is nothing in what we have seen that allows us to conclude that a non-normal regular or Hausdorff space can generate its own Urysohn separating function like a normal space can. In fact, as we shall soon see, Hausdorff and regular spaces cannot, in general, produce, by themselves, such a function. If we are presented with a regular space in which it is shown that, for any point and disjoint closed set, there exists a Urysohn separating function, then this topological space belongs to a separate class of topological spaces called "completely regular spaces". We formally define this class.

Definition 10.4 Suppose (S, τ_s) is a T_1-space. If, for any given point x and a non-empty closed subset F disjoint from $\{x\}$, there exists a continuous function, $f : S \to [0, 1]$, such that $\{x\} \subseteq f^{\leftarrow}\{0\}$ and $F \subseteq f^{\leftarrow}\{1\}$, then the space, S, is called a *completely regular space*, or a *Tychonoff space*[5] A completely regular space is often referred to as a $T_{3\frac{1}{2}}$*-space.*

We add a few remarks concerning this definition. In this text, our definition emphasizes that only T_1 spaces are completely regular. (Some authors may not require the T_1 condition.) Obviously, since a Urysohn separating function separates disjoint closed sets of a normal space, and since a regular space is T_1, then

$$\text{normal} \Rightarrow \text{completely regular}$$

Also, if S is completely regular, there exists a Urysohn separating function, f, such that $\{x\} \subseteq f^{\leftarrow}[\{0\}]$ and $F \subseteq f^{\leftarrow}[\{1\}]$. It follows that $\{x\} \subseteq f^{\leftarrow}[[0, 1/3)]$ and $F \subseteq f^{\leftarrow}[(2/3, 1]]$ where $[0, 1/3)$ and $(2/3, 1]$ are open in $[0, 1]$. Then (see implication chart on page 227),

$$\text{completely regular} \Rightarrow \text{regular} \Rightarrow \text{semiregular} \Rightarrow \text{Hausdorff}$$

$$\text{regular} \Rightarrow \text{completely Hausdorff} \Rightarrow \text{Hausdorff}$$

There are, however, regular spaces which are not completely regular. The standard example of a regular non-completely regular space is called the *Tychonoff corkscrew space.*[6] This space is rather involved and lengthy to describe. So, we will not describe it in this text. The *Tychonoff corksrew*, justifies,

$$\text{regular} \not\Rightarrow \text{completely regular}$$

Also, there are completely regular spaces which are *not* normal. The Moore plane (Niemytzki's plane) is such an example. At the

---

[5]Named after the Soviet and Russian mathematician Andreï Tychonoff (1906–1993).

[6]Interested readers will easily find a description of the Tychonoff corkscrew online.

end of this section (on page 263), we prove that the "Moore plane is completely regular but not normal". From which we deduce

$$\text{completely regular} \not\Rightarrow \text{normal}$$

This means, a $T_{3\frac{1}{2}}$-space is strictly in between T_3 and T_4 (which partially explains the tongue-in-cheek terminology, "$T_{3\frac{1}{2}}$-space", used by some authors).

The description of the *completely regular property* does not simply refer to the separation of closed sets by open sets. It is defined in terms of the existence of a continuous function. If we want to refer to the "completely regular property" as a "topological property" (or topological invariant) we should prove that it is one first. Fortunately, the proof is fairly straightforward.

"Completely regular" is a topological property: Suppose S is completely regular and $h : S \to T$ is a homeomorphism mapping S onto the space T. Suppose the singleton set, $\{x\}$, and the closed subset, F, are disjoint in T. Then there exists a function $f : S \to [0,1]$ such that $h^{\leftarrow}[\{x\}] \subseteq f^{\leftarrow}[\{0\}]$ and $h^{\leftarrow}[F] \subseteq f^{\leftarrow}[\{1\}]$. Let $g = f \circ h^{\leftarrow}$. Then $g[\{x\}] = \{0\}$ and $g[F] = \{1\}$. Then T is also completely regular. So, "completely regular" is indeed a topological property, as expected.

Readers who deal mostly with metrizable topological spaces will be glad to know that all such spaces are completely regular. We prove this as follows.

Theorem 10.5 *If (S, τ) is a metrizable space then it is completely regular.*

Proof. We are given that (S, τ) is metrizable. Then there is a metric, ρ, such that (S, ρ) and (S, τ) are equivalent topological spaces. Let F be a non-empty closed subset of S and x be a point in $S \backslash F$. We are required to construct a continuous real-valued function which separates F from x.

Consider the function, $d_F : S \to \mathbb{R}$ defined as

$$d_F(u) = \rho(u, F) = \inf \{\rho(u, y) : y \in F\}$$

The real-valued function, $d_F(u)$, numerically provides a reasonable measure of the smallest distance which separates u from F. **Claim:** That the function, d_F, is a continuous function on S.

Proof of claim: Let $u \in S$. Let $\varepsilon > 0$ and $\delta = \varepsilon/3$. It suffices to show that $\rho(x,u) < \delta$ implies $|d_F(x) - d_F(u)| < \varepsilon$.

Then with the variables, x, u and y, we write the two possible triangle inequalities with y on one side.

$$\begin{aligned} \rho(x,y) &\leq \rho(x,u) + \rho(u,y) \\ \rho(u,y) &\leq \rho(x,u) + \rho(x,y) \end{aligned}$$

implies

$$\begin{aligned} \inf\{\rho(x,y) : y \in F\} &\leq \rho(x,u) + \inf\{\rho(u,y) : y \in F\} \\ \inf\{\rho(u,y) : y \in F\} &\leq \rho(x,u) + \inf\{\rho(x,y) : y \in F\} \end{aligned}$$

implies

$$\begin{aligned} d_F(x) &\leq \delta + d_F(u) = \varepsilon/3 + d_F(u) \\ d_F(u) &\leq \delta + d_F(x) = \varepsilon/3 + d_F(x) \end{aligned}$$

implies

$$-\varepsilon < -\varepsilon/3 \leq d_F(x) - d_F(u) \leq \varepsilon/3 < \varepsilon$$

So, $|d_F(x) - d_F(u)| < \varepsilon$. Then d_F is continuous on S, *as claimed.*

Clearly $F \subseteq d_F^{\leftarrow}(0)$. We have a fixed point, x in $S \backslash F$. Since F is closed, then x is contained in an open ball $B_\varepsilon(x)$ disjoint from F. So $d_F(x)$ equals some $k \geq \varepsilon > 0$. Since $x \in Z(d_F(x) - k)$ and $F \subseteq Z(d_F[F])$, then F and $\{x\}$ are completely separated (see the fact proven on page 244). So, S is completely regular.

We conclude that every metrizable space is completely regular.

The following two theorems confirm that completely regular spaces are hereditary and also carry over from a family of completely regular factors to their products.[7]

---

[7] *Continuous images of completely regular spaces are not always completely regular.* A standard example of this fact is to produce a continuous function on the completely regular *Moore plane* which maps it onto a Hausdorff, but non-regular space.

Theorem 10.6 *Let S be a completely regular space and T be a subspace of S. Then T is completely regular.*

Proof. We are given that S is a completely regular space and T is a subspace of S.

First note that T is clearly T_1. Let $F \cap T$ be a closed subset of T, where F is closed in S, and let $u \in T \backslash F$. Since S is completely regular, there exists a continuous function $f : S \to [0, 1]$ such that $x \in f^{\leftarrow}[\{0\}]$ and $F \subseteq f^{\leftarrow}[\{1\}]$. Then the function, $f|_T : T \to [0, 1]$ separates x and $F \cap T$ and so T is completely regular.

Theorem 10.7 *Let $\{S_i : i \in I\}$ be a family of topological spaces and $S = \prod_{i \in I} S_i$ be a product space. Then S is completely regular if and only if each factor, S_i, is completely regular.*

Proof. We are given that $\{S_i : i \in I\}$ is a family of topological spaces and $S = \prod_{i \in I} S_i$ is a product space.

($\Rightarrow$) Suppose the product space, $S = \prod_{i \in I} S_i$, is completely regular. By Theorem 7.14, each S_i is homeomorphically embedded in the product space, S. By the preceding theorem, every subspace is completely regular, so each S_i is completely regular.

($\Leftarrow$) We are given that each S_i is completely regular. Let K be a closed subset of $S = \prod_{i \in I} S_i$ and $\langle x_i \rangle_{i \in I}$ be a point in $S \backslash K$.

Then

$$\langle x_i \rangle_{i \in I} \in \cap\{\pi_i^{\leftarrow}[U_i] : i \in F\} \subseteq S \backslash K$$

for some finite $F \subseteq I$ and open U_i's in S_i, where $x_i \in U_i$ for $i \in F$. Then, for each $i \in F$, there exists a continuous function, $f_i : S_i \to [0, 1]$, such that $x_i \in f_i^{\leftarrow}[\{1\}]$ and $S_i \backslash U_i \subseteq f_i^{\leftarrow}[\{0\}]$. We define $h : S \to [0, 1]$ as

$$h(\langle x_i \rangle_{i \in I}) = \inf\{f_i(x_i) : i \in F\}$$

Then $h = \inf\{f_i \circ \pi_i : i \in F\}$ is a continuous function. Note that $h(\langle x_i \rangle_{i \in I}) = 1$ and $S \backslash K \subseteq h^{\leftarrow}(0)$. So, S is completely regular.

10.4 Topic: On Zero-Sets in Normal Spaces

Recall that, in the proof of Urysohn's lemma (in Theorem 10.1), we showed that, for disjoint closed subsets F and W of the normal space, S, there is a continuous function, $f : S \to [0, 1]$, such that

$$\begin{aligned} F \subseteq U_0 &= \quad \cap\{U_i\}_{i\in J} &= f^{\leftarrow}(0) \\ W \subseteq U_1 &= \cap\{S \setminus \mathrm{cl}_S U_i\}_{i\in J} &= f^{\leftarrow}(1) \end{aligned}$$

The continuous real-valued function, f, on S was referred to as a *Urysohn separating function for the closed subsets F and W.*

The set, J, is a countably infinite indexing set so both U_0 and U_1 are G_δ's (countable intersection of open sets). Note that we did not prove that F and W are *equal* to U_0 and U_1. Usually, they are not. Although they appear in the proof, neither U_0 nor U_1 are referred to in the Urysohn lemma statement itself.

We will briefly continue our discussion of normal spaces (in relation to U_0 and U_1). First, we need the following definition.

Definition 10.8 A subset of a topological space, S, of the form, $f^{\leftarrow}[\{0\}]$, for some *continuous* real-valued function, f, on S is called a *zero-set*. Such a set is denoted by, $Z(f)$.

A *cozero-set* of S is a set of the form $S\backslash Z(f) = \{x \in S : f(x) \neq 0\}$ for some continuous real-valued function f. The cozero-set of f is denoted by $\mathrm{coz}(f)$ or $\mathrm{Cz}(f)$.

We note a few interesting facts about zero-sets.

When we speak of a zero-set in a topological space, S, we are referring to a specific subset, A, of S associated to some function $f : S \to \mathbb{R}$ such that $A = f^{\leftarrow}(0) = Z(f)$. A zero-set is sometimes described as the "fiber" of a continuous real-valued function $f : S \to \mathbb{R}$. To see this, if $f(u) = r \in \mathbb{R}$, then $u \in f^{\leftarrow}(r)$; if $g : S \to \mathbb{R}$ is the function defined as $g(x) = f(x) - r$ then

$$\begin{aligned} g(u) &= f(u) - r \\ &= r - r \\ &= 0 \end{aligned}$$

so, g is a continuous real-valued function.

Then the fiber, $f^{\leftarrow}(r)$, is the zero set, $Z(g)$, which contains the element u. So the fiber of a continuous real-valued function, f, on S is a zero-set.

Also see that, for any zero-set, $Z(f)$ of the continuous function, f,

$$Z(f) = \cap\{\, f^{\leftarrow}[(-1/n, 1/n)]\}_{n>0}$$

Since f is continuous,

any zero-set, $Z(f)$, is a G_δ

Countable intersections of zero-sets. We know that any zero-set is a countable intersection of open sets. What can we say about countable intersections of zero-sets? We claim: *The countable intersection of zero-sets is a zero-set.*

Proof of the claim: Let $\{Z(f_n) : n \in \mathbb{N}\}$ be a countable family of zero-sets with non-empty intersection. We are required to show that $\cap\{Z(f_n) : n \in \mathbb{N}\}$ is a zero-set. For each f_n, let[8]

$$h_n = |f_n| \wedge \frac{1}{2^n}$$

We see that $0 \leq h_n(x) \leq \frac{1}{2^n}$ on S. Since the series, $\sum_{n\in\mathbb{N}} \frac{1}{2^n}$, converges, $\sum_{n\in\mathbb{N}} h_n(x)$ converges uniformly to a continuous function $h(x)$ on S.[9] That is,

$$\sum_{n\in\mathbb{N}} h_n(x) = h(x)$$

Since convergence is uniform, then h is continuous.

$$\begin{aligned} x \in \cap\{Z(f_n) : n \in \mathbb{N}\} &\Leftrightarrow f_n(x) = 0 \text{ for all } n \\ &\Leftrightarrow \left(|f_n| \wedge \frac{1}{2^n}\right)(x) = 0 \text{ for all } n \\ &\Leftrightarrow h_n(x) = 0 \text{ for all } n \\ &\Leftrightarrow h(x) = 0 \\ &\Leftrightarrow x \in Z(h) \end{aligned}$$

[8] $(f \wedge g)(x) = \min\{f(x), g(x)\}$.
[9] By the Weierstrass M-test.

So,

$$Z(h) = \cap\{Z(f_n) : n \in \mathbb{N}\}$$

as required. So countable intersections of zero-sets are zero-sets, as claimed.

In the proof of Urysohn's lemma we showed that $F \subseteq f^{\leftarrow}(0)$ and $W \subseteq f^{\leftarrow}(1)$. This means $F \subseteq Z(f)$ and $W \subseteq Z(f-1)$. We can then confidently reformulate one direction of the Urysohn lemma statement as follows:

> If (S, τ_S) is a normal topological space and F and W are disjoint non-empty closed subsets of S, then F and W are contained in disjoint zero-sets.

The leads to an interesting question: Is the converse of this statement true? That is, does it hold true that "if F and W are respectively contained in disjoint zero-sets then S is normal"? We show that it is true, in general.

Theorem 10.9 *Let S be a T_1 topological space. The following are equivalent.*

(a) *The space S is a normal space.*
(b) *If A and B are disjoint closed subsets of S then they are contained in disjoint zero-sets.*
(c) *If A and B are disjoint closed subsets of S then they are completely separated.*

Proof. We are given that (S, τ_S) is normal.

(a $\Rightarrow$ b) This direction follows immediately from Urysohn's lemma stated in the paragraph above.

(b $\Leftarrow$ a) We are given that disjoint closed subsets F and W are contained in disjoint zero-sets, $Z(f)$ and $Z(g)$, respectively. Since the two zero-sets are disjoint, $0 \notin f[Z(g)]$ and $0 \notin g[Z(f)]$.[10]

We are required to show that S is normal. To do this we seek a Urysohn separating function, $h : S \to \mathbb{R}$, for F and W.

---

[10]Since, if $0 \in f[Z(g)]$, there is some $u \in Z(g)$ such that $f(u) = 0$ hence, $u \in Z(f) \cap Z(g)$; so $Z(f) \cap Z(g)$ is not empty, a contradiction.

Let the function $h :\to \mathbb{R}$ be defined as

$$h(x) = \frac{|f(x)|}{|f(x)| + |g(x)|}$$

Then h is continuous and real-valued on S with range in $[0, 1]$.

If $x \in F \subseteq Z(f)$ then $h(x) = \frac{0}{0+|g(x)|} = 0$, so $F \subseteq Z(f) \subseteq h^{\leftarrow}(0)$.

If $x \in W \subseteq Z(g)$ then $h(x) = \frac{|f(x)|}{|f(x)|+0} = 1$, so $W \subseteq Z(g) \subseteq h^{\leftarrow}(1)$.

So, h separates F and W. By Theorem 10.1 S is normal.

(a $\Leftrightarrow$ c) Is Urysohn's lemma 10.1.

(b $\Leftrightarrow$ c) Since(a $\Leftrightarrow$ b) and (a $\Leftrightarrow$ c), then (b $\Rightarrow$ c) and (c $\Rightarrow$ b).

From the above theorem we can then say that...

> *if A and B are completely separated and $r \neq k$ there exist $h \in C^*(S)$ such that if $h[A] = \{r\}$ and $h[B] = \{k\}$,*

by composing the function h with $t(x) = (k - r)x + r$.

10.5 On C^*-Embedded Subsets of Normal Spaces

The following statement refers to the notions of C^*-embedded subsets and C-embedded subsets:

> *A subset A of a topological space S is said to be a C^*-embedded subset (C-embedded subset) of S if for each $g \in C^*(A)$ (respectively, $g \in C(A)$) there exists $f \in C^*(S)$ (respectively, $C(S)$) such that $f|_A = g$.*

One of the most important theorem in basic point-set topology is the *Urysohn extension theorem.* It is often referred to as being a "deep theorem" since the proof involves innovative mathematical techniques introduced by Pavel Urysohn. We invoke the Urysohn extension theorem in the proof of the following result. But its proof is deferred to Theorem 21.8 on page 490.

> **Urysohn extension theorem:** *Let T be a subset of the completely regular space S. Then T is C^*-embedded in S if and only if pairs of sets which can be completely separated by some function in $C^*(T)$ can also be separated by some function in $C^*(S)$.*

The proof also invokes a statement proven in Theorem 23.2 whose proof is deferred to that chapter.

> **Theorem 23.2**: *Suppose U be C^*-embedded in S. The set U is also C-embedded in S if and only if U is completely separated from any disjoint zero-set in S.*

Theorem 10.10 *Let S be a T_1 topological space and A and B be any pair non-empty disjoint closed subsets of S. The following are equivalent.*

(a) *The space S is a normal space.*
(b) *Each closed subset of S is C^*-embedded in S.*
(c) *Each closed subset of S is C-embedded in S.*

Proof. We are given that (S, τ_s) is normal.

(a $\Rightarrow$ b) Suppose S is normal.

Let A be a closed subset of S. We are required to show that A is C^*-embedded in S.

Let B and C be completely separated subsets of A. Then, by 10.9, part (b) $\Leftrightarrow$ part (c), B and C are contained in disjoint zero sets, Z_B and Z_C in A. Since A is closed, both Z_B and Z_C are closed in S. Since S is normal, by Theorem 10.9, part (a) $\Leftrightarrow$ part (c), Z_B and Z_C are completely separated in S. Then B and C are completely separated subsets of S. By Urysohn's extension theorem stated above, A is C^*-embedded in S.

(b $\Rightarrow$ c) Suppose each closed subset of S is C^*-embedded in S. Let A be a closed subset of S. We are required to show that A is C-embedded in S.

By hypothesis, since A is closed in S it is C^*-embedded in S.

We claim that A is completely separated from any zero-set disjoint from A. Suppose Z is a zero-set in $Z[S]$ which is disjoint from A. Then the function, $h : A \cup Z :\rightarrow \{0, 1\}$, defined as

$$h[A] = \{1\}$$
$$h[Z] = \{0\}$$

is continuous on $A \cup Z$. Then the function h in $C(A \cup Z)$ completely separates Z and A. So, A is completely separated from any zero-set disjoint from A, as claimed.

By Theorem 23.2, A is C-embedded in S.

We mentioned above that zero-sets of a continuous function are closed G_δ's. Can we turn this phrase around and say that "*closed G_δ's are zero-sets*"? The answer is no. The statement "closed G_δ's are zero-sets" is however true in, at least, one class of topological spaces, namely, normal spaces (as we will now show).

Theorem 10.11 *Suppose F is a non-empty closed subset of the normal topological space, (S, τ). Then F is a G_δ if and only if F is a zero-set produced by some continuous function $f : S \rightarrow [0, 1]$.*

Proof. We are given that (S, τ_s) is normal.

($\Leftarrow$) This direction is obvious since, if $F = Z(f) \subseteq S$ where $f : S \rightarrow [0, 1]$ is continuous then $F = \cap\{f^{\leftarrow}[\,[0, 1/i)\,] : i \in \mathbb{N}\backslash\{0\}\,\}$, a G_δ in S.

($\Rightarrow$) Suppose F is a closed G_δ in the space S. If F is also open then it is clopen and so is easily seen to be a zero-set. Suppose the closed G_δ, F, is not open. Then, by definition, there exists a countably infinite family, $\{V_i : i = 1, 2, 3, \ldots,\ V_i \text{ is open in } S\ \}$ such that $F = \cap\{V_i\}$. Suppose

$$U_1 = V_1$$
$$U_2 = V_1 \cap V_2$$
$$\vdots$$
$$U_n = V_1 \cap V_2 \cap \ldots \cap V_n$$
$$\vdots$$

Then $\{U_i : i = 1, 2, 3, \ldots\} \subset \tau_s$, $F \subseteq U_{i+1} \subseteq U_i$, for all $i > 0$, and $F = \cap\{U_i\}$.

Normality of S implies that, for each $i > 0$, there is a continuous function, $f_i : S \to [0, 1]$, such that $F \subseteq f_i^{\leftarrow}\{0\}$ and $S \backslash U_i \subseteq f_i^{\leftarrow}[\{1\}]$.

So, we have a countable family, $\{f_i : i > 0\}$, of functions each mapping S into $[0, 1]$.

Consider the function, $f : S \to \mathbb{R}$, defined as

$$f(x) = \sum_{i=1}^{\infty} \frac{f_i(x)}{2^i} \qquad (*)$$

See that, since $0 \leq \frac{f_i(x)}{2^i} \leq \frac{1}{2^i}$ on S, and $\sum_{i=1}^{\infty} \frac{1}{2^i} = 1$, then, by the Weierstrass M-test, the series $(*)$ converges uniformly to the function $f(x)$ on S. Uniform convergence guarantees continuity of f on S.

Also see that, since $F = \cap\{V_i\}$, $F = f^{\leftarrow}[\{0\}]$. So, the G_δ, F, is the zero-set, $Z(f)$, produced by the continuous function, $f : S \to [0, 1]$, described in $(*)$.

Notation: If $C(S)$ denotes the set of all continuous functions on S, we let the subsets of S,

$$Z[S] = \{Z(f) : f \in C(S)\}$$
$$\text{coz}[S] = \{S \backslash Z(f) : f \in C(S)\}$$

denote the family of all zero-sets and cozero-sets of a space S, respectively. The cozero-set of f is also expressed as $\text{Cz}(f)$. The family of all zero-sets (cozero-sets) in a completely regular space provides us with another way of describing a base for closed (open) sets of S.

Theorem 10.12 *If (S, τ) is completely regular then,*

(a) *The family $Z[S] = \{Z(f) : f \in C(S)\}$ of all zero-sets forms a base for the closed sets in its topology.*
(b) *The family coz$[S] = \{S \backslash Z(f) : f \in C(S)\}$ of all cozero-sets forms a base for the open sets in its topology.*

Proof. (a) Let F be a non-empty proper closed subset in S. Given that S is completely regular, then for every $p \in S \backslash F$, there exists a function $f \in C(S)$ such that $F \subseteq Z(f)$ and $p \in Z(f-1)$. Then F is the intersection of zero-sets. So, by definition of "base for closed sets" (see Definition 5.2) $Z[S]$ forms a base for the closed sets in S.

The proof of part (b) is left as an exercise.

Example 1. Show that, if S is a topological space in which the family of all zero-sets, $Z[S]$ forms a base for closed sets, then S must be completely regular.

Solution: Suppose S is a space such that $Z[S]$ forms a base for closed sets. Suppose F is a closed subset of S and p is a point not in F.

By hypothesis, the closed set F is the intersection of zero-sets. Then, for every $p \in S \backslash F$, there exists a zero-set $Z(f)$ such that $F \subseteq Z(f) \subseteq S \backslash \{p\}$. Let $h : S \to \mathbb{R}$ be defined as

$$h(x) = \left(\frac{1}{f(p)}\right) f(x)$$

Then h is a continuous real-valued function which maps F to $\{0\}$ and p to 1. By definition, S is completely regular.

Example 2. Suppose that F is a closed subset of a metric space S. Show that F is a zero-set of some real-valued function. Show that F is a G_δ.

Solution: Suppose F is a closed subset of a metric space, (S, ρ). Define a real-valued function $f_F : S \to \mathbb{R}$ as $f_F(u) = \rho(u, F) = \inf\{\rho(u, y) : y \in F\}$. Then $f_F[F] = \{0\}$ and $f_F(u) \neq 0$ for $u \notin F$. To show that F is a zero-set it now suffices to show that f_F is a continuous function.

Claim: That the function, f_F, is a continuous function on S.

Proof of claim: Let $\varepsilon > 0$. It suffices to show that there is a $\delta > 0$ such that $f[B_\delta(x)] \subseteq B_\varepsilon(f(x))$. Let $\rho(x, u) < \delta = \varepsilon/3$.

Since,

$$\rho(x, y) \leq \rho(x, u) + \rho(u, y)$$
$$\rho(u, y) \leq \rho(x, u) + \rho(x, y)$$

then,

$$\inf\{\rho(x,y) : y \in F\} \leq \rho(x,u) + \inf\{\rho(u,y) : y \in F\}$$
$$\inf\{\rho(u,y) : y \in F\} \leq \rho(x,u) + \inf\{\rho(x,y) : y \in F\}$$

Hence,

$$f_F(x) \leq \varepsilon/3 + f_F(u)$$
$$f_F(u) \leq \varepsilon/3 + f_F(x)$$

Then

$$-\varepsilon < -\varepsilon/3 \leq f_F(x) - f_F(u) \leq \varepsilon/3 < \varepsilon$$

So, $|f_F(x) - f_F(u)| < \varepsilon$. Then f_F is continuous on S, *as claimed.*
So, the closed set F is a zero-set, $Z[f_F]$, of f_F.
Since a zero-set is a G_δ, then F is a G_δ.

Example 3. Let S be a completely regular space. Suppose A and B are two infinitely countable subsets such $\mathrm{cl}_S A \cap B = \varnothing = A \cap \mathrm{cl}_S B$. Then A and B are contained in disjoint cozero-sets.

Solution: We are given that S is completely regular. Then the family of all cozero-sets on S form a base for open sets in S (by Theorem 10.12).

Let $A = \{a_i : i \in \mathbb{N}\}$ and $B = \{b_i : i \in \mathbb{N}\}$. Since $\mathrm{cl}_S B \cap A = \varnothing$, there exists a cozero-set U_0 such that $a_0 \in U_0$ and $\mathrm{cl}_S U_0 \cap \mathrm{cl}_S B = \varnothing$.

Since $\mathrm{cl}_S A \cap B = \varnothing$, there exists a cozero-set V_0 such that $b_0 \in V_0$ and

$$\mathrm{cl}_S V_0 \cap [\mathrm{cl}_S A \cup \mathrm{cl}_S U_0] = \varnothing$$

Since $\mathrm{cl}_S V_0 \cap \mathrm{cl}_S U_0 = \varnothing$, $U_0 \cap V_0 = \varnothing$.

Suppose

$$\{a_0, a_1, \ldots, a_i\} \subseteq U_0 \cup \cdots \cup U_i \subseteq \mathrm{cl}_S U_0 \cup \cdots \cup \mathrm{cl}_S U_i \subseteq S \backslash [\mathrm{cl}_S B \cup \mathrm{cl}_S V_0 \cup \cdots \cup \mathrm{cl}_S V_i]$$
$$\{b_0, b_1, \ldots, b_i\} \subseteq V_0 \cup \cdots \cup V_i \subseteq \mathrm{cl}_S V_0 \cup \cdots \cup \mathrm{cl}_S V_i \subseteq S \backslash [\mathrm{cl}_S A \cup \mathrm{cl}_S U_0 \cup \cdots \cup \mathrm{cl}_S U_i]$$

There exists a cozero-sets U_{i+1} and V_{i+1} such that

$$a_{i+1} \in U_{i+1} \subseteq \mathrm{cl}_S U_{i+1} \subseteq S \backslash [\mathrm{cl}_S B \ \cup \ \mathrm{cl}_S V_0 \cup \cdots \cup \ \mathrm{cl}_S V_i]$$
$$b_{i+1} \in V_{i+1} \subseteq \mathrm{cl}_S V_{i+1} \subseteq S \backslash [\mathrm{cl}_S A \ \cup \ \mathrm{cl}_S U_0 \cup \cdots \cup \ \mathrm{cl}_S U_{i+1}]$$

implies

$$[U_0 \cup \cdots \cup U_{i+1}] \cap [V_0 \cup \cdots \cup V_{i+1}] = \varnothing$$

By mathematical induction,

$$\cup\{U_i : i \in \mathbb{N}\} \cap \cup\{V_i : i \in \mathbb{N}\} = \varnothing$$

We know that the countable intersection of zero-sets is a zero-set (as shown on page 250). So, $\cup\{U_i : i \in \mathbb{N}\}$ and $\cup\{V_i : i \in \mathbb{N}\}$ are disjoint cozero-set neighborhoods of A and B. As required.

10.6 Topic: Perfectly Normal Topological Spaces

In Theorem 10.11, we showed that in normal spaces closed G_δ's and zero-sets are one and the same. We should, however, be cautious and not deduce from this that closed subsets of normal spaces are necessarily G_δ's. We will soon exhibit a normal space which contains a non-G_δ singleton set. Those normal spaces in which all non-empty closed subsets are G_δ's form a special class of topological spaces which we will now define.

Definition 10.13 A normal topological space, (S, τ_S), is said to be *perfectly normal* if and only if every closed subset of S is a G_δ.

To say that every closed subset is a G_δ is equivalent to saying that every open set is an F_σ (a countable union of closed sets) as we shall now see.

Suppose F is closed in S and $\{U_i : i \in \mathbb{N}\}$ is a family of open sets in S.

Let $U = S \backslash F$ and $F_i = S \backslash U_i$ for each i.

$$\begin{aligned} F = \cap\{U_i : i \in \mathbb{N}\}, \text{ is a closed } G_\delta &\Leftrightarrow S\backslash F = S\backslash \cap\{U_i : i \in \mathbb{N}\}, \text{ is open} \\ &\Leftrightarrow U = \cup\{S\backslash U_i : i \in \mathbb{N}\}, \text{ is open} \\ &\Leftrightarrow U = \cup\{F_i : i \in \mathbb{N}\}, \text{ is an open } F_\sigma \end{aligned}$$

Example 4. Show that all second countable normal spaces are perfectly normal spaces.

Solution: Suppose S is a second countable normal space. If $x \in S$, let U_x denote an open base element of S which contains x. Since S is second countable, we can choose $\mathscr{B} = \{U_x : x \in S\}$ to be countable. Let F be a closed subset of S and $V = S \backslash F$. Since S is normal, if $x \in V$, there exist an open base element, U_x, of x such that

$$x \in \text{cl}_S U_x \subseteq V$$

Then we can find $\{\text{cl}_S U_x : x \in V\}$, a countable set of closed subsets each of which is contained in V. Then

$$V = \cup\{\text{cl}_S U_x : x \in V\}$$

an F_σ in S. This implies that $F = S \backslash V$ is a G_δ. So every closed set is a G_δ. Then S is perfectly normal.

Theorem 10.14 *Suppose (S, τ_S) is a normal topological space. Then S is perfectly normal if and only if every non-empty closed subset of S is a zero-set.*

Proof. We are given that (S, τ_S) is *normal*.

$(\Rightarrow)$ We are given that S is perfectly normal. Suppose F is a non-empty closed subset. By hypothesis, F is a G_δ. By Theorem 10.11, since F is a G_δ in a normal space then F is a zero-set, $Z(f)$, for some continuous $f : S \to [0, 1]$.

$(\Leftarrow)$ Suppose S is normal and every non-empty closed subset F is a zero-set. Zero-sets have been shown to be closed G_δ's. So, S is perfectly normal.

We know from Theorem 9.19 that metric spaces are normal. We raise the bar slightly higher to show that they are even perfectly normal.

Example 5. Show that a metric space is perfectly normal.

Solution: By Theorem 9.19, a metric space is normal. In an example found on page 256, it is shown that closed subsets of a metric space are zero-sets. By the above theorem a metric space is perfectly normal.

10.6.1 *Normal but not perfect*

Are there normal spaces which are not perfectly normal? We have shown in an example on page 259 that second countable normal spaces are perfect. If we seek a "non-perfect" normal space we investigate normal spaces which are not second countable.

Example 6. *A normal space which is not perfectly normal.* Let ω_1 denote the first uncountable ordinal. Show that the ordinal space, $S = [0, \omega_1]$, (previously discussed on page 110) shown previously to be normal (on page 234) is not perfectly normal.

Solution: Let (S, τ_ω) denote the ordinal space $[0, \omega_1]$ where ω_1 is the first uncountable ordinal. It will suffice to show that the closed singleton subset, $\{\omega_1\}$, of S is not a G_δ and hence, S is not perfect.

Recall that $(\mu, \omega_1]$ is an open neighborhood base element of ω_1. Suppose

$$\{\omega_1\} \subseteq \cap\{U_i : i \in \mathbb{N}\}$$

where $\{U_i : i \in \mathbb{N}\}$ is a countable family of open neighborhoods each containing ω_1. We will show that $\{\omega_1\} \neq \cap\{U_i : i \in \mathbb{N}\}$.

For each U_i, there exists μ_i such that $\omega_1 \in (\mu_i, \omega_1] \subseteq U_i$. Note that each μ_i is a countable ordinal (since $\mu_i < \omega_1$). Let $U = \{\mu_i : i \in \mathbb{N}\}$ and

$$\gamma = \sup U \leq \omega_1$$

We consider two cases based on the nature of γ.

Case 1. The ordinal γ *is a limit ordinal*: Then $\gamma = \cup\{\mu_i : i \in \mathbb{N}\}$, a countable union of countable ordinals. Countable unions of countable sets are countable. Hence, γ is a countable ordinal. Since ω_1 is uncountable, $\gamma < \omega_1$. Then $(\gamma, \omega_1] \neq \{\omega_1\}$.

Case 2. The ordinal γ is *not a limit ordinal*: Then $\gamma \in U = \cup\{\mu_i : i \in \mathbb{N}\}$. So again γ is a countable ordinal. That is $\gamma < \omega_1$. Then $(\gamma, \omega_1] \neq \{\omega_1\}$.

So in both cases, $(\gamma, \omega_1] \neq \{\omega_1\}$. Also,

$$(\gamma, \omega_1] \subseteq \cap\{(\mu_i, \omega_1] : i \in \mathbb{N}\} \subseteq \cap\{U_i : i \in \mathbb{N}\}$$

So, $\{\omega_1\} \neq \cap\{U_i : i \in \mathbb{N}\}$. This shows that $\{\omega_1\}$ cannot be a G_δ. So, S is not perfect, as required.

We provide another example.

Example 7. *A normal space which is not perfectly normal.* Let $S = \mathbb{R}^2$ and $p = (0, 0)$ and suppose S has a topology defined as follows:

$$\tau = \{T \subseteq S : S\backslash T \text{ is finite or } p \notin T\}$$

Show that (S, τ) is normal but not perfectly normal.

Solution: We claim that S is T_1. A set which does not contain p is declared to be open so the set $\{p\}$ is closed. Complements of finite sets are open so, if $x \neq p$, since $S \setminus \{x\}$ is open then $\{x\}$ is closed. So, S is T_1, as claimed.

Claim: That S is T_4. Suppose F and K are disjoint closed subsets of S. If $p \notin K \cup F$ then K and F are both open. So, F and K are clopen and so separated by open sets. Suppose $p \in F$. Then K is clopen and finite. Now, $S\backslash K$ is open since K is finite. So, F and K are contained in the disjoint open sets $S\backslash K$ and K. So, S is normal, as claimed.

Claim: That S is not perfect. We know that $\mathbb{R}^2$ is uncountable. Let $\{U_i : i \in \mathbb{N}\}$ be a countable family of sets in S each containing the point p. Then each U_i is closed in S and $p \in \cap\{U_i : i \in \mathbb{N}\}$. If each U_i is also open $S\backslash U_i$ if finite. So, each U_i is uncountable. Then $S\backslash\cap\{U_i : i \in \mathbb{N}\} = \cup\{S\backslash U_i : i \in \mathbb{N}\}$ a countable subset of S. So, $\cap\{U_i : i \in \mathbb{N}\}$ is uncountable. So, $p \neq \cap\{U_i : i \in \mathbb{N}\}$. So, p is not a G_δ. So, S is not perfectly normal, as claimed. The completes the problem.

So we have, the implications,

$$\text{normal} \not\Rightarrow \text{perfectly normal}$$

$$\text{perfectly normal} \Rightarrow \text{normal} \Rightarrow \text{completely regular}$$
$$\Rightarrow \text{regular} \Rightarrow \text{semiregular} \Rightarrow \text{Hausdorff}$$
$$\Downarrow$$
$$\text{completely Hausdorff} \Rightarrow \text{Hausdorff}$$

We have a particular characterization of the perfectly normal property.

Theorem 10.15 *Suppose S is a normal topological space. The space S is perfectly normal if and only if every closed subset of S is a continuous pre-image of some closed subset of a perfectly normal space.*

Proof. ($\Leftarrow$) Let F be a closed subset of S. Suppose that there exists a continuous function, $f : S \to T$, mapping S into the perfectly normal space, T, such that $F = f^{\leftarrow}[K]$ for some closed subset K of T. We are required to show that S is perfectly normal. It will suffice to show that F is a zero-set. Since K is a closed subset of a perfectly normal T, then by Theorem 10.14 K is a zero-set in T. Then $K = h^{\leftarrow}(0)$, for some continuous function $h : T \to \mathbb{R}$. Then $(h \circ f)^{\leftarrow}(0) = f^{\leftarrow}(h^{\leftarrow}(0)) = F$. So, F is a zero-set in S. Since F is an arbitrary closed subset, by Theorem 10.14, S is perfectly normal.

($\Rightarrow$) Suppose S is perfectly normal and F is closed subset of S. Then F is a zero-set and so $F = f^{\leftarrow}(\{0\})$, for some $f \in C(S)$. Since $\mathbb{R}$ is a metrizable it is perfectly normal. Let $K = \{0\}$, a closed subset of $\mathbb{R}$. Then $F = f^{\leftarrow}[K]$, as required.

10.7 Topic: Example of a "Completely Regular Non-Normal" Space

Up to now, we have not yet witnessed a completely regular space which is not normal. The *Moore plane* is such a space, as we now show.

Example 8. Show that the *Moore plane* is completely regular.

Solution: Let (S, τ_M) denote the Moore plane. We have already shown in a previous example (on page 228) that (S, τ_M) is regular. We have also seen on page 92, that if τ is the usual topology on S, $\tau \subset \tau_M$. We know that metrizable spaces are completely regular so (S, τ) is completely regular.

We proceed to show that (S, τ_M) is completely regular.

For what follows we let $F = \{(x, 0) \in S : x \in \mathbb{R}\}$ and $W = S \setminus F$.

Suppose K is a non-empty closed subset of S and (a, b) is a point which does not belong to K. We seek a continuous function which completely separates K and (a, b). We consider two cases.

Case 1: Suppose $b \neq 0$**.**

Now, F is closed with respect to both τ and τ_M.

Since (S, τ) is completely regular there exists an open neighborhood V (in τ) of F such that $(a, b) \notin \text{cl}_S V$. Then $K \cup \text{cl}_S V$ is closed with respect to both topologies. Then there exists a continuous function $g : S \to [0, 1]$ (with respect to τ) such that, $K \cup \text{cl}_S V \subseteq g^{\leftarrow}(0)$ and $(a, b) \in g^{\leftarrow}(1)$. Since g is continuous with respect to τ, it is continuous with respect to τ_M. Then the continuous function g with respect to τ_M completely separates K and (a, b).

Case 2: We consider the case where $(a, b) \in F$**.** That is, $(a, b) = (a, 0)$.

Without loss of generality suppose $K \cap W \neq \varnothing$. If $B_\delta = B_\delta(a, \delta) \cup \{(a, 0)\}$ is a basic open neighborhood of $(a, 0)$ such that $B_\delta \cap K = \varnothing$; then

$$F \cap S \backslash B_\delta = F \backslash \{(a, 0)\}$$

is a closed subset of S. So, $K \cup (F \backslash \{(a, 0)\})$ is a closed subset of S which misses B_δ and intersects W.

Since (S, τ_M) is regular, there exists a neighborhood, $B_\varepsilon = B_\varepsilon(a, \varepsilon) \cup \{(a, 0)\}$ of the point $(a, 0)$ such that,

$$B_\varepsilon \subseteq \text{cl}_S B_\varepsilon \subseteq B_\delta$$

Then $\mathrm{cl}_S B_\varepsilon$ is a closed neighborhood of $(a, 0)$ in S contained in the set B_δ, and $K \cup F \backslash \{(a, 0)\} \subseteq S \setminus B_\delta$. Then

$$\mathrm{cl}_S B_\varepsilon \cap W \quad \text{and} \quad W \cap (S \setminus B_\delta)$$

are disjoint closed subsets of W.

Since W is metrizable, it is normal (by Theorem 9.19) and so there exists a Urysohn separating function, $g : W \to [0, 1]$, such that

$$\mathrm{cl}_S B_\varepsilon \cap W \subseteq g^{\leftarrow}[\{0\}] \quad \text{and} \quad W \cap (S \setminus B_\delta) \subseteq g^{\leftarrow}[\{1\}]$$

Given $g[\mathrm{cl}_S B_\varepsilon \cap W] = \{0\}$, then we can extend g to a continuous function, $g_S : S \to [0, 1]$, on S where

$$g_S[\mathrm{cl}_S B_\varepsilon] = \{0\} \quad \text{and} \quad g_S[S \setminus B_\delta] = \{1\}$$

Since $K \subseteq S \backslash B_\delta$ and $(a, 0) \in \mathrm{cl}_S B_\varepsilon$, then g_S completely separates K and $(a, 0)$.

So, the Moore plane is completely regular.

Example 9. If (S, τ_M) denotes the Moore plane and $F = \{(x, 0) : x \in \mathbb{R}\}$ show that every subset of F is closed in (S, τ_M).

Solution: We are given that (S, τ_M) denotes the Moore plane and $F = \{(x, 0) : x \in \mathbb{R}\}$. Since $S \backslash F$ is easily seen to be open in S, then F is closed.

Suppose $T \subset F$ and $(a, 0) \in F \backslash T$. Then $(a, 0)$ belongs to some basic open neighborhood, $B_a = B_\varepsilon(a, \varepsilon) \cup \{(a, 0)\}$. If

$$U_{F \backslash T} = \cup \{B_a : (a, 0) \in F \backslash T\}$$

then $F \backslash T \subseteq U_{F \backslash T}$ where $U_{F \backslash T}$ is an open subset of S, and so $T = (S \backslash U_{F \backslash T}) \cap F$ is a closed subset of S. So, T is closed in S.

In the following example, we show that the Moore plane is not normal. To follow the solution presented, we need to remember that, if T is a set and $\mathscr{P}(T)$ is its power set (the set of all subsets of T), then the cardinality of $\mathscr{P}(T)$ is $2^{|T|}$ where $|T|$ is the cardinality of T. Also, it is important to know that if $c = |\mathbb{R}|$ and $\aleph_0 = |\mathbb{N}|$, then $2^c > 2^{\aleph_0} = c$.

Example 10. Show that the *Moore plane* is not a normal space.

Solution: Let (S, τ_M) denote the Moore plane. We claim S is not normal. To show this we will assume S is normal.

Let $F = \{(x, 0) : x \in \mathbb{R}\}$ and $W = S \setminus F$.

For each $(a, 0) \in F$, a basic open neighborhood of $(a, 0)$ is of the form

$$B_a = B_\varepsilon(a, \varepsilon) \cup \{(a, 0)\} \quad \text{for some } \varepsilon$$

If T is a non-empty subset of F, then, from the previous example, both T and $F \backslash T$ are closed in S.

Let

$$D = (\mathbb{Q} \times \mathbb{Q}) \cap W$$

a countable dense subset of S.

The cardinality of $\mathscr{P}(D)$ is $2^{|D|} = 2^{\aleph_0} = c$ and the cardinality of $\mathscr{P}(F)$ is $2^{|F|} = 2^c$.

The contradiction that we seek is that $|\mathscr{P}(D)| \geq |\mathscr{P}(F)|$.

We supposed that S is normal, so for each non-empty set T of F, there exists in τ_M a pair of disjoint open neighborhoods,

$$U_T = \cup\{B_a : (a, 0) \in T\}$$
$$U_{F\backslash T} = \cup\{B_a : (a, 0) \in F \backslash T\}$$

of T and $F \backslash T$, respectively.

For each $T \in \mathscr{P}(F)$, let

$$D_T = \cup\{B_a \cap D : B_a \subseteq U_T\} = U_T \cap D \in \mathscr{P}(D)$$
$$D_{F\backslash T} = \cup\{B_a \cap D : B_a \subseteq U_{F\backslash T}\} = U_{F\backslash T} \cap D \in \mathscr{P}(D)$$
$$\text{where } U_{F\backslash T} \cap U_T = \varnothing$$

Let $f : \mathscr{P}(F) \to \mathscr{P}(D)$ be defined as

$$f(T) = D_T, \ f(F \backslash T) = D_{F\backslash T}, \ D_{F\backslash T} \cap D_T = \varnothing$$

Claim: That f maps $\mathscr{P}(F)$ one-to-one into $\mathscr{P}(D)$:

Suppose T_1 and T_2 are distinct elements in $\mathscr{P}(F)$. Say, there exists $(b, 0)$, such that $(b, 0) \in T_2 \backslash T_1 \subseteq U_{F\backslash T_1}$. It suffices to show that $D_{T_2} \backslash D_{T_1} \neq \varnothing$.

Then $B_b \cap D \subseteq B_b \cap D_{T_2} \subseteq B_b \cap U_{F\backslash T_1}$. Since $D_{T_1} \subseteq U_{T_1}$ and $U_{T_1} \cap U_{F\backslash T_1} = \varnothing$, then $D_{T_2} \backslash D_{T_1} \neq \varnothing$. So, $D_{T_2} \neq D_{T_1}$.

We conclude that $f(T_1) \neq f(T_2)$, which means that $f : \mathscr{P}(F) \to \mathscr{P}(D)$ is one-to-one as claimed.

Then the cardinality of the domain of f, $\mathscr{P}(F)$, is less than or equal to the cardinality of its codomain, $\mathscr{P}(D)$. The cardinality of $\mathscr{P}(F)$ is $2^{|F|} = 2^c$ while the cardinality of $\mathscr{P}(D)$ is $2^{|D|} = 2^{\aleph_0} = c$. Then from the claim we obtain $2^c \leq c$. This contradicts the fact that $c < 2^c$. The source of our contradiction is our supposition that S is normal.

So, the Moore plane is *not* normal.

10.8 Topic: The Embedding Theorem Revisited

We remind the reader of the statement in Theorem 7.17, titled, The *Embedding theorem I*, presented earlier as an application of Cartesian products. Assuming that the set, and $\mathscr{F} = \{f_i : i \in J\}$, of functions separates points and closed sets of a space, S, and $e : S \to \prod_{i \in J} X_i$ is an evaluation function with respect to $\mathscr{F}$ on S, the *Embedding theorem I* states that:

> *the evaluation map e, with respect to $\mathscr{F}$, embeds a homeomorphic copy, e[S], of S in the product space, $\prod_{i \in J} X_i$ where each of the factors, X_i, of the product is the codomain of the function, f_i, in $\mathscr{F}$.*

We now have tools which will allow us to enhance the *Embedding theorem I* further. But first, we must do a bit of groundwork with a preliminary lemma.

In the following lemma, we prove an interesting characterization of a completely regular space. It shows that the topology of any completely regular space is precisely the "weak topology" induced by the family, $C^*(S)$, of all real-valued continuous bounded functions.

Lemma 10.16 *Suppose (S, τ) is a T_1 topological space and $\mathscr{S} = C^*(S)$, represents the set of all continuous bounded real-valued functions on S. The topological space, (S, τ), is completely regular with respect to τ if and only if its topology, τ, is the "weak topology" induced by $\mathscr{S}$.*

Proof. We are given that (S,τ) is a T_1 topological space. Recall that

$$\tau_{\mathscr{S}} = \{f^{\leftarrow}[U] : f \in \mathscr{S},\ U \text{ an open subset of } \mathbb{R}\}$$

is the weak topology generated by $\mathscr{S}$.

($\Rightarrow$) Suppose (S,τ) is a completely regular space. To show that τ is the weak topology induced by $\mathscr{S}$ it suffices to show that $\tau_{\mathscr{S}} = \tau$.

Claim 1: $\tau \subseteq \tau_{\mathscr{S}}$: Let $x \in V \in \tau$. We claim that $V \in \tau_{\mathscr{S}}$. Since S is completely regular, there exists $f : S \to [0,1]$ such that $x \in f^{\leftarrow}[\{0\}]$ and $S\backslash V \subseteq f^{\leftarrow}[\{1\}]$. Then $f^{\leftarrow}[\,[0,1/3)\,] \in \mathscr{S}$, an open neighborhood of x entirely contained in V. Then, $V \in \mathscr{S} \subseteq \tau_{\mathscr{S}}$. This means $\tau \subseteq \tau_{\mathscr{S}}$. As claimed.

Claim 2: $\tau_{\mathscr{S}} \subseteq \tau$: See that $\tau_{\mathscr{S}}$ is the weakest topology induced by $\mathscr{S} = C^*(S)$. Let $V \in \tau_{\mathscr{S}}$. Then there exists and open set, $U \in \mathbb{R}$, such that $f^{\leftarrow}[U] = V$ for some $f \in C^*(S)$. But for every function $f \in C^*(S)$, $f^{\leftarrow}[U]$ is open in S with respect to τ. Then f is continuous on S with respect to τ. So, $V \in \tau$. So, $\tau_{\mathscr{S}} \subseteq \tau$, as claimed.

Combining both claims we obtain $\tau_{\mathscr{S}} = \tau$.

So, τ is the smallest topology on S such that every function in $\mathscr{S}$ is continuous on S, as required. We are done with $\Rightarrow$.

($\Leftarrow$) Suppose τ is the weak topology on S, induced by the family of functions, $\mathscr{S}$. That is,

$$\tau = \{V \subseteq S : V = f^{\leftarrow}[U],\ \text{for some open } U \text{ in } \mathbb{R},\ f \in \mathscr{S}\,\}$$

We are required to show that (S,τ) is a completely regular space.

Let W be a non-empty open subset of S and $x \in W$. To show complete regularity we must construct a continuous function, $h : S \to [0,1]$, which separates $\{x\}$ from $S\backslash W$.

The following family of open sets

$$\mathscr{M} = \{f^{\leftarrow}[Z] :\ \text{some } f \in \mathscr{S},\ Z \text{ is a subbase element of } \mathbb{R}\}$$

forms a subbase for τ. (Note that subbase elements for $\mathbb{R}$ are of the form (a_i,∞) or $(-\infty,a_i)$.)

Since τ is generated by the subbase $\mathscr{M}$, there exists finitely many subbase elements, $\{V_i : i \in F = \{1, \ldots, n\}\} \subset \mathscr{M}$, such that

$$x \in \cap\{V_i : i \in F = \{1, \ldots, n\}\} \subseteq W \quad (*)$$

where $V_i = f^{\leftarrow}[U_i]$, then U_i is of the form (a_i, ∞) or $(-\infty, a_i)$. To simplify our expression, note that, if $U_j = (-\infty, a_j)$ then

$$f_j^{\leftarrow}[U_j] = f_j^{\leftarrow}[(-1)(-a_j, \infty)] = (-f)_j^{\leftarrow}[(-a_j, \infty)]$$

so we can assume that all U_i's are of the form $U_i = (a_i, \infty)$ after having made the required adjustments on the functions, f_i. So, $(*)$ can be expressed as

$$x \in \cap\{f_i^{\leftarrow}[(a_i, \infty)] : i \in F\} \subseteq W \quad (**)$$

For each i in $F = \{1, \ldots, n\}$, let $t_i : S \to \mathbb{R}$ be defined as: $t_i = f_i - a_i$.

See that $t_i > 0$ on $f_i^{\leftarrow}[(a_i, \infty)]$. So $x \in t_i^{\leftarrow}[(0, \infty]$.

Also, see that $t_i(x) \leq 0$ everywhere else, including on $S \backslash W$.

For each i in $F = \{1, \ldots, n\}$, let

$$t_i^* = t_i \vee 0^{11}$$

So, $t_i^* \geq 0$, and

$$x \in \cap\{t_i^{*\leftarrow}[(0, \infty)] : i \in F\ \} \subseteq W$$

We define $t : S \to \mathbb{R}$, as

$$t(x) = t_1^*(x) t_2^*(x) t_3^*(x) \cdots t_n^*(x)$$

We then have a real-valued function such that $x \in t^{\leftarrow}[(0, \infty)] \subseteq W$ and $S \backslash W \subseteq t^{\leftarrow}(0)$. We restrict the range of the function: Suppose $t(x) = \delta$. Then $h = (t/\delta) \wedge 1^{12}$ maps S into $[0, 1]$ and $x \in h^{\leftarrow}(1)$. So, h separates x and $S \backslash W$.

This allows us to conclude that S is completely regular.

---

[11] $(f \vee 0)(y) = \max\{f(y), 0(y)\}$.
[12] $(t/\delta) \wedge 1)(y) = \min\{(t/\delta)(y), 1(y)\}$.

In the *Embedding theorem II* statement, we will speak of a "cube". When one casually speaks of a "cube" a three dimensional space, $[a, b]^3$, comes to mind. In topology, unless it is otherwise specified, the word "cube" refers more generally to any finite or infinite Cartesian product,

$$\prod_{i\in J}[a_i, b_i]$$

of closed intervals, $[a_i, b_i]$ in $\mathbb{R}$. Since any closed and bounded interval $[a, b]$ is homeomorphic to the closed unit interval, $[0, 1]$, then, by Theorem 7.13, any cube, $\prod_{i\in J}[a_i, b_i]$, is homeomorphic to a product space, $\prod_{i\in J}[0, 1]$, of the unit interval $[0, 1]$.

Theorem 10.17 (The Embedding Theorem II). *Suppose S is a T_1 topological space. A space, S, is completely regular if and only if S can be homeomorphically embedded in some cube, $\prod_{i\in J}[a_i, b_i]$.*

Proof. We are given that (S, τ) is a T_1 topological space.

($\Leftarrow$) Suppose S is embedded in a cube, $\prod_{i\in J}[a_i, b_i]$. We are required to show that S is completely regular.

Since each $[a_i, b_i]$ is a metric space, then, by Theorem 10.5, each $[a_i, b_i]$ is completely regular. By Theorem 10.7, the cube, $\prod_{i\in J}[a_i, b_i]$, is completely regular. Since S can be viewed as a subspace of $\prod_{i\in J}[a_i, b_i]$, by Theorem 10.6, S is completely regular.

($\Rightarrow$) Suppose S is completely regular. We are required to show that S is embedded in a cube.

Two facts follow from S being completely regular.

— By Lemma 10.16, the topology on S, is the weak topology induced by $C^*(S)$.
— Since S is completely regular, $C^*(S)$ separates points and closed sets of S.

So, by the *Embedding theorem I*, the evaluation map,

$$e : S \to \prod_{i\in J}[a_i, b_i]$$

induced by $C^*(S)$, homeomorphically embeds S into the product space, $\prod_{i\in J} X_i$ where each of the factors, X_i, of the product is the codomain of the function, f_i, in $\mathscr{F}$.

Each function, f_i, in $C^*(S)$ is bounded and so maps S into some closed interval, $[a_i, b_i]$, of $\mathbb{R}$, so $X_i \subseteq [a_i, b_i]$.

So, the evaluation map, $e : S \to \prod_{i\in J}[a_i, b_i]$, induced by $C^*(S)$, homeomorphically embeds S into the cube, $\prod_{i\in J}[a_i, b_i]$.

Since $\prod_{i\in J}[a_i, b_i]$ and $\prod_{i\in J}[0, 1]$ are homeomorphic spaces (by Theorem 7.13), then the compact space, $\prod_{i\in J}[0, 1]$, contains a homeomorphic copy of S.

Notation: If S is a space and $f : S \to [0, 1]$ then $f \in [0, 1]^S = \prod_{x\in S}[0, 1]$.

10.9 Topic: On P-Points and P-Spaces

Suppose S is a completely regular space. We know that infinite intersections of open subsets need not be open. We have referred to such sets as G_δ's. In fact some G_δ sets are closed (but not open). Some G_δ-sets are points while others may have non-empty interior in S. We will be investigating completely regular spaces in which the following property is satisfied: If G is a G_δ which contains a point p then $p \in \text{int}_S G$. For example, $\mathbb{R}$ would not be such a space since $\{1\} = \cap\{(1 - 1/n, 1 + 1/n) : n \in \mathbb{N}\backslash\{0\}\}$ and $\text{int}_S\{1\} = \varnothing$. In relation to this concept we introduce the following definitions.

Definition 10.18 Let S be a completely regular topological space. We will say that a point p in S is a *P-point* in S if, for any G_δ-set B containing p, $p \in \text{int}_S B$. If every point in S is a P-point then we will say that S is a *P-space.*

We list a few properties by P-spaces.

Theorem 10.19 *Completely regular spaces P-spaces are zero-dimensional spaces.*

Proof. Suppose S is a completely regular P-space.

Let $x \in S$ and U_0 be some open neighborhood of x. Since S is regular, there exists an open subset, U_1, of S such that

$$x \in U_1 \subseteq \text{cl}_S U_1 \subseteq U_0$$

We now inductively construct the decreasing family, $\{U_i : i \in \mathbb{N}\}$, of open neighborhoods of x where $x \in U_i \subseteq \text{cl}_S U_i \subseteq U_{i-1}$. Let

$$B = \cap\{U_i : i \in \mathbb{N}\} = \cap\{\text{cl}_S U_i : i \in \mathbb{N}\}$$

a closed G_δ-set containing x. Since S is a P-space, then x belongs to the interior of B. Then, for any $x \in B$, $x \in \text{int}_S B$. This means $B \subseteq \text{int}_S B$. So, the G_δ, B, is a clopen set containing x which is itself contained in U_0. Then the set of all clopen subsets of S then forms a base for the open subsets.

Then S has a base of clopen sets. So, S is zero-dimensional.

Recall that any zero-set of S is a G_δ. So in a completely regular P-space, every point is in the interior of every zero-set which contains it. So, as shown in the proof above, every zero-set in such spaces are clopen. This is the case for all cozero-sets of S, also. As shown in Theorem 10.12, in a completely regular space, the family of all cozero-sets form a base for open sets. It follows that in completely regular P-space all zero-sets form a base for open sets.

On the other hand every clopen set is a G_δ equal to its interior. So, S would be a P-space.

So,

> *at least for completely regular spaces, P-spaces are precisely those spaces for which $Z[S]$ forms a base for open sets.*

Theorem 10.20 *A pseudocompact P-space is finite.*

Proof. The proof is differed to Theorem 23.7, where we will have at our disposal extra tools.

Concept Review

1. Define "dyadic rationals" used in the proof of Urysohn's lemma.
2. State Urysohn's lemma.
3. What are we referring to when we speak of a *Urysohn separating function for two closed sets.* Describe such a function.
4. Define a completely regular space (Tychonoff space).
5. How does a completely regular space differ from a $T_{3\frac{1}{2}}$-space?
6. Is a normal space necessarily completely regular? Is a completely regular space necessarily regular? Why?
7. Define a perfectly normal space.
8. What are we referring to when we speak of a zero-set in a topological space?
9. It was proven that, in a certain type of space, G_δ's and zero-sets mean the same thing. What type of space did we refer to?
10. Provide an example of a completely regular space that is not normal.
11. Provide an example of a perfectly normal topological space.
12. Provide an example of a normal space which is not perfectly normal.
13. Describe the *Moore plane* and its topology.
14. Describe an *ordinal space* and its base for open sets.
15. Is the completely regular property hereditary?
16. Suppose $S = \prod_{i\in J} S_i$. If S is completely regular does it follow that each S_i is completely regular?
17. Suppose $S = \prod_{i\in J} S_i$. If each S_i is completely regular does it follow that S is completely regular?
18. The completely regular topology is the weak topology induced by which family of functions?
19. The evaluation map, e, which is used embed S into a cube is with respect to which family of functions?
20. Are continuous images of a completely regular space always completely regular?

Exercises

1. Let (S, τ_S) be a normal topological space which contains a subset T which is an F_σ. Verify whether T is a normal subspace.

2. Are there any normal spaces which are not separable? If so, show one.
3. Is "perfectly normal" a hereditary property?
4. Is it true that a topological space, (S, τ_s), is normal if and only if any two non-empty disjoint closed subsets have disjoint closed neighborhoods? Prove it.
5. Let (S, τ_s) be a normal topological space which contains the non-empty closed subset F. If U is an open neighborhood of F, show that $F \subset V \subset U$ for some F_σ-set, V.
6. Let (S, τ_s) be a countable completely regular space. Show that S is normal.
7. Let (S, τ_s) be a completely regular topological space. Let V be an open neighborhood of a point x. Show that there exists a Urysohn separating function, $f : S \to [0, 1]$, for $\{x\}$ and $S \backslash V$ if and only if x is a G_δ.
8. We showed that closed subspaces of normal spaces are normal. A topological space (S, τ_s) is said to be *completely normal* if and only if *every one* of its subspaces is a normal space. Show that S is completely normal if and only if, whenever $\mathrm{cl}_S U \cap V = \varnothing = U \cap \mathrm{cl}_S V$, there exists disjoint open neighborhoods for U and V, respectively. Note: To prove "$\Rightarrow$" consider $S \backslash (\mathrm{cl}_S U \cap \mathrm{cl}_S V)$.

Part IV
Limit Points in Topological Spaces

Chapter 11

Limit Points in First Countable Spaces

Abstract
In this section, we will investigate whether the notion of limit points and their sequences can be applied to topological spaces. We know that, in metric spaces, closed subsets can be characterized by the limit points they possess. Also, continuous functions can be characterized in terms of how they act on sequences and their limits. We will show that first countable topological spaces have sequences and limits points which can be handled in the usual way.

11.1 Introduction

Limit points of sets play an important role in the study of normed and metric spaces. In particular, they are involved in characterizations of closed sets and the sequential definition of continuity. Up to this point limit points are assumed to be the limit of sets called *sequences*, all indexed by countably infinite linearly ordered sets. We will investigate *limit points* in a more general context of topological spaces.

11.2 Sequences and Limit Points in Topological Spaces

We will first start with the formal definition of a sequence of elements in a topological space. It is analogous to the definitions introduced in normed vector spaces and metric spaces.

Definition 11.1 Let S be any topological space.

(a) *Sequence*: A *sequence* in S is a function, $f : \mathbb{N} \to S$, mapping $\mathbb{N}$ into S. It can be denoted as $\{f(i) : i = 0, 1, 2, 3, \ldots\}$, $\{x_0, x_1, x_2, x_3, \ldots, x_n, \ldots\}$ where $x_i = f(i)$ for each $i \in \mathbb{N}$, or, $\{x_i : i = 0, 1, 2, 3, \ldots\}$ or, simply $\{x_i\}$.

For $T_j = \{i \in \mathbb{N} : i \geq j\}$, we will say that $f[T_j] = \{x_i : i \geq j\}$ is a "tail end" of the sequence $\{x_i\}$.

(b) *Convergence*: A sequence, $\{x_i\}$, is said to *converge to a point* p in S if and only if, for every open neighborhood U of p, there exists, N, such that, when $n > N$, x_n belongs to U. We also say "...if and only if every open neighborhood of p contains a tail end of the sequence".[1] The expression, $\{x_i\} \to p$, is used to indicate that the sequence converges to p.

(c) *Limit point of a sequence*: If $\{x_i\} \to p$, then p is a *limit point of the sequence* $\{x_i\}$.

(d) *Limit point of a set*: If F is a non-empty set, we say that p *is a limit point of the set*, F, if and only if p is the limit point of some sequence, $\{x_i\}$, which is entirely contained in F.

(e) *Accumulation point of a sequence*: We say that k is an *accumulation point* of the sequence, $A = \{f(i)\}$, if every open neighborhood of k has a non-empty intersection with every tail end, $f[T_j] = \{x_i : i \geq j\}$, of A.

The reader is no doubt familiar with the following characterization of "closed set" given in the study of metric spaces:

> *A non-empty subset, F, of the metric space, (M, ρ), is closed in M, if and only if, F contains the limit point of each and every sequence contained in F.*

Also, we have another definition which involves sequences, namely the sequential definition of continuity for functions mapping a metric space to a metric space.

> *A function, f, mapping a metric space, M_1, into a metric space, M_2, is said to be continuous at a point p if and only if, whenever p is the limit point of a sequence, $\{x_i\}$, in*

[1] The expressions "$\{x_i\}$ converges to p if the sequence eventually belongs to every neighborhood, U of p" is also used.

M_1 $f(p)$ is the limit point of the corresponding sequence, $\{f(x_i)\}$, in M_2.

Can we easily generalize the statements from "metric spaces" to non-metrizable topological spaces? We will show that the transition from sequences in metric spaces to sequences in topological spaces flows fairly smoothly whenever the topological spaces are first countable; that is, if every point has a countable neighborhood base.

We begin by providing the following definition.

Definition 11.2 Let S and T be topological spaces.

We say that a function $f : S \to T$ is *sequentially continuous* at p if and only if, whenever p is limit point of a sequence, $\{x_i\}$, in S, then $f(p)$ is the limit point of the corresponding sequence, $\{f(x_i)\}$ in T.

Theorem 11.3 *Let (S, τ_S) and (T, τ_T) be* first countable *topological spaces.*

(a) *A non-empty subset, F of S, is closed in S if and only if every limit point of F belongs to F.*
(b) *A function mapping S into T is continuous on S if and only if it is sequentially continuous at each point, p, in S.*

Proof. Let (S, τ_S) and (T, τ_T) be first countable topological spaces.

(a) ($\Rightarrow$) Suppose F is a closed subset of S and let p be a limit point of F.

Then, by definition, there is a sequence, $\{x_i\} \subseteq F$, which converges to p. Suppose $p \in S \backslash F$. Then, if U is any open neighborhood of p, $U \cap (S \backslash F)$ contains a tail end of the sequence $\{x_i\}$. Since this is impossible, then $p \in F$.

($\Leftarrow$) Suppose F contains all its limit points.

Suppose there exists a point $p \in \text{cl}_S F \backslash F$. By hypothesis, there exists, $\{U_i : i \in \mathbb{N}\}$, a countable open neighborhood base of p. If, for $n \in \mathbb{N}$, $V_n = U_1 \cap U_2 \cap \cdots \cap U_n$, then $\{V_n : n \in \mathbb{N}\}$ is a countable neighborhood base of open sets of p, such that $V_{n+1} \subseteq V_n$. Then $\{p\} = \cap\{V_n : n \in \mathbb{N}\}$.

Since p is a boundary point of F, we can choose, for each $i \in \mathbb{N}$, $x_i \in V_i \cap F$ (Choice function). Since every V_i contains a tail end of the sequence $\{x_i\}$, then p is a limit of the sequence in F; so it is a limit point of F. By hypothesis, p, being a limit of F, should belong to F. This contradicts, $p \in \text{cl}_S F \backslash F$. To avoid the contradiction we must have, $\text{cl}_S F \backslash F = \varnothing$, which implies $F = \text{cl}_S F$, and so F is closed.

(b) ($\Rightarrow$) Suppose $f : S \to T$ is continuous on S and let $p \in S$. Also, suppose that p is the limit point of a sequence, $\{x_i\}$, in S.

If V is an open neighborhood of $f(p)$, then there exists an open neighborhood U of p such that $f[U] \subseteq V$.

Then U contains some tail end, say $\{x_i : i > N\}$, of the sequence $\{x_i\}$. Then $\{f(x_i) : i > N\} \subseteq f[U] \subseteq V$. So, V contains a tail end of the sequence $\{f(x_i)\}$. Since V is an arbitrary open neighborhood of $f(p)$, then $\{f(x_i)\} \to f(p)$. This establishes the proof of ($\Rightarrow$).

($\Leftarrow$) Suppose $f : S \to T$ is such that $\{x_i\} \to p$ in S implies $\{f(x_i)\} \to f(p)$.

We are required to prove that f is continuous at p. By hypothesis, there exists, $\{U_i : i \in \mathbb{N}\}$, a countable open neighborhood base of p, where $U_{i+1} \subseteq U_i$. Then $\cap\{U_i : i \in \mathbb{N}\} = \{p\}$.

We will proceed by contradiction. That is, suppose f is not continuous at p. Then there exists an open neighborhood, V, of $f(p)$ such that, for each i, $f(x_i) \backslash V \neq \varnothing$. Then, for each j, there exists $y_j \in U_j$ such that $f(y_j) \notin V$. Then $\{y_j\} \to p$; but the tail end of the sequence, $\{f(y_j)\}$, is excluded from V and so can't converge to $f(p)$, a contradiction. So f is continuous at p.

The above theorem allows us to say that, in a first countable space S, if F is a subset of S, then the closure, $\text{cl}_S F$, of F is the set of all the limit points of F. Also, in first countable spaces, to say that a function, $f : S \to T$, is continuous on S is to say that f is "sequentially continuous" at each point p in S.

11.3 Subsequences of a Sequence

Sequences in a first countable topological space don't always converge to a point. But infinite subsets of a sequence are themselves sequences and might still "converge".

Definition 11.4 Let S be a first countable topological space. Let

$$A = \{f(i) : i = 1, 2, 3, \dots\} = \{x_i\}$$

be a sequence in S. Suppose $g : \mathbb{N}\backslash\{0\} \to \mathbb{N}\backslash\{0\}$ is a *strictly increasing function*. Then the sequence,

$$B = \{f(g(i)) : i = 1, 2, 3, \dots\} = \{x_{g(i)}\}$$

is called a *subsequence of the sequence* A. By strictly increasing we mean $i > j \Rightarrow g(i) > g(j)$.

Note that the notation, $x_{g(i)}$, means that $x_{g(i)}$ is the element of the sequence $\{x_i\}$ with the index number $g(i)$. The terms in a subsequence *always* respect the order in which they appear in the sequence.

Example 1. If $A = \{\frac{1}{1}, \frac{1}{2}, \frac{1}{3}, \frac{1}{4}, \dots, \frac{1}{n}, \dots\}$, where $a_n = 1/n$, and $g(n) = 2n$, then $B = \{a_{g(n)}\} = \{a_{2n}\}$ is the subsequence

$$B = \left\{\frac{1}{2}, \frac{1}{4}, \frac{1}{6}, \dots, \frac{1}{2n}, \dots\right\}$$

Every element of a subsequence must always be an element of the sequence from which it is derived. The key is that the index function must be strictly increasing. Also, remember that a sequence always has infinitely many terms (but not necessarily have infinitely many distinct elements). In this example, the sequence $\{\frac{1}{2}, \frac{1}{2}, \frac{1}{3}, \frac{1}{3}, \frac{1}{4}, \frac{1}{4}, \dots\}$ is not a subsequence of the sequence A since the indexing function used is not strictly increasing.

Example 2. Let

$$A = \{1, 1, 2, 1, 3, 1, 4, 1, \dots\}$$

Then $B = \{1, 2, 3, 4, 5, \dots\}$ and $C = \{1, 1, 1, 1, 1, 1, 1, \dots\}$ are both subsequences of A. The subsequence B does not converge and does not have a subsequence which converges. On the other hand, the subsequence C of A converges to 1.

So the element, 1, is, by definition, an accumulation point of the sequence A.

Note that, if we view the sequence A as simply the "set of all non-zero natural numbers" then 1 is not, by definition, a "cluster point"[1] of the set A even though 1 is an accumulation point of A. It will be understood in this textbook that "cluster point" refers to a point in relation to some set. While accumulation point is used in relation to sequences (or later, to nets).

Note. Some authors use "cluster point" or "accumulation point" interchangeably, which may cause some confusion in some theorem proofs. So readers should verify carefully how authors define both words.

Theorem 11.5 *Let* $A = \{f(i) : i \in \mathbb{N}\} = \{x_i\}$ *be a sequence in a* first countable *topological space* S. *Then* p *is an accumulation point of the sequence,* A, *if and only if it is a limit point of some subsequence,* $\{f(g(i)) : i \in \mathbb{N}\} = \{x_{g(i)}\}$, *of* A.

Proof. Let $A = \{f(i) : i \in \mathbb{N}\} = \{x_i\}$ be a sequence in a *first countable* topological space S.

($\Leftarrow$) Suppose A has a subsequence, T, which converges to the point p. Then every open neighborhood of p contains a tail end of T, and so intersects a tail end of A. So p is an accumulation of A.

($\Rightarrow$) Suppose p is an accumulation point of A. We are required to find a subsequence of A which converges to p.

Since S is first countable, p has a countable open neighborhood base,

$$\mathscr{B}_p = \{B_i : i = 1, 2, 3, \ldots\}$$

such that $B_{i+1} \subseteq B_i$ for all i. If $T_j = \{i : i \geq j\}$ is a tail end of $\mathbb{N}$, then for all i and j, $f[T_j] \cap B_i \neq \varnothing$.

We inductively construct a subsequence, B, of A which converges to p. Let $g(1)$ be the smallest number in T_1, such that, $f(g(1)) \in B_1$. Suppose that, for $g(1) < g(2) < \cdots < g(j)$, $f(g(i)) \in B_i$. Let $g(j+1)$ be the smallest number in $T_{g(j+1)}$ such that $f(g(j+1)) \in B_{j+1}$.

[1]Reminder: In the chapter on closures on page 62, we defined a *cluster point of a set* A as a point p such that any open neighborhood of p contains some other point x in A.

We thus obtain a subsequence

$$B = \{f(g(i)) : i = 1, 2, 3, \ldots\}$$

of A, which converges to p, as required.

We will later see that, if S is not first countable, the statement above will fail to be true, in general. So we will have to redefine our notions about "sequence".

Concept Review

1. If S is a first countable topological space what does it mean to say that $\{x_i\}$ is a sequence in S?
2. If S is a first countable topological space what does it mean to say that p is limit point of $\{x_i\}$, a sequence in S?
3. If S is a first countable topological space what does it mean to say that p is a limit of a subset, U, in S?
4. If S is a first countable topological space and the subset, U of S, contains all its limit points state a topological property possessed by U.
5. If S is a first countable topological space and F is a closed subset of S what can we say about the set of all limit points of F?
6. If S is a first countable topological space define "the closure of F" in terms of limit points.
7. Define a subsequence of a sequence in a first countable topological space.
8. Define an accumulation point of a sequence in a first countable topological space.
9. If p is and accumulation point of a sequence is p a limit point of the sequence? Explain.

Exercise

1. Let (S, τ_S) be a first countable topological space. Define the "interior of the subset F" in terms of limit points.

Chapter 12

Limit Points of Nets

Abstract
In this section, we will expand the definition of "limit point" so that it applies to arbitrary topological spaces, including those which are not first countable. The indexing set of all natural numbers used for sequences will be substituted by a more general indexing set called "directed set". The sets whose elements are ordered by a "directed set" will be called "nets". Nets will serve as a tool which will assist in recognizing closed subsets and continuous functions in arbitrary topological spaces. They will also, eventually, be used to identify sets which are "compact" once this concept has been properly defined.

12.1 Directed sets

Suppose "$\leq$" is a *partial order relation* on a set, S. That is,

— for all $x \in S$, $x \leq x$, (reflexive property)
— if $x \leq y$ and $y \leq x$ then $x = y$, (anti-symmetric property)
— if $x \leq y$ and $y \leq z$ then $x \leq z$ (transitive property)

The second condition (referred to as the "anti-symmetric property") is not absolutely required in what follows, but it is easier to set the stage in this way to introduce the notion of a "directed set" whose definition requires some sort of predefined order relation on a given set.

Definition 12.1 Let D be a non-empty infinite set on which is defined a partial ordering relation "$\leq$". Suppose the partially ordered set, $(D, \leq)$, also satisfies the property:

If $x, y \in D$, there exists $z \in D$ such that $x \leq z$ and $y \leq z$

(The directed set property)

Then we refer to $(D, \leq)$ as a *directed set.*

We already know of a directed set, namely, $(\mathbb{N}, \leq)$. Simply see that, if $m, n \in \mathbb{N}$, there exists $k \in \mathbb{N}$ such that $m \leq k$ and $n \leq k$. So, a directed set is a generalization of the linearly ordered countable sets we normally use as indexing sets for sequences. As we can see, the definition of directed set does not require that its order relation, "$\leq$", be linear. It is usually appears in the form of a partial ordering.

Example 1. Given a set, S, its power set, $(\mathscr{P}(S), \subseteq)$, is partially ordered by the inclusion relation, $\subseteq$. This order relation can be seen to satisfy the directed set property. Then $(\mathscr{P}(S), \subseteq)$ is a directed set where $\varnothing \subseteq T$ for all $T \in \mathscr{P}(S)$ and $T \subseteq S$ for all $T \in \mathscr{P}(S)$.

Example 2. Consider the set of all finite partitions, $\mathscr{P}_{art}$, of the closed interval $[0, 10]$ of real numbers. For example, $M = \{[0, 4] \cup [4, 7] \cup [7, 10]\} \in \mathscr{P}_{art}$. We define a partial order on $\mathscr{P}_{art}$ as follows: $M \leq K$ if and only if K is a finer partition then M. That is, each closed interval of M is a union of closed intervals in K. For example,

$$M = \{[0, 4] \cup [4, 7] \cup [7, 10]\} \leq \{[0, 2] \cup [2, 4] \cup [4, 7] \cup [7, 9] \cup [9, 10]\} = K$$

It is easy to verify that "$\leq$" respects the directed set property on $\mathscr{P}_{art}$. For example, suppose $M = \{[0, 5] \cup [5, 10]\}$ and $J = \{[0, 2] \cup [2, 10]\}$. If

$$K = \{[0, 2] \cup [2, 5] \cup [5, 7] \cup [7, 10] \cup [9, 10]\}$$

then $M \leq K$ and $J \leq K$. See that $[0, 10] \leq W$ for all $W \in \mathscr{P}_{art}$. Then $(\mathscr{P}_{art}, \leq)$ is a directed set.

We will now show how directed sets can be used to establish a direction in arbitrary subsets of a topological space. The definition of a net, in terms of a function, is similar to the definition of a sequence except that the index set is not necessarily linear and so can be "larger" than $\mathbb{N}$.

Definition 12.2 Let S be a set.

(a) A *net in* S is a function, $f : D \to S$, mapping a directed set, $(D, \leq)$, into S. It is usually denoted as

$$\{f(i) : i \in D\} = \{f(i)\}_{i \in D} = \{x_i\}_{i \in D}$$

or, simply, by $\{x_i\}$ whenever it is obvious that the index set is the directed set, D.

(b) *Tail-end of N*: Let $N = \{f(i)\}_{i \in D}$ be a net in S, $k \in D$ and $T_k = \{i \in D : i \geq k\}$. We will refer to a subset of the form, $f[T_k] = \{f(i) : i \geq k\}$ as a *tail end* of the net, N. We can obtain many tail ends of N by varying the value of k in D.

(c) *Convergence*: Suppose (S, τ) is a topological space and $N = \{x_i\}_{i \in D}$ is a net in S, ordered by a directed set, D. The net, N, is said to *converge to a point* p in S (with respect to τ and the given ordering) if and only if every open neighborhood, U of p, contains some tail end, $f[T_k]$, of the net N. That is, there exists $k \in D$ such that, whenever $j \geq k$, then $x_j \in U$. The expression,

$$\{x_i\} \to p$$

is used to indicate that this net converges to p.

(d) *Limit point*: If a net, $\{x_i\}_{i \in D}$ in a topological space, S, converges to the point p, we say that p is a *limit point of the net* $\{x_i\}_{i \in D}$.

(e) *Limit point of a set*: If F is a non-empty set in a topological space, S, we say that p *is a limit point of the set,* F, if and only if p is the limit point of some net, $\{x_i\}_{i \in D}$, which is entirely contained in F.

(f) *Accumulation point*: Suppose (S, τ) is a topological space. We will say that y is an *accumulation point of the net,* $N = \{f(i)\}_{i \in D}$ in S, or that A *accumulates at* y, if for *every* tail end $f[T_k] = \{f(i) : i \geq k\}$, of N and *every* open neighborhood B_y of y,

$$f[T_k] \cap B_y \neq \varnothing$$

Since $(\mathbb{N}, \leq)$ is a directed set, then a sequence, $\{x_i : i \in \mathbb{N}\}$, is a particular kind of net.

Example 3. Consider the net (sequence),

$$N = \{n + (-1)^n n : n \in \mathbb{N}\backslash\{0\}\} = \{0, 4, 0, 8, 0, 12, \ldots\}$$

with directed set $\mathbb{N}\backslash\{0\}$ in $\mathbb{R}$ (with the usual topology). This net does not have a limit point, since a sufficiently small open neighborhood, $(p - \varepsilon, p + \varepsilon)$, of any p cannot contain a tail end of N. However, any neighborhood, $(-\varepsilon, \varepsilon)$, of the number, 0, intersects every tail end of the net (since all tail ends of N contain zeroes). So, 0 is an accumulation point.[1]

It is important to remember that limit points or convergence of a net can only be discussed with respect to a particular topology in mind. Given a net, $N = \{f(i)\}_{i \in D}$, in a topological space, if we change the topology on the set, then we may be altering its convergence properties. However, the directed set, $(D, \leq)$, which indexes the net elements is *not* topologized.

By definition, nets are indexed subsets of a topological space. If N is a net in S we stated that N may have a limit point, p, and may have an accumulation point, y. If we say that the net, N, "has" a limit point, p, we do not imply that $p \in N$. Also, by definition, every neighborhood of a limit point p *contains* a tail end of N so every neighborhood of p *intersects* a tail end of N and so, by definition, every limit point, p, of N is an accumulation point of N. Of course, an accumulation point of N need not be a limit point, as the example above shows. So, accumulation points of a net are of interest only when the net does not converge to a point in S.

Example 4. Consider the product space, $S = [0, \omega_1) \times [0, \omega_1)$ and $S^* = [0, \omega_1] \times [0, \omega_1]$ where ω_1 is the first uncountable ordinal. Let $p = (\omega_1, \omega_1)$. The set,

$$\mathscr{B} = \{(\alpha, \omega_1] \times (\beta, \omega_1] : \alpha < \omega_1, \beta < \omega_1\}$$

---

[1] In some texts, authors use the word "cluster of a net" instead of "accumulation point of a net". This may lead to confusion since the word "cluster" is already defined in the context, "cluster point of a set" (see page 62). Note that the net N in this example, when viewed as a set, (according to definitions) has no cluster points but does have an accumulation point.

forms a neighborhood base of open sets for the point $p \in S^*$. Construct a net in S whose limit point is p.

Solution: We are given $\mathscr{B} = \{(\alpha, \omega_1] \times (\beta, \omega_1] : \alpha < \omega_1, \beta < \omega_1\}$ forms a neighborhood base of open sets for the point $p \in S^*$. For $U, V \in \mathscr{B}$ we define $U \leq V$ if and only if $V \subseteq U$. Then "$\leq$", thus defined on $\mathscr{B}$, is a partial ordering that satisfies the directed set property, and so forms a directed set, $(\mathscr{B}, \leq)$.

Let $f : \mathscr{B} \to S$ be a choice function which maps each $U \in \mathscr{B}$ to one point of our choice, say $f(U) = x_U$ in $U \cap S$. We thus obtain, by definition, a net

$$\{f(U) : U \in \mathscr{B}\} = \{x_U : U \in \mathscr{B}\}$$

an uncountable subset of the product space, $S = [0, \omega_1) \times [0, \omega_1)$, with directed set $\mathscr{B}$.

We claim that, $p = (\omega_1, \omega_1)$ is a limit point of the set, S. It suffices to show that, $\{x_U\} \to p$.

Let $K = (\alpha, \omega_1] \times (\beta, \omega_1]$ be an arbitrary open neighborhood of p in S^*. There exists γ, μ such that $V = (\gamma, \omega_1] \times (\mu, \omega_1] \subset K$. Then $V \geq K$ and so $x_V \geq x_K$. That is, a tail end, $f[T_K] = \{x_U : U \geq K\}$ is contained in the open neighborhood, K of p. Hence, the tail end, $f[T_K]$, of the net, $\{x_U : U \in \mathscr{B}\}$, in S is a subset of the open neighborhood, K, of p. Hence, $\{x_U\} \to p$.

Since S contains a net, $\{x_U : U \in \mathscr{B}\}$, which converges to p, then p is a limit point of the set, S, as claimed.

12.2 Closed Sets and Continuity in Terms of Nets

We now present the results which describe the closure of a set and the continuity of a function in terms net convergence.

Theorem 12.3 *Let (S, τ_S) and (T, τ_T) be topological spaces.*

(a) *Let E be a subset of S. A point p belongs to $cl_S E$ if and only if there is a net in E which converges to p.*

(b) *A function, $f : S \to T$, mapping S into T is continuous at p in S if and only if whenever p is limit point of a net, $\{x_i\}$, in S then $f(p)$ is the limit point of the corresponding net, $\{f(x_i)\}$ in T.*

Proof. Let (S, τ_S) and (T, τ_T) be topological spaces.

(a) ($\Rightarrow$) Suppose $E \subseteq S$ and p belongs to $\text{cl}_S E$. We can index all open neighborhoods of p and partially order them by reverse inclusion to form the directed set $\mathscr{D} = \{U : U$ is an open neighborhood of p. Then, for each open neighborhood, U, of p, we can, by a choice function, choose x_U in $U \cap E$. We obtain a net, $\{x_U : U \in \mathscr{D}\}$, in E, which converges to p.

($\Leftarrow$) Suppose $\{x_i\}$ is a net in E which converges to p. Then, by definition of "convergence", each neighborhood of p intersects E. Then p cannot belong to the open set $S \backslash \text{cl}_S E$. So, $p \in \text{cl}_S E$.

(b) ($\Rightarrow$) Suppose $f : S \to T$ is continuous at $p \in S$ and p is a limit point of a net, $A = \{x_i : i \in I\}$. Let U be an open neighborhood of $f(p)$ in T.
By definition of continuity, $f^{\leftarrow}[U]$ is an open neighborhood of p in S. Since the net A converges to p, then, by definition, $f^{\leftarrow}[U]$ contains a tail end, $\{x_i : i \geq k\}$, of A. Then $f[\{x_i : i \geq k\}] = \{f(x_i) : i \geq k\}$ is a tail end of $f[A]$ which belongs to U.
So, $f[A]$ converges to $f(p)$.

($\Leftarrow$) Left as an exercise for the reader.

The above result shows that, given a subset, F, of a topological space, S, we can use nets as a tool to determine the closure, $\text{cl}_S F$, of the set F.

(a) A set F in a topological space, S, is closed in S if and only if F contains the limit point of every net which is contained in F.
(b) Equivalently, p is a limit point of a closed subset of S if and only if every open neighborhood of p intersects F.

We choose sequences or nets depending on the nature of the topological space considered.

12.3 Subnet of a Net

We know that sequences have special subsets called "subsequences". Nets can have proper subsets which are analogous to subsequences. We formally define this concept in such a way that it generalizes the concept of a subsequence.

Definition 12.4 Let S be a set and $N = \{f(i) : i \in D\}$ be a net in S with directed set D.

Suppose $T_j = \{i : i \geq j\}$ denotes a tail end of D (with "least" element j).

(a) *Cofinal*: Suppose U is a subset of D such that for each $d \in D$ there is an element $u \in U$ such that $u \geq d$. Then we say that

$$U \text{ is } \textit{cofinal in } D.$$

Verify that, if U is cofinal in D, then U is also a directed set. By extension, we will say that $M = \{f(i) : i \in U\}$ is a cofinal net in N. In this case, M intersects every tail end of N. We can then say that, "if every neighborhood U of y, $U \cap N$ is *cofinal* in N then y is an accumulation point of N".

(b) *Subnet of a net*: Let $\mathscr{D}$ be a directed set (possibly different from D).
Suppose $g : \mathscr{D} \to D$ is an *increasing* function on $\mathscr{D}$ (i.e., $i \geq j \Rightarrow g(i) \geq g(j)$) so that $g[\mathscr{D}]$ is cofinal in D; then $(f \circ g)[\mathscr{D}]$ is cofinal in N.
Let

$$B = \{f(g(i)) : i \in \mathscr{D}\} = \{x_{g(i)} : i \in \mathscr{D}\}$$

be a cofinal subset of N. Then B is a called a *subnet of the net* N.

A few remarks on the notion of a subnet. In the above definition the function g points to a cofinal subset of the directed set D while the function f associates this subset to a cofinal subset (the subnet) of the net N. We alert the reader to the fact that, in the usual definition of a *subsequence* we require that the function g be "strictly" increasing. So g chooses not only a cofinal subset of $\mathbb{N}$ but a "subsequence" of $\mathbb{N}$.

The definition of "subnet" requires that g be only "increasing". So, g may be pointing to a cofinal subset of D, but some of the elements of this cofinal subset may be repeating.

We verify how this may make a difference. If $f(n) = 1/n$, so that the sequence, $A = \{f(n)\} = \{1, 1/2, 1/3, 1/4, \ldots\}$, and $\{g(n)\} = \{1, 1, 2, 2, 3, 3, \ldots\}$, then we end up with a subnet $B = \{f(g(n))\} = \{1, 1, 1/2, 1/2, 1/3, 1/3, \ldots\}$, a subset of A, but clearly not a subsequence of A (at least not as we viewed it previously). So, *B is a subnet of A, but B does not satisfy the definition of a subsequence of A.*

Given our experience with subsequences, we expect that, if a net, N, in the space S, has an accumulation point x in S, the net will have a subnet B which will converge to it. We confirm that this holds true in the next theorem.

Theorem 12.5 *Let S be a topological space containing a net, $N = \{f(i) : i \in D\}$ and $u \in S$. Then u is an accumulation point of N if and only if N has a subnet converging to u.*

Proof. Let S be a topological space containing a net $N = \{f(i) : i \in D\}$ with directed set D.

($\Leftarrow$) Suppose $N = \{f(i) : i \in D\} = \{x_i\}$ has a subnet $B = \{f(g(i)) : i \in \mathscr{D}\} = \{x_{g(i)}\}$ which converges to u (where $\mathscr{D}$ is a directed set). We must show that for any open neighborhood B_u of u B_u is cofinal in N. Let B_u be any open neighborhood of u. Since the subnet, B, converges to u, then B_u contains a tail end,

$$f[g[T_r]] = \{f(g(i)) : i \geq r\} = \{x_{g(i)} : i \geq r\}$$

of B, where $r \in \mathscr{D}$. Let

$$f[T_k] = \{f(i) : i \geq k\}$$

be some tail end of N. By definition, $g[\mathscr{D}]$ is cofinal in D. Then there exist $q \in \mathscr{D}$ such $g(q) > \max\{k, g(r)\}$. Then $f(g(q)) \in f[T_k] \cap f[g[T_r]] \subseteq B_u$. So every tail end of N intersects B_u. So, B_u is cofinal in N. So, by definition, u is an accumulation point of N.

($\Rightarrow$) Suppose u is an accumulation point of the net $N = \{f(i) : i \in D\}$. We are required to construct a subnet, B, of N which converges to u. For every $j \in D$, let $T_j = \{i \in D : i \geq j\}$. Let $\mathscr{B}_u$ be an open neighborhood base of u in S. By definition of accumulation point of N, if $U \in \mathscr{B}_u$, U is cofinal in N. (That is, $U \cap f[T_j]$ is non-empty, for all $j \in D$.)

Step 1: *Defining the set* $\mathscr{D}$. Consider the set

$$\mathscr{D} = \{(i, U) \in D \times \mathscr{B}_u : f(i) \in U\}$$

Step 2: *Ordering the elements of* $\mathscr{D}$. We will order the elements of $\mathscr{D}$ with, "$\leq$", as follows:

$$(i,\, U) \leq (j,\, V) \qquad \Leftrightarrow \qquad [i \leq j \text{ and } V \subseteq U]$$

Then $(\mathscr{D}, \leq)$ is a partially ordered set.

Step 3: We claim that $(\mathscr{D}, \leq)$ is a directed set. Suppose $(i,\, V)$, $(j,\, U) \in \mathscr{D}$ and $W = V \cap U$. Since D is directed, there exists $k \in D$ such that $i \leq k$ and $j \leq k$ and $u \in W$. Since u is an accumulation point of N, the tail end, $f[T_k]$, intersects W and so W contains some $f(q)$ for $q \in T_k$. Then $(q, W) \in \mathscr{D}$ where $(i,\, V) \leq (q,\, W)$ and $(j,\, U) \leq (q,\, W)$. We then have a directed set, $(\mathscr{D}, \leq)$.

Step 4: *Constructing the subnet* B. Define $g : \mathscr{D} \to D$ as $g(j,\, U) = j$, for each $(j, U) \in \mathscr{D}$. Then g assigns each (i, U) to i in D such that $f(i) \in U$.

To show that $(f \circ g)[\mathscr{D}]$ is a subnet of N it suffices to show that it is cofinal in N. Let T_k be a tail end of D. It suffices to show that $g[\mathscr{D}] \cap T_k$ is non-empty.

Let V be an open neighborhood of u. Since u is an accumulation point of N, there exists $q \in T_k$ such that $f(q) \in f[T_k] \cap V$. Then $(q, V) \in \mathscr{D}$. Then $q = g(q, V) \in g[\mathscr{D}] \cap T_k$. So, $g[\mathscr{D}]$ is cofinal in D, as claimed.

Then $f\,[g[\mathscr{D}]]$ is cofinal in N, so

$$B = \{f(g(i, U_u)) : (i, U_u) \in \mathscr{D}\}$$

satisfies the definition of a subnet of N.

Step 5: *Claim that the subnet,* B*, converges to* u. Let $U \in \mathscr{B}_u$. To show that the subnet B converges to u we are required to find a tail end of B which is entirely contained in U. Let T_z be a tail end of D.

Since u is an accumulation point of N, $U \cap f[T_z] \neq \varnothing$ so we can choose $f(j) \in U \cap f[T_z] \neq \varnothing$. Then $(j, U) \in \mathscr{D}$.

Suppose $(j, U) \leq (k, V)$. Then $j \leq k$ and $f(k) \in V \subseteq U$. So, $(f \circ g)(k, V) = f(k) \in U$. So, the tail end of all elements in B passed the point $(f \circ g)(j, U_u))$ belongs to U_u. So, the subnet B converges to u, as claimed.

Then the subnet B converges to the accumulation point, u, as required.

Remark. In the above theorem, the net N may be either a sequence or an uncountable net. The cardinality of N does not play a role in the proof, so the statement holds true for countable and uncountable nets.

12.4 A Few Examples

12.4.1 *A comment on a few infinite countable ordinals*

Ordinals can also serve as an indexing set for a net. Recall that there are two types of ordinals: Those ordinals, α, which have an immediate predecessor, say β, where $\beta + 1 = \alpha$ and limit ordinals, γ, where $\gamma = \sup\{\alpha : \alpha < \gamma\} = \cup\{\alpha : \alpha \in \gamma\}$ or, equivalently, γ is an ordinal which does not contain a maximal ordinal. The following represent countable limit ordinals.

$$
\begin{aligned}
\omega_0 &= \{0, 1, 2, 3, \ldots\} = [0, \omega_0) \\
2\omega_0 &= \{0, 1, 2, 3, \ldots, \omega_0, \omega_0 + 1, \omega_0 + 2, \ldots, \omega_0 + n, \ldots\} = [0, 2\omega_0) \\
3\omega_0 &= \{0, 1, 2, \ldots, \omega_0, \omega_0 + 1, \omega_0 + 2, \ldots, 2\omega_0, 2\omega_0 + 1, 2\omega_0 + 2, \ldots\} \\
&= [0, 3\omega_0) \\
&\vdots \\
n\omega_0 &= \{0, 1, 2, \ldots, \omega_0, \omega_0 + 1, \omega_0 + 2, \ldots, (n-1)\omega_0, (n-1)\omega_0 \\
&\qquad +1, \ldots\} = [0, n\omega_0) \\
&\vdots \\
\omega_0\omega_0 &= [0, \omega_0\omega_0) = \{0, \ldots, \omega_0, \ldots, 2\omega_0, \ldots, 99\omega_0 + 100, \ldots\}
\end{aligned}
$$

Those ordinals which are smaller than ω_0 are simply the natural numbers. Ordinals such as $\{\omega_0, 2\omega_0, 3\omega_0, \ldots, n\omega_0 \ldots\}$, are all countably infinite limit ordinals.

With these facts in mind we present a solution for the question in the following example.

Example 5. Consider the set $\{n + \frac{1}{m} : n \in \mathbb{N},\, m \in \mathbb{N}\backslash\{0\}\}$. We will use the set of ordinals, $D = [\omega_0, \omega_0\omega_0)$, as an index to a subset of $\mathbb{Q} \cap [0, \infty)$. The set, D, need not be topologized.

Define $f : [\omega_0, \omega_0\omega_0) \to \mathbb{Q} \cap [0, \infty)$ as follows where, for $n \geq 1$, $m \geq 0$:

$$f(n\omega_0 + m) = m + \frac{1}{n+1}$$

Then $N = \{f(i) : i \in [\omega_0, \omega_0\omega_0)\}$ is a net in $\mathbb{Q} \cap [0, \infty)$ with $[\omega_0, \omega_0\omega_0)$ as directed set.

With this definition, the elements of the net are described as follows:

$$\left\{\frac{1}{2}, 1\frac{1}{2}, 2\frac{1}{2}, 3\frac{1}{2}, \ldots, \frac{1}{3}, 1\frac{1}{3}, 2\frac{1}{3}, 3\frac{1}{3}, \ldots, \frac{1}{n}, 1\frac{1}{n}, 2\frac{1}{n}, 3\frac{1}{n}, \ldots, \frac{1}{n+1}, 1\frac{1}{n+1}, 2\frac{1}{n+1}, 3\frac{1}{n+1}, \ldots, \right\}$$

Show that each natural number is an accumulation point of the net N. Then construct a subnet which converges to each natural number.

Solution: A tail end of $[\omega_0,\ \omega_0\omega_0)$ is of the form

$$T_{a\omega_0+b} = [a\omega_0 + b,\ \omega_0\omega_0) = \{n\omega_0 + m : n\omega_0 + m \geq a\omega_0 + b\}$$

So, a tail end of N is of the form

$$\begin{aligned} f[T_{a\omega_0+b}] &= \{f(n\omega_0 + m) : n\omega_0 + m \geq a\omega_0 + b\} \\ &= \left\{m + \frac{1}{n+1} : n\omega_0 + m \geq a\omega_0 + b\right\} \end{aligned}$$

Let q be a natural number. We claim that q is an accumulation point of N.

Proof of claim: To see this, let $(q-\varepsilon, q+\varepsilon)$ be a small open neighborhood of q.

It suffices to show that $f[T_{a\omega_0+b}] \cap (q-\varepsilon, q+\varepsilon) \neq \varnothing$.

Then there exists $k > a$ such that $q + 1/k \in (q-\varepsilon, q+\varepsilon)$. Then $k\omega_0 + q > a\omega_0 + b$, so $f(k\omega_0 + q) \in f[T_{a\omega_0+b}]$. Since $f(k\omega_0 + q) = q + 1/k \in (q-\varepsilon, q+\varepsilon)$, then $f[T_{a\omega_0+b}] \cap (q-\varepsilon, q+\varepsilon) \neq \varnothing$. So q is an accumulation point of N, as claimed.

Since q is an accumulation point of N, then, by the preceding theorem, N must contain a subnet which converges to q. Define the function $g : [\omega_0, \omega_0\omega_0) \to [\omega_0, \omega_0\omega_0)$ as

$$g(n\omega_0 + m) = n\omega_0 + q \quad \text{(For all } m.)$$

We see that g is increasing (but not strictly increasing) on $[\omega_0, \omega_0\omega_0)$, since for any a and b such that $a < b$,

$$g(a\omega_0 + m) = a\omega_0 + q < b\omega_0 + q = g(b\omega_0 + m) \text{ (So increasing)}$$

$$g^{\leftarrow}(a\omega_0 + q) = \{a\omega_0 + m : m \in \mathbb{N}\} \text{ (So not strictly increasing)}$$

We also have that $f(g(n\omega_0 + m)) = f(n\omega_0 + q) = q + \frac{1}{n+1}$.

Then

$$(f \circ g)[D] = \{q + \frac{1}{n+1} : n \in \mathbb{N}\}$$

mapping D to a subnet $\{q + \frac{1}{n+1} : n \in \mathbb{N}\}$ of N which converges to q as $n \to \omega_0$. As required.

In the following example, we witness how in certain circumstances the theory governing sequences and their subsequences can have shortcomings.

Example 6. Let $S = [0,1) \times [0,1) = \{(a,b) : a, b \in [0,1)\}$. We will equip S with the usual topology inherited from $\mathbb{R}^2$. Let

$$D = \{(c,d) : c, d \in (0,1), \text{ both } c \text{ and } d \text{ are irrationals}\}$$

lexicographically ordered, untopologized. By "lexicographically ordered" we mean that $(e,k) \leq (c,d)$ if $e < c$, and, when $e = c$, then $k \leq d$.[2] Then, with this ordering, $(D, \leq)$ will serve as a directed

[2]This means, we order the ordered pairs by using the same principle as the one used to order words in the dictionary.

set for a net in S. The net,

$$N = \{f(c,d) : (c,d) \in D\}$$

in the space, S, is defined as follows: $f(c,d) = (c,d)$. Then N linearly orders the ordered pairs, (c,d), in the space S, where $(0,0) < (c,d) < (1,1)$.

(a) Is the point, $p = (1,1)$ in S, a limit point of the net, N?
(b) Is the point, $p = (1,1)$ in S, an accumulation point of the net, N?

Solution:

(a) Let $B_\varepsilon(1,1)$ be an open neighborhood of the point $p = (1,1)$ in S. Suppose $T_{(a,b)}$ is a tail end of D. Then $f[T_{(a,b)}] = T_{(a,b)}$ is a tail end of N. Suppose $(a,b) \in B_\varepsilon(1,1)$. See that $(a + \frac{1-a}{2}, b - 2\varepsilon) > (a,b)$ with respect to the lexicographic ordering since $b \neq 1 \Rightarrow \frac{1-a}{2} \neq 0$. But $(a + \frac{1-a}{2}, b - 2\varepsilon)$ does not belong to $B_\varepsilon(1,1)$. So no tail end of the net $N = \{f(c,d) : (c,d) \in D\}$ belongs entirely in $B_\varepsilon(1,1)$. So, p is not a limit point of the net, N.

(b) We claim that $p = (1,1)$ is an accumulation point of the net, N. By the theorem above, it suffices to show that N has a subnet converging to (1, 1).

Let $g : N \to N$ be defined as: $g(a,b) = (a,a)$ if $a \geq b$ and $g(a,b) = (b,b)$ if $a < b$. We see that g is an increasing function with respect to the lexicographic ordering of the ordered pairs. We claim that $g[D]$ is cofinal in D. For $(a,b) \in D$, if $a < b$, $(b,b) > (a,b)$, if $b < a$, $(a,a) > (a,b)$. Then for any (a,b) in D, there exists an element in $g[D]$ which is larger than (a,b). So, $g[D]$ is cofinal in D. So, $f \circ g : N \to \{(a,a) : a \in [0,1)\}$ defines a subnet of N. If $B_\varepsilon(1,1)$ is an open neighborhood of $(1,1)$ then $(f \circ g)[T_{(1-\frac{\varepsilon}{2}, 1-\frac{\varepsilon}{2})}] \subseteq B_\varepsilon(1,1)$. So, $(f \circ g)[D]$ converges to $(1,1)$.

Example 7. Consider the directed set, $(\mathscr{P}_{art}, \leq)$, of all finite partitions of the closed interval $[0,10]$ introduced in an example on page 286. Given two partitions P and Q in $\mathscr{P}_{art}$, $P \leq Q$ if Q is a finer partition than P. That is, each closed interval in P is a union of closed intervals in Q.

Let $f : [0,10] \to \mathbb{R}$ be any function on $[0,10]$. Let $P_{\alpha_n} = \{P_i : i = 1..n\}$ represent a partition of $[0,10]$. Let p_i be the width of the

ith interval, P_i, of the partition P_{α_n}. Let

$$\mathscr{P}_{art} = \{P_{\alpha_n} : P_{\alpha_n} \text{ is a partition of } [1, 10]\}$$

and $(\mathscr{P}_{art}, \leq)$ be its representation as a directed set of partitions.

For a given partition, P_{α_n}, let m_i, l_i denote the two points of the ith interval, P_i, such that, $f(m_i)$ is maximal at $m_i \in P_i$, and $f(l_i)$ is minimal at $l_i \in P_i$, respectively. Then

$$U(P_{\alpha_n}) = \sum_{i=1}^{n} f(m_i)p_i = x_{P_{\alpha_n}}$$

$$L(P_{\alpha_n}) = \sum_{i=1}^{n} f(l_i)p_i = y_{P_{\alpha_n}}$$

both approximate the area beneath the curve of f on $[0, 10]$. Then $U : \mathscr{P}_{art} \to \mathbb{R}$ and $L : \mathscr{P}_{art} \to \mathbb{R}$ define two nets,

$$N_U = \{U(P_{\alpha_n}) : P_{\alpha_n} \in \mathscr{P}_{art}\} = \{x_{P_{\alpha_n}} : P_{\alpha_n} \in \mathscr{P}_{art}\}$$
$$N_L = \{L(P_{\alpha_n}) : P_{\alpha_n} \in \mathscr{P}_{art}\} = \{y_{P_{\alpha_n}} : P_{\alpha_n} \in \mathscr{P}_{art}\}$$

If the two nets N_U and N_L have a common limit, say $q \in \mathbb{R}$, the theory of integration confirms that

$$\int_0^{10} f(x)\, dx = q$$

Example 8. Let the space, $\mathbb{R}^2$, be equipped with the radial topology τ_r. Recall that $(\mathbb{R}^2, \tau_r)$ is called the *radial plane*. This topology is described in Example 2 on page 93. In Example 11 on page 104, we show that the radial plane is not first countable. Consider the set

$$S = \mathbb{R}^2 \backslash (0, 0)$$

equipped with the subspace topology inherited from τ_r. Construct a net which converges to $p = (0, 0)$.

Solution: In what follows, (r, θ), represents the radial coordinates of a point in $\mathbb{R}^2$. Let $D = \{(r, \theta) : r > 0,\ \theta > 0\}$ be an uncountably infinite linearly ordered set where

$$(r_1, \theta_1) = (r_2, \theta_2) \quad \text{if } r_1 = r_2 \quad \text{and} \quad \theta_2 = \theta_1 + 2n\pi$$

and

$$(r_1, \theta_1) < (r_2, \theta_2) \text{ if } \begin{cases} r_2 < r_1 \\ \text{or} \\ r_1 = r_2 \quad \text{and} \quad \theta_1 + 2n\pi < \theta_2 + 2n\pi \end{cases}$$

Then D is a directed set. Suppose $f : D \to S$ is defined as $f(r, \theta) = (\frac{r}{\theta}, \theta)$. Then

$$N = \{f(r, \theta) : (r, \theta) \in D\} = \{(\tfrac{r}{\theta}, \theta) : (r, \theta) \in D\}$$

describes an uncountable net in S.

Claim: That the net, N, accumulates at $p = (0, 0)$.

Proof of claim: Suppose B is an open neighborhood of p, with respect to τ_r.

Let $(r_q, \theta_q) \in D$ and

$$T_{(r_q,\theta_q)} = \{(r, \theta) : (r, \theta) \geq (r_q, \theta_q)\}$$

be a tail end of D. Then

$$f[T_{(r_q,\theta_q)}] = \{f(r, \theta) : (r, \theta) \geq (r_q, \theta_q)\} = \{(\tfrac{r}{\theta}, \theta) : (r, \theta) \geq (r_q, \theta_q)\}$$

is a tail end of N.

By definition of *accumulation point* it suffices to show that $f[T_{(r_q,\theta_q)}]$ intersects B. If so, then N contains a subnet which converges to p.

Let (r_k, θ_k) be any point in B. We can choose m such that

$$r_s = \frac{r_k}{\theta_k + 2m\pi} < r_k$$

Let $\theta_s = \theta_k + 2m\pi$. See that (r_s, θ_s) is a point on the ray passing through (r_k, θ_k) and so belongs to B. Also $(r_s, \theta_s) > (r_k, \theta_k)$, so $(r_s, \theta_s) \in T_{(r_k,\theta_k)}$. Then

$$f(r_s, \theta_s) = \left(\frac{r_s}{\theta_s}, \theta_s\right) \in f[T_{(r_k,\theta_k)}] \cap B$$

So every neighborhood of p intersects a tail end of N. So, the net N accumulates at $p = (0, 0)$, as claimed.

Then by the above theorem, the net N has a subnet which converges to p.

Example 9. Let C denote the Cantor set. Recall from its formal definition 7.18 and its graphic representation given on page 175 that $C = \cap\{C_i : i \in \mathbb{N}\}$ where $C_{i+1} \subset C_i$ for all i. Also every element of C is the set of all elements in $[0, 1]$ of the form $\sum_{i=1}^{\infty} \frac{m_n}{3^n}$ where $m_n \in \{0, 2\}$. Let $W = [0, 1] \times \mathscr{P}([0, 1])$ be partially ordered by $\leq$ defined as

$$(x, A) \geq (y, B) \qquad \Leftrightarrow \qquad x \in A \subseteq B$$

Then $D = \{(x, C_i) : i = 0, 1, 2, \ldots,\ x \in C_i\}$ is a partially ordered set. If $m \geq n$ and $(x, C_m), (y, C_n) \in D$ where $x \in C_m$, then $(x, C_m) \geq (y, C_n)$. So, D is a directed set.

Define the function $f : D \to S$ as $f(x, C_i) = x$. Let

$$N = \{f(x, C_i) : (x, C_i) \in D\}$$

be a corresponding net in $[0, 1]$. Show that every element of C is an accumulation point of the net N.

Solution: Let $k \in C$. Then $k \in C_i$ for all $i \in \mathbb{N}$. Suppose B is a basic open neighborhood of k and $j \in \mathbb{N}$.

If $B \cap f(x, C_i) = \varnothing$ for all $i \geq j$ then

$$\begin{aligned} z \in B \qquad &\Rightarrow \qquad z \notin \{f(x, C_i) : i \geq j\} \\ &\Rightarrow \qquad z \notin C_i \quad \text{for all } i \geq j \\ &\Rightarrow \qquad k \notin C_i \quad \text{for all } i \geq j \\ &\Rightarrow \qquad k \notin C \end{aligned}$$

a contradiction. We can only conclude that the open neighborhood B of k is cofinal in the net $N = \{f(x, C_i) : C_i \in C\}$. So, k is an accumulation point of N.

Remark. In the example immediately above, we have shown that k is an accumulation point of the net,

$$N = \{f(x, C_i) : (x, C_i) \in D\}$$

Then N must have a subnet, say

$$M = \{(f \circ g)(x, C_i) : (x, C_i) \in D\}$$

which converges to k. We will produce such a subnet by defining an appropriate increasing function $g : D \to D$.

Suppose $k = \sum_{i=1}^{\infty} \frac{t_n}{3^n}$ where $t_n \in \{0, 2\}$ and let $k_m = \sum_{i=1}^{m} \frac{t_n}{3^n}$. Define

$$g(x, C_m) = (k_m, C_m)$$

for $m \in \mathbb{N}$. Then $(f \circ g)(x, C_m) = k_m$. So,

$$(f \circ g)[D] = [\{k_m : m \in \mathbb{N}\}]$$

is a subnet which clearly converges to k.

12.5 Convergence of Functions

We will now consider the notion of "convergence to a limit point" for the particular case of a set, S, whose elements are functions. This will be considered from a topological point of view. Since convergence to a limit point depends on the topology defined on the set, our first step is to decide which topology on S is the most suitable for our purposes.

Consider, for example, the set,

$$S = \mathbb{R}^{[0,1]}$$

representing the family of all functions, $f : [0, 1] \to \mathbb{R}$, mapping $[0, 1]$ into $\mathbb{R}$ (where the functions are not necessarily continuous). The standard topology on Cartesian products, $\prod_{i \in I} S_i$, can be described as being the set of all functions mapping the index set, $I = [0, 1]$, into $\cup_{i \in I} S_i$. We will examine convergence in $\prod_{i \in [0,1]} \mathbb{R}_i$ equipped with the product topology where each factor, $\mathbb{R}_i$ is $\mathbb{R}$. An element of the product space $\prod_{i \in [0,1]} \mathbb{R}$ can be viewed as an element of $\mathbb{R}^{[0,1]}$.

The product topology on $S = \mathbb{R}^{[0,1]}$.
We examine what the product topology on $S = \mathbb{R}^{[0,1]}$ when viewed as $\prod_{i\in[0,1]} \mathbb{R}$ looks like.

Suppose $f = \{y_i : i \in [0,1]\} = \{f(i) : i \in [0,1]\}$ is an element in S. Then, for $f \in S$ and $j \in [0,1]$, $\pi_j(f) = f(j)$ and, for $k \in \mathbb{R}$,

$$\pi_j^{\leftarrow}(k) = \{f \in S : f(j) = k\}$$

With the product topology on S, a basic open neighborhood, B_f, of f is of the form

$$B_f = \cap\{\pi_i^{\leftarrow}[U_i] : i \in F_{\text{finite}} \subseteq [0,1],\ U_i \subseteq \mathbb{R}\}$$

where F is a finite subset of $[0,1]$ and U_i is an open neighborhood of $y_i = f(i)$ in $\mathbb{R}$. The product topology declares that $\pi_i[B_f] = \mathbb{R}$ for all $i \notin F$ and $y_i \in \pi_i[B] = U_i$, for all $i \in F$. If $g = \{x_i : i \in [0,1]\} \in B_f$ (where $g(i) = x_i$) this means that $x_i \in U_i$ for each $i \in F$ and x_i is anything else in $\mathbb{R}$ for $i \notin F$. This means the element g is "close" to f if $g(i)$ and $f(i)$ both belong to U_i for $i \in F_{\text{finite}}$.

The basic open neighborhood, B_f, of $f = \{y_i\}$ establishes restrictions on convergence of a net only on finitely many index elements at a time, not on the whole set $\{y_i : i \in [0,1]\}$ simultaneously. So, a sequence or a net of functions, $\{g_\alpha : \alpha \in J\}$, will converge to a limit point, $f \in S = \mathbb{R}^{[0,1]} = \prod_{i\in[0,1]} \mathbb{R}$ provided $\{g_\alpha(i)\} \to f(i)$ at finitely may points $i \in [0,1]$ at a time. We will refer to this type of function convergence as "point-wise convergence".

The box topology on $S = \mathbb{R}^{[0,1]}$.

On the other hand, if we equip $S = \mathbb{R}^{[0,1]} = \prod_{i\in[0,1]} \mathbb{R}$ with the (stronger) box topology an open neighborhood of $f = \{y_i : i \in [0,1]\} = \{f(i) : i \in [0,1]\}$ is of the form

$$B_f = \cap\{\pi_i^{\leftarrow}[U_i] : i \in [0,1],\ U_i \subseteq \mathbb{R}\}$$

where U_i is an open neighborhood of y_i in $\mathbb{R}$. In this case, $g = \{x_i : i \in [0,1]\}$ belongs to B_f provided $x_i = g(i) \in U_i$ for each $i \in [0,1]$.

When equipped with the box topology, convergence in $S = \mathbb{R}^{[0,1]} = \prod_{i\in[0,1]} \mathbb{R}$ is referred to as "uniform convergence".

Note that a topology on $[0,1]$ is not relevant in the choice of topology on S. We formally define these in a more general context, $S = T^A$.

Definition 12.6 Let A be a set and T be a topological space. Let $S = T^A$ represent the set of all functions which map A into the topological space, T.

(a) *Point-wise convergence*: Suppose S is expressed as $S = \prod_{i\in A} T$ and is equipped with the product topology.

In this case a basic open neighborhood of $f \in S$, is of the form

$$B_f = \cap\{\pi_i^{\leftarrow}[U_i] : i \in F_{\text{finite}} \subseteq A,\ U_i \subseteq T\}$$

Then convergence of a sequence or net, $\{g_\alpha : \alpha \in J\}$ to f will occur if $\{g_\alpha(i)\}$ converges to $f(i)$ at finitely many values of $i \in A$ at a time. Convergence in $S = T^A$, equipped with the product topology, is called *point-wise convergence.*

(b) *Uniform convergence*: Suppose $S = T^A$ is equipped with the box topology.

In this case a basic open neighborhood of $f \in S$, is of the form

$$B_f = \cap\{\pi_i^{\leftarrow}[U_i] : i \in A,\ U_i \subseteq T\}$$

In this case, g is considered to be near f if $g(i) \in B_f$ for all $i \in A$. Then convergence of a sequence or net, $\{g_\alpha : \alpha \in J\}$ to f will occur if $\{g_\alpha(i)\}$ converges to $f(i)$ for all values of $i \in A$ simultaneously. In this case, we say that this sequence or net *converges to f uniformly.*

(c) *Uniform norm topology on $C(S)$*: If S is a topological space it is the custom for analysts to represent the set of all continuous real-valued functions on S by $C(S)$ (or $C(S,\mathbb{R})$). The uniform norm topology is equivalent to topology generated by the sup-norm $\|\ \|_\infty : C(S) \to \mathbb{R}$, defined as

$$\|f\|_\infty = \sup_{x\in S} |f(x)|$$

It is equivalent to the topology whose open sets are the basic open sets that are associated to the ones in the box topology.

In the following example, if f is a real-valued function on $[0, 1]$ (not necessarily continuous), let $Z(f) = f^{\leftarrow}(0)$ and

$$\kappa(f) = |Z(f)| \text{ (the cardinality of } Z(f))$$

with $\kappa(f) = c$ if $Z(f)$ is uncountably large, $\kappa(f) = \aleph_0 = |\mathbb{N}|$ if $Z(f)$ is countably infinite and $\kappa(f) \in \mathbb{N}$ otherwise.

Example 10. Let $S = [0,1]^{[0,1]}$ be the topological space of all functions mapping $[0,1]$ into $[0,1]$ equipped with the product topology (that is, the topology in which convergence is pointwise). Let

$$H = \{f \in S : f[0,1] \subseteq \{0,1\}\}$$

Let $z = 0(x)$ denote the zero-function on $[0,1]$.

(a) Show that there is a net, N, in the subset, H of S, which accumulates at $z = 0(x)$.
(b) Let $D^* = \{f \in H : \kappa(f) \in \mathbb{N}, \kappa(f) > 0\}$ (all uncountable strings of zeroes and ones which contain finitely many zeroes). Show that no sequence in the uncountable set D^* can converge to z.

Solution: We are given that $S = [0,1]^{[0,1]}$ is equipped with the product topology and

$$H = \{f \in S : f[0,1] \subseteq \{0,1\}\}$$

(a) If $f \in H$, $\kappa(f)$ represents the"cardinality of $Z(f) = f^{\leftarrow}(0)$". Let

$$D = \{f \in H : \kappa(f) \in \mathbb{N}\}$$

So, $\kappa(f)$ is a natural number for all $f \in D$. Since $[0,1]$ is uncountable, the set of all finite subsets of $[0,1]$ is uncountable, so D is an uncountable set of functions which map $[0,1]$ to finitely many zeroes. We will define "$\leq$" on D as follows:

$$f \leq p \text{ if and only if } \kappa(f) \leq \kappa(p)$$
$$f = p \text{ if and only if } \kappa(p) = \kappa(f)$$

If $s, t \in D$ are such that $\kappa(s) = m < k = \kappa(t)$ then there exists $r \in D$ such that $\kappa(r) = k+1$, so $r > s$ and $r > t$. So, D is a directed set. A tail end in D is as follows: If $g \in D$, $T_g = \{f : \kappa(f) \geq \kappa(g),\ \kappa(f) < \aleph_0\}$.

Let $h : D \to H$ be defined as $h(f) = f$.

Let

$$N = \{h(f) : f \in D\}$$

be a net in H.

Claim 1: That the net N accumulates at $0(x)$. Consider an arbitrary basic open neighborhood B of $0(x)$. Then, there is $F_{\text{finite}} \subset [0,1]$, where $|F_{\text{finite}}| = k$,

$$B = \prod_{i\in[0,1]} A_i$$

where, for $i \in F_{\text{finite}}$, $A_i = \{0\}$, $A_i = \{0,1\}$, otherwise.

Let $g \in N$ such that $\kappa(g) = m$. Then

$$h[T_g] = \{f \in N : \kappa(f) \geq \kappa(g), \kappa(f) < \aleph_0\}$$

is a tail end of N.

There exists $h \in N$ such that $\kappa(h) > \max\{m, k\}$, such that $h(i) = 0$ for $i \in F_{\text{finite}}$. Then $h \in B$. So $h[T_g] \cap B \neq \varnothing$. So, $0(x)$ is an accumulation point of the net N, as required for Claim 1.

Claim 2: That the net N does *not* converge to $0(x)$. Consider an arbitrary basic open neighborhood B of $0(x)$. Then, there is $F_{\text{finite}} \subset [0,1]$, where $|F_{\text{finite}}| = k$,

$$B = \prod_{i\in[0,1]} A_i$$

where, for $i \in F_{\text{finite}}$, $A_i = \{0\}$, $A_i = \{0,1\}$, otherwise.

It suffices to show that B does not contain a tail end of N. If $g \in B$, let $h[T_g] = \{f \in N : f \geq g\}$ be a tail end in N. Then for $i \in F$, $g(i) = 0$. Suppose $F^* \subset [0,1]$ such that $|F^*| = k+1$ and $F \backslash F^* \neq \varnothing$.

Suppose $h \in N$ such that, for $i \in F^*$, $h(i) = 0$ for $i \in F^*$ and $h(i) = 1$ otherwise. Then, since $\kappa(h) > \kappa(g)$, $h > g$, so $h \in h[T_g]$. But since $h(i) = 1$ for some $i \in F\backslash F^*$ then $h \notin B$. So B cannot contain a tail end of N. So, the net, N in H, does not converge to $0(x)$, as required. This proves Claim 2.

(b) Let $A = \{f_k : k \in \mathbb{N}\}$ be a sequence in D. For each $k \in \mathbb{N}$, let $Z(f_k) = f_k^{\leftarrow}(0)$. Since each $Z(f_k)$ is a finite subset of $[0,1]$, then

$$U = \cup\{Z(f_k) : k \in \mathbb{N}\}$$

is a countable proper subset of $[0,1]$. Let F be a finite non-empty subset of $[0,1]$ such that $F \cap U = \varnothing$. Let $B = \prod_{i\in[0,1]} A_i$ where, for $i \in F$, $A_i = \{0\}$ and $A_i = \{0,1\}$, otherwise.

Then $f_k(i) = 1$ for all $i \in F$ and $k \in \mathbb{N}$. Such f_k's cannot belong to the basic open neighborhood, B, of $0(x)$. S, no tail end of A is contained is contained in B. So, the sequence A cannot converge to $0(x)$.

Example 11. Let the set, $S = [0,1]^{[0,1]}$, of all functions mapping $[0,1]$ to $[0,1]$ be equipped with the product topology. Here, we view S as $\prod_{i\in[0,1]}[0,1]_i$. Show that S is not first countable.

Solution: We are given that the space $S = [0,1]^{[0,1]}$ viewed as $\prod_{i\in[0,1]}[0,1]_i$.

Suppose S is first countable and let

$$f = \langle f(i)\rangle_{i\in[0,1]} = \langle x_i\rangle_{i\in[0,1]}$$

be a point in S. Then S has a countable neighborhood base,

$$\mathscr{B}_f = \{B_n : n \in \mathbb{N}\}$$

at f. Then each open base neighborhood, B_n, of f in $\prod_{i\in[0,1]}[0,1]_i$, can be expressed as

$$\langle x_i\rangle_{i\in[0,1]} \in B_n = \cap\{\pi_i^{\leftarrow}[U_i] : i \in F_{B_n} \subseteq [0,1],\ U_i \subseteq [0,1]\}$$

where F_{B_n} is finite and U_i is open. Then $\cup\{F_{B_n} : \mathbb{N}\}$ is countable union of finite subsets and so is a countable proper subset of $[0,1]$. There must exist $k \in [0,1]\backslash\cup\{F_{B_n} : \mathbb{N}\}$ such that

$$\pi_k\left[\prod_{i\in[0,1]}[0,1]_i\right] = \pi_k[B_n] = [0,1]$$

for all $n \in \mathbb{N}$.

Now, let V be a proper open neighborhood of $\pi_k(f) = x_k \in \pi_k[S] = [0,1]$. Since V is proper in $[0,1]$, then

$$x_k \in \pi_k[B_n] = [0,1] \nsubseteq V$$

for all $B_n \in \mathscr{B}_f$. Then, for any B_n,

$$\begin{aligned} f &\in \pi_k^{\leftarrow}(x_k) \\ &\subseteq B_n \\ &\nsubseteq \pi_k^{\leftarrow}[V] \end{aligned}$$

This contradicts that $\mathscr{B}_f$ is a countable neighborhood base of f. So, S is not first countable.

Concept Review

1. Define a directed set.
2. Are the natural numbers a directed set?
3. Show that the set of all finite partitions of a closed interval of real numbers is a directed set.
4. Define a net of elements in a set.
5. Define a tail end of a net in a set.
6. What does it mean to say that a net converges to a point p in a topological space?
7. What is a limit point of a net.
8. What does it mean to say that a set is cofinal in a net?
9. What does it mean to say that a set B is a subnet of a net A?
10. If B is a subnet of the sequence A is B necessarily a subsequence of A?
11. Is it true that a point p belongs to the closure of a set B if and only if B contains a net which converges to p?
12. Define an accumulation point of a net.
13. If the sequence A has an accumulation p is there necessarily a subnet which converges to p?
14. If the sequence A has an accumulation p is there necessarily a subsequence which converges to p?
15. Complete the statement in terms of nets: A function $f : S \to T$ is continuous at p if and only if...
16. Complete the statement in terms of subnets: The point u is an accumulation point of the net A if and only if...
17. Describe two topologies used on sets of function T^A and describe the convergence properties with respect to these particular topologies.

Exercises

1. Show that the subnet of a subnet of a net A is also a subnet of A.
2. Show that a net, $\{x_i : i \in D\}$, in a product space, $S = \prod_{\alpha \in A} S_\alpha$, converges to p if and only if, for any $\mu \in A$, the net $\{\pi_\mu(x_i) : i \in D\}$ converges to $\pi_\mu(p)$ in S_μ.

3. Let $\mathbb{J}$ denote the set of all irrationals in $\mathbb{R}$ and let $\vec{p} = (\pi, \pi)$. If $S = \mathbb{J}^2 \backslash \{\vec{p}\}$ we will define a partial ordering, "$\leq$", on S as:

$$\vec{a} \leq \vec{b} \quad \text{whenever } ||\vec{b} - \vec{p}|| \leq ||\vec{a} - \vec{p}||$$

where "$||\ ||$" refers to the distance between points in $\mathbb{J}^2$.

(a) Verify that the pair, $(S, \leq)$, forms a directed set in S.

(b) Suppose $f : \mathbb{J}^2 \to \mathbb{R}$ is a real-valued function on $\mathbb{J}^2$. Then, when f is restricted to S, the set, $A = \{ f(\vec{x}),\ \vec{x} \in S \}$, defines a net in $\mathbb{R}$. Verify that the net, A, converges to a limit L if and only if $\lim_{\vec{x} \to \vec{p}} f(\vec{x}) = L$ (in the usual sense).

Chapter 13

Limit Points of Filters

Abstract
In this section, we will introduce a form of convergence by sets rather than the convergence by points. These families of sets are called filters. The construction of a filter is modeled closely on the structure of a neighborhood base at a point. We will formally define a "filter of sets" and explain how to determine whether it has a limit point or not. We will also define an "accumulation point" which belongs to the closure of all sets in a filter. Finally, we will introduce "ultrafilters" and a few of their characterizations; these will help us distinguish them from regular filters.

13.1 Filters and Some of Their Properties

Recall that a net is essentially a set whose elements are indexed by either a linearly or partially ordered set. The ordered elements may or may not converge to a particular element we called a "limit point". The type of convergence we will consider in this section is more topological in nature. Essentially, neighborhood bases serve as the main inspiration for this type of convergence.

Definition 13.1 Let S be a set. Let $\mathscr{F} \subseteq \mathscr{P}(S)$. If $\mathscr{F}$ satisfies the two properties:

(1) $\varnothing \notin \mathscr{F}$,
(2) For every $U, V \in \mathscr{F}$, there exists $F \in \mathscr{F}$ such that $F \subseteq U \cap V$.
then we will say that $\mathscr{F}$ is a *filter base* of sets.

If $\mathscr{F}$ is a filter base which also satisfies the condition:

(3) Whenever $U \in \mathscr{F}$ and there is $V \in \mathscr{P}(S)\backslash\{\varnothing\}$ such that $U \subseteq V$, then $V \in \mathscr{F}$.

Then $\mathscr{F}$ is called a *filter of sets.*[1]

A quick way of describing a filter of sets is to say: "A filter is a subset of $\mathscr{P}(S)\backslash\{\varnothing\}$ which is closed under finite intersections and supersets".[2] Note that the notion of a filter is a set-theoretic concept, not a topological one. Once we want to discuss "convergence" we will be in the context of a filter in some topological space.

A filter is, itself, a filter base, while a filter base, if not a filter, can be completed to become a filter. Given a filter base, $\mathscr{F}$, we define the larger set, $\mathscr{F}^*$, as follows:

$$\mathscr{F}^* = \{U \in \mathscr{P}(S) : F \subseteq U, \text{ for some } F \in \mathscr{F}\}$$

We have simply adjoined to $\mathscr{F}$ all its supersets. It is easily verified that $\mathscr{F}^*$ is a filter.

We will say that the filter base, $\mathscr{F}$, *generates the filter* $\mathscr{F}^*$.

Example 1. For a topological space, (S, τ), and a point $x \in S$, a neighborhood base of open sets, $\mathscr{B}_x$, is clearly a filter base of sets in S since it satisfies the filter base property:

> For any two open neighborhoods U and V of $\mathscr{B}_x$ there exists $W \in \mathscr{B}_x$ such that $x \in W \subseteq U \cap V$.

The following proposition states that a filter which is generated by a filter base is the intersection of all filters which contain the filter base in question.

---

[1]Not that there can be many types of filters: filters of sets, filters of closed sets, filters of zero-sets. The definition of "filters of sets" is independent of a topology on S.

[2]If $A \subseteq B$ then we say that B is a "superset" of A. It is worth noting that the set, $\mathscr{P}(S)$, is not a filter. Why?

Proposition 13.2 *If $\mathscr{F}^*$ is a filter generated by the filter base, $\mathscr{F}$ then,*

$$\mathscr{F}^* = \cap\{\mathscr{H} : \mathscr{H} \text{ is a filter and } \mathscr{F} \subseteq \mathscr{H}\}$$

Proof. Let $\mathscr{H}$ be a filter which contains the filter base $\mathscr{F}$. It suffices to show that $\mathscr{F}^* \subseteq \mathscr{H}$. Let $A \in \mathscr{F}^*$. Then, by definition, there is a non-empty set $U \in \mathscr{F}$ such that $U \subseteq A$. Since $U \in \mathscr{H}$ and $\mathscr{H}$ is a filter, $A \in \mathscr{H}$. Then $\mathscr{F}^* \subseteq \mathscr{H}$.

Definition 13.3 Let (S, τ) be a topological space and $\mathscr{F}$ be a filter base of sets in $\mathscr{P}(S)$.

(a) If $\cap\{\text{cl}_S F : F \in \mathscr{F}\} \neq \varnothing$ then we will say that $\mathscr{F}$ is *fixed* in S.
(b) Otherwise, we will say that $\mathscr{F}$ is *free* in S.

Example 2. Let (S, τ) be a topological space and K be a non-empty subset of S. Let $\mathscr{F}^*$ be the set of all subsets, U, of S such that $K \subseteq \text{int}_S U$.

(a) Show that $\mathscr{F}^*$ is a filter of sets in S.
(b) If $K = \{x\}$, describe a filter base, $\mathscr{F}$, of $\mathscr{F}^*$.
(c) Is the filter base $\mathscr{F}$ free or fixed in S?

Solution: We are given $\mathscr{F}^* = \{U \in \mathscr{P}(S) : K \subseteq \text{int}_S U\}$ for a fixed K.

Let $U, V \in \mathscr{F}^*$ and $M \in \mathscr{P}(S)$ such that $U \subseteq M$.

(a) Since $K \not\subseteq \text{int}_S\varnothing$, then $\varnothing \notin \mathscr{F}^*$.
Since $U, V \in \mathscr{F}^*$, then $K \subseteq \text{int}_S U$ and $K \subseteq \text{int}_S V$. Then

$$K \subseteq \text{int}_S U \cap \text{int}_S V = \text{int}_S(U \cap V)$$

So, $U \cap V \in \mathscr{F}^*$. Then $\mathscr{F}^*$ is closed under finite intersections, which implies that $\mathscr{F}^*$ is a filter base of sets in $\mathscr{P}(S)$.

Since $U \in \mathscr{F}^*$ and $U \subseteq M$, then $K \subseteq \text{int}_S U \subseteq \text{int}_S M$ then $M \in \mathscr{F}^*$. So, $\mathscr{F}^*$ is closed under supersets.

So, $\mathscr{F}^*$ is a filter of sets.

(b) Suppose $K = \{x\}$. Then $\mathscr{F}$ is the set, $\mathscr{B}_x$, of all neighborhoods of x. The set, $\mathscr{F}$, satisfies the property: "If A and B are two neighborhoods of x in $\mathscr{F}$ then there exists a neighborhood C in $\mathscr{F}$ such that $x \in C \subseteq A \cap B$". So, the filter base, $\mathscr{F} = \mathscr{B}_x$, is a neighborhood base of x.

(c) Since $K \subseteq \text{int}_S U$ and $\text{int}_S U \subseteq \text{cl}_S U$ for all $U \in \mathscr{F}$, then $K \subseteq \cap\{\text{cl}_S U : U \in \mathscr{F}\}$ so $\mathscr{F}$ is fixed.

13.2 Limit Points and Accumulation Points of Filters

As we have done for sequences and nets, we define the limit point and accumulation point of a filter base and of a filter. In what follows, filters are family of subsets of some topological space.[3]

Definition 13.4 Let $\mathscr{F}$ be a filter base of sets on the topological space S, and $x \in S$. Let $\mathscr{B}_x$ be a neighborhood base at the point x.

(a) We will say that the filter base of sets, $\mathscr{F}$, *converges to* x if and only if, for every open neighborhood, B of x, there is some $U \in \mathscr{F}$ such that $U \subseteq B$. This implies that, if $\mathscr{B}_x$ is an open neighborhood base of x, then $\mathscr{B}_x \subseteq \mathscr{F}^*$, the filter generated by $\mathscr{F}$. So the filter, $\mathscr{F}^*$, is "finer" than $\mathscr{B}_x$. We abbreviate the "convergence of $\mathscr{F}$ to x" by the expression, $\mathscr{F} \to x$. In this case, we will say that

$$x \textit{ is a limit point of the filter base } \mathscr{F}.$$

(b) We will say that x is an *accumulation point* for $\mathscr{F}$ if and only if every open neighborhood of x intersects each $F \in \mathscr{F}$. That is,

$$x \in \text{cl}_S F \text{ for all } F \in \mathscr{F}$$

(c) The set of all accumulation points of $\mathscr{F}$ is called that the *adherence of* $\mathscr{F}$. The adherence of $\mathscr{F}$ is denoted by $a(\mathscr{F})$. That is, $a(\mathscr{F}) = \cap\{\text{cl}_S F : F \in \mathscr{F}\}$.

---

[3]Later in the text we will discuss filters whose elements are zero-sets.

It is worth verifying that, if $\mathscr{F}$ is a filter base, then $a(\mathscr{F}) = a(\mathscr{F}^*)$. Also note that, by definition, a limit point of a filter base, $\mathscr{F}$, is an accumulation point of $\mathscr{F}$.

Observe that, for x to be a limit point of a filter base $\mathscr{F}$, x need not belong to of any of its elements. Witness the following example.

Example 3. Consider the set, $\mathscr{F} = \{(4, 4+\delta] : \delta > 0\}$. See that $\mathscr{F}$ doesn't contain $\varnothing$ and satisfies the filter base property and so qualifies as a filter base. The element 4 does not belong to any of its elements. But if we take any open neighborhood of 4, say $B = (4 - \varepsilon, 4 + \mu)$ there exists, say $U = (4, 4 + \mu/4)$ in $\mathscr{B}$, such that $U \subseteq B$. So 4 is a limit point of the filter base $\mathscr{F}$.[4]

Furthermore, distinct filter bases can converge to the same point. Witness such a case in the following example.

Example 4. Consider the sets, $\mathscr{F}_1 = \{(4, 4+\delta] : \delta > 0\}$ and $\mathscr{F}_2 = \{[4-\delta, 4) : \delta > 0\}$. See that both $\mathscr{F}_1$ and $\mathscr{F}_2$ are filter bases which converge to the limit point 4.

By definition, any filter base that has an accumulation point is fixed. Filter bases can have multiple accumulation points.

Example 5. Consider the set, $A = \{2\} \cup (3, 4]$ in $\mathbb{R}$, equipped with the usual topology. Let $\mathscr{F} = \{U \in \mathscr{P}(\mathbb{R}) : A \subseteq U\}$. Every point, $x \in A$, belongs to every $F \in \mathscr{F}$. So every point in A is an accumulation point of $\mathscr{F}$. But also, every open neighborhood of 3 will intersect every set which contains A, so 3 is also an accumulation point. The filter base, $\mathscr{F}$, has no limit point since any open interval of a point in A with radius less than 1 cannot be contained in a set which contains all of A.

However, verify that if, $\mathscr{F}_1 = \{U \in \mathscr{P}(\mathbb{R}) : 2 \in U\}$, is a fixed filter where $\mathscr{F} \subseteq \mathscr{F}_1$.

13.3 Filters Derived from Nets

Suppose we are given a net in a topological space S. We will show how a filter can be constructed from this net, simply by gathering

[4]However, by definition, the set, $B = (4 - \varepsilon, 4 + \mu)$, will belong to $\mathscr{F}^*$, the filter generated by $\mathscr{F}$.

together all tail-ends of the net. Eventually, we will show they have the same limit points.

Theorem 13.5 *Let S be a set and D be a directed set. Suppose the function, $f : D \to S$, defines a net, $A = \{f(i) : i \in D\}$, in S. If $u \in D$, let $T_u = \{i \in D : i \geq u\}$, a tail end of D. Then,*

$$f[T_u] = \{f(i) : i \geq u \text{ in } D\}$$

represents a tail end of the net, N, starting with $f(u)$. Let

$$\mathscr{F}_f = \{f[T_u] : u \in D\}$$

be the set of all tail ends of the net, N. Then $\mathscr{F}_f$ is a filter base of sets in S.

Proof. We are given $\mathscr{F}_f = \{f[T_u] : u \in D\}$ where

$$N = \{f(i) : i \in D\}$$

is a net in the set S. First note that $\varnothing \notin \mathscr{F}_f$.

We claim that $\mathscr{F}_f$ is closed under finite intersection: If $j, k \in D$, then there exist, $h \in D$, such that $j \leq h$ and $k \leq h$. Then $T_h \subseteq T_j \cap T_k$ and so,

$$f[T_h] \subseteq f[T_j] \cap f[T_k]$$

So, $\mathscr{F}_f$ is closed under finite intersections, as claimed. So, the set, $\mathscr{F}_f$, of all tail ends of the net, N, is a filter base of sets in S.

Definition 13.6 Let S be a set and $N = \{f(i) : i \in D\}$ be a net in S. Let $T_u = \{i \in D : i \geq u\}$. The filter base, $\mathscr{F}_f = \{f[T_u] : u \in D\}$, forming the set of tail ends, of the net, N, in S, is referred to as the *filter base in S which is determined by the net N.*

13.4 A Net Derived from a Filter

Having seen how we can construct a filter from a net, we now show how a net can be constructed given any filter.

Definition 13.7 Let S be a set and $\mathscr{F}$ be a filter of sets. We define the relation, $\leq$, on the set,

$$D_{\mathscr{F}} = \{(u, F) \in S \times \mathscr{F} : u \in F\}$$

as follows: $(s, F) \leq (u, H)$ if and only if $u \in H \subseteq F$. Then $(D_{\mathscr{F}}, \leq)$ forms a directed set.

With this directed set in hand, and the function, $f : D_{\mathscr{F}} \to S$, defined as $f(u, F) = u$, we define the net

$$N_{\mathscr{F}} = \{f(u, F) : (u, F) \in D_{\mathscr{F}}\}$$

in the set, S. The net $N_{\mathscr{F}}$ is called a *net generated by the filter* $\mathscr{F}$.

13.5 Corresponding limit points of a filter and its net

We will now confirm that the limit point or accumulation point of a net will always be a limit point or accumulation point of the filter corresponding to this net, and vice-versa.

Theorem 13.8 *Let S be a topological space. Suppose $\mathscr{F}$ is a filter base of sets in S and $A = \{h(i) : i \in D\}$ is a net.*

(a) *Suppose the net, A, is a net generated by $\mathscr{F}^*$. The point, u, in S is a limit point (accumulation point) of $\mathscr{F}$ if and only if u is a limit point (accumulation point) of A.*

(b) *Suppose the filter base, $\mathscr{F}$, is determined by the net, A. The point, u, in S is a limit point (accumulation point) of $\mathscr{F}$ if and only if u is a limit point (accumulation point) of A.*

Proof. We are given that S is a topological space.

(a) We are given that $\mathscr{F}$ is a filter base and the net,

$$A = \{h(u, F) : (u, F) \in D_{\mathscr{F}}, u \in F \in \mathscr{F}\}$$

is one which is generated by filter $\mathscr{F}^*$.

($\Rightarrow$) Case: *Limit point.* Suppose u is a limit point of the filter base $\mathscr{F}$. Then u is a limit point of $\mathscr{F}^*$. We are required to show that u is a limit point of the net A. Let U be an open neighborhood of u. To show that A converges to u we must show that U contains a tail end of A.

Since u is a limit point of $\mathscr{F}$ and U is a neighborhood u, there exists $F \in \mathscr{F}$ such that $F \subseteq U$. Then U belongs to the filter, $\mathscr{F}^*$. We can choose some point $v \in U$, so that the pair, (v, U) belongs to $D_{\mathscr{F}}$.

We claim that the tail end, $h[T_{(v,U)}]$ in A, is contained in U: Let $(x, V) \in D_{\mathscr{F}}$ such that $(v, U) \leq (x, V)$. Then $T_{(x,V)} \in T_{(v,U)}$ and $x \in V \subset U$. Then $h(x, V) = x \in U$. So neighborhood of u, U, contains the tail end of $h[T_{(v,U)}]$, as claimed. So the net, A, converges to u.

($\Rightarrow$) Case: *Accumulation point.* Suppose u is an accumulation point of the filter base $\mathscr{F}$ and $A = \{h(u, F) : (u, F) \in D_{\mathscr{F}}\}$, the net generated by $\mathscr{F}$. We are required to show that u is an accumulation point of A. Let U be an open neighborhood of u. By hypothesis, $U \cap F \neq \varnothing$ for all $F \in \mathscr{F}$ (equivalently, $u \in \mathrm{cl}_S F$ for each $F \in \mathscr{F}$).

To show that u is an accumulation point of the net, A, we must show that U intersects each tail end of A. Let $h[T_{(v,K)}]$ be a tail end in A and $M \in \mathscr{F}$. Then $V = M \cap K$ is an element of $\mathscr{F}$ which is a subset of K. Then, by hypothesis, there exists

$$t \in U \cap V \subseteq V \subseteq K$$

The expression, $t \in V \subseteq K$, translates to, $(v, K) \leq (t, V)$. Hence,

$$t = h(t, V) \in h[T_{(v,K)}] \cap U$$

So, U intersects the tail-end $h[T_{(v,K)}]$. We conclude that u is an accumulation point of the net corresponding to $\mathscr{F}$.

($\Leftarrow$) Case: *Limit point.* Suppose u is a limit point of the net A. We are required to show that the filter, $\mathscr{F}$, generated by A converges to u. Let U be an open neighborhood of u. To show that $\mathscr{F}$ converges to u we must show that U contains an element of $\mathscr{F}$.

By hypothesis, U contains a tail end of A. That is, $h[T_{(s,F)}] \subseteq U$ for some (s, F). Then $h(r, F) \in U$ for all $r \in F$. But $h(r, F) = r$. So $F \subseteq U$. Since F belongs to $\mathscr{F}$, then $\mathscr{F}$ converges to u.

($\Leftarrow$) Case: *Accumulation point.* The proof is left as an exercise for the reader.

(b) We are given that the filter base, $\mathscr{F}_h$, is determined by the net $A = \{h(i) : i \in D\}$. Recall that $\mathscr{F}_h = \{h[T_u] : i \geq u\}$ where $T_u = \{i \in D : i \geq u\}$. That is, it is the set of the tail ends of the net, $\{h(i) : i \in D\}$.

($\Rightarrow$) Case: *Limit point.* Suppose x is a limit point of the filter base, $\mathscr{F}_h$. We must show that the net A also converges to x. Suppose U is any open neighborhood of x. Then there is $a \in D$ such that $h[T_a] \subseteq U$. Then every open neighborhood contains a tail end of the net, $\{h(i) : i \in D\}$. So the net converges to x.

($\Leftarrow$) Proving the converse of the "Case, limit point", is left as an exercise.

($\Rightarrow$) Case: *Accumulation point.* Suppose y is an accumulation point of the filter base, $\mathscr{F}_h$. We must show that y is an accumulation point of the net A. Let U be any open neighborhood of y. Then $U \cap h[T_u] \neq \varnothing$ for all $u \in D$. That is, U contains an element of all tail ends of the net, $\{h(i) : i \in D\}$. Then y is an accumulation point of this net.

($\Leftarrow$) Proving the converse of the "Case, accumulation point" is left as an exercise.

13.6 Other Properties of Filters

In the following theorem, we see that "uniqueness of limits of filters (nets)" characterizes the Hausdorff property on a topological space.

Theorem 13.9 *Let S be a topological space. The space S is Hausdorff if and only if a convergent filter, $\mathscr{F}^*$, has only one limit point.*

Proof. We are given that S is a space.

($\Rightarrow$) Suppose S is Hausdorff. Suppose the points x and p are limits of the filter base of sets, $\mathscr{F}$, in S. Then neighborhood bases, $\mathscr{B}_x$ and $\mathscr{B}_p$, are subsets of the filter, $\mathscr{F}^*$, generated by $\mathscr{F}$. If $x \neq p$, then, by hypothesis, we can choose disjoint open neighborhoods U_x and V_p which separate p from x. Since $\mathscr{F}^*$ converges to p and x there exists, W and M, in $\mathscr{F}^*$ which are contained in U_x and V_p, respectively. But then $W \cap M = \varnothing$, contradicting the fact that $\mathscr{F}^*$ is a filter. So $x = p$.

($\Leftarrow$) We are given that limits of any filter, $\mathscr{F}^*$, of sets in S are unique. Suppose S is not Hausdorff. Then there is a pair of points, x and p, with open neighborhood bases, $\mathscr{B}_x$ and $\mathscr{B}_p$ such that, $U \cap V \neq \varnothing$ for all U and V in $\mathscr{B}_x$ and $\mathscr{B}_p$. It easily follows that $\mathscr{F} = \{U \cap V : U \in \mathscr{B}_x \text{ and } V \in \mathscr{B}_p\}$ is a filter base of sets in S.

We claim that $\mathscr{F}^*$ converges to both x and p. Let W_x and D_p be elements of $\mathscr{B}_x$ and $\mathscr{B}_p$, respectively. It suffices to show that there are elements, F_1 and F_2 in $\mathscr{F}^*$ such that $F_1 \subseteq W_x$ and $F_2 \subseteq D_p$, respectively. This would contradict convergence to a single element.

There exists $U_x \in \mathscr{B}_x$ and $V_p \in \mathscr{B}_p$ such that

$$x \in F_1 = U_x \cap V_p \subseteq W_x$$

and $T_p \in \mathscr{B}_p$ and $C_x \in \mathscr{B}_x$ such that,

$$p \in F_2 = T_p \cap C_x \subseteq D_p$$

By definition of $\mathscr{F}^*$ both F_1 and F_2 belong to $\mathscr{F}^*$. By definition of convergence, $\mathscr{F}^*$ converges to both x and p. This contradicts our hypothesis. So, S must be Hausdorff.

The statement in the above theorem translates to nets. That is, limit points of nets in a topological space, S, are unique if and only if S is Hausdorff.

In the following theorem, we show that, for any accumulation point, p, of a filter, $\mathscr{F}$, it is possible to increase the size of $\mathscr{F}$ to a larger filter which will converge to p. So every accumulation point of a filter is the limit point of a larger filter which contains it.

Theorem 13.10 *Suppose S is a topological space and p is an accumulation point for a filter base, $\mathscr{F}$, of sets in S. Then p is the limit point of some filter, $\mathscr{H}^*$, which contains $\mathscr{F}^*$.*

Proof. We are given that S is a space and that p is an accumulation point of a filter, $\mathscr{F}$.

Let $\mathscr{B}_p$ denote the set of all open neighborhoods of p. Since p is an accumulation point of $\mathscr{F}$, every open neighborhood of p intersects every element of $\mathscr{F}$. We can then consider the family

$$\mathscr{H} = \{U \cap F : U \in \mathscr{B}_p, F \in \mathscr{F}\}$$

itself a filter base. Let $\mathscr{H}^*$ be the filter generated by $\mathscr{H}$.

We claim that $\mathscr{H}$ converges to p. If V is an element of $\mathscr{B}_p$, for $F \in \mathscr{F}$, $V \cap F$ is non-empty so is an element of $\mathscr{H}$. Since $V \cap F$ is a subset of V, by definition of convergence, $\mathscr{H}$ converges to p, as claimed.

Suppose $F \in \mathscr{F}$. Since $U \cap F \subseteq F$, then $F \in \mathscr{H}^*$ so $\mathscr{F} \subseteq \mathscr{H}^*$. In fact, $\mathscr{F}^* \subseteq \mathscr{H}^*$.

We conclude that, if p is accumulation point of $\mathscr{F}$, then it is the limit point of $\mathscr{H}^*$, a filter which contains $\mathscr{F}^*$.

Theorem 13.11 *Suppose S is a topological space. Suppose $\mathscr{F}_1$ and $\mathscr{F}_2$ are two fixed filter bases in S. If $\mathscr{F}_1$ and $\mathscr{F}_2$ have the same adherence set then every element of the filter, $\mathscr{F}_1$, intersects every element of the filter, $\mathscr{F}_2$.*

Proof. We are given that $\mathscr{F}_1$ and $\mathscr{F}_2$ are two fixed filter bases in S.

Suppose $a(\mathscr{F}_1) = a(\mathscr{F}_2)$. Then, if $p \in a(\mathscr{F}_1)$, p is an accumulation point of both $\mathscr{F}_1$ and of $\mathscr{F}_2$. There exists a filter, $\mathscr{H}^*$, which contains both $\mathscr{F}_1$ and of $\mathscr{F}_2$ and which converges to p. (*Note*: There are some straightforward details involving Theorem 13.10 to work out for proving this; these are left as an exercise.) Since $\mathscr{H}^*$ is closed under finite intersections, every element of $\mathscr{F}_1$ intersects every element of $\mathscr{F}_2$.

13.7 Filter Bases Describe Closures of Sets and Continuity

Not surprisingly, given our experience with sequences and nets, the closure of a set can be described in terms of the limit points of its filter bases.

Theorem 13.12 *Let F be a non-empty subset of a topological space, S. Then a point u belongs to the closure, $cl_S F$, of F if and only if u is the limit point of some filter base in F.*

Proof. We are given that S is a space and F is a non-empty subset of S.

($\Rightarrow$) Suppose $u \in \text{cl}_S F$. Let $\mathscr{B}_u$ be a neighborhood base of open sets for u. Then

$$\mathscr{F} = \{U \cap F : U \in \mathscr{B}_u\}$$

forms a filter base of sets in F which converges to u.

($\Leftarrow$) Suppose $\mathscr{F}$ is a filter base in F converging to some $u \in S$. Let U be any open neighborhood of u. Then there exists some $V \in \mathscr{F}$ such that $V \subseteq U$. Since $V \subseteq F$, U intersects F and so $u \in \text{cl}_S F$.

Just as for nets, continuity of a function $f : S \to T$ can be expressed in terms of filters. But first we must determine whether the continuous image of a filter can itself be seen as a filter.

About the image, $f[\mathscr{F}^*]$, of a filter, $\mathscr{F}^*$. Given a function $f : S \to T$ and given that $\mathscr{F}^*$ is a filter of sets in S we define,

$$f[\mathscr{F}^*] = \{f[F] : F \in \mathscr{F}^*\}$$

We claim $f[\mathscr{F}^*]$ is a filter base in the range of the function, f. Let $f[U]$ and $f[V]$ be two elements of $f[\mathscr{F}^*]$ where U and V are elements of $\mathscr{F}^*$. Then there exists non-empty $W \in \mathscr{F}^*$ such that $W \subseteq U \cap V$. Then $f[W] \subseteq f[U \cap V] \subseteq f[U] \cap f[V] \neq \varnothing$. So...

$f[\mathscr{F}^*]$is a filter base which generates the filter $f[\mathscr{F}^*]^*$.

Theorem 13.13 *Let S and T be two topological spaces. Let $f : S \to T$ be a function and $u \in S$. Let $\mathscr{F}^*$ be a filter. The function f is* *continuous at* u *if and only if whenever* u *is a limit point of the filter,* $\mathscr{F}^*$*, then* $f(u)$ *is a limit point of the filter,* $f[\mathscr{F}^*]^*$.

Proof. We are given that S and T are two topological spaces and $f : S \to T$ is a function.

($\Rightarrow$) Suppose f is continuous at $u \in S$ and $\mathscr{F}^*$ converges to u. We are required to show that $f[\mathscr{F}^*]^*$ converges to $f(u)$.

Suppose V is an open neighborhood of $f(u)$. It suffices to show that an element of $f[\mathscr{F}^*]^*$ is contained in V. Since f is continuous at u, then there exists an open neighborhood U of u such that $f[U] \subseteq V$. Since $\mathscr{F}^*$ is a filter, then U belongs to the filter $\mathscr{F}^*$. Since $f[U] \in f[\mathscr{F}^*]$ and is contained in V, then $V \in f[\mathscr{F}^*]^*$. So, $f[\mathscr{F}^*]^*$ converges to $f(u)$.

($\Leftarrow$) Suppose that whenever $\mathscr{F}^*$ converges to u then $f[\mathscr{F}^*]^*$ converges to $f(u)$. Let $\mathscr{F}^*$ be the neighborhood filter of u. Since $f[\mathscr{F}^*]^*$ converges to $f(u)$, each neighborhood V of $f(u)$ contains some element of $f[\mathscr{F}^*]^*$. Then for some neighborhood U of u, $f[U] \subseteq V$. This shows that f is continuous at u.

The statement in the following example involves a subspace of $S = [0,1]^{[0,1]}$ similar to the one found in the example discussed in the chapter on nets (on page 306).

Example 6. Let $S = [0,1]^{[0,1]}$, the set of all functions mapping $[0,1]$ into $[0,1]$, be equipped with the product topology. Let H be a subspace of the subspace, $\{0,1\}^{[0,1]}$ of S which is defined as follows:

$$H = \{h \in S : h(x) = 0 \text{ for finitely many values, otherwise } h(x) = 1\}$$

Construct a filter in H, which accumulates at the zero element, z.

Solution: Given $S = [0,1]^{[0,1]}$, the set of all functions mapping $[0,1]$ into $[0,1]$, equipped with the product topology and

$$H = \{h \in S : h^{\leftarrow}(0) \text{ is finite, } \text{ otherwise } h(x) = 1\}$$

a subspace of $\{0,1\}^{[0,1]}$. Let z represent the zero function, $0(x)$.

Let $T_n = \{f \in H : n \leq |\, f^{\leftarrow}(0)\,| < \aleph_0\}$. Let

$$\mathscr{F} = \{T_n : n = 1, 2, 3, \ldots\}$$

where $T_n \subseteq H$ for each n. If $f \in T_m$, $f(i) = 0$ for m or more values of i in $[0, 1]$. So $T_m \subseteq T_{m-1}$. By finite induction, if $n > m$ then $T_n \subseteq T_m$. Hence, if $m > n$ then $T_m \cap T_n = T_m \neq \varnothing$. Since T_n is never empty, $\mathscr{F}$ is then a filter base of sets in S.

Let B_F be a basic open neighborhood of z in S where $|F| = q$, for some $q \in \mathbb{N}\backslash\{0\}$ and $f(x) = 0$ for the q values of x in $F \subset [0, 1]$.

To show z is an accumulation point of $\mathscr{F}$, we must show that, given $T_k \in \mathscr{F}$, $B_F \cap T_k \neq \varnothing$. If $k > q$, then, for some $f \in T_k$, $f \in B_F \cap T_k \neq \varnothing$. If $k < q$, since $T_q \subseteq T_k$ then, again, for some $f \in T_k$, $f \in B_F \cap T_k \neq \varnothing$. So, $B_F \cap T_k \neq \varnothing$. Then z is an accumulation point of the filter base, $\mathscr{F}$.

13.8 Partial Ordering the Family, $\mathcal{F}$, of All Filters in $\mathscr{P}(S)$

We will now discuss some properties possessed by families of filters of sets. The reader can assume that all filters discussed here are subsets of $\mathscr{P}(S)$ for some topological space, S. Let

$$\mathcal{F} = \{\mathscr{F} \subseteq \mathscr{P}(S) : \mathscr{F} \text{ is a filter of sets of } S\}$$

denote the family of all filters on a space S. We will define an ordering of the elements in $\mathcal{F}$ by inclusion, "$\subseteq$".

Given two filters, $\mathscr{F}$ and $\mathscr{H}$, in $\mathcal{F}$, if $\mathscr{F} \subseteq \mathscr{H}$, we will say that,

$$\mathscr{H} \text{ is } \textit{finer} \text{ than } \mathscr{F} \qquad \text{or} \qquad \mathscr{F} \text{ is } \textit{coarser} \text{ than } \mathscr{H}$$

For example, the sets $\mathscr{F}^* = \{(2 - 1/n, 2 + 1/n) : n \in \mathbb{N}\backslash\{0\}\}^*$ and $\mathscr{H}^* = \{(2 - \varepsilon, 2 + \varepsilon) : \varepsilon > 0\}^*$ are easily verified to be two filters both converging to 2. We see that since $\mathscr{F}^* \subseteq \mathscr{H}^*$, then $\mathscr{H}^*$ is finer than $\mathscr{F}^*$. This ordering is a partial ordering.[5]

13.9 Ultrafilters

We have seen that a filter base can contain an other smaller filter base and can often be itself contained in some other larger filter of sets.

[5]Of course, "$\subseteq$" is not a "linear ordering" of $\mathcal{F}$ since it is *not* always the case that one filter is a subset of another.

Filter bases with large adherence sets can be built up so that they eventually will converge to at most one of those points. But, as we shall soon see, at some point, a filter of sets will have attained its maximum size. We often call such filters, "maximal filters". Although the use of the words "maximal filter" are by themselves sufficiently descriptive, most writers have become accustomed to using the single term, "ultrafilter". We will formally define an "ultrafilter" and prove some of its characterizations. These characterizations will make it easier to recognize an ultrafilter when we see one. We will then discuss some ultrafilter properties. Ultrafilters play an important role in certain branches of general topology.

Definition 13.14 Let S be a topological space and $(\mathcal{F}, \subseteq)$ denote the set of all filters in $\mathscr{P}(S)$, partially ordered by "$\subseteq$". Suppose $\mathscr{F}$ is an element of $\mathcal{F}$.

We will say that the filter $\mathscr{F}$ is an *ultrafilter* of sets if and only if, for $\mathscr{H} \in \mathcal{F}$,

$$\mathscr{F} \subseteq \mathscr{H} \Rightarrow \mathscr{F} = \mathscr{H}$$

That is, $\mathscr{F}$ cannot be properly contained in some other distinct filter.

The most commonly used characterizations of ultrafilters of sets in $\mathscr{P}(S)$ are stated and proved below. These are extremely useful. It is a good exercise to try to prove them before consulting the proposed proofs.

Theorem 13.15 *Let S be a topological space and $\mathscr{F}$ be a filter of sets in $\mathscr{P}(S)$. The following statements are equivalent.*

(a) *The set $\mathscr{F}$ is an ultrafilter in $\mathscr{P}(S)$.*
(b) *If $U \in \mathscr{P}(S)$ is such that it intersects every element of $\mathscr{F}$, then $U \in \mathscr{F}$.*
(c) *For any $U \in \mathscr{P}(S)$, if $U \notin \mathscr{F}$ then $S \backslash U \in \mathscr{F}$.*

Proof. We are given that S is a topological space.

$(a \Rightarrow b)$ We are given that $\mathscr{F}$ is an ultrafilter. Suppose $U \in \mathscr{P}(S)$ is such that $U \cap F \neq \varnothing$ for all $F \in \mathscr{F}$. We are required to show that $U \in \mathscr{F}$.

Let

$$\mathscr{H} = \{U \cap F : F \in \mathscr{F}\}$$

The set, $\mathscr{H}$, is easily seen to be a filter base of sets in S. Then the filter, $\mathscr{H}^*$, which is generated by $\mathscr{H}$, contains both U and the set F. Then, $\mathscr{F} \subseteq \mathscr{H}^*$. But, since $\mathscr{F}$ is a maximal filter, then $\mathscr{F} = \mathscr{H}^*$. Then $U \in \mathscr{F}$, as required.

$(b \Rightarrow c)$ Suppose that, whenever a set U meets every element of $\mathscr{F}$, then $U \in \mathscr{F}$. Suppose $U \notin \mathscr{F}$. It suffices to show that $S \setminus U \in \mathscr{F}$.

Since $U \notin \mathscr{F}$, then $U \cap F = \varnothing$ for some $F \in \mathscr{F}$. Then $F \subseteq S \setminus U$. If $F_1 \in \mathscr{F}$ and $F_1 \neq F$, then $F_1 \cap F \subseteq S \setminus F$. Since $F_1 \cap F$ is non-empty and is a subset of $S \setminus U$, then $S \setminus U$ intersects every element of $\mathscr{F}$. By hypothesis, $S \setminus U \in \mathscr{F}$.

$(c \Rightarrow a)$ We are given that, if $U \notin \mathscr{F}$ then $S \setminus U \in \mathscr{F}$. We claim that $\mathscr{F}$ is an ultrafilter.

Let $\mathscr{H}^*$ be any filter such that $\mathscr{F} \subseteq \mathscr{H}^*$. If $V \in \mathscr{H}^*$ such that $V \notin \mathscr{F}$ then $S \setminus V \in \mathscr{F}$. This means that $\mathscr{H}^*$ contains both V and its complement. This would imply that $\mathscr{H}^*$ is not a filter. Then there can be no such V. So $\mathscr{H}^* \subseteq \mathscr{F}$. This means that $\mathscr{F}$ is an ultrafilter.

Example 7. Let u be an element of the space S. Let $\mathscr{F} = \{U \in \mathscr{P}(S) : u \in U\}$. We claim that $\mathscr{F}$ is an ultrafilter. To see this, let $U \in \mathscr{P}(S)$. If $u \in U$ then $U \in \mathscr{F}$; if $u \notin U$, then $u \in S \setminus U \in \mathscr{F}$. So, by the above theorem, $\mathscr{F}$ is an ultrafilter.

We have seen that $(\mathcal{F}, \subseteq)$ is partially ordered by "$\subseteq$". For every filter, $\mathscr{H}$, in $\mathcal{F}$ does there exist an ultrafilter which contains it? Many readers see this as an "existence statement". It suggests that, to prove it we may have to invoke Zorn's lemma.[6] We remind ourselves what Zorn's lemma formally states:

[6]Zorn's lemma is proven to be equivalent to the Axiom of choice. A proof appears in R. André, *Axioms and Set Theory*.

> *If every chain in a partially ordered set, $(S, \leq)$, has a maximal element then S has a maximal element.*

Theorem 13.16 *Let S be a topological space and $(\mathcal{F}, \subseteq)$, denote the set of all filters in $\mathscr{P}(S)$, partially ordered by inclusion. If $\mathscr{F} \in \mathcal{F}$, then there exists an ultrafilter, $\mathscr{U}$, in $\mathcal{F}$ which contains $\mathscr{F}$.*

Proof. We are given that $\mathscr{F} \in \mathcal{F}$. We wish to prove the existence of an ultrafilter which contains $\mathscr{F}$. We set up the problem so that Zorn's lemma applies.

For a given $\mathscr{F} \in \mathcal{F}$, consider the set,

$$\mathbb{S}_{\mathscr{F}} = \{\mathscr{H} \in \mathcal{F} : \mathscr{F} \subseteq \mathscr{H}\}$$

of all filters which contain $\mathscr{F}$. We are required to show that some filter in $\mathbb{S}_{\mathscr{F}}$ is an ultrafilter (i.e., $\mathbb{S}_{\mathscr{F}}$ has a maximal element). If $\mathbb{S}_{\mathscr{F}} = \{\mathscr{F}\}$ then $\mathscr{F}$ is an ultrafilter. Suppose $\mathbb{S}_{\mathscr{F}}$ contains filters other than $\mathscr{F}$. Now $\mathscr{F}$ can be the filter base element of many chains in $\mathbb{S}_{\mathscr{F}}$. In fact, $\mathbb{S}_{\mathscr{F}}$ can be viewed as the union of a family of filter-chains,

$$\{\mathscr{C}_\alpha : \alpha \in J\}$$

each with base element, $\mathscr{F}$. Pick an arbitrary filter-chain, $\mathscr{C}_\gamma = \{\mathscr{F}_i : i \in I\}$ in $\mathbb{S}_{\mathscr{F}}$. It looks like,

$$\mathscr{F} \subseteq \mathscr{F}_1 \subseteq \mathscr{F}_2 \subseteq \mathscr{F}_3 \subseteq \cdots$$

Let $\mathscr{T} = \cup\{\mathscr{F}_i : \mathscr{F}_i \in \mathscr{C}_\gamma\}$. Quite clearly, $\mathscr{F}_i \subseteq \mathscr{T}$ for each i. We claim that the family of sets, $\mathscr{T}$, is itself a filter in the chain $\mathscr{C}_\gamma$. If U and V are elements of $\mathscr{T}$, then U and V are elements of some $\mathscr{F}_j$ in the chain, $\mathscr{C}_\gamma$, so $U \cap V$ is an element of $\mathscr{F}_j$; so $U \cap V$ is also an element of $\mathscr{T}$. Then $\mathscr{T}$ is a filter base of sets which is a maximal element in the chain $\mathscr{C}_\gamma$.

So every chain, $\mathscr{C}_\alpha$, in $\mathbb{S}_{\mathscr{F}}$ has a maximal element. We can invoke Zorn's lemma, to conclude from this that $\mathbb{S}_{\mathscr{F}}$, must also have a maximal filter element, $\mathscr{U}$. Since $\mathbb{S}_{\mathscr{F}}$ contains every filter which contains $\mathscr{F}$, then $\mathscr{U}$ is the maximal filter which contains $\mathscr{F}$. Then, by definition, $\mathscr{U}$ is the unique ultrafilter which contains $\mathscr{F}$. This proves the statement.

Example 8. Show that an ultrafilter of sets in a Hausdorff space can have, at most, one accumulation point.

Solution: If $\mathscr{U}$ is a free ultrafilter then, then $\cap\{\text{cl}_S F : F \in \mathscr{U}\} = \varnothing$, so $\mathscr{U}$ has no accumulation point. Suppose, on the other hand that $\mathscr{U}$ is fixed and u and p are two accumulation points. Then, by Theorem 13.10, there exists a filter, $\mathscr{H}^*$, such that $\mathscr{U} \subseteq \mathscr{H}^*$, $\mathscr{H}^* \to p$ and $\mathscr{H}^* \to u$. Since $\mathscr{U}$ is an ultrafilter, then $\mathscr{U} = \mathscr{H}^*$. So $\mathscr{U}$ converges to both u and p. By Theorem 13.9, since S is a Hausdorff space the limit of a filter is unique. So, $u = p$. We conclude that an ultrafilter can have at most one accumulation point.

13.10 Convergence of Filters on Product Spaces

Suppose $\{S_i : i \in I\}$ is a family of topological spaces and $S = \prod_{i\in I} S_i$ is a product space.[7] Suppose there is a filter, $\mathscr{F}$, in S which converges to a point, $\langle x_i \rangle_{i\in I}$. Then, for each $i \in I$,

$$\pi_i[\mathscr{F}] = \{\pi_i[F] : F \in \mathscr{F}^*\} = \mathscr{F}_i$$

is a filter base in S_i.[8]

Since $\mathscr{F}$ converges to $\langle x_i \rangle_{i\in I}$, by continuity of π_i, $\pi_i[\mathscr{F}] = \mathscr{F}_i$ will converge to $\pi_i(\langle x_i \rangle_{i\in I}) = x_i$ (by Theorem 13.13). So if a filter converges to a particular point in a product space, $\prod_{i\in I} S_i$, then the corresponding filter in each factor converges to the corresponding point under π_i. The following theorem shows that the converse holds true.

Theorem 13.17 *Let $\{S_i : i \in I\}$ be a family of topological spaces and $p = \langle x_i \rangle_{i\in I}$ be a point in the product space, $S = \prod_{i\in I} S_i$. Suppose*

[7]We are assuming that the product space is non-empty. This is a consequence of Axiom of choice.

[8]To see this simply note that if $\pi_i[F_1]$ and $\pi_i[F_2]$ belong to $\mathscr{F}_i$ then $\pi_i[F_1 \cap F_2] \subseteq \pi_i[F_1] \cap \pi_i[F_2]$ where $F_1 \cap F_2 \neq \varnothing$ and $\pi_i[F_1]$ is never empty.

that $\mathscr{F}^*$ is a filter of sets in S such that, for each $i \in I$, the filter,

$$\pi_i[\mathscr{F}^*] = \{f[F] : F \in \mathscr{F}^*\} = \mathscr{F}_i$$

converges to x_i. Then $\mathscr{F}$ converges to the point, $p = \langle x_i \rangle_{i \in I}$.[9]

Proof. Given: A filter, $\mathscr{F}^*$, in the product space, $S = \prod_{i \in I} S_i$, and that $p = \langle x_i \rangle_{i \in I}$ is a point in S. For each i, the filter, $\pi_i[\mathscr{F}^*] = \mathscr{F}_i$, converges to x_i. We are required to show that $\mathscr{F}^*$ converges to p.

Let F be a finite subset of I. For each $i \in F$, let U_i be an open neighborhood of $x_i \in S_i$. Then

$$B = \cap\{\pi_i^{\leftarrow}[U_i] : i \in F\}$$

forms a *basic* open neighborhood of $p = \langle x_i \rangle_{i \in I} \in S$. To show that $\mathscr{F}^*$ converges to p we must find an element, A of $\mathscr{F}^*$ which is contained in B.

Since $\mathscr{F}_i$ converges to x_i, $U_i \in \mathscr{F}_i$ (since $\mathscr{F}_i$ is a filter), so $\pi_i^{\leftarrow}[U_i] \in \mathscr{F}$. For each $i \in F$, since π_i is continuous there exists an open neighborhood, A_i, of p, in $\mathscr{F}$ such that $x_i \in \pi_i[A_i] \subseteq U_i$. By the filter base property, there exists $A \in \mathscr{F}$ such that $p \in A \subseteq \cap\{A_i : i \in F\}$. Now, for each $i \in F$,

$$\begin{aligned} \pi_i[A] &\subseteq \pi_i[\cap\{A_i\}] \\ &\subseteq \cap\{\pi_i[A_i]\} \\ &\subseteq \pi_i[A_i] \\ &\subseteq U_i \end{aligned}$$

So, we have found $A \in \mathscr{F}$ such that $p \in A \subseteq \cap\{\pi_i^{\leftarrow}[U_i] : i \in F\}$. So, $\mathscr{F}$ converges to $p = \langle x_i \rangle_{i \in I}$, as required.

Corollary 13.18 *Let $\{S_j : j \in J\}$ be a family of topological spaces and*

$$S = \prod_{j \in J} S_j$$

[9]This result will be useful to show that the product of compact spaces is compact in the Section 13.11.

be a corresponding product space. Let $B = \{\beta(i) : i \in \mathbb{N}\}$ *be a sequence in* S*, where* $\beta(i) = \langle \beta_j(i) \rangle_{j \in J}$*. For each* $j \in J$*, let*

$$\pi_j(\beta(i)) = \pi_j(\langle \beta_j(i) \rangle_{j \in J}) = \beta_j(i)$$

Then, for each $j \in J$*,*

$$\pi_j[B] = \{\beta_j(i) : i \in \mathbb{N}\}$$

the corresponding sequence in S_j*.*

Suppose that, for each $j \in J$*, the sequence* $\{\beta_j(i) : i \in \mathbb{N}\}$ *converges to* p_j *in* S_j*.*

Then the sequence $\{\beta(i) : i \in \mathbb{N}\}$ *converges to* $\langle p_j \rangle_{j \in J}$ *in* S*.*

Proof. We are given the product space, $S = \prod_{j \in J} S_j$, and a sequence, $B = \{\beta(i) : i \in \mathbb{N}\}$ in S. Also, for each $j \in J$,

$$\pi_j[\{\beta(i) : i \in \mathbb{N}\}] = \{\beta_j(i) : i \in \mathbb{N}\}$$

Suppose that, for each $j \in J$, $\{\beta_j(i) : i \in \mathbb{N}\} \to p_j \in S_j$.

Let $\mathscr{F}$ be the filter base generated by the sequence B and, for each $j \in J$, $\mathscr{F}_j = \pi_j[\mathscr{F}]$.

Then, by Theorem 13.8, the filter, $\mathscr{F}_j$, in S_j generated by the sequence $\{\beta_j(i) : i \in \mathbb{N}\}$ converges to p_j.

By Theorem 13.17, the filter $\mathscr{F}$ converges to $p = \langle p_j \rangle_{j \in J} \in \prod_{j \in J} S_j$. Then the sequence, $\{\beta(i) : i \in \mathbb{N}\}$, in $\prod_{j \in J} S_j$, converges to $p = \langle p_j \rangle_{j \in J}$, as required.

13.11 Ultranets: A Close Relative of Ultrafilters

We will now formally define what is also known as maximal nets.

Definition 13.19 Let S be a topological space and $A = \{f(i)\}_{i \in D}$ be a net in S. We will say that A is an *ultranet* if, for any subset, B, in S, a tail end of A either belongs to B and, if not, belongs to its complement, $S \backslash B$.

A net which is constant on its tail end, is an example of an ultranet. It is not immediately obvious from this definition that non-constant ultranets exist. We will be able to address this question below.

Theorem 13.20 *Let $A = \{f(i) : i \in D\}$ be an ultranet in a space S. If E is a subnet of A then E is a also an ultranet.*

Proof. Let S be a topological space which contains an ultranet, $A = \{f(i) : i \in D\}$.

Suppose E is a subnet of A. Let B be a non-empty subset of S. Then B or $S\setminus B$ contains a tail end of A. Suppose, without loss of generality, that B contains a tail end of A. Then B must contain the tail end of any subnet of A. Then every tail end of E belongs to B. By definition of "ultranet" E is an ultranet.

For what follows recall, from Definition 13.19, that "ultranets are those nets, $\{x_i\}$, such that, for each $B \in \mathscr{P}(S)$, either B or $S\setminus B$ contains a tail end of $\{x_i\}$".

Example 9. Let S be a topological space. Suppose $A = \{f(i) : i \in D\}$ is a net in S and $\mathscr{F}$ is a filter of sets in S.

(a) Show that, if A is an ultranet and $\mathscr{F}$ is the filter generated by A, then $\mathscr{F}$ is an ultrafilter.
(b) Show that, if $x \in S$ is the limit point of the ultranet A, then a tail end of A is constant. So a convergent ultranet must have a constant tail end.
(c) Show that, if $\mathscr{F}$ is an ultrafilter and A is a net determined by $\mathscr{F}$, then A is an ultranet.
(d) Show that a non-constant ultranet exists.

Solution: We are given that S is a topological space and $A = \{f(i) : i \in D\}$ is a net in S. Also, $\mathscr{F}$ is a filter of sets in S.

(a) Suppose A is an ultranet and $\mathscr{F} = \{f[T_u] : u \in D\}$, where $T_u = \{i \in D : i \geq u\}$, is the filter generated by the net A. We are required to show that $\mathscr{F}$ is an ultrafilter.

Let $B \in \mathscr{P}(S)\setminus\varnothing$. By Theorem 13.15, either B or $S\setminus B$ belongs to an ultrafilter of sets. Suppose $S\setminus B$ does not belong to $\mathscr{F}$.

Then $S \setminus B$ does not contain a tail end, say $f[T_j]$, of A (for, if it did, then $S \setminus B$ would belong to $\mathscr{F}$). Then, by definition of "ultranet", $f[T_j] \subseteq B$. Since $\mathscr{F}$ is a filter, then $B \in \mathscr{F}$. So $\mathscr{F}$ is an ultrafilter.

(b) We are given that the ultranet, A, converges to x and that $\mathscr{F} = \{f[T_u] : u \in D\}$ is the ultrafilter generated by A. Then $\mathscr{F}$ must also converge to x.

We are required to show that some tail end of A is constant. Suppose no tail end of A is constant. Then, for $k \in D$, there exists two non-empty non-intersecting cofinal sets, C and E of D, such that $f[D_k] = C \cup E$ and both C and E converge to the net limit point x. Since $\mathscr{F}$ is an ultrafilter, either C or $S \setminus C$ belongs to $\mathscr{F}$. Suppose $C \in \mathscr{F}$. Then $S \setminus C$ does not belong to $\mathscr{F}$. Since $\mathscr{F}$ is an ultrafilter, $S \setminus C$ does not intersect some element F of $\mathscr{F}$. Then, since $S \setminus C$ contains E, the cofinal set, E, does not intersect some $F \in \mathscr{F}$, that is some tail end of the ultranet, A. But this can't be since E is cofinal in A. So, A must be constant on a tail end. Since x is a limit point of A, then the constant is x.

(c) We are given that $\mathscr{F}$ is an ultrafilter and $A = \{f(y, F) : F \in \mathscr{F}$ and $y \in F\}$ is the net determined by $\mathscr{F}$. Recall that $(u, F) \leq (v, H)$ if and only if $v \in H \subseteq F$ and $f(u, F) = u$. We are required to show that A is an ultranet.

Let $B \in \mathscr{P}(S) \setminus \{0\}$. Then $B \in \mathscr{F}$ or $S \setminus B \in \mathscr{F}$. Suppose $s \in B \in \mathscr{F}$. Suppose $(u, H) \in D_{\mathscr{F}}$ such that $(u, H) \geq (s, B)$. Then $u \in H \subseteq B$. Then $f(u, H) = u \in B$. Then the tail end, $\{f(y, F) : (y, F) \geq (s, B)\}$, of A belongs to B. By definition, A is an ultranet.

(d) Since convergent ultranets have a constant tail end, we will only investigate non-converging ultranets. We begin with a free ultrafilter. Let $\mathscr{F}$ be a *free* ultrafilter of sets in S. Let $\kappa : \mathscr{F} \to S$ be a choice function defined as $\kappa(F) = x_F$, where $x_F \in F$. Let $A = \{f(x_F, F) : F \in \mathscr{F}\}$ be the free ultranet determined by $\mathscr{F}$. Since A is free, it cannot be constant on any tail end, otherwise it would converge to that constant element. So, A is a free ultranet with no constant tail end. As required.

Theorem 13.21 *Let $A = \{f(i) : i \in D\}$ be a net in a topological space, S. Then A has a subnet U which is an ultranet.*

Proof. We are given that $A = \{f(i) : i \in D\}$ is a net in a space, S. Let $\mathscr{F} = \{f[T_u] : u \in D\}$ be the filter generated by the net A, where $T_u = \{i \in D : i \geq u\}$ and $f[T_u] = \{f(i) : i \in T_u\}$. Then there exists an ultrafilter, $\mathscr{U}$, which contains the filter $\mathscr{F}$.

We now construct a net derived from the ultrafilter, $\mathscr{U}$. Let

$$(D_{\mathscr{U}}, \leq) = \{(i, F) \in D \times \mathscr{U} : x_i \in F\}$$

where $(i, F) \leq (j, H)$ if and only if $j \geq i$ and $x_j \in H \subseteq F$. So, $D_{\mathscr{U}}$ is a set directed by $\leq$.

We define the function, $g : D_{\mathscr{U}} \to D$, as $g(j, F) = j$ if $x_j \in F$.

Then $(f \circ g)(j, F) = f(j) \in A$ where $i \geq j$ implies $(f \circ g)(i, F) = f(i) \geq f(j)$. So, $(f \circ g)$ maps $D_{\mathscr{U}}$ in A, respecting the order, and $A_{\mathscr{U}} = (f \circ g)[D_{\mathscr{U}}]$ is cofinal. We have shown above that the net,

$$A_{\mathscr{U}} = \{(f \circ g)(u, F) : (u, F) \in D_{\mathscr{U}}\}$$

generated by an ultrafilter is an ultranet. So, $A_{\mathscr{U}}$ is an ultranet which is a subnet of A.

Concepts Review

1. Define filter base and filter.
2. How does a filter base generate a filter?
3. What does it mean to say that a filter base is fixed? When is it free?
4. If A is a net describe a filter base of sets determined by the net, A.
5. What does it mean to say that a filter base of sets converges to a point x?
6. How do we define an accumulation point of a filter base of sets?
7. Given a filter base, $\mathscr{F}$, how do we define the adherence of $\mathscr{F}$?
8. In what kind of topological space do *convergent* filter bases have precisely one limit point?
9. If a filter base, $\mathscr{F}$, has a non-limit accumulation point, p, describe a filter which contains $\mathscr{F}$ and converges to p.
10. How do we recognize the points in the closure, $\mathrm{cl}_S F$, of a set, F, in terms of filters?

11. Give a characterization of a continuous function, $f: S \to T$, at a point in terms of filters.
12. What does it mean to say that the filter, $\mathscr{F}$, is finer than the filter, $\mathscr{H}$?
13. Define an ultrafilter of sets, in $\mathscr{P}(S)$.
14. Give two characterizations of an ultrafilter of sets.
15. If S is a Hausdorff space, how many accumulation points can an ultrafilter in $\mathscr{P}(S)$ have?
16. Define an ultranet.

Exercises

1. Suppose $\mathscr{H}$ is a filter base of sets which is finer than the filter base $\mathscr{F}$. If $\mathscr{F}$ converges to p does $\mathscr{H}$ necessarily converge to p?
2. Suppose p is an accumulation point of the ultrafilter of sets, $\mathscr{U}$. Does $\mathscr{U}$ necessarily converge to p?
3. Let $f: S \to T$ be a function mapping the topological space S onto the topological space T. If $\mathscr{F}$ is an ultrafilter of sets in $\mathscr{P}(S)$ and $f[\mathscr{F}] = \{f[F] : F \in \mathscr{F}\}$ is $f[\mathscr{F}]$ necessarily a filter?
4. If V is a non-empty open subset of the topological space S and the filter of sets, $\mathscr{F}$, converges to some point $p \in V$, does V necessarily belong to $\mathscr{F}$?
5. Suppose U is a non-empty subset of the space S. Suppose U belongs to every filter, $\mathscr{F}$, which converges to a point p of U, is U necessarily open in S?
6. Let u be a point in a topological space S. Let $\mathscr{F} = \{F \in \mathscr{P}(S) : u \in F \text{ and } S \backslash F \text{ is finite}\}$.

   (a) Show that $\mathscr{F}$ is a filter of sets in $\mathscr{P}(S)$.
   (b) Show that $\mathscr{F}$ is a neighborhood system of u.

7. Let S be a topological space and $\mathscr{F}$ and $\mathscr{H}$ be two filters in $\mathscr{P}(S)$. Let

$$\mathscr{F} \times \mathscr{H} = \{F \times H : F \in \mathscr{F} \text{ and } H \in \mathscr{H}\}$$

   Is $\mathscr{F} \times \mathscr{H}$ a filter base?

8. Let S be a topological space an $\mathscr{F}$ is a filter of sets in $\mathscr{P}(S)$. Let $a(\mathscr{F})$ be the adherence set of $\mathscr{F}$. Show that $a(\mathscr{F})$ is a closed subset of S.
9. Suppose $\mathscr{U}$ is an ultrafilter in $\mathscr{P}(S)$. If U and V are disjoint non-empty subsets of S such that $U \cup V$ belongs to $\mathscr{U}$ show that either U or V belongs to $\mathscr{U}$.

Part V
Compact Spaces and Relatives

Chapter 14

Compactness: Definition and Basic Properties

Abstract

In this section we define and describe properties of compactness in a topological space. The concept of compactness is a generalization of the "closed and bounded" property in the Euclidean space $\mathbb{R}$. It was introduced by Pavel Urysohn in 1929. We give the most general definition of compactness in terms of covers and subcovers. The few characterizations given for this property will simplify the task of determining whether a space is compact. These characterizations also provide a deeper understanding of what it means for a set to be "compact". We also show that the compact property is carried over from one topological space to another by continuous functions. Furthermore, the compact property carries over from the factors of a product space to the product space itself. Subsets of a compact space, S, will be seen to be closed only if S is Hausdorff. But closed subsets of any compact space are always compact.

14.1 Introduction

Many readers may already be familiar with the notion of "compactness". It is often encountered early in a course of real analysis, in a form that is somewhat different from the one we will encounter here. Initially, some readers will wonder whether we are talking about the same property. If viewed in this context a possible definition could have been: "A subset, F, of $\mathbb{R}^n$ is compact if and only if every sequence in F has a convergent subsequence with its limit point inside F". Or, in the context of normed vector spaces, the reader may have encountered a theorem which states: "If V is a finite dimensional vector space, the compact subsets of V are precisely the closed and bounded subsets". One of many reasons the notion of compactness would have been introduced in a previous course is to access a statement commonly called the "extreme value theorem" (EVT) which says: "Given that $f : V \to \mathbb{R}$ is a continuous function mapping a normed vector into $\mathbb{R}$, if F is a compact subset of V, then f attains a maximum value and a minimum value on F". We will be studying precisely the same notion of compactness in the more general context of a topological space. Compact spaces are encountered in numerous other fields of mathematics. In general, most students will feel that spaces which are both compact and Hausdorff are more intuitive; they seem to be "well-behaved" since they come with many useful tools that can be used to solve various problems.

In what follows, we will be referring to an "open cover" of a subset F of a space, S. We say that $\mathscr{U} = \{U_i : i \in I\}$ is an *open cover of* F if each U_i is an open subset of S and $F = \cup\{U_i : i \in I\}$. A "finite subcover", $\{U_i : i \in K\}$, is a finite subset of an open cover, $\mathscr{U}$, of F, which also covers F.

Definition 14.1 Let F be any non-empty subspace of the topological space, S. We say that F is *compact* if and only if every S-open cover of F has a finite subcover.

Since the reader may have encountered other definitions of "compact", we will refer to 14.1 above as the *topological definition of compactness.* We provide a few examples of compact and non-compact spaces to help the reader develop a better feeling for what "every open cover has an a finite subcover" means.

Example 1. Note that the cover, $\mathscr{V} = \{(i - 2/3, i + 2/3) : i \in \mathbb{Z}\}$, forms an open cover of each of the four spaces, $\mathbb{R}$, $\mathbb{Q}$, $\mathbb{J}$, (the irrationals) $\mathbb{Z}$ with the usual topology. But removing just one set from $\mathscr{V}$ will leave a point which is not covered by the other sets in $\mathscr{V}$. So $\mathscr{V}$ has no finite subcover. So *none* of these four spaces are compact.

Example 2. Consider the set, $F = \{1/n : n = 1, 2, 3, \ldots\} \cup \{0\}$ with the subspace topology inherited from $\mathbb{R}$. Let $\mathscr{V}$ be an $\mathbb{R}$-open cover of F. There exists $U \in \mathscr{V}$ such that $0 \in U$. Then there are at most finitely many points in $F \backslash U$. For each of these points we can choose precisely one set from $\mathscr{V}$ which contains it. The set U along with each one of these sets covers F. So F is compact. If we remove the point, 0, from F then the resulting space is not compact. (Verify this!)

It will be useful to remember the particular technique exhibited in the following example, since it will be called upon to help solve various questions in the next few chapters.

Example 3. Consider the bounded closed interval, $S = [3, 7]$, viewed as a subspace of $\mathbb{R}$. Let $\mathscr{V} = \{V_i : i \in I\}$ be an open covering of S. Let

$$\mathscr{U} = \{u \in [3, 7] : \text{where } [3, u] \text{ is covered by finitely many sets from } \mathscr{V}\}$$

Let $k = \sup\{u : u \in \mathscr{U}\}$. Then $3 \leq k \leq 7$ and $[3, 3] \subseteq [3, k] \subseteq [3, 7]$.

There exists $V_j \in \mathscr{V}$ such that $k \in V_j$.

Suppose $k < 7$. Then there exists $\varepsilon > 0$ such that $k \in (k - \varepsilon, k + \varepsilon) \subseteq V_j \subseteq [3, 7]$. By definition of k, $[3, k - \frac{\varepsilon}{2}]$ has a finite subcover, say $\mathscr{V}_F$. Then $\mathscr{V}_F \cup \{V_j\}$ is a finite subcover of $[3, k]$. Also, $[3, k + \frac{\varepsilon}{2}]$ has a finite subcover, $\mathscr{V}_F \cup \{V_j\}$. This contradicts the definition of k. The source of our contradiction is our supposition that $k < 7$. So, $k = 7$. Hence, S has a finite subcover so S is compact.

14.2 Characterizations of the Compact Property

We will be needing a few efficient ways of determining whether a subspace is compact (or not). The following characterizations will be useful.

But first we should develop some familiarity with the following concept. It is often abbreviated by the acronym, FIP.

> A family, $\mathscr{F}$, of sets is said to satisfy the *finite intersection property* if "every finite subfamily of $\mathscr{F}$ has non-empty intersection".

A "filter of sets" is a prime example of a family of sets which satisfies the finite intersection property.

We present the following various ways of recognizing a compact set.[1]

Theorem 14.2 *Let S be a topological space. The following are equivalent.*

(a) *The space S is compact.*
(b) *Any family, $\mathscr{F}$, of closed subsets of S which satisfies the finite intersection property has non-empty intersection.*
(c) *Every filter of sets in S has an accumulation point.*
(d) *Every ultrafilter of sets in S has a limit point.*
(e) *Every net in S has an accumulation point.*
(f) *Every ultranet in S has a limit point.*

Proof. We are given that S is a topological space.

(a $\Rightarrow$ b) Let S be a compact topological space. Suppose the family, $\mathscr{F}$, of closed subsets of S satisfies the finite intersection property. That is, for every finite subset, $\mathscr{W}$, of $\mathscr{F}$, $\cap\{F : F \in \mathscr{W}\} \neq \varnothing$.

We are required to show that $\cap\{F : F \in \mathscr{F}\} \neq \varnothing$.

Suppose $\cap\{F : F \in \mathscr{F}\} = \varnothing$. Then, $\{S\backslash F : F \in \mathscr{F}\}$ is an open cover of S. Then, by hypothesis, there exists a finite subset, $\mathscr{W}$, of $\mathscr{F}$, such that $S = \cup\{S\backslash F : F \in \mathscr{W}\}$. Equivalently, $\cap\{F : F \in \mathscr{W}\} = \varnothing$, contradicting the given fact that $\mathscr{F}$ satisfies the finite intersection property. So, $\cap\{F : F \in \mathscr{F}\} \neq \varnothing$, as required.

(b $\Rightarrow$ a) Suppose that, whenever a family, $\mathscr{F}$, of closed sets satisfies the finite intersection property, then $\cap\{F : F \in \mathscr{F}\} \neq \varnothing$. Let $\mathscr{U}$ be an open cover of S. If, for any finite $\mathscr{W}$ of $\mathscr{U}$, $\cup\{F : F \in \mathscr{W}\} \neq S$, then the set $\{S\backslash F : F \in \mathscr{U}\}$ satisfies the finite intersection property, and so $\cap\{S\backslash F : F \in \mathscr{U}\} \neq \varnothing$. This can only mean that, $\mathscr{U}$ is not an open cover of S, a contradiction. So, $\mathscr{U}$ has a finite subcover of S.

(b $\Rightarrow$ c) We are given that if a family of closed sets satisfies the FIP then it has non-empty intersection. Let $\mathscr{F}$ be a filter base of subsets in S. We are required to show that $\mathscr{F}$ has an accumulation point.

---

[1]Those readers who have skipped studying the chapters on nets, will find properties (b), (d) the most useful. These are the characterizations that are the most often called upon, in future proofs.

That is, there is a point x such that $x \in \mathrm{cl}_S F$ for all $F \in \mathscr{F}$. Then, by definition of filter base, $\mathscr{F}$ satisfies the finite intersection property. Then, if $\mathscr{G}$ is a finite subset, of $\mathscr{F}$, $\cap\{F : F \in \mathscr{G}\} \neq \varnothing$. Then $\cap\{\mathrm{cl}_S F : F \in \mathscr{G}\} \neq \varnothing$. By hypothesis, there exists $x \in \cap\{\mathrm{cl}_S F : F \in \mathscr{F}\} \neq \varnothing$. Then $\mathscr{F}$ has an accumulation point.

(c $\Rightarrow$ d) Suppose every filter of sets has an accumulation point. Let $\mathscr{U}$ be an ultrafilter of sets in $\mathscr{P}(S)$. We must show that $\mathscr{U}$ has a limit point. By hypothesis, $\mathscr{U}$ must have an accumulation point, say p. By Theorem 13.10, p is a limit point of some filter, $\mathscr{H}^*$ containing $\mathscr{U}$. Since $\mathscr{U}$ is a maximal filter, then $\mathscr{U} = \mathscr{H}^*$. So, p is a limit point of $\mathscr{U}$.

(d $\Rightarrow$ c) We are given that every ultrafilter has a limit point. We must show that every filter of sets has an accumulation point. Let $\mathscr{F}$ be a filter base. Then, by Theorem 13.16, $\mathscr{F}$ is contained in some ultrafilter, $\mathscr{U}$. Let p be the limit point of $\mathscr{U}$. Then, for any open neighborhood, $U \in \mathscr{B}_p$, of p, there is $F \in \mathscr{U}$ such that $F \subseteq U$. This means every open neighborhood of p intersects every $F \in \mathscr{U}$. Then $p \in \cap\{\mathrm{cl}_S F : F \in \mathscr{F}\}$. So, p is an accumulation point of $\mathscr{F}$.

(c $\Rightarrow$ b) We are given that every filter base in $\mathscr{P}(S)$ has an accumulation point.

Suppose the family, $\mathscr{F}$, of closed subsets of S satisfies the finite intersection property. By hypothesis, $\mathscr{F}$ has an accumulation point, say p.

We are required to show that $\cap\{F : F \in \mathscr{F}\} \neq \varnothing$.

Note that, $\mathscr{F}$, satisfies the main filter base property and so $\mathscr{F}$ a "filter base of closed sets" which has an accumulation point, p. Then, by definition, $p \in \cap\{\mathrm{cl}_S F : F \in \mathscr{F}\} \neq \varnothing$. Since each F is closed in S, $p \in \cap\{F : F \in \mathscr{F}\} \neq \varnothing$, as required.

(f $\Leftrightarrow$ d) By Theorem 13.8 and the example on page 331, an ultrafilter has limit point p if and only if its corresponding ultranet has limit p.

(e $\Leftrightarrow$ c) By Theorem 13.8, a filter has accumulation point p if and only if its corresponding net has accumulation limit p. By Theorem 12.5, if a net has an accumulation point p then it has a subnet which converges to p.

The reader will soon discover that, compact spaces are more intuitive and richer in interesting properties when equipped with the

Hausdorff separation axiom. It is often assumed that "compact sets are always closed" since, in many books, to remove clutter from the main body of the text, a blanket assumption is initially made that all spaces are hypothesized to be Hausdorff. But a compact space may not be closed if the space is not Hausdorff. When referring to various texts, readers should keep this in mind, and check at the begin of the text to see if any blanket assumptions are made.

14.3 Properties of Compact Subsets

We will now verify whether the compactness property is carried over by continuous functions from its domain to its codomain. Also, we will show that closed subspaces of compact spaces are compact. These results will be familiar to readers with some experience in real analysis. The proofs confirm that these properties generalize nicely, from $\mathbb{R}^n$ to topological spaces.

Theorem 14.3 *Let S and T be topological spaces.*

(a) *Suppose $f : S \to T$ is a continuous function mapping S into T. If F is a compact subset of S then $f[F]$ is compact in T. So continuous images of compact sets are compact.*
(b) *Suppose S is compact. If H is a closed subset of S, then H is compact.*
(c) *The union of a finite collection of compact sets is compact.*
(d) *Suppose S is Hausdorff.*

(i) *If H is a compact subset of S then H is closed in S.*
(ii) *If S is also compact then S is regular.*
(iii) *If S is also compact then S is normal.*
(iv) *The intersection of any collection of compact subsets is compact. (Note: The Hausdorff property is required in the proof of this statement.)*

Proof. We are given that S is a topological space.

(a) We are given that the function, $f : S \to T$, is continuous and that F is a compact subset of S.

Let $\mathscr{U}$ be an open cover of $f[F]$. Then $\{f^{\leftarrow}[U] \cap F : U \in \mathscr{U}\}$ is an open cover of F. Since F is compact, there is a finite subcover, $\{f^{\leftarrow}[U] \cap F : U \in \mathscr{F}\}$, of open subsets of F. Since,

$$F \subseteq \cup\{f^{\leftarrow}[U] \cap F : U \in \mathscr{F}\} \Rightarrow f[F] \subseteq f\left[\cup\{f^{\leftarrow}[U] \cap F : U \in \mathscr{F}\}\right]$$

then,

$$\begin{aligned} f[F] &\subseteq \cup\{f\left[\{f^{\leftarrow}[U] \cap F\right] : U \in \mathscr{F}\} \\ &\subseteq \cup\{f[\{f^{\leftarrow}[U]] \cap f[F] : U \in \mathscr{F}\} \\ &= \cup\{U \cap f[F] : U \in \mathscr{F}\} \end{aligned}$$

So, $\{U \cap f[F] : U \in \mathscr{F}\}$ is a finite subcover of $f[F]$. So continuous images of compact sets are compact.

(b) Suppose S is a compact space. We are given that H is a closed subset of S. Let $\mathscr{F} = \{F_i : i \in I\}$ be a family of closed subsets in the subspace, H, such that $\mathscr{F}$ satisfies the finite intersection property. Note that, since H is closed, then $\text{cl}_S F_i = F_i$ for all $i \in I$. So that $\mathscr{F}$ is a family of closed subsets of S such that $\mathscr{F}$ satisfies the FIP. Since S is compact, there is a point $p \in \cap\{F : F \in \mathscr{F}\}$. Since $p \in F \subseteq H$, $\cap\{F : F \in \mathscr{F}\}$ is a non-empty subset of H. So, H is compact.[2]

(c) Let $\{K_1, K_2, \ldots, K_n\}$ represent a finite collection of compact subsets of a space, S. Let $K = \cup_{i=1 \text{ to } n} K_i$. Let $\mathscr{U} = \{U_\alpha : \alpha \in A\}$ be an open cover of K. Then for each i, K_i has a finite subcover $\{U_{\alpha_i} : \alpha_i \in F_i\}$. Then $\cup_{i=1 \text{to } n}[\cup_{\alpha_i \in F_i} U_{\alpha_i}]$ forms a finite subcover of K. So, K is compact.

(d) For what follows, suppose S is Hausdorff.

   (i) Let H be a compact subset of S (where S need not be compact). We are required to show that H is closed in S. It will suffice to show that $S \setminus H$ is open in S. Let $u \in S \setminus H$.

[2]Note that the Hausdorff property for S is not required for this to hold true. That is, *closed subsets of compact spaces are always compact.*

Since S is Hausdorff, then for each $x \in H$, there exists disjoint open neighborhoods, U_x and V_u^x, of x and u respectively. Since H is compact, the open cover, $\{U_x : x \in H\}$, will have a finite subcover, say $\{U_{x_i} : i \in F\}$ and finite open neighborhoods of u, say $\{V_{x_i}^u : i \in F\}$, where $U_{x_i} \cap V_{x_i}^u = \varnothing$, for $i \in F$. Then, for each $i \in F$,

$$U_{x_i} \cap [\cap\{V_{x_i}^u : i \in F\}] = \varnothing$$

Let $W_u = \cup\{U_{x_i} : i \in F\}$ and $M_u = \cap\{V_{x_i}^u : i \in F\}$.

Then $W_u \cap M_u = \varnothing$. So $M_u \cap H = \varnothing$. Then $M_u \cap H = \varnothing$ holds true for any choice of $u \in S \setminus H$. We have then shown that every point, u, in $S \setminus H$ has an open neighborhood, M_u, which has empty intersection with H. So $S \setminus H$ is open; hence, H is closed. (Note how the Hausdorff property plays a role in the proof.)

(ii) We are given that S is compact Hausdorff. We are required to show that S is regular. Let H be closed and $u \in S \setminus H$. Then, since S is compact, H is compact. In part (ii) we constructed disjoint open neighborhoods, W_u and M_u, of H and $\{u\}$, respectively. So, S is regular.

(iii) We are given that S is compact Hausdorff. Let H and K be disjoint closed subsets of S. Then, since S is compact, H and K are compact. Let $u \in K$. We showed in part (ii) that S is regular and so there exists disjoint open neighborhoods, W_u and M_u of H and $\{u\}$, respectively. Then $\{M_u : u \in K\}$ forms an open cover of K which has a finite subcover, say $\{M_{u_i} : i \in F\}$. Then $\cap\{W_{u_i} : i \in F\}$ and $\cap\{M_{u_i} : i \in F\}$ form disjoint open neighborhoods of H and K. So, S is normal.

(iv) Let $\{K_\alpha : \alpha \in A\}$ be a collection of compact subsets of a Hausdorff space, S. Let $K = \cap_{\alpha \in A} K_\alpha$. Since S is Hausdorff and each K_α is closed, then so is K. Let $\alpha_i \in A$. Since, by hypothesis, K_{α_i} is compact and $K \subseteq K_{\alpha_i}$, then K is compact.

Some readers may find it useful to "think" of (or manipulate) compact subsets of Hausdorff spaces as if they are "points".

Example 4. Construct a topological space, S, with two compact subspaces, A and B, such that $A \cap B$ is non-compact.

Solution: Given Theorem 14.3, part (d) we need only consider spaces which are not Hausdorff. Consider the set, $\mathbb{N}$, equipped with topology generated by the base,

$$\mathscr{B} = \{\varnothing, \mathbb{N}, \mathbb{N}\backslash\{0\}, \mathbb{N}\backslash\{1\}\} \cup \{\{n\} : n \geq 2\}$$

Let $A = \mathbb{N}\backslash\{0\}$ and $B = \mathbb{N}\backslash\{1\}$. Let $\mathscr{U}_A$ be an open cover of A. Then the only element of $\mathscr{B}$ which will cover $\{1\}$ is $A = \mathbb{N}\backslash\{0\}$. So A is compact. Similarly, B is compact. But any cover of $A \cap B$ contains infinitely many isolated points. So even though both A and B are compact, $A \cap B$ is not compact.

Example 5. Let F be a compact subset of a topological space S. Suppose $f : S \to T$ is a continuous one-to-one function mapping F into a Hausdorff space, T. Show that $f[F]$ is a homeomorphic copy of F.

Solution: By Theorem 6.9, it will suffice to show that $f : F \to T$ is a closed function on F. Let K be a closed subset of F. Then, by the above theorem, K is compact. Then $f[K]$ is compact in T. Since T is Hausdorff, then $f[F]$ is Hausdorff. Since $f[K]$ is a compact subset of $f[F]$, then it is closed. Then f is a closed mapping on F. So, $f : F \to f[F]$ is homeomorphism.

The following very important theorem titled the *Tychonoff theorem* will often be referred to in the proofs that will follow. It is surprising how often this theorem is invoked in proofs of topology and real analysis. It shows that the Cartesian product of any number of compact spaces is compact.

Theorem 14.4 The Tychonoff theorem. *Let $\{S_i : i \in I\}$ be a family of topological spaces.*

Then the product space, $S = \prod_{i\in I} S_i$, is compact if and only if S_i is compact for each $i \in I$.

Proof. Let $\{S_i : i \in I\}$ be a family of topological spaces.

($\Rightarrow$) We are given that $S = \prod_{i\in I} S_i$ is compact. Then, since the projection map, $\pi_i : S \to S_i$, is continuous for each $i \in I$, and continuous images of compact sets are compact, then $\pi_i[S] = S_i$ is compact, for each $i \in I$.

($\Leftarrow$) We are given that S_i is compact for each $i \in I$.

Let

$$\mathscr{U} = \{\pi_i^{\leftarrow}[U] : U \text{ open in } S_i, i \in I\}$$

be a cover of open subbase sets of S. Let

$$\mathscr{U}_i = \{U : U \subseteq S_i, \pi_i^{\leftarrow}[U] \in \mathscr{U}\}$$

be an open cover of S_i corresponding to $\mathscr{U}$.

Let

$$\mathscr{U}_F = \{\pi_i^{\leftarrow}[U] : \text{finitely many open } U\text{'s in each } S_i, i \in F\}$$

be a finite subset of $\mathscr{U}$. For each $i \in F$, let $\mathscr{U}_i^*$ be a finite subset of $\mathscr{U}_i$ which covers S_i.

Then

$$\mathscr{U}_F^* = \{\pi_i^{\leftarrow}[U] : U \in \mathscr{U}_i^*, i \in F\}$$

is a finite subset of $\mathscr{U}$.

If $\mathscr{U}$ doesn't not have a finite subcover for S then there must be a point $p = \langle x_i \rangle_{i\in I} \in S$ such that $p \notin \pi_j^{\leftarrow}[U]$ for all $j \in F$ and all $U \in \mathscr{U}_j^*$. This means $x_j \notin U \in \mathscr{U}_j^*$ for all $j \in F$. This contradicts that $\mathscr{U}_i^*$ covers S_i for each $i \in F$. So $\mathscr{U}$ has a finite subcover for S.

The proof for the case where $\mathscr{U}$ is a cover of basic open sets mimics the one provided above for the case of the cover of open subbase sets of S.

Alternate approach for the proof of ($\Leftarrow$). We are given that S_i is compact for each $i \in I$. Let $\mathscr{U}$ be an ultrafilter in $S = \prod_{i\in I} S_i$. By Theorem 14.2, compact spaces are precisely those spaces in which every ultrafilter of sets converges to a point in that set. So, it will suffice to show that $\mathscr{U}$ has a limit point in S. Then

$$\mathscr{U}_i = \pi_i[\mathscr{U}] = \{\pi_i[U] : U \in \mathscr{U}\}$$

is a filter base in S_i.[3]

[3]Let $\pi_i[U]$ and $\pi_i[V]$ be two elements of $\pi_i[\mathscr{U}]$ where U and V are elements of $\mathscr{U}$. Then there exists non-empty $W \in \mathscr{U}$ such that $W \subseteq U \cap V$. Then $\pi_i[W] \subseteq f[U \cap V] \subseteq f[U] \cap f[V] \neq \varnothing$. So, $\pi_i[\mathscr{U}]$ is a filter base.

Since $\mathscr{U}$ is a maximal filter, then $\mathscr{U}_i = \pi_i[\mathscr{U}]$ must be a maximal filter for each $i \in I$.[4]

Since S_i is compact, each $\mathscr{U}_i$ converges to a point x_i. By Theorem 13.17, $\mathscr{U}$ converges to $\langle x_i \rangle_{i \in I}$. Then every ultrafilter of sets in S converges to a point of S. So, the product space S is compact.

We will now see that, in Euclidean spaces, $\mathbb{R}^n$, the compact subsets are precisely the "closed and bounded" subsets of $\mathbb{R}^n$.

In the following theorem, we consider the Euclidean topological space $\mathbb{R}^n$. We will say that a subset, U, is *bounded in* $\mathbb{R}^n$ if there exists $k \in \mathbb{R}^+$ such that

$$U \subseteq [-k, k]^n \subseteq \mathbb{R}^n$$

The following theorem states that...

> *in $\mathbb{R}^n$, compactness is equivalent to the closed and bounded properties combined.*

Theorem 14.5 *Let $\mathbb{R}^n$ be the topological space equipped with the product topology. The non-empty subset, T of $\mathbb{R}^n$, is a compact subset if and only if T is both closed and bounded in $\mathbb{R}^n$.*

Proof. When we invoke the Tychonoff theorem, the proof is straightforward and so is assigned as an exercise question.

Theorem 14.6 *Let F be a compact subset of a topological space, S. If f is a continuous real-valued function on F then f attains its maximum and minimum values inside F.*

Proof. We are given that F is compact in S and $f[F]$ is the continuous image of F in $\mathbb{R}$. Then $f[F]$ is compact in $\mathbb{R}$ (by 14.3) and so is closed and bounded (by 14.5). Then boundedness of $f[F]$ guarantees that $f[F] \subseteq [-k, k]$, for some k. Since $f[F]$ is closed in $\mathbb{R}$, then

[4]To see this, suppose U_i is a subset of S_i which intersects every element of $\pi_i[\mathscr{U}]$. Then $\pi_i^{\leftarrow}[U_i]$ intersects every element of $\mathscr{U}$. Since $\mathscr{U}$ is an ultrafilter, $\pi_i^{\leftarrow}[U_i] \in \mathscr{U}$. Then $\pi_i[\pi_i^{\leftarrow}[U_i]] = U_i \in \pi_i[\mathscr{U}]$.

$\sup f[F]$ and $\inf f[F]$ must both belong to $f[F]$. So, f attains both its maximum and minimum values on F.

Example 6. Suppose S and T are topological spaces. We know that projection maps on product spaces are open maps. (See theorem on page 149.)

Show that, in the case where the space S is Hausdorff and the space T is compact then the projection map, $\pi_1 : S \times T \to S$, is a closed map.

Solution: Let K be a closed subset of $S \times T$. We are required to show that $\pi_1[K]$ is closed. It then suffices to show that $S \backslash \pi_1[K]$ is open in S.

Let $u \in S \backslash \pi_1[K]$. Then $(\{u\} \times T) \cap K = \varnothing$. Since T is compact, then we easily see that $\{u\} \times T$ is compact. Since S is Hausdorff, for each $(u, x) \in \{u\} \times T$, there is an open neighborhood, $V_u^x \times U_x$, which does not meet K. In this way we obtain an open cover

$$\{V_u^x \times U_x : x \in T\}$$

of $\{u\} \times T$ which then has a finite subcover

$$\{V_u^{x_i} \times U_{x_i} : x_i \in F \subseteq T\}$$

Then $\cap\{V_u^{x_i} \times U_{x_i} : x_i \in F \subseteq T\}$ forms an open neighborhood of u which does not intersect $\pi_1[K]$. So, $S \backslash \pi_1[K]$ is open in S, as claimed. We conclude that $\pi_1 : S \times T \to S$ is a closed projection map.

Example 7. For the Hausdorff topological space, S, and the compact space, T, let $f : S \to T$ be a function mapping S into T. The set, G, will represent the graph of f, defined as

$$G = \{(x, f(x)) : x \in S\} \subseteq S \times T$$

Show that, if G is a closed subspace of the product space, $S \times T$, then $f : S \to T$ must be a continuous function.

Solution: Given: S is Hausdorff and T is compact. Also, we are given that $G = \{(x, f(x)) : x \in S\}$ is a closed subset of $S \times T$. Let K be a closed subset of T. To show continuity of f, it will suffice to show that $f^{\leftarrow}[K]$ is a closed subset of S.

By the result stated in the example immediately above, $\pi_1 : S \times T \to S$ is a closed map. See that $\pi_2^{\leftarrow}[K] \cap G$ is the intersection of two closed sets and so is a closed subset of $S \times T$. Since

$$\pi_2^{\leftarrow}[K] \cap G = \{(x, f(x)) : f(x) \in K\}$$

then $\pi_1[\pi_2^{\leftarrow}[K] \cap G] = f^{\leftarrow}[K]$.

Since π_1 is a closed map, $f^{\leftarrow}[K]$ is closed. We have shown that f pulls back closed sets to closed sets; so f is continuous.

14.4 Topic : The Embedding Theorem III

Recall the statement of the *Embedding theorem II* (Theorem 10.16) which says that a space S is completely regular if and only if the evaluation map with respect to $C^*(S)$ embeds S inside a cube, $T = \prod_{i\in I}[a_i, b_i]$.

Now, we know something else about cubes: Since they are the product of compact sets, the Tychonoff theorem guarantees that they are compact. Since $\prod_{i\in I}[a_i, b_i]$ and $\prod_{i\in I}[0, 1]$ are homeomorphic, then S can be embedded in the compact space, $\prod_{i\in I}[0, 1]$. With this in mind, we promote the *Embedding theorem II* to *Embedding theorem III*, in the form of the following theorem.

Theorem 14.7 (Embedding theorem III). *If a topological space S is completely regular then S can be* densely embedded *in a compact Hausdorff space.*

Proof. We are given a completely regular topological space S.

Then, by Theorem 10.16, the evaluation map with respect to $C^*(S)$ embeds S into a cube, $T = \prod_{i\in I}[a_i, b_i]$. Each factor of the product T is Hausdorff. Since a product of Hausdorff spaces was shown to be Hausdorff, T is Hausdorff. By the Tychonoff theorem, T is a compact Hausdorff space.

Since $\text{cl}_T e[S]$ is a closed subset of a Hausdorff compact space T, it is Hausdorff compact. Then the homeomorphic image, $e[S]$, of S is dense in $\text{cl}_T e[S]$, a Hausdorff compact set, as required.

Theorem 14.8 *Let S be a topological space. Then S is completely regular if and only if S is a subspace of a Hausdorff compact space.*

Proof. We are given that S is any topological space.

($\Rightarrow$) Suppose S is a subspace of a compact Hausdorff space, T. By Urysohn's lemma, there is a continuous bounded function which separates any two closed subsets (so one that separates a point and a closed set). So, T is completely regular. By Theorem 10.5, since S is a subspace of a completely regular space, it is completely regular, as required.

($\Leftarrow$) Suppose S is completely regular. Then, by Theorem 14.7, S is densely embedded in a compact Hausdorff space T. That is, a topological copy of S is a subspace of T. So, S is a subspace T.

14.5 Topic: Completely Normal Spaces

With the few results involving compact spaces behind us we can now delve a bit deeper on the topic of complete separation of sets. We first remind ourselves of some previously proven results.

— The Cartesian product (equipped with the product topology) of compact spaces is compact. (Tychonoff theorem 14.4)
— Any compact Hausdorff space is normal. (Theorem 14.3(c))
— Metrizable spaces are normal. (Theorem 9.19)
— Since subspaces of metrizable spaces are metrizable (Example on page 107), those normal spaces which are metrizable have normal subspaces.

We have never proven that subspaces of normal spaces are always normal. The main reason is that it is impossible to do so. The Embedding theorem III (Theorem 14.7) confirms that every completely regular space can be (densely) embedded in a compact Hausdorff space. We know, however, that there are completely regular spaces which are not normal. The Moore plane is such an example. (See the Moore plane on page 265). Then...

the Moore plane is a non-normal subspace of the compact (normal) space in which it is embedded.

Those normal spaces whose subspaces are normal form a particular class of topological spaces.

Definition 14.9 A *completely normal space* is a space in which every subspace is normal.

Then every metrizable space is completely normal.

By the given definition, since a topological space is a subspace of itself,

completely normal spaces are normal spaces.

Recall that a *perfectly normal space* is a normal space whose closed subsets are G_δ's. We investigate how "completely normal" spaces compare with "perfectly normal" spaces.

Suppose S is perfectly normal. We claim that S must then be completely normal. Then, by definition, S is normal. Let T be a subspace of S. To show that S is completely normal it suffices to show that T is a normal subspace. Let F be a closed subset of T. Then $F = F^* \cap T$ for some closed subset, F^*, of S. Since S is perfectly normal, F^* is a G_δ in S, then F is a G_δ of T. So, T is perfectly normal and so is a normal subspace of S. We can then say that a perfectly normal space, S, is completely normal. So we have the chain,

perfectly normal $\Rightarrow$ *completely normal* $\Rightarrow$ *normal*

$\Rightarrow$ *completely regular*

In the next example, we have a characterization of a completely normal space.

Example 8. Show that S is completely normal if and only if for any pair of subsets A and B such that $A \cap \mathrm{cl}_S B = \varnothing = \mathrm{cl}_S A \cap B$ there exists disjoint open subsets U and V such that $A \subseteq U$ and $B \subseteq V$.[5]

---

[5]This is an exercise question which appears in 15B of S. Willard's *General Topology* (1970).

Solution: ($\Leftarrow$) Let S be a topological space. Suppose that for any pair of subsets A and B such that $A \cap \text{cl}_S B = \varnothing = \text{cl}_S A \cap B$ there exists *disjoint* open subsets U and V such that $A \subseteq U$ and $B \subseteq V$. If C and D are disjoint closed subsets of S then $C \cap \text{cl}_S D = \varnothing = \text{cl}_S C \cap D$. Then, by hypothesis, there exists *disjoint* open subsets U and V such that $C \subseteq U$ and $D \subseteq V$. So, S is normal.

We now claim that S is completely normal. Let T be a subspace of S. It suffices to show that T is normal. Suppose F and K are disjoint non-empty closed subsets of T. Then it suffices to show that F and K are contained in disjoint open sets. Let F^* and K^* are closed subsets of S such that $F = F^* \cap T$ and $K = K^* \cap T$. Then $F^* \cap \text{cl}_S K^* = \varnothing = \text{cl}_S F^* \cap K^*$. By hypothesis, there exists disjoint open subsets of S, U^* and V^*, such that $F^* \subseteq U^*$ and $K^* \subseteq V^*$. Then $F \subseteq U^* \cap T$ and $K \subseteq V^* \cap T$. Then T must be normal. We can conclude that, by definition, S is completely normal.

The direction ($\Rightarrow$) is left as an exercise.

A normal space which is not perfectly normal. The following example illustrates a completely normal space which is not perfectly normal.

Example 9. Suppose S is an uncountable set and $p \in S$. We equip S with a topology, τ, defined as follows:

$$\tau = \{T \subseteq S : S \backslash T \text{ is finite or } p \notin T\}$$

Show that S is completely normal but not perfectly normal.

Solution: We claim that S is T_1. The set $\{p\}$ is closed and if $x \neq p$ then $S \backslash \{x\}$ is open so $\{x\}$ is closed. So, S is T_1, as claimed.

We claim that S is completely normal. Suppose F and K are disjoint closed subsets of S. If $p \notin K \cup F$ then K and F are both open. So, F and K are clopen and so $F \cap \text{cl}_S K = \varnothing = \text{cl}_S F \cap K$. Suppose $p \in F$. Then K is clopen and finite. That is $K = \text{cl}_S K$ is open. Now $S \backslash K$ is open since K is finite. So, $\text{cl}_S F$ and K are contained disjoint open sets, $S \backslash K$ and K as well as F and $\text{cl}_S K$. So, by the statement proven in example immediately above, S is completely normal, as claimed.

We claim that S is not perfectly normal. Let $\{U_i : i \in \mathbb{N}\}$ be a countable family of sets in S each containing the point p. Then each U_i is closed in S and $p \in \cap\{U_i : i \in \mathbb{N}\}$. If each U_i is also open $S \backslash U_i$

if finite. So each U_i is uncountable. Then $S\backslash\cap\{U_i : i \in \mathbb{N}\} = \cup\{S\backslash U_i : i \in \mathbb{N}\}$, a countable subset of S. So, $\cap\{U_i : i \in \mathbb{N}\}$ is uncountable. So, $p \neq \cap\{U_i : i \in \mathbb{N}\}$. So, p is not a G_δ. So, S is not perfectly normal, as claimed.

So, we can confidently state that

$$\textit{completely normal} \quad \not\Rightarrow \quad \textit{perfectly normal}$$

14.6 Topic: Compactness in Terms of Filters of Zero-Sets

For a given topological space, (S, τ), the symbols $C(S)$ and $C^*(S)$ denote the set of all real-valued continuous functions and the set of real-valued continuous bounded functions, respectively, on S. We recall a few concepts presented in an earlier chapter on separation of closed subsets with functions.

— If S is a topological space and $f \in C(S)$ then $Z(f) = f^{\leftarrow}(0)$ is referred to as a "zero-set" in S. We represent the family of all zero-sets in S by

$$Z[S] = \{Z(f) : f \in C(S)\}$$

— Recall that zero-sets are closed G_δ's.
— On the other hand, given a closed G_δ, F, of S, F need not necessarily be a zero-set.
— However, we have shown that, in a normal space, S, every closed G_δ is a zero-set in S.

Even though $Z[S]$ is just a particular type of subset of $\mathscr{P}(S)$, the expression $Z[S]$ makes sense only if S is a known topological space. After all, by definition, $Z(f)$ is the zero-set associated to a *continuous* real-valued function on S.

We would like to discuss subfamilies of $Z[S]$ which are "filters". To distinguish a filter in $Z[S]$ from a filter of sets in $\mathscr{P}(S)$ we refer to filters restricted to $Z[S]$ as

z-filters.

Even though z-filters are, in most ways, analogous to filters of sets, for the sake of completeness, we include the following definitions.

Definition 14.10 Let S be a topological space. Let $\mathscr{F} \subseteq Z[S]$. If $\mathscr{F}$ satisfies the two properties:

(1) $\varnothing \notin \mathscr{F}$.
(2) For every $U, V \in \mathscr{F}$, there exists $F \in \mathscr{F}$ such that $F \subseteq U \cap V$. then we will say that $\mathscr{F}$ is a *z-filter base.*

If $\mathscr{F}$ is a z-filter base which also satisfies the condition:

(3) Whenever $U \in \mathscr{F}$ and there is $V \in Z[S]\backslash\{\varnothing\}$ such that $U \subseteq V$, then $V \in \mathscr{F}$.

Then $\mathscr{F}$ is called a *z-filter.*

(4) If $\mathscr{F}$ is a z-filter which is not contained in any other strictly larger z-filter then we say that $\mathscr{F}$ is a *maximal z-filter* or, more commonly, a *z-ultrafilter.*

A quick way of describing a z-filter is to say:

> *A z-filter is a subset of $Z[S]\backslash\{\varnothing\}$ which is closed under finite intersections and supersets both in $Z[S]$.*

Suppose $\mathscr{F} = \{Z(f) : f \in \mathscr{U}\}$ is known to be a z-filter where f and g both belong to $\mathscr{U}$. Then, by definition of z-filter, $Z(f) \cap Z(g)$ and $Z(f) \cup Z(g)$ must be zero-sets which belong to $\mathscr{F}$. We must then assume that both $Z(f) \cap Z(g)$ and $Z(f) \cup Z(g)$ are zero-sets. This should nevertheless be verified. Verify that

$$Z(f^2 + g^2) = Z(f) \cap Z(g)$$
$$Z(fg) = Z(f) \cup Z(g)$$

A z-filter is, itself, a z-filter base, while a z-filter base, if not a z-filter, can be completed to become a z-filter. Given a z-filter base, $\mathscr{F}$, we define the larger set, $\mathscr{F}^*$, as follows:

$$\mathscr{F}^* = \{U \in Z[S] : F \subseteq U, \text{ for some } F \in \mathscr{F}\}$$

We have simply adjoined to $\mathscr{F}$ all its supersets which belong to $Z[S]$. It is easily verified that $\mathscr{F}^*$ is a z-filter. We will say that the z-filter base, $\mathscr{F}$, *generates the z-filter* $\mathscr{F}^*$.

Every z-filter is usually expected to be a proper subfamily of a filter of sets. In the event that every subset is a zero-set (such as in a discrete space, for example) a z-filter is simply a filter of sets.

The use of the notation, $Z(f)$, for the zero-set generated by f suggests that Z can be viewed as a function

$$Z : C(S) \to Z[S]$$

where $C(S)$ is the algebraic ring, $(C(S), +, \cdot)$, of continuous functions on S, and $Z[S]$ is a subset of $\mathscr{P}(S)$. So, Z maps f to $Z(f) = f^{\leftarrow}(0)$.

A subset $(U, +, \cdot)$ of $C(S)$ which is also a ring is called a *subring.* Those subrings, $(U, +, \cdot)$, of $C(S)$ satisfying the property,

$$\text{For every } f \in U,\ gf \in U, \quad \text{for all } g \in C(S)$$

are called

ideals in $C(S)$.

For obvious reasons we tend to represent an ideal in $\mathscr{P}(C(S))$ by I.[6] If J is an ideal in $C(S)$ such that for any ideal I, $J \subseteq I$ implies $I = J$ we say that J is a *maximal ideal* in $C(S)$. Ideals in $C(S)$ are relevant in our study since the image $Z[I]$ of an ideal I under the map $Z : C(S) \to Z[S]$, will be seen to be z-filters. Also, the function Z will pull back a z-filter, U, in $Z[S]$ to a corresponding ideal, $I = Z^{\leftarrow}[U]$.[7]

The following theorem formally describes the association between ideals in the ring of continuous functions, $C(S)$, and the z-filters in the family of all zero-sets, $Z[S]$, of the topological space, S.

Theorem 14.11 *Suppose S is a topological space. Let I be an ideal in $C(S)$ and $\mathscr{F}$ be a z-filter in $Z[S]$ (where $Z : C(S) \to Z[S]$ is viewed as a function mapping f in $C(S)$ to $Z(f)$ in $Z[S]$).*

(a) *Then the subset, $Z[I]$, of $Z[S]$ is a z-filter.*

---

[6]When we will say that I is an ideal in $C(S)$, we will always mean a "proper ideal".

[7]Note that $(C^*(S), +, \cdot)$ also constitutes a ring, and so we can speak of *ideals* and *maximal ideals* in $C^*(S)$ also.

(b) *Then the subset, $Z^{\leftarrow}[\mathscr{F}]$, of $C(S)$ is an ideal in $C(S)$.*
(c) *If J is a maximal ideal then $Z[J]$ is a z-ultrafilter. Conversely, if $\mathscr{F}$ is a z-ultrafilter then $Z^{\leftarrow}[\mathscr{F}]$ is maximal ideal.*
(d) *If $\mathscr{M}$ is the set of all maximal ideals in $\mathscr{P}(C(S))$ and $\mathscr{Z}$ is the set of all z-ultrafilters in $\mathscr{P}(Z[S])$, then $Z : \mathscr{M} \to \mathscr{Z}$ is one-to-one and onto.*
(e) *If $\mathscr{Z} = Z[\mathscr{M}]$ is a z-ultrafilter in $\mathscr{P}(Z[S])$ and $Z(f)$ is a zero-set which meets every member of $\mathscr{Z}$ then $Z(f) \in \mathscr{Z}$.*

Proof. We are given that I is a (proper) ideal in $C(S)$ and $\mathscr{F}$ is a z-filter in $Z[S]$.

(a) Given: I is a proper ideal. We are required to show that $Z[I]$ is a z-filter.

Since I is proper, then the unit function 1 is not an element of I. Then the empty-set, in the form of $Z(1)$, is not in $Z[I]$. Suppose $Z(f)$ and $Z(g)$ belong to $Z[I]$. Since $f, g \in I$ and I is an ideal, $f^2 + g^2 \in I$ and so

$$Z(f^2 + g^2) = Z(f) \cap Z(g) \in Z[I]$$

Suppose $f \in I$ and $Z(g) \in Z[S]$ such that $Z(f) \subseteq Z(g)$. It suffices to show that $Z(g) \in Z[I]$. Then $gf \in I$ (since I is an ideal). See that

$$Z(g) = Z(f) \cup Z(g) = Z(fg) \in Z[I]$$

So, $Z[I]$ is a z-filter, as claimed.

(b) Given: $\mathscr{F}$ is a z-filter. We consider $Z^{\leftarrow}[\mathscr{F}]$. We claim that $Z^{\leftarrow}[\mathscr{F}]$ is an ideal.

Since $\varnothing \notin \mathscr{F}$, the unit function 1 is not in $Z^{\leftarrow}[\mathscr{F}]$. Suppose f and g belong to $Z^{\leftarrow}[\mathscr{F}]$. To show that $Z^{\leftarrow}[\mathscr{F}]$ is an ideal we must show it is closed under addition and $fk \in Z^{\leftarrow}[\mathscr{F}]$ for all $k \in C(S)$.

Then $Z(f)$, $Z(g)$ and $Z(-g)$ belong to $\mathscr{F}$. So, $Z(f) \cap Z(-g) \in \mathscr{F}$. Now, if $u \in Z(f) \cap Z(-g)$, then $f(u) = 0 = -g(u)$ then $(f+g)(u) = 0$, so $Z(f) \cap Z(g) \subseteq Z(f+g)$. Since $\mathscr{F}$ is a z-filter, then $Z(f+g) \in \mathscr{F}$. This implies $f + g \in Z^{\leftarrow}[\mathscr{F}]$. So, $Z^{\leftarrow}[\mathscr{F}]$ is closed under addition.

Let $k \in C(S)$. Since $Z(f) \subseteq Z(fk)$, then $Z(fk) \in \mathscr{F}$ so $fk \in Z^{\leftarrow}[\mathscr{F}]$. So, $Z^{\leftarrow}[\mathscr{F}]$ is an ideal, as claimed.

(c) Suppose J is a maximal ideal. Then $Z[J]$ is a z-filter (by part a).
We are required to show that $Z[J]$ is a z-ultrafilter. Suppose $\mathscr{F}$ is a z-filter such that $Z[J] \subseteq \mathscr{F}$. It suffices to show that $\mathscr{F} = Z[J]$.

$$\begin{aligned} Z[J] \subseteq \mathscr{F} &\Rightarrow J \subseteq Z^{\leftarrow}[Z[J]] \subseteq Z^{\leftarrow}[\mathscr{F}] \\ &\Rightarrow J \subseteq Z^{\leftarrow}[\mathscr{F}] \text{ (where } J \text{ is maximal ideal)} \\ &\Rightarrow J = Z^{\leftarrow}[\mathscr{F}] \text{ (Since } Z^{\leftarrow}[\mathscr{F}] \text{ is an ideal (by part b))} \\ &\Rightarrow Z[J] = Z[Z^{\leftarrow}[\mathscr{F}]] = \mathscr{F} \end{aligned}$$

So, $Z[J]$ is a z-ultrafilter, as claimed.

On the other hand, if $\mathscr{F}$ is a z-ultrafilter, then $Z^{\leftarrow}[\mathscr{F}]$ is an ideal (by part b). We are required to show that $Z^{\leftarrow}[\mathscr{F}]$ is a maximal ideal. Suppose K is an ideal such that $Z^{\leftarrow}[\mathscr{F}] \subseteq K$. It will suffice to show that $K \subseteq Z^{\leftarrow}[\mathscr{F}]$.

See that $\mathscr{F} = Z[Z^{\leftarrow}[\mathscr{F}]] \subseteq Z[K]$. Then $Z[K] = \mathscr{F}$ (since $Z[K]$ is a z-filter and $\mathscr{F}$ is a maximal z-filter). So, $K \subseteq Z^{\leftarrow}[Z[K]] = Z^{\leftarrow}[\mathscr{F}]$. Then $K \subseteq Z^{\leftarrow}[\mathscr{F}]$. So $Z^{\leftarrow}[\mathscr{F}]$ is a maximal ideal.

(d) The result follows immediately from the definition of "maximal".

(e) The proof is analogous to the similar statement referring to an ultrafilter of sets proven Theorem 13.15.

In the following theorem, we guarantee that, given any z-filter, $\mathscr{F}$, in $Z[S]$ there exists in $Z[S]$ a z-ultrafilter which contains it.

Theorem 14.12 *Let S be a topological space and $(\mathcal{Z}, \subseteq)$, denote the set of all z-filters in $\mathscr{P}(S)$, partially ordered by inclusion. If $\mathscr{F}$ is any z-filter belonging to $\mathcal{Z}$ then there exists a z-ultrafilter, $\mathscr{U}$, in $\mathcal{Z}$ which contains $\mathscr{F}$.*

Proof. We are given that $\mathscr{F} \in \mathcal{Z}$. We wish to prove the existence of a z-ultrafilter which contains $\mathscr{F}$. We set up the problem so that Zorn's lemma applies.

For a given $\mathscr{F} \in \mathcal{Z}$, consider the set $\mathbb{S}_{\mathscr{F}} = \{\mathscr{H} \in \mathcal{Z} : \mathscr{F} \subseteq \mathscr{H}\}$ of all z-filters which contain $\mathscr{F}$. We are required to show that some z-filter in $\mathbb{S}_{\mathscr{F}}$ is a z-ultrafilter (i.e., $\mathbb{S}_{\mathscr{F}}$ has a maximal element).

If $\mathbb{S}_{\mathscr{F}} = \{\mathscr{F}\}$ then $\mathscr{F}$ is a z-ultrafilter. Suppose $\mathbb{S}_{\mathscr{F}}$ contains other filters. Now, $\mathscr{F}$ can be the base element of many chains in $\mathbb{S}_{\mathscr{F}}$. In fact, $\mathbb{S}_{\mathscr{F}}$ can be viewed as the union of a family of filter-chains, $\{\mathscr{C}_\alpha : \alpha \in J\}$, each with base element, $\mathscr{F}$. Pick an arbitrary filter-chain, $\mathscr{C}_\gamma = \{\mathscr{F}_i : i \in I\}$ in $\mathbb{S}_{\mathscr{F}}$. It looks like,

$$\mathscr{F} \subseteq \mathscr{F}_1 \subseteq \mathscr{F}_2 \subseteq \mathscr{F}_3 \subseteq \cdots$$

Let $\mathscr{T} = \cup\{\mathscr{F}_i : \mathscr{F}_i \in \mathscr{C}_\gamma\}$. Quite clearly, $\mathscr{F}_i \subseteq \mathscr{T}$ for each i. We claim that the family of sets, $\mathscr{T}$, is itself a z-filter in the chain $\mathscr{C}_\gamma$. If U and V are elements of $\mathscr{T}$, then U and V are elements of some $\mathscr{F}_j$ in the chain, $\mathscr{C}_\gamma$, so $U \cap V$ is an element of $\mathscr{F}_j$; so $U \cap V$ is also an element of $\mathscr{T}$. Then $\mathscr{T}$ is a z-filter base of sets which is a maximal element in the chain $\mathscr{C}_\gamma$.

So every chain, $\mathscr{C}_\alpha$, in $\mathbb{S}_{\mathscr{F}}$ has a maximal element. We can invoke Zorn's lemma, to conclude from this that $\mathbb{S}_{\mathscr{F}}$, must also have a maximal z-filter element, $\mathscr{U}$. Since $\mathbb{S}_{\mathscr{F}}$ contains every z-filter which contains $\mathscr{F}$, then $\mathscr{U}$ is the maximal z-filter which contains $\mathscr{F}$. Then, by definition, $\mathscr{U}$ is the unique z-ultrafilter which contains $\mathscr{F}$.

Definition 14.13 Let S be a topological space. Let I be an ideal in $C(S)$ or $C^*(S)$ and $\mathscr{F} = Z[I]$ be the corresponding z-filter.

(a) If $\cap\{Z(f) : f \in I\} \neq \varnothing$ then we will say that $\mathscr{F}$ is a *fixed* z-filter and that I is a *fixed* ideal.
(b) Otherwise, we will say that $\mathscr{F}$ is a *free* z-filter and I is a *free* ideal.

Theorem 14.14 *Suppose S is a completely regular topological space. Then the following are equivalent.*

(a) *The space S is a compact space.*
(b) *Every z-filter in $Z[S]$ is fixed. Every ideal in $C(S)$ is fixed. Every ideal in $C^*(S)$ is fixed.*
(c) *Every z-ultrafilter is fixed. Every maximal ideal in $C(S)$ is fixed. Every maximal ideal in $C^*(S)$ is fixed.*

Proof. We are given that S is a completely regular topological space.

(a $\Rightarrow$ b) Suppose S is compact. Since S is compact, then every function in $C(S)$ is bounded; so $C(S) = C^*(S)$. Suppose $\mathscr{F} = \{Z(f) : f \in I\}$ is a z-filter in $Z[S]$ corresponding to an ideal I in $C(S)$.

Since S is compact and $\mathscr{F}$ satisfies the finite intersection property, then $\cap\{Z(f) : f \in I\}$ is non-empty. So by definition, $\mathscr{F}$ is fixed. Equivalently I is a fixed ideal in $C(S)$.

(b $\Rightarrow$ a) Suppose every z-filter in $Z[S]$ is fixed. We are required to show that S is compact. Suppose $\mathscr{F}$ is a family of closed subsets with the FIP. It suffices to show that $\mathscr{F}$ has non-empty intersection.

If $F \in \mathscr{F}$ and $p \in S\backslash F$ then, since S is completely regular, there exists $f \in C(S)$ such that $F \subseteq Z(f)$ and $p \in f^{\leftarrow}(1)$. Then the set $F = \cap\{Z(f) : f \in U\}$, the intersection of zero-sets. Then

$$\cap\{F : F \in \mathscr{F}\} = \cap\{Z : Z \in \mathscr{G} \subseteq Z[S]\}$$

the intersection of zero-sets in $Z[S]$. Since $\{Z : Z \in \mathscr{G} \subseteq Z[S]\}$ is a fixed z-filter, then $\cap\{F : F \in \mathscr{F}\}$ is non-empty so S is compact.

(b $\Rightarrow$ c) Suppose every z-filter in $Z[S]$ is fixed. Then every z-ultrafilter in $Z[S]$ is fixed. As well every ideal is fixed and so every maximal ideal is fixed.

(c $\Rightarrow$ b) Suppose every z-ultrafilter in $Z[S]$ is fixed. Then every maximal ideal is fixed. Suppose $\mathscr{F}$ is a z-filter in $Z[S]$. Then, by Theorem 14.12, $\mathscr{F}$ is contained some z-ultrafilter, $\mathscr{G}$. By hypothesis, $\mathscr{G}$ is fixed, so there exists $p \in \cap\{F : F \in \mathscr{G}\}$. Then $p \in \cap\{F : F \in \mathscr{F}\}$. So, $\mathscr{F}$ is fixed. Also every corresponding ideal in $C(S)$ is fixed.

Concepts Review

1. Define an open cover of a space. What does it mean to say that it has a finite subcover?
2. State the formal topological definition of compact subspace.
3. What does it mean to say that a family of sets satisfies the "finite intersection property"?

4. State a characterization of compact space involving the "finite intersection property".
5. What kind of properties do nets in a compact space satisfy?
6. What kind of properties do filters in a compact space satisfy?
7. What can we say about ultrafilters of sets in a compact set?
8. What can we say about continuous images of compact sets?
9. What can we say about closed subsets of compact sets?
10. Under what conditions on S does the statement "Compact subsets of S are closed" hold true, in general?
11. Complete the statement: A one-to-one continuous function, $f : F \to T$, mapping a compact space, F, into a Hausdorff space, T, is...
12. Suppose $S = \prod_{i\in I} S_i$ is a product space with compact factors, S_i. Is this statement true or false? "The product space S maybe compact but it depends on the size of the index I".
13. What separation axioms will compact Hausdorff spaces satisfy?
14. Define the separation property referred to as *completely normal.*
15. Give a characterization of "completely normal".
16. How does a completely normal space compare with a normal space and a perfectly normal space.
17. Define a z-filter base and a z-filter.
18. Define a z-ultrafilter.
19. Define an ideal and a maximal ideal in $C^*(S)$.
20. Describe a relationship between ideals in $C^*(S)$ and z-filters.
21. Define a fixed filter and a free filter. How are they recognized?
22. Give a characterization of a compact space in terms of z-filters.

Exercises

1. Show that a closed and bounded subset of $\mathbb{R}$ is compact.
2. Show that disjoint compact sets in a Hausdorff space are respectively contained in disjoint open subsets.
3. Let F and H be disjoint closed subsets of $[0,5]^n$. Show that they are contained in disjoint open neighborhoods.
4. Suppose $g : S \to T$ is a function mapping the topological space S into the compact Hausdorff space T. Show that g is continuous

on S if and only if the graph of g, $\{(x, g(x)) : x \in S\}$ is a closed subset of $S \times T$.

5. Show that the finite union of compact subsets of a topological space is compact.
6. The set T is a compact subset of $\mathbb{R}^n$ if and only if T is both closed and bounded in $\mathbb{R}^n$.
7. Suppose $\{F_i : i \in I\}$ is a family of compact sets in a Hausdorff space. Show that $\cap\{F_i : i \in I\}$ is compact.

Chapter 15

Countably Compact Spaces

Abstract
In this section, we discuss a slightly weaker version of the compactness property called "countable compactness". We formally define it and discuss its properties. We will see that there are countably compact spaces which are not compact spaces. We also show that, in some topological spaces (such as metric spaces) the countably compact and compact properties are equivalent.

15.1 The Countably Compact Property

The compact property imposes strong restrictions on a topological space. Sometimes, a weaker version of, "Every open cover has a finite subcover" will suit our purposes. A slightly weaker version, can be expressed as "Every countable open cover has a finite subcover". We will investigate properties which are specific to countable compactness.

Definition 15.1 Let F be any non-empty subset of the topological space, S. We say that F is *countably compact* if and only if every *countable* open cover of F has a finite subcover.

Just as for compact subsets, it will be useful to obtain various ways of recognizing those spaces which are countably compact. It will also allow us to better see how countably compact spaces distinguish themselves from those which satisfy the compact property. Recall that p is said to be a "*cluster point* of the set A if every open neighborhood of p contains some other point of A".

Theorem 15.2 *Let S be a topological space and T be a non-empty subset of S. Then the following are equivalent.*

(a) *The space S is a countably compact set.*
(b) *If T is a countably infinite subset of S, then T has at least one cluster point.*
(c) *If $A = \{x_i : i \in \mathbb{N}\}$ is a sequence with an infinite number of distinct points, then A has at least one accumulation point.*
(d) *A* countable *family, $\mathscr{F} = \{F_i : i \in \mathbb{N}\}$, of closed sets in S satisfying the finite intersection property has non-empty intersection in S.*

Proof. Given: S is a topological space.

($a \Rightarrow b$) Given: S is countably compact and T is countably infinite subset of S.

We are required to show that T has a cluster point. Suppose T has *no* cluster point.

Claim 1: That $T = \text{cl}_S T$. If not, $T \subset \text{cl}_S T$. Then there is a point, p, in $\text{cl}_S T \cap (S \setminus T)$. So, p is a boundary point of T. This would mean that every open neighborhood of p intersects $T \setminus \{p\}$; this would make of p a cluster point,a contradiction. So, $T = \text{cl}_S T$, as claimed.

By Claim 1, we can say that $S \setminus T$ is open in S. Since T has no cluster point, then every point x in T is contained in an open neighborhood V_x of S which does not intersect $T \setminus \{x\}$. Then $V_x \cap T = \{x\}$. Then $\{V_x : x \in T\}$ is a countably infinite family of open subsets of S which covers T. Clearly, it has no finite subset which covers T. Then,

$$\{(S \setminus T)\} \cup \{V_x : x \in T\}$$

is a countably infinite open cover of S with no finite subcover, contradicting the fact that S is countably compact. So, T must have a cluster point.

($b \Rightarrow c$) We are given that, a countably infinite subset of S must have a cluster point. Let $A = \{f(i) : i \in \mathbb{N}\}$ be an infinite sequence. We are required to show that the sequence, A, has an accumulation point.

By hypothesis, A has a cluster point, say z, in T. Then, for any neighborhood, B_z, of z, $B_z \cap A$ contains infinitely many distinct points. So, $B_z \cap \{f(i) : i > q\}$ is non-empty for all $q \in \mathbb{N}$. This means that any neighborhood of z intersects a tail end of A. So by definition, z is an accumulation point of the sequence, A.

($c \Rightarrow a$) Suppose that any sequence in any subset, T, of S has an accumulation point. Recall that p is an accumulation point of a sequence if every open neighborhood of p intersects the tail end of the sequence. We are required to show that S is countably compact.

Let $\mathscr{U} = \{U_i : i \in \mathbb{N}\}$ be a countable family of open sets which covers S. We are required to show that $\mathscr{U}$ has a finite subcover. Suppose $\mathscr{U}$ has no finite subcover. Then, for every n, it is possible to choose,

$$f(n) \in T \setminus \cup\{U_i : i = 1 \text{ to } n\}$$

Then $\{f(n) : n \in \mathbb{N}\setminus\{0\}\}$ forms a sequence in T.

Then, for each k,

$$f[\,[k, \infty)\,] \subseteq T \setminus \cup\{U_i : i = 1 \text{ to } k\}$$

So no element of $\mathscr{U}$ can intersect a tail end of the sequence, $\{f(n)\}$, in T. This contradicts our hypothesis stating that every sequence in T has an accumulation point. So, $\mathscr{U}$ must have a finite subcover.

($a \Rightarrow d$) We are given that S is countably compact. Let $\mathscr{F} = \{F_i : i \in \mathbb{N}\}$ be countable family of closed sets in S with the finite intersection property. We are required to show that $\cap\{F_i : i \in \mathbb{N}\} = \varnothing$.

Suppose $\cap\{F_i : i \in \mathbb{N}\}$ is empty. Then $\{S \setminus F_i : i \in \mathbb{N}\}$ forms an open cover of S with no finite subcover, contradicting the property of countable compactness on S.

($d \Rightarrow a$) We are given that, if $\mathscr{F}$ is a countable family of closed sets in S with the finite intersection property, then $\cap\mathscr{F}$ is not empty. We are required to show that S is countably compact.

Suppose $\mathscr{U} = \{U_i : i \in \mathbb{N}\}$ is an open cover of S. We are required to show that $\mathscr{U}$ contains a finite subcover of S. Suppose $\mathscr{U}$ has no finite subcover. Then $\{S\backslash U_i : i \in \mathbb{N}\}$ forms a family of closed sets satisfying the finite intersection property which has empty intersection. This contradicts our hypothesis. So, S is countably compact.

15.2 Bolzano–Weierstrass Property

We will examine a bit more closely the countable compactness characterization given in part (c) of the characterization theorem stated above. It states that, in countably compact spaces:

> *The space S is countably compact if and only if every sequence,*
>
> $$A = \{x_i : i \in \mathbb{N}\}$$
>
> *in S has an accumulation point, p, in S.*

The statement is quite similar to the one found in the compactness characterization theorem 14.2:

> *A space S is compact if and only if every net in S has an accumulation point, p, in S.*

By Theorem 12.5, "p is an accumulation point of a net N if and only if N has a subnet converging to p". We can then say:

> *The space S is a compact space if and only if every net N in S has a subnet which converges to some accumulation point, p, of N.*

And,

> *the space S is a countably compact space if and only if every sequence A in S has a subnet which converges to some accumulation point, p, of A.*

To some readers, this may be reminiscent of a real analysis statement called the *Bolzano–Weierstrass theorem* (BWT). It states that

every bounded sequence in $\mathbb{R}$ has a convergent subsequence.

There is, however, a fundamental difference between the statement of BWT and this particular characterization of the countable compactness. We see that the BWT refers only to $\mathbb{R}$, not an arbitrary topological space, and only to those sequences which are "bounded". (Note that the property of "boundedness" has not yet been defined for arbitrary topological spaces.)

For future reference we will formally provide a name for the property described in the above countably compact characterization theorem.

Definition 15.3 A topological space, S, is said to satisfy the

Bolzano–Weierstrass property

(BWP) if and only if every sequence, $A = \{x_i : i \in \mathbb{N}\}$, in S has an accumulation point in S.

By the characterization theorem above, countably compact topological spaces are precisely those spaces which satisfy the Bolzano–Weierstrass property.

15.3 Properties of Countably Compact Spaces

Before presenting some examples, it will help to list a few properties of countably compact spaces. The first statement says that a countably compact space is "closed-hereditary".

Theorem 15.4 *Closed subspaces of countably compact spaces are countably compact.*

Proof. Given: The space, S, is a countably compact topological space and F is a non-empty closed subset of S.

Suppose $\mathscr{F}$ is a countable family of closed subsets of F which satisfies the finite intersection property. It suffices to show that $\mathscr{F}$ has non-empty intersection. Then every set in $\mathscr{F}$ is also a closed subset in S. Since S is countably compact, the family of sets in $\mathscr{F}$ has non-empty intersection inside F. So, F is also countably compact.

We know that continuous images of compact sets are compact. The following theorem shows that an analogous statement for countably compact spaces holds true.

Theorem 15.5 *Suppose S is countably compact and the continuous function, $f : S \to T$, maps S into the topological space, T. Then $f[S]$ is a countably compact subspace of T.*

Proof. Given: The space S is countably compact and T is a topological spaces. Suppose $f : S \to T$ is continuous.

Let $\mathscr{U} = \{U_i : i \in \mathbb{N}\}$ be a countable open cover of $f[S]$. Then $\mathscr{V} = \{f^{\leftarrow}[U_i] : i \in \mathbb{N}\}$ is a countable open cover of S. Since S is countably compact, $\mathscr{V}$ has a finite subcover $\mathscr{V}_F = \{f^{\leftarrow}[U_i] : i \in F\}$ of S. Then

$$\begin{aligned} f[S] &\subseteq f[\cup\{f^{\leftarrow}[U_i] : i \in F\}] \\ &= \cup\{f[f^{\leftarrow}[U_i]] : i \in F]\} \\ &= \{U_i : i \in F\} \end{aligned}$$

So, $\mathscr{U}$ has a finite subcover $\mathscr{U}_F = \{U_i : i \in F\}$ of $f[S]$. So, $f[S]$ is also countably compact.

We now consider how the countable compactness property carries over products.

Theorem 15.6 *Let $\{S_j : j \in \mathbb{N}\}$ be a family of topological spaces. We consider the corresponding product space, $S = \prod_{j \in \mathbb{N}} S_j$, with countably many factors.*

(a) *If the product space, S, is countably compact then so is each one of its S_j factors.*
(b) *If each S_j factor of the product space, S, is both countably compact and first countable then the infinite product space, S, is countably compact.*

Proof. Given: The set, $S = \prod_{j \in \mathbb{N}} S_j$, is a product space.

(a) Suppose the product space, S, is countably compact. Since each projection map, π_j, is continuous, then, for each factor S_j, $S_j = \pi_j[S]$, the continuous image of a countably compact space. So by Theorem 15.5 each factor, S_j, is countably compact. We are done with part (a).

(b) We are given that $\{S_j : j \in \mathbb{N}\}$ is a family of first countable and countably compact topological spaces. We are required to show that $S = \prod_{j \in \mathbb{N}} S_j$ is countably compact.

Let $B = \{\beta(i) : i \in \mathbb{N}\}$ be a sequence in S. To show that S is countably compact it suffices to show that B has an accumulation point.

Step 0: Since $\pi_0(\beta(i)) = \pi_0(\langle \beta_j(i) \rangle_{j \in \mathbb{N}}) = \beta_0(i)$, then

$$\pi_0[B] = \{\beta_0(i) : i \in \mathbb{N}\}$$

is a sequence in S_0. Since S_0 is countably compact, then, by Theorem 15.2, $\pi_0[B]$ has an accumulation point, say p_0. By Theorem 11.5, since S_0 is first countable, $\pi_0[B]$ has a subsequence,

$$\{\beta_0(g_0(i)) : i \in \mathbb{N}\}$$

in S_0 converging to p_0.

Let

$$B_0 = \{\beta(g_0(i)) : i \in \mathbb{N}\}$$

Then B_0 is a subsequence of B in $S = \prod_{j \in \mathbb{N}} S_j$.

Step 1: We consider the sequence $\pi_1[B_0]$ in S_1 where

$$\pi_1(\beta(g_0(i))) = \pi_1(\langle \beta_j(g_0(i)) \rangle_{j \in \mathbb{N}}) = \beta_1(g_0(i))$$

Then

$$\pi_1[B_0] = \{\beta_1(i) : i \in \mathbb{N}\}$$

is a sequence in S_1.

Since S_1 is first countable, $\pi_1[B_0]$ has a subsequence,

$$\{\beta_1(g_1(g_0(i))) : i \in \mathbb{N}\}$$

in S_1 converging to p_1.

Let $B_1 = \{\beta(g_1(g_0(i))) : i \in \mathbb{N}\}$. Then B_1 is a subsequence of B in $S = \prod_{j\in\mathbb{N}} S_j$.

Step n: Eventually we obtain a sequence $\{\beta_n(g_n(\cdots(g_0(i)))\cdots)):$ $i \in \mathbb{N}\}$ in S_n, converging to $p_n \in S_n$.

From this, we obtain the subsequence

$$B_n = \{\beta(g_n(g_{n-1}(\cdots(g_0(i)))\cdots) : i \in \mathbb{N}\}$$

of B.

Let $\gamma : \mathbb{N} \to B$ be defined as

$$\gamma(n) = \beta(g_n(g_{n-1}(\cdots(g_0(n)))\cdots)$$

That is,

$$\begin{aligned}
\gamma(0) &= \beta(g_0(0)) \\
\gamma(1) &= \beta(g_1(g_0(1))) \\
\gamma(2) &= \beta(g_2(g_1(g_0(2)))) \\
&\vdots \\
\gamma(n) &= \beta(g_n(g_{n-1}(\cdots(g_0(n)))\cdots) \\
&\vdots
\end{aligned}$$

Then $A = \{\gamma(i) : i \in \mathbb{N}\}$ is a subsequence of B.

We claim that the subsequence, A, converges to $p = \langle p_j\rangle_{j\in\mathbb{N}}$.

See that, for each $j \in \mathbb{N}$, $\{\pi_j(\gamma(i)) : i \in \mathbb{N}\}$ converges to p_j. By Corollary 13.18,

$$\{\gamma(i) : i \in \mathbb{N}\} \subset B \subset S = \textstyle\prod_{j\in\mathbb{N}} S_j$$

converges to $p = \langle p_j\rangle_{j\in\mathbb{N}}$, as claimed.

Since B has a converging subsequence, $A = \{\gamma(i) : i \in \mathbb{N}\}$, then B has an accumulation point and so the product space, S, is countably compact. As required.

15.4 Examples

Example 1. Consider the set $S = [0, \omega_1]$, of all countable ordinals union the singleton set, $\{\omega_1\}$, where ω_1 is the first uncountable ordinal. The Hausdorff topology on S is generated by the base, $\mathscr{B}$, for open sets defined as

$$\mathscr{B} = \{(\alpha, \beta] : \alpha < \beta\}$$

(a) Show that $S = [0, \omega_1]$ is a compact space.
(b) Show that the subspace, $T = [0, \omega_1)$, of S is not compact but is, nevertheless, countably compact.

Solution: Given: $S = [0, \omega_1]$ and $T = [0, \omega_1)$.

(a) We will use the formal definition of compactness. The technique mimics the one used to show that closed and bounded intervals of $\mathbb{R}$ are compact. Let $\mathscr{U} = \{B_i : i \in I\}$ be an open cover of S. It suffices to show that S has a finite subcover.

Let $V = \{u : [0, u]$ has a finite subcover$\}$.[1] Suppose $\omega_1 \notin V$. Since the ordinal space, S, is well-ordered, every non-empty subset has a least element. Let k be the least element in $S \backslash V$. Then $[0, k]$ does not have a finite subcover. There is some element, $B_j \in \mathscr{U}$, which contains k. Then there exists some $m \in S$ such that the basic open neighborhood of k, $(m, k] \subseteq B_j$. But $[0, m]$ has a finite subcover, say $\mathscr{U}_F \subset \mathscr{U}$. Then $\mathscr{U}_F \cup \{B_j\}$ is a finite subcover of $[0, k]$. This is a contradiction. So, $V = S$. Hence, $S = [0, \omega_1]$ has a finite subcover and so is compact.

(b) Consider $T = [0, \omega_1)$, a subspace of compact S. It is not closed seeing that it does not contain the boundary point, ω_1 (since every basic open neighborhood of ω_1 intersects T). Since compact subsets of Hausdorff spaces are closed, then $T = [0, \omega_1)$ is *not compact.*

We claim that T is countably compact. Suppose $A = \{x_i : i \in \mathbb{N}\}$ is a sequence in T. By a characterization of "countably compact" it suffices to show that the sequence A has an accumulation point.

[1]The set S is well-ordered by $\leq$.

As explained on page 108, if $\sup A = \omega_1$, then $\omega_1 = \cup\{x_i : i \in \mathbb{N}\}$. Then ω_1 would be the countable union of countable ordinals and so would be countable, a contradiction. So, $\sup A < \omega_1$. Then there must exist $k \in T$ such that $A \subseteq [0, k]$. Since $[0, k]$ is a closed subset of the compact space, $[0, \omega_1]$, then $[0, k]$ is itself compact. We know that, since $[0, k]$ is compact, every net in $[0, k]$ has an accumulation point in $[0, k]$ (this is a characterization of the compact property). Then the sequence A must have an accumulation point, say p in $[0, k]$. Since $p \in [0, k] \subset [0, \omega_1)$, then every sequence in T has an accumulation point in T. Then T is countably compact.

15.5 On Countably Compact Spaces Which Are Compact

In the following results, we see how the countably compact property compares with the compact property in certain spaces. Recall that *second countable* spaces are those spaces which have a countable base for open sets.

Theorem 15.7 *Suppose S is a second countable topological space. Then S is compact if and only if S is countably compact.*

Proof. Given: The space S is a second countable topological space.

($\Rightarrow$) It is always true that, if S is compact then S is countably compact.

($\Leftarrow$) Suppose S is countably compact. Let $\mathscr{U}$ be an open cover of S. It suffices to show that $\mathscr{U}$ has a finite subcover.

Since S is second countable, then S has a countable open base, $\mathscr{B}$.

Let x be a point in S. Since $\mathscr{U}$ is an open cover, there exist $U_x \in \mathscr{U}$ which contains x. Since $\mathscr{B}$ is a base for open sets, there exists in $\mathscr{B}$, at least one B_x such that $x \in B_x \subseteq U_x$. For each $p \in S$, let

$$\mathscr{A}_p = \{B_p \in \mathscr{B} : p \in B_p \subseteq U_p\}$$

Let $\mathscr{A} = \cup\{\mathscr{A}_p : p \in S\}$. Then $\mathscr{A} \subseteq \mathscr{B}$.

Since S is second countable, then $\mathscr{A}$ is a countable set of open neighborhoods, say $\{B_i : i \in \mathbb{N}\backslash\{0\}\}$. For each $i \in \mathbb{N}\backslash\{0\}, p \in B_i \subseteq U_p$, for some p. For each i, choose exactly one U_{p_i} such that $B_i \subseteq U_{p_i}$. Then $\mathscr{V} = \{U_{p_i} : i \in \mathbb{N}\backslash\{0\}\}$ forms a countable subcover of S.

Since S is countably compact, then $\mathscr{V}$ contains a finite subcover $\mathscr{W} = \{U_{p_i} : i \in F\}$ of S. We have shown that $\mathscr{U}$ contains a finite subcover, $\mathscr{W}$, of S. So, the second countable property combined with the countably compact property on S implies S is compact. As required.

Theorem 15.8 *Those metric spaces which are countably compact are compact metric spaces.*

Proof. Suppose (S, ρ) is a countably compact metric space. It suffices to show that S is separable since, by Theorem 5.11, this implies S is second countable and, by Theorem 15.7, second countable countably compact spaces are compact.

Let $\varepsilon > 0$. Recall that since S is countably compact every infinite subset V has a cluster point.

Claim 1: There exists a finite set $\{x_i : i \in F\} \subseteq S$ such that $S \subseteq \cup\{B_\varepsilon(x_i) : i \in F\}$.

Proof of claim: Suppose not. Then there is some $\delta > 0$ such that $S \nsubseteq \cup\{B_\delta(x_i) : i \in F\}$ for all finite sets F. Suppose $y_0 \in S$. If $B_\delta(y_0)$ does not cover S, then there is $y_1 \in S$ such that $\rho(y_0, y_1) \geq \delta$. Inductively, suppose $\{y_i : i = 1, 2, 3, \ldots, k\}$ are such that $S \nsubseteq \cup\{B_\delta(y_i) : i = 0, 1, 2, 3, \ldots, k\}$. then there exists

$$y_{k+1} \in S\backslash\cup\{B_\delta(y_i) : i = 0, 1, 2, 3, \ldots, k\}$$

Then, we can construct in this way, the infinite subset, $V = \{y_i : i \in \mathbb{N}\}$, of S. Then V has no cluster point, contradicting the countable compactness property. This establishes the claim.

Claim 2: The metric space (S, ρ) is separable.

Proof of claim: Let $n \in \mathbb{N}\backslash\{0\}$. By Claim 1, for this value, n, we can construct, $\{y_{(i,n)} : i = 1, 2, 3, \ldots, k_n\}$ such that, if $S \subseteq B_n = \cup\{B_{1/n}(y_{(i,n)}) : i = 1, 2, 3, \ldots, k_n\}$. By constructing in this way the sets, $\{B_n : n \in \mathbb{N}\backslash\{0\}\}$, we extract the subset,

$$D = \{y_{(i,n)} : i = 1, 2, 3, \ldots k_n, n = 1, 2, 3, \ldots\}$$

where $S \subsetneq B_n$ for each $n = 1, 2, 3, \ldots$. Then D is easily seen to be dense in S. So, S is separable, as claimed.

By Theorems 15.7 and 5.10, S is compact.

Example 2. Show that the space $S = [0, \omega_1)$ is non-metrizable.

Solution: We have shown in the previous example that $[0, \omega_1)$ is countably compact. By Theorem 15.8, if S was metrizable it would be compact. But we showed in the example that S is not compact. So, $S = [0, \omega_1)$ is a non-metrizable topological space.

Many of the results above allow us to confirm that the real line, equipped with the usual topology, is not countably compact. One of the characterization states that any sequence will have an accumulation point. We already mentioned that the sequence constructed from the natural numbers has no accumulation point. One could also raise the argument that, since $\mathbb{R}$ is known to be second countable, then, if $\mathbb{R}$ was countably compact, it would have to be compact. In Theorem 15.9, we show that continuous real-valued functions on countably compact spaces are bounded functions. If $\mathbb{R}$ was countably compact, then every real-valued continuous function on $\mathbb{R}$ would have to be bounded, which is not the case.

Theorem 15.9 *Let S be a countably compact topological space. If f is a real-valued continuous function on S, then $f[S]$ is compact.*[2]

Proof. Given: The space S is a countably compact topological space and $f[S]$ is the continuous image of S in $\mathbb{R}$, equipped with the usual topology.

Then, by Theorem 15.5, $f[S]$ is countably compact in $\mathbb{R}$. In the example on page 103, it is shown that the real line, $\mathbb{R}$, equipped with

[2]Later in the text, we will refer to those spaces, S, in which all continuous real-valued functions are bounded on S as being *pseudocompact spaces.* This theorem can then correctly be paraphrased as "All countably compact Hausdorff spaces are pseudocompact".

the usual topology is second countable. In Theorem 5.13, it is shown that the second countable property is hereditary. So, $f[S]$ is second countable in $\mathbb{R}$. Then by Theorem 15.7, $f[S]$ is compact.

Concepts Review

1. Define a countably compact space.
2. Give three characterizations of the countable compactness property.
3. State the Bolzano–Weierstrass property.
4. Are closed subsets of countably compact spaces necessarily countably compact?
5. What can we say about continuous images of countably compact spaces?
6. Suppose S is the product of countably compact spaces. If we want some guarantee that S be countably compact, what conditions must be satisfied?
7. Give an example of a space which is countably compact but not compact.
8. Identify topological spaces in which the compact property is equivalent to the countable compactness property.
9. What can we say about continuous real-valued functions on a countably compact space.

Exercises

1. Show that a space S is countably compact if and only if whenever a family, $\mathscr{F} = \{F_i : i \in \mathbb{N}\}$ of closed non-empty sets in S, is such that $F_{i+1} \subseteq F_i$ for all i, then $\cap\{F_i : i \in \mathbb{N}\}$ is non-empty.
2. Suppose $f : S \to T$ is a one-to-one continuous function mapping a countably compact space, S, onto a first countable space, T. Show that T is a homeomorphic image of S.

Chapter 16

Lindelöf Spaces

Abstract
We will now investigate a property which can be seen as a weak relative of the compact property. It is called the Lindelöf property. We will formally define it and study its characterizations and basic properties. We compare its characteristics to those of the countable compactness property. Finally, we will provide a few examples.

16.1 Introduction

There are various ways to weaken the "open cover... finite subcover" property used to described compactness of a space. The "countably compact" property, for example, refers to those spaces for which every *countable* open cover has a finite subcover. There is another approach we can use to weaken the compactness property. We can require that arbitrarily large open covers have a countable subcover (rather than the stricter "finite subcover"). This property is referred to as the *Lindelöf* property. Although, some readers may not necessarily view the Lindelöf property as a relative of compactness, we included it in this part of the text since it often appears in the literature combined with other weaker compact properties. Furthermore, it shares with the "compact relatives" a property which allows for the reduction of a "large" open cover to one of a "smaller" size. We will see that this property may be of interest when combined with the properties of other non-compact topological spaces.

16.2 The Definition

The following definition applies to arbitrary topological spaces. Some text define the *Lindelöf property* only for regular spaces, possibly because it is in such spaces that the Lindelöf property is most commonly studied.

Definition 16.1 A topological space, S, is said to be *Lindelöf*, or satisfies the *Lindelöf property*, if every open cover of S has a countable subcover.[1]

Obviously, since finite sets are by definition countable, if an open cover has a finite subcover then this cover has a countable subcover; so we can say that

every compact space is Lindelöf.

Also, by combining the Lindelöf property with the countably compact property, we see that the Lindelöf property reduces an open cover to a countable one, and the countable compact property reduces the countable open cover to a finite one. So,

every countably compact Lindelöf space is a compact space.

There are other properties which guarantee that an open cover will have a countable subcover.

Theorem 16.2 *If S is second countable then S must be Lindelöf.*

Proof. Given: We are given that S is a second countable space and that $\mathscr{U}$ is an open cover on S.

If $p \in S$, there exist $V_p \in \mathscr{U}$ which contains p. By hypothesis, S has a countable open base, $\mathscr{B}$, for S. Then, we can choose, from $\mathscr{B}$, at least one B_p such that $p \in B_p \subseteq V_p$. Let

$$\mathscr{A}_p = \{B_p \in \mathscr{B} : p \in B_p \subseteq V_p\}$$

Let $\mathscr{A} = \cup\{\mathscr{A}_p : p \in S\}$.

[1]Named after the Finnish mathematician Ernst Lindelöf (1870–1946). He made contributions to the fields of real analysis, complex analysis, and general topology.

Since S is second countable, then $\mathscr{A}$ can be reduced to a countable subfamily, $\mathscr{A}^*$, say $\mathscr{A}^* = \{B_i : i \in \mathbb{N}\backslash\{0\}\}$, without sacrificing its open base property. For each $i \in \mathbb{N}\backslash\{0\}$, $p \in B_i \subseteq V_p$, for some p. For each i, choose exactly one V_{p_i} such that $B_i \subseteq V_{p_i}$. Then $\mathscr{V} = \{V_{p_i} : i \in \mathbb{N}\backslash\{0\}\}$ forms a countable subcover of S. So, if S is second countable, then S must be Lindelöf.

16.3 Characterizations

Those who only work with metrizable topological spaces will soon notice that the Lindelöf property provides nothing new in their topological region of the universe. This is due to the fact that, in metric spaces, the Lindelöf property is equivalent to the second countable property. We will say that the family of subsets, $\mathscr{F}$, of a space S satisfies the *countable intersection property* (CIP) if, for any countable subfamily, $\mathscr{F}_{\mathbb{N}}$, of $\mathscr{F}$,

$$\cap\{F : F \in \mathscr{F}_{\mathbb{N}}\} \neq \varnothing$$

Theorem 16.3 *Let S be a topological space.*

(a) *The space, S, is Lindelöf if and only if each filter of closed sets with the* countable intersection property *has non-empty intersection.*

(b) *Suppose (S, ρ) is a metric space. Then the following are equivalent.*

   (i) *The space S is Lindelöf.*
   (ii) *The space S is second countable.*
   (iii) *The space S is separable.*

(c) *Suppose S is a regular space. Then the following are equivalent.*

   (i) *The space S is Lindelöf.*
   (ii) *If each open cover, $\mathscr{U} = \{U_i : i \in I\}$, of S has a subfamily, $\{U_{i_j} : j \in \mathbb{N}\}$, such that $\{cl_S U_{i_j} : j \in \mathbb{N}\}$ covers S, then $\{U_{i_j} : j \in \mathbb{N}\}$ covers S.*
   (iii) *Every filter of open sets in S with the countable intersection property has an accumulation point.*

Proof. We are given that S is a topological space.

(a) We are given that $\mathscr{F} = \{F_i : i \in I\}$ represents a filter of closed sets in the space S.

($\Rightarrow$) Suppose S is Lindelöf and $\mathscr{F}$ satisfies the CIP. That is, countable subfamilies of $\mathscr{F}$ have non-empty intersection. We are required to show that $\cap\{F_i : i \in I\} \neq \varnothing$.

Suppose not. That is, suppose $\cap\{F_i : i \in I\} = \varnothing$. Then $\{S\backslash F_i : i \in I\}$ is an open cover of S. Since S is Lindelöf, $\{S\backslash F_{i_j} : j \in \mathbb{N}\}$ forms a countable subcover. This means, $\cap\{F_{i_j} : j \in \mathbb{N}\} = \varnothing$. This contradicts the fact that $\mathscr{F}$ satisfies the CIP.

So, $\cap\{F_i : i \in I\} \neq \varnothing$, as required.

($\Leftarrow$) Given: Every filter of closed sets with the *countable intersection property* has non-empty intersection. Suppose $\mathscr{U} = \{U_i : i \in I\}$ is an open cover of S.

We are required to show that $\mathscr{U}$ has a countable subcover.

Suppose not. That is, suppose that, for any countable subfamily, $\{U_{i_j} : j \in \mathbb{N}\}$, $\cap\{S \backslash U_{i_j} : j \in \mathbb{N}\} \neq \varnothing$. Then $\mathscr{F}$ satisfies the CIP. Then $\cap\{S \backslash U_i : i \in I\} \neq \varnothing$. This contradicts that $\mathscr{U}$ is an open cover. So, $\mathscr{U}$ has a countable subcover. This means that S is Lindelöf.

(b) We are given that S a metric space.

(i $\Leftarrow$ ii) Given that S is second countable. By Theorem 16.2, second countable spaces are Lindelöf spaces, always. So S is Lindelöf. We are done with (i $\Leftarrow$ ii).

(i $\Rightarrow$ ii) Suppose the metric space, S, is a Lindelöf space. We are required to show that S has a countable base for open sets.

For each $i \in \mathbb{N}\backslash\{0\}$, we construct an open cover, $\mathscr{U}_i = \{B_{1/i}(u) : u \in S\}$, of S. By hypothesis, each $\mathscr{U}_i$ has a countable subcover, say

$$\mathscr{B}_i = \{B_{1/i}(u_{i_j}) : u_{i_j} \in S,\ j \in \mathbb{N}\}$$

Claim: That the countable family of open sets, $\mathscr{B} = \cup\{\mathscr{B}_i : i \in \mathbb{N}\backslash\{0\}\}$, is a base for open sets of S. That is, every open subset is the union of elements from $\mathscr{B}$.

Proof of claim: Suppose U is an open neighborhood of $p \in S$. We are required to show that p belongs to some element, B, of

$\mathscr{B}$ such that $B \subseteq U$. Then there exists some natural number, k, such that $B_{1/k}(p) \subseteq U$. Now, $\mathscr{B}_{3k}$ has been hypothesized to be an open cover of S. So, there is some $v \in S$, such that $p \in B_{1/3k}(v)$. Then

$$x \in B_{1/3k}(v) \;\Rightarrow\; \rho(x,p) < \frac{2}{3k} < \frac{1}{k}$$

So, $x \in B_{1/k}(p) \subseteq U$. Then $p \in B_{1/3k}(v) \subseteq U$ where $B_{1/3k}(v)$ is an element of $\mathscr{B}$. Then $\mathscr{B}$ is a countable base for open sets, as claimed.

So, S is second countable, as required. We are done with (i $\Rightarrow$ ii).

(ii $\Leftarrow$ iii) If S a separable metric space, by Theorem 5.11, S is second countable.

(ii $\Rightarrow$ iii) Suppose S is second countable. Then, by Theorem 5.10, S is separable, always.

(c) We are given that S is a regular topological space.

(i $\Rightarrow$ ii) Suppose S is Lindelöf. Then S has a countable subcover, $\{U_{i_j} : j \in \mathbb{N}\}$ and so $\{\text{cl}_S U_{i_j} : j \in \mathbb{N}\}$ is a countable subcover. So, (ii) holds true.

(i $\Leftarrow$ ii) We are given that, if $\mathscr{V} = \{U_i : i \in I\}$ is an open cover, then $\{\text{cl}_S U_{i_j} : j \in \mathbb{N}\}$ covers S for some countable subfamily, $\{U_{i_j} : j \in \mathbb{N}\}$, of $\mathscr{V}$.

We claim that $\{U_{i_j} : j \in \mathbb{N}\}$ must also be a subcover.

Proof of claim: Let $y \in S$. Then there exist U_y in the open cover, $\mathscr{V}$, which contains y. By regularity, there exists an open neighborhood, B_y, of y such that $y \in B_y \subseteq \text{cl}_S B_y \subseteq U_y$. Then $\{B_y : y \in S\}$ forms an open cover of S. By hypothesis, we have $\{\text{cl}_S B_{y_j} : j \in \mathbb{N}\}$ covers S. Since $\text{cl}_S B_{y_j} \subseteq U_{y_j}$ for each j, then $\{U_{y_j} : j \in \mathbb{N}\}$ covers S, as claimed.

We conclude that S is Lindelöf. We are done for $\Leftarrow$.

(i $\Rightarrow$ iii) Suppose S is Lindelöf.

Let $\mathscr{U} = \{U_i : i \in I\}$ be a filter base of open sets in S which satisfies the countable intersection property.

We are required to show that $\cap\{\text{cl}_S U_i : i \in I\}$ is non-empty.

Consider the family $\mathscr{V} = \{\mathrm{cl}_S U_i : i \in I\}$. Since $\mathscr{U}$ satisfies the countable intersection property, then so does $\mathscr{V}$. Then $\mathscr{V}$ is a filter of closed sets in S. Since S is Lindelöf, by part (a), $\cap\{\mathrm{cl}_S U_i : i \in I\}$ is non-empty. If $p \in \cap\{\mathrm{cl}_S U_i : i \in I\}$, then by definition, p is an accumulation point of the filter of open sets, $\mathscr{U}$.

(i $\Leftarrow$ iii) We are given that filters of open sets in S which satisfy the countable intersection property have an accumulation point. Suppose $\mathscr{U} = \{U_i : i \in I\}$ is an open cover of S. We are required to show that $\mathscr{U}$ has a countable subcover, $\{U_{i_j} : j \in \mathbb{N}\}$.

To obtain a contradiction, let us suppose $\mathscr{U}$ does not have a countable subcover.

Let $x \in S$. Then there exists U_x in $\mathscr{U}$ such that $x \in U_x$. By regularity, there exists W_x such that $x \in W_x \subseteq \mathrm{cl}_S W_x \subseteq U_x$. Then $\mathscr{W} = \{W_x : x \in S\}$ as well as

$$\mathscr{W}^* = \{\mathrm{cl}_S W_x : x \in S\}$$

cover S. Then, for any countable subfamily, $\{\mathrm{cl}_S W_{x_j} : j \in \mathbb{N}\}$, of $\mathscr{W}^*$,

$$\cap\{S \backslash \mathrm{cl}_S W_{x_j} : j \in \mathbb{N}\}$$

is non-empty. Then, $\{S \backslash \mathrm{cl}_S W_x : x \in S\}$ is a family of open sets which satisfies the countable intersection property. By hypothesis, it must have an accumulation point p. This means $p \in \cap\{\mathrm{cl}_S[S \backslash \mathrm{cl}_S W_x] : x \in S\}$. But $\mathrm{cl}_S[S \backslash \mathrm{cl}_S W_x] = S \backslash W_x$. Then $p \in \cap\{S \backslash W_x] : x \in S\}$. Then p is not in any W_x, contradicting that $\{W_x : x \in S\}$ is an open cover of S. The source of the contradiction is the assumption that $\mathscr{U}$ does not have a countable subcover.

So, $\mathscr{U}$ has a countable subcover. We conclude that S is Lindelöf.

Example 1. We have shown in Chapter 5 (page 103) that the metrizable set of all real numbers, equipped with the usual topology, is second countable and so is separable. Then, by part (b) of the above characterization theorem, $\mathbb{R}$ is a Lindelöf space.

Example 2. We have seen that the real numbers, $(\mathbb{R}, \tau_S)$, equipped with the upper limit topology (Sorgenfrey line) is not second countable (see page 102). Show that it is Lindelöf.

Solution: Let $\mathscr{U}$ be an open cover of $\mathbb{R}$. We will fix $n \in \mathbb{N}$. Let $S_n = \{[u, n] : [u, n] \text{ has a countable subcover}\}$.

Let $k_n = \inf\{u : [u, n] \in S_n\}$. We claim that $k_n \notin \mathbb{R}$. Suppose $k_n \in \mathbb{R}$. Then $[k_n, n]$ has a countable subcover, $\mathscr{U}_n = \{U_i \in \mathscr{U} : i \in \mathbb{N}\}$. This means that $k_n \in U_j$, for some $j \in \mathbb{N}$. Since U_j is open, there is some $\varepsilon > 0$ such that $k_n \in (k_n - \varepsilon, k_n] \subseteq U_j$. Then $[k_n - \varepsilon/2, n]$ has a countable subcover contradicting the definition of k_n. So, $(-\infty, n]$ has a countable subcover, $\mathscr{U}_n$, as claimed. This is independent of the value of n. Then $\cup\{\mathscr{U}_n : n \in \mathbb{N}\}$ is a countable subcover of $\cup\{(-\infty, n] : n \in \mathbb{N}\} = \mathbb{R}$. So $\mathbb{R}$ is Lindelöf.

Note that $(\mathbb{R}, \tau_S)$ has been found to be Lindelöf in spite of it not being second countable. Does this contradict Theorem 16.3, part (b)? The answer is "No it doesn't". It simply guarantees that $(\mathbb{R}, \tau_S)$ is not metrizable.

Example 3. In the example on page 373, we showed that the ordinal space, $S = [0, \omega_1)$, is countably compact but not compact. Then it cannot be Lindelöf, for, if it was both countably compact and Lindelöf, it would have to be compact.

16.4 Two Invariance Theorems for the Lindelöf Property

Arbitrary subspaces generally do not inherit the Lindelöf property from its superset. Also, products of Lindelöf spaces need not be Lindelöf. But closed subspaces do inherit the Lindelöf property from its superset.

We will also show that the Lindelöf property is carried over by continuous functions.

Theorem 16.4 *If S is Lindelöf and F is a closed subset of S then F is Lindelöf.*

Proof. Given: That S is Lindelöf and F is closed in S.

Let $\mathscr{U} = \{U_i : i \in I\}$ be a family of open sets in S which covers F. Then $\mathscr{U}^* = \mathscr{U} \cup \{S\backslash F\}$ covers S. Since S is Lindelöf, then $\mathscr{U}^*$ has a countable subcover $\{U_{i_j} : j \in \mathbb{N}\} \cup \{S\backslash F\}$. Then $\{U_{i_j} : j \in \mathbb{N}\}$ is a countable subcover of F. So, F is Lindelöf.

Theorem 16.5 *If S is Lindelöf and $f : S \to T$ is continuous and onto T, then T is Lindelöf.*

Proof. Given: That S is Lindelöf and $f : S \to T$ is continuous and onto T.

Let $\mathscr{U} = \{U_i : i \in I\}$ be a family of open sets which covers T. Then $\mathscr{V} = \{f^{\leftarrow}[U_i] : i \in I\}$ covers S. Then $\mathscr{V}$ has a countable subcover, $\{f^{\leftarrow}[U_{i_j}] : j \in \mathbb{N}\}$, of S. Then $\{f[f^{\leftarrow}[U_{i_j}]] : j \in \mathbb{N}\} = \{U_{i_j} : j \in \mathbb{N}\}$ is a countable subcover of T. So, T is Lindelöf.

16.5 Regular Lindelöf Spaces are normal

We now show that regular spaces that are equipped with the Lindelöf property are in fact normal spaces.

Theorem 16.6 *If S is both regular and Lindelöf then it is normal.*

Proof. Given: That S is regular and Lindelöf.

Let F and K be disjoint closed sets. If $x \in F$ and $y \in K$, then, by regularity, there exists an open B_x such that $\text{cl}_S B_x \cap K = \varnothing$ and open D_y such that $\text{cl}_S D_y \cap F = \varnothing$. Then

$$\mathscr{B}_F = \{B_x : x \in F\}$$

$$\mathscr{D}_K = \{D_y : y \in K\}$$

form open covers of F and K, respectively. Since S is Lindelöf, then so are the closed subsets F and K. So, we can obtain

$$\mathscr{B}_F{}^* = \{B_{x_i} : i \in \mathbb{N}\}$$

$$\mathscr{D}_K{}^* = \{D_{y_i} : i \in \mathbb{N}\}$$

as countable subcovers of F and K, respectively.

We now inductively construct disjoint open neighborhoods of F and K:

For	$F_1 = B_{x_1}$	let $K_1 = D_{y_1} \backslash \text{cl}_S F_1$
For $i \in \mathbb{N}$ if	$F_i = B_{x_i} \backslash \text{cl}_S(K_1 \cup \cdots \cup K_{i-1})$	let $K_i = D_{y_i} \backslash \text{cl}_S(F_1 \cup \cdots \cup F_i)$

Then $\cup\{F_i : i \in \mathbb{N}\}$ and $\cup\{K_i : i \in \mathbb{N}\}$ form disjoint open neighborhoods of F and K, respectively. So, S is normal.

Concept Review

1. Define the Lindelöf property on a topological space.
2. What kind of space is obtained if we combine countable compactness with the Lindelöf property?
3. How does a space with the second countable property compare with a Lindelöf space?
4. Give a characterization of the Lindelöf property in terms of a filter of closed sets.
5. In a metric space, state two properties each of which is equivalent to the Lindelöf property.
6. In a regular space, state two properties which are equivalent to the Lindelöf property.
7. Is there a quick argument you can use to conclude that $\mathbb{R}$ is Lindelöf, based on the theory developed in the chapter.
8. What can you say about subspaces of Lindelöf spaces?
9. What about continuous images of Lindelöf spaces.
10. In what way does the Lindelöf property enhance a regular space?
11. Provide examples of Lindelöf and non-Lindelöf spaces.

Exercises

1. Suppose that the space S is such that it can be expressed as a countable union of compact spaces. Show that S is Lindelöf.
2. Suppose S is a topological space. We say that the open cover, $\mathscr{U} = \{U_i : i \in I\}$, refines (or is a refinement of) the open cover, $\mathscr{V} = \{V_i : i \in I\}$, if for every V_i there is some U_j such that $U_j \subseteq V_i$.

Show that S is Lindelöf if and only if every open cover of S has a countable refinement.

3. Recall that a T_1-space, S, is perfectly normal if and only if, for any pair of non-empty disjoint close subsets, F and K, there exists a continuous function $f : S \to [0,1]$ such that $F = f^{\leftarrow}(0)$ and $K = f^{\leftarrow}(1)$. Suppose S is a regular space such that each of its open subsets is Lindelöf. Show that S is perfectly normal.

Chapter 17

Sequentially and Feebly Compact Spaces

Abstract
In this section, we define the sequentially compact property and determine in which precise circumstances it is just another way of referring to the compactness property. At the same time, we outline conditions under which a compact space is not sequentially compact. To illustrate this, we provide various examples. We also introduce two closely related classes of spaces: feebly compact and pseudocompact spaces.

17.1 Sequentially Compact Spaces

For many readers, sequentially compact is just another way of saying "compact" in two words instead of one. In many cases, this is fine. As long as we keep in mind that the two properties are, in general, not equivalent. We will begin with a formal definition and then examine specific cases where the sequential compact spaces are equivalent to compact ones. Based on our experience with countably compact spaces we expect that countability properties will, somehow, be relevant in many arguments.

Definition 17.1 Let S be a topological space. We say that a subspace, T, of S is *sequentially compact* if and only if every sequence in T has a subsequence which converges to a point in T.

Theorem 17.2 *Let S be a topological space.*

(a) *If S is second countable, then the following are equivalent.*
  (i) *The space S is compact.*
  (ii) *The space S is countably compact.*
  (iii) *The space S is sequentially compact.*

(b) *If S is metric, then the following are equivalent.*
  (i) *The space S is compact.*
  (ii) *The space S is countably compact.*
  (iii) *The space S is sequentially compact.*

(c) *If $T \subseteq S = \mathbb{R}^n$ the following are equivalent.*
  (i) *The subspace T is compact.*
  (ii) *The subspace T is closed and bounded.*
  (iii) *The subspace T is sequentially compact.*

Proof. We are given that S is a topological space.

(a) We are given that S is second countable.

(i $\Leftrightarrow$ ii) In second countable spaces, the equivalence of compactness and countable compactness is proved in Theorem 15.7.

(ii $\Leftrightarrow$ iii) We must show "countably compact if and only if sequentially compact".

Recall that second countable spaces are first countable, always.

S ctbly cpct $\Leftrightarrow$ every seq in S has an accum'tion pt. (Theorem 15.2)

In a first countable space S,

$p =$ accum'tion point of a seq $\Leftrightarrow p =$ limit point of a subseq. (Theorem 11.5)

So in a second countable space,

$$S \to \text{sequentially compact} \quad \Leftrightarrow \quad S \to \text{countably compact}$$

(b) We are given that S is a metric space.

(i $\Leftrightarrow$ ii) In Theorem 15.8, it is shown that in a metric space, compactness and countable compactness are equivalent.

(ii $\Leftrightarrow$ iii) Recall that metric spaces are first countable. As shown in (ii $\Leftrightarrow$ iii) of part (a), in first countable spaces, countable compactness and sequential compactness are equivalent.

(c) We are given that $T \subseteq S = \mathbb{R}^n$ with the usual topology.

(i $\Leftrightarrow$ iii) Since S is a metric space, S is compact if and only if S is sequentially compact. (By part (b).)

(i $\Leftrightarrow$ ii) For $\Rightarrow$. We are given that T is compact. Since S is Hausdorff, compact subspaces are closed in S, so T is closed in S.

We now show boundedness of T. For $u \in T$, let $f : T \to \mathbb{R}$ be defined as $f(x) = \text{distance}(u, x)$ (known to be continuous, see page 258). Since f is continuous on the compact set T, f attains a maximum value, k, at some element $v \in T$, so $k = \text{dist}(u, v) \geq \text{dist}(u, x)$ independently of the choice of x in T. So, for any $a, b \in T$, $\text{dist}(a, b) \leq \text{dist }(a, u) +$ $\text{dist}(u, b) \leq k + k = 2k$. So, T is bounded, as required.

For $\Leftarrow$. We are given that T is closed and bounded in $\mathbb{R}^n$. Since T is bounded in $\mathbb{R}^n$, there exists $k \in \mathbb{R}^+$ such that $T \subseteq [-k, k]^n$. Since $[-k, k]$ is compact then so is its product, $\prod_{i=1}^{n}[-k, k]_i$. Then, since T is a closed subset of a compact space, T is compact.

Many readers may already be familiar with the following result about compact subsets of metric spaces.

Theorem 17.3 *If S is a metric space and T is compact in S, then T is closed and bounded.*

Proof. The proof is left as an exercise for the reader.

Note: The converse fails. Consider an infinite discrete space with metric $\rho(a, b) = 1$ if $a \neq b$.

Theorem 17.4 *If S is a sequentially compact space then it must necessarily be countably compact.*

Proof. We are given that S is a sequentially compact space. Then every sequence in S has a subsequence which converges to a point in S. We are required to show that it is countably compact.

Suppose S is not a countably compact space. Then S has a countable open cover, $\mathscr{U} = \{U_i : i \in \mathbb{N}\}$, with no finite subcover. For each $n \in \mathbb{N}$, we can then choose

$$p_n \in S \backslash \cup \{U_i : i = 0, 1, 2, 3, \ldots, n\}$$

Let $p \in S$. We claim that p cannot be the limit of a subsequence of $A = \{p_i : i \in \mathbb{N}\}$. Since $\mathscr{U}$ covers S, then $p \in U_k$ for some $k \in \mathbb{N}$. If $m > k$ then

$$p_m \notin \cup \{U_i : i = 0, 1, 2, 3, \ldots, m\}$$

Then the tail end, $\{p_i : i \geq k\}$, of A, does not intersect the open neighborhood, U_k, of p. Then p cannot be the limit of a subsequence of A, as claimed.

This contradicts the fact that S is sequentially compact. So spaces which are sequentially compact are countably compact.

We now verify that sequential compactness is preserved by continuous functions.

Theorem 17.5 *Let S be a topological space and T be a Hausdorff topological space. If $f : S \to T$ is a continuous function and S is sequentially compact then $f[S]$ is sequentially compact in T.*

Proof. We are given that S is a sequentially compact space, T is Hausdorff and $f : S \to T$ is continuous. We are required to show that $f[T]$ is sequentially compact.

The proof follows directly from the definition of sequentially compact. The details are left as an exercise.

17.2 Example of a Compact Space Which Is Not Sequentially Compact

We have seen that, if S is either second countable or is metrizable then

$$S \text{ sequential compactness} \Leftrightarrow S \text{ countable compactness}$$

We have seen in 15.2 that "countably compact spaces are precisely those spaces in which every sequence has an accumulation point". We can then also say:

If S is either second countable or is metrizable then

Every sequence in S has a convergent subsequence in S

$$\Leftrightarrow$$

Every sequence in S has a convergent subnet in S

This leads us to wonder whether there are any spaces in which sequences have a convergent subnet but not a convergent subsequence.

We studied convergence properties of $S = [0,1]^{[0,1]}$ on page 306. In that example, we saw that S is a compact space in which a sequence has an accumulation point but no subsequence converging to that point. We will revisit the example, $S = [0,1]^{[0,1]}$, to show that a compact space need not be sequentially compact.

Example 1. Let $S = [0,1]^{[0,1]}$ be equipped with the *product topology.* That is, we view S as $\prod_{i\in[0,1]}[0,1]_i$. Show that S is compact, hence, countably compact.

(a) Show that S is compact, hence, countably compact.
(b) Show that, in spite of its compactness, S is not sequentially compact.

Solution: We are given that the space $S = [0,1]^{[0,1]}$ viewed as $\prod_{i\in[0,1]}[0,1]_i$ is equipped with the product topology.

(a) Since $[0,1]$ is compact and the any product space of compact sets is compact (by Tychonoff theorem), then $S = \prod_{i\in[0,1]}[0,1]_i$ is compact. Since any compact set is countably compact, then S is also countably compact.
(b) We are given that $S = [0,1]^{[0,1]}$ is equipped with the product topology. That is, we view S as $\prod_{i\in[0,1]}[0,1]_i$. We will construct a sequence in S which has no converging subsequence. Suppose each element, x, of $[0,1]$ is expressed in its binary expansion form. For each $n \in \mathbb{N}\setminus\{0\}$ we define the one-to-one function

$f_n : [0,1] \to \{0,1\}$ as

$$f_n(x) = \text{“the } n\text{th digit in the binary expansion of } x\text{”}$$

For example, $f_3(0.10101010\ldots) = 1$.[1] This allows us to view S as an uncountable family,

$$T = \{\{f_n(x) : n = 1, 2, 3, \ldots\} : x \in [0,1]\}$$

of sequences of 0's and 1's.

We define a sequence $\{x_k : k = 1, 2, 3, \ldots\}$ in S, such that

$$N = \{\{f_n(x_k) : n = 1, 2, 3, \ldots\} : x_k \in S\}$$

where $f_k(x_k) = 1$ for all values of k, and $f_n(x_k) = 0$ if $n \neq k$. That is, the diagonal of N in T is made only of 1's and all other entries are 0's.

We claim that N cannot have a convergent subsequence. To see why, suppose N has a convergent subsequence, say

$$M = \{\{f_n(x_{k_n}) : n = 1, 2, 3, \ldots\} : x_{k_n} \in \{x_k : k = 1, 2, 3, \ldots\}\}$$

Then every column of M is constant on a tail end and so must converge (to zero). If so, what can we say about the point (1, 0)? This particular point becomes a limit point of a sequence of both a string of 0's (above it) and a string of 1's (on the diagonal). A contradiction. The source of the contradiction is our supposition that M converges.

Since M was an arbitrary subsequence of N, N cannot have a convergent subsequence. So, S cannot be sequentially compact.

The example does not contradict Theorem 17.2. It simply shows that S is not metrizable.

17.3 Topic: A Compact Characterization for Metric Spaces

Completeness is a concept which is defined only in the context of metric spaces.

[1] One-to-one if we avoid infinite tail ends of zeroes. For example, $0.1111111\ldots = 1.000000\ldots$

A sequence $\{x_n : n \in \mathbb{N}\}$ in a metric space (M, ρ) is said to be a *Cauchy sequence* if, for any $\varepsilon > 0$, there exists N such that, whenever $n, m > N$, $\rho(x_n, x_m) < \varepsilon$. If every Cauchy sequence in (M, ρ) converges to a point in M, then (M, ρ) it is said to be a *complete* metric space.

For example the metric space, $\mathbb{R}^n$, is known to be complete. But we can still associate completeness in a metric space to the topological notion of compactness. For example, it is not difficult to show that all compact metric spaces are complete metric spaces, as shown below.

Example 2. Verify that a compact metric space is a complete metric space.

Solution: Suppose (M, ρ) is a compact metric space and $S = \{x_n : n \in \mathbb{N}\}$ is a Cauchy sequence in M. Then, for any $k \in \mathbb{N}\backslash\{0\}$ there exists N_k such that, if $\{x_n : n > N_k\} \subseteq B_{1/k}(x_n) \neq \varnothing$. So for every $1/k$, there is some n such that $B_{1/k}(x_n)$ contains a tail-end $\{x_n, x_n + 1, \ldots\}$ of S. Then the set of closed subsets,

$$F = \{\ \mathrm{cl}_M \left[B_{1/k}(x_n) \cap \{x_n : n > N_k\}\right] : k \in \mathbb{N}\backslash\{0\}\}$$

of M satisfies the FIP. Since M is compact, $\cap F$ is non-empty and so contains a point, say $z \in M$. Since every open neighborhood of z will intersect every element of F. Then every open neighborhood of z contains a tail-end of S. It follows that z is the limit point of the sequence, S. So, the Cauchy sequence, S, converges. Then M is complete.

Since the metric space, $\mathbb{R}$, is known to be complete and non-compact, the converse obviously does not hold true. A compact metric space would also have to satisfy some other property. We introduce the following notion for metric spaces:

Definition 17.6 A metric space, (M, ρ), is said to be *totally bounded* if for any $\varepsilon > 0$ there exists a *finite* subset F of M such that $\{B_\varepsilon(x) : x \in F\}$ which covers M.

Theorem 17.7 *A metric space is compact if and only if it is both complete and totally bounded.*

Proof. Suppose (M, ρ) is a metric space.

($\Rightarrow$) We have already shown that compact metric spaces are complete. To show totally bounded, let $U = \{B_\varepsilon(x) : x \in I\}$ be an open cover of M. Since M is compact, U has a finite subcover and so M is totally bounded.

($\Leftarrow$) Suppose (M, ρ) is both a complete and totally bounded metric space. We are required to show that M is compact.

To show compactness of M it suffices to show sequential compactness of M. That is, that any sequence must have a convergent subsequence.

Let $\kappa_0 = \{x_i : i \in J_0\}$ be a sequence in M. Since M is complete, it will suffice to show that κ_0 has a subsequence which is Cauchy.

Suppose $\mathscr{M}_1$ is an open cover of M made only of balls of radius 1. Since M is totally bounded, $\mathscr{M}_1$ has a finite covering, $\mathscr{F}_1$. Let B_1 be an element of $\mathscr{F}_1$ which contains infinitely many points, say $\kappa_1 = \{x_i : i \in J_1\} \subseteq \kappa_0$, for some countable $J_1 \subseteq J_0$.

Suppose $\mathscr{M}_2$ is an open cover of M made only of balls of radius 1/2. Since M is totally bounded, $\mathscr{M}_2$ has a finite covering, $\mathscr{F}_2$. Let B_2 be an element of $\mathscr{F}_2$ which contains infinitely many points, say $\kappa_2 = \{x_i : i \in J_2\} \subseteq \kappa_1$, for some countable $J_2 \subseteq J_1$.

More generally, suppose we are given $\kappa_n = \{x_i : i \in J_n\} \subseteq \kappa_{n-1}$ and suppose $\mathscr{M}_{n+1}$ is an open cover of M made only of balls of radius $1/(n+1)$. Then $\mathscr{M}_{n+1}$ has a finite subcover, $\mathscr{F}_{n+1}$, of M. Let B_{n+1} be an element of $\mathscr{F}_{n+1}$ which contains infinitely many points $\kappa_{n+1} = \{x_i : i \in J_{n+1}\} \subseteq \kappa_n$ of M where $J_{n+1} \subseteq J_n$.

Applying this process, we have a nested family of sequences $\{\kappa_n : n = 1, 2, 3, \ldots\}$ where $\kappa_{n+1} \subseteq \kappa_n$. Then we can choose $x_j \in \kappa_j$ for each $j \in \mathbb{N}$ to form a subsequence $\kappa = \{x_j : j \in \mathbb{N}\}$ of κ_0 where for any $\varepsilon > 0$, κ has a tail-end contained in a ball of radius ε. So, κ is Cauchy.

By hypothesis, M is complete so κ converges to, say, a point $y \in M$. Then every sequence has a convergent subsequence. That is, M is sequentially compact. So, M is compact.

17.4 Feebly Compact Spaces

We introduce a new class of spaces called *feebly compact* spaces. Feebly compact spaces have a few very interesting characterizations. We will see that they form a proper subclass of the all pseudocompact spaces (those spaces for which $C^*(S) = C(S)$). We will also see that, for completely regular spaces (Tychonoff spaces), pseudocompact spaces are always feebly compact. The definition refers to "locally finite collections of open subsets".

Definition 17.8 A *locally finite* collection, $\mathscr{C}$, of subsets of a topological space, S, is one for which every point in S has at least one neighborhood which meets only finitely many elements of $\mathscr{C}$.

Definition 17.9 Let S be a topological space. A space S for which there can be no infinite locally finite collection of open sets is called a *feebly compact* space.

The formal definition is not particularly enlightening. The following characterizations will provide different perspectives on this property.

Theorem 17.10 *Let S be any topological space. Then the following are equivalent.*

(1) *Any locally finite collection of pairwise disjoint open subsets of S is finite.*
(2) *The space S is feebly compact.*
(3) *For any decreasing collection, $\{U_i : i \in \mathbb{N}\}$ of open subsets of S,*

$$\cap\{cl_S U_i : i \in \mathbb{N}\} \neq \varnothing$$

(4) *Any countable open cover, $\{U_i : i \in \mathbb{N}\}$, of open subsets of S has a finite subcollection, $\{U_i : i \in F\}$, such that $\cup\{U_i : i \in F\}$ is dense in S.*

Proof. We are given that S is a topological space.

(1 $\Rightarrow$ 2) Suppose any locally finite collection of pairwise disjoint open subsets of S is finite. We are required to show that S is feebly compact.

Suppose S is not feebly compact. Then there exists an infinite locally finite collection of open sets. Let $\mathscr{U} = \{U_i : i \in \mathbb{N}\}$ be such a collection. Choose $x_0 \in U_0$. There exists an open neighborhood, V_0 of x_0 which intersects at most finitely many elements of $\mathscr{U}$. Then there is a $n_1 \in \mathbb{N}$ that if $n \geq n_1$, $U_n \cap V_0 = \varnothing$.

We proceed inductively. Let $\{n_i : i = 0 \ldots m\}$ be a strictly increasing set of natural numbers, and let $\{V_i : i = 0 \ldots m-1\}$ be a sequence of open sets such that $U_{n_i} \cap V_i \neq \varnothing$ but, for $i < j$ implies $V_i \cap U_{n_j} = \varnothing$. We can then choose $x_m \in U_{n_m}$ for which we can find an open set V_m and a natural number $n_{m+1} > n_m$ such that $x_m \in U_{n_m} \cap V_m$ and $V_m \cap U_j = \varnothing$ whenever $j \geq n_{m+1}$.

Then $\{V_i \cap U_{n_i} : i \in \mathbb{N}\}$ forms and infinite pairwise disjoint locally finite collection of open sets in S. This is contrary to our hypothesis. So, S must be feebly compact.

(2 $\Rightarrow$ 3) Suppose S is feebly compact. Then any locally finite collection of open subsets of S is finite. Suppose $\mathscr{U} = \{U_i : i \in \mathbb{N}\}$ is a collection of non-empty open subsets of S such that $U_{i+1} \subseteq U_i$ for all i.

We are required to show that $\cap\{\text{cl}_S U_i : i \in \mathbb{N}\} \neq \varnothing$.

Suppose $\cap\{\text{cl}_S U_i : i \in \mathbb{N}\} = \varnothing$. Then, if $x \in S$ there exists some k_x such that $x \notin \text{cl}_S V_i$ for all $i \geq k_x$. Then, if $i \geq k_x$, $S \setminus \text{cl}_S V_i$ is an open neighborhood of x which intersects only finitely many elements of $\{U_i : i \in \mathbb{N}\}$. So, $\mathscr{U}$ is an infinite locally finite collection of open sets contrary to our hypothesis about the properties of $\mathscr{U}$. So, $\cap\{\text{cl}_S U_i : i \in \mathbb{N}\} \neq \varnothing$.

(3 $\Rightarrow$ 4) Suppose S is a space which satisfies the property: Whenever $\mathscr{U} = \{U_i : i \in \mathbb{N}\}$ is a collection of non-empty open subsets of S such that $U_{i+1} \subseteq U_i$ for all i, then $\cap\{\text{cl}_S U_i : i \in \mathbb{N}\} \neq \varnothing$.

Let $\mathscr{V} = \{V_i : i \in \mathbb{N}\}$ be any countable open cover of S. We are required to show that $\mathscr{V}$ has a subfamily whose union is dense in S.

For each $n \in \mathbb{N}$ we define

$$B_n = S \setminus \text{cl}_S[\cup\{V_i : i \in \{0, n\}\}]$$

Suppose B_n is non-empty for all $n \in \mathbb{N}$. Then $\{B_n : n \in \mathbb{N}\}$ is a collection of non-empty open sets such that $B_{n+1} \subseteq B_n$.

By hypothesis, $\cap\{\text{cl}_S B_n : n \in \mathbb{N}\} \neq \varnothing$. But we also have,

$$\cap\{\text{cl}_S B_n : n \in \mathbb{N}\} \subseteq S \backslash \cup\{\text{cl}_S V_i : n \in \mathbb{N}\} = \varnothing$$

We have a situation where $\cap\{\text{cl}_S B_n : n \in \mathbb{N}\}$ is both empty and non-empty. The fact that $\cap\{\text{cl}_S B_n : n \in \mathbb{N}\} \neq \varnothing$ followed from our assumption that B_n is non-empty for all $n \in \mathbb{N}$. So, this assumption is not valid. There must be some $k \in \mathbb{N}$ such that $B_k = \varnothing$. So $\cup\{\text{cl}_S V_i : i \in \{0, k\}\} = \text{cl}_S \cup \{V_i : i \in i \in \{0, k\}\} = S$. Then $\cup\{V_i : i \in i \in \{0, k\}\}$ is dense in S.

($4 \Rightarrow 1$) Suppose that, if $\mathscr{V} = \{V_i : i \in \mathbb{N}\}$ be a countable open cover of S, then $\mathscr{V}$ has a subfamily whose union is dense in S. We are required to show that all locally finite families of pairwise disjoint open subsets are finite.

We will suppose that an infinite locally finite family, $\mathscr{U} = \{U_i : i \in \mathbb{N}\}$, of pairwise disjoint open subsets exists. For each $x \in S$, let M_x be an open neighborhood of x and F_x be a finite subset of $\mathbb{N}$ such that $M_x \cap U_i \neq \varnothing$ for $i \in F_x$. Then $\{M_x : x \in S\}$ forms an open cover of S.

Let $\mathscr{F} = \{F : F$ is a finite subset of $\mathbb{N}\}$. Note that $\mathscr{F}$ is a countably infinite subset of $\mathscr{P}(\mathbb{N})$.[2]

Let $W : \mathscr{F} \to S$ be defined as

$$W(F) = \cup\{M_x : x \in S, F = F_x\}$$

See that, for each F, $W(F)$ is a non-empty open neighborhood of x in S. Since $S \subseteq \cup\{W(F) : F \in \mathscr{F}\}$,

$$\mathscr{W} = \{W(F) : F \in \mathscr{F}\} \text{ is an open cover of } S$$

Since $\mathscr{F}$ is countable, then so is the open cover, $\mathscr{W}$.

[2]The number of finite subsets of $\mathbb{N}$ is countable: For each $n \in \mathbb{N}$, let $\mathscr{U}_n = \{A \in \mathscr{P}(\mathbb{N}) : \max A \leq n\}$, the set of all subsets of $\{0, 1, 2, \ldots, n\}$. Then, for each n, $|\mathscr{U}_n| = 2^{n+1}$. The countable union of countable sets is countable (see 19.3 of Appendix A.2), so $\cup_{n \in \mathbb{N}}\{\mathscr{U}_n\}$ is countable. If F is a finite subset of $\mathbb{N}$, $F \in \mathscr{U}_m$ for some m. Then $\cup_{n \in \mathbb{N}}\{\mathscr{U}_n\}$ contains all finite subsets of $\mathbb{N}$. Then the number of finite subsets of $\mathbb{N}$ is countable.

We will show that no finite subfamily of $\mathscr{W}$ can be dense in S. To see this, suppose $\{W(F_i) : F_i \in \{F_1, F_2, \ldots, F_n\}\}$ is any finite subfamily of $\mathscr{W}$.

Choose any $F_k \in \mathscr{W} \backslash \{F_i : i \in \{0, n\}\}$.

We claim that $U_k \cap \cup\{W(F_i) : F_i \in \{F_1, F_2, \ldots, F_n\}\} = \varnothing$.

For, if there is a y in $\cup\{W(F_i) : i \in \{0, n\}\} \cap U_k$, then $y \in M_y$ and $F_y = F_k$. Since $F_k \notin \{F_i : i \in \{0, n\}\}$, we have a contradiction. The source of the contradiction is our supposition that $\cup\{W(F_i) : i \in \{0, n\}\} \cap U_k$ contains an element y.

So, $U_k \cap \cup\{W(F_i) : F_i \in \{F_1, F_2, \ldots, F_n\}\} = \varnothing$, as claimed.

Then we can conclude that for any finite collection $\{W(F_i) : F_i \in \{F_1, F_2, \ldots, F_n\}\} \subseteq \mathscr{W}$, $S \backslash \cup\{W(F_i) : F_i \in \{F_1, F_2, \ldots, F_n\}\}$ contains a non-empty open set. So, no finite subfamily of $\mathscr{W}$ can be dense in S. This contradicts our hypothesis.

The assumption that there is an infinite locally finite family, $\mathscr{U} = \{U_i : i \in \mathbb{N}\}$, of pairwise disjoint open subsets cannot possibly hold true. So, $\mathscr{U}$ must be finite. We are done with $(4 \Rightarrow 1)$.

The fourth statement in the above characterization theorem confirms for us that compact spaces are most certainly feebly compact.

Example 3. Let $S = [0, 1]$ (the closed unit interval). Let $\mathscr{B} = \{U_i : i \in \mathbb{N}\}$ denote a partition of S where each U_i is a dense subset of S (with respect to the usual topology).

We show that such a partition of S exists. Suppose we represent each element of S in its binary expansion form, $0.x_1x_2x_3\cdots x_n\cdots$. Let the function $f : S \to \mathbb{R}$ be defined as

$$\begin{aligned} f(x) &= f(0.x_1x_2x_3\cdots x_n\cdots) \\ &= \lim_{n\to\infty} \sup_{m\geq n} \frac{x_1 + x_2 + x_3 + \cdots + x_m}{m} \\ &= \lim_{n\to\infty} \left\{ \sup \left\{ \frac{x_1 + x_2 + x_3 + \cdots + x_m}{m} : m > n \right\} : n \geq 0 \right\} \end{aligned}$$

Then f is well-defined and the set of fibers, $\{f^{\leftarrow}[\{y\}] : y \in \mathbb{R}\}$, of f forms a partition of S.

If $s_m = \frac{x_1+x_2+x_3+\cdots+x_m}{m}$, the terms of the sequence,

$$\{\sup\{s_m : m > n\}\}_{n\in\mathbb{N}}$$

decrease to the value $f(x)$ as n increases.

Let $u \in \mathbb{R}$. We claim that $f^{\leftarrow}[\{u\}]$ is dense in S.

There exists $z = 0.z_1 z_2 z_3 \cdots z_n \cdots \in f^{\leftarrow}(u)$.

See that,

$$\begin{aligned}
u &= f(z) \\
&= \lim_{n\to\infty} \sup_{m>n} \frac{z_1 + z_2 + z_3 + \cdots + z_m}{m} \\
&= \lim_{n\to\infty} \sup_{m>n} \left[\frac{z_1 + z_2 + z_3 + \cdots + z_n}{m} + \frac{z_{n+1} + z_{n+2} + \cdots + z_m}{m}\right] \\
&= 0 + \lim_{n\to\infty} \sup_{m>n} \left[\frac{z_{n+1} + z_{n+2} + \cdots + z_m}{m}\right] \qquad (^*)^3 \\
&= \lim_{n\to\infty} \sup_{m>n} \left[\frac{z_{n+1} + z_{n+2} + \cdots + z_m}{m}\right] \\
&= f(0.z_{n+1} z_{n+2} z_{n+3} \cdots)
\end{aligned}$$

Suppose

$$y = 0.y_1 y_2 y_3 \cdots y_n \text{ (in binary expansion form)}$$

and $\varepsilon > 0$. We claim that there exists $q = 0.q_1 q_2 q_3 \cdots \in f^{\leftarrow}(u)$ such that $|y - q| < \varepsilon$.

There exists $N_1 > 0$ so that, for $n > N_1$, for $q = 0.y_1 y_2 \cdots y_n q_{n+1} q_{n+2} \cdots$,

$$|q - y| = |0.y_1 y_2 \cdots y_n q_{n+1} q_{n+2} \cdots - 0.y_1 y_2 \cdots y_n y_{n+1} y_{n+2} \cdots| < \varepsilon$$

We can choose

$$q = 0.y_1 y_2 y_3 \cdots y_n z_{n+1} z_{n+2} z_{n+3} \cdots$$

[3]The step $(^*)$ is an application of the rule: If $\{a_m\}$ converges to a then

$$\lim_{n\to\infty} \sup_{m>n} (a_m + b_m) = a + \lim_{n\to\infty} \sup_{m>n} (b_m)$$

Then $f(q) = f(0.z_{n+1}z_{n+2}z_{n+3}\cdots) = f(z) = u$. So, $q \in f^{\leftarrow}(u)$ as claimed. So, $q \in B_\varepsilon(y)$. Then $f^{\leftarrow}(u)$ is dense in S.

So, the closed unit interval $S = [0, 1]$ can be partitioned into a countably infinite family,

$$\mathscr{B} = \{U_i : i \in \mathbb{N}\backslash\{0\}\}$$

of dense subsets.

Suppose τ denotes the usual topology on S.

For each $i \in \mathbb{N}\backslash\{0\}$, let

$$V_i = U_{2i-1} \cup U_{2i} \cup U_{2i+1}$$

So, $\{V_i : i = 1, 2, 3, \ldots\}$ covers S.

Define the collection γ of subsets of S as

$$\gamma = \{U_{2i-1} : i \in \mathbb{N}\backslash\{0\}\} \cup \{V_i : i \in \mathbb{N}\backslash\{0\}\}\dagger^2$$

The subset

$$\mathscr{S} = \tau \cup \gamma$$

of $\mathscr{P}(S)$ generates a topology on S, denote it by σ, for which it is a subbase. The topology σ is stronger than the usual topology, τ.

Since $\tau \subseteq \sigma$ and (S, τ) is Hausdorff, then (S, σ) is Hausdorff.

In what follows we show that (a) (S, σ) is not regular and (b) (S, σ) is not feebly compact.

(a) We will now prove that (S, σ) is not regular.

Proof. Let $x \in S$. Then $x \in U_j$ for some j. Suppose j is odd.

Say, for example, that the point x belongs to U_3. The set U_3 is a subbase element and so is an open subset in (S, σ). Then the set U_4 is a closed subset of S with respect to σ, disjoint from U_3.

For any τ-basic open neighborhood, (r, s), of x,

$$(r, s) \cap U_3$$

is a σ-basic open neighborhood of x.

[2]The set γ is, $\gamma = \{U_1, U_3, U_5, \ldots,\} \cup \{U_1 \cup U_2 \cup U_3,\ U_3 \cup U_4 \cup U_5,\ U_5 \cup U_6 \cup U_7, \ldots\}$.

Claim: We claim that every σ-open neighborhood of U_4 intersects the σ-basic open neighborhood, $(r, s) \cap U_3$, of x. Note that $V_2 = U_3 \cup U_4 \cup U_5$ is a σ-basic open neighborhood of U_4.

Proof of claim: Since U_4 is dense in S with respect to τ, then the τ-basic open neighborhood, (r, s) of x in U_3 must intersect U_4. Say $q \in (r, s) \cap U_4$. Let (w, z) be any τ-basic open neighborhood of q. Then $(w, z) \cap (r, s)$ is an open interval containing q. Since U_3 is dense in S with respect to τ, $(w, z) \cap (r, s)$ intersects U_3. So, $[(w, z) \cap (r, s)] \cap U_3$ is non-empty. To prove the claim, it suffices to show that

$$[(w, z) \cap V_2] \cap [(r, s) \cap U_3] \neq \varnothing$$

See that

$$\begin{aligned} [(w, z) \cap (r, s)] \cap U_3 &= [(w, z) \cap (r, s)] \cap [U_3 \cup U_4 \cup U_5] \cap U_3 \\ &= [(w, z) \cap V_2] \cap [(r, s) \cap U_3] \end{aligned}$$

Then $[(w, z) \cap V_2] \cap [(r, s) \cap U_3] \neq \varnothing$. So every σ-open neighborhood of U_4 will intersect $(r, s) \cap U_3$. This establishes claim A.

So, (S, σ) is not a regular space.

(b) We now claim that (S, σ) is not feebly compact.

Proof. Let

$$\begin{aligned} A_1 &= \{U_3 \cup U_5 \cup U_7 \cup \cdots\} \\ A_2 &= \{U_5 \cup U_7 \cup U_9 \cup \cdots\} \\ A_3 &= \{U_7 \cup U_9 \cup U_{11} \cup \cdots\} \\ &\vdots \\ A_n &= \{U_{2n+1} \cup U_{2n+3} \cup U_{2n+5} \cup \cdots\} \\ &\vdots \end{aligned}$$

Then, for $\{A_n : n = 1, 2, 3, \ldots\}$,

$$A_{n+1} \subseteq A_n \quad \text{and} \quad \mathrm{cl}_S A_{n+1} \subseteq \mathrm{cl}_S A_n$$

Also,

$$\begin{aligned}
\mathrm{cl}_S A_2 &\subseteq S\backslash(U_1 \cup U_2 \cup U_3) = S\backslash V_1 \\
\mathrm{cl}_S A_3 &\subseteq S\backslash(U_1 \cup U_2 \cup U_3 \cup U_4 \cup U_5) = S\backslash(V_1 \cup V_2) \\
&\vdots \\
\mathrm{cl}_S A_n &\subseteq S\backslash(V_1 \cup V_2 \cup \cdots \cup V_{n-1}) \\
&\vdots
\end{aligned}$$

Then

$$\begin{aligned}
\cap\{\mathrm{cl}_s A_n : n=1,2,3,\ldots\} &\subseteq \cap\{S\backslash(V_1 \cup V_2 \cup \cdots \cup V_{n-1}) : n=1,2,3,\ldots\} \\
&= S\backslash\cup\{V_n : n = 1,2,3,\ldots\} \\
&= S\backslash S \\
&= \varnothing
\end{aligned}$$

By Theorem 17.10, (S, σ) is not feebly compact.

17.4.1 *On pseudocompact and feebly compactness*

Feebly compact and countably compact spaces share a common property. It is that neither topological space admits unbounded continuous real-valued function. The next theorem confirms that...

every feebly compact space is pseudocompact.

But there are spaces which are pseudocompact but not feebly compact. (Such a space would have to be non-"completely regular".)

Theorem 17.11 *Let S be a feebly compact topological space. Then every continuous real-valued function on S is bounded. (So every feebly compact space is a pseudocompact space.)*

Proof. We are given that S is a feebly compact topological space. We are required to show that any continuous real-valued function on S is bounded. Suppose $f \in C(S)\backslash C^*(S)$. Note that we can assume, without loss of generality, that $f \geq 0$. We can then construct in $f[S]$

a sequence, $\{x_n : n \in \mathbb{N}\}$, of real numbers where $x_{n+1} \geq x_n + 1$. See that

$$\mathscr{U} = \{f^{\leftarrow}[(x_n - \tfrac{1}{3},\ x_n + \tfrac{1}{3})] : n \in \mathbb{N}\}$$

is an infinite locally finite collection of open subsets of S. So, S cannot be feebly compact, a contradiction of our hypothesis.

17.5 Pseudocompact Spaces

We know that the continuous image of a compact space is always compact. This means that continuous real-valued functions on a compact space are always bounded. However, a space S in which the property "all real-valued continuous functions are bounded on S" need not be compact. A countably compact space is an example. A space which satisfies this property is said to be *pseudocompact.* We formally define this concept.

Definition 17.12 Let S be a topological space. If every continuous real-valued function on S is bounded in $\mathbb{R}$ then S is *pseudocompact.*

Using the newly introduced terminology above, we can say that all compact spaces are pseudocompact spaces. The set of all real numbers, $\mathbb{R}$, is, of course, not pseudocompact (as witnessed by $f(x) = x^3$ on $\mathbb{R}$). We have seen in Theorem 15.9, that

> *every continuous real-valued function on a countably compact space, S, is bounded on S.*

So countably compact spaces are pseudocompact.

What about sequentially compact spaces? Consider a continuous real-valued function, $f : S \to \mathbb{R}$, on a sequentially compact space S. By Theorem 17.4, S must also be countably compact. We have seen in Theorem 15.9 that f must be bounded on S. So, sequentially compact spaces are pseudocompact.

We have already encountered an example of a space which is pseudocompact but not compact. We saw that the ordinal space,

$$S = [0, \omega_1)$$

is countably compact and not compact. So $S = [0, \omega_1)$ is pseudocompact but not compact.

The next theorem illustrates how closely related the feebly compact and pseudocompact properties are. We see that, in the particular case where a space is completely regular, the pseudocompact property and the feebly compact property simply different ways of representing the same topological property.

Various characterizations of the pseudocompact property will be discussed in chapters to come. But we can present immediately characterizations of the pseudocompact property for completely regular spaces.

The following theorem also confirms that if we seek

a pseudocompact space which is not feebly compact,

we only need investigate those spaces which are not completely regular.

So, we know that pseudocompact spaces are not feebly compact. But, under the right conditions, they are equivalent.

Theorem 17.13 *Let S be a completely regular space. Then the following are equivalent.*

(1) *The space S is pseudocompact.*
(2) *If $\{U_i : i \in \mathbb{N}\}$ is a nested collection of non-empty open sets in S (i.e., $U_{i+1} \subset U_i$ for all $i \in \mathbb{N}$) then*

$$\cap\{cl_S U_i : i \in \mathbb{N}\} \neq \varnothing$$

(3) *The space S is feebly compact.*

Proof. We are given that S is a completely regular space.

($1 \Rightarrow 2$) Suppose S is pseudocompact. Let $\{U_i : i \in \mathbb{N}\}$ be a collection of non-empty open subsets in S such that $U_0 \supset U_1 \supset U_2 \supset \cdots$. We are required to show that $\cap\{\text{cl}_S U_i : i \in \mathbb{N}\} \neq \varnothing$.

Suppose $\cap\{\text{cl}_S U_n : n \in \mathbb{N}\} = \varnothing$.

For each n, choose a point $x_n \in U_n$ to form a closed infinite sequence $\{x_n : n \in \mathbb{N}\}$. For each x_n we can inductively choose a closed neighborhood, V_n, of x_n such that $V_n \subseteq U_n$ and $V_m \cap V_n = \varnothing$ whenever $m \neq n$.

Since S is completely regular, there is, for each n, a continuous function $g_n : S \to [0,1]$ such that $g_n(x_n) = 1$ and $g_n[S \backslash V_n] = \{0\}$.

Let $f_n = ng_n$. Then f_n maps S into $[0, n]$. We define the function,

$$h(x) = \sum_{n \in \mathbb{N}} f_n(x)$$

on S.

We investigate properties of this function carefully. First see that h is an unbounded function.

Also see that, for any $u \in S$, then $f_n(u) \neq 0$ for no more than one n. Then $h(u) = f_m(u)$, for some m.

We now proceed to show that h is a continuous function. Let $u \in S$. We claim that u has an open neighborhood W_u which intersects at most a single V_n.

Proof of claim: Suppose every open neighborhood in $h[S]$ of u intersects infinitely many V_n's. Then u would belong to $\cap\{\text{cl}_S U_n : n \in \mathbb{N}\}$ which has been hypothesized to be empty. So u has an open neighborhood, say W, which intersects finitely many V_n's, say $\{V_{n_1}, V_{n_2}, \ldots, V_{n_k}\}$. Since each V_{n_i} is closed, u has an open neighborhood W_u which intersects only V_{n_1}, as claimed.

Then, for some m, h agrees with f_m on a neighborhood W_u of u.

Let $h(y) \in h[S]$ and U be an open neighborhood of $h(y)$ in $\mathbb{R}$. Then $h(y) = f_m(y)$ for precisely one $m \in \mathbb{N}$. We have shown that $h = f_m$ on some open neighborhood, W_y, of y in S. Since f_m is continuous on W_y, then there exists an open neighborhood B of y such that $h(x) = f_m(x)$ on B and $f_m[B] \subseteq U$ So $h[B] \subseteq U$. We conclude that h is continuous at each point $y \in S$. Then h is continuous on S, as desired.

Since h is unbounded, we have a contradiction. So $\cap\{\text{cl}_S U_i : i \in \mathbb{N}\} \neq \varnothing$.

$(2 \Rightarrow 3)$ Follows from $2 \Leftrightarrow 3$ of Theorem 17.10.

$(3 \Rightarrow 1)$ This follows from Theorem 17.11. This does not require complete regularity.

Concept Review

1. Define a sequentially compact space.
2. If S is a second countable or metrizable topological spaces which satisfies the sequentially compact property, name two other properties which are equivalent properties in such spaces.
3. Describe a topological space in which sequentially compact subsets are precisely the closed and bounded ones.
4. Is it true that, in all metric spaces, the compact subsets are always closed and bounded?
5. Is it true that, in all metric spaces, the the closed and bounded subsets are always compact?
6. Given a sequentially compact property and the countably compact property, one always implies the other. Which one? Provide a counter example for the one that doesn't.
7. Define a "pseudocompact space".
8. Is a sequentially compact space pseudocompact?
9. Is a countably compact space necessarily pseudocompact?
10. Are there any spaces which are pseudocompact and not compact? If so, give an example.
11. Are there any spaces which are compact and not pseudocompact? If so, give an example.

Exercises

1. Show that, if F is a non-empty closed subset of a sequentially compact space, then F is sequentially compact.
2. Let $f : S \to T$ denote a function mapping S into a Hausdorff space T. Show that, if f is continuous on S and S is a sequentially compact space, then $f[S]$ is sequentially compact.
3. Suppose $f : S \to \mathbb{R}$ is a continuous function mapping the sequentially compact space into $\mathbb{R}$, equipped with the usual topology. Can the function f be "onto" $\mathbb{R}$?

Chapter 18

Locally Compact Spaces

Abstract
We will discuss another way of weakening the compact property without sacrificing many of the properties we find desirable in a topological space. It is called the "locally compact" property. Rather than have the compact property assigned to the whole space, we assign it on the elements of a certain neighborhood base of each point. After defining this property formally we examine what are its main characteristics. Even though local compactness is usually seen combined with the Hausdorff property our formal definition will be in its most general form. We have seen that all Hausdorff compact spaces are normal. But locally compact Hausdorff spaces will be proven to be only completely regular. All metrizable spaces will also be seen to be locally compact and Hausdorff. These concepts will set the stage for the future study of "compactifications of locally compact Hausdorff spaces".

18.1 The Locally Compact Property

We now present a formal definition of the locally compact property. We define this property for arbitrary topological spaces. We alert the reader to the fact that many authors define local compactness for Hausdorff spaces only, probably because it simplifies its definition, and, more often then not, it is mostly referred to in the context of Hausdorff spaces. We prefer to define it in its most general form.

Definition 18.1 Let S be a topological space. We say that S is *locally compact* if, for every point, x, in S there is a compact set, K, such that $x \in \text{int}_S K \subseteq K$.

When a particular topological property is applied only to some unspecified neighborhood of each point of a space rather than on the whole space, we traditionally precede the name of this property with the adverb "locally". In this particular case, we are referring to a *compact* neighborhood.

18.2 A Characterization of Local Compactness

Note that, in the above definition, in the particular case where the space S is *not* Hausdorff, it may occur that the compact neighborhood is not a closed subset of S. The following characterization referring specifically to *locally compact spaces which are also Hausdorff* allows us to say, "there is an open set U with compact closure such that $x \in U \subseteq \text{cl}_S U$".

Theorem 18.2 *Let S be a Hausdorff topological space. The space, S, is locally compact if and only if each point, x, in S has an open neighborhood base, $\mathscr{B}_x = \{B_i : i \in I\}$, such that $\text{cl}_S B_i$ is compact.*

Proof. We are given that S is a topological space.

($\Leftarrow$) If B is an open neighborhood of a point, x in S, such that $\text{cl}_S B$ is compact then, by definition, x has a compact neighborhood. So, S is locally compact.

($\Rightarrow$) We are given that S is locally compact and Hausdorff. Let $p \in S$ and U be an open neighborhood of p and K be any compact neighborhood of p. Then $p \in (\text{int}_S K) \cap U \subseteq K$. Since K is compact and S is Hausdorff, then, by Theorem 14.3(b) (iii), K is a regular subspace. This means that there exists an open set V such that

$$p \in V \subseteq \text{cl}_K V \subseteq (\text{int}_S K) \cap U \subseteq K$$

Since $\text{cl}_K V$ is closed in the compact set K, it must be compact. So, $p \in \text{int}_S \text{cl}_K V \subseteq \text{cl}_K V \subseteq U$. We can conclude that each point of S has an open neighborhood base whose elements have a compact closure.

In the proof of the above theorem, note why the Hausdorff property on S is required.

The above characterization of local compactness on Hausdorff spaces allows us to quickly see that every compact Hausdorff space, S, is locally compact: Since compact Hausdorff spaces are regular, we can construct a compact neighborhood of a point, p, inside any open neighborhood of p.

Example 1. Consider the space $\mathbb{R}$ with its usual topology. Show that $\mathbb{R}$ is locally compact.

Solution: Let $\mathscr{B}_p = \{(a,b) : a < p < b\}$ be an open neighborhood base for the point p. For each $(a,b) \in \mathscr{B}_p$, there exists $\varepsilon > 0$, such that,

$$p \in \left(p - \frac{\varepsilon}{3}, p + \frac{\varepsilon}{3}\right) \subseteq \left[p - \frac{\varepsilon}{3}, p + \frac{\varepsilon}{3}\right] \subseteq (p - \varepsilon, p + \varepsilon) \subseteq (a,b)$$

Since the closed interval in this chain of containments is the closure of the first interval and is compact, then $\mathbb{R}$, with the usual topology, is locally compact.

The space $\mathbb{R}^2$ *is locally compact.* Note that $\mathbb{R}^2$ is also locally compact since, for $(a,b) \in \mathbb{R}^2$, $[a+\varepsilon, a-\varepsilon] \times [b+\varepsilon, b-\varepsilon]$ is compact and

$$\begin{aligned}(a,b) &\in \text{int}_{\mathbb{R}^2}[a+\varepsilon, a-\varepsilon] \times [b+\varepsilon, b-\varepsilon] \\ &= (a+\varepsilon, a-\varepsilon) \times (b+\varepsilon, b-\varepsilon)\end{aligned}$$

Example 2. Consider the space, $\mathbb{Z}$, of all integers with the discrete topology. Show that $\mathbb{Z}$ is locally compact.

Solution: If $n \in \mathbb{Z}$, and U is an open set containing n, $\text{cl}_{\mathbb{Z}}\{n\}$ is a compact neighborhood of n, so,

$$n \in \{n\} = \text{int}_{\mathbb{Z}}\{n\} = \text{cl}_{\mathbb{Z}}\{n\} = \{n\} \subseteq U$$

Then $\mathbb{Z}$, with the discrete topology, is locally compact. In fact, every discrete space is locally compact.

18.3 Some Basic Properties of Local Compactness

We now examine some invariance properties. A subset of a locally compact space does not always inherit the locally compact property from its superset, as we shall soon see. The locally compact property is carried over from the domain of a continuous function to its range provided the *function is an open mapping.* For a product space, the locally compact property is carried over from the factors to the product, provided all factors are locally compact and *at most* finitely many of those factors are *not* compact. We now immediately prove these facts.

Theorem 18.3 *Let S be a locally compact Hausdorff topological space.*

(a) *If F is a closed subspace of S, then F is locally compact. So a* closed *subspace inherits the locally compact property from its Hausdorff superset.*
(b) *If U is an open subspace of S, then U is locally compact. So an* open *subspace inherits the locally compact property from its Hausdorff superset.*
(c) *In any Hausdorff topological space, the intersection of two locally compact subspaces of S is locally compact. In particular, if S is locally compact and Hausdorff, any subset which is the intersection of an open set and a closed one is a locally compact subspace.*
(d) *Let S be any Hausdorff topological space which contains a non-empty subspace, W. If W is locally compact, then W is the intersection of an open subset with a closed subset of S.*

Proof. We are given that S is a topological space.

(a) We are given that F is a closed subset of the locally compact Hausdorff space S. Let $x \in F$ and V be an open neighborhood of x in S. Since S is locally compact and Hausdorff, there exist an open neighborhood U of x such that $x \in U \subseteq \mathrm{cl}_S U \subseteq V$.

    Then $x \in U \cap F \subseteq \mathrm{cl}_S U \cap F \subseteq V \cap F$, where $\mathrm{cl}_S U \cap F$ is a closed subset of the compact set $\mathrm{cl}_S U$, hence is a compact neighborhood of x in F. So, F is locally compact.
(b) Let V be an open subset of the locally compact Hausdorff space S and U be an open neighborhood of $x \in V$. Then $U \cap V$ is

an open neighborhood of x in S. Since S is locally compact and Hausdorff, there exist an open neighborhood, W, such that $x \in W \subseteq \text{cl}_S W \subseteq U \cap V$, where $\text{cl}_S W$ is compact in V. So, V is locally compact, as required.

(c) We are given that S is a Hausdorff space containing the two locally compact subspaces U and V with non-empty intersection. Since S is Hausdorff, so are U, V, and $U \cap V$. Let $x \in U \cap V$ and Z be an open subset of S such that $x \in Z$. Then there exists an open subset, A, in U such that

$$x \in A \subseteq \text{cl}_U A \subseteq Z \cap U \subseteq U$$

and an open subset, B, in V such that

$$x \in B \subseteq \text{cl}_V B \subseteq Z \cap V \subseteq V$$

where $\text{cl}_U A$ and $\text{cl}_V B$ are compact; because of the Hausdorff property, they are closed. Then $U \cap V$ is locally compact, as required.

Combining parts a, b, and c the second part of the statement quickly follows.

(d) Suppose W is a subset of a Hausdorff topological space. Suppose W is locally compact. We are required to show that W is the intersection of an open set and a closed set in S. If we can show that W is open in S then $W = W \cap \text{cl}_S W$, an intersection of an open and closed subset of S, and we will be done. To show W is open we will prove that every point in W belongs to an open subset of S, contained in W.

Let $x \in W$. Since W is locally compact and Hausdorff, S contains an open neighborhood, A, of x such that

$$x \in A \cap W \subseteq \text{cl}_W(A \cap W) = \text{cl}_S(A \cap W) \cap W$$

where $\text{cl}_W(A \cap W)$ is a closed and compact set in W and so $\text{cl}_S(A \cap W) \cap W$ is also closed and compact in S. Since $A \cap W$ is contained in the closed set, $\text{cl}_S(A \cap W) \cap W$, it must follow that $\text{cl}_S(A \cap W) \subseteq \text{cl}_S(A \cap W) \cap W$, and so, $\text{cl}_S(A \cap W) \subseteq W$. So, $x \in \text{int}_S(\text{cl}_S(A \cap W)) \subseteq W$. Then each point of W is contained in an open neighborhood of S and W is open. So, $W = W \cap \text{cl}_S W$, as required.

Corollary 18.4 *Let S be a compact Hausdorff topological space and D be a dense subset of S. Then the subset, D, of S is locally compact if and only if D is an open subset of S.*

Proof. We are given that S is a compact Hausdorff topological space and D is a dense subset of S.

($\Rightarrow$) Suppose the dense subset, D, is a locally compact subspace of S. Then, by Corollary 18.4, D is an open subset of $\text{cl}_S D = S$. So, D is open in S.

($\Leftarrow$) Suppose D is a an open dense subset of the compact space, S. Since S is compact, it is locally compact. As shown in Theorem 18.3, open subsets of locally compact sets are locally compact. So, D is locally compact. We are done.

Example 3. Determine whether the set, $\mathbb{Q}$, of all rationals is locally compact or not.

Solution: Suppose $\mathbb{Q}$ is locally compact. We know that $\mathbb{Q}$ is Hausdorff. Suppose $x, y \in \mathbb{Q}$ where $x < y$. Then $(x, y) \cap \mathbb{Q}$ is open in $\mathbb{Q}$. Let $a \in \mathbb{Q}$ where $x < a < y$. Then there exists a compact neighborhood F of a such that

$$a \in (c, d) \cap \mathbb{Q} \subseteq F \subseteq (x, y) \cap \mathbb{Q}$$

Let r be an irrational such that $c < r < d$. Then there is a sequence in $(c, d) \cap \mathbb{Q}$ which converges to $r \notin F$. Then F is not closed in $\mathbb{Q}$ and so cannot be compact. Then $\mathbb{Q}$ cannot be locally compact.

We know that the compact property is carried over from the domain to the codomain of a continuous function. A similar result holds for the locally compact property *but only if the function is both continuous and open.*

Theorem 18.5 *Let S be a locally compact topological space and suppose T is any topological space. If $f : S \to T$ is a continuous open function mapping S into T then $f[S]$ is locally compact.*

Proof. We are given that S is a locally compact topological space and $f : S \to T$ is a continuous open function.

We are required to show that $f[S]$ is locally compact. Let $u \in f[S]$ and U be any open neighborhood of u. Suppose $v \in f^{\leftarrow}[\{u\}] \subseteq f^{\leftarrow}[U]$. Then, by hypothesis, v has a compact neighborhood, say K, such that $v \in \text{int}_S K \subseteq K \subseteq f^{\leftarrow}[U]$. Since f is open and continuous and

$$u = f(v) \in f[\text{int}_S K] \subseteq f[K] \subseteq f[f^{\leftarrow}[U]] = U$$

where $f[\text{int}_S K]$ is open and $f[K]$ is compact, then every point in $f[S]$ has a compact neighborhood. So, $f[S]$ is locally compact.

Note that the Hausdorff property is not required in the above statement.

Example 4. Show that the closed continuous image of a locally compact space need not be locally compact.

Solution: Recall that in the usual topology, $\mathbb{R}^2$ is locally compact. Let $A = \{(x, 0) : x \in \mathbb{R}\}$. Let $\mathscr{D}$ denote the decomposition of $\mathbb{R}^2$,

$$\mathscr{D} = \{A\} \cup \{\, \{(x, y)\} : y \neq 0\,\}$$

Given the quotient map, $q : \mathbb{R}^2 \to \mathscr{D}$, $\mathscr{D}$ is equipped with the quotient topology induced by q.

Claim 1: We claim that function q is a closed map. Suppose B is a closed subset of $\mathbb{R}^2$. Case 1: Suppose $B \cap A = \varnothing$. Then $q^{\leftarrow}[q[B]] = B$. So, $q^{\leftarrow}[\mathscr{D} \setminus q[B]] = \mathbb{R}^2 \setminus q^{\leftarrow}[q[B]] = \mathbb{R}^2 \setminus B$ which is open in $\mathbb{R}^2$. Thus, $\mathscr{D} \setminus q[B]$ is open in $\mathscr{D}$ with respect to the quotient topology. So, $q[B]$ is closed. Case 2: Suppose $B \cap A \neq \varnothing$. Then $q^{\leftarrow}[q[B]] = A \cup B$. So,

$$q^{\leftarrow}[\mathscr{D} \setminus q[B]] = \mathbb{R}^2 \setminus q^{\leftarrow}[q[B]] = \mathbb{R}^2 \setminus (A \cup B)$$

an open subset of $\mathbb{R}^2$. Then $\mathscr{D} \setminus q^{\leftarrow}[B]$ is open in $\mathscr{D}$ (by definition of quotient topology); so $q[B]$ is closed. Claim 1 is proved.

Claim 2: We claim that $\mathscr{D}$ is not locally compact. Suppose $\mathscr{S}$ is an open neighborhood of A. It suffices to show that $\text{cl}_{\mathscr{D}} \mathscr{S}$

is not compact. Then $V = q^{\leftarrow}[\mathscr{S}]$ is an open subset of $\mathbb{R}^2$.

$$\begin{aligned} q[\text{cl}_{\mathbb{R}^2} V] &= q[\text{cl}_{\mathbb{R}^2} q^{\leftarrow}[\mathscr{S}]] \\ &= \text{cl}_{\mathscr{D}}(q[q^{\leftarrow}[\mathscr{S}]]) \\ &= \text{cl}_{\mathscr{D}} \mathscr{S} \end{aligned}$$

Consider the open neighborhood, $V \cup \{(x, y) \in \mathbb{R}^2 : -n < x < n\} \backslash A$, of A in $\mathbb{R}^2$.

Then $q\left[V \cup \{(x, y) \in \mathbb{R}^2 : -n < x < n\} \backslash A\right]$ is an open neighborhood of A in $\mathscr{D}$, for each $n \in \mathbb{N}$.

Then the collection

$$\{q\left[V \cup \{(x, y) \in \mathbb{R}^2 : -n < x < n\} \backslash A\right] : n \in \mathbb{N}\}$$

forms an open cover of $\text{cl}_{\mathscr{D}} \mathscr{S}$, with no finite subcover. So $\text{cl}_{\mathscr{D}} \mathscr{S}$, is not compact. We conclude that $\mathscr{D}$ is not locally compact.

In the following statement, we show that finite products of locally compact spaces are locally compact, but one has to be cautious when considering infinite products of locally compact spaces.

Theorem 18.6 *Let $\{S_i : i \in I\}$ be a family of topological spaces and $S = \prod_{i \in I} S_i$ be a product space. Then S is locally compact if and only if every factor, S_i, is locally compact and at most, finitely many of the factors are* not *compact.*

Proof. We are given that $\{S_i : i \in I\}$ is a family of topological spaces and $S = \prod_{i \in I} S_i$ is a product space.

($\Rightarrow$) Suppose $S = \prod_{i \in I} S_i$ is locally compact. Since every projection map is open (see page 149), then, by Theorem 18.5, every factor is locally compact. We now claim that, at most, finitely many of the factors are not compact spaces.

If $u \in S$, then, since S is locally compact, there exists a compact set K in S, such that $u \in \text{int}_S K \subseteq K$. Then there exists a finite subset, F, of I such that

$$u \in U = \cap\{\pi_i^{\leftarrow}[U_i] : i \in F\} \subseteq \text{int}_S K$$

where U_i is an open subset of S_i. We claim that, if $i \notin F$ then S_i is compact. Consider $i \notin F$: Then $\pi_i[U] = S_i \subseteq \pi_i[K]$ (since $U \subseteq K$).

So $S_i = \pi_i[K]$. Since π_i is continuous and K is compact, then S_i is compact, for all $i \notin F$, as claimed. Done.

($\Leftarrow$) Suppose every factor, S_i, is locally compact and at most finitely many are not compact. We are required to show that the product space, S, is locally compact.

Let $\langle x_i \rangle_{i\in I} \in S = \prod_{i\in I} S_i$ and $U = \cap\{\pi_i^{\leftarrow}[U_i] : i \in F\}$ be a basic open neighborhood of $\langle x_i \rangle_{i\in I}$. Suppose F_c is the largest subset of I, such that, if $i \in F_c$, S_i is not compact. Then, by hypothesis, F_c is finite. Let $F_1 = F \cup F_c$. Then F_1 is finite. Let $U_1 = \cap\{\pi_i^{\leftarrow}[U_i] : i \in F_1\} \subseteq U$.

We claim that $\langle x_i \rangle_{i\in I}$ has a compact neighborhood, K, which is contained in U_1.

Given that, for each $i \in F_1$, S_i is locally compact, there must be a compact set, K_i, such that $x_i \in \text{int}_{S_i} K_i \subseteq K_i \subseteq U_i$.

Then

$$\langle x_i \rangle_{i\in I} \in \cap\{\pi_i^{\leftarrow}[\text{int}_S K_i] : i \in F_1\} \subseteq \cap\{\pi_i^{\leftarrow}[K_i] : i \in F_1\} = K \subseteq U_1$$

where K is a neighborhood of $\langle x_i \rangle_{i\in I}$. See that, for each $i \notin F_1$, X_i is a compact factor of K, and, for each $i \in F$, K_i, is a compact factor of K. So, K is a compact neighborhood of $\langle x_i \rangle_{i\in I}$, as claimed. So, S is locally compact.

18.4 More on the Hausdorff Locally Compact Property

In the Embedding theorem part III, we arrived at the conclusion that "any completely regular space can be densely embedded in a Hausdorff compact space". We would like to state a similar result involving Hausdorff locally compact spaces. That is, we will show that "any locally compact Hausdorff space, S, can be embedded in a Hausdorff compact space".

But in this particular context the compact space in question is *not* the product of closed intervals of $\mathbb{R}$, as one might expect. This compact space, denoted by, S_ω, is constructed by adding a single point to S and then defining an appropriate topology on it so that it is compact.

Theorem 18.7 *Let (S,τ) be a Hausdorff topological space and let ω be a point which does not belong to S. The set ωS is defined as*

$$\omega S = S \cup \{\omega\}$$

The topology, τ_ω, on ωS is defined as follows: Firstly, $\tau \subseteq \tau_\omega$. Secondly, if $\omega \in U \subseteq \omega S$ and $\omega S \backslash U$ is a compact subset of S, then $U \in \tau_\omega$.

(a) *Then the family of sets, τ_ω, thus defined, is a valid topology on ωS.*
(b) *The topological space, $(\omega S, \tau_\omega)$, is a compact space.*
(c) *The compact space, $(\omega S, \tau_\omega)$, densely contains a homeomorphic copy of S.*
(d) *The compact space, $(\omega S, \tau_\omega)$, is Hausdorff if and only if S is locally compact.*

Proof. We are given that (S,τ) is a Hausdorff topological space and that ω is a point which does not belong to S. Also, $\omega S = S \cup \{\omega\}$.

(a) The proof is routine and so is left as an exercise.
(b) We are given $(\omega S, \tau_\omega)$. Let $\mathscr{U} = \{U_i : i \in I\}$ be an open cover of ωS.

    By hypothesis, there exists, U_k, such that $\omega \in U_k$ and $S \backslash U_k$ is compact. Then $S \backslash U_k$ has a finite subcover, $\{U_i : i \in F\}$. Then $\{U_i : i \in F\} \cup \{U_k\}$ is a finite subcover of ωS and so ωS is compact.
(c) The identity map, $i : S \to S \cup \{\omega\}$, is easily seen to be a continuous one-to-one open map onto $i[S] = S$.
(d) We are given that ωS is compact.

    ($\Rightarrow$) Suppose ωS is Hausdorff. Then S is a Hausdorff subspace (since the Hausdorff property is hereditary). Also, for any $x \in S$, there exists disjoint open neighborhoods U and V such that $\omega \in U$ and $x \in V$. Then S is open and dense in ωS. By Corollary 18.4, S is locally compact.

    ($\Leftarrow$) We are given that S is a Hausdorff locally compact dense subspace of the compact space, ωS. We are required to show that ωS is Hausdorff. It follows immediately that distinct pairs of points in S are contained in disjoint open neighborhoods of ωS.

Suppose $p \in S$. Since S is Hausdorff and locally compact, there exists an open neighborhood U of p such that $p \in U \subseteq \text{cl}_S U \subseteq S$, where $\text{cl}_S U$ is a compact neighborhood of p. By hypothesis, $S \backslash \text{cl}_S U$ is an open neighborhood of ω disjoint from U. So, ωS is Hausdorff, as required.

We now show that the local compactness property on a space S, when combined with the Hausdorff property satisfies the completely regular separation property.

Theorem 18.8 *A locally compact Hausdorff space is completely regular.*

Proof. We are given that S is a Hausdorff locally compact topological space. Then S is embedded in the compact Hausdorff topological space $\omega S = S \cup \{\omega\}$. Then by Theorem 14.3, ωS is normal and so is completely regular. The completely regular property is hereditary, so S is completely regular.

From the above result we again argue that if S is locally compact and Hausdorff it can be densely embedded in at least two essentially different Hausdorff compact spaces.

We now know that S is embedded in $\omega S = S \cup \{\omega\}$; secondly, from the Embedding theorem 14.7, we know that...

> *a locally compact Hausdorff space, S, can be densely embedded in a compact Hausdorff subspace of a cube.*

18.5 On Spaces Which are σ-Compact

Some spaces can be described as being the union of a countably infinite number of compact sets. Such spaces need not be compact but still have properties that are worth discussing. In particular, we will see that a locally compact Hausdorff space which is a countable union of compact sets is a Lindelöf Hausdorff space.

Definition 18.9 Let S be a topological space. We say that S is σ-compact if S is the union of countably many compact sets.

Theorem 18.10 *Suppose S is a locally compact Hausdorff space. Then S is σ-compact if and only if S is Lindelöf.*

Proof. We are given that S is a locally compact Hausdorff topological space.

($\Leftarrow$) Suppose S is Lindelöf. We are required to show that S is σ-compact. Since S is locally compact and Hausdorff, we can construct an open covering of

$$\mathscr{B} = \{U_x : x \in S\}$$

of sets with compact closures,

$$\mathscr{B}^* = \{\text{cl}_S U_x : x \in S\}$$

Let $\{U_{x_i} : i \in \mathbb{N}\}$ be a countable subcover of $\mathscr{B}$. Then

$$S = \cup\{U_{x_i} : i \in \mathbb{N}\} \subseteq \cup\{\text{cl}_S U_{x_i} : i \in \mathbb{N}\}$$

So, S is σ-compact, as required.

($\Rightarrow$) Suppose S is σ-compact. We are required to show that S is Lindelöf.

Then $S = \cup\{C_i : i \in \mathbb{N},\ C_i \text{ is compact}\}$, a countable union of compact sets.

Consider the compact subset, C_0.

Claim: That there is an open neighborhood, U_0, with compact closure, $\text{cl}_S U_0$, such that $C_0 \subseteq U_0 \subseteq \text{cl}_S U_0$.

Proof of claim: Since S is locally compact and Hausdorff and C_0 is compact, any open cover, $\mathscr{U}$, of C_0 has a finite cover $\{V_i : i = 1 \ldots k\}$, such that each $\text{cl}_S V_i$ is compact for $i = 1$ to k. Let $U_0 = \cup\{V_i : i = 1 \ldots k\}$. Then

$$C_0 \subseteq U_0 \subseteq \text{cl}_S U_0 = \cup\{\text{cl}_S V_i : i = 1 \ldots k\}$$

is compact in S, as claimed.

We can inductively construct $\{U_i : i \in \mathbb{N}\}$, such that each closure, $\text{cl}_S U_i$, is compact where

$$\begin{aligned}
C_0 \quad &\subseteq U_0 \subseteq \text{cl}_S U_0 \\
C_1 \cup \text{cl}_S U_0 &\subseteq U_1 \subseteq \text{cl}_S U_1 \\
C_2 \cup \text{cl}_S U_1 &\subseteq U_2 \subseteq \text{cl}_S U_2 \\
&\vdots \qquad \vdots \\
C_i \cup \text{cl}_S U_{i-1} &\subseteq U_i \subseteq \text{cl}_S U_i \\
&\vdots \qquad \vdots
\end{aligned}$$

Set

$$K_i = \text{cl}_S U_i \subseteq U_{i+1} \subseteq \text{cl}_S U_{i+1} = K_{i+1}$$

we obtain a family of compact sets, $\{K_i : i \in \mathbb{N}\}$ where $K_{i-1} \subseteq K_i$ and $S = \cup\{K_i : i \in \mathbb{N}\}$.

Suppose $\mathscr{V} = \{V_j : j \in J\}$ is an open cover of S. To show Lindelöf it suffices to show that $\mathscr{V}$ has a countable subcover.

Since each K_i is compact, there exist a finite subset, $F_i \subseteq J$ such that

$$K_i \subseteq \cup\{V_{j_k} : k \in F_i\}$$

Then $\mathscr{V}^* = \{V_{j_k} : k \in F_i,\ i \in \mathbb{N}\}$ is a countable subfamily of $\mathscr{V}$.

Since $S = \cup_{i,k}\{V_{j_k} : k \in F_i,\ i \in \mathbb{N}\}$, $\mathscr{V}^*$ is a countable subcover of S. So, S is Lindelöf, as required.

Example 5. Verify if the Moore plan (Niemytzki's plane), Γ, is locally compact or not.

Solution: The Moore plane, Γ, has $\{(x, y) \in \mathbb{R}^2 : y \geq 0\}$ as underlying set. Basic neighborhoods of points in $\mathbb{R}^2$, and, if $z = (x, 0)$, then $\{z\} \cup A$ is a basic neighborhood of z in $\mathbb{R}^2$ where A is an open disc tangent to the x-axis at z.

We claim that the Moore plane is not locally compact. If it was, then $(0,0)$ would have a compact basic open neighborhood with compact closure. If

$$V = \{(0,0)\} \cup \{(x,y) : x^2 + (y - \delta^2) < \delta^2\}$$

is such a basic neighborhood then

$$\text{cl}_\Gamma V = \{(x,y) : x^2 + (y - \delta^2) \leq \delta^2\}$$

We claim that $\text{cl}_\Gamma V$ is not compact. To see this consider the open cover

$$\{V\} \cup \{\, \{(x,y) \in \Gamma : y > 1/n\} : n \in \mathbb{N}, n > 0\,\}$$

of $\text{cl}_\Gamma V$. It has no finite subcollection which covers $\text{cl}_\Gamma V$.

Example 6. Is the set $\mathbb{R}$ line equipped with the lower limit topology, $(\mathbb{R}, \tau_S)$, locally compact?

Solution: The answer is no. Recall that an open base for τ_S, is $\mathscr{B} = \{[a,b) : a < b\}$. Let $r \in \mathbb{R}$. Suppose there is a $\delta > 0$, such that $\text{cl}_\mathbb{R}[r, r+\delta)$ is a compact neighborhood of r. See that

$$\mathbb{R}\backslash[r, r+\delta) = \cup_{r=1}^{\infty}[r-n, r) \cup [r+\delta, r+\delta+n]$$

is open, so $\text{cl}_\mathbb{R}[r, r+\delta) = [r, r+\delta)$. But $[r, r+\delta)$ is not compact since

$$\{[r, r+\delta - 1/n) : n \in \mathbb{N}, n > 0\}$$

is an open cover of $[r, r+\delta)$ with no finite subcover.

Concept Review

1. Define the locally compact property on a topological space.
2. State a characterization of Hausdorff locally compact spaces.
3. If S is locally compact and Hausdorff what kind of subsets are guaranteed to inherit the locally compact property?

4. If T is a locally compact subspace of the Hausdorff space, S, what property does T satisfy?
5. If D is a dense subset of a compact Hausdorff space, S, what can we say about D?
6. Provide an easy example of a non-locally compact space.
7. Are continuous images of locally compact spaces necessarily locally compact? Explain.
8. If S is a product space which is locally compact. What can say about its factors?
9. What conditions must be satisfied if we want the locally compact factors to be carried over to their product.
10. Describe the Hausdorff compact space, ωS, which contains a dense homeomorphic copy of a locally compact Hausdorff space, S.
11. If S is a Hausdorff locally compact space, what other separation axiom does it satisfy?
12. If S is a Hausdorff locally compact space, is there another compact space, other than ωS, which contains a dense copy of S?
13. Define a σ-compact space.
14. If S is locally compact and Hausdorff. In such a case provide a characterization of the σ-compact property.

Exercises

1. Suppose $f : S \to T$ is a continuous open function mapping a locally compact space, S, onto a space T. Show that any compact subspace, K, of T is the image under f of some compact subspace, F, of S.
2. Show that the real numbers equipped with the upper limit topology (Sorgenfrey line) is not locally compact.
3. A *perfect function*, $f : S \to T$, is a continuous closed and onto function which pulls back each point in T to a compact set in S. Suppose S is Hausdorff and T is locally compact. Show that $f : S \to T$ is a perfect function if and only if f pulls back compact sets in T to compact sets in S.

4. Let S be a locally compact regular space and K be a closed and compact subset of S. Suppose K is a subset of an open set, U, in S. Show that there is some compact set V, such that $F \subseteq \text{int}_S V \subseteq V \subseteq U$.
5. Show that the Moore plane is not locally compact.
6. Suppose S is a Hausdorff space which contains a dense subset, D. Suppose $x \in D$. Show that, if V is compact neighborhood of x in D, then V is a neighborhood of x in S.

Chapter 19

Paracompact Topological Spaces

Abstract
We will develop in this section some familiarity with the paracompact property. After giving a formal definition, we provide a few examples and discuss its invariance properties. Finally we show that all metrizable spaces are paracompact.

19.1 Paracompact Topological Spaces

We now introduce a last important class of topological spaces closely related to the family of compact spaces. Its importance was discovered in the role it played in characterizations of metrizable spaces. We begin by introducing some new terminology. Note that some of what we present here may appear somewhat familiar since we have already (informally) introduced the concept of a "locally finite" family of sets before (on page 136). For convenience we formally reintroduce the concept here along with associated terms.

Definition 19.1 Let S be a topological space.

(a) Let $\mathscr{U} = \{U_i : i \in I\}$ be a family of subsets of S. We say that $\mathscr{U}$ is a *locally finite family of sets* if each point in S has a neighborhood which intersects only finitely elements of $\mathscr{U}$.[1]

[1] The word "neighborhood-finite" is sometimes used instead of "locally finite".

(b) We say that $\mathscr{V} = \{V_j : j \in J\}$ *is a refinement of*, (or refines) the family, $\mathscr{U} = \{U_i : i \in I\}$, if every element of $\mathscr{V}$ is contained in some element of $\mathscr{U}$. If the elements of $\mathscr{V}$ are open, then we will more specifically say that $\mathscr{V}$ *is an open refinement* of $\mathscr{U}$.

(c) Let $\mathscr{U}$ be a family of subsets of the topological space S. We say that $\mathscr{U}$ *has a locally finite open refinement* if there is a family, $\mathscr{V}$, of open subsets in S which both refines $\mathscr{U}$ and satisfies the locally finite property in S.

Whether a family of sets is locally finite or not in the space, S, depends on the topology of S (since the definition refers to neighborhoods of S). Whether $\mathscr{V}$ refines $\mathscr{U}$ or not refers to a set-theoretic property which characterizes a particular relationship between between the elements of $\mathscr{V}$ and those of $\mathscr{U}$. *Open refinement* adds a particular topological characteristic to the elements of the refinement.

Note that if, for any open cover $\mathscr{U}$ of a space S, there is a finite subfamily, $\mathscr{F}$, of $\mathscr{U}$ which covers the space S, then by definition, $\mathscr{F}$ refines $\mathscr{U}$ and, again by definition, $\mathscr{F}$ is locally finite in S.

We can then say that...

> *if S is compact and $\mathscr{U}$ is an open cover then $\mathscr{U}$ has a locally finite open refinement $\mathscr{V}$.*

In the particular case described above, $\mathscr{V}$ simply turns out to be a finite subfamily of $\mathscr{U}$. We now define the main subject of this section.

Definition 19.2 Let S be a topological space. The space S is said to be a *paracompact* space if, for any open cover $\mathscr{U}$ of S, there exists a locally finite open cover, $\mathscr{V}$, of S which refines $\mathscr{U}$.[2]

We alert the reader to the fact that, in some books, the Hausdorff property is incorporated into the formal definition of the paracompact property, in the sense that, for these authors, all paracompact

[2]Note that $\mathscr{V}$ satisfies four conditions: (1) $\mathscr{V}$'s elements must be open sets, (2) $\mathscr{V}$ must refine $\mathscr{U}$, (3) $\mathscr{V}$ must be locally finite, (4) $\mathscr{V}$ covers S.

spaces are hypothesized to be Hausdorff. In this book, a paracompact space is Hausdorff only when we explicitly state it as such.

Theorem 19.3 *Any compact space is a paracompact space.*

Proof. This statement has been proven in the paragraph immediately preceding the definition above.

Since all compact spaces are paracompact, we then know of a large family of topological spaces which are paracompact. We consider the following example of a non-compact paracompact space.

Example 1. Let S be an infinite space equipped with the discrete topology. Since $\mathscr{V} = \{\{x\} : x \in S\}$ is an open cover with no subcover, the space, S, is not compact. Show that S is, nevertheless, paracompact.

Solution: Let $\mathscr{U} = \{U_i : i \in I\}$ be an arbitrary open cover of S. To show that S is paracompact it suffices to show that $\mathscr{U}$ has a locally finite open refinement. From the definition, the family $\mathscr{V} = \{\{x\} : x \in S\}$ is an open refinement of $\mathscr{U}$ (since every one of its elements, $\{x\}$, is a subset of some set, U_i, in $\mathscr{U}$). It then suffices to show that $\mathscr{V}$ is a locally finite family of sets. Let $x \in S$. Consider the open neighborhood, $B = \{x\}$, of x. The neighborhood B intersects only one element, $\{x\} \in \mathscr{V}$. So, $\mathscr{V}$ is a locally finite family of sets in S.

So, $\mathscr{V}$ is a locally finite open refinement of $\mathscr{U}$. We can conclude that S is paracompact.

Example 2. Consider the space $S = \mathbb{R}$ equipped with the usual topology. We know that this metrizable space is unbounded and so is not compact. Show that S is paracompact.

Solution: Suppose we have an open cover, $\mathscr{U} = \{U_i : i \in I\}$, of S. We are required to construct an open refinement of $\mathscr{U}$ which is locally finite.

Firstly, we will first construct an open refinement, $\mathscr{V}$, of $\mathscr{U}$ which covers S.

For each $n \in \mathbb{N}\backslash\{0\}$, let $(-n, n)$ denote an open interval centered at 0 of radius n. Since, for each n, $[-n, n]$ is closed and bounded, it is a compact subset of S. So, for each n, $[-n, n]$ has a finite open cover, say, $\mathscr{U}_n = \{U_i : i \in F_n\} \subseteq \mathscr{U}$. That is, $[-n, n] \subseteq \cup\mathscr{U}_n = \cup\{U_i : i \in F_n\}$. For each $i \in F_n$, let

$$V_i = U_i \backslash [-n+1, n-1] \subseteq U_i$$

and let

$$\mathscr{V}_n = \{V_i : i \in F_n\}$$

So each V_i in $\mathscr{V}_n$ is open in S and is contained in some U_j. Then, for each n, $\mathscr{V}_n$ is an open cover of $[-n, n] \backslash [-n+1, n-1]$. So,

$$\mathscr{V} = \cup\{\mathscr{V}_n : n \in \mathbb{N}\backslash\{0\}\}$$

is an open refinement of $\mathscr{U}$ which covers all of S, as required.

Secondly, we will show that $\mathscr{V}$ is locally finite in S. If $p \in S$, then $p \in [-m, m] \backslash [-m+1, m-1]$, for some m, and so p has a neighborhood which intersects at most finitely elements of $\mathscr{V}_m \subseteq \mathscr{V}$. So, $\mathscr{V}$ is locally finite.

We have constructed the family, $\mathscr{V}$, which is both an open refinement of the open cover, $\mathscr{U}$, and is locally finite. So, $S = \mathbb{R}$ is paracompact. We are done.

Example 3. Consider the space $S = \mathbb{R}^n$ equipped with the usual topology. We know that, for any n, this metrizable space is not compact. Show that S is paracompact.

Solution: To solve this we mimic the procedure in the previous example, replacing the interval $(-n, n)$ with the open ball, $B_n(0)$, centered at 0.

We know that Hausdorff compact spaces are normal. We have a similar result for Hausdorff paracompact spaces. We prove that Hausdorff paracompact spaces are guaranteed to be "normal" topological spaces.

In the proof of the following theorem, we invoke a lemma (6.16 on page 136) which we restate here, for convenience.

> *If $\mathscr{U} = \{V_i : i \in I\}$ is a locally finite collection of sets in S then $\mathscr{U}^* = \{\text{cl}_S V_i : i \in I\}$ is also locally finite. Furthermore,*
>
> $$\text{cl}_S[\cup\{V_i : i \in I\}] = \cup\{\text{cl}_S V_i : i \in I\}$$

Theorem 19.4 *Let S be a Hausdorff paracompact topological space.*

(a) *The space, S, is a regular space.*
(b) *The space, S, is a normal space.*

Proof. We are given that S is a Hausdorff paracompact topological space.

(a) We are required to show that S is regular.

Suppose H is a closed subset of S and $u \in S\backslash H$. It suffices to show that there is an open set, W, such that $H \subseteq W \subseteq \text{cl}_S W \subseteq S\backslash\{u\}$.

Since S is Hausdorff and H is closed, then, for each $x \in H$, there exists an open neighborhood, U_x, of x such that $u \notin \text{cl}_S U_x$. Now,

$$\mathscr{U} = \{U_x : x \in H\} \cup \{S\backslash H\}$$

forms an open cover of S. Since S is paracompact and $\mathscr{U}$ is an open cover of S, there exists a locally finite collection of open sets

$$\mathscr{V} = \{V_i : i \in I\} \cup \{V\}$$

which covers S and refines the open cover, $\mathscr{U}$ (where each $V_i \subseteq U_{y_i}$, for some $y_i \in H$, and $V \subseteq S\backslash H$).

Then, for each i,

$$V_i \subseteq \text{cl}_S V_i \subseteq \text{cl}_S U_{y_i} \subseteq S\backslash\{u\} \quad \text{(for some } y_i \in H\text{)}$$

Let $W = \cup\{V_i : i \in I\}$. Then by Lemma 6.16,

$$\text{cl}_S W = \cup\{\text{cl}_S V_i : i \in I\}$$

It follows that

$$H \subseteq W = \cup\{V_i : i \in I\} \subseteq \text{cl}_S W = \cup\{\text{cl}_S V_i : i \in I\} \subseteq S\backslash\{u\}$$

So, S is regular. As required.

(b) We are now required to prove that S is a normal space. Let H be a closed set in S. To prove the normal property holds we replace the point u in the proof of part (a) with a closed set, F, and invoke the regular property and mimic the steps in the proof above. Suppose H is a closed subset of S and F is a closed subset such that $F \in S \backslash H$. For each $x \in H$, there exists an open neighborhood, U_x such that $F \cap \mathrm{cl}_S U_x = \varnothing$ (since we have shown that S satisfies the regular property).

Now,

$$\mathscr{U} = \{U_x : x \in H\} \cup \{S \backslash H\}$$

forms an open cover of S (with $F \subseteq S \backslash H$). By hypothesis, there exists a locally finite collection of open sets

$$\mathscr{V} = \{V_i : i \in I\} \cup \{V\}$$

which refines $\mathscr{U}$, where each V_i is a subset of some U_x and $V \subseteq S \backslash H$.

Then, for each i,

$$V_i \subseteq \mathrm{cl}_S V_i \subseteq \mathrm{cl}_S U_x \subseteq S \backslash F \quad \text{(for some } x \in H\text{)}$$

Again, by Lemma 6.16, $\cup\{\mathrm{cl}_S V_i : i \in I\} = \mathrm{cl}_S(\cup\{V_i : i \in I\})$.

Since $\mathrm{cl}_S(\cup\{V_i : i \in I\}) = \cup\{\mathrm{cl}_S V_i : i \in I\} \subseteq (\cup\{\mathrm{cl}_S U_x : x \in H\}) \subseteq S \backslash F$, S is normal, the desired property.

We know that the compact property is carried over from the domain to the codomain by arbitrary continuous functions. This is not the case for the paracompact property. But, if the function is one which is closed and continuous, it does, as the following result shows.

Theorem 19.5 *Let S be a paracompact topological space and T be any space. If $f : S \to T$ is a closed and continuous function then $f[S]$ is a paracompact subspace of T.*

Proof. The proof is lengthy and involved. A reasonably complete proof is found in [Engelking].

Theorem 19.6 *Let S be a paracompact topological space and F be a closed subset of S. Then F inherits the paracompact property from S.*

Proof. We are given that S is a paracompact topological space and F is closed in S. We are required to show that F is also paracompact.

Let $\mathscr{U} = \{U_i : i \in I\}$ be an open cover of F. Then, for each i, $U_i = V_i \cap F$, for some open subset V_i of S. We then obtain an open cover, $\mathscr{V} = \{V_i : i \in I\} \cup \{S \backslash F\}$ of S. Since S is paracompact, then there is a locally finite open refinement, $\mathscr{W} = \{W_i : i \in I\}$, of $\mathscr{V}$.

The family $\{W_i \cap F : W_i \cap F \neq \varnothing\}$ forms a locally finite open refinement of $\mathscr{U}$. Then F is paracompact.

19.2 A Non-Paracompact Space

We now present an example of a space which is not paracompact.

Example 4. A standard, non-trivial, example of a non-paracompact space is the deleted Tychonoff plank, $S = [0, \omega_1] \times [0, \omega_0] \backslash \{(\omega_1, \omega_0)\}$. On page 235, we showed that the space S is not normal. Since paracompact spaces have been shown to be normal, then S cannot be paracompact.

Also, the Moore plane was shown to be completely regular but not normal. So, this is another example of a non-paracompact space.

19.3 Topic: Metrizability and Paracompactness

Any result which can bring us a step closer to a characterization of metrizable spaces is considered to be important and worth the effort required to study the proofs of related statements. The following definitions and technical results will lead to the threshold of a proof of the statement, "Metrizable spaces are paracompact spaces".

Definition 19.7 Let S be a topological space and $\mathscr{U}$ be a subfamily of $\mathscr{P}(S)$. If

$$\mathscr{U} = \cup\{\mathscr{U}_n : n \in \mathbb{N}\}$$

where each $\mathscr{U}_n$ is a locally finite subset of $\mathscr{P}(S)$ then we say that $\mathscr{U}$ is σ-locally finite.

Lemma 19.8 *Suppose S is a metrizable space and $\mathscr{U}$ is an open cover of S. Then there is an open cover, $\mathscr{E}$, which both refines $\mathscr{U}$ and is σ-locally finite.*[3]

Proof. Let S be a metrizable topological space. Then there is a metric, ρ, which allows us to express S as a metric space, (S, ρ). Suppose $\mathscr{U}$ is an open cover of S.

We will construct a subfamily, $\mathscr{E}$, of $\mathscr{P}(S)$ such that

(1) $\mathscr{E}$ is an open cover of S,
(2) $\mathscr{E}$ refines $\mathscr{U}$,
(3) $\mathscr{E} = \cup\{\mathscr{E}_n : n \in \mathbb{N}\backslash\{0\}\}$ where each $\mathscr{E}_n$ is a locally finite collection of open sets.

We will index the elements of the open cover, $\mathscr{U}$, with ordinals Ω: $\mathscr{U} = \{U_\alpha : \alpha \in \Omega\}$. That is,

$$\mathscr{U} = \{U_0, U_1, \ldots, U_{\omega_0}, U_{\omega_0+1} \ldots, U_{\omega_1}, U_{\omega_1+1}, U_{\omega_1+2}, \ldots\}^4$$

Step 1: Construction of the collection of sets, $\mathscr{E}$.

For each $U_\gamma \in \mathscr{U}$ and a fixed n in $\mathbb{N}\backslash\{0\}$ let

$$S_n(U_\gamma) = \{x \in S : B_{1/n}(x) \subseteq U_\gamma\}$$

Let

$$T_n(U_\gamma) = S_n(U_\gamma)\backslash\cup\{U_\alpha : \alpha < \gamma\}$$

Now, see that $T_n(U_\gamma) \subseteq U_\gamma$.[5] If we repeat this for each element, U_α, of $\mathscr{U}$, then

$$\mathscr{T}_n = \{T_n(U_\alpha) : \alpha \in \Omega\}$$

is a refinement of $\mathscr{U}$.

---

[3]Note that $\mathscr{E}$ is "σ-locally finite" (not locally finite) so we cannot conclude immediately that metrizable spaces are paracompact. This will come later.

[4]Here we are invoking the *Well-ordering theorem* which permits such an ordering. It is a statement which is equivalent to the *Axiom of choice*.

[5]For, if $x \in T_n(U_\gamma)$, then $x \in B_{1/n}(x) \subseteq U_\gamma$.

We claim that the elements of $\mathscr{T}_n$ are pairwise disjoint.

Proof of claim: Consider $T_n(U_\beta)$ and $T_n(U_\gamma)$ where $\beta < \gamma$. Choose element $c \in T_n(U_\gamma)$. Then for any point $b \in T_n(U_\beta)$, $b \in S_n(U_\beta)$. So, $b \in B_{1/n}(b) \subseteq U_\beta$. Since $c \in T_n(U_\gamma)$ and $\beta < \gamma$, $c \notin U_\beta$. So, $c \notin B_{1/n}(b)$. So, $c \notin T_n(U_\beta)$. We have shown that,

$$b \in T_n(U_\beta) \quad \text{and} \quad c \in T_n(U_\gamma) \Rightarrow c \notin B_{1/n}(b) \quad (*)$$

So the elements of $\mathscr{T}_n$ are pairwise disjoint, *as claimed.*

Even if $\mathscr{T}_n$ is a refinement of $\mathscr{U}$, its elements may not be open, so we have more work to do.

We will construct an open set $E_n(U_\gamma)$ such that

$$T_n(U_\gamma) \subseteq E_n(U_\gamma) \subseteq U_\gamma$$

To construct $E_n(U_\gamma)$, let

$$E_n(U_\gamma) = \cup\{B_{1/3n}(x) : x \in T_n(U_\gamma)\}$$

We repeat this for each element, U_α, of $\mathscr{U}$, to form the collection of sets,

$$\mathscr{E}_n = \{E_n(U_\alpha) : \alpha \in \Omega\}$$

Step 2: Proof that the family of sets,

$$\mathscr{E} = \cup\{\mathscr{E}_n : n \in \mathbb{N}\backslash\{0\}\}$$

satisfies all the required properties.

(1) The collection $\mathscr{E}$ is an open refinement of $\mathscr{U}$:
Each $E_n(U_\alpha)$ is easily seen to be open so, for each n, $\mathscr{E}_n$ is a family of open sets, the same then holds true for $\mathscr{E}$.

We *claim* that $\mathscr{E}_n$ refines $\mathscr{U}$.

$$u \in E_n(U_\gamma) \Rightarrow u \in B_{1/3n}(c) \text{ for some } c \in T_n(U_\gamma)$$

$$T_n(U_\gamma) = S_n(U_\gamma)\backslash\cup\{U_\alpha : \alpha < \gamma\} \Rightarrow c \in S_n(U_\gamma)$$

$$S_n(U_\gamma) = \{x \in S : B_{1/n}(x) \subseteq U_\gamma\} \Rightarrow B_{1/n}(c) \subseteq U_\gamma$$

$$u \in B_{1/3n}(c) \subseteq B_{1/n}(c) \Rightarrow u \in U_\gamma$$

Then $E_n(U_\gamma) \subseteq U_\gamma$. Hence, $\mathscr{E}_n$ is an open refinement of $\mathscr{U}$, *as claimed.*

This satisfies one of the required properties of $\mathscr{E}$.

(2) The collection $\mathscr{E}$ is σ-locally finite. That is, $\mathscr{E}$ is the countable union of locally finite families of sets:

Claim 1: We first *claim* that the elements of $\mathscr{E}_n = \{E_n(U_\alpha) : \alpha \in \Omega\}$ are pairwise disjoint.

Proof of claim: Consider the pair $E_n(U_\beta)$ and $E_n(U_\gamma)$ where $\beta < \gamma$. Let $v \in E_n(U_\gamma)$ and u be any element in $E_n(U_\beta)$. We will show that $v \neq u$.

$$u \in E_n(U_\beta) = \cup\{B_{1/3n}(x) : x \in T_n(U_\beta)\} \Rightarrow u \in B_{1/3n}(b) \text{ for some } b \in T_n(U_\beta)$$
$$v \in E_n(U_\gamma) = \cup\{B_{1/3n}(x) : x \in T_n(U_\gamma)\} \Rightarrow v \in B_{1/3n}(c) \text{ for some } c \in T_n(U_\gamma)$$
$$b \in T_n(U_\beta) \text{ and } c \in T_n(U_\gamma) \Rightarrow c \notin B_{1/n}(b) \quad (\text{By } (^*).)$$

Then

$$\begin{aligned} 1/n &\leq \rho(b, c) \\ &\leq \rho(b, v) + \rho(c, v) \\ &< [\rho(b, u) + \rho(u, v)] + 1/3n \\ &< 1/3n + \rho(u, v) + 1/3n \\ &= \rho(u, v) + 2/3n \end{aligned}$$

$$\text{implies} \quad \rho(u, v) > 1/3n \qquad (^{**})$$

So, $v \notin B_{1/3n}(u)$.

We have shown that,

$$u \in E_n(U_\beta) \text{ and } v \in E_n(U_\gamma) \Rightarrow v \notin B_{1/3n}(u) \quad (***)$$

So, $v \neq u$ and $u \in E_n(U_\beta)$ then $v \notin E_n(U_\beta)$. We conclude, if $\beta < \gamma$, $E_n(U_\beta) \cap E_n(U_\gamma) = \varnothing$; this proves *Claim* 1.

Claim 2: We now claim that each $\mathscr{E}_n$ is locally finite.

Proof of claim: Let $z \in S$. By Claim 1, z belongs to at most one element of $\mathscr{E}_n$, say $E_n(U_\gamma) = \cup\{B_{1/3n}(x) : x \in T_n(U_\gamma)\}$. Say $z \in B_{1/3n}(y)$. By (**) the distance between points in different sets in $\mathscr{E}_n$ is larger than $1/3n$. So, the ball $B_{1/6n}(z)$ intersects only $E_n(U_\gamma)$. Then $\mathscr{E}_n$ is locally finite, as claimed.

We are done with the second property.

(3) The collection $\mathscr{E}$ is an open cover of S: Let y be any point in S. We must show that y must belong to some member, $E_m(U_\mu) \in \mathscr{E}_m$, of $\mathscr{E}$.

Let μ be the least ordinal such that $y \in U_\mu \in \mathscr{U}$.[6] Since U_μ is open, we can choose a natural number m such that $B_{1/m}(y) \subseteq U_\mu$. Then $y \in S_m(U_\mu) = \{x \in S : B_{1/m}(x) \subseteq U_\mu\}$. Since μ is the least ordinal such that $y \in U_\mu$, then $y \in T_m(U_\mu) = S_m(U_\mu) \backslash \cup\{U_\alpha : \alpha < \mu\}$. By definition, $B_{1/3m}(y) \subseteq E_m(U_\mu) = \cup\{B_{1/3m}(x) : x \in T_n(U_\mu)\}$, so $y \in E_m(U_\mu) \in \mathscr{E}_m \subseteq \mathscr{E}$. Then each element of S belongs to some element of $\mathscr{E}$. So, $\mathscr{E}$ is an open cover of S.

So, $\mathscr{E}$ is the countable union of locally finite collections of open sets which both covers S and refines $\mathscr{U}$.

We are done.

In the next theorem, takes us a step closer to the statement "Metrizable spaces are paracompact spaces". The reader will recognize the use of techniques similar to the ones used in the proof of Lemma 19.8.

Theorem 19.9 *Suppose the space S is metrizable and $\mathscr{U}$ is an open cover of S. Then $\mathscr{U}$ has a refinement, $\mathscr{C}$, which both covers S and is locally finite.*[7]

Proof. Suppose S is a metrizable space and $\mathscr{U}$ is an open cover of S. By Lemma 19.8 there is an open cover, $\mathscr{E}$, which both refines $\mathscr{U}$ and is σ*-locally finite*. Let

$$\mathscr{E} = \cup\{\mathscr{E}_n : n \in \mathbb{N}\backslash\{0\}\}$$

be such an open cover where each $\mathscr{E}_n = \{E_{n(i)} : i \in I_n\}$ is a locally finite collection of open sets which refines $\mathscr{U}$. For each $m \in \mathbb{N}\backslash\{0\}$, let

$$V_m = \cup\{E_{m(i)} : i \in I_m\} = \cup\,\mathscr{E}_m$$

---

[6]The ordinal μ exists since the ordinals are "well-ordered".

[7]Note that we are not hypothesizing that the members of $\mathscr{C}$ are open in S.

For $n \in \mathbb{N}\backslash\{0\}$, and $i \in I_n$ let

$$\begin{aligned} S_n(E_{n(i)}) &= E_{n(i)}\backslash\cup\{V_m : m < n\} \\ &= E_{n(i)}\backslash\cup\left\{\cup\{E_{m(i)} : i \in I_m\} : m < n\right\} \end{aligned}$$

We are required to construct a collection of sets, $\mathscr{C}$, which covers S, refines $\mathscr{E}$ and is locally finite. (Note that the members of $\mathscr{C}$ need not be open.)

Let

$$\mathscr{C}_n = \{S_n(E_{n(i)}) : i \in I_n\}$$

Since $S_n(E_{n(i)}) \subseteq E_{n(i)} \in \mathscr{E}_n$, then $\mathscr{C}_n$ refines $\mathscr{E}_n$.

Let

$$\mathscr{C} = \cup\{\mathscr{C}_n : n \in \mathbb{N}\backslash\{0\}\}$$

Since each $\mathscr{C}_n$ refines $\mathscr{E}_n$ and each $\mathscr{E}_n$ refines $\mathscr{U}$, then $\mathscr{C}$ is a refinement of $\mathscr{U}$.

Claim 1: It is claimed that $\mathscr{C}$ covers S.

Proof of claim: Let $p \in S$. We know that $\mathscr{E} = \cup\{\mathscr{E}_n : n \in \mathbb{N}\setminus\{0\}\}$ is an open cover of S. Let $k = \min\{n : p \text{ belongs to some element of } \mathscr{E}_n\}$. Let j be such that $p \in E_{k(j)} \in \mathscr{E}_k$. Then $p \in S_k(E_{k(j)}) \in \mathscr{C}_k \subseteq \mathscr{C}$. Then p belongs to some element of $\mathscr{C}$. So, $\mathscr{C}$ covers S. This establishes Claim 1.

Claim 2: It is claimed that for $p \in S$, p belongs to a neighborhood $B \subseteq E_{k(j)}$ which meets at most finitely many elements of $\mathscr{C}$. That is, $\mathscr{C}$ is a locally finite collection of sets in S.

Proof of claim: Let $p \in S$ and $k = \min\{n : p \text{ belongs to some element of } \mathscr{E}_n\}$ (as defined in the proof of Claim 1. Then $p \in S_k(E_{k(i)}) \subseteq E_{k(i)}$, for some $i \in I_k$.

Recall that, for each n, $\mathscr{E}_n$ is locally finite. So, for each n, there exists a neighborhood, B_n, of p which intersects only finitely many elements in $\mathscr{E}_n = \{E_{n(i)} : i \in I_n\}$.

Because of this, this B_n can only intersect finitely may elements of $\mathscr{C}_n = \{S(E_{n(i)}) : i \in I_n\}$ for, whenever it intersects $S(E_{n(i)}) \subseteq E_{n(i)}$, it intersects $E_{n(i)}$. Also, if $n > k$, $E_{k(j)} \cap S_n(E_{n(i)}) = \varnothing$ (by definition of $S_n(E_{n(i)})$).

Let $B = \cap\{B_n : n \in \{1, 2, \ldots, k\}\}$. Then $p \in B$ and so $B \cap E_{k(i)}$ is an open neighborhood of p which meets at most finitely many elements of $\mathscr{C}$ (these are in $\cup\{\mathscr{C}_1, \mathscr{C}_2, \ldots, \mathscr{C}_k\}$). So, $\mathscr{C}$ is locally finite.

Then the collection, $\mathscr{C}$, refines $\mathscr{U}$, is locally finite and covers S. As required.

Lemma 19.10 *Suppose the space S is metrizable and $\mathscr{U}$ is an open cover of S. Then $\mathscr{U}$ has a refinement of* closed *sets which both covers S and is locally finite.*

Proof. Suppose S is a metrizable space and $\mathscr{U}$ is an open cover of S. Since metrizable spaces are regular (by Theorem 9.19), then S is regular.

Then, for each $U \in \mathscr{U}$ and for each $x \in S$, there is some open neighborhood, B_x, of x such that $\text{cl}_S B_x \subseteq U$. Then $\mathscr{B} = \{B_x : x \in S\}$ is an open cover of S which refines $\mathscr{U}$ where $B_x \subseteq U$ implies $\text{cl}_S B_x \subseteq U$.

By Theorem 19.9, there is a collection of sets, $\mathscr{C} = \{C : C \in \mathscr{C}\}$, which refines $\mathscr{B}$, covers S and is locally finite. By Lemma 6.16, if $\mathscr{C}$ locally finite then $\mathscr{C}^* = \{\text{cl}_S C : C \in \mathscr{C}\}$ is also locally finite. Then the family $\mathscr{C}^*$ covers S (since $\mathscr{C}$ covers S) and is a refinement of $\mathscr{U}$ (since $\mathscr{C}$ refines $\mathscr{U}$).

So, $\mathscr{C}^*$ is the desired refinement of $\mathscr{U}$ whose members are closed and which covers S.

Example 5. Suppose $\mathscr{F}$ is a locally finite collection of closed subsets in $\mathscr{P}(S)$. Show that $\cup\{F : F \in \mathscr{F}\}$ is a closed subset of S.

Solution: We are given that each $F \in \mathscr{F}$ is closed where $\mathscr{F}$ is locally finite. Let $M = \cup\{F : F \in \mathscr{F}\}$. From Lemma 6.16, we know that, if $\mathscr{F}$ is locally finite, $\{\text{cl}_S F : F \in \mathscr{F}\}$ is locally finite and $\text{cl}_S M = \cup\{\text{cl}_S F : F \in \mathscr{F}\}$. Since

$$\cup\{\text{cl}_S F : F \in \mathscr{F}\} = \cup\{F : F \in \mathscr{F}\}$$

then $M = \text{cl}_S M$. So, M is closed in S.

We are finally set for the main result of this section.

Theorem 19.11 *If a space is metrizable then it is paracompact.*

Proof. Suppose S is a metrizable space and $\mathscr{U}$ is an open cover of S. By Theorem 19.9, there exists a family of sets, $\mathscr{H} = \{H : H \in \mathscr{H}\}$, which refines $\mathscr{U}$, covers S and is locally finite.

Since $\mathscr{H}$ is locally finite, for each $x \in S$ we can find open a neighborhood, V_x, of x which intersects at most finitely many elements of $\mathscr{H}$. Let

$$\mathscr{V} = \{V_x : x \in S\}$$

be the collection of all such sets.

Since $\mathscr{V}$ is an open cover of S, Lemma 19.10 applies: There is a collection of *closed* sets,

$$\mathscr{C} = \{C : C \in \mathscr{C}\}$$

which refines $\mathscr{V}$, covers S and is locally finite.

For each $C \in \mathscr{C}$,

$$C \text{ intersects with finitely many elements of } \mathscr{H} (^*)$$

(since for each C, $C \subseteq V_x$, for some x, and V_x itself intersects with finitely many elements of $\mathscr{H}$).

Since $\mathscr{H}$ refines the open cover, $\mathscr{U}$, for each $H \in \mathscr{H}$, we can choose an element, U_H, in $\mathscr{U}$, such that

$$H \subseteq U_H \in \mathscr{U}$$

For each $H \in \mathscr{H}$. let

$$E_H = S \backslash \cup\{C \in \mathscr{C} : C \subseteq S \backslash H\}$$

See that E_H is open, since $\mathscr{C}$ is locally finite, $\cup\mathscr{C}$ is closed (see the example immediately preceding this theorem). We claim that $H \subseteq E_H$. For, if $y \in H$, we can argue,

$$\begin{aligned} y \notin E_H \quad &\Rightarrow \quad y \in \cup\{C \in \mathscr{C} : C \subseteq S \backslash H\} \\ &\Rightarrow \quad y \in C, \text{ for some } C \in \mathscr{C} \text{ such that } C \cap H = \varnothing \\ &\Rightarrow \quad y \notin H \end{aligned}$$

So, $H \subseteq E_H$, as claimed.

Then H is a subset of the open subset, $E_H \cap U_H$. Then we define

$$\mathscr{D} = \{E_H \cap U_H : H \in \mathscr{H}\}$$

Then,

(1) $\mathscr{D}$ is an open cover of S: Since $\mathscr{H}$ covers S, and $H \subseteq E_H \cap U_H$.
(2) $\mathscr{D}$ refines $\mathscr{U}$: Since, $[y \in E_{H_0} \cap U_{H_0}$ some $H_0] \Rightarrow [y = U_{H_0} \in \mathscr{U}]$.
(3) $\mathscr{D}$ is locally finite. (If so, $\mathscr{D}$ is the family of sets we seek.)

Proof of (3): Let $p \in S$. Since $\mathscr{C}$ is locally finite, p has a neighborhood, B, which intersects finitely many elements, $\{C_i : i = 1, \ldots, k\} \subseteq \mathscr{C}$. Since $\mathscr{C}$ is a cover of S, then

$$B \subseteq \cup\{C_i : i = 1, \ldots, k\}$$

It suffices to show that each element of $\mathscr{C}$ intersects $E_H \cap U_H$ for, at most, finitely many H; for, if so, B intersects only finitely many elements of $\mathscr{D}$.

Suppose that, for some j, C_j intersects $E_{H_1} \cap U_{H_1}$; say, $y \in C_j \cap [E_{H_1} \cap U_{H_1}]$. Then $y \in C_j \subseteq U_{H_1} \in \mathscr{U}$. Also,

$$\begin{aligned} y \in C_j \cap E_{H_1} \quad &\Rightarrow \quad y \in C_j \cap S \backslash \cup\{C \in \mathscr{C} : C \cap H_1 = \varnothing\} \\ &\Rightarrow \quad y \notin \cup\{C \in \mathscr{C} : C \cap H_1 = \varnothing\} \\ &\Rightarrow \quad y \text{ belongs to } C_j \text{ such that } C_j \cap H_1 \neq \varnothing \\ &\Rightarrow \quad C_j \text{ intersects some } U_{H_1} \in \mathscr{U} \end{aligned}$$

Recall that C in $\mathscr{C}$ can intersect at most finitely many H's in $\mathscr{H}$ (see (*)). Then each C_j can intersect at most finitely many elements of $\mathscr{D}$.

We conclude that B intersects at most finitely many elements of $\mathscr{D}$.

We conclude that $\mathscr{D}$ is locally finite. This proves statement (3).

We conclude that $\mathscr{D}$ is an open refinement of $\mathscr{U}$ which both covers S and is locally finite in S. So, the metrizable space, S, is paracompact.

Concepts Review

1. What does it mean to say that $\mathscr{U}$ is a *locally finite* family of subsets of a space S?
2. What does it mean to say that the family of subsets, $\mathscr{V}$, is an *open refinement* of the family $\mathscr{U}$?
3. What does it mean to say that the family of subsets, $\mathscr{V}$, is a *locally finite open refinement* of the family $\mathscr{U}$?
4. Define the paracompact property of a topological space.
5. What class of topological spaces has elements which are guaranteed to be paracompact?
6. Provide an example of a paracompact space and briefly summarize arguments which confirms your answer.
7. Provide an example of a non-paracompact space.
8. If a paracompact space is Hausdorff what other separation axioms does it satisfy?
9. If $f : S \to T$ is a function mapping the paracompact space S into T, what properties must be satisfied by f if we want to guarantee that $f[S]$ is paracompact?
10. If S is a paracompact space what kind of subsets of S will share this property?

Exercises

1. Show that, if a paracompact space S is countably compact, then S is compact.
2. Suppose S is a Lindelöf space and $\mathscr{U}$ is a family of subsets of S. Show that $\mathscr{U}$ is a countable family of subsets.
3. Suppose S is a regular space and $\mathscr{U} = \{F_i : i \in I\}$ is a family of paracompact closed subsets of S. Show that if $\mathscr{U}$ is locally finite then $\cup\{F_i : i \in I\}$ is paracompact.
4. Let S be a perfectly normal paracompact space. Show that any non-empty subspace is also paracompact.

Part VI
The Connected Property

Chapter 20

Connected Spaces and Properties

Abstract
In this section, we formally define the topological property of connectedness while providing a few examples of connected spaces, and of spaces which are not. Continuous images of connected spaces are proven to be connected. Arbitrary unions of families of connected sets are proven to be connected provided they have at least one point in common. If the set A is connected and is dense in a set, B, then B is shown to be connected. In particular, closures of connected sets are connected. The connected property will be seen to be invariant over arbitrary products of connected spaces, without conditions on the number of factors. A connected component is defined as being the largest connected subspace containing a given point. Components will be seen to be closed, but not necessarily open. Properties of spaces that have an open base of connected neighborhoods will be briefly discussed. We will also define the pathwise connected property; we will see that sets which satisfy this property are connected. But the converse does not hold true. Finally, we will introduce the totally disconnected property.

20.1 Definition

The property of *connectedness* is one that is, more or less, independent of the topological properties that we have studied until now, in the sense that we cannot characterize this property with some combination of the other properties. It is fairly easy to construct mental images of "connected spaces".

A good way to start is to think of a single string and ask yourself "Is this string connected?" to which you would answer, "...of course!". Then take a pair of scissors and cut it, and ask, "what about now?" to which you would intuitively answer "well no, not anymore". A bystander might add: "Both represent the same string, but the second one differs topologically from the the first".

All this is stated based on an intuitive understanding of the word "connected". But intuition, as a guiding tool, precedes the process of *providing a rigourous mathematical definition.* That is, producing a definition which can be interpreted in only one way while simultaneously satisfying our intuitive understanding of this property.

A rigourous definition should allow us to answer more difficult questions about objects which possess this property but whose complexity defies our imagination.

Definition 20.1 A topological space S is said to be *connected* if it is **not** the *disjoint union* of two non-empty *open* sets.

Recall that, in this book, those subsets of a space that are both open and closed are referred to as being *clopen.* So, if S is the disjoint union of two open sets A and B if and only if both A and B are complements of an open sets and so are also closed. So, we can say that a space

> *S is* not *connected if and only if S contains a proper non-empty clopen subset.*

Example 1. The topological space, $\mathbb{R}$, with the usual topology is connected since no non-empty proper subset in $\mathbb{R}$ is clopen. However, the subspace of all rationals, $\mathbb{Q}$, is not connected since $\mathbb{Q}$ the disjoint union of the two open sets,

$$U = (-\infty, \pi) \cap \mathbb{Q} \quad \text{and} \quad \mathbb{Q} \backslash U = (\pi, \infty) \cap \mathbb{Q}$$

So, the connected property is not hereditary.

It is not difficult to prove that the only proper non-empty subsets of $\mathbb{R}$ which are *connected subspaces* are those subsets which are either a single element or those which are intervals (either open, closed or half open).

Example 2. The topological space $\mathbb{R}$ with the upper limit topology is not connected since $\mathbb{R}$ is the disjoint union of $U = (-\infty, 3]$ and $\mathbb{R}\backslash U = (3, \infty)$ both of which are open in this topology. (Verify this.)

20.2 Properties of Connected Spaces

We now investigate some invariance properties for connectedness.

Theorem 20.2 *The continuous image of a connected space is connected.*

Proof. We are given that S and T are topological spaces, that S is connected and the function, $f : S \to T$, is continuous. We are required to show that $f[S]$ is connected.

Suppose U is clopen subset of $f[S]$ where $U \neq f[S]$. We claim that U must be empty.

To see this suppose,

$$h : f[S] \to \{0, 1\}$$

is defined as $h[U] = \{0\}$ and $h[\, f[S]\backslash U\,] = \{1\}$. Then $h[f[S]] \subseteq \{0, 1\}$, so h is continuous on $f[S]$. Since f is continuous on S,

$$h{\circ}f : S \to \{0, 1\}$$

is also continuous on S. Now, $\{0\}$ and $\{1\}$ are clopen subsets of $h[f[S]]$, hence,

$$(h{\circ}f)^{\leftarrow}[\{0\}] = f^{\leftarrow}[h^{\leftarrow}(0)] = f^{\leftarrow}[U] \neq S \quad \text{(Since } U \neq f[S].\text{)}$$

where $f^{\leftarrow}[U]$ must be a clopen subset of S (but not equal to S). Since S is connected, the only clopen subset in S other than S itself is $\varnothing$. So $f^{\leftarrow}[U] = \varnothing$ which implies

$$U = f[f^{\leftarrow}[U]] = f[\varnothing] = \varnothing$$

So $U = \varnothing$, as claimed. So the continuous image, $f[S]$, contains no non-empty proper clopen set. So, $f[S]$ must be connected.

Example 3. Suppose $(\mathbb{R}, \tau)$ is the set of reals equipped with the usual topology and $(\mathbb{R}, \tau_s)$ is the set of reals with the upper limit topology. Show that the identity map

$$i : (\mathbb{R}, \tau) \to (\mathbb{R}, \tau_s)$$

is not continuous.

Solution: We have seen that $(\mathbb{R}, \tau)$ is connected and that $(\mathbb{R}, \tau_s)$ is not. Since the continuous image of a connected set is connected, then $i : (\mathbb{R}, \tau) \to (\mathbb{R}, \tau_s)$ cannot be continuous, as required.

Does the union of connected sets preserve the connected property? Our instinct states that they must, a least, have non-empty intersection. We should check this out.

Suppose

$$S = \cap\{U_i : i \in I\}$$

where each U_i is connected in S. Suppose $q \in \cap\{U_i : i \in I\}$.

If there exists a function, $h : \cup\{U_i : i \in I\} = \{0, 1\}$, which is continuous on S then, since each U_i is connected, h must be constant on each U_i. There is no other option. It must then follow that, for each i, $h(q)$ and $h[U_i]$ have the same value. So, h must be constant on the union of these sets. So, $\cup\{U_i : i \in I\}$ cannot be the union of two non-empty clopen sets and so must be connected.

It is worth recording this as a theorem, for future reference.

Theorem 20.3 *The arbitrary union of a family of connected sets which have at least one point in common is connected.*

Proof. The statement is proved in the paragraph above.

Example 4. Suppose $S = \mathbb{R}^2$ equipped with the usual topology. Show that S is a connected space.

Solution: For each point $x \in S$, let L_x denote the infinite line containing both x and 0.

Then $S = \cup\{L_x : x \in S\}$. For each $x \in S$, L_x is homeomorphic to $\mathbb{R}$ (previously shown to be connected) and contains $(0, 0)$; hence, S is the union of a family of connected spaces which contain one point in common. Then S is connected.

In the following theorem, we see that the closure operation preserves connectedness. We then verify that arbitrarily large products preserve connectedness, provided every factor is connected. Also that connected product spaces must have connected factors.

Theorem 20.4 *The closure of a connected set is connected.*

Furthermore, any space, S, which has a connected dense subset is connected.

Proof. We are given that T is a connected subspace of the topological space S.

We are required to show that $\text{cl}_S T$ is connected. Suppose $h : \text{cl}_S T \to \{0, 1\}$ is a continuous function. To show that $\text{cl}_S T$ is connected it suffices to show that h is constant on $\text{cl}_S T$. Since h is continuous, then h must be constant on the connected set T. Suppose $q \in \text{cl}_S T \backslash T$. If $h(q) \neq h[T]$, then $h^{\leftarrow}[h(q)]$ is the pull-back of a clopen set in $\{0, 1\}$ and so must intersect T, contradicting the fact that h is constant on T. So, h must be constant on $\text{cl}_S T$ and so is connected.

By applying a similar reasoning, if S has a connected dense subset then S is connected.

Theorem 20.5 *Let $\{S_i : i \in I\}$ be a family of topological spaces and $S = \prod_{i \in I} S_i$ be a product space. Then S is connected if and only if each S_i is connected.*

Proof. We are given that $S = \prod_{i \in I} S_i$ is a product space.

($\Rightarrow$) If the product space, S, is connected then each factor, S_i, is the continuous image of S under the projection map, $\pi_i : S \to S_i$, and so is connected.

($\Leftarrow$) Suppose S_i is connected for each $i \in I$.
Let's first consider the case where S has only two connected factors, S_1 and S_2.

We *claim* that $S_1 \times S_2$ is connected. Suppose (q_1, q_2) is a fixed point in $S_1 \times S_2$ and let (x, y) be any other point. Then $(q_1, q_2) \in S_1 \times \{q_2\}$

and $(x, y) \in \{x\} \times S_2$. Then the set,

$$M_{(x,y)} = (S_1 \times \{q_2\}) \cup (\{x\} \times S_2)$$

is the union of two connected. Hence, $M_{(x,y)}$ is connected and contains the fixed point (q_1, q_2), for any choice of (x, y). More generally, every element of $\{M_{(x,y)} : (x, y) \in S_1 \times S_2\}$ is a connected set and contains (q_1, q_2). So, $\cup\{M_{(x,y)} : (x, y) \in S_1 \times S_2\}$ is connected. Since $S_1 \times S_2 = \cup\{M_{(x,y)} : (x, y) \in S_1 \times S_2\}$, then $S_1 \times S_2$ is connected, as claimed.

By finite induction, the product space of finitely many connected factors is connected.

Let $\mathscr{F}$ denote the family of all finite subsets of the index set, I. Suppose $q = \langle q_i \rangle_{i \in I}$ is an element of the product space, S, and let $F \in \mathscr{F}$. Let

$$\begin{aligned} T_i &= S_i \quad \text{if } i \in F \\ T_i &= \{q_i\} \quad \text{otherwise} \end{aligned}$$

Let $K_F = \prod_{i \in I} T_i$. By the claim established above, K_F is a connected subspace of the product space, S. Hence, $\{K_F : F \in \mathscr{F}\}$ is a family of connected subspaces of S, where

$$\{q\} = \{\, \langle q_i \rangle_{i \in I} \,\} = \cap\{K_F : F \in \mathscr{F}\}$$

By Theorem 20.3, $V = \cup\{K_F : F \in \mathscr{F}\}$ is a connected subset of S.

We claim that V is dense in S. Let $u = \langle u_i \rangle_{i \in I} \in S$. Let $U = \prod_{i \in I} U_i$ be a basic open neighborhood of u. Then there is a finite subset, F^* of I, such that, $U_i = \pi_i^{\leftarrow}[U]$ for $i \in F^*$ and equals S_i, otherwise. It suffices to show that, $U \cap V \neq \varnothing$. See that $K_{F^*} \cap U = \prod_{i \in I} V_i$ such that, $V_i = U_i$ for $i \in F^*$ and $V_i = \{q_i\}$, otherwise. So $K_{F^*} \cap U$ is non-empty. Hence, $V \cap U$ is non-empty. So, V is dense in S. Since V is dense and connected, then, by Theorem 20.4, so is S.

Given a function $f : T \to T$, the point x in T is called a *fixed point* of f if $f(x) = x$. If $T \subseteq S$ and $f : S \to T$ is a continuous function on S such that every point of T is a fixed point of f, then we say that *T is retract of f*. The following theorem shows that connectedness of $[a, b]$ guarantees that a continuous function $f : [a, b] \to [a, b]$ on $[a, b]$

will have at least one fixed point of f. This well-known statement is commonly referred to as the *Brouwer's fixed point theorem.*

Theorem 20.6 *For a closed interval $[a, b]$ in $\mathbb{R}$ a continuous function $f : [a, b] \to [a, b]$ mapping $[a, b]$ onto $[a, b]$ will have at least one fixed point.*

Proof. We are given that $f : [a, b] \to [a, b]$ is continuous. Let

$$h(x) = \frac{f(x) - x}{|f(x) - x|}$$

Suppose f has no fixed points in $[a, b]$.

See that the function $h(x)$ is well-defined and continuous on $[a, b]$.

Then $h(x)$ maps $[a, b]$ into $\{1, -1\}$. Since f is continuous on the connected set $[a, b]$, $h[a, b]$ is connected. So, h is constant on it's domain.

This leaves two possibilities: Either $h[a, b] = \{1\}$ or $h[a, b] = \{-1\}$.

If $h(x) = 1$, then $|f(x) - x| = f(x) - x$ for each $x \in [a, b]$. Then $f(x) - x \geq 0$. This implies that $f(x) \geq x$ on $[a, b]$. Since f has no fixed point in $[a, b]$, $f(x) > x$ on $[a, b]$ So, $f(b) > b$. Since f maps $[a, b]$ into $[a, b]$, we have a contradiction. So, $h[a, b] \neq \{1\}$.

If $h[a, b] = \{-1\}$, by a similar reasoning, we deduce that $f(a) < a$. This is also impossible.

We then must conclude that, if $f : [a, b] \to [a, b]$ is continuous, it has at least one fixed point in $[a, b]$.

20.3 The Connected Components of a Topological Space

A topological space, S, which is not connected can always be partitioned in a family of subspaces which are connected. A trivial example could be the partitioning of $S = \{\{x\} : x \in S\}$, into singleton sets, since each singleton set is connected. However, a point x may be contained in a larger connected subspace of S which contains it.

One might want to partition S as

$$S = \cup\{C_x : x \in S\}$$

where C_x denotes "the largest connected subspace of S which contains x". To remove any ambiguity about what we mean by "C_x is the largest connected set in S," we might prefer to say "C_x is the union of all the connected subspaces of S which contain x". This leads us to the following definition.

Definition 20.7 If S is a topological space and $u \in S$, then we define the *connected component of u in S* as being the union of all connected subspaces of S that contain the point u.[1]

Clearly, the only component of a connected space is the space itself.

The definition confirms our perception that the component of a point u in S, is the *unique* largest connected subset of S which contains the point u. Also, two components, C_1 and C_2, in S cannot intersect, for if $z \in C_1 \cap C_2$, then $C_1 \cup C_2$ is connected; hence, neither C_1 nor C_2 could be components of z. Since every point of a space S belongs to precisely one connected component, then the family of all components partitions the space, S.

If C is a "connected component" in S, we mean that C is a connected as a subspace of S. That is, C cannot contain a non-empty clopen sets in C. This means that,

— if U and V are both open in S such that

$$(U \cap C) \cap (V \cap C) = \varnothing \quad \text{and} \quad (U \cap C) \cup (V \cap C) = C$$

then one $(U \cap C)$ or $(V \cap C)$ must be empty.

— if $x \in S \backslash C$, and C_x is the component containing x and $T = C \cup C_x$ then T cannot be a connected subspace of S, and so T must contain a non-empty clopen subset of T, say B. Since C and C_x are connected in T, $C \subseteq B$ or $C \subseteq T \backslash B$. Then C and C_x are clopen in T.

[1]When there can be (by context) no confusion we often drop the adjective "connected" and only say "component".

Example 5. Consider the subspace, $T = B_1(0,0) \cup B_1(0,2)$, the union of two open balls of radius one in $\mathbb{R}^2$ with center $(0,0)$ and $(0,2)$, respectively. Since T is the union of two open disjoint subsets of T, then, by definition, T is not a connected space. The space, T, has precisely two components, $B_1(0,0)$ and $B_1(0,2)$.

Example 6. Let $U = B_1(0,0)$, $V = B_1(0,2)$ in $\mathbb{R}^2$. Consider the subspace,

$$T^* = U \cup V \cup \{(0,1)\}$$

of $\mathbb{R}^2$ (see Figure 20.1). It is clearly the case that T^* is the union of three non-intersecting connected subsets, the two connected sets U and V and the connected set $\{(0,1)\}$. But we cannot conclude from this that T^* is not connected. We can view T^* as the union of the two connected sets, $U \cup \{(0,1\}$ and $V \cup \{0,1)\}$ with non-empty intersection, $\{(0,1)\}$. So, T^* is a connected set with itself as the only component.

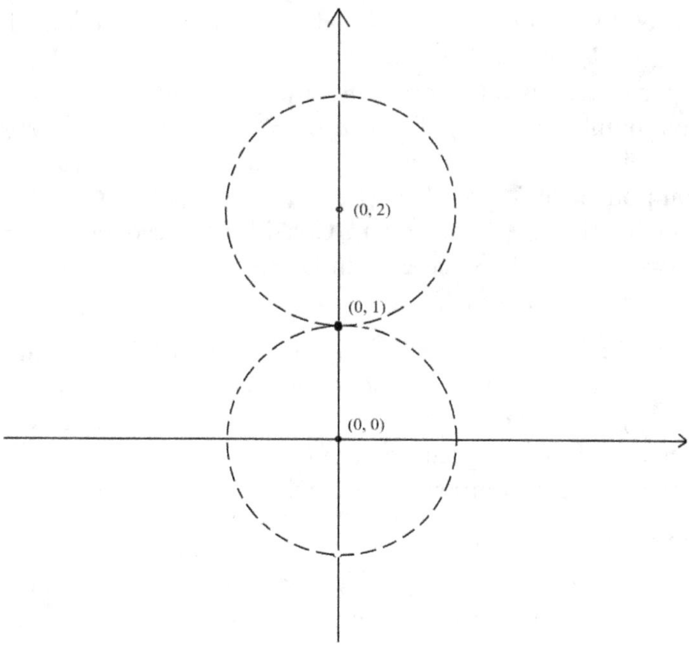

Figure 20.1. Example 6: $B_1(0,0) \cup B_1(0,2) \cup \{(0,1)\}$ a connected set.

In the following theorem, we will verify that the connected components of a topological space are always closed subsets.

Theorem 20.8 *Suppose C is a component of a space, S. Then C is a closed subset of S.*

Proof. We are given that C is a component of the topological space S.

Since C is, by definition, connected, then $\text{cl}_S C$ is also connected. Since C is the largest connected set containing one of its points, then $\text{cl}_S C = C$. So, C is closed.

The definition of the "connected space" may lead one to surmise that components in S must also be open in S. If the reader is tempted by this assertion, the following example may incite this reader to reconsider.

Example 7. Consider the subspace, $\mathbb{Q}$ with the subspace topology. The subset $\{5\}$ is a connected closed subset of $\mathbb{Q}$. We claim that the component, C_5, of 5 is $\{5\}$.

Suppose (u, v) is an interval in $\mathbb{Q}$ which contains 5. Let i be an irrational number in (u, v). Then $(u, v) = [(u, i) \cup (i, v)] \cap \mathbb{Q}$ represents the union of two clopen sets in $\mathbb{Q}$ which contains 5. Then (u, v) cannot be a component of 5. So, $C_5 = \{5\}$.

We claim that C_5 is not open. Consider the sequence $T = \{5 + 1/n : n = 1, 2, 3, \ldots\}$. Since T is in $\mathbb{Q} \backslash C_5$ but does not converge in $\mathbb{Q} \backslash C_5$, C_5 is not open in $\mathbb{Q}$.

Example 8. Let $\{S_i : i \in I\}$ be a family of topological spaces and $S = \prod_{i \in I} S_i$ be the corresponding product space. Suppose that $u = \langle u_i \rangle_{i \in I} \in S$. For each $i \in I$, let C_i be the unique connected component in S_i which contains the point u_i. Suppose $C = \prod_{i \in I} C_i$. Show that C is the component in S which contains the point u.

Solution: We are given $u = \langle u_i \rangle_{i \in I} \in S = \prod_{i \in I} S_i$. If $u_i \in C_i$ for each $i \in I$, then $u \in C = \prod_{i \in I} C_i$. By Theorem 20.5 (which states that products of connected spaces are connected) C is a connected subset of S containing u. We claim that C is a component in S. Suppose not. Suppose D is the component in S which contains u

and suppose $v = \langle v_i \rangle_{i \in I} \in D \backslash C$. Then there must be some $j \in I$ such that $v_j \notin C_j$. Since $\pi_j : S \to S_j$ is continuous, it must be that $\pi_j[D]$ is a connected subset of S_j where $C_j \cup \{v_j\} \subseteq \pi_j[D]$. Since C_j is the largest connected set in S_j which contains u_j, then $\pi_j[D]$ cannot be connected. We have a contradiction due to our supposition that $D \backslash C$ is non-empty. So, C is the component in S which contains u, as claimed.

The previous example shows that

> *the product of connected components is a connected component of the product.*

Example 9. Let $\{S_i : i \in I\}$ be an infinite family of discrete topological spaces, none of which contain only one point. Then the connected components of each S_i must be singleton sets. Let $S = \prod_{i \in I} S_i$ be the corresponding product space. Show that the connected components of S are also singleton sets. Show that the product space, S, is not a discrete space.

Solution: Let $u = \langle u_i \rangle_{i \in I} \in S$. Since S_i is discrete, $C_i = \{u_i\}$ is a connected component, for each i. Then, by the example above (where it is shown that the product of connected components is a connected component),

$$C = \prod_{i \in I} C_i = \prod_{i \in I} \{u_i\} = \{u\}$$

is the connected component in S containing u. So the connected components of S are singleton sets, as required.

Are sets whose components are singletons discrete? If S is the disjoint union of components each of which is a singleton set, S need not be a discrete space. Consider $S = \prod_{i \in I} S_i$ and $u = \langle u_i \rangle_{i \in I} \in S$, as in the above example (where S is equipped with the product topology). Consider a basic open neighborhood,

$$B = \cap \{\pi_i^{\leftarrow}(u_i) : i \in F\}$$

of u where F is finite in I. Suppose each S_i is discrete. Then $C_u = \{u\}$ is a connected component in S. Since B cannot be contained in $\{u\}$, C_u is not open.

Hence,

the product of discrete spaces need not be a discrete space.

Example 10. Show that the components of the Cantor set are all singleton sets.

Solution: Recall that, in Example 2 of page 174 and Theorem 7.19, the Cantor set was defined as being the image of the infinite product, $\prod_{n\in\mathbb{Z}^+}\{0,2\}$, of the discrete space, $\{0,2\}$, under the homeomorphism function $\varphi : \prod_{n\in\mathbb{Z}^+}\{0,2\} \to [0,1]$. By the above example, the components of $\prod_{n\in\mathbb{Z}^+}\{0,2\}$ are singleton sets. The homeomorphic image of a component must be a component. So, the components of the Cantor set are singleton sets.

Example 11. *Decomposition of a compact space into components.* Let S be a compact Hausdorff topological space which is not connected. We will partition the space S by its connected components. That is,

$$\mathscr{D}_S = \{C_x : x \in S,\ C_x \text{ a component containing } x\}$$

denotes the family of all connected components of S where $x \in C_x$. Let $\theta : S \to \mathscr{D}_S$ be defined such that $\theta(u) = C_u$ if $u \in C_u$ in $\mathscr{D}_S$. Then we obtain a decomposition space $(\mathscr{D}_S, \tau)$ where U is open in $\mathscr{D}_S$ if and only if $\theta^{\leftarrow}[U]$ is open in S and is the union of connected components of S.

Show that the only connected components of $\mathscr{D}_S$ are points.[2]

Solution: We are given that S is compact Hausdorff and the function

$$\theta : S \to \mathscr{D}_S$$

defined as

$$\theta(u) = C_u \quad \text{if } u \in C_u$$

Let $p, q \in S$, contained in distinct components of S. Then $\theta(p)$ and $\theta(q)$ are distinct points in $\mathscr{D}_S$.

---

[2]We will later refer to such spaces as being *totally disconnected.*

Each connected component, $\theta^{\leftarrow}(C_x)$, of S has been shown to be closed in S. Since S is compact, then $\theta^{\leftarrow}(C_x)$ is compact. Now "compact and Hausdorff" $\Rightarrow$ "normal", so, for each $x \neq p$ in S, there exists a pair of *disjoint* open sets, V_x and P_x, such that $\theta^{\leftarrow}(C_x) \subseteq V_x$ and $\theta^{\leftarrow}(C_p) \subseteq P_x$. Then the family open sets,

$$\mathscr{V} = \{V_x : x \in S, x \neq p\} \cup \{P_x : x \in S, x \neq p\}$$

forms an open cover of S. Since S is compact, then $\mathscr{V}$ has a finite subcover (of S)

$$\mathscr{F} = \{V_{x_i} : x_i \in S, x_i \neq p, i \in F\} \cup \{P_{x_i} : x_i \in S, x_i \neq p, i \in F\}$$

where F is a finite indexing set and $V_{x_i} \cap P_{x_i} = \varnothing$. Then S is the disjoint union of the two open subsets,

$$V = \cup\{V_{x_i} : x_i \in S, x_i \neq p, i \in F\}$$
$$P = \cap\{P_{x_i} : x_i \in S, x_i \neq p, i \in F\}$$

where $\theta^{\leftarrow}(C_q) \subseteq V$ and $\theta^{\leftarrow}(C_p) \subseteq P$. Since P and V are both open and disjoint such that $S = P \cup V$, then V and $S \backslash V = P$ are clopen in S and so are both compact. So, $\mathscr{D}_S$ is the disjoint union of the compact sets $\theta[V]$ and $\theta[P]$. Each of these must then be clopen in $\mathscr{D}_S$. Given that the set, $\theta[V]$, is clopen, then

$$\theta(q) = C_q \subseteq \theta[V] \subseteq S \backslash \theta(p)$$

as claimed.

It follows that no two points in $\mathscr{D}_S$ belong to the same connected component and so the only connected components of $\mathscr{D}_S$ are its points.

20.4 Locally Connected: Spaces with an Open Base of Connected Sets

Even when a space is not connected it may have a base for open sets whose elements are connected subspaces. For example, any discrete space, S, can easily be seen to have an open base of connected subspaces, $\mathscr{B} = \{\{x\} : x \in S\}$. So does the connected space of real numbers, $\mathbb{R}$, with the usual topology since it has an open base, $\mathscr{B} = \{(a, b) : a, b \in \mathbb{R}, a < b\}$.

We formally define this particular notion.

Definition 20.9 We say that the space S is *locally connected* if it has an open base, $\mathscr{B} = \{B_i : i \in I\}$, where each element, B_i, is a connected subspace.

What would a non-locally connected space look like? We construct a connected subspace of $\mathbb{R}^2$ which is not locally connected.

Example 12. For each $n \in \mathbb{N}\backslash\{0\}$, let

$$g_n(x) = \left(\frac{\sin\left(\frac{\pi}{4n}\right)}{\cos\left(\frac{\pi}{4n}\right)} \right) x$$

For each n, let L_n denote the line,

$$L_n = \{(x, y_n) : y_n = g_n(x),\ x \geq 0\}$$

Let

$$T = \cup\{L_n : n \in \mathbb{N}\backslash\{0\}\} \cup \{(1, 0)\}$$

be a subspace of $\mathbb{R}^2$ with the usual topology. Show that T is a connected subspace which is not locally connected.

Solution: For each $n \in \mathbb{N}\backslash\{0\}$, L_n is a line in $\mathbb{R}^2$ which is homeomorphic to $\mathbb{R}^+$ and so is connected. Since each L_n contains the point, $(0, 0)$, then $\cup\{L_n : n \in \mathbb{N}\backslash\{0\}\}$ is connected. The connected set $\cup\{L_n : n \in \mathbb{N}\backslash\{0\}\}$ is dense in T and so T is connected.

Consider the open neighborhood base,

$$\mathscr{B} = \{B_{1/k}(1, 0) \cap T : k \geq 2\}$$

of the point $(1, 0) \in T$. For each k, the open "ball", $B = B_{1/k}(1, 0) \cap T$, in T contains countably many line segments (belonging to the L_n's). If $j \neq m$ there is a line L^* whose slope is strictly in between the slopes of L_j and L_m, so the line of L^* does not appear in T. Then B is disconnected at the line L^*. So, T is not locally connected at the point (1, 0).

20.5 Which Spaces Have Clopen Connected Components?

Suppose S is a locally connected space and U is an open subspace of S. Then S has an open base (Figure 20.2),

$$\mathscr{B} = \{B_i : i \in I\}$$

of connected subsets. If $u \in U$, let C_u denote a connected component of the subspace, U, which contains u. We have already proven that C_u must be closed in U.

We claim that C_u must be a clopen subset in U. To see this simply note that there is some connected B_j in $\mathscr{B}$ such that $u \in B_j \subseteq U$. Since $u \in C_u$ and B_j is connected and intersects C_u, then $B_j \cup C_u$ is connected so $u \in B_j \subseteq C_u$. Since B_j and U are open in S, B_j is open in U and so C_u is open in U and so is clopen in U, as claimed.

This leads to an interesting characterization of the locally connected property on a space.

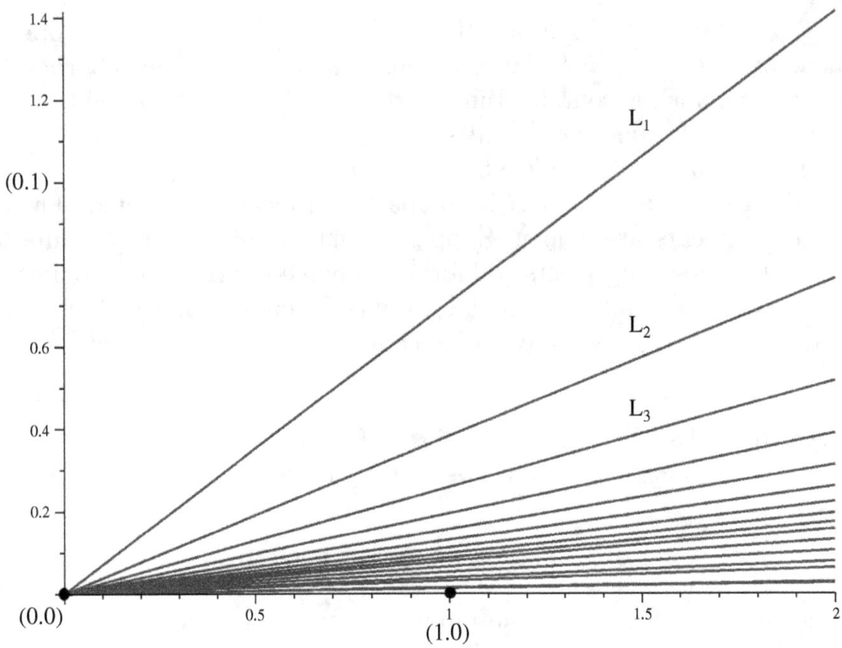

Figure 20.2. A connected but not locally connected subspace of $\mathbb{R}^2$.

Theorem 20.10 *Suppose S is a topological space.*

(a) *The space S is a locally connected space if and only if every open subspace of S (including S itself) has clopen components.*
(b) *If the space S is both locally connected and compact then it has at most finitely many connected components.*

Proof. For part (a): The direction ($\Rightarrow$) of part (a) is proven in the paragraph above.

We prove the direction ($\Leftarrow$): Suppose that every open subspace of S has clopen components. We are required to show that S has an open base of connected sets.

Let V be an open neighborhood of a point x in S. Let C_x be a connected component of V which contains x.

We claim that C_x is a connected subset of S. If not, then there is a clopen subset, U, of S which intersects only a part of C_x. Then $U \cap V$ is a clopen subset of V which intersects only a part of C_x, contradicting the fact that C_x is a connected subset of V. So, C_x is connected in S, as claimed.

By hypothesis, C_x is an open subset of V. So, C_x is an open neighborhood of x in S. So, C_x is an open connected neighborhood of x in S which is contained in V. So every open neighborhood V of x contains an open connected set C_x in S such that $x \in C_x \subseteq V$. This implies that S is a locally connected space.

For part (b): Suppose S is compact and locally connected. Then its components are clopen. Suppose it has infinitely many components then its components will form an open cover of S with no finite subcover, a contradiction. So a locally connected compact space can have at most finitely many components.

Example 13. Let $g_n(x) = [\sin(\pi/4n)/(\cos(\pi/4n)]x$ and $L_n = \{(x, y_n) : y_n = g_n(x)\}$ where $n \in \mathbb{N}\backslash\{0\}$. Consider the space S defined as

$$S = \cup\{L_n : n \in \mathbb{N}\backslash\{0\}\} \setminus \{(0,0)\}$$

The subspace S of $\mathbb{R}^2$ is the infinite union of disjoint straight lines in $\mathbb{R}^2$. Each point has a neighborhood base of connected sets (line segments).

Consider any two distinct sets $S_j = L_j \backslash \{(0,0)\}$ and $S_k = L_k \backslash \{(0,0)\}$ of slope s_j and s_k, respectively, in S. These can easily seen to be separable by two disjoint open subsets of $\mathbb{R}^2$ and so can be seen as four disjoint open subsets of S. It can also be shown that each S_i is a connected component of S and so is closed in S. (Note that the closure of S in $\mathbb{R}^2$ adds the line $0(x) = 0$ to S. So, S is the union of infinitely many clopen components of S.)

This does not contradict part (b) of the previous theorem since S is not a compact subspace of $\mathbb{R}^2$.

Theorem 20.11 *Let $\{S_i : i \in I\}$ be a family of topological spaces and $S = \prod_{i \in I} S_i$ be the corresponding product space. Then S is locally connected if and only if every factor, S_i, is locally connected and, at most, finitely many of the factors are* not *connected.*

Proof. We are given that $\{S_i : i \in I\}$ is a family of topological spaces and $S = \prod_{i \in I} S_i$ is a product space.

($\Rightarrow$) Suppose $S = \prod_{i \in I} S_i$ is locally connected.

Claim 1: That all but finitely many of the factors are connected spaces. Let $u \in S$. Since S is locally connected, then, by Theorem 20.10, $u \in C$ for some clopen connected component, C, of S. Let U be a basic open neighborhood of u in S such that $u \in U \subseteq C$. That is, for a finite subset, F, of I,

$$u \in U = \cap\{\pi_i^{\leftarrow}[U_i] : i \in F\} \subseteq C$$

where U_i is an open subset of S_i.

Consider $i \notin F$. Then $\pi_i[U] = S_i \subseteq \pi_i[C]$ (since $U \subseteq C$). So, $\pi_i[U] = S_i = \pi_i[C]$. Since C is a connected component of S and π_i is continuous, S_i is connected for all i's except possibly the ones in F. This establishes the Claim 1.

Claim 2: That each S_i is locally connected. For $i \in I$ let $v_i \in S_i$. Let W be an open neighborhood of v_i in S_i and $v \in S$ such that $\pi_i(v) = v_i$. It suffices to show that there is a connected open neighborhood of v_i which is entirely contained in W.

Since S is locally connected, and $\pi_i^{\leftarrow}[W]$ is an open neighborhood of v in S, there exists an open connected subspace, V, in S such that $v \in V \subseteq \pi_i^{\leftarrow}[W]$. Since π_i is both continuous and open, then $\pi_i[V]$

is a connected open neighborhood of v_i contained in W. So, v_i has a connected neighborhood base. Then S_i is locally connected. This establishes Claim 2.

We are done with this direction.

($\Leftarrow$) Suppose every factor, S_i, is locally connected and at most finitely many of these factors are not connected. We are required to show that the product space, S, is locally connected.

Suppose F_c is the largest subset of the indexing set, I, such that, if $i \in F_c$, S_i is not connected. Then, by hypothesis, F_c is finite.

Let

$$x = \langle x_i \rangle_{i \in I} \in S = \prod_{i \in I} S_i$$

and let $V = \cap\{\pi_i^{\leftarrow}[U_i] : i \in F\}$ be any open base element containing x. Let $F_1 = F \cup F_c$ and

$$U_1 = \cap\{\pi_i^{\leftarrow}[U_i] : i \in F_1\}$$

Clearly, since $F \subseteq F_1$, then $U_1 \subseteq V$.

We *claim* that the point, x, has a connected neighborhood, W, which is contained in U_1.

Given that every S_i is locally connected, for each $i \in F_1$, there must be an open connected set, C_i in S_i, such that $x_i \in C_i \subseteq U_i$.

Then

$$x \in \cap\{\pi_i^{\leftarrow}[C_i] : i \in F_1\} = W \subseteq U_1 \subseteq V$$

exhibits an open neighborhood, W, of x in S which is contained in V. If $j \notin F_1$, the jth factor of W is the connected space, S_j. If $j \in F_1$, the jth factor of W is the connected, C_j. By Theorem 20.5, W a connected subspace in S.

So, x has a connected open neighborhood, W, contained in U_1, as claimed. Since $U_1 \subseteq V$, S is locally connected, as required.

20.6 Pathwise Connected Spaces

There are different ways of perceiving the connected property on a set. The formal definition of the *connected space* given earlier was "...a space that is not the union of two disjoint open sets" appeals to its proponents because of its simplicity. It is a clear and unambiguous definition. However, some critics might have argued that this definitions sounds to much like "a connected space is one that is not disconnected; and we all know what *disconnected* means". The formal definition, along with a few examples, and some discussion of the properties derived from it, allowed users to develop a better understanding of what a connected space is. But, in spite of this, the space, T, described in the example found on page 454, may not appear to be connected, by some participants of a random survey. We define below, the

pathwise connected property,

a slightly stronger definition than that of the *connected property* in the sense that *pathwise connected spaces* are always connected, but not conversely. To some, it is a more appealing form of connectedness even though it is sightly more difficult to work with.

Definition 20.12 Let S be a topological space. A *path* in S is a *continuous* function, $f : [0,1] \to S$, mapping the closed unit interval into S. If, for a path, $f : [0,1] \to S$, $a = f(0)$ and $b = f(1)$, we say that...

f is a path from a to b in S

where a is referred to as being the *initial point of the path f* and b is called the *terminal point of the path f*. These terms prescribe a *direction* for the image, $f[0,1]$, of f.

The space, S, is said to be a

pathwise connected space

if, for any pair of points, a and b in S, there is a path, $f : [0,1] \to S$, joining $a = f(0)$ to $b = f(1)$ (Figure 20.3).[3]

---

[3]If the continuous function, f, from $a = f(0)$ to $b = f(1)$ is a *homeomorphism* then f is called an *arc*. If every path on S is a homeomorphism then the space is said to be *arcwise connected.*

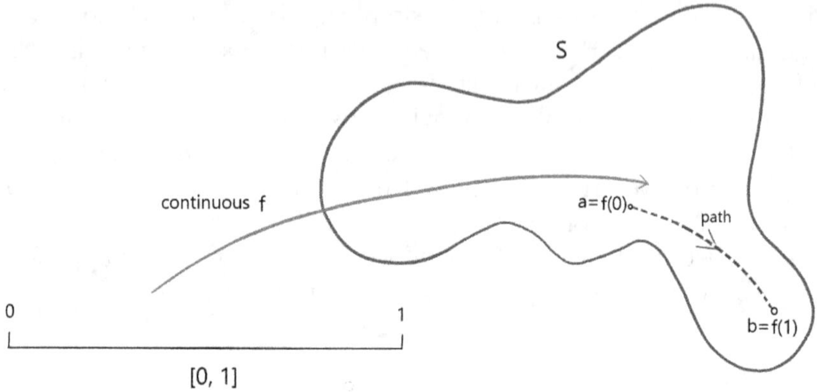

Figure 20.3. A pathwise connected space with a path $f[0,1]$ in S.

Since $[0,1]$ is connected in $\mathbb{R}$ and f is continuous on $[0,1]$, then the image $f[0,1]$ is a connected subspace of S. The set of all real numbers, $\mathbb{R}$, is of course pathwise connected since for any a and b in $\mathbb{R}$

$$f(x) = (b-a)x + a$$

is a path connecting $a = f(0)$ to $b = f(1)$. This particular perception of "connected" is also quite intuitive. It suggests that if one can draw a curve joining any two points in S, without lifting the pencil off the page, then the space is connected. The following theorem guarantees that the pathwise connected spaces are connected as described by the formal definition.

Theorem 20.13 *If the topological space, S, is pathwise connected then it is connected.*

Proof. Suppose S is pathwise connected. If $a \in S$, then, for any $b \in S$, there is a path, $f_b : [0,1] \to S$, such that $a = f_b(0)$ is its initial point and $b = f_b(1)$ is its terminal point.

Since f_b is continuous, then $f_b[0,1]$ is connected. Then, given the point a, the family of subsets of S,

$$\mathscr{C} = \{f_b[0,1] : b \in S, b \neq a\}$$

is a family of connected sets which entirely covers S each of which contains the point a as initial point.

Since S is the union of connected subsets of S each containing a common element, then S is connected.

We have then confirmed that

"Pathwise connected space" $\Rightarrow$ "Connected space"

It is worth noting that, if there is a path, f, from the point a to the point b in a space S, then there is a path, $g : [0,1] \to S$, defined as $g(x) = f(1-x)$ which goes in the opposite direction, from b to a.

20.6.1 *A connected space which is not pathwise connected*

Example 14. We reconsider the example presented earlier described as follows: For each $n \in \mathbb{N}\backslash\{0\}$, let $L_n \subseteq \mathbb{R}^2$ denote the set,

$$L_n = \left\{(x, y_n) : y_n = g_n(x) = \left(\frac{\sin\left(\frac{\pi}{4n}\right)}{\cos\left(\frac{\pi}{4n}\right)}\right)x, \quad x \geq 0\right\}$$

Let

$$T = \cup\{L_n : n \in \mathbb{N}\backslash\{0\}\} \cup \{(1,0)\}$$

be a subspace of $\mathbb{R}^2$ with the usual topology. We have shown (in an example on page 454) that T is a connected subspace but is not locally connected. Show that T is not pathwise connected.

Solution: Suppose there is a path joining the point $a = (1,0)$ in T to a point, $b = (u,v) \neq a$ in $T\backslash\{(0,0)\}$. That is, suppose there is a continuous function $f : [0,1] \to T$, such that

$$f(0) = (1,0)$$
$$f(1) = b = (u,v)$$

Note that, any natural number > 0 the family $\{L_n\backslash\{(0,0)\} : n > m\} \cup \{(1,0)\}$ is the union of an infinite family of connected components of $T\backslash\{(0,0)\}$.

Continuity of $f : [0,1] \to T$ guarantees that

$$0 < \varepsilon < 1/10 \quad \Rightarrow \quad f[0,\delta) \subseteq B_\varepsilon(1,0) \text{ for some } \delta > 0$$

Since f is continuous and $[0,\delta)$ is connected, then $f[0,\delta)$ is connected.

Suppose $q \in L_k \cap f[0, \delta)$ for some $k \in \mathbb{N}\backslash\{0\}$. Then $f[0, \delta)$ intersects the two components, $L_k\backslash\{(0,0)\}$ and $\{(1,0)\}$, of $T\backslash\{(0,0)\}$; so $f[0, \delta)$ cannot be connected, contradicting continuity of f. Then $f[0, \delta) = \{(1,0)\}$.

So continuity of f holds true only for $b = (1,0)$. So, T is not pathwise connected.

In the following theorem, we confirm that pathwise connectedness is carried over by continuous functions.

Theorem 20.14 *If $g : S \to T$ is a continuous function mapping a pathwise connected space S to a space T, then $g[S]$ is a pathwise connected subspace of T.*

Proof. Suppose S is a pathwise connected space. Then if u and v are distinct points in S there is a continuous function $f : [0, 1] \to S$ such that $f(0) = u$ and $f(1) = v$.

Suppose now that $g : S \to T$ is a continuous function mapping S onto $g[S] \subseteq T$. We claim that $g[S]$ is pathwise connected. To see this let a and b be distinct points in $g[S]$. Then, we can choose distinct points, $x \in g^{\leftarrow}(a)$ and $y \in g^{\leftarrow}(b)$. Since S is pathwise connected, there is a continuous function $f : [0, 1] \to S$ such that $f(0) = x$ and $f(1) = y$. Then the continuous function, $g \circ f : [0, 1] \to g[S]$, maps 0 to a and 1 to b.

So, $g \circ f$ is a path from a to b.

We conclude that $g[S]$ is pathwise connected, as required.

We now consider invariance of the pathwise connected property with respect to products.

Theorem 20.15 *Let $\{S_i : i \in I\}$ be a family of topological spaces. Then the product space, $S = \prod_{i \in I} S_i$, is pathwise connected if and only if every factor, S_i, is pathwise connected.*

Proof. We are given that $\{S_i : i \in I\}$ is a family of topological spaces and that $S = \prod_{i \in I} S_i$ is the corresponding product space.

For ($\Rightarrow$), if S is pathwise connected, since each projection map, π_i, is continuous, then each S_i is pathwise connected.

For ($\Leftarrow$), suppose S_i is pathwise connected for each $i \in I$.

Let $\langle a_i \rangle_{i \in I}$ and $\langle b_i \rangle_{i \in I}$ be distinct points in S. Since each S_i is pathwise connected, then, for each $i \in I$, there is a continuous function, $f_i : [0, 1] \to S_i$, such that $f_i(0) = a_i$ and $f_i(1) = b_i$. Let $f : [0, 1] \to S$ be defined as

$$f(u) = \langle f_i(u) \rangle_{i \in I}$$

Then $f_i(u) = (\pi_i \circ f)(u)$. By Lemma 7.11, the function, $f : [0, 1] \to S$, is continuous on $[0, 1]$ if and only if each f_i is continuous on $[0, 1]$. Since each f_i maps 0 to a_i and 1 to b_i, then

$$f(0) = \langle f_i(0) \rangle_{i \in I} = \langle a_i \rangle_{i \in I}$$
$$f(1) = \langle f_i(1) \rangle_{i \in I} = \langle b_i \rangle_{i \in I}$$

So, S is pathwise connected, as required.

Example 15. It follows from the previous theorem that the space, $S = \mathbb{R}^{\mathbb{R}}$, equipped with the product topology is pathwise connected since each factor, $\mathbb{R}$, is pathwise connected.

Example 16. Let u be a point in a topological space, S. Show that S is pathwise connected if and only if there is a path joining each point, $x \in S$, to u (Figure 20.4).

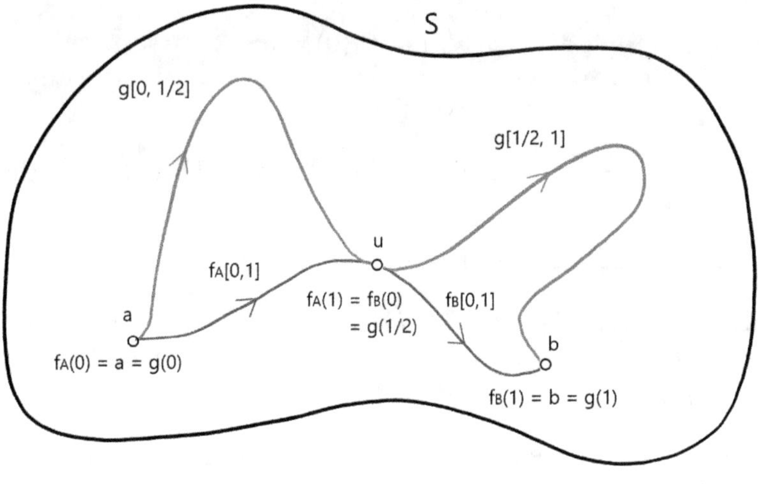

Figure 20.4. A pathwise connected theorem with f_A, f_B, and g.

Solution: For the direction ($\Rightarrow$) the statement is obviously true.

For the direction ($\Leftarrow$), we are given that every point in S can be joined by a path to u. Suppose a and b are distinct points in S. We are required to show that there is a path linking a to b.

By hypothesis, there are two paths, $f_A : I \to S$ and $f_B : I \to S$ such that

$$f_A(0) = a$$

$$f_A(1) = u$$

$$f_B(1) = b$$

$$f_B(0) = u$$

Consider the function

$$g(x) = \begin{cases} f_A(2x) & \text{if } x \in [0, 1/2] \\ f_B(2x-1) & \text{if } x \in [1/2, 1] \end{cases}$$

Then g is continuous on $[0, 1]$ where

$$g(0) = a$$

$$g(1/2) = f_A(1) = u$$

$$g(1/2) = f_B(0) = u$$

$$g(1) = b$$

So, $g : [0, 1] \to S$ forms a path from a to b. Then S is pathwise connected.

Theorem 20.16 *Let $\{S_i : i \in I\}$ be a family of pathwise connected topological spaces which have the point u in common. Then space $S = \cup\{S_i : i \in I\}$, is pathwise connected.*

Proof. The statement follows immediately from the argument presented in the previous example.

Definition 20.17 If S is a topological space and $u \in S$, then we define the

pathwise component of u

as being the union of all pathwise connected spaces that contain the point u. Hence, if C is the pathwise component containing u, it is the largest subspace containing u which is pathwise connected.

Example 17. Consider the subspace, $T = \cup\{L_n : n \in \mathbb{N}\backslash\{0\}\} \cup \{(1,0)\}$ of $\mathbb{R}^2$ as presented in the example on page 461 where we showed that T is connected but not pathwise connected. Show that a pathwise component in T need not be closed.

Solution: Consider the subspace, $S = T\backslash\{(1,0)\}$. The space, S, is an infinite union of straight lines in $\mathbb{R}^2$ each of which is pathwise connected and contains the point $(0,0)$. From this we deduce that S is pathwise connected. It can also be seen to be a pathwise component of T. But since S is missing the limit point $(1,0)$ in T, it is not closed in T. So a pathwise components of a space need not be closed.

Under particular conditions, it does occur that a pathwise component is a clopen subset. This condition is described in the following theorem.

Theorem 20.18 *Let S be a topological space. The pathwise components of S are clopen if and only if every point in S has a pathwise connected neighborhood.*

Proof. We are given that S is a topological space.

($\Rightarrow$) Suppose each pathwise component is clopen. Let $x \in S$. Then there exists be a pathwise component C containing x. We are required to show that C is a neighborhood of x. This follows immediately from our hypothesis.

($\Leftarrow$) Let $x \in S$ and C be the (unique) pathwise component containing x. By hypothesis, C is a neighborhood of x. We are required to show that C is clopen.

Then since C is the largest pathwise connected neighborhood which contains x, then $x \in \text{int}_S U \subseteq C$. So, C is open. We claim it is also closed: Since the components partition the space S, then C is the complement of the union of all other open components of S and so is also closed. So all pathwise components are clopen.

The equivalent conditions, "pathwise components are clopen" and "every point in S has a pathwise connected neighborhood" described in the above theorem, when combined to the connected property characterizes pathwise connectedness, as shown in the next theorem.

Theorem 20.19 *Let S be a topological space. The space, S, is pathwise connected if and only if S is both connected and each of its points, has a pathwise connected neighborhood.*

Proof. We are given that S is a topological space.

($\Rightarrow$) We are given that S is a pathwise connected space. The fact that S is a connected space follows from Theorem 20.13. That every point has a pathwise connected neighborhood follows from the the fact that the pathwise connected space is a neighborhood of every point in S.

($\Leftarrow$) Suppose S is connected and every one of its points has a pathwise connected neighborhood. We are required to show that S is pathwise connected.

Let $x \in S$. By hypothesis, x has a pathwise connected neighborhood, C. Then, by Theorem 20.18, C is clopen. Since pathwise connected spaces are connected, the clopen set, C, cannot be a proper subset. So, the pathwise component, C, must be S itself. So, S is pathwise connected.

In $\mathbb{R}^n$, any connected open subspace is a pathwise connected subspace as the following theorem shows.

Theorem 20.20 *Let $\mathbb{R}^n$ be equipped with the usual topology. An open subspace, U, of $\mathbb{R}^n$ is connected if and only if U is a pathwise connected subspace of $\mathbb{R}^n$.*

Proof. The space $\mathbb{R}^n$ be equipped with the usual topology and U is an open subspace of $\mathbb{R}^n$.

($\Rightarrow$) Suppose U is a connected subspace of $\mathbb{R}^n$. We are required to show that U is a pathwise connected subspace. By Theorem 20.19 it suffices to show that each point in U has a pathwise connected neighborhood.
Let $x \in U$. Since U is open in $\mathbb{R}^n$, there is an $\varepsilon > 0$ such that $x \in B_\varepsilon(x) \subseteq U$. Let $a = \langle a_i \rangle_{i=1\ldots n}$ and $b = \langle b_i \rangle_{i=1\ldots n}$ be distinct points in $B_\varepsilon(x)$. If $f(x) = (b-a)(x) + a$ then $f[0,1]$ is a line segment entirely contained in $B_\varepsilon(x)$. So, $B_\varepsilon(x)$ is pathwise connected. From this we can conclude that every point in U has a pathwise connected neighborhood, so by the above theorem, U is pathwise connected, as required.

($\Leftarrow$) Suppose U is pathwise connected. By Theorem 20.13, U is a connected subspace. Done.

20.7 Totally Disconnected Spaces

For many spaces, the only connected subspaces are singleton sets. We briefly discuss some basic properties of these types of spaces.

Definition 20.21 A topological space, S, is said to be *totally disconnected* if and only if the only connected components of S are its points.

Discrete spaces, the subspace $\mathbb{Q}$, the Cantor set (viewed as a subspace of $[0,1]$) and $\mathbb{R}\backslash\mathbb{Q}$ all previously discussed sets in this chapter are standard examples of totally disconnected spaces.

We verify that the totally disconnected property is preserved over arbitrary products.

Theorem 20.22 *The product space of totally disconnected spaces is itself totally disconnected.*

Proof. Let $\{S_i : i \in I\}$ be a family of totally disconnected spaces and $S = \prod_{i\in I} S_i$ be the corresponding Cartesian product space. We are required to show that S is totally disconnected.

Let C be a connected component in S. It suffices to show that C is a singleton set. Suppose $a = \langle a_i \rangle_{i\in I}$ and $b = \langle b_i \rangle_{i\in I}$ are two points in C. If $a \neq b$ then $a_j \neq b_j$ for some j. Then $\{a_j, b_j\} \subseteq \pi_j[C]$, a continuous image of the connected set, C, in S. Since the only connected sets in S_j are singleton sets, then $a_j = b_j$. So, $a = b$ in C. So, C is a singleton set. So, S is totally disconnected, as required.

Example 18. Show that if S is totally disconnected and $f : S \to T$ is continuous then $f[S]$ need not be totally disconnected.

Solution: In Theorem 7.20, we showed that there is a function $\psi : C \to [0, 1]$ which maps the totally disconnected Cantor set C onto the connected set, $[0, 1]$. So the continuous image of a totally disconnected set need not by totally disconnected.

Total disconnectedness versus the zero-dimensional property. Recall from Definition 5.16 that zero-dimensional spaces are those spaces that have an open base of clopen sets. Equivalently, each point has a neighborhood base of clopen sets. We know that a non-empty clopen set allows us to express a space as a union of two open sets. So, we surmise that zero-dimensional spaces and totally disconnected spaces are similar in many ways. For certain types of topological spaces they may even be equivalent, as we shall soon see.

Example 19. Show that a T_1 zero-dimensional space, S, is totally disconnected.

Solution: Since S is T_1, any point in S is closed. Given distinct points p and y the set $S\backslash\{y\}$ is an open set containing p. Since p has a neighborhood base of clopen sets, then there exists a clopen set U such that $p \in U \subset S\backslash\{y\}$. So, p and y cannot belong to the same component. So the only connected component which contains p is $\{p\}$. This means that S is totally disconnected.

Example 20. Suppose S is a compact T_2 space satisfying the property: "For distinct pairs of points, p and x, there is a clopen set U_x such that $x \in U_x \subset S\backslash\{p\}$". Show that S must then be zero-dimensional.

Solution: Given: S is a compact T_2 space. Suppose V is a proper open subset containing the point p. It suffices to show that there is a clopen set W such that $p \in W \subseteq V$.

By hypothesis, for each $x \in S\backslash V$, there is a clopen set U_x such that $x \in U_x \subset S\backslash\{p\}$. Now, $S\backslash V$ is closed and so is a compact subset of S. Then

$$\mathscr{U} = \{U_x : x \in S\backslash V\}$$

covers $S\backslash V$ with clopen sets each of which misses p. Let

$$\mathscr{U}_F = \{U_{x_i} : i \in F\}$$

be a finite subfamily of $\mathscr{U}$ which covers $S\backslash V$. If $M = \cup\{U_{x_i} : i \in F\}$ and $W = S\backslash M$, W is a clopen subset of S such that $p \in W \subseteq V$. So, S is zero-dimensional, as required.

Lemma 20.23 *Let S be a compact Hausdorff space, $x \in S$ and C_x be a connected component containing x. Then $C_x = \cap\{D : D$ is clopen in $S, x \in D\}$* (Figure 20.5).

Proof. Given: That S is compact Hausdorff and C_x is a connected component which contains x. If U is a clopen neighborhood of x then $C_x \subseteq U$ (since C_x is connected).

Let

$$\mathscr{D} = \{D : D \text{ is clopen in } S, x \in D\}$$

and let

$$M = \cap\{D : D \text{ is clopen in } S, x \in D\}$$

Then C_x is a subset of M, a closed subset of the compact space, S.

Claim: We claim that M is connected in S.

Proof of claim: Suppose A and B are subsets of M, such that $M = A \cup B$. To show that M is connected it suffices to show that either A or B is empty. Since A and B are closed in M and M is closed in S, then A and B are closed in S; so they are both compact.

Since S is Hausdorff, there exists *disjoint* open subsets U and V of S which contain A and B, respectively. Then

$$M = A \cup B \subseteq U \cup V$$

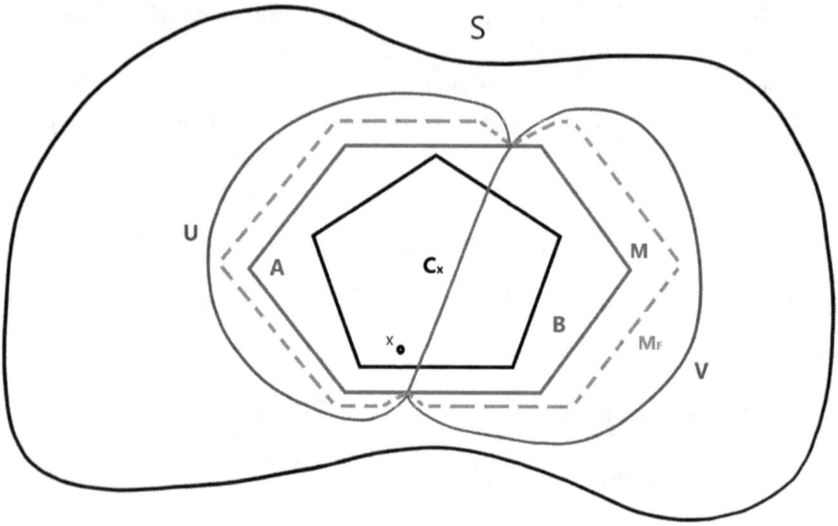

Figure 20.5. On a connected component C_x in a compact T_2 space: $M = C_x$.

Without loss of generality suppose $x \in A \subseteq U$. Then $\mathscr{D}$ contains a finite subset, $\{D_i : D_i \in \mathscr{D},\ i \in F\}$, such that

$$M \subseteq M_F = \cap\{D_i : D_i \in \mathscr{D},\ i \in F\} \subseteq U \cup V$$

for, if $(S \backslash D) \cap [S \backslash (U \cup V)] \neq \varnothing$ for all $D \in \mathscr{D}$, given that S is compact, M would intersect $S \backslash (U \cup V)$.

See that M_F is a clopen neighborhood of $M = A \cup B$.

Then $M_F \cap U$ and $M_F \cap V$ are (disjoint) open subsets of S.

Since $M_F \cap V = M_F \cap S \backslash V$ is closed in S and

$$M_F \backslash V = M_F \cap U$$

the subset, $M_F \cap U$, is also closed in S, so $M_F \cap U$ is a clopen subset of S.

Since $x \in A \subseteq M_F \cap U$, $M_F \cap U$ is a clopen neighborhood of x and so $M_F \cap U$ belongs to $\mathscr{D}$.

Then, by definition of M (which is the disjoint union of A and B), $M \subseteq M_F \cap U$. Since $B \subseteq M_F \cap V$ and $U \cap V = \varnothing$, then $B = \varnothing$. So, M must be connected, as claimed.

Since C_x is the largest connected set containing x, then $C_x = M$. We are done.

Theorem 20.24 *A locally compact Hausdorff space is totally disconnected if and only if it is zero-dimensional.*

Proof. ($\Leftarrow$) We have shown in the example above that T_1 zero-dimensional spaces are totally disconnected. Since Hausdorff spaces are T_1, we are done.

($\Rightarrow$) Suppose S is a locally compact Hausdorff totally disconnected space.

Let $p \in S$ and V be an open neighborhood of p in S. Then $C_p = \{p\}$. We are required to show that S is zero-dimensional. To do this it suffices to show the existence of a clopen neighborhood D, in S such that $p \in D \subseteq V$.

Since S is locally compact, there is an open subset M whose closure, $\mathrm{cl}_S M$, is compact such that

$$p \in M \subseteq \mathrm{cl}_S M \subseteq V$$

Let

$$\mathscr{D} = \{D \in S : D \text{ is clopen in } S, p \in D\}$$

By the above lemma, $C_p = \cap\mathscr{D}$.

Since $\mathrm{cl}_S M$ is compact, there exists a finite subset, $\mathscr{D}_F$, of $\mathscr{D}$ such that

$$C_p \subseteq \cap\mathscr{D}_F \subseteq \mathrm{cl}_S M \subseteq V$$

Then $\cap\mathscr{D}_F$ is a clopen neighborhood of p which is contained in V.

We conclude that S is zero-dimensional.

Example 21. Show that a subspace of a locally compact totally disconnected Hausdorff space is also totally disconnected.

Solution: Suppose T is a subspace of a totally disconnected locally compact Hausdorff space S. Then S is zero-dimensional (by Theorem 20.24). Let C be a connected component of T. To show that T is totally disconnected it suffices to show that C is a singleton set.

If a and b are points in C then, since S is zero-dimensional Hausdorff, there exists a clopen neighborhood U (in S) of a which does not contain b. So, the clopen subset $U \cap T$ (of T) separates a and b

and so, if $a \neq b$, C cannot be connected. So, $a = b$. Then connected components of T are singleton sets. So, T is totally disconnected.

Example 22. Let S be a compact Hausdorff topological space which is not connected. We showed in an example on page 452, that if we collapse the connected components of S to points by the quotient map $\theta : S \to \theta[S]$ we obtain a decomposition space $\theta[S]$ whose points are connected components. This means that the compact space $\theta[S]$ is totally disconnected and hence zero-dimensional.

Concepts Review

1. What does it mean to say that a topological space is connected?
2. Is the connected property invariant with respect to the continuous functions?
3. Under what conditions, if any, are unions of connected sets connected?
4. If U is a connected dense subset of V what can say about V?
5. Under what conditions, if any, is the product space, $S = \prod_{i \in I} S_i$, of connected, S_i's, connected?
6. Define a connected component of a topological space.
7. Briefly argue that a connected component of a topological space is closed.
8. Provide a simple example which illustrates that a connected component need not be open.
9. Describe the connected components of a discrete space.
10. Describe the connected components of an infinite product of discrete spaces all of which have more than one point.
11. Is an infinite product of discrete spaces (all of which have more than one point) discrete?
12. Define a locally connected space.
13. What kind of locally connected spaces are guaranteed to have clopen connected components.
14. Describe the conditions under which the locally connected property carries over on a product space, in both directions.
15. Suppose a locally connected space is shown to be compact. What can we say about its connected components?

16. If a person speaks of a path joining a point, a, to a point, b, in a space S, what is this person talking about?
17. If you are familiar with the notion of a "curve" in a space S, is there a subtle difference between a curve and a path?
18. What does it mean to say that a space S is pathwise connected?
19. When is a connected space a pathwise connected space?
20. Are continuous images of pathwise connected spaces necessarily pathwise connected?
21. Under what conditions is the pathwise connected property carried over products in both directions?
22. Under what conditions, if any, are unions of pathwise connected spaces pathwise connected?
23. Define the pathwise component of a point in a space S.
24. Are pathwise components necessarily closed?
25. Describe a particular condition under which a pathwise component is clopen.
26. Describe a condition under which connected spaces are pathwise connected.
27. Define the property called totally disconnected.
28. Give a characterization of the totally disconnected property and describe the conditions under which it holds true.
29. What can we say about the product of totally disconnected spaces?
30. What can we say about the subspace of a totally disconnected space?

Exercises

1. Suppose U is subset of $\mathbb{R}$ which is not connected and $V \subseteq U$. Is it possible for V to be connected?
2. Let (S, τ_S) and (T, τ_T) be two topological spaces where T is connected. If T has a strictly stronger topology then S, determine whether S is necessarily connected.
3. Consider the subset

$$T = \{(x \sin(1/x)) : x \in [0, 2\pi]\} \cup \{(0,0)\}$$

of $\mathbb{R}^2$. Is T a connected set?

4. Is the set T described in the previous question locally connected?
5. Is the set T described in the previous question pathwise connected?
6. Let $\{S_i : i \in I\}$ be an infinite family of connected sets. Suppose that, if $i \neq k$, then $S_i \cap S_k \neq \varnothing$. Determine whether the space $S = \cup\{S_i : i \in I\}$ is connected or not.
7. Let S be an infinite set equipped with the cofinite topology (i.e., $\{\varnothing\}$ union the family of all sets with a finite complement). Determine whether S is a connected set or not.
8. Suppose the space (S, τ) is the topological space where

$$\tau = \{\{\varnothing\} \cup \{U : S \backslash U \text{ is finite }\}\}$$

Let the function, $f : [0, 1] \to S$, be continuous. Show that $f[\,[0, 1]\,]$ in S is a singleton set.
9. Let $S = \mathbb{R}^n$ be the space equipped with the usual topology. Suppose U is a non-empty open subspace of S. How many connected components can S have? If the possible cardinality of all components is infinite, is it countable or uncountable?
10. Are open subspaces of locally connected spaces necessarily locally connected?
11. Suppose S is a T_1-space which has, at each point, a neighborhood base of clopen sets. Show that the connected components of S are all singleton sets.

Part VII

Topics

Chapter 21

Compactifications of Completely Regular Spaces

Abstract
In this section, we discuss those spaces, S, which can be densely embedded in a compact Hausdorff space. Only completely regular spaces can possess this property. The process by which we determine such a compact space, αS, for S, is called compactifying S. The space, αS, is called the compactification of S. The family of all compactifications of a completely regular space can be partially ordered. The maximal compactification of S with respect to the chosen partial ordering is called the Stone–Čech compactification. We discuss methods for its construction. We will show that only locally compact spaces have a minimal compactification with respect to the chosen partial ordering. It is called the one-point compactification.

21.1 Compactifying a Space

In this section, we will briefly talk about methods for "compactifying a space (S, τ_S)". This essentially means adding a set, F, of points to S, to obtain a larger set, $T = S \cup F$, and topologizing T so that (T, τ_T) is a compact Hausdorff space in which a homeomorphic copy of S appears as a dense subspace of T.

With certain bounded subspaces of $\mathbb{R}^n$, this can, sometimes, be quite easy to do. For example, if $S = [-1, 3) \cup (3, 7)$ is equipped with the subspace topology, then by simply adding the points $\{3, 7\}$ to S we obtain the set $T = S \cup \{3, 7\} = [-1, 7]$ which, when equipped with the subspace topology, is a compact Hausdorff space which densely contains a homeomorphic copy of S. In such a case, we will say that T is a *compactification* of S. A compactification of a bounded subset of $\mathbb{R}^n$ can always be obtained by taking its closure. This does not mean, however, that there are not others. If we are given a space such as $\mathbb{N}$ or $\mathbb{Q}$, it is not at all obvious how one would go about compactifying such spaces. We will show techniques which allow us to achieve this objective.

For what follows, recall that the evaluation map $e : S \to \prod_{i\in J}[a_i, b_i]$ induced by $C^*(S)$ (the set of all continuous bounded real-valued functions on S) is defined as

$$e(x) = \langle f_i(x)\rangle_{i\in J} \in \prod_{i\in J}[a_i, b_i]$$

where $f_i \in C^*(S)$ and $f_i[S] \subseteq [a_i, b_i]$.

Definition 21.1 Let (S, τ_S) be a topological space and (T, τ_T) be a compact Hausdorff space. We will say that T *is a compactification of* S if S is densely embedded in T.[1]

If S is a compact Hausdorff space, then S can be viewed as being a compactification of itself. Recall that a compact Hausdorff space is normal, and so is completely regular. Since subspaces of completely regular spaces are completely regular, then...

only a completely regular space can have a compactification.

In Theorem 14.7, we showed how any completely regular space can be compactified. In the proof of that theorem, we witnessed how an evaluation map, $e : S \to T$, induced by $C^*(S)$ embeds S into a cube,

$$T = \prod_{i\in J}[a_i, b_i]$$

There may be different ways of describing this type of compactification of S. Since each interval $[a_i, b_i]$ is homeomorphic to $[0, 1]$, then

[1]When we say "compactification of S" we always mean a Hausdorff compactification.

there is a homeomorphism, $h : T \to \prod_{i\in J}[0,1]$, which maps T onto $P = \prod_{i\in J}[0,1]$. By Tychonoff's theorem, P is guaranteed to be compact. So the function, $q : S \to P$, defined as, $q = h \circ e$, embeds S into $\prod_{i\in J}[0,1]$. Hence, $\text{cl}_P q[S]$ is a compact subspace of the product space, P, which densely contains the homeomorphic image, $q[S]$, of S.[2] So, even common topological spaces such at $\mathbb{R}$, $\mathbb{Q}$, and $\mathbb{N}$ have at least the compactification obtained by the method just described.

21.2 The Stone–Čech Compactification

We have described only one of the various ways to obtain a homeomorphic copy of the compactification, $\text{cl}_T e[S]$, of S. This particular compactification has a special name.

Definition 21.2 Let (S, τ_s) be a completely regular topological space. Let

$$e : S \to \prod_{i\in I}[a_i, b_i]$$

be the evaluation map induced by $C^*(S)$ which embeds S in the product space, $T = \prod_{i\in I}[a_i, b_i]$.

The subspace,

$$\text{cl}_T e[S]$$

is called the *Stone–Čech compactification of* S. The Stone–Čech compactification of S is uniquely (and universally) denoted by, βS.

So we see that a non-compact completely regular space, S, always has at least one compactification, namely, it's Stone–Čech compactification, βS. We will soon see that there can be more than one compactification, for the same space.

[2]This is just one small example which shows why the Tychonoff theorem deserves to be titled and why it is such an important theorem in topology.

21.2.1 *Equivalent compactifications*

Given a completely regular space S, there is nothing stated up to now which would lead us to conclude that S has only one compactification. In fact most spaces we will consider will have many compactifications. Suppose we are given two compactifications for a space, S, say αS and γS. If there is a homeomorphism

$$h : \alpha S \to \gamma S$$

mapping αS onto γS such that $h(x) = x$ for all $x \in S$, then αS and γS will be considered to be *equivalent compactifications* of S. Two compactifications of the same space which have been determined to be "equivalent" in this way will be assumed to be the same compactification, topologically speaking. This equivalence is often expressed by the symbol, $\alpha S \equiv \gamma S$. For convenience, it is sometimes expressed by, $\alpha S = \gamma S$, even though, strictly speaking, αS and γS may not necessarily be equal sets.

21.2.2 *The outgrowth of a topological space*

Given a topological space, S, and a compactification, αS, the set

$$\alpha S \backslash S$$

is referred to as being the *outgrowth of* S with respect to this particular compactification or as the *remainder* of αS. Equivalent compactifications will be considered to have the same outgrowth. We will sometimes be interested in determining whether an outgrowth satisfies certain topological properties.

Example 1. *Compactifications of* $\mathbb{R}$. Construct two compactifications of the space, $\mathbb{R}$, one whose outgrowth contains a single point and another one whose outgrowth contains two points. We will refer to these as a "one-point compactification" and "two-point compactification" of $\mathbb{R}$.

Solution: *A two-point compactification of* $\mathbb{R}$. We know that arctan: $\mathbb{R} \to (-\pi/2, \pi/2)$ maps $\mathbb{R}$ homeomorphically onto $(-\pi/2, \pi/2)$. Hence, $\mathbb{R}$ is embedded in the compact space

$$\gamma\mathbb{R} = \text{cl}_{\mathbb{R}}\arctan[\mathbb{R}] = \text{cl}_{\mathbb{R}}(-\pi/2, \pi/2) = [-\pi/2, \pi/2]$$

with outgrowth $\{-\pi/2, \pi/2\}$.

A one-point compactification of $\mathbb{R}$. We know that there is a homeomorphism $h : \mathbb{R} \to (0, 2\pi)$ which maps $\mathbb{R}$ onto $(0, 2\pi)$. The function,

$$h(x) = 2[\arctan(x) + \pi/2]$$

is an example. Define the homeomorphic function $f : (0, 2\pi) \to \mathbb{R}^2$ as

$$f(x) = (\sin(x), \cos(x))$$

Then $f \circ h : \mathbb{R} \to \mathbb{R}^2$ densely embeds $\mathbb{R}$ into the (closed and bounded) compact set

$$\alpha\mathbb{R} = K = \{(\sin(x), \cos(x)) : x \in (0, 2\pi)\} \;\cup\; \{(1, 0)\}$$

with the single point, $\{(1, 0)\}$, in its outgrowth.

Example 2. Let $\mathbb{R}^+ = \{x \in \mathbb{R} : x \geq 0\}$. Is there a two-point compactification of $\mathbb{R}^+$?

Solution: No. Suppose $\alpha\mathbb{R}^+ = \mathbb{R}^+ \cup \{a, b\}$. Then, since $\alpha\mathbb{R}^+$ is Hausdorff, there must be disjoint open neighborhoods B_a and B_b in $\alpha\mathbb{R}^+$ of a and b, respectively. Then $\alpha\mathbb{R}^+\backslash\mathbb{R}^+ \subseteq B_a \cup B_b$. Since $B_a \cup B_b$ is open in $\alpha\mathbb{R}^+$ containing $\alpha\mathbb{R}^+\backslash\mathbb{R}^+$, then

$$K = \alpha\mathbb{R}\backslash(B_a \cup B_b)$$

is a compact subset of $\mathbb{R}^+$. Then there exists $k \in \mathbb{R}^+$ such that $K \subseteq [0, k]$. Then $[k, \infty) = [(B_a \cup B_b) \cap \mathbb{R}^+] \backslash [0, k)$. We see that $[k, \infty)$ is a connected subspace of $\mathbb{R}^+$ while $[(B_a \cup B_b) \cap \mathbb{R}^+] \backslash [0, k)$ is not. So, we have a contradiction. So, $\mathbb{R}^+$ cannot have a two-point compactification.

Example 3. *There is no three-point compactification of* $\mathbb{R}$. Prove this statement.

Solution: Suppose that

$$\alpha\mathbb{R} = \mathbb{R} \cup \{a, b, c\}$$

represents a compactification of $\mathbb{R}$ with outgrowth $\alpha\mathbb{R}\backslash\mathbb{R} = \{a, b, c\}$. Let $\mathbb{R}^+ = \{x \in \mathbb{R} : x \geq 0\}$ and $\mathbb{R}^- = \{x \in \mathbb{R} : x \leq 0\}$. Then

$$\begin{aligned}\alpha\mathbb{R} &= \text{cl}_{\alpha\mathbb{R}}\mathbb{R} \\ &= \text{cl}_{\alpha\mathbb{R}}(\mathbb{R}^+ \cup \mathbb{R}^-) \\ &= \text{cl}_{\alpha\mathbb{R}}\mathbb{R}^+ \cup \text{cl}_{\alpha\mathbb{R}}\mathbb{R}^- \\ &= \mathbb{R}^+ \cup \mathbb{R}^- \cup \{a, b, c\}\end{aligned}$$

By the previous example neither $\mathbb{R}^+$ nor $\mathbb{R}^-$ can have a two-point (nor a three-point) outgrowth. So, there is no three-point compactification of $\mathbb{R}$.

We can easily generalize the statement in the previous example to "If the compactification $\alpha\mathbb{R}$ has a finite outgrowth then $\alpha\mathbb{R}\backslash\mathbb{R}$ must be either a singleton set or a doubleton set".

21.3 A Partial Ordering of Hausdorff Compactifications of a Space

Let's gather together all compactifications of a completely regular space, S, as follows:

$$\mathscr{C} = \{\alpha_i S : i \in I\}$$

We will partially order the family, $\mathscr{C}$, by defining "$\preceq$" in the following way:

> *If $\alpha_i S$ and $\alpha_j S$ belong to $\mathscr{C}$, we will write*
>
> $$\alpha_i S \preceq \alpha_j S$$
>
> *if and only if there is a continuous function $f : \alpha_j S \to \alpha_i S$, mapping $\alpha_j S\backslash S$ onto $\alpha_i S\backslash S$ which fixes the points of S.*

Notation: Suppose two compactifications αS and γS are such that $\alpha S \preceq \gamma S$, then by definition, there is a continuous function $f : \gamma S \to \alpha S$, mapping $\gamma S\backslash S$ onto $\alpha S\backslash S$ which fixes the points of S. We will represent this continuous function f as

$$\pi_{\gamma\to\alpha} : \gamma S \to \alpha S$$

The function $\pi_{\gamma \to \alpha}$ explicitly expresses which compactification is larger than (or equal to) the other. Note that a pair of compactifications need not necessarily be comparable in the sense that one need not necessarily by "less than" the other.

21.4 On C^*-Embedded Subsets of a Topological Space

Given a subset T of a topological space, S, and a continuous bounded real-valued function $f : T \to \mathbb{R}$, it is not guaranteed that there is a bounded continuous function $g : S \to \mathbb{R}$ such that $g|_T = f$ on T. If there is, then we will say that

g is a continuous extension of f from T to S.

This motivates the following definition.

Definition 21.3 Let (S, τ) be a topological space and U be a proper non-empty subset of S. We say that U is C^*-*embedded* in S if every real-valued bounded function, $f \in C^*(U)$, extends to a function, $g \in C^*(S)$, in the sense that $g|_U = f$.[3] In this case we will say that...

$g : S \to \mathbb{R}$ is a *continuous extension* of $f : U \to \mathbb{R}$ from U to S

The notion of "C^*-embedding" is closely related to completely regular spaces and their Stone–Čech compactification. For this reason, we will discuss this property in depth now (even though C^*-embeddings may be discussed in other contexts). The following theorem shows that, for any completely regular space S, S is C^*-embedded in its Stone–Čech compactification, βS. That is,

every function, $f \in C^*(S)$, extends continuously to a function, $f^\beta \in C(\beta S)$.

---

[3]There is an analogous definition for "C-embedded" studied later: We say that U is C-*embedded* in S if every real-valued function, $f \in C(U)$, continuously extends to a function, $f^* \in C(S)$.

Theorem 21.4 *Let S be a completely regular space. Then S is C^*-embedded in βS.*

Proof. If $f \in C^*(S)$, let I_f be the range of f. Let

$$T = \prod_{f \in C^*(S)} \text{cl}_{\mathbb{R}} I_f$$

Then the evaluation map, $e : S \to T$ embeds S in T. Recall that, by definition,

$$\beta S = \text{cl}_T e[S] \subseteq \prod_{f \in C^*(S)} \text{cl}_{\mathbb{R}} I_f$$

Suppose $g \in C^*(S)$.

We are required to show that $g : S \to \mathbb{R}$ extends continuously to some function, $g^\beta : \beta S \to \mathbb{R}$.

If π_g is the gth-projection map then

$$\pi_g : \prod_{f \in C^*(S)} \text{cl}_{\mathbb{R}} I_f \to \text{cl}_{\mathbb{R}} I_g$$

where $\beta S = \text{cl}_T e[S]$ and so,

$$\pi_g|_{\beta S} : \beta S \to \text{cl}_{\mathbb{R}} I_g$$

maps βS into $\text{cl}_{\mathbb{R}} I_g$. Let $g^\beta = \pi_g|_{\beta S}$. Then $g^\beta[\beta S] = g^\beta[\text{cl}_T e[S]] = \text{cl}_{\mathbb{R}} g[S] \subseteq \text{cl}_{\mathbb{R}} I_g$.

It follows that $g^\beta : \beta S \to \text{cl}_{\mathbb{R}} I_g$ and, since $g[S] \subseteq I_g$, for $x \in S$, $g^\beta|_S(x) = g(x)$.

So, g^β is a continuous extension of g from S to βS.

Note that, in the case where S is a compact space, $e[S]$ is a compact space densely embedded in βS and so $\beta S \backslash S = \varnothing$. Then S and βS are homeomorphic.

The above theorem guarantees that every real-valued continuous bounded function, f, on a completely regular space, S, extends to a continuous function $f^\beta : \beta S \to \mathbb{R}$. The function f^β is the *extension of* f from S to βS. Recall (from Theorem 9.8) that continuity guarantees that two continuous functions which agree on a dense subset D of a Hausdorff space, S, must agree on all of S. So, there can only be one extension, f^β, of f from S to βS.

We will soon show (in Theorem 21.9) that, if αS is a compactification of S and S is C^*-embedded in αS then αS must be the compactification, βS. That is,

βS is the only compactification in which S is C^*-embedded.

21.4.1 *A generalization of the extension, $f \to f^\beta$*

The above theorem can be generalized a step further. Suppose $C(S, K)$ denotes all continuous functions mapping S into a compact set K. We show that every function f in $C(S, K)$ extends to a function $f^{\beta(K)} \in C(\beta S, K)$. Note that neither f or $f^{\beta(K)}$ need be real-valued. The space, K, represents any compact set which contains the image of S under f.

Theorem 21.5 *Let S be a completely regular (non-compact) space and $g : S \to K$ be a continuous function mapping S into a compact Hausdorff space, K. Then g extends uniquely to a continuous function, $g^{\beta(K)} : \beta S \to K$.*

Proof. We are given that $g : S \to K$ continuously maps the completely regular space, S, into the compact Hausdorff space K. We are required to show that g extends to $g^{\beta(K)} : \beta S \to K$.

Since K is compact Hausdorff, it is completely regular; hence, there exists a function (the evaluation map) which embeds the compact set K in $V = \prod_{i\in J}[0,1]$. Since V contains a topological copy of K, let us view K as a subset of V.

Since $g : S \to K$, then, for every $x \in S$, $g(x) = \langle\!\langle g_i(x)\rangle\!\rangle_{i\in J} \in K \subseteq [0,1]^J$.

Since S is C^*-embedded in βS, then, for each $i \in J$, $g_i : S \to [0,1]$ extends to $g_i^\beta : \beta S \to [0,1]$.

We define the function $g^{\beta(K)} : \beta S \to V$ as

$$g^{\beta(K)}(x) = \langle\!\langle g_i(x)\rangle\!\rangle_{i\in J} \in V = \textstyle\prod_{i\in J}[0,1]$$

Since g_i^β is continuous on βS, for each i, then $g^{\beta(K)} : \beta S \to \prod_{i\in J}[0,1]$ is continuous on βS.

See that $g^{\beta(K)}[\beta S] = g^{\beta(K)}[\text{cl}_{\beta S}(S)] = \text{cl}_V(g[S]) \subseteq \text{cl}_V(K) = K$.

Since $g^{\beta(K)}|_S = g$, then $g : S \to K$ continuously extends to $g^{\beta(K)} : \beta S \to K$ on βS. As required.

Given a completely regular topological space S, and $T = \prod_{i\in I}[a_i, b_i]$ we now see that the evaluation function $e_{C^*(S)} : S \to \prod_{i\in I}[a_i, b_i]$ (which homeomorphically embeds a copy of S into T) then continuously extends to βS as follows:

$$\begin{aligned} e^{\beta}_{C^*(S)}[\beta S] &= e^{\beta}_{C^*(S)}[\text{cl}_{\beta S} S] \\ &= \text{cl}_T e_{C^*(S)}[S] \end{aligned}$$

where

$$e^{\beta}_{C^*(S)}(x) = \langle f^{\beta}(x)\rangle_{f\in C^*(S)}$$

So, we can represent the Stone–Čech compactification, βS, of S as being equivalent to

$$e^{\beta}_{C^*(S)}[\beta S]$$

21.4.2 *The maximal compactification, βS*

Suppose αS is any compactification of S possibly distinct from βS. We will now show that, in the partially ordered family, $\mathscr{C}$, of all compactifications of S, $\alpha S \preceq \beta S$. Showing this requires that we produce a continuous function $\pi_{\beta\to\alpha} : \beta S \to \alpha S$ such that $\pi_{\beta\to\alpha}(x) = x$, for all $x \in S$. If we can prove this, then we will have shown that βS is the unique maximal compactification of a completely regular space, S.

Theorem 21.6 *If αS is a compactification of S, then $\alpha S \preceq \beta S$.*

Proof. By Theorem 21.5, the identity map $i : S \to \alpha S$, extends to a continuous function,

$$i^* : \beta S \to \alpha S$$

Then $S \subseteq i^*[\beta S] \subseteq \alpha S$, where $i^*[\beta S]$ is compact, hence, closed in αS. Since S is dense in αS, then the open set $\alpha S \backslash i^*[\beta S]$ must be empty.

So, the continuous function,

$$i^*[\beta S] = i^*[\mathrm{cl}_{\beta S} S] = \mathrm{cl}_{\alpha S} i^*[S] = \mathrm{cl}_{\alpha S} S = \alpha S$$

maps βS onto αS. So, $\alpha S \preceq \beta S$, as required.

Then for any compactification αS, there is the continuous function

$$\pi_{\beta \to \alpha} : \beta S \to \alpha S$$

which maps βS onto αS where $\pi_{\beta \to \alpha}$ fixes the points of S.

21.4.3 *More on C^*-embedded subsets of a space S*

We present a miscellany of results which will help us more easily recognize a C^*-embedded subset, T of a space S. We will return to our discussion of compactification immediately following this.

The simplest example of a C^*-embedded subset of $\mathbb{R}$ is a compact subset of $\mathbb{R}$.

Theorem 21.7 *If K is a compact subset of $\mathbb{R}$ then K is C^*-embedded in $\mathbb{R}$.*

Proof. Let K be a compact subset of $\mathbb{R}$ and $f : K \to \mathbb{R}$ be a continuous function on K. Since every continuous real-valued function is bounded on a compact subset, then $f \in C^*(K) = C(K)$. Suppose

$$u = \sup\{K\} \text{ and } v = \inf\{K\}$$

Since K is closed and bounded in $\mathbb{R}$, then u and v belong to K. Suppose $g : \mathbb{R} \to \mathbb{R}$ is a function such that

$$g = f \text{ on } K$$
$$g(x) = f(u), \text{ if } x \geq u$$
$$g(x) = f(v), \text{ if } x \leq v$$

It is easily verified that g is a continuous extension of f from K to $\mathbb{R}$. Then K is C^*-embedded in $\mathbb{R}$.

Example 4. The set $\mathbb{N}$ is C^*-embedded in $\mathbb{R}$. One way of visualizing this is to plot the points of $\{(n, f(n)) : n \in \mathbb{N}\}$ of a function $f \in C^*(\mathbb{N})$ in the Cartesian plane $\mathbb{R}^2$ and join every pair of successive points $(n, f(n))$ and $(n+1, f(n+1))$ by a straight line. This results in a continuous curve representing a continuous function g on $\mathbb{R}$ which extends f.

21.4.4 *Urysohn's extension theorem*

The following theorem often referred to as *Urysohn's extension theorem* provides an important and useful tool for recognizing C^*-embedded sets.[4]

Theorem 21.8 *Urysohn's extension theorem: Let T be a subset of the completely regular space S. Then T is C^*-embedded in S if and only if pairs of sets which can be completely separated by some function in $C^*(T)$ can also be separated by some function in $C^*(S)$.*

Proof. ($\Rightarrow$) Suppose T is C^*-embedded in S and U and V are completely separated subsets of T. Then there exists $f \in C^*(T)$ such that $U \subseteq f^{\leftarrow}(0)$ and $V \subseteq f^{\leftarrow}(1)$. Then by hypothesis f extends to $f^* \in C^*(S)$. Then $U \subseteq f^{\leftarrow *}(0)$ and $V \subseteq f^{\leftarrow *}(1)$. So, U and V are completely separated in S.

($\Leftarrow$) Suppose that pairs of sets which can be completely separated by some function in $C^*(T)$ can also be separated by some function in $C^*(S)$.

Let f_1 be a function in $C^*(T)$. We are required to show that there exists a function $g \in C^*(S)$ such that $g|_T = f_1$.

---

[4]Urysohn's extension theorem should not be confused with *Urysohn's lemma.* Urysohn's lemma states that "The topological space (S, τ_S) is *normal* if and only if given a pair of disjoint non-empty closed sets, F and W, in S there exists a continuous function $f : S \to [0,1]$ such that, $F \subseteq f^{\leftarrow}[\{0\}]$ and $W \subseteq f^{\leftarrow}[\{1\}]$".

Since f_1 is bounded on T, then there exists, $k \in \mathbb{R}$, such that $|f_1(x)| \leq k$ for all $x \in T$. Then $f_1 \leq k = 3r_1 = 3 \cdot \left[\frac{k}{2} \cdot \left(\frac{2}{3}\right)^1\right] = k$ (where $r_1 = \frac{k}{2} \cdot \left(\frac{2}{3}\right)^1$).

We now inductively define a sequence of functions $\{f_n\} \subseteq C^*(T)$. For $n = 1, 2, 3, \ldots$, there exists $f_n \in C^*(T)$ such that $-3r_n \leq f_n(x) \leq 3r_n$ where,

$$3r_n = 3 \cdot \left[\frac{k}{2} \cdot \left(\frac{2}{3}\right)^n\right] = k \cdot \left(\frac{2}{3}\right)^{n-1} \quad \text{(where } r_n = \tfrac{k}{2} \cdot \left(\tfrac{2}{3}\right)^n\text{)}$$

For this n, let $U_n = f^{\leftarrow}{}_n[[-3r_n, -r_n]]$ and $V_n = f^{\leftarrow}{}_n[[r_n, 3r_n]]$.

We see that U_n and V_n are completely separated in T.[5]

By hypothesis, U_n and V_n are completely separated in S. This means there exists $g_n \in C^*(S)$ such that, $g_n[S] \subseteq [-r_n, r_n]$, $g_n[U_n] = \{-r_n\}$ and $g_n[V_n] = \{r_n\}$ (where $r_n = \left[\frac{k}{2} \cdot \left(\frac{2}{3}\right)^n\right]$). So the sequence, $\{g_n : n = 1, 2, 3, \ldots\}$, thus constructed is well-defined in $C^*(S)$.

We now inductively define the sequence $\{h_n\} \subseteq C^*(T)$ initiating the process with $h_1 = f_1$ and continuing with

$$h_{n+1} = h_n - g_n|_T$$

Then for each n,

$$|h_{n+1}| \leq 2r_1 = 2 \cdot \frac{k}{2} \cdot \left(\frac{2}{3}\right)^n = 3 \cdot \frac{k}{2} \cdot \left(\frac{2}{3}\right)^{n+1} = 3r_{n+1}$$

So, $g_n|_T = h_n - h_{n+1}$. Define $g : S \to \mathbb{R}$ as the series

$$g(x) = \sum_{n \in \mathbb{N} \setminus \{0\}} g_n(x)$$

See that $g(x)$ is continuous on S: Since $|g_n(x)| \leq \frac{k}{2}\left(\frac{2}{3}\right)^n$, and $\sum_{n \in \mathbb{N} \setminus \{0\}} \frac{k}{2}\left(\frac{2}{3}\right)^n$ is a converging geometric series, then $\sum_{n \in \mathbb{N} \setminus \{0\}} g_n(x)$ converges uniformly to $g(x)$. Since each $g_n(x)$ is continuous on S, then $g \in C^*(S)$. So, g is a continuous on S.

[5]To see this, the function $h_n = (-r_n \vee f_n) \wedge r_n)$ has $U_n \subseteq Z(h_n - (-r_n))$ and $V_n \subseteq Z(h_n - r_n)$.

Also see that, $g|_T = f_1$:

$$\begin{aligned} g|_T(x) &= \lim_{m\to\infty} \sum_{n=1}^{m} g_n|_T(x) \\ &= \lim_{m\to\infty} (h_1(x) - h_2(x)) + (h_2(x) - h_3(x)) + \cdots \\ &\quad + (h_m(x) - h_{m+1}(x)) \\ &= \lim_{m\to\infty} h_1(x) - h_{m+1}(x) \\ &= h_1(x) = f_1(x) \quad (\text{Since } \lim_{m\to\infty} 3r_{m+1} = 0) \end{aligned}$$

Then, $g|_T = f_1$ so f_1 extends continuously from T to S. We are done.

Example 5. Use Urysohn's extension lemma to show that any compact subset, K, of a completely regular space, S, is C^*-embedded in S.

Solution: Let U and V be disjoint subsets of the compact set, K, which are completely separated in K. Then there is a function $f \in C(K)$ such that $U \subseteq A = f^{\leftarrow}[\{0\}]$ and $V \subseteq B = f^{\leftarrow}[\{1\}]$. Both A and B are disjoint closed subsets of compact K and so are compact sets. Then A and B are compact in the completely regular space, S. Then A and B are completely separated in S. So, U and V are completely separated in S. By Urysohn's extension lemma, K is C^*-embedded in S. We conclude that...

> *any compact subset, K, of a completely regular space, S, is C^*-embedded in S.*

21.4.5 *Uniqueness of βS*

We are now able to prove that, up to equivalence, the Stone–Čech compactification of S is the only compactification in which S is C^*-embedded. By this we mean that, if S is C^*-embedded in the compactification, γS, of S, then γS is equivalent to βS. So the symbol, βS, is strictly reserved for the Stone–Čech compactification of S.

Theorem 21.9 *The completely regular space S is C^*-embedded in the compactification, γS, if and only if $\gamma S \equiv \beta S$.*

Proof. We are given that S is completely regular.

($\Leftarrow$) Suppose $\gamma S \equiv \beta S$. Then there is a homeomorphism, $h : \gamma S \to \beta S$, mapping γS onto βS such that $h(x) = x$, for all $x \in S$. We are required to show that S is C^*-embedded in γS.

If $f \in C^*(S)$ then f extends to $f^\beta : \beta S \to \mathbb{R}$. Then $f^\beta{\circ}h : \gamma S \to \mathbb{R}$. Define $f^\gamma = f^\beta{\circ}h$. We see that $f^\gamma : \gamma S \to \mathbb{R}$ is the continuous extension of f from S to γS. Then S is C^*-embedded in γS.

($\Rightarrow$) Suppose S is C^*-embedded in γS. We are required to show that γS and βS are equivalent compactifications.

Let $i : S \to S$ be the identity map. Then, by Theorem 21.5, i extends to $i^* : \beta S \to \gamma S$. Also, just as shown in the proof of Theorem 21.5, $i^{\leftarrow}$ extends to $i^{\leftarrow\hat{}} : \gamma S \to \beta S$. Then $i^{\leftarrow}{\circ}i$ and $i{\circ}i^{\leftarrow}$ are both identity maps on S and, since S is dense in both βS and γS, respectively, then $i^*{\circ}i^{\leftarrow\hat{}}$ and $i^{\leftarrow\hat{}}{\circ}i^*$ are identity maps on γS and on βS, respectively. Then both i^* and $i^{\leftarrow\hat{}}$ are homeomorphisms. Hence, γS and βS are equivalent compactifications.

Example 6. Show that if F is a closed subset of a metric space S then F is C^*-embedded in S.

Solution: Let F be a closed subset of the metric space S. We will set up the solution so that we can invoke the Urysohn extension theorem.

Let A and B be completely separated in F. Then, by definition, there is a function f in $C^*(F)$ such that $A \subseteq Z(f)$ and $B \subseteq Z(f-1)$. Then $\mathrm{cl}_F A \subseteq Z(f)$ and $\mathrm{cl}_F B \subseteq Z(f-1)$. Since F is closed in S, then so are the disjoint sets $\mathrm{cl}_F A$ and $\mathrm{cl}_F B$. It is shown on page 258 that in metric spaces closed subsets are the same as zero-sets. So, $\mathrm{cl}_F A$ and $\mathrm{cl}_F B$ are disjoint zero-sets in S, say $\mathrm{cl}_F A = Z(k)$ and $\mathrm{cl}_F B = Z(g)$ in S. If

$$h = \frac{|k|}{|k| + |g|}$$

on S, $\mathrm{cl}_F A = Z(h)$ and $\mathrm{cl}_F B = Z(h-1)$ in S. So, A and B are completely separated in S. By Urysohn's extension lemma every closed subset of a metric space is C^*-embedded.

Because of this, it is useful to remember that

any closed subset of $\mathbb{R}$ *is* C^**-embedded in* $\mathbb{R}$.

21.5 Associating Compactifications to Subalgebras of $C^*(S)$

To each compactification αS we can associate a subset, $C_\alpha(S)$, of $C^*(S)$ defined as

$$C_\alpha(S) = \{f|_S \in C^*(S) : f \in C(\alpha S)\}$$

That is, $f \in C_\alpha(S)$ if and only if f extends to $f^\alpha \in C(\alpha S)$. If αS and γS are two compactifications of S such that $\alpha S \preceq \gamma S$ it is normal to wonder how $C_\alpha(S)$ compares to $C_\gamma(S)$ in $C^*(S)$.

Theorem 21.10 *Let αS and γS be two compactifications of S such that $\alpha S \preceq \gamma S$. Let $C_\alpha(S)$ denote the set of all real-valued continuous bounded functions on S that extend to αS. Let $C_\gamma(S)$ denote the set of all real-valued continuous bounded functions on S that extend to γS. Then $C_\alpha(S) \subseteq C_\gamma(S)$.*

Proof. We are given that $\alpha S \preceq \gamma S$. Then there is a continuous function $\pi_{\gamma\to\alpha} : \gamma S \to \alpha S$ such that $\pi_{\gamma\to\alpha}(x) = x$ on S.

Suppose $t \in C_\alpha(S)$. It suffices to show that $t \in C_\gamma(S)$. Then there a function $t^\alpha : \alpha S \to \mathbb{R}$ is such that $t^\alpha|_S = t$. Define the function $g : \gamma S \to \mathbb{R}$ as

$$g = t^\alpha \circ \pi_{\gamma\to\alpha}$$

Since $\pi_{\gamma\to\alpha} : \gamma S \to \alpha S$ and $t^\alpha : \alpha S \to \mathbb{R}$ are both continuous, then g is continuous on γS and

$$g|_S(x) = (t \circ \pi_{\gamma\to\alpha})(x) = t(x)$$

So, $t = g|_S \in C_\gamma(S)$. Hence,

$$\alpha S \preceq \gamma S \qquad \Rightarrow \qquad C_\alpha(S) \subseteq C_\gamma(S)$$

as required.

21.5.1 Equivalent functions in $C^*(S)$

In what follows we will refer to pairs of functions in $C^*(S)$ which are "equivalent".

We define a particular subset, $C_\omega(S)$, of $C^*(S)$ as follows:

$$C_\omega(S) = \{f \in C^*(S) : f^\beta \text{ is constant on } \beta S \backslash S\}$$

The subset $C_\omega(S)$ is easily seen to be closed under addition, multiplication, scalar multiplication and contains all constant functions. We say that it is a *subalgebra* of $C^*(S)$.

We will say that two functions f and g in $C^*(S)$ are *equivalent functions in* $C^*(S)$ if

$$f - g \in C_\omega(S)$$

Equivalent functions f and g are sometimes expressed by the notation

$$f \cong g$$

The next theorem shows that there are as many compactifications of S as there are subalgebras of $C^*(S)$ of a certain type.

Suppose (S, τ) is a completely regular space and $\mathscr{F}$ is a *subalgebra* of $C^*(S)$ which contains $C_\omega(S)$ and separates points and closed sets. We can invoke Theorems 7.17 and 10.16 to obtain a compactification,

$$\alpha S = e^\beta_{\mathscr{F}}[\beta S] = \text{cl}_T e_{\mathscr{F}}[S]$$

of S, generated by $\mathscr{F}$. We can associate to the compactification αS, the set $C_\alpha(S) \subseteq C^*(S)$ of real-valued continuous functions. In the following theorem, we show conditions that $\mathscr{F}$ must satisfy to guarantee that $\mathscr{F} = C_\alpha(S)$.

Theorem 21.11 *Let (S, τ) be a completely regular space and $\mathscr{F}$ be a* subalgebra *of $C^*(S)$ which separates points and closed sets in S. Suppose $\mathscr{F}$ generates the compactification*

$$\alpha S = e^\beta_{\mathscr{F}}[\beta S] = cl_T e_{\mathscr{F}}[S]$$

If $\mathscr{F}$ satisfies the properties:

(1) *$\mathscr{F}$ contains $C_\omega(S)$,*
(2) *every $f \in C_\alpha(S)$ is equivalent to some function $g \in \mathscr{F}$,*

then $\mathscr{F} = C_\alpha(S)$.

Proof. We are given that $\mathscr{F}$ separates points and closed sets of S and $C_\omega(S) \subseteq \mathscr{F}$. Then

$$e^\beta_{\mathscr{F}}[\beta S] = \text{cl}_T e_{\mathscr{F}}[S] = \alpha S$$

is a compactification of S (generated by $\mathscr{F}$) where $C_\alpha(S) = \{f|_S : f \in C(\alpha S)\}$.

We are also given that every function in $C_\alpha(S)$ is equivalent to some function in $\mathscr{F}$.

We are required to show that $\mathscr{F} = C_\alpha(S)$.

Claim 1: $\mathscr{F} \subseteq C_\alpha(S)$. Note that $e^{\beta\leftarrow}_{\mathscr{F}} : \alpha S \to \beta S$. Let $f \in \mathscr{F}$. We define the real-valued function $f^\alpha : \alpha S \to \mathbb{R}$ as

$$f^\alpha = f^\beta \circ e^{\beta\leftarrow}_{\mathscr{F}}$$

We see that $f^\alpha \in C(\alpha S)$ and so is a continuous extension of $f \in \mathscr{F}$. We then conclude that $\mathscr{F} \subseteq C_\alpha(S)$, as claimed.

Claim 2: $C_\alpha(S) \subseteq \mathscr{F}$. Suppose $g \in C_\alpha(S)$. Then, by hypothesis, there is a function $f \in \mathscr{F}$ such that $g - f = h \in C_\omega(S)$. Then $g = f + h$. Since both f and h belong to the subalgebra $\mathscr{F}$, then $g \in \mathscr{F}$. We can then conclude that $C_\alpha(S) \subseteq \mathscr{F}$, as claimed.

Then $C_\alpha(S) = \mathscr{F}$, as required.

The converse is easily seen to be true. That is, if $\mathscr{F} = C_\alpha(S)$, then $\mathscr{F}$ satisfies the two given properties.

The reader can expect to encounter, a bit later, similar theorem statements, one of which is called the *Stone–Weierstrass theorem* in Theorem 30.3, the other is Theorem 30.4.[6]

[6]The Stone–Weierstrass theorem states: "Let S be a compact topological space. Let $\mathscr{F}$ be a complete subring of $C^*(S)$ which contains the constant functions. If $\mathscr{F}$ separates the points of S then $\mathscr{F} = C^*(S)$".

Suppose γS is a compactification of S and $C_\gamma(S) = \{f|_S \in C(S) : f \in C(\gamma S)\}$. That is, $C_\gamma(S)$ is the set of all function, f, in $C^*(S)$ which extend to $f^\gamma : \gamma S \to \mathbb{R}$. We have shown that $\gamma S \preceq \beta S$ and $C_\gamma(S)$ is a subalgebra of $C^*(S)$. We have also seen that there is a continuous map $\pi_{\beta\to\gamma} : \beta S \to \gamma S$ which fixes the points of S.

In the following lemma, we show that we can express the function $\pi_{\beta\to\gamma} : \beta S \to \gamma S$ in a form which better describes the mechanism behind the function itself.

Lemma 21.12 *Let γS be a compactification of the space S. Let $\mathscr{G} = C_\gamma(S)$. Then*

$$\pi_{\beta\to\gamma} = e_{\mathscr{G}}{}^{\gamma\leftarrow} \circ e_{\mathscr{G}}{}^{\beta}$$

where $e_{\mathscr{G}}$ is the evaluation map generated by $\mathscr{G}$.

Proof. If $f \in C_\gamma(S)$, for $x \in \beta S$, $f^\beta(x) = (f^\gamma \circ \pi_{\beta\to\gamma})(x)$. Then, for $x \in \beta S$,

$$\begin{aligned}
e_{\mathscr{G}}{}^{\beta}(x) &= \langle f^\beta(x)\rangle_{f\in C_\gamma(S)}\\
&= \langle (f^\gamma \circ \pi_{\beta\to\gamma})(x)\rangle_{f\in C_\gamma(S)}\\
&= \langle f^\gamma(\pi_{\beta\to\gamma}(x))\rangle_{f\in C_\gamma(S)}\\
&= e_{\mathscr{G}}{}^{\gamma}(\pi_{\beta\to\gamma}(x))\\
&= (e_{\mathscr{G}}{}^{\gamma} \circ \pi_{\beta\to\gamma})(x)
\end{aligned}$$

Then $e_{\mathscr{G}}{}^{\beta} = (e_{\mathscr{G}}{}^{\gamma} \circ \pi_{\beta\to\gamma})$ on βS. Then $\pi_{\beta\to\gamma} = e_{\mathscr{G}}{}^{\gamma\leftarrow} \circ e_{\mathscr{G}}{}^{\beta}$.

21.6 Limits of z-Ultrafilters in βS

For what follows, recall the definitions of terms related to z-filters in 14.10.

A consequence of the Stone–Weierstrass statement is Theorem 30.4 which roughly states that: "If the set, $C^*(S)$ contains a subring, $\mathscr{F}$, which is complete and contains $C_\omega(S)$ then $\mathscr{F} = C_\alpha(S)$ for some compactification, αS, of S".

Suppose $\mathscr{Z} = \{Z(f) : f \in M \subseteq C^*(S)\}$ is a free z-ultrafilter in the locally compact Hausdorff space S, where M is the corresponding free maximal ideal in $C^*(S)$. Let

$$\mathscr{Z}^* = \{\text{cl}_{\beta S} Z(f) : f \in M\}$$

denote a family of closures of the elements in $\mathscr{Z}$. Since βS is compact Hausdorff and $\mathscr{Z}$ is a filter, $\mathscr{Z}^*$ satisfies the finite intersection property. Then $\mathscr{Z}^*$ must have non-empty intersection in βS. Then it is fixed and so must have a unique limit point,

$$\{p\} = \cap\{\text{cl}_{\beta S} Z(f) : f \in M\}$$

in $\beta S \backslash S$. We clearly have $\text{cl}_{\beta S} Z(f) \subseteq Z(f^\beta)$ for each $f \in M \subseteq C^*(S)$. Since $f^\beta|_{Z(f)}$ agrees with f^β on $\text{cl}_{\beta S} Z(f)$, then its extension to $Z(f^\beta)$ agrees with f^β on $Z(f^\beta)$. So,

$$\text{cl}_{\beta S} Z(f) = Z(f^\beta) \quad (*)$$

So,

> *for any free z-ultrafilter, $\mathscr{Z} = \{Z(f) : f \in M \subseteq C^*(S)\}$ in $Z[S]$, we can write,*
>
> $$\{p\} = \cap\{cl_{\beta S} Z(f) : f \in M\} = \cap\{Z(f^\beta) : p \in Z(f^\beta)\}$$
>
> *where $p \in \beta S \backslash S$. So the points in $\beta S \backslash S$ are precisely the unique limits of a unique free z-ultrafilter in $Z[S]$.*

Of course, if $\mathscr{Z}$ is a fixed z-ultrafilter in $Z[S]$ then

$$\{p\} = \cap\{Z(f) : f \in M\}$$

for some p in S.

Theorem 21.13 *Suppose T is a dense subset of the completely regular space S. Then the following are equivalent:*

(a) *The subset, T, is C^*-embedded in the space S.*
(b) *Disjoint zero-sets in T have disjoint closures in S.*
(c) *Every point of S is the limit of a unique z-ultrafilter on T.*

Proof. We are given that S is completely regular.

(a $\Rightarrow$ b) Suppose T is a C^*-embedded dense subset of S. Let $h, g \in C^*(T)$ be such that $Z(h)$ and $Z(g)$ are disjoint zero-sets of T. Then $Z(h)$ and $Z(g)$ are completely separated in T.[7]

By Urysohn's extension theorem, $Z(h)$ and $Z(g)$ are completely separated in S. Then there exists $t \in C^*(S)$ such that $Z(h) \subseteq Z(t)$ and $Z(g) \subseteq Z(t-1)$. Then $\mathrm{cl}_S Z(h) \subseteq Z(t)$ and $\mathrm{cl}_S Z(g) \subseteq Z(t-1)$. We can then conclude that $\mathrm{cl}_S Z(h) \cap \mathrm{cl}_S Z(g) = \varnothing$, as required.

(b $\Rightarrow$ c) Suppose that disjoint zero-sets in T have disjoint closures in S. Let Z_1 and Z_2 be disjoint zero-sets in T.

Claim 1: We claim that $\mathrm{cl}_S(Z_1 \cap Z_2) = \mathrm{cl}_S Z_1 \cap \mathrm{cl}_S Z_2$.

Proof of claim: Of course, $LHS \subseteq RHS$ is always true. Suppose, on the other hand, that $x \in \mathrm{cl}_S Z_1 \cap \mathrm{cl}_S Z_2$.

We are required to show that $x \in \mathrm{cl}_S(Z_1 \cap Z_2)$. For any zero-set neighborhood, Z, of x, $Z \cap Z_1$ is dense in $Z \cap \mathrm{cl}_S Z_1$ so

$$x \in \mathrm{cl}_S(Z \cap Z_1) \cap \mathrm{cl}_S(Z \cap Z_2) \neq \varnothing \quad (*)$$

Now both $Z \cap Z_1$ and $Z \cap Z_2$ are zero-sets so, by hypothesis, $(Z \cap Z_1) \cap (Z \cap Z_2)$ cannot be empty. So, $Z \cap (Z_1 \cap Z_2) \neq \varnothing$. Since Z is any zero-set neighborhood of x, this means that $x \in \mathrm{cl}_S(Z_1 \cap Z_2)$.

We conclude $\mathrm{cl}_S(Z_1 \cap Z_2) = \mathrm{cl}_S Z_1 \cap \mathrm{cl}_S Z_2$, as claimed.

Claim 2: We claim that each point in $S \backslash T$ is the limit of a unique z-ultrafilter on T.

Proof of claim: Let $y \in T$. Then y belongs to the closure of a zero-set, Z, in T. Hence, y is the limit point of a z-ultrafilter, $\mathscr{Z}$, in T. Now, if y is also the limit of another z-ultrafilter, $\mathscr{Z}_1$, in T then $\mathscr{Z}_1$, will contain a zero-set Z_1 which will not intersect some zero-set, Z_2, of $\mathscr{Z}$. Then $y \in \mathrm{cl}_S Z_1 \cap \mathrm{cl}_S Z_2 = \mathrm{cl}_S(Z_1 \cap Z_2) = \varnothing$. We can only conclude that every point in $S \backslash T$ is the limit of a unique z-ultrafilter in T. As claimed.

(c $\Rightarrow$ a) We are given that $S \backslash T$ is a set of limits of free z-ultrafilters on T. Since $\beta T \backslash T$ is the set of all limit points of free z-ultrafilters

[7] To see this, note that $Z(|h| + |g|) = \varnothing$ in T and so for the function, $k(x) = |h(x)|/[|h(x)| + g(x)|]$, $Z(h) \subseteq Z(k)$ and $Z(g) \subseteq Z(k-1)$.

on T, we can say that

$$S \backslash T \subseteq \beta T \backslash T$$

Now, since S is dense in βS and T is dense in S, then T is dense in βS so we can view βS as a compactification, say γT, of T with outgrowth

$$\gamma T \backslash T = (\beta S \backslash S) \cup (S \backslash T)$$

Then for the function $\pi_{\beta \to \gamma} : \beta T \to \gamma T$, we have, $\pi_{\beta \to \gamma}[\beta T \backslash T] = \gamma T \backslash T$ where,

$$\pi_{\beta \to \gamma}|_S(x) = x$$

Then, for $f \in C^*(T)$ define $f^* : S \to \mathbb{R}$ as

$$f^*(x) = f^\beta \circ \pi_{\beta \to \gamma}|_S^{\leftarrow}(x)$$

where f^* is seen to be continuous on S and $f^*|_S = f$ on T. So, f^* is a continuous extension of f from T to S. We conclude that T is C^*-embedded in S.

21.7 Pseudocompact Spaces Revisited

Recall (from Definition 17.10) that a topological space is said to be *pseudocompact* if every continuous real-valued function on S is bounded. That is, if $C(S) = C^*(S)$.

Although the pseudocompact property has a simple and easily understood definition, it turns out that, when not compact, such spaces are not easily recognizable. It will be helpful to obtain a few characterizations. In the following theorem, we show that, for locally compact Hausdorff spaces, pseudocompact spaces are precisely those spaces, S, where $\beta S \backslash S$ does not contain a zero-set.

Theorem 21.14 *A locally compact Hausdorff space S is pseudocompact if and only if no zero set Z in $Z[\beta S]$ is entirely contained in $\beta S \backslash S$.*

Proof. Let S be a locally compact Hausdorff space.

($\Rightarrow$) Suppose S is a pseudocompact space and $Z(f^\beta) \in Z[\beta S]$. We are required to show that $Z(f^\beta) \cap S \neq \varnothing$.

Suppose not. That is, suppose, $Z(h) \subseteq \beta S \backslash S$ (where $h \in C(\beta S)$. Then, since $h|_S$ is not zero in S, we can define the function $g = 1/h|_S$ and so $g \in C(S)$. Let $z \in Z(h)$. Since z belongs to $\text{cl}_{\beta S}S$, then there is a sequence, $\{x_i\}$ in S, which converges to z. By continuity, the corresponding sequence, $\{h(x_i)\}$ in $\mathbb{R}$, must converge to $h(z) = 0$. So, g is unbounded on S, which contradicts the hypothesis which states that all real-valued continuous functions on S are bounded. So, $Z(h)$ must intersect with S, as required.

($\Leftarrow$) Suppose now that, for any $Z \in Z[\beta S]$, $Z \cap S \neq \varnothing$. We are required to show that S is pseudocompact.

Suppose S is not pseudocompact. Then $C(S)$ contains an unbounded function g. Let

$$f = |g| \wedge k$$

where $k > 0$. Then f is continuous real-valued unbounded on S. Then for each $n \in \mathbb{N}$, there exists $x_n \in S$ such that $f(x_n) \in (n, \infty)$. Then $h = 1/f$ is a continuous well-defined real-valued bounded function on S. Since βS is compact, $\{x_n : n \in \mathbb{N}\}$ has a converging subsequence $\{x_{n_i} : i \in \mathbb{N}\}$ with limit, say $x \in \beta S \setminus S$. Since h is continuous and bounded on S, h extends to $h^\beta : \beta S \to \mathbb{R}$ with $h^\beta(x) = 0$. Then $Z(h^\beta) \subseteq \beta S \backslash S$, a contradiction of our hypothesis. Then S is pseudocompact.

It is interesting to note that, in the above theorem, a property of the outgrowth, $\beta S \backslash S$ characterizes a property of the space S.

We point out one easy consequence of the above theorem.

Suppose S is locally compact and completely regular. If S is pseudocompact and $k = f^\beta(x) \in f^\beta[\beta S \setminus S]$, then there exists $y \in Z(f^\beta - k) \cap S$, so, $f^\beta(y) = k \in f[S]$. So, $f^\beta[\beta S \backslash S] \subseteq f[S]$.

Conversely, suppose $f^\beta[\beta S \backslash S] \subseteq f[S]$. Suppose $Z(f^\beta - t) \cap \beta S \backslash S \neq \varnothing$. Then there exists $y \in \beta S \backslash S$ such that $f^\beta(y) = t$ and $x \in S$ such that $f(x) = t$. Then $x \in Z(f^\beta - t) \cap S \neq \varnothing$. So, S is pseudocompact.

> *The locally compact completely regular space, S, is pseudocompact if and only if, for every $f \in C^*(S)$, $f^\beta[\beta S \backslash S] \subseteq f[S]$.*

21.8 The One-Point Compactification

It was shown in Theorem 18.8, that every locally compact Hausdorff space is completely regular. Hence, a locally compact Hausdorff space, S, has at least one compactification, namely, βS. This compactification is maximal when compared to all others in the family, $\mathscr{C}$, of all compactifications. Does the family $\mathscr{C}$ contain a minimal element? That is, does $\mathscr{C}$ have a compactification, γS, such that $\gamma S \preceq \alpha S$, for all compactifications, αS in $\mathscr{C}$? The answer will depend on the space, S. In a previous chapter of the book, we, in fact, provided an answer to this question, as we shall soon see.

In Theorem 18.7, we showed that given a locally compact Hausdorff space (S, τ) and a point $\omega \notin S$, we can construct a larger set,

$$\omega S = S \cup \{\omega\}$$

By first defining, $\mathscr{B}_\omega = \{U \cup \{\omega\} : U \in \tau \text{ and } S \backslash U \text{ is compact}\}$, we then define a topology, τ_ω, on ωS as follows:

$$\tau_\omega = \tau \ \cup \ \mathscr{B}_\omega$$

We then showed that $(\omega S, \tau_\omega)$ is a compact space which densely contains S. Then $(\omega S, \tau_\omega)$ satisfies the definition of a compactification of S. Furthermore, and quite importantly, we show in Theorem 18.7 that ωS is Hausdorff if and only if S is locally compact.

So a non-compact, locally compact Hausdorff space, S, has a compactification, ωS, which may be different from βS. We formally define it.

Definition 21.15 Let (S, τ) be a locally compact Hausdorff topological space, ω be a point not in S and $\omega S = S \cup \{\omega\}$. If

$$\mathscr{B}_\omega = \{U \cup \{\omega\} : U \in \tau \text{ and } S \backslash U \text{ is compact}\}$$

and $\tau_\omega = \tau \ \cup \ \mathscr{B}_\omega$ then $(\omega S, \tau_\omega)$ is called

the one-point compactification of S.[8]

[8]The one-point compactification of S is also referred to as the *Alexandrov compactification* of S, named after the soviet mathematician, Pavel Alexandrov (1896–1982).

We will, more succinctly, denote the one-compactification of S by,

$$\omega S = S \cup \{\omega\}$$

Since we have chosen the symbol ωS as notation, then, to be consistent with the notation used up to now, we should let

$$C_\omega(S) = \{f|_S : f \in C(\omega S)\}$$

But the symbol, $C_\omega(S)$, has already been used in another sense on page 493 where $C_\omega(S)$ is used to represent the set of all $f \in C^*(S)$ such that f^β is constant on $\beta S \backslash S$. We verify that these are the same set. For $f \in C_\omega(S)$ if and only if,

$$f^\beta[\beta S \backslash S] = f^\beta[\pi^{\leftarrow}_{\beta \to \omega}[\{\omega\}]] = \{f^\omega(\omega)\} \subset \mathbb{R}$$

Then $C_\omega(S)$ separates points and closed sets of S.[9]

It is also worth emphasizing the fact that

> *for a space S to have a one-point compactification which is Hausdorff, S must be locally compact.*

We can even say more. Amongst all the the completely regular spaces, S, the only ones that are open in any compactification are the ones where S is locally compact. We will prove this now.

Theorem 21.16 *Let S be a completely regular topological space and αS be any compactification of S. Then S is open in αS if and only if S is locally compact.*

Proof. We are given that S is completely regular and αS is a compactification of S.

($\Rightarrow$) Suppose S is open in αS. Since the space S is the intersection of the open set S and the closed set αS, by Theorem 18.3, S is locally compact.

---

[9] Had we known this fact before we would has seen that, in the statement of Theorem 21.11, the condition (1) is redundant.

($\Leftarrow$) Suppose S is locally compact in αS. By Theorem 18.3, part (d), S is the intersection of an open subset, U, and a closed subset, F. Since $F = \alpha S$ and S is dense in αS, then S is open in αS, as required.

If S is locally compact then the map, $\pi_{\beta\to\omega} : \beta S \to \omega S$, collapses the set $\beta S \setminus S$ down to the singleton set $\{\omega\}$. More generally, if αS is any compactification of S, then $\pi_{\alpha\to\omega}$ continuously collapses the outgrowth $\alpha S \setminus S$ down to $\{\omega\}$ and fixes the points of S. So for any compactification αS,

$$\omega S \preceq \alpha S \quad \text{and} \quad C_\omega(S) \subseteq C_\alpha(S)$$

Theorem 21.17 *Uniqueness of ωS. Let S be a locally compact completely regular topological space. Suppose αS and γS are both compactifications of S which contain only one point in their compact extension. Then they are equivalent compactifications.*

Proof. We are given that S is locally compact completely regular. Suppose $\alpha S \setminus S = \{\omega_\alpha\}$ and $\gamma S \setminus S = \{\omega_\gamma\}$, both singleton sets.

Consider the map, $h : \alpha S \to \gamma S$, where $h(x) = x$ on S and $h(\omega_\alpha) = \omega_\gamma$. Then h is one-to-one and onto. It will suffice to show that h maps open neighborhoods to open neighborhoods.

Let U_α be any open neighborhood of a point in αS. If $U_\alpha \subseteq S$ then U_α is open in S. Then $h[U_\alpha] = U_\alpha$. Since S is locally compact, it is open in γS and $U_\alpha = U_\alpha \cap S$ is open in γS.

Suppose $\omega_\alpha \in U_\alpha$. Since $\alpha S \setminus U_\alpha$ is compact and $h|_S$ is continuous, then $h[\alpha S \setminus U_\alpha]$ is a compact. Since h is one-to-one, $h[\alpha S \setminus U_\alpha] = h[\alpha S] \setminus h[U_\alpha] = \gamma S \setminus h[U_\alpha]$, a compact set which doesn't contain ω_γ. So, $h[U_\alpha]$ is open. Hence, $h : \alpha S \to \gamma S$ is a homeomorphism. We can conclude that $\alpha S \equiv \gamma S$.

From this theorem, we can conclude that the one-point compactification is unique, up to equivalence. We now consider a few examples involving compactifications of a space.

Example 7. Suppose that S is locally compact and its one-point compactification, ωS, of S is metrizable. Show that S must be second countable.

Solution: Suppose ωS is metrizable. In the proof of Theorem 15.8, it is shown that countably compact metric spaces are separable. By Theorem 5.11, a separable metric space is second countable. Since ωS is compact and so is countably compact, then, by combining these two results, we obtain that ωS is second countable. By Theorem, 5.13, subspaces of second countable spaces are second countable. So, S is second countable. As required.[10]

It may happen that βS and ωS are the same compactification. We provide an example where they are not equivalent.

Example 8. Consider the set $S = (0, 1]$ equipped with the usual subspace topology. Determine ωS. Show that the one-point compactification, $\omega S = [0, 1]$, of S is not equivalent to βS.

Solution: Since [0, 1] is a compact set which densely contains S, and the one-point compactification is unique, then $\omega S = [0, 1]$. Note that, since the function $f(x) = \sin \frac{1}{x}$ is a bounded continuous function on S, then it extends to βS. (Plot $f(x) = \sin \frac{1}{x}$.) Verify that f does not extend continuously to $[0, 1]$.

Then $[0, 1]$ cannot be the Stone–Čech compactification of S.

The following theorem provides an example of a space, S, such that $\beta S = \omega S$.

Theorem 21.18 *If S is the ordinal space $[0, \omega_1)$ (where ω_1 is the first uncountable ordinal) then $\beta S \backslash S = \{\omega_1\}$ and so $\beta S = \omega S = [0, \omega_1]$, the one-point compactification of S.*

Proof. Given: The space, S, is the ordinal space $[0, \omega_1)$. Then $\omega S = [0, \omega_1]$ is its one-point compactification. So, S is completely regular and locally compact.

We are required to show that $\beta S = \omega S = [0, \omega_1]$.

In the example on page 373, it is shown that $S = [0, \omega_1)$ is countably compact but non-compact. In Theorem 15.9, it is shown that, if S is countably compact every function in $C(S)$ is bounded and so has a compact image in $\mathbb{R}$.

---

[10]The converse of the statement in this example is true. That is, "Locally compact second countable spaces have a metrizable one-point compactification" has been proven. But its proof is fairly involved. So, we will not show it here.

Let $f \in C^*(S) = C(S)$. Then

$$f^\beta[\beta S] = f^\beta[\mathrm{cl}_{\beta S} S] = \mathrm{cl}_{\mathbb{R}} f[S] = f[S] \subseteq \mathbb{R}$$

To show that $\beta S = \omega S$ it suffices to show that $f^\beta[\beta S \backslash S]$ is a singleton set in $f[S]$. Suppose q and q^* are two points in $f^\beta[\beta S \backslash S] \subseteq f[S] \subseteq \mathbb{R}$.

We claim that $q = q^*$.

Proof of claim: Express $f[S]$ as a net

$$N = \{f(\alpha) : \alpha \in S = [0, \omega_1)\}$$

Both q and q^* are accumulation points of the net $N = f[S]$. So, for each $n \in \mathbb{N}$, both $B_{1/2n}(q)$ and $B_{1/2n+1}(q^*)$ each contain a cofinal subset of the tail end of the net N. We can then choose, $\{\alpha_n : n \in \mathbb{N}\}$ strictly increasing in S such that $f(\alpha_{2n}) \in B_{1/2n}(q)$ and $f(\alpha_{2n+1}) \in B_{1/2n+1}(q^*)$. If $\sup\{\alpha_n\} = \kappa$ then $\sup\{\alpha_{2n}\} = \kappa = \sup\{\alpha_{2n+1}\}$. Hence,

$$\lim_{n\to\infty} f(\alpha_{2n}) = q = f(\kappa) = q^* = \lim_{n\to\infty} f(\alpha_{2n+1})$$

So, $q = q^*$ as claimed.

From this we can conclude that, for all $f \in C^*(S)$, f^β is constant on $\beta S \backslash S$. Then $\beta S = [0, \omega_1] = \omega S$, the one-point compactification of S.

In the following statement, an *n-sphere* , S^n, refers to a subset of $\mathbb{R}^{n+1}$ defined as $\{\vec{x} \in \mathbb{R}^{n+1} : \|\vec{x}\| = 1\}$. For example, S_2 is a subset of $\mathbb{R}^3$ where $(x, y, z) \in S^2$ if and only if $\|(x, y, z)\| = 1$. We will show that, the 2-sphere, S_2 is homeomorphic to $\omega\mathbb{R}^2$.

Theorem 21.19 *Let S^n denote the n-sphere in $\mathbb{R}^{n+1}$. Then the "punctured n-sphere", S^n minus a single point, $(S^n \backslash \{p\})$, is homeomorphic to $\mathbb{R}^n$. Then, S^n is homeomorphic to the one-point compactification, $\omega\mathbb{R}^n$, of $\mathbb{R}^n$.*

Proof. We are given that S^n is the n-sphere which is a subset of $\mathbb{R}^{n+1}$ and that $p = (0, 0, 0, \ldots, 0, 1)$ is a point in $\mathbb{R}^{n+1}$ which belongs to S^n. We will first show that $S^n \backslash \{p\}$ is homeomorphic to $\mathbb{R}^n$. We achieve this by showing that the function $g : S^n\backslash\{p\} \to \mathbb{R}^n$ defined as

$$g(x_1, x_2, \ldots, x_n, x_{n+1}) = \frac{1}{1 - x_{n+1}}(x_1, x_2, \ldots, x_{n-1}, x_n)^{11}$$

maps $S^n\backslash\{p\}$ homeomorphically onto $\mathbb{R}^n$.

Since $p = (0, 0, \ldots, 1)$, then $1 - x_{n+1} \neq 0$, so g is well-defined.

Let $a = (a_1, a_2, \ldots, a_{n+1})$ and $b = (b_1, b_2, \ldots, b_{n+1})$ be distinct points in $S^n\backslash\{p\}$. Then $a_k \neq b_k$ for at least one $k \in \{1, 2, \ldots, n+1\}$. Then

$$\frac{1}{1 - a_{n+1}}(a_1, a_2, \ldots, a_{n-1}, a_n) \neq \frac{1}{1 - b_{n+1}}(b_1, b_2, \ldots, b_{n-1}, b_n)$$

So, g is one-to-one. The function g is also verified to be onto $\mathbb{R}^n$.

To prove continuity of $g : S^n\backslash\{p\} \to \mathbb{R}^n$ it suffices to show that $\pi_i \circ g$ is continuous for each $i \in \{1, \ldots, n\}$, and invoke 7.11. See that

$$\begin{aligned}(\pi_i \circ g)(x_1, x_2, \ldots, x_n, x_{n+1}) &= \pi_i \left[\frac{1}{1 - x_{n+1}}(x_1, x_2, \ldots, x_{n-1}, x_n)\right] \\ &= \frac{x_i}{1 - x_{n+1}} \\ &= \frac{\pi_i(x_1, x_2, \ldots, x_n)}{1 - x_{n+1}} \\ &= \left(\frac{\pi_i}{1 - x_{n+1}}\right)(x_1, x_2, \ldots, x_n)\end{aligned}$$

so $\pi_i \circ g$ is continuous for each i.

Also, since π is an open map, the function g is an open map.

Then g maps $S^n\backslash\{p\}$ homeomorphically onto $\mathbb{R}^n$, as we claimed earlier.

We now define a function $g^* : S_n \to \omega\mathbb{R}^n$ and show it is a homeomorphism.

Let $g^* : S_n \to \omega\mathbb{R}^n$ be a function mapping the one-point compactification, S^n, of $S^n\backslash\{p\}$ onto the one-point compactification,

---

[11] The function, g, is known as a *stereographic projection.*

$\omega\mathbb{R}^n = \mathbb{R} \cup \{\omega\}$, of $\mathbb{R}^n$ which is defined as

$$g^*|_{S^n\backslash\{p\}}(x) = g(x)$$
$$g^*(p) = \omega$$

If U is an open subset of $\omega\mathbb{R}^n$ which doesn't contain ω then $g^{*\leftarrow}[U]$ is open in $S^n\backslash\{p\}$, and so is open in S^n.

If U is an open subset of $\omega\mathbb{R}^n$ which contains ω then $g^{*\leftarrow}[U] = \{p\} \cup g^{\leftarrow}[U \cap \mathbb{R}^n]$. Now, $\omega\mathbb{R}^n\backslash U$ is compact in $\omega\mathbb{R}^n$ and so is compact in $\mathbb{R}^n$. Then $g^{\leftarrow}[\omega\mathbb{R}^n \backslash U]$ is compact in S^n. So, $g^{*\leftarrow}[U]$ is an open neighborhood of p in S^n. So, g^* is continuous on S^n.

Similarly, g^* maps open subsets of S^n to open subsets of $\omega\mathbb{R}^n$.

So, g^* is a homeomorphism between the compact sets S^n and $\omega\mathbb{R}^n$.

We also showed along the way that $g^*|\backslash_{S_n\backslash\{p\}} = g$ maps the punctured n-sphere, $S_n\backslash\{p\}$, homeomorphically onto $\mathbb{R}^n$.

21.9 Topic: Cardinality of Some Common Stone–Čech Compactifications

We know the cardinality of the most common sets we encounter (such as $\mathbb{N}$, $\mathbb{Q}$ and $\mathbb{R}^n$). We can sometimes determine the cardinality of associated sets such as their Stone–Čech compactification. We know the cardinality, $|\mathbb{N}|$, of the set $\mathbb{N}$ is $\aleph_0$ while $|\mathbb{R}| = c = 2^{\aleph_0}$. In the following theorem, we compute the cardinalities of $\beta\mathbb{N}$, $\beta\mathbb{Q}$, and $\beta\mathbb{R}$. This is good practice in working with these particular compactifications.

Theorem 21.20 *The cardinality, $|\beta\mathbb{N}|$, of the set $\beta\mathbb{N}$ is 2^c.*

Proof. We are given the compactification, $\beta\mathbb{N}$, of $\mathbb{N}$.

Claim 1: We first claim that $|\beta\mathbb{N}| \geq 2^c$.

Proof of Claim 1: In Theorem 7.10, it is shown that, since $[0, 1]$ is separable, then the product space, $K = \prod_{i\in\mathbb{R}}[0, 1]$, with $|\mathbb{R}| = c$ factors, is also separable. This means that there is a countably infinite set, D, contained in K. There then exist a function

$$g : \mathbb{N} \to D \subset K$$

which indexes the elements of D. Note that g is continuous on $\mathbb{N}$ and densely embeds D in K. By Tychonoff's theorem, K is compact. See that g extends continuously from $\mathbb{N}$ to

$$g^{\beta(K)} : \beta\mathbb{N} \to K$$

Then

$$\begin{aligned} g^{\beta(K)}[\beta\mathbb{N}] &= g^{\beta(K)}[\mathrm{cl}_{\beta\mathbb{N}}\mathbb{N}] \\ &= \mathrm{cl}_K g^{\beta(K)}[\mathbb{N}] \\ &= \mathrm{cl}_K D \\ &= K \end{aligned}$$

Now, $|K| = |\prod_{i\in\mathbb{R}}[0,1]| = c^c = 2^c$.[12] Since $\beta\mathbb{N}$ is the domain of the function $g^{\beta(K)}$ (which could possibly not be one-to-one), then

$$|\beta\mathbb{N}| \geq |K| = 2^c$$

as claimed.

Claim 2: That $|\beta\mathbb{N}| \leq |K| = 2^c$.

Proof of Claim 2: We know that each function in $C^*(\mathbb{N})$ can be seen as a sequence $\{x_i : i \in \mathbb{N}\}$ in $\mathbb{R}^{\mathbb{N}}$. Then

$$|C^*(\mathbb{N})| = |\mathbb{R}^{\mathbb{N}}| = c^{\aleph_0} = c^{13}$$

That is, if $I = |C^*(\mathbb{N})|$, then

$$C^*(\mathbb{N}) = \{f_i : \mathbb{N} \to \mathbb{R} : i \in I\}$$

contains $I = c$ distinct functions.

---

[12]The proof of $|\prod_{i\in\mathbb{R}}[0,1]| = c^c = 2^c$ is shown in an example of Section 24.2 of *Axioms and Set theory*, R. André, in which we compute the cardinality of $\mathbb{R}^{\mathbb{R}}$.
[13]The proof of $|\mathbb{R}^{\mathbb{N}}| = c$ is shown in Theorem 25.2 of *Axioms and Set theory*, R. André.

Let $T = \prod_{i \in I}[a_i, b_i] \subseteq \prod_{i \in I}\mathbb{R}$ where, by Tychonoff theorem, T is compact.

Recall that $e_{C^*(\mathbb{N})} : \mathbb{N} \to \prod_{i \in I} \mathbb{R}$ is the evaluation map generated by $C^*(\mathbb{N})$, explicitly defined as

$$e_{C^*(\mathbb{N})}(n) = \langle f_i(n) \rangle_{f_i \in C^*(\mathbb{N})} \in e_{C^*(\mathbb{N})}[\mathbb{N}] \subseteq T \subseteq \prod_{i \in I}\mathbb{R}$$

Then $e_{C^*(\mathbb{N})}[\mathbb{N}] \subseteq T = \prod_{i \in I}[a_i, b_i]$, and $|T| \leq |\mathbb{R}|^I = c^c = 2^c$. So,

$$\begin{aligned} e^{\beta}_{C^*(\mathbb{N})}[\beta\mathbb{N}] &= e^{\beta}_{C^*(\mathbb{N})}[\mathrm{cl}_{\beta S}\mathbb{N}] \\ &= \mathrm{cl}_T e_{C^*(\mathbb{N})}[\mathbb{N}] \\ &\subseteq T \end{aligned}$$

So, $|\beta\mathbb{N}| \leq |T| \leq 2^c$. This establishes our second claim.

Since $|\beta\mathbb{N}| \leq 2^c$ and $|\beta\mathbb{N}| \geq 2^c$, then $|\beta\mathbb{N}| = 2^c$. We are done.

Now, $\mathbb{N}$ contains countably many points and so $|\beta\mathbb{N}\backslash\mathbb{N}| = 2^c$. Since we can associate to each point in $\beta\mathbb{N}\backslash\mathbb{N}$ a unique free z-ultrafilter in $Z[\mathbb{N}]$, the above theorem confirms that there are 2^c free z-ultrafilters in $Z[\mathbb{N}]$.

The cardinality of $\beta\mathbb{N}$ will help us determine the cardinalities of $\beta\mathbb{R}$ and $\beta\mathbb{Q}$.

Theorem 21.21 *The sets $\beta\mathbb{N}$, $\beta\mathbb{R}$ and $\beta\mathbb{Q}$ each have a cardinality equal to 2^c.*

Proof. We have already shown that $|\beta\mathbb{N}| = 2^c$.

Claim 1: $|\beta\mathbb{Q}| \leq 2^c$.

Proof of Claim 1: We know $\mathbb{N}$ and $\mathbb{Q}$ are countable so both have cardinality $\aleph_0$. Then there exists a one-to-one function, $f : \mathbb{N} \to \beta\mathbb{Q}$, mapping $\mathbb{N}$ onto $\mathbb{Q} \subseteq \beta\mathbb{Q}$. Since $\mathbb{N}$ is discrete, f is continuous on $\mathbb{N}$. By Theorem 21.5, $f : \mathbb{N} \to \beta\mathbb{Q}$ extends to

$$f^{\beta(\beta\mathbb{Q})} : \beta\mathbb{N} \to \mathrm{cl}_{\beta\mathbb{Q}} f[\mathbb{N}] = \mathrm{cl}_{\beta\mathbb{Q}}\mathbb{Q} = \beta\mathbb{Q}$$

Then

$$f^{\beta(\beta\mathbb{Q})}[\beta\mathbb{N}] = f^{\beta(\beta\mathbb{Q})}[\mathrm{cl}_{\beta\mathbb{N}}\mathbb{N}] = \mathrm{cl}_{\beta\mathbb{Q}}f[\mathbb{N}] = \mathrm{cl}_{\beta\mathbb{Q}}\mathbb{Q} = \beta\mathbb{Q}$$

So, $|\beta\mathbb{Q}| \leq |\beta\mathbb{N}| = 2^c$. This establishes Claim 1.

Claim 2: $|\beta\mathbb{R}| \leq |\beta\mathbb{Q}|$.

Proof of Claim 2: Since $\mathbb{Q}$ is dense in $\mathbb{R}$ and $\mathbb{R}$ is dense in $\beta\mathbb{R}$, then it is dense in $\beta\mathbb{R}$ then $\mathrm{cl}_{\beta\mathbb{R}}\mathbb{Q} = \beta\mathbb{R}$. Consider the continuous inclusion function

$$i : \mathbb{Q} \to \mathrm{cl}_{\beta\mathbb{R}}\mathbb{Q} = \beta\mathbb{R}$$

By Theorem 21.5, $i : \mathbb{Q} \to \mathrm{cl}_{\beta\mathbb{R}}\mathbb{Q} = \beta\mathbb{R}$ extends to

$$i^{\beta(\beta\mathbb{R})} : \beta\mathbb{Q} \to \beta\mathbb{R}$$

Then

$$i^{\beta(\beta\mathbb{R})}[\beta\mathbb{Q}] = i^{\beta(\beta\mathbb{R})}[\mathrm{cl}_{\beta\mathbb{Q}}\mathbb{Q}] = \mathrm{cl}_{\beta\mathbb{R}}i[\mathbb{Q}] = \mathrm{cl}_{\beta\mathbb{R}}\mathbb{Q} = \beta\mathbb{R}$$

So, $|\beta\mathbb{R}| \leq |\beta\mathbb{Q}|$. This establishes Claim 2.

Up to now we have shown that $|\beta\mathbb{R}| \leq |\beta\mathbb{Q}| \leq |\beta\mathbb{N}| = 2^c$.

Claim 3: $|\beta\mathbb{R}| \geq |\beta\mathbb{N}|$.

Proof of Claim 3: We know that $\mathbb{N}$ is C^*-embedded in $\mathbb{R}$. (See example on page 488 or Theorem 21.8.) Then, if $i : \mathbb{N} \to \beta\mathbb{N}$ is the continuous inclusion map, since $\mathbb{N}$ is C^*-embedded in $\mathbb{R}$, $i : \mathbb{N} \to \beta\mathbb{N}$ extends continuously to

$$i^* : \mathbb{R} \to \beta\mathbb{N}$$

Also, $i^* : \mathbb{R} \to \beta\mathbb{N}$ extends continuously to $i^{*\beta} : \beta\mathbb{R} \to \beta\mathbb{N}$. Then

$$\begin{aligned}
\beta\mathbb{N} &= \mathrm{cl}_{\beta\mathbb{N}}\mathbb{N} \\
&= \mathrm{cl}_{\beta\mathbb{N}}i[\mathbb{N}] \\
&\subseteq \mathrm{cl}_{\beta\mathbb{N}}i^*[\mathbb{R}] \quad (\text{Since } i \text{ embeds } \mathbb{N} \text{ in } i^*[\mathbb{R}]) \\
&= i^{*\beta}[\mathrm{cl}_{\beta\mathbb{R}}\mathbb{R}] \\
&= i^{*\beta}[\beta\mathbb{R}]
\end{aligned}$$

Since $i^{*\beta}[\beta\mathbb{R}]$ contains $\beta\mathbb{N}$, then $|\beta\mathbb{N}| \leq |\beta\mathbb{R}|$. This establishes Claim 3.

Combining the results in the three claims above we conclude that $|\beta\mathbb{R}| = |\beta\mathbb{Q}| = |\beta\mathbb{N}| = 2^c$ as required.

21.10 Compactifying a Subset T of $S \subseteq \beta S$

If T is a non-compact subset of S, it is interesting to reflect on how $\text{cl}_{\beta S}T$ compares with βT. Does it make sense to say that $\beta T \subseteq \beta S$? We examine this question in the following example.

Example 9. Let T be a non-empty subspace of a completely regular space, S. Show that

$$\text{cl}_{\beta S}T \text{ is equivalent to } \beta T.$$

Solution: We are given that $T \subseteq S$. Since subspaces of completely regular spaces are completely regular, then T is completely regular. Let $i : T \to \beta S$ be the identity function which embeds T into βS. By Theorem 21.5, $i : T \to \beta S$ extends continuously to $i^{\beta(\beta S)} : \beta T \to \beta S$. Then

$$\begin{aligned}\beta T &= i^{\beta(\beta S)}[\beta T]\\ &= i^{\beta(\beta S)}[\text{cl}_{\beta T}T]\\ &= \text{cl}_{\beta S}i[T]\\ &= \text{cl}_{\beta S}T\end{aligned}$$

So, $\text{cl}_{\beta S}T$ is equivalent to βT.

Suppose now that F is a *closed* non-empty subset of the completely regular space S. The statement in the above example will allow us to say something more about F.

To see this, suppose $f \in C^*(S)$. As stated in the example,

$$\text{cl}_{\beta S}F = \beta F$$

So, F is C^*-embedded in $\mathrm{cl}_{\beta S}F$. Since F is closed in S, then

$$(\mathrm{cl}_{\beta S}S)\backslash S \subseteq \beta S\backslash S$$

So, $f : F \to \mathbb{R}$ extends to $f^{\beta} : \beta F \to \mathbb{R}$. Since βF is compact in βS, then f^{β} extends to $f^{\beta^*} : \beta S \to \mathbb{R}$. So,

$$f^{\beta^*}|_S : S \to \mathbb{R}$$

is a continuous extension of f on F to $f^{\beta^*}|_S$ on S (via f^{β^*} on βS).

Then F is C^*-embedded in S. We conclude that:

> *If F is a closed non-empty subset of a completely regular space S then F is C^*-embedded in S.*

21.11 The Zero-Sets of $\beta\mathbb{N}$ Are Clopen

Zero-sets in $\beta\mathbb{N}$ play a role in the solution of the following example. If Z is a zero-set in $\mathbb{N}$ then it is clopen in $\mathbb{N}$. It is easily seen that any zero-set in $\mathbb{N}$ is a zero-set of a characteristic function, $g : \mathbb{N} \to \{0, 1\}$, on $\mathbb{N}$. Since g extends to $g^{\beta} : \beta\mathbb{N} \to \{0, 1\}$, $g^{\beta^{\leftarrow}}(0) = Z(g^{\beta}) = \mathrm{cl}_{\beta\mathbb{N}}Z$ is a clopen zero-set in $\beta\mathbb{N}$. As well, $g^{\beta^{\leftarrow}}(1) = \beta\mathbb{N} \setminus Z(g^{\beta})$ is a clopen zero-set in $\beta\mathbb{N}$. So, for every zero-set $Z(g)$ in $\mathbb{N}$, $Z(g^{\beta})$ is clopen in $\beta\mathbb{N}$. Hence, zero-sets of $\beta\mathbb{N}$ are clopen in $\beta\mathbb{N}$.

Example 10. The compactification, $\beta\mathbb{N}$, is easily seen to be separable ($\mathbb{N}$ is a dense subset of $\beta\mathbb{N}$). Show that $\beta\mathbb{N}\backslash\mathbb{N}$ is *not* separable.

Solution: Suppose that $\beta\mathbb{N}\backslash\mathbb{N}$ is separable. Then, $\beta\mathbb{N}\backslash\mathbb{N}$ contains a dense countable subset

$$D = \{x_i : i \in \mathbb{N}\}$$

Since the cardinality of $\beta\mathbb{N}$ is 2^c (shown above), we can fix two points

$$n^* \in (\beta\mathbb{N}\backslash\mathbb{N})\backslash D \quad \text{and} \quad n \in \mathbb{N}$$

Now, for each $i \in \mathbb{N}$, $\{x_i\}$ and $\{n^*, n\}$ form disjoint closed subsets of $\beta\mathbb{N}$.

Since $\beta\mathbb{N}$ is normal, for each $x_i \in D$, there is a clopen zero-set, $Z_i = Z(g^\beta)$ such that

$$x_i \in Z_i \quad \text{and} \quad \{n^*, n\} \subseteq \beta\mathbb{N}\backslash Z_i$$

Then $\{\beta\mathbb{N}\backslash Z_i : i \in \mathbb{N}\}$ is a family of zero-set clopen neighborhoods of $\{n^*, n\}$.

If $W = \cap\{\beta\mathbb{N}\backslash Z_i : i \in \mathbb{N}\}$ then W is a countable intersection of zero-sets and so is, itself, a zero-set which contains $\{n^*, n\}$ (see page 252) which does not intersect D. Since $n^* \in \beta\mathbb{N}\backslash\mathbb{N} \cap W$, then $W \cap \beta\mathbb{N}\backslash\mathbb{N}$ is a non-empty clopen subset of $\beta\mathbb{N}\backslash\mathbb{N}$ which does not intersect D. Since D is dense in $\beta\mathbb{N}\backslash\mathbb{N}$, we have a contradiction. So, $\beta\mathbb{N}\backslash\mathbb{N}$ is not separable.

Concepts Review

1. Suppose S is a topological space and T is a compact Hausdorff space. What does it mean to say that T is a compactification of S?
2. If S has a compactification, αS, what separation axiom is guaranteed to be satisfied by S?
3. Given a completely regular space S let $e : S \to \pi_{i\in I}[a_i, b_i]$ be the evaluation map on S induced by $C^*(S)$. Give a definition of the Stone–Čech compactification of S which involves this evaluation map.
4. What does it mean to say that the two compactifications of S, αS and γS, are equivalent compactifications?
5. If $\mathscr{C} = \{\alpha_i S : i \in I\}$ denotes the family of all compactifications of S. Define a partial ordering of $\mathscr{C}$.
6. If U is a subset of the topological space S, what does it mean to say that U is C^*-embedded in S?
7. If S is C^*-embedded in the compactification, αS, of S what can we say about αS?
8. Suppose S is completely regular and $g : S \to K$ is a continuous function mapping S into a compact Hausdorff space K. For which compactifications, αS, does the following statement hold true: "the function g extends to a continuous function $g^* : \alpha S \to K$"?

9. Suppose S is locally compact and Hausdorff. Define the one-point compactification, ωS, of S.
10. What can we says about those subspaces of a compactification, αS, which are locally compact? What can we says about those subspaces of a compactification, αS, which are open in αS?
11. What is the Stone–Čech compactification of the ordinal space $[0, \omega_1)$?
12. State a characterization of the pseudocompact property stated in this chapter.

Chapter 22

Singular Sets and Singular Compactifications

Abstract
In this chapter, we introduce an alternate method to construct a compactification of a locally compact Hausdorff space. We define the notion of the singular set, $S(f)$, of a function, $f : S \to T$. We will show that we can always use $S(f)$ to construct a compactification, $\alpha S = S \cup S(f)$, by applying a suitable topology on αS. If the singular set, $S(f)$, contains the image, $f[S]$, of f, we refer to f as a singular map. When f is singular the resulting compactification, $S \cup_f S(f)$, is called a singular compactification.

22.1 Singular Sets and Functions: Definitions

We begin by formally defining a "singular set" of a continuous function on a locally compact non-compact Hausdorff space. We also define what is meant when a function is referred to as being a "singular map".

Definition 22.1 Let (S, τ) be a locally compact non-compact Hausdorff topological space and $f : S \to T$ be a continuous function mapping S into some *compact* space T. We define the *singular set*,

$S(f)$, of f as follows:

$$S(f) = \{x \in \mathrm{cl}_T f[S] : \mathrm{cl}_S f^{\leftarrow}[U] \text{ is not compact } \forall\ \mathrm{cl}_T f[S]\text{-open nbhd}, U, \text{ of } x\}$$

If $S(f) = \mathrm{cl}_T f[S]$ then $f : S \to T$ is said to be a *singular function* or *singular map.*

We make the following few remarks about the two concepts we have just introduced.

(1) *If S is a non-compact locally compact Hausdorff, the singular set $S(f)$ is never empty in the compact codomain, T.* This is so, whether f is a singular map or not. To see this, recall that, by Theorem 21.5, $f : S \to T$, extends to $f^{\beta(T)} : \beta S \to T$. Suppose $u \in \beta S \setminus S$ and

$$x = f^{\beta(T)}(u) \in f^{\beta(T)}[\beta S \setminus S]$$

If U is an open neighborhood of x in T, then $u \in f^{\beta(T)\leftarrow}[U]$ an open subset of βS. If $\mathrm{cl}_S f^{\leftarrow}[U]$ is a compact subset of S, then $f^{\beta(T)\leftarrow}[U] \setminus \mathrm{cl}_S f^{\leftarrow}[U]$ is an open neighborhood of u contained in $\beta S \setminus S$, a contradiction. So, $\mathrm{cl}_S f^{\leftarrow}[U]$ is not compact. By definition, $x \in S(f)$. So, $S(f)$ is non-empty.

(2) *The singular set $S(f)$ is always closed, and hence compact, in T.* This is so whether f is a singular map or not. To see this, suppose $x \in T \setminus S(f)$. Then there exists an open neighborhood U of x in T, such that $\mathrm{cl}_S f^{\leftarrow}[U]$ is compact in S. Then every point $p \in U$ also belongs to $T \setminus S(f)$. So, $x \in U \subseteq T \setminus S(f)$. Hence, $S(f)$ is a closed (and so is a compact) subset of the compact space T.

(3) If $f : S \to T$ is a *singular map*, then $f[S]$ is a dense subset of $S(f)$ (since, by definition of singular function, $S(f) = \mathrm{cl}_T f[S]$).

22.2 Singular Compactifications: Definitions

We now show how the singular set, $S(f)$, of a function, $f : S \to T$, can be used as the outgrowth of a compactification of S.

Definition 22.2 Let (S,τ) be a locally compact non-compact Hausdorff topological space and $f : S \to T$ be a continuous function mapping S into a compact space T. Then $S(f)$ denotes its singular set. If f is a singular map then, by definition, $S(f) = \text{cl}_T f[S] \subseteq T$. We construct a new set by adjoining $S(f)$ to S to obtain a larger set,

$$\gamma S = S \cup_f S(f)$$

The basic open neighborhoods, $\mathscr{B}_1$, of points in S will be the same as the ones in S when viewed as a topological space on its own.[1] If $x \in S(f)$, F is a compact subset of S and U an open neighborhood of x in $S(f)$, we define

$$B_x = U \cup f^{\leftarrow}[U]\backslash F$$

as a basic open neighborhood of x. Let $\mathscr{B}_2 = \{B_x : x \in S(f)\}$. Let $\mathscr{B} = \mathscr{B}_1 \cup \mathscr{B}_2$. This defines a base for a topology on γS which is easily seen to be a Hausdorff compactification of S. We will refer to γS as the *singular compactification* induced by the singular map $f : S \to T$.

The definition of a singular compactification is not a simple one to grasp nor to visualize. It invites many questions. For example, if given a singular compactification γS, one may want to know which function induces it. Can there be more than one such function that induces it? Are there compactifications which are non-singular?

To help us answer such questions it will be helpful to find simpler characterizations of a singular compactification.

Retractions and retracts. We remind the reader that, for a topological space $S, \ldots$

> *if $A \subset S$ and $r : S \to A$ is a* continuous *function which fixes the points of A, then r is referred to as a "retraction" of S onto A. In such a case, A is called a "retract" of S.*

[1]Remember that, if S is locally compact Hausdorff, then S is open in any compactification of S.

We will see that those compactifications, αS, whose outgrowth, $\alpha S \backslash S$, is a retract of αS characterize singular compactifications.

Theorem 22.3 *Let S be locally compact and Hausdorff. Let αS be a Hausdorff compactification of S. Then αS is a singular compactification of S if and only if $\alpha S \backslash S$ is a retract of αS.*

Proof. We are given that αS is a Hausdorff compactification of S.

($\Rightarrow$) Suppose $\alpha S = S \cup_f S(f)$ is a singular compactification of S induced by the continuous function,

$$f : S \;\rightarrow\; \mathrm{cl}_T f[S] = S(f) = \alpha S \backslash S$$

We are required to show that $\alpha S \backslash S$ is a retract of αS.

By definition, $f[S]$ is dense in $S(f)$. Let

$$f^{\alpha} : \alpha S \rightarrow S(f)$$

be a function which agrees with f on S and fixes the points of $S(f)$.

We claim, that f^{α} is continuous on αS: Let $x \in S(f)$ and U be an open neighborhood of x. Then $f^{\alpha\leftarrow}[U] = U \cup f^{\leftarrow}[U]$, by definition of a basic open neighborhood in αS. So, f^{α} is continuous on αS, as claimed.

Hence, $\alpha S \backslash S$ is a retract of αS, as required.

($\Leftarrow$) Suppose $\alpha S \backslash S$ is a retract of αS. We are required to show that αS is a singular compactification.

By hypothesis, there is a continuous function $r : \alpha S \rightarrow \alpha S \backslash S$ which fixes the points of $\alpha S \backslash S$. By definition of retract, $r[S] \subseteq \alpha S \backslash S$.

We claim that $r|_S$ is a singular function on S: Let $x \in \alpha S \backslash S$ and U be an open neighborhood of x in $\alpha S \backslash S$. Now, U must intersect $r|_S[S]$, for if not, $r^{\leftarrow}[U] = U \subseteq \alpha S \backslash S$ which is not open in αS. So, $r^{\leftarrow}[U] = U \cup (r^{\leftarrow}[U] \cap S)$. Then, since $\mathrm{cl}_S r|_S{}^{\leftarrow}[S]$ is not compact in S, then $\alpha S \backslash S = S(r|_S)$ is the singular set of $r|_S$, and so

$$\alpha S = S \cup_r S(r|_S)$$

a singular compactification induced by the map $r|_S$. So,

$$r|_S : S \rightarrow \alpha S \backslash S$$

is a singular, map as claimed.

We now have another way of recognizing a singular compactification:

> *Singular compactifications are precisely those compactifications, αS, whose outgrowth, $\alpha S \setminus S$, is a retract of the whole space.*

In the following example, we verify that, if we have one singular compactification of S, then every compactification "below" it in the partially ordered family of all compactifications will also be a singular compactification.

Example 1. Show that if αS is a singular compactification and γS is another compactification such that $\gamma S \preceq \alpha S$ then γS is also a singular compactification.

Solution: Suppose αS is a singular compactification and γS is another compactification such that $\gamma S \preceq \alpha S$. Then $\alpha S = S \cup_f S(f)$ where $f : S \to T$ is continuous and $f[S] \subseteq S(f) = \mathrm{cl}_T f[S]$. Also, there exists $\pi_{\alpha \to \gamma} : \alpha S \to \gamma S$ such that $\pi_{\alpha \to \gamma}$ is continuous and onto and fixes the points of S. If $g = \pi_{\alpha \to \gamma} \circ f$ then

$$\begin{aligned}
\mathrm{cl}_{\gamma S} g[S] &= \mathrm{cl}_{\gamma S}(\pi_{\alpha \to \gamma} \circ f)[S] \\
&= \pi_{\alpha \to \gamma}[\mathrm{cl}_{\alpha S}(f[S])] \\
&= \pi_{\alpha \to \gamma}[\alpha S \setminus S] \\
&= \gamma S \setminus S
\end{aligned}$$

If U is an open subset of $\gamma S \setminus S$

$$\begin{aligned}
\mathrm{cl}_S g^{\leftarrow}[U] &= \mathrm{cl}_S(\pi_{\alpha \to \gamma} \circ f)^{\leftarrow}[U] \\
&= \mathrm{cl}_S f^{\leftarrow}(\pi^{\leftarrow}_{\alpha \to \gamma}[U]) \\
&= \mathrm{cl}_S f^{\leftarrow}[V] \quad (V \text{ open in } \alpha S \setminus S) \\
&\quad \text{a non-compact set in } S.
\end{aligned}$$

So, $\gamma S \setminus S = S(g)$. This means that $\gamma S \setminus S$ is a retract of γS and so γS is a singular compactification. We are done with the solution.

Does every locally compact Hausdorff space, S, have at least one singular compactification? Well, we know that such a space S, has a one-point compactification, $\omega S = S \cup \{\omega\}$. Consider the constant function $r : \omega S \to \{\omega\}$. It is a retraction on ωS. So, $r|_S : S \to \{\omega\}$ is a singular map on S. So, we can answer this question in the affirmative.

22.3 Compactifications Induced by Non-Singular Functions

Suppose (S, τ) is a topological space and $f : S \to T$ is a continuous function mapping a non-compact locally compact Hausdorff space S into a compact space T. For what follows, f may or may not be a singular map.

22.3.1 *Definition of a non-singular compactification induced by f*

Suppose f is *not* a singular map on S. Recall that

$$S(f) = \{x \in \text{cl}_T f[S] : \text{cl}_S f^{\leftarrow}[U] \text{ is not compact for any neighborhood } U \text{ of } x\}$$

Since f is not singular, then $S(f)$ does not satisfy the condition, $f[S] \subseteq S(f)$, required to construct a singular compactification. Despite this, we can still adjoin the singular set $S(f)$ to S to form a larger set, $K = S \cup S(f)$.

We define a topology on K as follows: The set, $\mathscr{B}_S$, of basic open neighborhoods of points in S will be the same as the ones in S when viewed as a topological space on its own. For $x \in S(f)$, F a compact subset of S and U an open neighborhood of x in T, we define

$$B_x = (U \cap S(f)) \cup f^{\leftarrow}[U] \backslash F$$

as a basic open neighborhood of x. This defines a topology on $K = S \cup S(f)$ generated by the basic open sets,

$$\mathscr{B} = \mathscr{B}_S \cup \{B_x : x \in S(f)\}$$

The set, K, will be seen to be a compact set which contains S as a dense subset. These facts will be verified following 22.4. In the case where f is not singular (that is, $f[S] \cap T \backslash S(f) \neq \varnothing$), the set $S(f)$

will however not be a retract of the compactification $K = S \cup S(f)$.[2] We would like to adopt notation which will help distinguish those compactifications (induced by a function) which are singular from those that are not.

Notation 22.4 Suppose $S(f)$ is a singular set of a function $f : S \to T$ where T is compact. If f is not a singular map we will represent the compactification γS induced by f as

$$\gamma S = S \cup S(f)$$

Note how we distinguish it from those compactifications where f is a singular map:

$$\gamma S = S \cup_f S(f)$$

We first justify a few statements made in the introductory paragraph above. All properties are for a continuous function $f : S \to T$ mapping the locally compact Hausdorff space into the compact space T.

(1) *We verify that $\gamma S = S \cup S(f)$ is a well-defined topological space.* To do this we will show that the given family, $\mathscr{B}$, of sets satisfies the "base property"(See page 91). If so it generates a topology on γS. Clearly $\gamma S = \cup\{B \in \mathscr{B}\}$. Given

$$A \cap B = (U_x \cup f^{\leftarrow}[U_x] \backslash F_U) \,\cap\, (V_x \cup f^{\leftarrow}[V_x] \backslash F_V)$$

we have,

$$\begin{aligned} x \in A \cap B &= (U_x \cap V_x) \,\cup\, [(f^{\leftarrow}[U_x] \backslash F_U) \,\cap\, (f^{\leftarrow}[V_x] \backslash F_V)] \\ &= (U_x \cap V_x) \cup [f^{\leftarrow}[U_x] \cap f^{\leftarrow}[V_x]] \backslash (F_U \,\cup\, F_V) \end{aligned}$$

implies

$$x \in (U_x \cap V_x) \cup [f^{\leftarrow}[U_x \cap V_x]] \backslash (F_U \,\cup\, F_V) = A \cap B$$

So, $\mathscr{B}$ satisfies the base property and so generates a topology on γS.

---

[2]If f is not singular, $f[S]$ may still intersect $S(f)$, or may not even intersect $S(f)$ at all; it is just that $f[S]$ cannot be entirely contained in $S(f)$.

(2) *The set S is dense in $\gamma S = S \cup S(f)$.* See that very open neighborhood, $U \cup f^{\leftarrow}[U]\backslash F$, of a point in $S(f)$ intersects S.

(3) **Fact:** Suppose $f : S \to T$ is a continuous function, where T is compact.

Case 1: If $f[S]$ is compact, $S(f) \subseteq f[S]$.

Case 2: If $f[S]$ *is non-compact*, f also induces a compactification of $f[S]$:

$$f^{\beta(T)}[\beta S] = f[S] \cup S(f) = \mathrm{cl}_T f[S]$$

with compactification outgrowth, $S(f)\backslash f[S]$.

Proof. We are given that $f : S \to T$ is a continuous function mapping S into the compact space T. The function f extends to $f^{\beta(T)} : \beta S \to T$.

By definition of $S(f)$, $f[S] \cup S(f) \subseteq \mathrm{cl}_T f[S] = f^{\beta(T)}[\beta S]$.

Case 1: Suppose $f[S]$ is compact. Then $f[S] \cup S(f) \subseteq \mathrm{cl}_T f[S] = f[S]$, so $S(f) \subseteq f[S]$, as required.

Case 2: Suppose $f[S]$ is non-compact. We claim that $f^{\beta(T)}[\beta S] \subseteq f[S] \cup S(f)$. Suppose not. Suppose there is a point $u \in f^{\beta(T)}[\beta S]\backslash S(f))$. Then, since $u \notin S(f)$, there is a $\mathrm{cl}_T f[S]$-open neighborhood, U, of u such that $\mathrm{cl}_S f^{\leftarrow}[U]$ is a compact subset of S.

Then $f^{\beta(T)^{\leftarrow}}(u) \subseteq \mathrm{cl}_S f^{\leftarrow}[U] \subseteq S$. So, $u \in f[S]$. Then $f^{\beta(T)}[\beta S] \subseteq f[S] \cup S(f)$ as claimed.

We conclude that $f^{\beta(T)}[\beta S] = f[S] \cup S(f)$.

So, $f[S] \cup S(f)$ is a compactification of $f[S]$ with compactification outgrowth $S(f) \setminus f[S]$.

(4) **Fact:** For $f : S \to T$, $f^{\beta(T)}[\beta S\backslash S] \subseteq S(f)$.

Proof. Let $f^{\beta(T)}(u) \in f^{\beta(T)}[\beta S\backslash S]$ and U be an open neighborhood of $f^{\beta(T)}(u)$ in $\mathrm{cl}_T f[S]$. Then

$$u \in f^{\beta(T)^{\leftarrow}}[U] \cap \beta S\backslash S \neq \varnothing$$

Then $\mathrm{cl}_S f^{\leftarrow}[U]$ cannot be compact.[3] So $f^{\beta(T)}(u) \in S(f)$.

We conclude $f^{\beta(T)}[\beta S\backslash S] \subseteq S(f)$.

[3]For if $\mathrm{cl}_S f^{\leftarrow}[U]$ is compact and V is an open neighborhood of u which misses $\mathrm{cl}_S f^{\leftarrow}[U]$ then $V \cap f^{\beta(T)^{\leftarrow}}[U]$ is an open subset of βS contained in $\beta S\backslash S$.

(5) *We verify that $\gamma S = S \cup S(f)$ is indeed compact.*

Proof. Given: $\gamma S = S \cup S(f)$ and $f : S \to T$ a continuous function mapping completely regular, S, into the compact space T. Then f extends to $f^{\beta(T)} : \beta S \to T$.

Define $\pi_{\beta(\gamma)} : \beta S \to \gamma S = S \cup S(f)$ as:

$$\pi_{\beta(\gamma)}|_{\beta S \backslash S} = f^{\beta(T)}|_{\beta S \backslash S} \quad \text{and} \quad \pi_{\beta(\gamma)}|_S(x) = x$$

We claim that $\pi_{\beta(\gamma)} : \beta S \to S \cup S(f)$ is continuous. Let $U \cup f^{\leftarrow}[U]\backslash F$ be an open neighborhood of $y \in S(f) \subseteq \gamma S$ (where F is compact in S). Then

$$\begin{aligned}
\pi^{\leftarrow}_{\beta(\gamma)}[U \cup f^{\leftarrow}[U]\backslash F] &= \pi^{\leftarrow}_{\beta(\gamma)}[U] \cup \pi^{\leftarrow}_{\beta(\gamma)}[f^{\leftarrow}[U]\backslash F] \\
&= \pi_{\beta(\gamma)}|^{\leftarrow}_{\beta S \backslash S}[U] \cup \pi_{\beta(\gamma)}|^{\leftarrow}_{S}[f^{\leftarrow}[U]\backslash F] \\
&= f^{\beta(T)}|^{\leftarrow}_{\beta S \backslash S}[U] \cup f^{\leftarrow}[U]\backslash F \\
&= f^{\beta(T)}|^{\leftarrow}_{\beta S \backslash S}[U] \cup f^{\beta(T)}|^{\leftarrow}_{S}[U]\backslash F \\
&= f^{\beta(T)\leftarrow}[U]\backslash F
\end{aligned}$$

an open subset of βS. So, $\gamma S = S \cup S(f)$ is the continuous image of βS. Then $\gamma S = S \cup S(f)$ is compact.
Then $S \cup S(f)$ is a compactification of S.

(6) **Fact:** For $f : S \to T$, $f^{\beta(T)}[\beta S \backslash S] = S(f)$.

Proof. We have shown above that $f^{\beta(T)}[\beta S \backslash S] \subseteq S(f)$. It suffices to show that $S(f) \subseteq f^{\beta(T)}[\beta S \backslash S]$. Let $x \in S(f)$. To show that $x \in f^{\beta(T)}[\beta S \backslash S]$ it suffices to show $f^{\beta(T)}(y) = x$ for some $y \in \beta S \backslash S$. Then

$$\pi^{\leftarrow}_{\beta(\gamma)}(x) = f^{\beta(T)\leftarrow}(x) \cap \beta S \backslash S \subseteq \beta S \backslash S$$

So, for $y \in f^{\beta(T)\leftarrow}(x) \cap \beta S \backslash S$,

$$f^{\beta(T)}(y) = x$$

So,

$$x \in f^{\beta(T)}[\beta S \backslash S]$$

Then $S(f) \subseteq f^{\beta(T)}[\beta S \backslash S]$. Then

$$S(f) = f^{\beta(T)}[\beta S \backslash S]$$

(7) Suppose $f : S \to T$ is continuous (not necessarily singular) where T is compact and $\gamma S = S \cup S(f)$ is the compactification of S induced by f.

Fact: $f : S \to T$ extends continuously to $f^\gamma : \gamma S \to \mathrm{cl}_T f[S]$.

Proof of Fact: Let $\pi_{\beta(\gamma)} : \beta S \to \gamma S$ be the continuous function defined as above. That is, $\pi_{\beta(\gamma)}|_{\beta S \backslash S} = f^{\beta(T)}|_{\beta S \backslash S}$ and $\pi_{\beta(\gamma)}|_S(x) = x$.

We define $f^\gamma : \gamma S \to \mathrm{cl}_T f[S] = f^{\beta(T)}[\beta S]$ as

$$f^\gamma(x) = f^{\beta(T)}[\pi^{\leftarrow}_{\beta(\gamma)}(x)]$$

Then for $x \in S(f)$, $f^\gamma(x) = f^{\beta(T)}[\pi^{\leftarrow}_{\beta(\gamma)}(x)] = f^{\beta(T)}[f^{\beta(T)\leftarrow}(x)] = x$.

Then $f^\gamma : \gamma S \to \mathrm{cl}_T f[S]$ is easily seen to be continuous on its domain.

We can then say, f extends to $f^\gamma : \gamma S \to \mathrm{cl}_T f[S]$ where

$$\begin{aligned} f^\gamma[\gamma S] &= f^\gamma[S \cup S(f)] \\ &= f^\gamma[S(f)] \cup f[S] \\ &= S(f) \cup f[S] \\ &= \mathrm{cl}_T f[S] \end{aligned}$$

22.3.2 *Summary of the above results*

We are given a continuous function $f : S \to T$ mapping S into a compact space T (not necessarily a singular map). We obtain a compactification of S,

$$\gamma S = S \cup S(f)$$

(not necessarily a singular compactification). Then, if $f[S]$ is non-compact,

$$\begin{aligned} f^{\beta(T)}[\beta S] &= f[S] \cup S(f) = \mathrm{cl}_T f[S] \\ f^\gamma[\gamma S] &= f[S] \cup S(f) = \mathrm{cl}_T f[S] \\ f^{\beta(T)}[\beta S \backslash S] &= S(f) \ = \ f^\gamma[S(f)] \end{aligned}$$

where $f[S] \cup S(f)$ may be disjoint, but not necessarily. If f is a singular map then

$$f[S] \cup S(f) = S(f)$$

If $f[S]$ is compact, $f^{\beta(T)}[\beta S] = f[S] = f[S] \cup S(f)$.

22.4 A Few Examples

The above definition shows that any continuous function, $f : S \to T$ from S into a compact space, T, can be used to construct a compactification of a locally compact Hausdorff space S. We consider a few examples to better visualize how this is done.

Example 2. Consider the space $\mathbb{R}$ equipped with the usual topology. The space $\mathbb{R}$ is known to be locally compact non-compact Hausdorff. Consider the continuous functions

$$\sin : \mathbb{R} \to [-1, 1] \quad \text{and} \quad \cos : \mathbb{R} \to [-1, 1]$$

both mapping $\mathbb{R}$ into the compact subspace $[-1, 1]$. Show that sin and cos are both singular maps on $\mathbb{R}$ and so each induce a singular compactification of $\mathbb{R}$.

Solution: Case sine: See that, for any $x \in [-1, 1]$ and open neighborhood, U of x, in $[-1, 1]$, $\sin^{\leftarrow}[U]$ is unbounded and so its closure, $\mathrm{cl}_{\mathbb{R}} \sin^{\leftarrow}[U]$, in $\mathbb{R}$ is not compact. Then $x \in S(\sin)$, and so $[-1, 1] \subseteq S(\sin)$. By definition, $S(\sin) \subseteq \mathrm{cl}_{[-1,1]} \sin[\mathbb{R}]$, so

$$S(\sin) = \mathrm{cl}_{\mathbb{R}}[\sin[\mathbb{R}]] = [-1, 1]$$

Then $\sin : \mathbb{R} \to [-1, 1]$ is a singular map. We can then use the sine function to construct the singular compactification

$$\mathbb{R} \cup_{\sin} S(\sin) = \mathbb{R} \cup_{\sin} [-1, 1]$$

of $\mathbb{R}$ with outgrowth $[-1, 1]$.

Case cosine: Proceed similarly to show that $\mathbb{R} \cup_{\cos} S(\cos) = \mathbb{R} \cup_{\cos} [-1, 1]$ is also a singular compactification of $\mathbb{R}$.

The above example produces two compactifications of $\mathbb{R}$ with identical outgrowth. It may be tempting to conclude that they are equivalent compactifications. We will later show (on page 527) that this is not the case.

Example 3. Consider the space $\mathbb{R}$ equipped with the usual topology. The space S is known to be a locally compact non-compact Hausdorff. Let $T = [-\pi/2, \pi/2]$. Show that $\mathbb{R} \cup S(\arctan)$ is *not* a singular compactification of $\mathbb{R}$. Then find the compactification induced by arctan.

Solution: We consider the function, $\arctan : \mathbb{R} \to T$. We then verify which points in $\mathrm{cl}_T[\arctan[\mathbb{R}]] = [-\pi/2, \pi/2]$ belong to the singular set $S(\arctan)$. We see that arctan, pulls back open intervals of the form (a, b) in $[-\pi/2, \pi/2]$, to intervals whose closure is compact in $\mathbb{R}$. Then $\arctan[\mathbb{R}] = (-\pi/2, \pi/2) \not\subseteq S(\arctan)$. So arctan is not a singular map. Thus $\mathbb{R} \cup S(\arctan)$ is *not* a singular compactification of $\mathbb{R}$.

We now determine $S(\arctan)$. Since the curve of $y = \arctan(x)$ is asymptotic to the horizontal lines $y = -\pi/2$ and $y = \pi/2$, the function arctan pulls backs open intervals of the form $(a, \pi/2]$ and $[-\pi/2, b)$ to unbounded sets and so the "pull backs" of these have non-compact closures in $\mathbb{R}$. So, $S(\arctan) = \{-\pi/2, \pi/2\}$. So, even though arctan is not a singular map on $\mathbb{R}$ it does induce a "two-point compactification"

$$\mathbb{R} \cup S(\arctan) = \mathbb{R} \cup \{-\pi/2, \pi/2\}$$

of $\mathbb{R}$.

Not surprisingly,

$$\arctan[\mathbb{R}] \cup S(\arctan) = (-\pi/2, \pi/2) \cup \{-\pi/2, \pi/2\} = [-\pi/2, \pi/2]$$

is the two-point compactification of $\arctan[\mathbb{R}]$.

Example 4. Show that $\beta\mathbb{R}$ is not a singular compactification of $\mathbb{R}$.

Solution: We have shown that $\mathbb{R} \cup S(\arctan) = \mathbb{R} \cup \{-\pi/2, \pi/2\}$ is a compactification of $\mathbb{R}$ but not a singular one. Since $\mathbb{R} \cup S(\arctan) \preceq \beta\mathbb{R}$, then, by the example on page 519, $\beta\mathbb{R}$ cannot be singular.

In the following example, we show that two compactifications of the same set with the same outgrowth need not be equivalent compactifications.

Example 5. Given the two singular compactifications,

$$\alpha\mathbb{R} = \mathbb{R} \cup_{\cos} S(\cos)$$
$$\gamma\mathbb{R} = \mathbb{R} \cup_{\sin} S(\sin)$$

show that, in spite of $S(\sin) = [-1, 1] = S(\cos)$, $\alpha\mathbb{R}$ and $\gamma\mathbb{R}$ are *not* equivalent compactifications.

Solution: Suppose $\alpha\mathbb{R}$ and $\gamma\mathbb{R}$ are equivalent compactifications. We will show that this will lead to a contradiction.

Then, by definition of "equivalent compactifications", there exists a homeomorphism,

$$\pi_{\gamma\to\alpha} : \gamma\mathbb{R} \to \alpha\mathbb{R}$$

such that for $x \in \mathbb{R}$, $\pi_{\gamma\to\alpha}(x) = x$.

By definition, the function, $\pi_{\gamma\to\alpha}|_{\gamma\mathbb{R}\backslash\mathbb{R}}$, is a homeomorphism mapping $[-1, 1]$ onto $[-1, 1]$. This means that $\pi_{\gamma\to\alpha}|_{\gamma\mathbb{R}\backslash\mathbb{R}}$ is monotone and maps endpoints to endpoints. Suppose, without loss of generality, that $\pi_{\gamma\to\alpha}$ is increasing and so $\pi_{\gamma\to\alpha}|_{\gamma\mathbb{R}\backslash\mathbb{R}}(-1) = -1$. Then $\pi_{\gamma\to\alpha}|_{\gamma\mathbb{R}\backslash\mathbb{R}}(1) = 1$.

If $U = (a, 1] \subseteq S(\sin)$ and $V = \pi_{\gamma\to\alpha}|_{\gamma\mathbb{R}\backslash\mathbb{R}}U \subseteq S(\cos)$ see that can choose a small enough so that $\sin^{\leftarrow}[U] \cap \cos^{\leftarrow}[V] \cap [-\pi, \pi]$ is empty. Then $\sin^{\leftarrow}[U] \cap \cos^{\leftarrow}[V] = \varnothing$.

Now, $\cos : \mathbb{R} \to [-1, 1]$ extends to $\cos^{\alpha} : \alpha\mathbb{R} \to [-1, 1]$.

Then $U \cup \sin^{\leftarrow}[U]$ is open in $\gamma\mathbb{R}$. Then by continuity, the two sets

$$\pi_{\gamma\to\alpha}[U \cup \sin^{\leftarrow}[U]] = V \cup \sin^{\leftarrow}[U]$$
$$\cos^{\alpha\,\leftarrow}[V] = V \cup \cos^{\leftarrow}[V]$$

are both open subsets of $\alpha\mathbb{R}$.

So $(V \cup \sin^{\leftarrow}[U]) \cap (V \cup \cos^{\leftarrow}[V]) = V$, a non-empty open subset of $\alpha\mathbb{R}$ which is entirely contained in $\alpha\mathbb{R}\backslash\mathbb{R} = [-1, 1]$. Since $\mathbb{R}$ is dense in $\alpha\mathbb{R} = \mathbb{R} \cup [-1, 1]$, this is impossible. The source of our contradiction is our supposition that $\alpha\mathbb{R}$ and $\gamma\mathbb{R}$ are equivalent. We are done.

Example 6. Determine whether $f(x) = \sin(1/x)$ with domain $\mathbb{R}\backslash\{0\}$ and range $T = [-1, 1] = f[\mathbb{R}\backslash\{0\}]$ induces a singular compactification. If not determine the non-singular compactification it induces (Figure 22.1).

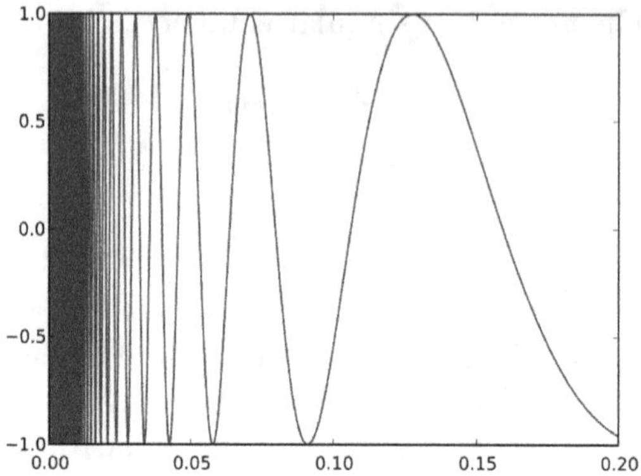

Figure 22.1. The graph of $y = \sin(1/x)$.

Solution: We are given $f(x) = \sin(1/x)$ where $f : S \to T$, maps $S = \mathbb{R}\backslash\{0\}$ onto the compact set $T = [-1, 1]$.

Suppose $a \in [-1, 0) \cup (0, 1]$ and $y \in M = [-1, a) \cup (-a, 1]$. Then, for an open interval U in M, $f^{\leftarrow}[U]$ produces a bounded sequence of open intervals converging to zero. If we take the closure, $\mathrm{cl}_S f^{\leftarrow}[U]$, we obtain a bounded sequence of closed intervals converging to zero. Since $0 \notin S$, then $\mathrm{cl}_S f^{\leftarrow}[U]$ is not compact. If $y \in U = (-a, a)$ then $f^{\leftarrow}[U]$ is unbounded on both ends of $\mathbb{R}$ so $\mathrm{cl}_S f^{\leftarrow}[U]$ is not compact. So, $f : S \to T$ is a singular map and, for $S(f) = [-1, 1]$,

$$S \cup_f S(f) = \mathbb{R}\backslash\{0\} \ \cup_f \ [-1, 1]$$

is a singular compactification of $\mathbb{R}\backslash\{0\}$ induced by $f(x) = \sin(1/x)$.

22.5 More on Equivalent Singular Compactifications

We will now produce a characterization of pairs of singular compactifications which are equivalent. But first we must present a few lemmas involving singular sets $S(f)$.

Lemma 22.5 *Let $f : S \to K$ be a continuous function mapping a locally compact Hausdorff space into a compact Hausdorff space, K, and $Y = cl_K f[S]$. Then*

$$S(f) = \cap\{cl_Y f[S\backslash F] : F \text{ is compact in } S\}$$

Proof. We are given that $f : S \to K$ is a continuous function mapping S into the compact space K and $Y = \text{cl}_K f[S]$. In the proof, F will always represent a compact set in S.

Claim 1: We claim that $S(f) \subseteq \text{cl}_Y f[S\backslash F]$ for all compact F. To prove this, it suffices to show that $Y\backslash\text{cl}_Y f[S\backslash F] \cap S(f) = \varnothing$.

Let $p \in Y\backslash\text{cl}_Y f[S\backslash F]$. Then there exists an open neighborhood U of p in Y such that $f^{\leftarrow}[U] \subseteq F$. Then $\text{cl}_S f^{\leftarrow}[U]$ is compact so $p \notin S(f)$. Then $Y\backslash\text{cl}_Y f[S\backslash F] \cap S(f) = \varnothing$; so $S(f) \subseteq \text{cl}_Y f[S\backslash F]$, as claimed. We can deduce that

$$S(f) \subseteq \cap\{\text{cl}_Y f[S\backslash F] : F \text{ is compact in } S\}$$

Claim 2: $\cap\{\text{cl}_Y f[S\backslash F] : F$ is compact in $S\} \subseteq S(f)$.

Let $p \in \cap\{\text{cl}_Y f[S\backslash F] : F$ is compact in $S\}$. Suppose $p \notin S(f)$. Then there is an open neighborhood U_1 of p in $Y = \text{cl}_K f[S]$ such that $\text{cl}_S f^{\leftarrow}[U_1]$ is compact. But

$$\begin{aligned} p &\in \cap\{\text{cl}_Y f[S\backslash F] : F \text{ is compact in } S\} \\ &\subseteq \text{cl}_Y f[S\backslash \text{cl}_S f^{\leftarrow}[U_1]] \\ &\subseteq \text{cl}_Y f[S\backslash f^{\leftarrow}[U_1]] \\ &\subseteq \text{cl}_Y f\circ f^{\leftarrow}[Y\backslash U_1]] \\ &= Y\backslash U \end{aligned}$$

The statement "$p \in Y\backslash U$" contradicts the fact that U is a neighborhood of p. Consequently, $\cap\{\text{cl}_Y f[S\backslash F] : F$ is compact in $S\} \subseteq S(f)$ as claimed.

So, $S(f) = \cap\{\text{cl}_Y f[S\backslash F] : F$ is compact in $S\}$. This completes the proof of the lemma.

Lemma 22.6 *Let S be a locally compact Hausdorff space and αS be a compactification of S. Let K be a compact space. If $f \in C_\alpha(S,K)$*[4] *then $f^\alpha[\alpha S\backslash S] = S(f)$.*

[4]Where $C_\alpha(S,K)$ is the set of continuous functions mapping S into K which extend to $f^\alpha C(\alpha S, K)$.

Proof. If F is compact in S then, since $\alpha S\backslash S \subseteq \alpha S\backslash K$, $\alpha S\backslash S \subseteq \mathrm{cl}_{\alpha S}(S\backslash F)$, and so

$$f^{\alpha}[\alpha S\backslash S] \subseteq f^{\alpha}[\mathrm{cl}_{\alpha S}(S\backslash F)] \subseteq \mathrm{cl}_{\mathrm{cl}_K f[S]} f[S\backslash F]$$

Then $f^{\alpha}[\alpha S\backslash S] \subseteq \cap\{\mathrm{cl}_{\mathrm{cl}_K f[S]} f[S\backslash F] : F \text{ is compact in } S\}$. By the previous lemma $S(f) = \cap\{\mathrm{cl}_{K f[S]} f[S\backslash F] : F \text{ is compact in } S\}$, so

$$f^{\alpha}[\alpha S\backslash S] \subseteq S(f)$$

On the other hand, if $p \notin f^{\alpha}[\alpha S\backslash S]$ and U is an open neighborhood of p in K such that $\mathrm{cl}_K U$ misses $f^{\alpha}[\alpha S\backslash S]$ then $\mathrm{cl}_S f^{\leftarrow}[U] \subseteq f^{\leftarrow}[\mathrm{cl}_K U]$, a compact subset of S. Hence, $\mathrm{cl}_S f^{\leftarrow}[U]$ is compact and so $p \notin S(f)$. So

$$S(f) \subseteq f^{\alpha}[\alpha S\backslash S]$$

We conclude that, if f extends to $f^{\alpha} : \alpha S \to K$, then $f^{\alpha}[\alpha S\backslash S] = S(f)$. This proves the lemma.

Lemma 22.7 *Let $f : S \to K$ be a continuous function in $C(S, K)$ mapping the locally compact Hausdorff space S into a compact Hausdorff space, K. Suppose f extends to $f^{\alpha(K)} \in C(\alpha S, K)$ for some compactification αS such that $f^{\alpha(K)}$ is one-to-one on $\alpha S\backslash S$. Then*

$$\alpha S \equiv S \cup S(f)$$

Proof. Let $S(f)$ be the singular set of the continuous map $f : S \to K$ (not necessarily a singular function). We are given a compactification, αS, such that f extends continuously to $f^{\alpha(K)} : \alpha S \to K$ in such a way that f^{α} is one-to-one on $\alpha S\backslash S$.

By Lemma 22.6, $f^{\alpha(K)}[\alpha S\backslash S] = S(f)$.

If $\gamma S = S \cup S(f)$, we define $\pi_{\alpha\to\gamma} : \alpha S \to S \cup S(f)$ as follows:

$$\pi_{\alpha\to\gamma}(x) = \begin{cases} f^{\alpha}(x) & \text{if } x \in \alpha S\backslash S \\ x & \text{if } x \in S \end{cases}$$

Claim: That $\pi_{\alpha\to\gamma}$ is continuous on αS. It suffices to show that $\pi_{\alpha\to\gamma}$ pulls back open neighborhoods in $S \cup S(f)$ to open sets in αS.

Suppose $p \in S \cup S(f)$. Clearly, if $p \in S$ and $p \in U$ is open in S, then $\pi^{\leftarrow}_{\alpha\to\gamma}[U]$ is open in αS. If $p \in S(f)$ then an open neighborhood of p is, by definition, of the form $[U \cap S(f)] \cup f^{\leftarrow}[U]\backslash F$ where U is an open subset of K and F some compact set in S. See that,

$$\begin{aligned}
\pi^{\leftarrow}_{\alpha\to\gamma}[[U \cap S(f)] \cup f^{\leftarrow}[U]\backslash F] &= \pi^{\leftarrow}_{\alpha\to\gamma}[U \cap S(f)] \cup \pi^{\leftarrow}_{\alpha\to\gamma}[f^{\leftarrow}[U]\backslash F]] \\
&= \pi^{\leftarrow}_{\alpha\to\gamma}[U] \cap \pi^{\leftarrow}_{\alpha\to\gamma}[S(f)] \cup f^{\leftarrow}[U]\backslash F \\
&= (f^{\alpha\leftarrow}[U] \cap f^{\alpha\leftarrow}[S(f)]) \cup f^{\leftarrow}[U]\backslash F \\
&= (f^{\alpha\leftarrow}[U \cap S(f)] \cup f^{\leftarrow}[U]\backslash F \\
&= f^{\alpha\leftarrow}[U]\backslash F \\
&= f^{\alpha\leftarrow}[U] \cap S\backslash F
\end{aligned}$$

We see that $\pi_{\alpha\to\gamma}$ pulls back open neighborhoods of points in $S(f)$ to open sets. So, $\pi_{\alpha\to\gamma} : \alpha S \to S \cup S(f)$ is continuous, as claimed.

By definition, αS and $S \cup S(f)$ are equivalent compactifications.

Theorem 22.8 *Let S be a completely regular topological space. Suppose $f : S \to K$ and $g : S \to K$ are two continuous* singular *functions mapping S into a compact space K. Suppose $S(f)$ and $S(g)$ are homeomorphic. Then the two induced singular compactifications,*

$$\alpha S = S \cup_f S(f)$$
$$\gamma S = S \cup_g S(g)$$

are equivalent if and only if the singular function $f : S \to S(f)$ of αS extends continuously to $f^\gamma : \gamma S \to S(g)$ such that f^γ separates the points of $\gamma S\backslash S = S(g)$.

Proof. We are given that $f : S \to K$ and $g : S \to K$ are two continuous *singular* functions mapping S into a compact space K inducing the two singular compactifications, $\alpha S = S \cup_f S(f)$ and $\gamma S = S \cup_g S(g)$. Also, $S(f)$ and $S(g)$ are seen to be homeomorphic.

($\Rightarrow$) Suppose $\alpha S = S \cup_f S(f)$ and $\gamma S = S \cup_g S(g)$ are equivalent.

We are required to show that the singular function $f : S \to S(f)$ extends continuously to $f^\gamma : \gamma S \to S(g)$ such that f^γ separates the points of $\gamma S\backslash S = S(g)$.

Recall that $f^\alpha : \alpha S \to \alpha S\backslash S$ acts as the identity map on $\alpha S\backslash S$. Also, since αS and γS are equivalent, then there is a continuous map

$\pi_{\gamma\to\alpha} : \gamma S \to \alpha S$ such that $\pi_{\gamma\to\alpha}(x) = x$ on S and $\pi_{\gamma\to\alpha}$ maps $\gamma S \backslash S$ homeomorphically onto $\alpha S \backslash S$.

Let $f^\gamma : \gamma S \to S(f)$ be defined as follows:

$$f^\gamma = f^\alpha \circ \pi_{\gamma\to\alpha}$$

Then f^γ is continuous, $f^\gamma|_S = f$ and $f^\gamma|_{S(g)} = \pi_{\gamma\to\alpha}|_{S(g)}$. This shows that $f : S \to S(f)$ extends continuously to the function $f^\gamma : S \cup_g S(g) \to S(f)$. Since $\pi_{\gamma\to\alpha}$ is a homeomorphism on $S(g)$ and f^α is the identity function on $S(f)$, then f^γ separates points of $S(g)$, as required.

($\Leftarrow$) We are given that both f and g are singular maps on S and $\alpha S = S \cup_f S(f)$ and $\gamma S = S \cup_g S(g)$. We are also given that the singular function $f : S \to S(f)$ extends continuously to $f^\gamma : S \cup_g S(g) \to S(g)$ such that f^γ separates the points of $S(g)$.

Then by Lemma 22.7, $S \cup_g S(g)$ is a compactification which is equivalent to $S \cup S(f)$.

Since f is singular, $S \cup S(f) = S \cup_f S(f)$. So $S \cup_f S(f)$ and $S \cup_g S(g)$ are equivalent.

Theorem 22.9 *Two continuous functions, $f : S \to K$ and $g : S \to K$, will be said to be* homeomorphically related *if there exists a homeomorphic function $h : cl_K g[S] \to cl_K h[S]$ such that $h(g(x)) = f(x)$ for all $x \in S$. Suppose that $f : S \to K$ and $g : S \to K$ are two singular maps such that $S(f) = S(g)$. If f and g are homeomorphically related then $S \cup_f S(f)$ and $S \cup_g S(g)$ are equivalent compactifications.*

Proof. We are given two continuous functions, $f : S \to K$ and $g : S \to K$, mapping S into the compact space K.

Suppose f and g are homeomorphically related singular maps which, respectively, induce the singular compactifications,

$$\alpha S = S \cup_f S(f)$$
$$\gamma S = S \cup_g S(g)$$

where $S(f) = S(g)$. By definition, there exists a homeomorphism $h : S(g) \to S(f)$ such that $h(g(x)) = f(x)$.

See that $g : S \to S(g)$ extends continuously to $g^\gamma : S \cup_g S(g) \to S(g)$ where g^γ is the identity map when restricted to $S(g)$. Then $h_\circ g : S \to S(f)$ extends to

$$(h_\circ g)^\gamma : S \cup_g S(g) \to S(f)$$

where $(h_\circ g^\gamma)|_{S(g)}(x) = h(x)$. Since $(h_\circ g)(x) = f(x)$ on S, f extends to $(h_\circ g)^\gamma = f^\gamma$ where

$$f^\gamma : S \cup_g S(g) \to S(f)$$

and $f^\gamma = h$ on $S(g)$. So f^γ separates points of $S(g)$.

By Theorem 22.8, $S \cup_f S(f)$ and $S \cup_g S(g)$ are equivalent, as required.

We can now present an example where this particular concept plays a key role.

Example 7. The two compactifications $\mathbb{R} \cup_{\sin^2} S(\sin^2)$ and $\mathbb{R} \cup_{\cos^2} S(\cos^2)$ are easily seen to be singular compactifications with outgrowth $[0, 1]$. Show that they are equivalent compactifications.

Solution: Consider the function $h(x) = 1 - x$ mapping $[0, 1]$ onto $[0, 1]$. It is a one-to-one continuous function. Also, note that

$$h(\sin^2(x)) = 1 - \sin^2(x) = (1 - \sin^2)(x) = \cos^2(x)$$

on $[0, 1]$. Then $\sin^2$ and $\cos^2$ are homeomorphically related. By the above theorem $\mathbb{R} \cup_{\sin^2} S(\sin^2)$ and $\mathbb{R} \cup_{\cos^2} S(\cos^2)$ are equivalent compactifications. This is what we were required to show.

There can be various ways of showing that

the compactification $\beta\mathbb{N}$ is not a singular compactification.

In the next example, we propose one method.

Example 8. Show that $\beta\mathbb{N}$ cannot be a singular compactification.

Solution: Suppose $\beta\mathbb{N}$ is a singular compactification, Then there is a retraction function $r : \beta\mathbb{N} \to \beta\mathbb{N}\backslash\mathbb{N}$ which maps $\beta\mathbb{N}$ onto $\beta\mathbb{N}\backslash\mathbb{N}$.

We know that $\beta\mathbb{N}$ is separable, but on page 513, we showed that $\beta\mathbb{N}\backslash\mathbb{N}$ is not a separable space. By Theorem 6.11, we know that continuous images of separable spaces are separable. So, $\beta\mathbb{N}\backslash\mathbb{N}$ cannot be the continuous image of $\beta\mathbb{N}$. So, $\beta\mathbb{N}$ is not a singular compactification.

22.6 What Kind of Space Has Only Singular Compactifications?

We have already shown a compactification "less than" a singular compactification must be singular. So, if βS is a singular compactification of S, then all compactifications of S are singular. We wonder what class of topological spaces satisfies this property?

Theorem 22.10 *Let S be locally compact, non-compact, and Hausdorff. If βS is a singular compactification then S is pseudocompact.*

Proof. The proof of the statement is differed to the end of the next section in Theorem 23.8.

22.7 Compact spaces as a β-outgrowth, $\beta S\backslash S$, for some S

If we are given a compact Hausdorff space, T, we may wonder whether it is the outgrowth of some topological space. We can answer is question in the affirmative. Moreover, we can show that all such spaces will appear as the outgrowth, $\beta S\backslash S$, of at least one space S. We prove this statement now. The proof involves a few notions involving uncountably infinite limit ordinals, $\omega_1, \omega_2, \omega_3, \ldots, \omega_n, \ldots$

Theorem 22.11 *Every compact Hausdorff topological space, T, is the β-outgrowth of some locally compact non-compact Hausdorff topological space, S. That is, βS is equivalent to some compactification, $S \cup T$, of S, with $\beta S\backslash S$ homeomorphic to T.*

Proof. Let T be a compact Hausdorff space. Let ω_n be a limit ordinal such that $n \in \mathbb{N}$ and $|\omega_n| > |T|$. Let

$$S = [0, \omega_n) \times T^5$$

See that $[0, \omega_n] = \omega_n + 1$ is compact and, since ω_n is a limit ordinal, $[0, \omega_n)$ is dense in $[0, \omega_n]$ (see example on page 373). Since both T and $[0, \omega_n] = \omega_n + 1$ are compact, then so is $[0, \omega_n] \times T$. Also

$$S = \omega_n \times T = [0, \omega_n) \times T \quad \subseteq_{\text{dense}} \quad (\omega_n + 1) \times T = [0, \omega_n] \times T$$

See that $\{\omega_n + 1\} \times T$ is a homeomorphic copy of T.

We claim that $\beta S = [0, \omega_n] \times T$. If so, then T can be seen at the β-outgrowth of S.

Proof of claim: Let $i : S \to [0, \omega_n] \times T$ denote the identity map which densely embeds S into the compact set $[0, \omega_n] \times T$. It is easily verified that $\{\omega_n + 1\} \times T \subseteq S(i)$. Then, when equipped with the topology described earlier

$$\alpha S = S \cup S(i) = [0, \omega_n] \times T$$

is the compactification whose topology is induced by i.

To establish the claim it suffices to show that αS is equivalent to βS. To do this we need to show that every bounded function, f, in $C^*(S)$ extends to f^α on αS.

Let $f \in C^*(S)$.

If $p \in T$, see that $f|_{[0,\omega_n)\times\{p\}}$ must be constant on some tail end, say

$$[\alpha_p, \omega_n) \times \{p\}^6$$

of $[0, \omega_n) \times \{p\}$.

---

[5]The limit ordinal (equipped with the interval topology), ω_n is the set of all ordinals less than ω_n. It is also expressed as $\omega_n = [0, \omega_n)$. For example, $\omega_0 = [0, \omega_0) = \mathbb{N}$, is the set of all finite ordinals and $[0, \omega_0]$ is the successor, $\omega_0 + 1$, of ω_0. The first uncountable ordinal, $\omega_1 = [0, \omega_1)$, is the set of all countable ordinals and $[0, \omega_1]$ is the successor $\omega_1 + 1$ of ω_1.

If $|T| \geq |\omega_n|$ for all $n \in \mathbb{N}$, we can always choose a non-limit ordinal, α, such that $|\omega_\alpha| > |T|$. Things would then follow identically.

[6]Since $|[0, \omega_n) \times \{p\}| = |\omega_n| > |\omega_1| = |\mathbb{R}| \geq |f[[0, \omega_n) \times \{p\}]|$. Then $\sup |f^{\leftarrow}[f[[0, \omega_n) \times \{p\}]]| < |\omega_n|$.

Since $|\omega_n| \geq |T|$, then

$$|T| = |\{[\alpha_p, \omega_n) : p \in T\}| < |\omega_n|$$

So, $\sup\{\alpha_p : p \in T\} = \kappa < \omega_n$.

So, for this ordinal, κ, $f : S \to \mathbb{R}$ is constant on $\{[\kappa, \omega_n) \times \{p\}\}$ for all $p \in T$.

For $p \in T$, say $f[[\kappa, \omega_n) \times \{p\}] = k_p$.

For each $p \in T$ we set $f^\alpha(\{\omega_n + 1\} \times p) = k_p$.

Then $f^\alpha : \alpha S \to \mathbb{R}$ is continuous. We can conclude that every function f in $C^*(S)$ extends to $f^\alpha \in C(\alpha S)$. So $\alpha S = \beta S$ as claimed.

So every compact Hausdorff space T is the β-outgrowth, $\beta S \backslash S$, of some locally compact non-compact Hausdorff space, S.

Concepts Review

1. Given a continuous function $f : S \to T$ from the completely regular set S into the compact set T, define the singular set $S(f)$.
2. Given a continuous function $f : S \to T$ from the completely regular set S into the compact set T, what does it mean to say that f is a singular map?
3. Given a continuous function $f : S \to T$ from the completely regular set S into the compact set T, define a compactification induced by f.
4. Given a continuous function $f : S \to T$ from the completely regular set S into the compact set T, define a singular compactification induced by f.
5. Show that arctan: $\mathbb{R} \to [-\pi/2, \pi/2]$ is not a singular map. Find a compactification of $\mathbb{R}$ induced by arctan.
6. Produce an example of a pair of singular compactifications with the same singular sets but which are not equivalent.
7. Produce an example of a Stone–Čech compactification which is not singular.
8. State one way of recognizing a pair of singular compactifications which are equivalent.

Chapter 23

On C-Embeddings and Pseudocompactness

Abstract
In this section, we define the C-embedded property. We describe a characterization associated to the existence of a C-embedded subset. We then present a few applications associated to these types of subsets. We then discuss the pseudocompact property in relation to C-embeddings.

23.1 Definition: C-Embedded Subsets

The notion of C^*-embedded subsets was introduced in the last section in the context of an important characterization of the Stone–Čech compactification. By "C^*-embedded subset U of a space S" we mean that all bounded real-valued functions in $C^*(U)$ can be continuously extended to a function in $C^*(S)$. This allowed us to characterize βS as:

> *The compactification αS is the Stone–Čech compactification of S if and only if S is C^*-embedded in αS.*

We presented a characterization of a C^*-embedded set T in S in the form of the Urysohn extension theorem:

> *The set T is C^*-embedded in $S \Leftrightarrow$ completely separated sets in T are completely separated in S.*

From this we deduced that "compact subsets of a completely regular set, S, are C^*-embedded in S" and "closed subsets of a metric space S are C^*-embedded in S".

We now want to extend our study of C^*-embedded subsets to C-embedded subsets.

Definition 23.1 A subset U of topological space S is *C-embedded* in S if *every* continuous function f in $C(U)$ can be continuously extended to a function f^* in $C(S)$.

The notion of a C-embedded subset will play an important role in the study of another extension of the space S denoted by, υS, (in the chapter called "realcompact spaces").

Clearly, it is always true that, for any subset U of S, $C^*(U) \subseteq C(U)$. But one may wonder: Is a C^*-embedded subset, U, in S necessarily C-embedded in S, and vice-versa? To answer such a question we must compare the definitions of each of these two types of subsets carefully.

We first verify the easier part of this question. We show that a C-embedded set in S is C^*-embedded in S. Suppose U is C-embedded in S. Consider a function, $f \in C^*(U)$. Then there exists a number, M, such that $f[M] \in [-M, M]$. Since U is C-embedded, then f extends to $f^* \in C(S)$. Define the function $g : S \to \mathbb{R}$ as

$$g = (f^* \vee -M) \wedge M$$

We see that g is both a continuous and bounded function on S such that $g|_U = f$. So f in $C^*(U)$ will extend to $g \in C^*(S)$. So,

> *whenever U is C-embedded in S then it is also C^*-embedded in S.*

That settles this particular part of the question.

But it is *not* true in general that a C^*-embedded set U in S is C-embedded in S (unless certain conditions on U and S are satisfied). That is, there are subsets U of S, for which every function f in $C^*(U)$ extends to a function $f^* \in C^*(S)$, but not every function $f \in C(U)$ extends to a function $f^* \in C(S)$.

The following theorem shows under what conditions *if U is C^*-embedded in S then U is C-embedded in S.*

Theorem 23.2 *Suppose U is a non-empty subset of the topological space, S.*

(a) *If U is C-embedded in S then U is C^*-embedded in S.*
(b) *Let U be C^*-embedded in S. The set U is also C-embedded in S if and only if U is completely separated from any disjoint zero-set in S.*

Proof. We are given that U is a subset in S.

(a) That "U is C-embedded in S" $\Rightarrow$ "U is C^*-embedded in S" is shown above.
(b) We are given that U is a subset of S.

($\Leftarrow$) Suppose that U is C^*-embedded in S and is completely separated from any disjoint zero-set in S.

We are required to show that U is also a C-embedded subset of S.

Let $f : U \to \mathbb{R}$ be a function in $C(U)$. To prove this direction of the statement, it suffices to find a function $t \in C(S)$ such that $t|_U = f$.

Step 1: We first construct a zero-set, Q, in S which is disjoint from U. Using the given function, $f \in C(U)$, we consider the continuous function, $g : U \to (-\pi/2, \pi/2)$ defined as

$$g(x) = (\arctan \circ f)(x)$$

Then $|g| : U \to [0, \pi/2)$ is a bounded continuous function on U. Since U is C^*-embedded in S, $|g|$ extends to $|g|^* : S \to \mathbb{R}$ in $C^*(S)$.

The set $[\frac{\pi}{2}, \infty)$ is a closed G_δ in the normal space, $\mathbb{R}$, so by 10.10, it is a zero-set in $\mathbb{R}$, say $Z(h)$. We thus obtain the zero-set

$$Q = Z(h_\circ |g|^*) = (h_\circ |g|^*)^{\leftarrow}(0) = |g|^{*\leftarrow}[h^{\leftarrow}(0)] = |g|^{*\leftarrow}\left[\frac{\pi}{2}, \infty\right)$$

in S. For the zero-set, $Q = Z(h_\circ |g|^*)$, $Q \cap U = \varnothing$, since

$$x \in U \Rightarrow |g|^*(x) \in [0, \pi/2) \Rightarrow x \notin |g|^{*\leftarrow}\left[\frac{\pi}{2}, \infty\right) = Q$$

So, Q is the desired zero-set disjoint from U.

Step 2: Given a zero-set $Q = Z(h_\circ|g|^*)$ in S disjoint from U, we now apply the hypothesis stating that U and Q are completely separated. That is, there is a function, $k : S \to [0, 1]$ such that $k[Q] = \{0\}$ and $k[U] = \{1\}$.

Let $t(x) = \tan(g(x)k(x))$ be a continuous function in $C(S)$. We claim that $t : S \to \mathbb{R}$ is such that $t|_U = f$.

Proof of claim: See that, if $x \in U$,

$$\begin{aligned} t(x) &= \tan[(gk)(x)] \\ &= \tan[g(x)(1)] \\ &= \tan[\arctan(f(x))] \\ &= f(x) \end{aligned}$$

So $t|_U = f$, as claimed. Then $t : S \to \mathbb{R}$ is a continuous extension of $f : U \to \mathbb{R}$.

We are done with this direction.

($\Rightarrow$) Suppose that U is a C^*-embedded set which is also C-embedded in S. Suppose that $Z(f)$ is the zero-set of a function $f \in C(S)$ which is disjoint from U. We are required to show that U and $Z(f)$ are completely separated in S.

Let

$$h = \frac{1}{f}$$

on U. Then $h \in C(U)$. Since U is C-embedded in S, then $h : U \to \mathbb{R}$ extends to $h^* \in C(S)$. Then $t = h^*f \in C(S)$ where $U \subseteq Z(h^*f - 1)$ and $Z(f) \subseteq Z(h^*f)$. So, U and $Z(f)$ are completely separated by t, as required.

We are done with part (b).

Example 1. Show that every closed subset of $\mathbb{R}$ is C-embedded.

Solution: Let F be a closed subset of $\mathbb{R}$.

By the previous example on page 493 F is C^*-embedded in $\mathbb{R}$. Let $Z(f)$ be a zero-set which does not intersect F. Since $\mathbb{R}$ is known to be normal, F and $Z(f)$ are completely separated (by 10.1). By the above theorem, F is C-embedded in $\mathbb{R}$.

Theorem 23.3 *Suppose U is a subspace of a topological space S and let $f \in C(S)$. If f maps U homeomorphically onto a closed subset, $f[U]$, of $\mathbb{R}$ then U is C-embedded in S.*

Proof. We are given a function, f, in $C(S)$ which maps U homeomorphically onto a closed subset, $A = f[U]$, in $\mathbb{R}$. We are required to show that U is C-embedded in S.

Let $h \in C(U)$. It suffices to show that there is a function $t \in C(S)$ such that $t|_U = h$.

See that, the function $h \circ f|_U^{\leftarrow} : A \to \mathbb{R}$ is continuous on A and that

$$[h \circ f|_U^{\leftarrow}][A] = h\,[f|_U^{\leftarrow}[A]] = h[U]$$

in $\mathbb{R}$. So, $h \circ f|_U \in C(A)$. Since A is a closed subset of $\mathbb{R}$, it is C-embedded in $\mathbb{R}$ (as shown in the example above). This means there is a function $g \in C(S)$ such that $g|_A = h \circ f|_U^{\leftarrow}$. Let $t = g \circ f \in C(S)$. Then for $x \in U$

$$\begin{aligned} t|_U(x) &= (g \circ f)|_U(x) \quad \text{(By definition of } t) \\ &= g[f(x)] \quad (x \in U) \\ &= (h \circ f|_U^{\leftarrow})(f(x)) \quad (f(x) \in A) \\ &= h(x) \quad (f \text{ is a homeomorphism on } U) \end{aligned}$$

So, $t|_U = h$. Then $t \in C(S)$ is a continuous extension of $h \in C(U)$. We conclude U is C-embedded in S.

23.2 C-Embeddings and the Pseudocompact Property

One consequence of the previous theorem is the statement proven in the following example.

Example 2. Suppose S is a non-pseudocompact space. Show that S contains a C-embedded copy of $\mathbb{N}$.

Solution: We are given that S is non-pseudocompact and so there is a real-valued continuous function, $f \in C(S)$, which is unbounded.

For each $i \in \mathbb{N}$, choose $m_i \in [2i - 1, 2i + 1] \cap f[S]$, if non-empty and not equal to m_{i-1}. Since f is unbounded, we can choose infinitely many such m_i's to construct the subset

$$N = \{m_i : i \in \mathbb{N}\}$$

in $f[S] \subseteq \mathbb{R}$. See that...

— N is an unbounded subset of $\mathbb{R}$,
— Each $\{m_i\}$ is a clopen subset of N,
— The set N is a closed subset of $\mathbb{R}$ since N contains all its limit points.

So, N is a closed unbounded copy of $\mathbb{N}$ in $f[S]$.

We claim that, S contains a C-embedded copy, A, of $\mathbb{N}$.

Note that $f^{\leftarrow}[N] \subseteq S$. For each $i \in \mathbb{N}$, choose $n_i \in f^{\leftarrow}(m_i)$ in S. If $A = \{n_i : i \in \mathbb{N}\}$, then the function, $f|_A : A \to N$, maps A homeomorphically onto the copy, N, of $\mathbb{N}$ in $f[S]$, as claimed.

Since f maps A onto a closed subset of $\mathbb{R}$, then, by Theorem 23.3, A is C-embedded copy of $\mathbb{N}$ in S, as required.

It follows from this statement that...

> *if S does* not *contain a C-embedded copy of $\mathbb{N}$ then $C(S)$ cannot contain an unbounded function and so S must be pseudocompact.*

The converse also holds true as we shall show in the form of a theorem.

Theorem 23.4 *The space S is pseudocompact if and only if S does not contain a C-embedded copy of $\mathbb{N}$.*

Proof. ($\Leftarrow$) We have shown in the example above that if S does not contain a C-embedded copy of $\mathbb{N}$ then S must be pseudocompact.

($\Rightarrow$) Suppose S is pseudocompact. We are required to show that S does not contain a C-embedded copy of $\mathbb{N}$. Suppose S contains a C-embedded copy, A, of $\mathbb{N}$. Let $g : A \to \mathbb{N} \subseteq \mathbb{R}$ be one-to-one and onto $\mathbb{N}$. Then $g \in C(A)$ and is unbounded. Since A is C-embedded in S, there exists a function $f \in C(S)$ such that $f|_A = g$. Since such an f is unbounded in $C(S)$, this contradicts the fact that S is pseudocompact. So, S does not contain a C-embedded copy of $\mathbb{N}$, as required.

We present a few consequences of the previous theorems.

Corollary 23.5 *Let S be a locally compact Hausdorff space and N be a C-embedded copy of $\mathbb{N}$ in S. Let U be an open neighborhood of N in S. Then $cl_{\beta S}(S\backslash U) \cap cl_{\beta S}N = \varnothing$.*

Proof. We are given that $N = \{n_i : i \in \mathbb{N}\}$ is a C-embedded copy of $\mathbb{N}$ in S and U is an open neighborhood of N in S.

By local compactness, S is open in βS, so we can construct the family, $\{\mathrm{cl}_S V_i : i \in \mathbb{N}\}$, of pairwise disjoint *compact sets* such that V_i is an open neighborhood of n_i and $V_i \subseteq \mathrm{cl}_S V_i \subseteq U$.

Then $S\backslash U \subseteq S\backslash \mathrm{cl}_S V_i \subseteq S\backslash V_i$, for each i. So

$$S\backslash U \subseteq \cap\{S\backslash V_i : i \in \mathbb{N}\}$$

Since S is completely regular, for each i, there a function $g_i \in C^*(S)$ such that

$$n_i \in Z(g_i - 1)$$
$$S\backslash V_i \subseteq Z(g_i)$$
$$S\backslash U \subseteq \cap\{Z(g_i) : i \in \mathbb{N}\}$$

If $Z = \cap\{Z(g_i) : i \in \mathbb{N}\}$, it is a countable intersection of zero-sets such that, for all i, $n_i \notin Z$. So, $Z \cap N = \varnothing$. Since Z is itself a zero-set,[1] then $Z = Z(t)$ for some $t \in C(S)$.

So,

$$Z(t) \cap N = \varnothing$$

By Theorem 23.2, since N is C-embedded in S, then N is completely separated from any zero set disjoint from N.

---

[1]See justification on page 252 where it is shown that a countable intersection of zero-sets is a zero-set.

Then there is a continuous function $h : S \to [0, 1]$ such that

$$N \subseteq Z(h) \quad \text{and} \quad Z(t) \subseteq Z(h-1)$$

It follows that,

$$\mathrm{cl}_{\beta S}N \subseteq \mathrm{cl}_{\beta S}Z(h) = Z(h^{\beta})$$
$$\mathrm{cl}_{\beta S}(S\backslash U) \subseteq \mathrm{cl}_{\beta S}Z(h-1) = Z(h^{\beta}-1)$$

Since $Z(h^{\beta}-1) \cap Z(h^{\beta}) = \varnothing$, then $\mathrm{cl}_{\beta S}(S\backslash U) \cap \mathrm{cl}_{\beta S}N = \varnothing$, as required.

In the following statement, we show that, given a C-embedded copy, N, of $\mathbb{N}$ in a locally compact Hausdorff space, S, its outgrowth, $\mathrm{cl}_{\beta S}N\backslash N$, in $\beta S\backslash S$ is contained in the $\beta S\backslash S$-interior of any zero-set which contains it.[2]

Corollary 23.6 *Let S be a locally compact Hausdorff space and N be a C-embedded copy of $\mathbb{N}$ in S. Suppose $Z(f^{\beta})$ is a zero-set in βS, such that $(cl_{\beta S}N)\backslash N \subseteq Z(f^{\beta}) \cap \beta S\backslash S$. Then*

$$(\mathrm{cl}_{\beta S}N)\backslash N \subseteq \mathrm{int}_{\beta S\backslash S}(Z(f^{\beta}) \cap \beta S\backslash S)$$

Proof. We are given a locally compact space, S, which contains a C-embedded copy, $N = \{n_i : i \in \mathbb{N}\}$, of $\mathbb{N}$. Also, we are given that, for some function $f \in C^*(S)$,

$$(\mathrm{cl}_{\beta S}N)\backslash N \subseteq Z(f^{\beta}) \cap \beta S\backslash S$$

We are required to show that $(\mathrm{cl}_{\beta S}N)\backslash N \subseteq \mathrm{int}_{\beta S\backslash S}(Z(f^{\beta}) \cap \beta S\backslash S)$.

For each $i \in \mathbb{N}$, we can choose an open neighborhood V_i of n_i such that $V_i \cap V_j = \varnothing$ for $j < i$. We can choose these V_i's such that

[2]This statement expresses a property related to the concept of "p-points".

> A *p-point* in a space S, is a point $p \in S$ such that p belongs to the interior of any G_δ which contains it. It is easily verified that, if S is completely regular and p is a p-point in S, then $p \in \mathrm{int}_S Z$ for any zero-set Z in $Z[S]$ such that $p \in Z$.

$\{\text{cl}_{\beta S}V_i : i \in \mathbb{N}\}$ forms a family of pairwise disjoint compact subsets of S. We define,

$$U = \cup\{V_i : i \in \mathbb{N}\}$$

as a βS-open neighborhood of the C-embedded set N in S.

Fact 1: Suppose $x_i \in V_i$, for each $i \in \mathbb{N}$. Then $\{f^\beta(x_i) : i \in \mathbb{N}\}$ must converge to zero.

Proof of this fact: Let $p \in (\text{cl}_{\beta S}N)\backslash N$. Since $(\text{cl}_{\beta S}N)\backslash N \subseteq Z(f^\beta) \cap \beta S \backslash S$, $f^\beta(p) = 0$ and p is an accumulation point of N. Then there is some subsequence $\{n_{j(i)} : i \in \mathbb{N}\} \subseteq N$, converging to p. Since f^β is continuous on βS,

$$\{f^\beta(n_{j(i)}) : i \in \mathbb{N}\} \longrightarrow f^\beta(p) = 0$$

For each $i \in \mathbb{N}$ and open ball $B_{1/i+1}(f^\beta(n_{j(i)}))$ in $\mathbb{R}$, by virtue of continuity of f^β on βS (and that S is open in βS), there exists a βS-open neighborhood, V_i, of $n_{j(i)}$, contained in S such that

$$f^\beta(n_{j(i)}) \in f^\beta[V_i] \subseteq f^\beta[\text{cl}_{\beta S}V_i] \subseteq B_{1/i+1}(f^\beta(n_{j(i)}))$$

For each $i \in \mathbb{N}$ and $x_i \in V_i$, then,

$$\begin{aligned} |f^\beta(x_i)| &= |f^\beta(x_i) - f^\beta(n_{j(i)}) + f^\beta(n_{j(i)})| \\ &\leq |f^\beta(x_i) - f^\beta(n_{j(i)})(x)| + |f^\beta(n_{j(i)})| \\ &< 1/(i+1) + |f^\beta(n_{j(i)})| \end{aligned}$$

Since $\{f^\beta(n_{j(i)}) : i \in \mathbb{N}\} \longrightarrow 0$, then $\{f^\beta(x_i) : i \in \mathbb{N}\} \longrightarrow 0$. This proves Fact 1.

Corollary 23.5 states that

$$\text{cl}_{\beta S}(S\backslash U) \cap \text{cl}_{\beta S}N = \varnothing \quad (\dagger)$$

Fact 2: There exists a function a function $h^\beta : \beta S \to [0,1]$ such that $\text{cl}_{\beta S}N \subseteq Z(h^\beta - 1)$ and $\text{cl}_{\beta S}(S\backslash U) \subseteq Z(h^\beta)$.

This follows immediately from (†) where it is shown that $\text{cl}_{\beta S}(S\backslash U)$ and $\text{cl}_{\beta S}N$ are disjoint compact sets in the normal space βS, and so are completely separated. So a function, $h^\beta : \beta S \to [0,1]$, exists such that such that $\text{cl}_{\beta S}(S\backslash U) \subseteq Z(h^\beta)$ and $\text{cl}_{\beta S}N \subseteq Z(h^\beta - 1)$, as claimed.

We are then guaranteed that there exists a continuous function $h : S \to [0, 1]$ such that,

$$\text{cl}_{\beta S}N \subseteq Z(h^\beta - 1) \subseteq \text{Cz}(h^\beta) \subseteq \beta S\backslash Z(h^\beta) \subseteq \beta S\backslash \text{cl}_{\beta S}(S\backslash U). \quad (*)$$

Fact 3: $\text{Cz}(h^\beta)\backslash S \subseteq Z(f^\beta)\backslash S$.

Proof of this fact: Let $q \in \text{Cz}(h^\beta)\backslash S$. To show that $q \in Z(f^\beta)\backslash S$ it suffices to show that $f^\beta(q) = 0$.

By (∗), $\text{Cz}(h^\beta) \subseteq \beta S\backslash \text{cl}_{\beta S}(S\backslash U)$.

Let W be any βS open neighborhood of q.

Claim: We claim that W must intersect infinitely many of the compact $\text{cl}_S V_i$'s.

Suppose not. Suppose $K = \cup\{\text{cl}_{\beta S}V_{n_i} : i = 1 \text{ to } m\}$ is the union of finitely many $\text{cl}_{\beta S}V_i$'s, where $W \cap N\backslash K = \varnothing$. Then K is compact. From (*), we have $q \in \beta S\backslash \text{cl}_{\beta S}(S\backslash U) \cap W$. Verify that

$$[W \cap \beta S\backslash \text{cl}_{\beta S}(S\backslash U)]\backslash K$$

is a non-empty open subset of βS which is contained in $\beta S\backslash S$. This is impossible. So, W must intersect infinitely many of the compact $\text{cl}_S V_i$'s, as claimed.

So, W must must meet infinitely many V_i's.

We construct an infinite family of open neighborhoods

$$\{f^{\beta\leftarrow}[B_{1/j}(f(q))] \ : \ j = 1, 2, 3, \ldots\}$$

of q. The proven claim guarantees that each of these intersects infinitely many V_i's.

Then, for each j, we can choose u_j such that

$$u_j \in f^{\beta\leftarrow}[B_{1/j}(f(q))] \cap V_j$$

By Fact 1, $\{f^\beta(u_j) : j \in \mathbb{N}\} \longrightarrow f^\beta(q) = 0$. So, $q \in Z(f^\beta)$.

We conclude that,

$$\text{Cz}(h^\beta)\backslash S \subseteq Z(f^\beta)\backslash S$$

This completes the proof of the Fact 3.

By (*), $\text{cl}_{\beta S}N \subseteq \text{Cz}(h^\beta)$ so

$$\text{cl}_{\beta S}N \backslash N \subseteq \text{Cz}(h^\beta) \cap \beta S \backslash S \subseteq Z(f^\beta) \cap \beta S \backslash S$$

Since $\text{Cz}(h^\beta) \cap \beta S \backslash S$ is open in $\beta S \backslash S$, then $\text{cl}_{\beta S}N \backslash N \subseteq \text{int}_{\beta S \backslash S}(Z(f^\beta) \cap \beta S \backslash S)$, as required.

Recall that (from Definition 10.17), for a completely regular space, a *P-point* in S is a point which belongs to the interior of any G_δ-set that contains it. Also, every space in which all points are P-points is a *P-space*. In Theorem 10.18, we proved that completely regular spaces P-spaces are zero-dimensional spaces. We also stated without proof that pseudocompact P-spaces are finite. We prove that particular statement now.

Theorem 23.7 *All pseudocompact P-spaces are finite.*

Proof. Suppose S is a pseudocompact P-space.

Then every continuous real-valued function on S is bounded.

If S is infinite then we can inductively construct a sequence $A = \{x_i : i \in \mathbb{N}\}$ for which there is a clopen neighborhood B_i of x_i which does not intersect a clopen neighborhood B_j of x_j for $j < i$. So A is a discrete subset. Also, $S \backslash A = \cap\{S \backslash \{x_i\} : i \in \mathbb{N}\}$ so $S \backslash A$ is a G_δ of S. Then both $S \backslash A$ and A are clopen in S. Then A is C^*-embedded in S.

Let $f : A \to \mathbb{R}$ be the continuous function defined as $f(x_i) = i$.

Suppose Z is a zero-set in S which is disjoint from A. Since zero-sets in a P-space are clopen, then Z and A are completely separated. By Theorem 23.2, part (b) A is C-embedded in S. So, $f : A \to \mathbb{R}$ extends to $f^* : S \to \mathbb{R}$. But f^* is unbounded; this contradicts pseudocompactness of S. Then S cannot be infinite.

23.3 When Is βS a Singular Compactification?

With the tools presented above, we are now able to prove an interesting statement (in Theorem 22.10) presented in the last chapter whose proof was differed to this time. We stated that spaces, S, such that βS is a singular compactification must be pseudocompact. We prove this in the next theorem.

Theorem 23.8 *Let S be locally compact, non-compact, and Hausdorff. If βS is a singular compactification then S is pseudocompact.*

Proof. We are given that S is locally compact and Hausdorff and that βS is a singular compactification.

Since βS is singular, there exists a continuous function $r : \beta S \to \beta S \backslash S$ where $r[\beta S] = r[\beta S \backslash S] = \beta S \backslash S$ and $r(x) = x$ for $x \in \beta S \backslash S$.

We are required to show that S is pseudocompact.

Suppose S is *not* pseudocompact.

By Theorem 23.4, since the space S is not pseudocompact, it must contain a C-embedded copy, N, of $\mathbb{N}$.

From the Example 9 found on page 512, we know that $\mathrm{cl}_{\beta S} N = \beta N$.

Then $r[\beta N] = (\beta N \setminus N) \cup r[N]$. Since βN is separable, then $r[\beta N]$ is separable.[3] Since $\beta N \setminus N$ was shown to be non-separable (see page 513), the subset $r[\beta N] \backslash (\beta N \backslash N)$ of $\beta S \backslash S$ must contain countably infinite elements say,

$$T = \{t_i : i \in \mathbb{N}\}^4$$

where $T \subseteq r[\beta N]$. For each $i \in \mathbb{N}$, choose in N, $m_i \in r^{\leftarrow}(t_i)$, to obtain the subset,

$$M = \{m_i : i \in \mathbb{N}\}$$

of N. So, $r[M] = T \subseteq \beta S \backslash S$. Note that $M \subseteq N \subseteq S$; also T and M can be viewed as disjoint subsets of βS.

---

[3]By Theorem 6.11, the continuous image of a separable set is separable.
[4]For if $r[\beta N] \backslash (\beta N \backslash N)$ is finite, then $\beta N \backslash N$ is separable.

Claim: That $(\mathrm{cl}_{\beta S\backslash S}T)\backslash T = (\mathrm{cl}_{\beta S}M)\backslash M$.

Proof of claim: First see we have two representations of $r[\mathrm{cl}_{\beta S}M]$:

$$r[\mathrm{cl}_{\beta S}M] = \mathrm{cl}_{\beta S\backslash S}(r[M]) = \mathrm{cl}_{\beta S\backslash S}T \quad (\text{Since } r(m_i) = t_i)$$
$$r[\mathrm{cl}_{\beta S}M] = r[(\mathrm{cl}_{\beta S}M)\backslash M] \cup r[M] = (\mathrm{cl}_{\beta S}M)\backslash M \ \cup\ T$$

So,

$$\mathrm{cl}_{\beta S\backslash S}T = (\mathrm{cl}_{\beta S}M)\backslash M \ \cup\ T \qquad (*)$$

Verify that $\mathrm{cl}_{\beta S}M\backslash M \cap T = \varnothing$.[5] So,

$$(\mathrm{cl}_{\beta S\backslash S}T)\backslash T = (\mathrm{cl}_{\beta S}M)\backslash M \qquad (\dagger)$$

as claimed.

We now define the function $g : \mathrm{cl}_{\beta S\backslash S}T \to \mathbb{R}$ as

$$g(t_i) = \frac{1}{i+1}$$
$$g[\,(\mathrm{cl}_{\beta S\backslash S}T)\backslash T\,] = \{0\}$$

So, $\{g(t_i) : i = 1, 2, 3, \ldots\} \longrightarrow \{0\}$. The function g is continuous and bounded on the compact domain, $\mathrm{cl}_{\beta S\backslash S}T$, and so is the function $(g{\circ}r) \in C^*(M)$ where

$$(g{\circ}r)(m_i) = g(t_i) = 1/(i+1)$$

Then $(g{\circ}r)$ extends to $(g{\circ}r)^* \in C^*(S)$.[6] So, $(g{\circ}r)$ extends continuously to $(g{\circ}r)^\beta \in C(\beta S)$.

---

[5] $T \subseteq r[\beta N]\backslash(\beta N\backslash N) \ \Rightarrow\ T \cap \beta N\backslash N = \varnothing$

and

$$\begin{aligned} M \subseteq N \ &\Rightarrow\ \mathrm{cl}_{\beta S}M \subseteq \mathrm{cl}_{\beta S}N = \beta N \\ &\Rightarrow\ \mathrm{cl}_{\beta S}M\backslash M \subseteq \mathrm{cl}_{\beta S}N\backslash N = \beta N\backslash N \\ &\Rightarrow\ \mathrm{cl}_{\beta S}M\backslash M \subseteq \beta N\backslash N \\ &\Rightarrow\ T \cap \mathrm{cl}_{\beta S}M\backslash M = \varnothing \ \ (\text{Since } T \cap \beta N\backslash N = \varnothing) \end{aligned}$$

[6] Every function is continuous on N so M is C-embedded in N. Now, N was declared to be C-embedded in S so M is C-embedded in S. Then M is C^*-embedded in S.

So,

$$(\text{cl}_{\beta S}M)\backslash M = (\text{cl}_{\beta S\backslash S}T)\backslash T \subseteq Z((g_\circ r)^\beta) \cap \beta S\backslash S \text{ (By } \dagger)$$

By Corollary 23.6,

$$(\text{cl}_{\beta S}M)\backslash M \subseteq \text{int}_{\beta S\backslash S}[Z((g_\circ r)^\beta) \cap \beta S\backslash S]$$

Since, by (†), $(\text{cl}_{\beta S\backslash S}T)\backslash T = (\text{cl}_{\beta S}M)\backslash M$,

$$(\text{cl}_{\beta S\backslash S}T)\backslash T \subseteq \text{int}_{\beta S\backslash S}[Z((g_\circ r)^\beta) \cap \beta S\backslash S]$$

But this is impossible since it implies there exists infinitely many i's such that $g(r(m_i)) = g(t_i) = 1/(i+1) = 0$.

The contradiction is the result of our assumption that S is not pseudocompact.

So, S is pseudocompact, as required. This completes the proof of the theorem.

23.4 Another Characterization of Pseudocompactness

A space, S, was originally said to satisfy the "pseudocompact property" if and only if every continuous real-valued function on S is bounded. Any compact Hausdorff space is easily seen to be pseudocompact. But a pseudocompact space need not be compact: We showed, for example, that any countably compact space is pseudocompact (15.9). For example, the ordinal space, $S = [0, \omega_1)$, was shown to be pseudocompact but not compact. We have also seen that every sequentially compact space satisfies the pseudocompact property (17.4).

We saw other ways of describing those spaces which are pseudocompact.

— In Theorem 21.14, we saw that (for locally compact Hausdorff spaces) pseudocompact spaces are precisely those spaces for which zero-sets, Z, in $Z[\beta S]$ *cannot* be entirely contained in $\beta S\backslash S$. That is, for any zero-set, $Z \in Z[\beta S]$, $Z \cap S \neq \varnothing$. Equivalently, for any continuous, $f : S \to \omega\mathbb{R}$,

$$f^{\beta(\omega)}[\beta S \backslash S] \subseteq f[S]$$

This is interesting, since it shows that we can characterize a property of S in terms of a property of its β-outgrowth. By itself, this can motivate the study of compactifications of S.

— Also, a space S is pseudocompact if and only if S does not contain a C-embedded copy of $\mathbb{N}$ (23.4).[7]

We also have the following results associated to the pseudocompact property.

— Let S be locally compact, non-compact, and Hausdorff. If βS is a singular compactification then S is pseudocompact (23.8).[8]

We add to the list of characterizations for the pseudocompactness property, the following statement.

Theorem 23.9 *Let S be a locally compact Hausdorff space. Then the following statements are equivalent:*

(a) *The space S is pseudocompact.*
(b) *The space S* does not *admit an infinite family, $\mathscr{U}$, of non-empty open sets which satisfies the property "every point has an open neighborhood which will intersect only finitely many elements of $\mathscr{U}$".*[9]

Proof. We are given S is a locally compact Hausdorff space. Then S is completely regular (by 18.8).

(a $\Rightarrow$ b) Suppose S is pseudocompact. Suppose S admits an infinite family, $\mathscr{U}$, of non-empty open sets such that every point has an open neighborhood which will intersect at most finitely many elements of $\mathscr{U}$.

[7]We will add later to this list the following characterization: Pseudocompact spaces, S, are precisely those spaces whose realcompactification, υS, is βS (24.8).

[8]We will add later: A Hausdorff topological space, S, is compact if and only if S is both realcompact and pseudocompact (24.7).

[9]When $\mathscr{U}$ satisfies this property we say that $\mathscr{U}$ is "locally finite".

Step 1: We inductively construct sequences $\{p_i : i = 1, 2, 3, \ldots\}$ and open subsets, $\{V_i : i = 1, 2, 3, \ldots\}$, where $p_i \in V_i$ and the V_i's are open and pairwise disjoint.

Let $U_1 \in \mathscr{U}$. Then there is a point p_1 and open V_1 such that $p_1 \in V_1 \subseteq U_1$ where V_1 intersects at most finitely many elements of $\mathscr{U}$.

There exists $U_2 \in \mathscr{U}$ such that $U_2 \cap V_1 = \varnothing$. Then there is a point p_2 and open V_2 such that $p_2 \in V_2 \subseteq U_2$ where V_2 intersects at most finitely many elements of $\mathscr{U}$.

There exists $U_3 \in \mathscr{U}$ such that $U_3 \cap (V_1 \cup V_2) = \varnothing$. Then there is a point p_3 and open V_3 such that $p_3 \in V_3 \subseteq U_3$ where V_3 intersects at most finitely many elements of $\mathscr{U}$.

$\vdots$

Inductively, there exists $U_n \in \mathscr{U}$ such that $U_n \cap (V_1 \cup \cdots \cup V_{n-1}) = \varnothing$. Then there is a point p_n and open V_n such that $p_n \in V_n \subseteq U_n$ where V_n intersects at most finitely many elements of $\mathscr{U}$.

We thus obtain the sequences $\{p_i : i = 1, 2, 3, \ldots\}$ and $\{V_i : i = 1, 2, 3, \ldots\}$ where $p_i \in V_i$ and the V_i's are open and pairwise disjoint.

Step 2: By complete regularity, we can find, for each $i \in \mathbb{N}\backslash\{0\}$, a continuous $f_i : S \to [0, i]$ such that $f_i(p_i) = i$ and $f_i[S \backslash V_i] = \{0\}$. Since the V_i's are pairwise disjoint, then the function $g : S \to \mathbb{R}$ defined as

$$g(x) = \sum_{i=1}^{\infty} f_i(x)$$

is continuous and unbounded on S. This contradicts our hypothesis. So, if S is pseudocompact, it *cannot* admit an infinite family, $\mathscr{U}$, of open sets such that every point has an open neighborhood which will intersect at most finitely many elements of $\mathscr{U}$. This concludes a $\Rightarrow$ b.

(b $\Rightarrow$ a) Suppose S *does not* admit an infinite family, $\mathscr{U}$, of open sets such that every point has an open neighborhood which will intersect at most finitely many elements of $\mathscr{U}$. We are required to show that S is pseudocompact.

Suppose not. Then S contains a C-embedded copy, $N = \{n_i : i \in \mathbb{N}\}$, of $\mathbb{N}$. Let $g : N \to \mathbb{N}$ be a function defined as $g(n_i) = i$. Then g is continuous on N. Since N is C-embedded, g extends to a continuous function, $f : S \to \mathbb{R}$, where $f|_N = g$. Let $\{V_i : i \in \mathbb{N}\}$ be a pairwise disjoint family of open sets such that $f(n_i) \in V_i$.

Then $\mathscr{U} = \{f^{\leftarrow}[V_i] : i \in \mathbb{N}\}$ forms a pairwise disjoint family of open sets in S such that $n_i \in f^{\leftarrow}[V_i]$.

We claim that every point of S has an open neighborhood which intersects at most finitely many elements of $\mathscr{U}$ (thus contradicting our hypothesis).

Proof of claim: Let $p \in S$, $q = f(p)$ and $B = B_{1/2}(q)$. Then $f^{\leftarrow}[B]$ is an open neighborhood of p in S which intersects at most two elements of $\mathscr{U}$. Then $\mathscr{U}$ is an infinite family of open subsets of S such every point has an open neighborhood which intersects at most finitely many points. Since this contradicts our hypothesis, S must be pseudocompact. This concludes b $\Rightarrow$ a.

23.5 The Stone–Čech Compactification of a Product

Suppose we are given two completely regular spaces S and T and we consider the Stone–Čech compactification of their corresponding product space, $S \times T$. By the Tychonoff theorem, $\beta S \times \beta T$ is a compactification of $S \times T$. Under what conditions is it equivalent to $\beta(S \times T)$? We investigate the question below.

In the proof of the following theorem, we invoke in various steps the statements 18.8, 23.2, 21.14, and 23.5.

Theorem 23.10 *Let S and T be locally compact Hausdorff infinite spaces. If $S \times T$ is C^*-embedded in $\beta S \times \beta T$, then $S \times T$ must be pseudocompact.*

Proof. We are given that S and T are locally compact Hausdorff infinite spaces such $S \times T$ is C^*-embedded in $\beta S \times \beta T$. We want to show that $S \times T$ is pseudocompact.

Notation: We start by introducing some notation. Let

$$X = S \times T$$

$$A = S \times \beta T \backslash T$$

$$B = \beta S \backslash S \times T$$

$$C = \beta S \backslash S \times \beta T \backslash T \quad \text{(compact)}$$

Let $\alpha X = X \cup A \cup B \cup C$ where $B \cup C$ and $A \cup C$ are compact. Since $S \times T$ is C^*-embedded in $\beta S \times \beta T$,

$$\alpha X = \beta S \times \beta T \equiv \beta X$$

Claim 1: We first show that $\beta(X \cup A \cup B) \equiv \alpha X$.

Proof of claim: Since

$$\beta X \equiv \alpha X$$

if $f \in C^*(X \cup A \cup B)$, then $f|_X \in C^*(X)$, so $f|_X$ extends to $(f|_X)^\alpha \in C(\alpha X) = C(X \cup A \cup B \cup C)$. So,

$$X \cup A \cup B \text{ is } C^*\text{-embedded in } \alpha X \qquad (*)$$

Then $\mathrm{cl}_{\beta X} X = \beta(X \cup A \cup B) \equiv \alpha X$, as claimed.

We prove the theorem statement in two steps: In step one we prove that $X \cup A \cup B$ is pseudocompact. In step two that X is pseudocompact.

Step 1. We show that $X \cup A \cup B$ is pseudocompact.
Suppose $X \cup A \cup B$ is not pseudocompact.
Then C contains a zero-set $Z(f^\alpha)$ for some $f \in C^*(X \cup A \cup B)$ (by Theorem 21.14).

We know that a zero-set is a G_δ. Then $Z(f^\alpha) = \cap\{B_n : n \in \mathbb{N}\}$ where $\{B_n : n \in \mathbb{N}\}$ is a countable family of open sets in αX. For each $i \in \mathbb{N}$, let $V_i = \cap\{B_n : n = 0 \text{ to } i\}$, to obtain a decreasing collection $\{V_i : i \in \mathbb{N}\}$ of open neighborhoods of $Z(f^\alpha)$ such that

$$Z(f^\alpha) = \cap\{V_i : i \in \mathbb{N}\}$$

Since $Z(f^\alpha)$ is a zero-set in $\alpha X \setminus (X \cup A \cup B)$, then the function $g : X \cup A \cup B \to \mathbb{R}$ defined as

$$g = 1/f$$

is continuous on $X \cup A \cup B$.

Suppose $(u, v) \in Z(f^\alpha)$, a subset of C.

Claim 2: The function g is unbounded on $S \times \{v\}$ and on $\{u\} \times T$.

Proof of Claim 2: We show this for the case, $g : S \times \{v\} \to \mathbb{R}$. Suppose $g|_{S\times\{v\}}$ is bounded. Then there exists $M > 0$ such that $|g|_{S\times\{v\}}|[S \times \{v\}] \subseteq [0, M]$. For a neighborhood $B_{1/M+1}(0)$ of 0 there exists an open neighborhood $U \times V$ of (u, v) in X such that $f^\alpha[U \times V] \subseteq B_{1/M+1}(0)$. Now, $(U \times V) \cap (S \times \{v\})$ is non-empty. Then there exists $(a, v) \in S \times \{v\}$ such that $g(a, v) = 1/f(a, v) > 1/(1/M) = M$. Since M was assumed to be a bound for $g|_{S\times\{v\}}$, $g|_{S\times\{v\}}$ is unbounded on $S \times \{v\}$, as claimed.

The proof which shows that $g|_{\{u\}\times T}$ is unbounded on $\{u\} \times T$ is similar. This establishes Claim 2.

Claim 3: We now claim that $S \times \{v\}$ contains a closed C-embedded copy of $\mathbb{N}$.

Proof of Claim 3: For each $i \in \mathbb{N}$, choose $m_i \in [2i - 1, 2i + 1] \cap g[S \times \{v\}]$, if non-empty. Since g is unbounded on $S \times \{v\}$, we can choose infinitely many such m_i's to construct the subset

$$M = \{m_i : i \in \mathbb{N}\}$$

in $g[S \times \{v\}] \subseteq \mathbb{R}$. See that...

- M is an unbounded subset of $\mathbb{R}$.
- Each $\{m_i\}$ is a clopen subset of M.
- The set M is a closed subset of $\mathbb{R}$ since it contains all its limit points.

So, M is a closed unbounded copy of $\mathbb{N}$ in $\mathbb{R}$.

Note that $g^{\leftarrow}[M] \subseteq X \cup A \cup B$. For each $i \in \mathbb{N}$, choose $(n_i, v) \in g^{\leftarrow}(m_i)$ to form the set $N_1 = \{(n_i, v) : i \in \mathbb{N}\}$ in $X \cup A \cup B$. Then the function, $h : M \to N_1$ defined as $h(m_i) = (n_i, v)$ maps the copy, M, of $\mathbb{N}$ in $g[X \cup A \cup B]$ homeomorphically onto N_1. Then N_1 is a closed copy of $\mathbb{N}$ in $X \cup A \cup B$.

Since g maps N_1 onto a closed subset, M, of $\mathbb{R}$ then, by Theorem 23.3, N_1 is C-embedded copy of $\mathbb{N}$ in $S \times \{v\}$. This establishes the Claim 3.

By applying an analogous reasoning, $\{u\} \times T$ contains a closed C-embedded copy, N_2, of $\mathbb{N}$.

Since every open neighborhood $U \times V$ of (u, v) intersects both N_1 and N_2, then (u, v) is an accumulation point of both N_1 and N_2. This implies that,

$$(u, v) \in \mathrm{cl}_{\alpha X} N_1 \cap \mathrm{cl}_{\alpha X} N_2$$

See that $S \times \beta T = (X \cup A \cup B) \backslash (\beta S \backslash S \times T)$ is an open neighborhood of N_1 in $X \cup A \cup B$. By Theorem 23.5,

$$\mathrm{cl}_{\alpha X}[(X \cup A \cup B) \backslash (S \times \beta T)] \quad \cap \quad \mathrm{cl}_{\alpha X} N_1$$

must be empty. Since

$$(u, v) \in \mathrm{cl}_{\alpha X} N_2 \cap \mathrm{cl}_{\alpha X} N_1 \subseteq \mathrm{cl}_{\alpha X}[(X \cup A \cup B) \backslash (S \times \beta T)] \cap \mathrm{cl}_{\alpha X} N_1$$

we have a contradiction.

So, $X \cup A \cup B$ must be pseudocompact. We are done with step 1.

Step 2. In this part we conclude the proof by showing that X is pseudocompact.

We are given that αX and βX are equivalent compactifications and so X is C^*-embedded in αX. Also we know (from step 1) that $X \cup A \cup B$ is pseudocompact.

Suppose X is *not* pseudocompact.

Then the space X contains a copy, $N = \{(n_i, m_i) : i \in \mathbb{N}\}$, of $\mathbb{N}$ which is C-embedded in X. That is, there is a continuous function $d : N \to \mathbb{R}$ such that $d[N]$ is closed and unbounded in $\mathbb{R}$ and $\{d(n_i, m_i)\}$ is clopen in $d[N]$, for each i. Suppose, without loss of generality, that $Z = \{n_i : (n_i, m_i) \in N\}$ is infinite. Let

$$Y = \{n_{i_j} : n_{i_j} \in Z,\ j \in \mathbb{N},\quad n_{i_j} \neq n_{i_k} \text{ if } k < j\}$$

Then $\{d(n_{i_j}, m_{i_j}) : j \in \mathbb{N}\}$ is an unbounded copy of $\mathbb{N}$ in $\mathbb{R}$.

Let $(u, v) \in C$. Let

$$W = \{(n_{i_j}, v) : n_{i_j} \in Y\}$$

a subset of $S \times \{v\} \subseteq A$.

Let $r : W \to d[N]$ be defined as $r(n_{i_j}, v) = d(n_{i_j}, m_{i_j})$.

Then $r[W]$ is a C-embedded copy of $\mathbb{N}$. So W is a C-embedded copy of $\mathbb{N}$ in $X \cup A \cup B$. Since $X \cup A \cup B$ is pseudocompact, we have a contradiction.

We conclude that X is pseudocompact, what is required for Step 2.

We have a simple example illustrating the above result.

Example 3. Since $\mathbb{R} \times \mathbb{R}$ not pseudocompact, we can invoke the above theorem to deduce that $\beta(\mathbb{R} \times \mathbb{R}) \not\equiv \beta\mathbb{R} \times \beta\mathbb{R}$.

In the following theorem, we show that pseudocompactness of $S \times T$ is sufficient to guaranty that $\beta(S \times T) \equiv \beta S \times \beta T$.

23.6 Tietze's Extension Theorem

The proof of the following theorem invokes the Urysohn lemma in an important way to deduce that closed subspaces of a normal space are C-embedded.

Theorem 23.11 *Let S be a normal topological space which contains a non-empty closed subset, A. If $f : A \to [a, b]$ is a continuous real-valued function then f extends to a continuous function which maps all of S into $[a, b]$.*

Proof. We are given that S is normal and contains a non-empty closed subset A.

Fact 1. If f is a continuous function which maps A into $I = [-k, k]$ there is a function $g \in C^*(S)$ such that $g[S] \subseteq [-\frac{1}{3}k, \frac{1}{3}k]$ and

$$|\, g|_A(x) - f(x)\,| \leq \tfrac{2}{3}k$$

for $x \in A$. That is, there is a function $g \in C(S)$ which "approximates" f on A.

Proof of Fact 1: Subdivide the interval $[-k, k]$ into

$$I_1 = [-k, -\frac{1}{3}k], \quad I_2 = [-\frac{1}{3}k, \frac{1}{3}k], \quad I_3 = [\frac{1}{3}k, k].$$

Then $f^{\leftarrow}[I_1]$ and $f^{\leftarrow}[I_3]$ are disjoint closed subsets of S (since f is continuous on A and A is closed in S). We can then invoke the Urysohn lemma to justify the existence of a function $g : S \to [-\frac{1}{3}k, \frac{1}{3}k]$ such that $g[f^{\leftarrow}[I_1]] = \{-\frac{1}{3}k\}$ and $g[f^{\leftarrow}[I_3]] = \{\frac{1}{3}k\}$.

Then $g[S] \subseteq [-\frac{1}{3}k, \frac{1}{3}k]$ and $|g|_A(x) - f(x)| \leq \frac{2}{3}k$. This establishes Fact 1.

Without loss of generality, suppose $f : A \to [-1, 1]$ is continuous. We apply the general principle illustrated in Fact 1 replacing k with 1.

We then know of the existence of a continuous function $g_1 \in C(S)$:

$$g_1[S] \subseteq \left[-\tfrac{1}{3}, \tfrac{1}{3}\right], \text{ where } |\, f(x) - (g_1)|_A(x)\,| \leq \tfrac{2}{3} \text{ on } A$$

Applying the same principle, replacing f with $f - (g_1)|_A$, we know of the existence of a continuous function $g_2 \in C(S)$:

$$g_2[S] \subseteq \left[-\tfrac{1}{3}\left(\tfrac{2}{3}\right), \tfrac{1}{3}\left(\tfrac{2}{3}\right)\right] \text{ where}$$
$$|f(x) - (g_1)|_A(x) - (g_2)|_A(x)| \leq [2/3]^2 \text{ on } A$$

We can inductively construct a sequence of continuous functions $\{g_1, g_2, g_3, \ldots, g_n, \ldots\}$ where, if given $\{g_1, g_2, \ldots, g_n\} \subseteq C(S)$ where,

$$|g_n(x)| \leq \tfrac{1}{3}\left(\tfrac{2}{3}\right)^{n-1} \quad \text{and}$$
$$|f(x) - (g_1)|_A(x) - \cdots - (g_n)|_A(x)| \leq [2/3]^n \text{ on } A$$

there exists a continuous function $g_{n+1} \in C(S)$ such that

$$|g_{n+1}(x)| \leq \tfrac{1}{3}\left(\tfrac{2}{3}\right)^{n} \quad \text{and}$$
$$|f(x) - (g_1)|_A(x) - \cdots - (g_{n+1})|_A(x)| \leq [2/3]^{n+1} \text{ on } A$$

We can then consider the series

$$g(x) = \sum_{n=1}^{\infty} g_n(x)$$

on S. This series converges uniformly to $g(x)$,[10] so g is continuous on S.

[10] By comparison with a geometric series and the Weierstrass M-test. Note that, since $\frac{1}{3}\sum_{n=1}^{\infty}\frac{2}{3}^n$ converges to 1 the range of g is contained in $[-1, 1]$.

Claim: We claim that $g|_A = f$.

Proof of claim: To see this simply witness the inequality,

$$|\, f(x) - (g_1)|_A(x) - \cdots - (g_{n+1})|_A(x)\,| \leq [2/3]^{n+1}$$

on A. Taking the limit as n goes to ∞ on each side produces 0. Then the sequence of partial sums $\{\sum_{n=1}^{m} g_n(x) : m \in \mathbb{N}\}$ converges uniformly to $f(x)$ at each $x \in A$.

We can conclude that, if S is normal, a continuous, $f \in C(A)$, mapping a closed subset, A, into $[-1,1]$ extends to a continuous function, $g : S \to [-1,1]$.

The above theorem states that in normal spaces, S, those continuous real-valued functions f on a closed subset A extend to a *bounded* continuous function on S. In the following statement, we generalize the statement so as to include unbounded continuous functions.

Corollary 23.12 *Let S be a normal topological space which contains a non-empty closed subset, A. If $f : A \to \mathbb{R}$ is a continuous real-valued function then f extends to a continuous function $h : S \to \mathbb{R}$, on all of S.*

Proof. We are given that S is normal and contains a non-empty closed subset A.

We know that the open interval $(-1,1)$ is homeomorphic to $\mathbb{R}$, so we will more simply consider a continuous function $f : A \to (-1,1)$ mapping the closed subset, A, into $(-1,1)$. Then we will show that it extends to a continuous function $h : S \to (-1,1)$ mapping S inside $(-1,1)$.

The theorem above guarantees that f extends to a function $g : S \to [-1,1]$ mapping S into $[-1,1]$. But we need a continuous function $h : S \to (-1,1)$ mapping S into $(-1,1)$ such that $h|_A = f$. We will use the function $g : S \to [-1,1]$ to construct h.

We define the subset, D, of S as

$$D = g^{\leftarrow}[\{-1\}] \cup g^{\leftarrow}[\{1\}]$$

We see that D is a closed subset of S which entirely misses the closed subset, A (since $f[A] \subseteq (-1,1)$).

Since $D \cap A = \varnothing$, by the Urysohn lemma, there exists a continuous function $k : S \to [0, 1]$ such that

$$D \subseteq Z(k) \quad \text{and} \quad A \subseteq Z(k-1)$$

We let

$$h(x) = k(x)g(x)$$

where h is seen to be continuous on S. Since

$$\begin{aligned} A \subseteq Z(k-1) \quad &\Rightarrow \quad h[A] = k[A]f[A] = 1 \cdot f[A] \;\; (\text{So } h = f \text{ on } A) \\ D \subseteq Z(k) \quad &\Rightarrow \quad h[D] = k[D]g[D] = 0 \cdot g[D] = \{0\} \\ a \notin D \quad &\Rightarrow \quad |g(a)| < 1 \end{aligned}$$

So, $h[S] \subseteq (-1, 1)$. We are done.

Concepts Review

1. Define a C-embedded subset of a space S.
2. Describe a property which characterizes those C^*-embedded which are C-embedded.
3. Show that closed subsets of $\mathbb{R}$ are C-embedded.
4. State a characterization of pseudocompactness involving a copy N of $\mathbb{N}$.
5. Give a characterization of those spaces S such that βS is singular.
6. State a characterization of pseudocompactness involving a locally finite family of sets.
7. State Tietze's theorem.

Chapter 24

Realcompact Spaces

Abstract
In this section, we introduce the set of all points in βS called the "real points" of a space S. The set of all real points of S is denoted by υS. Given a space S, the extension of S, υS, is referred to as the Hewitt–Nachbin realcompactification of S. If $\upsilon S = S$ then S is said to be realcompact. We provide a characterization as well as a few examples of realcompact spaces.

24.1 Realcompact Space: Definitions and Characterizations

Suppose S is a locally compact Hausdorff topological space and $f : S \to \mathbb{R}$ is a function in the set, $C(S)$, of all real-valued continuous functions (including unbounded ones) on S. Let $i : \mathbb{R} \to \omega\mathbb{R}$ be the inclusion function (identity map) which embeds $\mathbb{R}$ into the one-point compactification,

$$\omega\mathbb{R} = \mathbb{R} \cup \{\infty\}$$

defined as $i(x) = x$ (where ∞ represents a point not in $\mathbb{R}$). We can combine the two functions f and i to define the function $f_+ = i_\circ f$, where $f_+ : S \to \omega\mathbb{R}$ maps S into the compact set $\omega\mathbb{R}$. In this case, we have practically identical functions, f_+ and f, except that the codomain of f_+ is $\omega\mathbb{R}$. We have shown in Theorem 21.5 that any continuous function, $g : S \to K$, mapping a locally compact Hausdorff space, S, into a compact space K extends to a function, $g^{\beta(K)} : \beta S \to K$. We will denote the extension of $f_+ : S \to \omega\mathbb{R}$ to βS as

$$f^{\beta(\omega)} : \beta S \to \omega\mathbb{R}$$

That is, if $f \in C(S)$ (possibly unbounded), f extends to $f^{\beta(\omega)} \in C(\beta S, \omega\mathbb{R})$.

In the case where f is bounded, the extension $f^{\beta(\omega)} : \beta S \to \omega\mathbb{R}$ maps the compact space βS onto the compact space

$$f^{\beta(\omega)}[\beta S] = f^{\beta(\omega)}[\mathrm{cl}_{\beta S} S] = \mathrm{cl}_{\omega\mathbb{R}} f[S] = \mathrm{cl}_{\mathbb{R}} f[S]$$

a compact set in $\mathbb{R} \subset \omega\mathbb{R}$. So in the case of a bounded function, f, $f^{\beta(\omega)} : \beta S \to \omega\mathbb{R}$ and the usual, $f^{\beta} : \beta S \to \mathbb{R}$, both represent the same function.

Of course, if $f : S \to \mathbb{R}$ is unbounded,

$$\begin{aligned} f^{\beta(\omega)}[\beta S] &= f^{\beta(\omega)}[\beta S \backslash S \ \cup \ S] \\ &= f^{\beta(\omega)}[\beta S \backslash S] \cup f[S] \\ &\subseteq f^{\beta(\omega)}[\beta S \backslash S] \ \cup \ \mathbb{R} \end{aligned}$$

where $f[S]$ is non-compact in $\mathbb{R}$ (since $f[S]$ is unbounded). So, $f^{\beta(\omega)}$ must map at least one point in $\beta S \backslash S$ to ∞. We stress *at least* one point, since $f^{\beta(\omega)\leftarrow}(\infty)$ need not be a singleton set in βS. We then have the disjoint union,

$$\beta S = f^{\beta(\omega)\leftarrow}[\mathbb{R}] \ \cup \ f^{\beta(\omega)\leftarrow}(\infty)$$

So, if $C(S)$ has some unbounded functions,

$$C(\beta S, \omega\mathbb{R}) \nsubseteq C(\beta S)$$

24.1.1 *Real points in βS*

Given $f \in C(S)$, we will want to distinguish those points in βS which belong to $f^{\beta(\omega)\leftarrow}(\infty)$ from those which belong to $f^{\beta(\omega)\leftarrow}[\mathbb{R}]$. For a given $f \in C(S)$ we then define $\upsilon_f S$[1] as

$$\upsilon_f S = \beta S \backslash f^{\beta(\omega)\leftarrow}(\infty) = f^{\beta(\omega)\leftarrow}[\mathbb{R}]$$

Clearly, $S \subseteq \upsilon_f S \subseteq \beta S$, for all f.

For a bounded function $f \in C^*(S)$,

$$\begin{aligned}\upsilon_f S &= f^{\beta(\omega)\leftarrow}[\mathbb{R}]\\ &= f^{\beta(\omega)\leftarrow}[\mathrm{cl}_{\mathbb{R}} f[S]]\\ &= f^{\beta\leftarrow}[\mathrm{cl}_{\mathbb{R}} f[S]]\\ &= \beta S\end{aligned}$$

So, in the particular case, $f \in C^*(S)$

$$\upsilon_f S = \beta S$$

The function, $f : S \to \mathbb{R}$, in $C(S)$ is one which continuously extends to a real-valued function, $f^{\beta(\omega)} : \upsilon_f S \to \mathbb{R}$ in $C(\upsilon_f S)$. The points in $\upsilon_f S$ are commonly referred to as being the set of all *real points of* f (in the sense that $f^{\beta(\omega)}(x)$ has a real number value at each of point x in $\upsilon_f S$). To consider the "real points" associated to all functions, $f \in C(S)$, we define.

$$\upsilon S = \cap\{\upsilon_f S : f \in C(S)\} = \cap\{f^{\beta(\omega)\leftarrow}[\mathbb{R}] : f \in C(S)\}$$

Note that every function, $f \in C(S)$, extends continuously to a real-valued function,

$$f^{\beta(\omega)}|_{\upsilon S}$$

on υS. More specifically, each function in $C(S)$ is associated to a unique function in $C(\upsilon S)$.

With these facts in mind, we can now formally define the following related concepts.

[1]The Greek letter, υ, is pronounced *upsilon*.

Definition 24.1 Let S be locally compact and Hausdorff.

(a) If $f \in C(S)$, the function $f^{\beta(\omega)} : \beta S \to \omega\mathbb{R}$ represents the extension of f to βS with range $\omega\mathbb{R}$. We define the subset, $\upsilon_f S$, of βS as

$$\upsilon_f S = f^{\beta(\omega)\leftarrow}[\mathbb{R}]$$

If $p \in \beta S \backslash \upsilon_f S$, then $f^{\beta(\omega)}(p) = \infty$. The points in $\upsilon_f S$ are referred to as *real points of* f.

(b) We define, the subspace, υS of βS, as

$$\upsilon S = \cap\{\upsilon_f S : f \in C(S)\} = \cap\{f^{\beta(\omega)\leftarrow}[\mathbb{R}] : f \in C(S)\}$$

The elements of $\upsilon S \subseteq \beta S$ are called *real points of* S. Every element of S is a real point. If $p \in \beta S \backslash \upsilon S$, p is not *a real point of* S and so there must be at least one function $f \in C(S)$ such that $f^{\beta(\omega)}(p) = \infty$.

(c) If the space S is such that $S = \upsilon S$, then we refer to S as being a *realcompact space*. A space S is realcompact if and only if it contains all real points. If S is realcompact $\beta S \backslash S$ contains no real points.

Based on the above definition, for any locally compact Hausdorff space S,

$$S \subseteq \upsilon S \subseteq \beta S$$

Furthermore, if f is any function in $C(S)$, f can be viewed as belonging to $C(S, \omega\mathbb{R})$ and so extends to a continuous function $f^{\beta(\omega)}|_{\upsilon S} : \upsilon S \to \omega\mathbb{R}$. In fact,

$$f \in C(S) \qquad \Rightarrow \qquad f^{\beta(\omega)}|_{\upsilon S} \in C(\upsilon S)$$

We will more succinctly denote the function $f^{\beta(\omega)}|_{\upsilon S}$ by

$$f^{\upsilon} = f^{\beta(\omega)}|_{\upsilon S}$$

At one end of the spectrum we have, for example, the class of countably compact spaces. These spaces have been shown to have only bounded functions. So for a countably compact space, S, every function in $C(S)$ is real-valued on βS; therefore, for any f, every point in $\beta S\backslash S$ is a real point of f. In this case $\upsilon S = \beta S$.

In fact, if S is any pseudocompact space, then $\upsilon S = \beta S$.

At the other end of the spectrum, we have the class of all realcompact spaces S where $S = \upsilon S$. In this case, for every point $p \in \beta S\backslash S$, there is some function $f \in C(S)$ such that $f^{\beta(\omega)}(p) = \infty$. That is, only S contains real points. We eventually show that $\mathbb{R}$ (equipped with the usual topology) is such a space.

We will require the following concept for our first characterization of a "realcompact space". It refers to a slightly stronger condition on z-ultrafilters.

Definition 24.2 Let S be locally compact and Hausdorff. A z-ultrafilter, $\mathscr{U}$, in $Z[S]$ is called a

real z-ultrafilter

if and only if $\mathscr{U}$ is closed under countable intersections. That is, all countable subfamilies of a real z-ultrafilter, $\mathscr{U}$, have non-empty intersection.

There are z-ultrafilters in $Z[S]$ which are not "real".

Theorem 24.3 *Let S be a locally compact non-compact Hausdorff space.*

(a) *If $p \in \beta S\backslash \upsilon S$ then $p \in Z(h^\beta) \subseteq \beta S\backslash \upsilon S$ for some $h \in C^*(S)$.*
(b) *A real z-ultrafilter in $Z[S]$ can only converge to a real point in υS.*
(c) *The following three statements are equivalent.*

  (i) *The space S is realcompact. That is, $S = \upsilon S$.*
  (ii) *For any $p \in \beta S \setminus S$, there is a function, $h \in C^*(S)$, such that $p \in Z(h^\beta) \subseteq \beta S\backslash S$.*
  (iii) *Every* real z-ultrafilter *in $Z[S]$ converges to a point $p \in S$.*

Proof. We are given that S is a locally compact non-compact Hausdorff space.

(a) Let $p \in \beta S \backslash \upsilon S$.
We claim that $p \in Z(h^\beta) \subseteq \beta S \backslash \upsilon S$ for some $h \in C^*(S)$.
Since $p \in \beta S \backslash \upsilon S$, then p is *not* a real point of S, in the sense that, for some function, say $g \in C(S)$, $g^{\beta(\omega)}(p) = \infty$. Then g^υ is unbounded on υS. For, if g^υ was bounded on υS,

$$\begin{aligned} g^{\beta(\omega)}(p) &\in g^{\beta(\omega)}[\mathrm{cl}_{\beta S} S] = g^{\beta(\omega)}[\mathrm{cl}_{\beta S} \upsilon S] = \mathrm{cl}_{\omega \mathbb{R}} g^\upsilon[\upsilon S] \\ &= \mathrm{cl}_{\mathbb{R}} g^\upsilon[\upsilon S] \subseteq \mathbb{R} \end{aligned}$$

contradicting $g^{\beta(\omega)}(p) = \infty$.
Let $f^\upsilon = (g^\upsilon)^2 \vee 1$. Then $f^\upsilon : \upsilon S \to [1, \infty)$ is continuous, unbounded and *nowhere zero* on υS. Furthermore, f^υ extends to $f^{\beta(\omega)} : \beta S \to \omega \mathbb{R}$.
Define the function $h^\upsilon : \upsilon S \to \mathbb{R}$ as

$$h^\upsilon(x) = (1/f^\upsilon)(x) = 1/f^\upsilon(x)$$

Then $h^\upsilon \in C^*(\upsilon S)$ is bounded by 0 and 1.
Now, $h^\upsilon : \upsilon S \to \mathbb{R}$ extends continuously to $h^\beta : \beta S \to \mathbb{R}$ where

$$h^\beta[\beta S] = \mathrm{cl}_{\mathbb{R}} h^\upsilon[\upsilon S] \subseteq \mathrm{cl}_{\mathbb{R}}(0, 1] = [0, 1]$$

Since $p \in \beta S \backslash \upsilon S$, there is sequence (net) $\{x_i : i \in N\}$ in υS converging to p. Then

$$\{h^\beta(x_i) : i \in N\} \quad \longrightarrow \quad h^\beta\left(\lim_{x_i \to p} x_i\right) = h^\beta(p) \notin (0, 1]$$

So, $h^\beta(p) = 0$. That is, $p \in Z(h^\beta)$. Since h^β is not zero on υS, then

$$p \in Z(h^\beta) \subseteq \beta S \backslash \upsilon S$$

as required.

(b) We are given that every point in $\beta S \backslash \upsilon S$ belongs to a zero-set entirely contained in $\beta S \backslash \upsilon S$. Suppose p is a point in $\beta S \backslash \upsilon S$. We will show that no real z-ultrafilter converges to p; so a real z-ultrafilter can only converge to a point in υS.

In the particular case where S is pseudocompact $\upsilon S = \beta S$ so every real z-ultrafilter can only converge to a point in υS.
We now consider the case where $\beta S \setminus \upsilon S$ is non-empty. If $p \in \beta S \setminus \upsilon S$ then, by hypothesis, $p \in Z(h^{\beta}) \subseteq \beta S \setminus \upsilon S$, for some $h \in C^*(S)$. We know there exists a z-ultrafilter, $\mathscr{Z}$, in $Z[S]$ which converges to p. It suffices to prove that, if $\mathscr{Z}$ is a z-ultrafilter which converges to $p \notin \upsilon S$, then $\mathscr{Z}$ does not satisfy the "CIP", and so is not "real".
See that,

$$Z(h^{\beta}) = \cap\{h^{\beta\leftarrow}[-1/n, 1/n] : n \in \mathbb{N}\setminus\{0\}\}$$

where each $h^{\beta\leftarrow}[-1/n, 1/n]$ is a zero-set in $Z[\beta S]$.[2]
Therefore, $\{h^{\leftarrow}[-1/n, 1/n] : n \in \mathbb{N}\setminus\{0\}\}$, is a countable family of (non-empty) sets in the z-ultrafilter $\mathscr{Z}$ (which converges to p). Since $Z(h^{\beta}) \subseteq \beta S \setminus \upsilon S \subseteq \beta S \setminus S$, then

$$\cap\{h^{\leftarrow}[-1/n, 1/n] : n \in \mathbb{N}\setminus\{0\}\} = \varnothing$$

So a real z-ultrafilter in $Z[S]$ can only converge to a real point in υS. As required.

(c) Let S be a locally compact non-compact Hausdorff space.
(i $\Rightarrow$ ii) We are given that S is realcompact. Then $\upsilon S = S$.

Let $p \in \beta S \setminus \upsilon S = \beta S \setminus S$.
By part (a) $p \in Z(h^{\beta}) \subseteq \beta S \setminus S$ for some $h \in C^*(S)$.

(ii $\Rightarrow$ i) We are given that every point in $\beta S \setminus S$ belongs to a zero-set entirely contained in $\beta S \setminus S$. We are required to show that S is realcompact. Let $p \in \beta S \setminus S$. It suffices to show that p is not a real point.
To say that p is not a real point means that there is a function $h \in C(S)$ such that $h^{\beta(\omega)}(p) = \infty$. By hypothesis, $p \in Z(f^{\beta}) \subseteq \beta S \setminus S$ for some $f \in C(S)$. Then f is never 0 on S.
We can then define $h : S \to \omega\mathbb{R}$ as $h = 1/f$. Then h extends to $h^{\beta(\omega)} : \beta S \to \omega\mathbb{R}$. There exists a sequence (net), $\{x_i : i \in N\}$,

---

[2]See that $h^{\beta\leftarrow}[\,[-a, b]\,] = \cap\{h^{\beta\leftarrow}[\,(a - 1/n, b + 1/n)\,]\}$ is a closed G_{δ}-set in βS. In a normal space, closed G_{δ}'s are zero-sets (by Theorem 10.10). Then $h^{\beta\leftarrow}[-1/n, 1/n]$ is a zero-set, for each n.

in S which converges to $p \in Z(f^\beta)$. Then $\{f(x_i) : i \in N\}$ approaches zero as x_i approaches p. By continuity of $h^{\beta(\omega)}$,

$$\begin{aligned} h^{\beta(\omega)}(p) &= h^{\beta(\omega)}\left(\lim_{x_i \to p}\{x_i\}\right) \\ &= \lim_{x_i \to p}\{h(x_i)\} \\ &= \lim_{x_i \to p}\{1/f(x_i)\} \\ &= \infty \in \omega\mathbb{R} \end{aligned}$$

So, p is not a real point. Then, if every point in $\beta S \backslash S$ belongs to a zero-set entirely contained in $\beta S \backslash S$ the subset, $\beta S \backslash S$, cannot contain any real points. So, $\upsilon S = S$. By definition, S is realcompact. This proves (ii $\Rightarrow$ i).

(ii $\Rightarrow$ iii) We are given that every point in $\beta S \backslash S$ belongs to a zero-set entirely contained in $\beta S \backslash S$. We are required to show the limit of any real z-ultrafilter belongs to S.

Since (ii) $\Rightarrow$ (i), our hypothesis implies that no point in $\beta S \backslash S$ is real. So, $\upsilon S = S$. By part b) no real z-ultrafilter converges to a point in $\beta S \backslash \upsilon S = \beta S \backslash S$. Then a real z-ultrafilter can only converge to point in S, as required.

(iii $\Rightarrow$ ii) We are given that every *real* z-ultrafilter in $Z[S]$ converges to a point in S. Let $p \in \beta S \backslash S$. We are required to produce a zero-set $Z(h) \in Z[S]$ such that $p \in Z(h^\beta) \subseteq \beta S \backslash S$.

Suppose $\mathscr{Z}_p^\beta = \{Z(f^\beta) : f \in C^*(S), p \in Z(f^\beta)\}$ is the unique z-ultrafilter in $Z[\beta S]$ which converges to p. By hypothesis, the corresponding z-ultrafilter, $\mathscr{Z} = \{Z(f) : f \in C^*(S), p \in Z(f^\beta)\}$, in $Z[S]$ is not real. Then, $\mathscr{Z}$ contains a countable subfamily $\{Z(f_n) : f_n \in D, p \in Z(f_n^\beta\}$ such that

$$\cap\{Z(f_n) : f_n \in D, p \in Z(f_n^\beta\} = \varnothing$$

Let

$$W^\beta = \cap\{Z(f_n^\beta) : f_n \in D, p \in Z(f_n^\beta)\}$$

Then W^β is a zero-set, say $W^\beta(h^\beta)$, in $Z[\beta S]$ such that $p \in W^\beta(h^\beta) \subseteq \beta S \backslash S$, as required.

We add a few remarks in reference to non-compact spaces S.

(1) Compare statement (b)

> *"$S = \upsilon S$ if and only if, for any $p \in \beta S \setminus S = \beta S \setminus \upsilon S$, there is a function, $h \in C^*(S)$, such that $p \in Z(h^\beta) \subseteq \beta S \setminus S$".*

above to the theorem statement in 21.14 which states:

> *"S is pseudocompact if and only if no zero set Z in $Z[\beta S]$ is entirely contained in $\beta S \setminus S$".*

When viewed together, see what this implies.[3]

(2) The statement above also says that $S = \upsilon S$ if and only if every *real* z-ultrafilter in $Z[S]$ is fixed (that is, it converges in $S = \upsilon S$). By definition of *real* z-ultrafilter, this means that a space S is realcompact if and only if every z-ultrafilter in $Z[S]$ which satisfies the countable intersection property must converge in $S = \upsilon S$.

Further on in the book, we will be presenting still another characterization of realcompact spaces (in Corollary 25.6).

Example 1. Show that $\mathbb{R}$ equipped with the usual topology is a realcompact space.

Solution: Consider the function $f : \mathbb{R} \to \mathbb{R}$ defined as $f(x) = 1/(1+|x|)$. The function f is continuous, bounded and nowhere zero on $\mathbb{R}$, with range, $(0, 1]$. Recall from Theorem 10.10 that...

closed G_δ's in a normal space are zero-sets.

Then, since $\mathbb{R}$ is normal, for each $n \in \mathbb{N}\setminus\{0\}$, $f^{\beta\leftarrow}[0, 1/n]$ is a zero-set neighborhood of $\beta\mathbb{R}\setminus\mathbb{R}$. Then

$$\beta\mathbb{R}\setminus\mathbb{R} = \cap\{f^{\beta\leftarrow}[0, 1/n] : n > 0\}$$

Being a countable intersection of G_δ's in $\beta\mathbb{R}$, $\beta\mathbb{R}\setminus\mathbb{R}$ is a $\beta\mathbb{R}$ zero-set disjoint from $\mathbb{R}$. By Theorem 24.3, $\mathbb{R}$ is realcompact.[1]

---

[3]If S is pseudocompact then every function in $C(S)$ is bounded and so $\beta S\setminus\upsilon S$ is empty. Being empty it cannot contain a zero-set of $Z[\beta S]$. On the other hand, if a non-compact space, S, is realcompact, then $\beta S\setminus S = \beta S\setminus\upsilon S$ is non-empty. So, $C(S)$ contains unbounded functions. So a non-compact realcompact space cannot be pseudocompact. In this case, $\beta S\setminus S$ contains a zero-set.

[1]In Theorem 25.7 of the next chapter where better tools are available, we show that any product of $\mathbb{R}$'s is realcompact.

24.2 Extending Functions to υS

Let S and T be locally compact and Hausdorff. Then, being completely regular, S and T can be densely embedded into βS and βT, respectively.

Suppose $f : S \to T$ is a continuous function which maps S onto T. By Theorem 21.5, the function $f : S \to \beta T$ extends to $f^{\beta(\beta T)} : \beta S \to \beta T$. We know that υS is a subspace of βS, hence the restriction, $f^{\beta(\beta T)}|_{\upsilon S} : \upsilon S \to \beta T$, maps υS into βT.

In the following theorem, we will show a little bit more. We show that $f^{\upsilon}[\upsilon S] \subseteq \upsilon T$. That is, if $f : S \to T$ is any continuous function mapping S to T, (both locally compact and Hausdorff) f will extend to a continuous function f^{υ} which maps υS *into* υT.[4]

Theorem 24.4 *If S and T are locally compact Hausdorff spaces and $f : S \to T$ is a continuous function then f extends continuously to a unique continuous function*

$$f^{\upsilon} : \upsilon S \to \upsilon T$$

Proof. We are given that S and T are locally compact Hausdorff spaces and $f : S \to T$ is a continuous function. Since f extends to $f^{\beta(\beta T)} : \beta S \to \beta T$ and $S \subseteq \upsilon S \subseteq \beta S$, then f extends to $f^{\upsilon} : \upsilon S \to \beta T$ (where $f^{\upsilon} = f^{\beta(\beta T)}|_{\upsilon S}$). We are required to show that $f^{\upsilon}[\upsilon S] \subseteq \upsilon T$.

Suppose $h \in C(T)$. Then, if $\omega\mathbb{R} = \mathbb{R} \cup \{\infty\}$,

$$h : T \to \mathbb{R} \text{ will continuously extend to } h^{\omega} : \beta T \to \omega\mathbb{R}$$

Furthermore,

$$(h_{\circ} f) : S \to \mathbb{R} \text{ will continuously extend to } (h_{\circ} f)^{\omega} : \beta S \to \omega\mathbb{R}$$

For $x \in S$, $(h_{\circ} f)^{\omega}|_S(x) = h(f(x)) = (h^{\omega}{}_{\circ} f^{\beta})|_S(x)$. Since S is dense in βS, then

$$(h_{\circ} f)^{\omega} : \beta S \to \omega\mathbb{R}$$
$$(h^{\omega}{}_{\circ} f^{\beta}) : \beta S \to \omega\mathbb{R}$$

are equal functions on βS.

---

[4]Where $f^{\upsilon} = f^{\beta(\beta T)}|_{\upsilon S}$.

Let $r \in \upsilon S$. By definition, r is a real point of the function $h \circ f : S \to \omega\mathbb{R}$. This means that $(h \circ f)^{\omega}(r) = h^{\omega}[f^{\beta}(r)] \neq \infty$. Then $f^{\beta}(r)$ is a real point of h. That is,

$$f^{\upsilon}(r) \in \upsilon_h T$$

But h was an arbitrarily chosen function in $C(T)$. The point r was also arbitrarily chosen in υS. We can then conclude that

$$f^{\upsilon}[\upsilon S] \subseteq \cap\{\upsilon_h T : h \in C(T)\} = \upsilon T$$

So, $f^{\upsilon}[\upsilon S] \subseteq \upsilon T$. This is what we were required to prove.

Corollary 24.5 *Suppose S and T are locally compact Hausdorff spaces. If T is realcompact and $f : S \to T$ is a continuous function then f extends continuously to a unique continuous function $f^{\upsilon} : \upsilon S \to T$. In particular, if $f \in C(S)$, f extends continuously to $f^{\upsilon} \in C(\upsilon S)$.*

Proof. This is a special case of the theorem where $T = \upsilon T$. The second part follows from the fact that $\mathbb{R}$ is realcompact as shown in the above example.

From the above corollary, we see that, if S is completely regular, every continuous function $f : S \to \mathbb{R}$ in $C(S)$ extends to $f^{\upsilon} : \upsilon S \to \mathbb{R}$. We know that S is C^*-embedded in βS, but now we can confidently say that

S is C-embedded in υS.

The function

$$\phi : C(S) \to C(\upsilon S)$$

defined as $\phi(f) = f^{\upsilon}$ is then an isomorphism from the ring $C(S)$ onto $C(\upsilon S)$.

24.3 The *Hewitt–Nachbin Real Compactification* of a Space *S*

We have seen that $S \subseteq \upsilon S \subseteq \beta S$. We have also seen that, for every function $f \in C(S)$, the function, $f^{\beta(\omega)}|_{\upsilon S}$, continuously maps υS into $\mathbb{R}$. That is, every function $f \in C(S)$ extends to a function, f^{υ}, in $C(\upsilon S)$.

We also make the following observation. Since $\upsilon S \subseteq \beta S$, then $\text{cl}_{\beta S}\upsilon S \subseteq \beta S$. Given that $\text{cl}_{\beta S}S \subseteq \text{cl}_{\beta S}\upsilon S$, then

$$\beta(\upsilon S) = \beta S$$

Theorem 24.6 *Suppose S is locally compact and Hausdorff. Then υS is a realcompact subspace of βS.*

Proof. To show that υS is realcompact, it suffices to show that $\upsilon(\upsilon S) = \upsilon S$. To do this, it suffices to show that the set of all real points of υS is υS.

First note that $\upsilon S \subseteq \upsilon(\upsilon S)$. Then it suffices to show that $\upsilon(\upsilon S) \subseteq \upsilon S$. Suppose $p \in \upsilon(\upsilon S)$. Then p is a real point of υS. Let $f \in C(S)$. Then there is $g \in C(\upsilon S)$ such that $g = f^{\upsilon}$. Since p is a real point of υS, then $g^{\beta(\omega)}(p) \in \mathbb{R}$. Since g and f^{υ} agree on υS, $g^{\beta(\omega)} = f^{\beta(\omega)}$ on βS. So, $f^{\beta(\omega)}(p) \in \mathbb{R}$. So, p is a real point of S. That is, $p \in \upsilon S$. We have shown that $\upsilon(\upsilon S) \subseteq \upsilon S$.

So,

$$\upsilon(\upsilon S) = \upsilon S$$

Then we can view υS as a *realcompact space which densely contains the locally compact Hausdorff space S.*

Given a space S, the topological space, υS, is normally referred to as the

Hewitt–Nachbin realcompactification of S.[5]

[5]Named after the American mathematician Edwin Hewitt (1920–1999) and the Jewish-Brazilian mathematician Leopoldo Nachbin (1922–1993).

So, instead of referring to υS as being "the set of all real points of S", we will now speak of υS as being "the *realcompactification of* S". If a space, S, is realcompact then it is its own realcompactification.

If S is pseudocompact then $C(S) = C^*(S)$ and so every point of βS is a real point of S. Hence, if S is pseudocompact, βS is both the Stone–Čech compactification and the Hewitt–Nachbin realcompactification of S.

24.4 Another Characterization of Compactness and of Pseudocompactness

The statements which appear above are mostly in reference to non-compact spaces. The realcompact property provides us with another characterization of the compact property.

Theorem 24.7 *A Hausdorff topological space, S, is compact if and only if S is both realcompact and pseudocompact.*

Proof. ($\Rightarrow$) If S is compact then $S = \beta S$; hence, every real-valued function on S is bounded. Then, by definition, S is pseudocompact. Since $S \subseteq \upsilon S \subseteq \beta S$, then $S = \upsilon S$, so S is realcompact.

($\Leftarrow$) Suppose S is both pseudocompact and realcompact. We are required to show that S is compact.

Since S is pseudocompact, every $f \in C(S)$ is bounded. Then, for such an f,

$$f^{\beta(\omega)}[\beta S] \subseteq \text{cl}_{\mathbb{R}} f[S] \subseteq \mathbb{R}$$

Then

$$\beta S \subseteq \cap\{f^{\beta(\omega)\leftarrow}[\mathbb{R}] : f \in C(S)\} = \upsilon S$$

So, $\upsilon S = \beta S$. Since S is realcompact, then $S = \upsilon S$. So, $S = \beta S$. So, S is compact, as required.

Theorem 24.8 *Let S be a non-compact locally compact Hausdorff space. Then the following are equivalent.*

(a) *The space S is pseudocompact.*
(b) *The equality, $\beta S = \upsilon S$, holds true.*
(c) *Every z-ultrafilter is a real z-ultrafilter.*

Proof. We are given that S is a non-compact locally compact Hausdorff space.

(a ⇒ b) We are given that, S is pseudocompact. We are required to show that $\beta S = \upsilon S$.

Then, for any $f \in C(S)$, $f^{\beta(\omega)} : \beta S \to \omega\mathbb{R}$ maps βS onto $\mathrm{cl}_{\omega\mathbb{R}} f[S] \subseteq \mathrm{cl}_{\mathbb{R}} f[S] \subseteq \mathbb{R}$. Then, for any $f \in C(S)$, we have $\upsilon_f S = \beta S$. So,

$$\beta S = \cap\{\upsilon_f S : f \in C(S)\} = \upsilon S$$

as required.

(b ⇒ a) We are given that, $\beta S = \upsilon S$. We are required to prove that S is pseudocompact.

Suppose not. Then there is an unbounded function, f, in $C(S)$. Then $f : S \to \mathbb{R}$ extends to $f^{\beta(\omega)} : \beta S \to \omega\mathbb{R}$. If $f^{\beta(\omega)}[\beta S] \subseteq \mathbb{R}$ then $f^{\beta(\omega)}[\beta S]$ is non-compact since it is unbounded. So, there is an $x \in \beta S \backslash S$ such that $f^{\beta(\omega)}(x) = \infty$. In such a case, $\beta S \neq \upsilon S$. Contradiction. So, S must be pseudocompact.

(b ⇔ c) Given that S is a non-compact locally compact Hausdorff space.

See that $\beta S = \upsilon S$ if and only if every point in βS is a real point. A z-ultrafilter converges to a real point if and only if it is a real z-ultrafilter. So, $\beta S = \upsilon S$ if and only if every z-ultrafilter is a real z-ultrafilter.[6]

24.5 A Few Examples of Real Compact Spaces

In Theorem 16.3, we showed that in any Lindelöf space, S, every filter of *closed* sets in $\mathscr{P}(S)$ which satisfies the "countable intersection property" has a non-empty intersection (in S). In fact, this property characterizes Lindelöf spaces. Now every real z-ultrafilter in $Z[S]$ is a

[6]We can also conclude that "if and only if every z-ultrafilter satisfies the countable intersection property".

collection of closed subsets of S which satisfies the CIP. Then if S is lindelöf, every real z-ultrafilter in $Z[S]$ has a non-empty intersection. Then, by Theorem 24.3, Lindelöf spaces are realcompact. We state this as a formal result.

Theorem 24.9 *If S is a Lindelöf space then S is realcompact.*

Proof. Given in the paragraph above.

Example 2. By Definition 18.9, a σ-compact space is the countable union of compact sets. State conditions under which a σ-compact space is guaranteed to be realcompact.

Solution: In Theorem 18.10, it is shown that in a locally compact Hausdorff space the σ-compact property is equivalent to the Lindelöf property. So any locally compact Hausdorff σ-compact space is realcompact.

Example 3. Show that $\mathbb{N}$ is realcompact.

Solution: Any countable space is the union of singleton sets and hence is σ-compact. Since $\mathbb{N}$ is locally compact Hausdorff and countable, $\mathbb{N}$ is realcompact.

Example 4. Show that any *separable metric space* is realcompact.

Solution: By definition, a space that has a countable open base is Lindelöf. By Theorem 16.3, for metric spaces, the following are equivalent: (1) the second countable property, (2) the separable property, and (3) the Lindelöf property. By Theorem 24.9, a separable metric space is realcompact.

Example 5. Show that every subspace of $\mathbb{R}^n$ equipped with the usual topology is realcompact.

Solution: The Euclidean space $\mathbb{R}^n$ is a finite product of the metric separable space $\mathbb{R}$ and hence is both metrizable and separable (see Theorem 7.10 on products of separable spaces and Lemma 31.1 on countable products of metrizable spaces). Then it is second countable (see Theorem 5.11 where it is shown that metrizable spaces which are separable are also second countable). So every subspace of $\mathbb{R}^n$ is both second countable and metrizable (see the example after Definition 5.12 of hereditary property and Theorem 5.13 where we see that both metrizability and second countability are hereditary). In metrizable spaces the Lindelöf property is equivalent to the separable property and the second countable property (see Theorem 16.3). Then every subspace of $\mathbb{R}^n$ is Lindelöf. Since Lindelöf spaces are realcompact (by Theorem 24.9), then every subspace of $\mathbb{R}^n$ is realcompact.

24.6 Properties of Realcompact Spaces

We now examine when a particular space inherits the realcompact property from other spaces known to be realcompact. We show, in particular, that the realcompact property is closed under arbitrary intersections and arbitrary products. We also show that a closed subspace inherits the realcompact property from its superset.

Theorem 24.10 *Let $\{S_\alpha : \alpha \in I\}$ be a family of non-empty completely regular spaces each of which is realcompact.*

(a) *If S is a realcompact space which contains each S_α and $T = \cap\{S_\alpha : \alpha \in I\}$ is non-empty in S, then T is a realcompact subspace of S.*
(b) *The product space, $\prod_{\alpha \in I} S_\alpha$, is realcompact.*

Proof. We are given that $\{S_\alpha : \alpha \in I\}$ is a family of completely regular non-empty *realcompact* spaces.

(a) Suppose the S_α's are all subspaces of a realcompact space S and

$$T = \cap\{S_\alpha : \alpha \in I\}$$

is non-empty. We are required to show that T is realcompact.

Since each S_α is realcompact, $S_\alpha = \upsilon S_\alpha$. Since subspaces of completely regular spaces are completely regular, $T = \cap\{S_\alpha : \alpha \in I\}$ is a completely regular subspace. As shown in an example on page 512, $\text{cl}_{\beta S} T \equiv \beta T \subseteq \beta S_\alpha$. Since $\upsilon T \subseteq \beta T$, $\upsilon T \subseteq \beta S_\alpha$ for each α.

We claim that $\upsilon T \subseteq \upsilon S_\alpha$. For each α, let $i_\alpha : \upsilon T \to \beta S_\alpha$ denote the identity map embedding the subspace υT into βS_α. By Theorem 24.4, for each α, $i_\alpha|_T : T \to S$ extends continuously to the function

$$(i_\alpha|_T)^\upsilon : \upsilon T \to \upsilon S_\alpha = S_\alpha$$

Since T is dense in υT, $(i_\alpha|_T)^\upsilon = i_\alpha$ on υT, so $(i_\alpha|_T)^\upsilon$ embeds υT into υS_α, establishing the claim.

Since $(i_\alpha|_T)^\upsilon[\upsilon T] = \upsilon T \subseteq \upsilon S_\alpha = S_\alpha$ for all $\alpha \in I$, then $\upsilon T \subseteq \cap\{S_\alpha : \alpha \in I\} = T$.

Given that $T \subseteq \upsilon T$, we can only conclude that $T = \upsilon T$ and so T is realcompact.

(b) Suppose $S = \prod_{\alpha\in I} S_\alpha$ where each S_α is realcompact. For $\gamma \in I$, the projection map, $\pi_\gamma : S \to S_\gamma$ is continuous and hence extends continuously to

$$\pi_\gamma^\upsilon : \upsilon S \to \upsilon S_\gamma$$

where $S_\gamma = \upsilon S_\gamma$.

Let $\mathscr{G} = \{\pi_\alpha : \alpha \in I\}$ where $\pi_\alpha : S \to S_\alpha$. Then the evaluation map, $e_{\mathscr{G}} : S \to \prod_{\alpha\in I} S_\alpha$ extends to $e_{\mathscr{G}}{}^\upsilon : \upsilon S \to \prod_{\alpha\in I} S_\alpha = S$ a function which embeds υS into S.

Then $\upsilon S \subseteq S$. We conclude that $S = \upsilon S$ and so S is realcompact.

The subspace, T, of a realcompact space, S, need not, generally, be realcompact. It is if T is closed in S.

Theorem 24.11 *If F is a closed non-empty subspace of a realcompact space S then F is realcompact.*

Proof. We are given that F is closed in realcompact S. Then $S = \upsilon S$ hence,

$$\mathrm{cl}_{\upsilon S} F = \mathrm{cl}_S F = F$$

(since F is closed in S). We are required to show that $F = \upsilon F$.

Let $i : F \to S$ denote the inclusion map embedding F into S. Then (by Theorem 24.4) $i : F \to F \subseteq S$ extends continuously to $i^\upsilon : \upsilon F \to \upsilon S$ where

$$i^\upsilon[\upsilon F] = \upsilon F \subseteq \upsilon S = S$$

Now, F is dense in υF hence,

$$\upsilon F \subseteq \mathrm{cl}_{\upsilon S} F = \mathrm{cl}_S F = F$$

Then $\upsilon F \subseteq F$. So $F = \upsilon F$.

24.7 Example of a Non-Realcompact Space

The above few examples may lead the reader to suspect that non-realcompact spaces are uncommon. This motivates us to construct such a space. We do so in the following example.

Example 6. Recall that ω_1 denotes the first uncountable ordinal while ω_0 denotes the first countable infinite ordinal. Let $X = [0, \omega_1)$ and $Y = [0, \omega_0)$ be equipped with the ordinal topology, and $T = X \times Y$ be the corresponding product space.[7]

In the example on page 373, it is shown that $X = [0, \omega_1)$ is countably compact. By Theorem 15.9, since X is countably compact, for any $f \in C(X)$, $f[X]$ is compact in $\mathbb{R}$. This implies that $[0, \omega_1)$ is pseudocompact. But the space X is non-compact since the open cover $\{[0, \gamma) : \gamma \in S\}$ has no finite subcover. By Theorem 24.7, pseudocompact spaces which are realcompact must be compact. So,

$$[0, \omega_1) \textit{ is not realcompact.}$$

But the space X, when viewed as a homeomorphic image of the subspace, $[0, \omega_1) \times \{0\}$, of the product, T, is closed in T. By Theorem 24.11, closed subspaces of realcompact spaces are realcompact. So,

the product space, $[0, \omega_1) \times [0, \omega_0)$, cannot be realcompact.

Concepts Review

1. If f is a function in $C(S)$ define set, $\upsilon_f S$, of all the real points of f.
2. Define υS in terms of all the sets $\upsilon_f S$.

---

[7]In Theorem 21.18, it is shown that $\beta X = [0, \omega_1] = \omega X$.

3. What does it mean to say that the space, S, is realcompact?
4. Define a real z-ultrafilter in $Z[S]$.
5. Give a characterization of the realcompactness property in terms of zero-sets in $Z[\beta S]$.
6. Give a characterization of the realcompactness property in terms of z-ultrafilters of zero-sets in $Z[S]$.
7. If $f : S \to T$ is a continuous function mapping S into T what can be said about a particular continuous extension of f?
8. What does it mean to say that $f \in C(S)$ is C-embedded in υS?
9. What can we say about a space that is both realcompact and pseudocompact?
10. What can we say about Lindelöf spaces in the context discussed in this chapter?
11. What can we say about arbitrary intersections of realcompact space?
12. What can we say about arbitrary products of realcompact space?
13. What kind of subsets of a realcompact space are guaranteed to be realcompact?
14. Give an example of a realcompact space.
15. Give an example of a space which is not realcompact.

Chapter 25

Perfect Functions

Abstract
In this section, we introduce the notion of a "perfect function". After providing a formal definition we derive two of its most well-known characterizations. Our brief discussion of perfect functions will involve notions discussed in our study of singular functions and of realcompact spaces.

25.1 Introduction

Our study of realcompact spaces started with a brief discussion of a continuous function, $f : S \to \mathbb{R}$ mapping a completely regular space, S, into $\mathbb{R}$. So, f could be seen as a map, $f_+ : S \to \omega\mathbb{R}$, continuously mapping S into the one-point compactification denoted as $\omega\mathbb{R} = \mathbb{R} \cup \{\infty\}$, where $f_+ = f$ on S. By Theorem 21.5, $f_+ : S \to \omega\mathbb{R}$ extends continuously to a function, $f^{\beta(\omega)} : \beta S \to \omega\mathbb{R}$. Even if it was clear that $f^{\beta(\omega)}[S] \subseteq \mathbb{R}$, there was no reason to assume that, in the case where f is unbounded, $f^{\beta(\omega)}$ would map all of $\beta S \setminus S$ onto $\{\infty\}$. In fact, $f^{\beta(\omega)}$ could still map some points, x, in $\beta S \setminus S$ to $f^{\beta(\omega)}(x) \in \mathbb{R}$.

We defined $\upsilon_f S$ as

$$\upsilon_f S = \{x \in \beta S : f^{\beta(\omega)}(x) \in \mathbb{R}\} = f^{\beta(\omega)\leftarrow}[\mathbb{R}]$$

The points in $\upsilon_f S$ were referred to as the "real points of f". When $\upsilon_f S$ was not entirely contained in S this was viewed as a property of

S rather than as a property of f. The set

$$\upsilon S = \cap\{\upsilon_f S : f \in C(S)\}$$

was called the set of "all real points of S". When $\upsilon S \cap \beta S \backslash S = \varnothing$, we referred to S as satisfying the "realcompact" property and referred to S as being a realcompact space.

In this section, we will generalize the procedure used to construct $\upsilon_f S$ and study those functions $f : S \to T$ such that $f_+ : S \to \alpha T$, *for any compactification* αT of T (instead of only for $\omega\mathbb{R}$).

25.2 Perfect Function: Definition

Recall that, for a function, $f : S \to T$, the *fibers of* f refers to the elements of the set

$$\{f^{\leftarrow}(y) : y \in f[S]\} \subseteq \mathscr{P}(S)$$

Definition 25.1 Let S and T be completely regular spaces and $f : S \to T$ be a continuous function mapping S into T. We say that f is a *compact mapping* if its fibers are all compact (subsets of S). That is, $f^{\leftarrow}(y)$ is compact for each y in the range of f.

We say that the continuous function, $f : S \to T$, is a *perfect function* if f is both a closed function and a compact mapping.[1]

In our introductory paragraph, we considered a continuous function $f : S \to \mathbb{R}$ to construct the function $f^{\beta(\omega)} : \beta S \to \omega\mathbb{R}$. Here we will start with a function, $f : S \to T$, to construct, by using a similar procedure, a function $f^{\beta(\alpha)} : \beta S \to \alpha T$.

Suppose $f : S \to T$ continuously maps S into T (both completely regular). Let $i : T \to \alpha T$ be the identity map embedding T into αT. Let $f_+ : S \to \alpha T$ be defined as

$$f_+(x) = i(f(x))$$

[1]Some authors may not require that perfect functions be continuous.

Then, by Theorem 21.5, $f_+ : S \to \alpha T$ extends continuously to

$$f^{\beta(\alpha)} : \beta S \to \alpha T$$

With this in mind we will describe a perfect function, $f : S \to T$ by referring to a property of, $f^{\beta(\alpha)} : \beta S \to \alpha T$.

Theorem 25.2 *Let $f : S \to T$ be a continuous function mapping a non-compact completely regular space, S, into a non-compact completely regular space T. Let αT be any Hausdorff compactification of T. Then the following are equivalent:*

(a) *The function $f : S \to T$ is perfect.*
(b) *If $f : S \to T$, and $f^{\beta(\alpha)} : \beta S \to \alpha T$ is its continuous extension to βS, then*

$$f^{\beta(\alpha)}[\beta S \backslash S] \subseteq cl_{\alpha T} f[S] \backslash f[S]$$

(c) *If $f : S \to T$, and $f^{\beta(\alpha)} : \beta S \to \alpha T$ is its continuous extension to βS, then*

$$f^{\beta(\alpha)}[\beta S \backslash S] \cap f[S] = \varnothing$$

Proof. Let $f : S \to T$ be a continuous function mapping a non-compact completely regular space, S, *into* a non-compact completely regular space T.

(a $\Rightarrow$ b) We are given that $f : S \to T$ is perfect, αT is any Hausdorff compactification of T and $f^{\beta(\alpha)} : \beta S \to \alpha T$ is the continuous extension of f to βS where $f^{\beta(\alpha)}[\text{cl}_{\beta S} S] = \text{cl}_{\alpha T} f[S] \subseteq \alpha T$.

We are required to show that $f^{\beta(\alpha)}[\beta S \backslash S] \subseteq \text{cl}_{\alpha T} f[S] \backslash f[S]$.

Suppose $f^{\beta(\alpha)}[\beta S \backslash S] \nsubseteq \text{cl}_{\alpha T} f[S] \backslash f[S]$. Therefore $f^{\beta(\alpha)}[\beta S \backslash S] \cap f[S] \neq \varnothing$. There must then exist a $u \in \beta S \backslash S$ such that $f^{\beta(\alpha)}(u) \in f[S]$.

Let

$$K = f^{\leftarrow}(f^{\beta(\alpha)}(u))$$

Then, given that $f : S \to T$ is perfect, K is a compact subset of the completely regular subspace $S \cup \{u\}$ of βS where $f^{\beta(\alpha)}[S \cup \{u\}] = f[S]$.

Then there is an open neighborhood, U of K in $S \cup \{u\}$ such that

$$K = \text{cl}_{S\cup\{u\}} K \subseteq U \subseteq \text{cl}_{S\cup\{u\}} U \subset (S \cup \{u\})\backslash\{u\} \qquad (*)$$

So, $u \notin \text{cl}_{S\cup\{u\}} U$. Then $u \in \text{cl}_{S\cup\{u\}}(S\backslash U)$.[2]

We then have,

$$\begin{aligned} f^{\beta(\alpha)}(u) &\in f^{\beta(\alpha)}[\text{cl}_{S\cup\{u\}}[S\backslash U]] \\ &\subseteq \text{cl}_T f[S\backslash U] \quad \text{(By continuity and Theorem 6.4)} \\ &= f[S\backslash U] \quad \text{(Since } f \text{ is a closed function.)} \end{aligned}$$

Then, there is a point t in $S\backslash U$ such that $f^{\beta(\alpha)}(u) = f(t)$. Then $t \in f^{\leftarrow}(f^{\beta(\alpha)}(u)) = K$.

So, $t \in K \cap S\backslash U$, contradicting $K \subseteq U$ (see $(*)$).

Then $f^{\beta(\alpha)}[\beta S\backslash S] \subseteq \text{cl}_{\alpha T} f[S]\backslash f[S]$ as required.

We are done with (a $\Rightarrow$ b).

(b $\Rightarrow$ a) We are given that $f : S \to T$ is a continuous function which extends to $f^{\beta(\alpha)} : \beta S \to f^{\beta(\alpha)}[\beta S] = \text{cl}_{\alpha T} f[S] \subseteq \alpha T$ such that

$$f^{\beta(\alpha)}[\beta S\backslash S] \subseteq \text{cl}_{\alpha T} f[S]\backslash f[S] \qquad (\dagger)$$

We are required to show that f is both a compact and closed function.

Step 1: *The function $f : S \to T$ is compact*: Let $y \in f[S] \subseteq T$. Then $f^{\beta(\alpha)\leftarrow}(y)$ is closed in βS.

Case (a): $f^{\beta(\alpha)\leftarrow}(y) \subseteq S$. Then it is a closed subset of βS so it is a compact subset of S.

Case (b): $f^{\beta(\alpha)\leftarrow}(y) \cap \beta S\backslash S \notin \varnothing$. Then, by hypothesis,

$$f^{\beta(\alpha)}[[f^{\beta(\alpha)\leftarrow}(y)] \cap \beta S\backslash S] = \{y\} \subseteq \text{cl}_{\alpha T} f[S]\backslash f[S]$$

Since $y \in f[S]$ (by hypothesis), this case cannot occur. So, $f^{\beta(\alpha)\leftarrow}(y) \subseteq S$. Then $f^{\leftarrow}(y)$ is compact in S. So, $f : S \to T$ is a compact function. This concludes Step 1.

---

[2]Since $u \in \text{cl}_{S\cup\{u\}} S = \text{cl}_{S\cup\{u\}}(S\backslash U \cup U) = \text{cl}_{S\cup\{u\}}(S\backslash U) \cup \text{cl}_{S\cup\{u\}} U$ and $u \notin \text{cl}_{S\cup\{u\}} U$ then $u \in \text{cl}_{S\cup\{u\}}(S\backslash U)$.

Step 2: *The function $f : S \to T$ is a closed function*: Let F be a closed subset of S.

See that $\mathrm{cl}_{\beta S}F$ is compact in βS and so $f^{\beta(\alpha)}[\mathrm{cl}_{\beta S}F] \cap T$ is a closed subset of T.

We claim that $f^{\beta(\alpha)}[\mathrm{cl}_{\beta S}F] \cap T = f[F]$.

$$\begin{aligned}
f^{\beta(\alpha)}[\mathrm{cl}_{\beta S}F] \cap T &= f^{\beta(\alpha)}[(\mathrm{cl}_{\beta S}F \cap \beta S\backslash S) \cup F] \cap T \\
&= [f^{\beta(\alpha)}[(\mathrm{cl}_{\beta S}F \cap \beta S\backslash S)] \cup f^{\beta(\alpha)}[F]] \cap T \\
&= [f^{\beta(\alpha)}[(\mathrm{cl}_{\beta S}F \cap \beta S\backslash S)] \cap T] \cup [f^{\beta(\alpha)}[F] \cap T] \\
&\subseteq [[\mathrm{cl}_{\alpha T}f[S]\backslash f[S]] \cap T] \cup [f^{\beta(\alpha)}[F] \cap T] \quad \text{(By (†))}. \\
&= \varnothing \cup [f[F] \cap T] \text{ (By our hypothesis)} \\
&= f[F]
\end{aligned}$$

So, $f^{\beta(\alpha)}[\mathrm{cl}_{\beta S}F] \cap T \subseteq f[F]$. Since $f[F] \subseteq f^{\beta(\alpha)}[\mathrm{cl}_{\beta S}F] \cap T$, then

$$f^{\beta(\alpha)}[\mathrm{cl}_{\beta S}F] \cap T = f[F]$$

a closed subset of T, as claimed.

So, f is a closed map.

By definition, f is a perfect function. We are done with (b $\Rightarrow$ a).

(b $\Leftrightarrow$ c) Proof is left as an exercise.

25.3 Relating a Perfect Function, f to Its Singular Set, $S(f)$

Recall (from page 526) that, if we are given a continuous function $f : S \to T$ mapping S into a compact space T (not necessarily a singular map), we obtain a compactification of S,

$$\gamma S = S \cup S(f)$$

(not necessarily a singular compactification). We showed that,

$$f^{\beta(T)}[\beta S] = f[S] \cup S(f) = \text{cl}_T f[S] \quad (\dagger\dagger)$$
$$f^{\gamma}[\gamma S] = f[S] \cup S(f) = \text{cl}_T f[S] \quad (\dagger)$$
$$f^{\beta(T)}[\beta S \backslash S] = S(f) = f^{\gamma}[S(f)] \quad (\dagger\dagger\dagger)$$

where $f[S]$ and $S(f)$ may be disjoint, but not necessarily.

In the case where f is a singular map (that is, $f[S]$ is dense in $S(f)$) then

$$f[S] \cup S(f) = S(f)$$

So should f be singular and $f[S]$ be a compact subset of T,

$$S(f) = f[S]$$

In the case where f is not singular $f[S]$ will intersect $T\backslash S(f)$. It may even happen that $f[S] \cap S(f)$ is empty. Whether f is singular or not (and $f[S]$ is not compact), the set

$$S(f) \backslash f[S]$$

is the outgrowth of the compactification, $\text{cl}_T f[S]$, of the image, $f[S]$, of S under f.

We now show that the function $f : S \to T$ is perfect depending, specifically, on how its range, $f[S]$, intersects its singular set, $S(f)$.

Theorem 25.3 A characterization of perfect functions. *Let S and T be* non-compact *locally compact Hausdorff spaces and $f : S \to T$ be a continuous function mapping S* into *T. Then f is perfect if and only if*

$$f[S] \cap S(f) = \varnothing$$

Proof. We are given that $f : S \to T$ is a continuous function mapping S *into* T, where S and T are both locally compact, non-compact and Hausdorff. Let αT be a compactification of T so that $f[S]$ is a subset of αT.

Then the function $f : S \to \alpha T$, extends to

$$f^{\beta(\alpha)} : \beta S \to \text{cl}_{\alpha T} f[S] = f[S] \cup S(f) \subseteq \alpha T$$

where $f^{\beta(\alpha)}[\beta S \backslash S] = S(f)$ (see (†) above).

Let

$$\gamma S = S \cup S(f)$$

The function $f : S \to f[S] \cup S(f)$, extends to

$$f^{\gamma(\alpha)} : S \cup S(f) \to f[S] \cup S(f) = \mathrm{cl}_{\alpha T} f[S] = f^{\beta(\alpha)}[\beta S]$$

where $f^{\gamma(\alpha)}[S(f)] = S(f)$ appears in the role of the identity map on $S(f)$.

($\Rightarrow$) Suppose f is perfect. We are required to show that $f[S] \cap S(f) = \varnothing$.

Then, $f^{\beta(\alpha)}[\beta S \setminus S] \cap f[S] = \varnothing$ ((By Theorem 25.2, part (c)). Therefore,

$$\begin{aligned} f^{\beta(\alpha)}[\beta S \backslash S] &\subseteq f^{\beta(\alpha)}[\beta S] \backslash f[S] \quad \text{(By Theorem 25.2, part (c)} \\ &= [\, f[S] \cup S(f)\,] \backslash f[S] \quad \text{(See (††))} \end{aligned}$$

Since $f^{\beta(\alpha)}[\beta S \backslash S] = S(f)$ (see († † †), above), and $[\, f[S] \cup S(f)\,] \backslash f[S] = S(f) \backslash f[S]$, we obtain

$$S(f) \subseteq S(f) \backslash f[S] \text{ which implies } f[S] \cap S(f) = \varnothing$$

So, if f is a perfect function $f[S] \cap S(f) = \varnothing$. This is the desired result.

($\Leftarrow$) Conversely, suppose $f[S] \cap S(f) = \varnothing$.

$$\begin{aligned} f^{\beta(\alpha)}[\beta S \backslash S] &= S(f) \text{ (see († † †), above)} \\ &= S(f) \backslash f[S] \quad \text{(Since } f[S] \cap S(f) = \varnothing) \\ &\subseteq [\, S(f) \cup f[S]\,] \backslash f[S] \\ &= \mathrm{cl}_{\alpha T} f[S] \backslash f[S] \quad \text{(see (†), above)} \end{aligned}$$

This implies $f^{\beta(\alpha)}[\beta S \backslash S] \cap f[S] = \varnothing$.

By Theorem 25.2, (c) $\Rightarrow$ (a), f is perfect, as required.

So intersection of $S(f)$ with $f[S]$ helps recognize both perfect functions and singular functions. Since a function $f : S \to T$ is singular when $f[S] \subseteq S(f)$, a singular function can never be perfect. The above theorem confirms that a perfect function is one such that $f[S] \cap S(f)$ is empty. So perfect functions and singular functions are at opposite ends of a "spectrum of functions" each characterized by the contents of the set $f[S] \cap S(f)$. At one end we have $f[S] \subseteq S(f)$ for a singular function, f, and at the other end we have $f[S] \cap S(f) = \varnothing$ for a perfect function, f. For all other cases of $f[S] \cap S(f) \neq \varnothing$, f is perfect neither nor singular.

Example 1. Suppose S is non-compact completely regular and *connected*. Let $f : S \to \mathbb{R}$ be a continuous real-valued bounded function on S. Verify that f a perfect function if and only if $f[S]$ is not closed.

Solution: We are given that non-compact S is connected and $f : S \to \mathbb{R}$ belongs to $C^*(S)$. Since f is bounded, the function $f : S \to \mathbb{R}$ extends to $f^\beta : \beta S \to \text{cl}_{\mathbb{R}} f[S]$ where $f^\beta[\beta S] = \text{cl}_{\mathbb{R}} f[S]$. Since S is connected, then $f[S]$ is connected and $\text{cl}_{\mathbb{R}} f[S]$ is a closed bounded interval.

Recall that $f^\beta[\beta S] = \text{cl}_{\mathbb{R}} f[S] = f[S] \cup S(f)$ (see (††)). This implies

$$\text{cl}_{\mathbb{R}} f[S] = f[S] \cup S(f) = [a, b]$$

for some a and b.

Case $f[S] = [a, b]$: Then $f^\beta[\beta S] = f[S] = [a, b]$. So $S(f) \subseteq f[S]$. Since $S(f) \cap f[S] \neq \varnothing$, then f is not perfect.

Other cases, $f[S] = [a, b)$ or $f[S] = (a, b]$ or $f[S] = (a, b)$:

If $f[S] = [a, b)$ then $f^\beta[\beta S] = \text{cl}_{\mathbb{R}} f[S] = f[S] \cup S(f) = [a, b) \cup \{b\}$ where $\{b\} = S[f]$. Then $S(f) \cap f[S] = \varnothing$ and so f is perfect. Similarly, if $f[S] = (a, b]$ or $f[S] = (a, b)$ then $S(f) \cap f[S] = \varnothing$ and so f is perfect.

Example 2. Let $f = (\frac{2}{\pi})\text{arctan}$. We then obtain the function $f : \mathbb{R} \to [-1, 1]$. Show that f is perfect.

Solution: The function, f, maps the connected interval $(-\infty, \infty)$ one-to-one onto $T = (-1, 1)$. Since $f : \mathbb{R} \to [-1, 1]$ is real-valued and bounded then, by the statement the previous example, f is perfect.

25.4 The Realcompact Property and Perfect Evaluation Maps

We will review a few facts related to the realcompact property. Suppose S is completely regular and $\mathscr{F} = C(S)$. If $f \in \mathscr{F}$, we have seen that $f : S \to \mathbb{R}$ extends continuously to $f^{\beta(\omega)} : \beta S \to \omega\mathbb{R}$ (where $\omega\mathbb{R} = \mathbb{R} \cup \{\infty\}$).

Furthermore,

$$\upsilon_f S = \{x \in \beta S : f^{\beta(\omega)}(x) \in \mathbb{R}\} = f^{\beta(\omega)\leftarrow}[\mathbb{R}]$$

represents all the "real points in βS associated to f". If a point p in βS is a real point associated *to all* $f \in C(S)$ then we say that p is a "real point of βS". Then the set

$$\upsilon S = \cap\{f^{\beta(\omega)\leftarrow}[\mathbb{R}] : f \in \mathscr{F}\} = \cap\{\upsilon_f S : f \in \mathscr{F}\}$$

represents the set of all real points of βS.

So, if $x \in \beta S \backslash \upsilon S$ there is some $g \in C(S)$ such that $g^{\beta(\omega)}(x) = \infty$. If $\upsilon S = S$ then S is said to be "realcompact". That is, for every point, p, in $\beta S \backslash S$ there is some $f \in C(S)$ such that $f^{\beta(\omega)}(p) = \infty$.

In the special case where $f \in C^*(S)$, then

$$\begin{aligned} f^{\beta(\omega)}[\beta S] &= f^{\beta(\omega)}[\mathrm{cl}_{\beta S} S] \\ &= \mathrm{cl}_{\omega\mathbb{R}} f[S] \\ &= \mathrm{cl}_{\mathbb{R}} \mathbb{R} \\ &\subseteq \mathbb{R} \end{aligned}$$

So, in this case, $\upsilon_f S = f^{\beta(\omega)\leftarrow}[\mathbb{R}] = \beta S$. Then $\cap\{f^{\beta(\omega)\leftarrow}[\mathbb{R}] : f \in C^*(S)\} = \beta S$. When S is not pseudocompact[3] the functions, f, in $C^*(S)$ play no role in distinguishing the real points in $\beta S \backslash S$ from the "non-real points". Only unbounded functions, f, will extend to a function, $f^{\beta(\omega)}$, so that $f^{\beta(\omega)\leftarrow}(\infty)$ is non-empty.

So, if S is *not* pseudocompact and $\mathscr{F} = C(S)$ we will denote the non-empty set of all unbounded real-valued functions on S by

[3] A pseudocompact space is one for which $C(S) = C^*(S)$.

$\mathscr{H} = \mathscr{F} \backslash C^*(S)$. Then

$$\begin{aligned} \upsilon S &= \cap\{f^{\beta(\omega)\leftarrow}[\mathbb{R}] : f \in \mathscr{H}\} \\ &= \cap\{f^{\beta(\omega)\leftarrow}[\mathbb{R}] : f \in \mathscr{F}\} \end{aligned}$$

If $g \in \mathscr{H}$, then $g^{\beta(\omega)}[\mathrm{cl}_{\beta S} S] = \mathrm{cl}_{\omega\mathbb{R}} g[S]$. For this function g, the image, $g[S]$, of S is unbounded in $\mathbb{R}$ and so is not a compact subset of $\mathbb{R}$; this means that there must be some point q in $\beta S \backslash S$ such that $g^{\beta(\omega)}(q) = \infty$.

We will now consider these few notions from a slightly different point of view.

First, some notation.

If $\mathscr{F} = C(S)$, $\mathscr{H} = \mathscr{F} \backslash C^*(S)$, and $\omega\mathbb{R}_f$ is the codomain of $f : S \to \omega\mathbb{R}$.

$$\begin{aligned} H &= \textstyle\prod_{f\in\mathscr{F}} \omega\mathbb{R}_f \\ K &= \textstyle\prod_{f\in\mathscr{F}} \mathrm{cl}_{\omega\mathbb{R}} f[S] \\ T &= e_{\mathscr{F}}[S] \\ \alpha T &= \mathrm{cl}_K T \end{aligned}$$

By Tychonoff's theorem, H and K are compact with T dense in $\alpha T \subseteq K \subseteq H$.

For each $f : S \to \mathbb{R}$, we obtain $f_+ : S \to \omega\mathbb{R}$ which extends to $f^{\beta(\omega)} : \beta S \to \mathrm{cl}_{\omega\mathbb{R}} f[S]$.

The evaluation function, $e_{\mathscr{F}} : S \to T$, maps S onto T, where

$$e_{\mathscr{F}}(x) = \langle\!\langle f(x) \rangle\!\rangle_{f\in\mathscr{F}} \in T$$

It extends to

$$e_{\mathscr{F}}^{\beta(\alpha)} : \beta S \to \alpha T$$

where $e_{\mathscr{F}}^{\beta(\alpha)}$ maps βS onto $\mathrm{cl}_K T = \alpha T$.

In the following theorem, we establish a fundamental relationship between the realcompact property and properties of $e_{\mathscr{F}}$.

Theorem 25.4 *Let S be both a completely regular and non-pseudocompact space and $\mathscr{F} = C(S)$. Let $\mathbb{R}_f$ denote the codomain* *of* $f : S \to \mathbb{R}_f$. *If* S *is realcompact then the evaluation map,*

$$e_{\mathscr{F}} : S \to \prod_{f \in \mathscr{F}} \mathbb{R}_f$$

is a perfect function *which maps* S *onto* $e_{\mathscr{F}}[S]$ *in* $\prod_{f \in \mathscr{F}} \mathbb{R}_f$.

Proof. We are given that S is both a completely regular and non-pseudocompact space and $\mathscr{F} = C(S)$.

Let $\mathscr{H} = \mathscr{F} \backslash C^*(S)$ and $K_{\mathscr{H}} = \prod_{f \in \mathscr{H}} \text{cl}_{\omega\mathbb{R}} f[S]$.

Since S is non-pseudocompact, then $\mathscr{H}$ is non-empty.

Let function, $e_{\mathscr{H}} : S \to e_{\mathscr{H}}[S]$, be the evaluation map generated by $\mathscr{H}$ defined as

$$e_{\mathscr{H}}(x) = \langle f(x) \rangle_{f \in \mathscr{H}} \in e_{\mathscr{H}}[S]$$

The function $e_{\mathscr{H}}$ extends to $e_{\mathscr{H}}^{\beta(\alpha)} : \beta S \to e_{\mathscr{H}}^{\beta(\alpha)}[\beta S]$ defined as

$$e_{\mathscr{H}}^{\beta(\alpha)}(x) = \langle f^{\beta(\omega)}(x) \rangle_{f \in \mathscr{H}}$$

Suppose S is realcompact.

We are required to show that $e_{\mathscr{F}} : S \to e_{\mathscr{F}}[S] \subseteq \prod_{f \in \mathscr{F}} \mathbb{R}$ is a perfect function. To do this it will suffice to show that $e_{\mathscr{F}}^{\beta(\alpha)}[\beta S \backslash S] = (\text{cl}_K T) \backslash T$ and invoke Theorem 25.2, part (b).

Since S is realcompact, $\beta S \backslash S = \beta S \backslash \upsilon S$, so, for every $x \in \beta S \backslash S$, there is some $f \in \mathscr{H}$ such that $f^{\beta(\omega)}(x) = \infty$.

Then

$$\cap \{ f^{\beta(\omega)\leftarrow}[\mathbb{R}] : f \in \mathscr{H} \} = S$$

Then, for every $x \in \beta S \backslash S$, $e_{\mathscr{H}}^{\beta(\alpha)}(x) = \langle f^{\beta(\omega)}(x) \rangle_{f \in \mathscr{H}}$ has an entry which is ∞.

So,

$$e_{\mathscr{H}}^{\beta(\alpha)}[\beta S \backslash S] \cap \prod_{f \in \mathscr{H}} \mathbb{R}_f = \varnothing$$

hence,

$$e_{\mathscr{F}}^{\beta(\alpha)}[\beta S \backslash S] \cap \prod_{f \in \mathscr{F}} \mathbb{R}_f = \varnothing \quad (*)$$

Let $T = e_{\mathscr{F}}[S]$. Since $T \subseteq \prod_{f\in\mathscr{F}} \mathbb{R}_f$, then (*) implies

$$e_{\mathscr{F}}^{\beta(\alpha)}[\beta S\backslash S] \cap T = \varnothing \qquad (**)$$

For $K = \prod_{f\in\mathscr{F}} \text{cl}_{\omega\mathbb{R}} f[S]$.

$$\begin{aligned} e_{\mathscr{F}}^{\beta(\alpha)}[\beta S] &= e_{\mathscr{F}}^{\beta(\alpha)}[\beta S\backslash S] \ \cup\ e_{\mathscr{F}}[S] \\ &= e_{\mathscr{F}}^{\beta(\alpha)}[\beta S\backslash S] \ \cup\ T \quad \text{(By (**), disjoint union.)} \\ &= \text{cl}_K T \end{aligned}$$

Since $[e_{\mathscr{F}}^{\beta(\alpha)}[\beta S\backslash S] \ \cup\ T]\backslash T = e_{\mathscr{F}}^{\beta(\alpha)}[\beta S\backslash S]$, then

$$e_{\mathscr{F}}^{\beta(\alpha)}[\beta S\backslash S] = (\text{cl}_K T)\backslash T$$

So, if S is realcompact, then, by 25.2, $e_{\mathscr{F}} : S \to \prod_{f\in\mathscr{F}} f[S]$ is perfect.

We now show that the converse of the above theorem statement holds true. That is, if $e_{\mathscr{F}} : S \to \prod_{f\in\mathscr{F}} f[S]$ is perfect then S must be realcompact.

Theorem 25.5 *Let S be both a completely regular non-pseudocompact space and $\mathscr{F} = C(S)$. If $e_{\mathscr{F}} : S \to \prod_{f\in\mathscr{F}} \mathbb{R}_f$ is a perfect function then S is realcompact.*

Proof. Let $R = \prod_{f\in\mathscr{F}} \mathbb{R}_f$, $K = \prod_{f\in\mathscr{F}} \omega\mathbb{R}_f$ and $T = e_{\mathscr{F}}[S]$. Then $T \subseteq R \subseteq K$.

Given: $e_{\mathscr{F}} : S \to T$ is perfect.

We are required to show that S is realcompact. Let $\alpha T = \text{cl}_K T$. Then $e_{\mathscr{F}}$ extends to

$$e^{\beta(\alpha)}{}_{\mathscr{F}} : \beta S \to \alpha T$$

To prove S is realcompact it suffices to show that, if $y \in \beta S \setminus S$, then

$$e^{\beta(\alpha)}{}_{\mathscr{F}}(y) = \langle\!\langle f^{\beta(\omega)}(y) \rangle\!\rangle_{f \in \mathscr{F}}$$

in $\prod_{f \in \mathscr{F}} \omega\mathbb{R}_f$ has at least one entry, $g^{\beta(\omega)}(y) = \infty$.

Since $e_{\mathscr{F}} : S \to \prod_{f \in \mathscr{F}}, \mathbb{R}_f$ is perfect it is, by definition, a closed function. Then $e_{\mathscr{F}}[S] = T$ is a closed subset of R. This means that $\mathrm{cl}_R T = T$. So, $\mathrm{cl}_K T \cap R = \mathrm{cl}_R T = T$. Then

$$\begin{aligned} (\mathrm{cl}_K T) \setminus T \cap R &\subseteq (\mathrm{cl}_K T \cap R) \setminus T \\ &= T \setminus T \\ &= \varnothing \end{aligned}$$

implies

$$(\mathrm{cl}_K T) \setminus T \cap R = \varnothing \qquad (\dagger)$$

Since $e_{\mathscr{F}}$ is perfect, by Theorem 25.2,

$$\begin{aligned} e^{\beta(\alpha)}{}_{\mathscr{F}}[\beta S \setminus S] &\subseteq (\mathrm{cl}_K T) \setminus T \\ &= [(\mathrm{cl}_K T) \setminus T] \setminus R \quad \text{(By } (\dagger).) \\ &\subseteq K \setminus R \end{aligned}$$

So,

$$e^{\beta(\alpha)}{}_{\mathscr{F}}[\beta S \setminus S] \subseteq \left(\textstyle\prod_{f \in \mathscr{F}} \omega\mathbb{R} \right) \Big\backslash \left(\textstyle\prod_{f \in \mathscr{F}} \mathbb{R} \right)$$

If $y \in \beta S \setminus S$ then $e^{\beta(\alpha)}{}_{\mathscr{F}}(y)$ has at least one entry which is ∞. So, y is not a real point of S. So the only real points of βS belong to S. So, S is realcompact, as required.

Corollary 25.6 *Let S be both a completely regular non-pseudocompact space and $\mathscr{F} = C(S)$. The space S is realcompact if and only if the evaluation map, $e_{\mathscr{F}} : S \to \prod_{f \in \mathscr{F}} f[S]$, is a perfect function which homeomorphically maps S into $\prod_{f \in \mathscr{F}} \mathbb{R}_f$.*

Proof. This simply summarizes the two previous theorems into a single "if and only if" statement.

From the previous results we restate the general statement:

Realcompact spaces are precisely those spaces, S, such that

$$e_{C(S)} : S \to \prod_{f\in C(S)} \mathbb{R}_f$$

is a perfect function.

We will use the techniques illustrated in the proof of the above characterization of realcompactness to prove the following statement about arbitrary products of $\mathbb{R}$'s.

Theorem 25.7 *Let $R = \prod_{j\in J} \mathbb{R}_j$ denote a product of $\mathbb{R}$'s. The product space, R, is realcompact.*

Proof. Let $R = \prod_{j\in J} \mathbb{R}_j$ and $K = \prod_{j\in J} \omega\mathbb{R}_j$.

We are required to show that R is realcompact.

To do this, it suffices to show that $e_{C(R)} : \prod_{j\in J} \mathbb{R}_j \to \prod_{f\in C(R)} \mathbb{R}_f$ is perfect and invoke Corollary 25.6.

Consider the identity function $i : R \to R$ expressed as $i : R \to K$, where i is seen as an inclusion map which embeds $R = \prod_{j\in J} \mathbb{R}$ into $K = \prod_{j\in J} \omega\mathbb{R}$.

Then i extends to $i^{\beta(K)} : \beta R \to \text{cl}_K i[R] \subseteq K$. See that

$$\text{cl}_K i[R] = i^{\beta(K)}[\beta R] = i^{\beta(K)}[\beta R \backslash R] \,\cup\, i[R]$$

Now i pulls back points to points and so is a compact function on R. If F is closed in R then $i[F]$ is closed in R and so it is a closed function. So, i is a perfect function.

Also, see that

$$\begin{aligned} i^{\beta(K)}[\beta R\backslash R] &\subseteq \text{cl}_K i[R]\backslash i[R] \quad \text{(Since } i : R \to R \text{ is a perfect map)} \\ &\subseteq K\backslash R \\ &= \prod_{j\in J}\omega\mathbb{R}\backslash\prod_{j\in J}\mathbb{R} \end{aligned}$$

Let $\pi_i : \prod_{j\in J} \mathbb{R}_j \to \mathbb{R}_i$ denote the function defined as $\pi_i(\langle x_j\rangle_{j\in J}) = x_i \in \mathbb{R}_i$. Essentially π_i is the real-valued continuous function which projects R into $\mathbb{R}$. So,

$$\mathscr{P} = \{\pi_j : j \in J\} \subseteq C(R)$$

The family $\mathscr{P}$ generates the evaluation map $e_{\mathscr{P}} : \prod_{j\in J} \mathbb{R}_j \to \prod_{j\in J} \mathbb{R}_j$ defined as

$$\begin{aligned} e_{\mathscr{P}}(\langle x_j\rangle_{j\in J}) &= \langle\, \pi_j(\langle x_i\rangle_{i\in J})\,\rangle_{j\in J} \\ &= \langle x_j\rangle_{j\in J} \\ &\in \textstyle\prod_{j\in J}\mathbb{R}_j \end{aligned}$$

Then $e_{\mathscr{P}}(\langle x_j\rangle_{j\in J}) = \langle x_j\rangle_{j\in J}$. The function, $e_{\mathscr{P}}$, is simply another way of representing the identity function mapping

R onto R.

Since $i^{\beta(K)}[\beta R\backslash R] \subseteq \prod_{j\in J}\omega\mathbb{R}_j\backslash\prod_{j\in J}\mathbb{R}_j$, if $y \in \beta R\backslash R$, then

$$e_{\mathscr{P}}{}^{\beta(K)}(y) = \langle \pi_j^{\beta(\omega)}(y)\rangle_{\pi_j\in\mathscr{P}} \in \textstyle\prod_{\pi_j\in\mathscr{P}}\omega\mathbb{R}_{\pi_j}\backslash\prod_{j\in J}\mathbb{R}_{\pi_j}$$

has at least one entry which is ∞.

Since $\mathscr{P} \subseteq C(R)$, for $y \in \beta R\backslash R$,

$$e_{C(R)}{}^{\beta(K)}(y) = \langle f^{\beta(\omega)}(y)\rangle_{f\in C(R)} \in \textstyle\prod_{f\in C(R)}\omega\mathbb{R}_f\backslash\prod_{j\in J}\mathbb{R}_j$$

has at least one entry which is ∞.

So, y is not a real point of R. Then,

$$e_{C(R)}{}^{\beta(K)}[\beta R\backslash R] \subseteq \textstyle\prod_{f\in C(R)}\omega\mathbb{R}_f\backslash\prod_{f\in C(R)}\mathbb{R}_f$$

This means that $e_{C(R)} : R \to \prod_{f\in C(R)}\mathbb{R}_f$ is a perfect function. By Corollary 25.6, R is realcompact.

We have shown that R is realcompact.

Concept Review

1. Define a compact function.
2. Define a perfect function.
3. Provide two characterizations of a perfect function.
4. Provide a characterization of the realcompact property involving a perfect function.
5. Provide a characterization of a perfect function $f : S \to K$ in terms of a singular set.
6. If $e_{C(S)} : S \to \prod_{f \in C(S)} f[S]$ is an evaluation map what can we say about S if $e_{C(S)}$ is perfect?
7. If $S = \prod_{i \in J} S_i$ what property must S satisfy if we want the projection map to be perfect?

Chapter 26

Perfect and Freudenthal Compactifications

Abstract
In this chapter, we define both the Freudenthal compactification and the family of "perfect compactifications". In spite of its name, a perfect compactification has very little to do with "perfect functions". A perfect function is one which is closed and whose fibers are compact, while a perfect compactification, αS, is one for which the fibers of $\pi_{\beta \to \alpha}$ are connected. We produce an algebraic characterization of those compactifications we call "perfect". We also present examples of both the Freudenthal compactification of a space and a perfect compactification. This chapter can be studied immediately after the chapter introducing compactifications without loss of continuity.

26.1 The Freudenthal Compactification

We introduce a particular compactification for locally compact Hausdorff spaces called the *Freudenthal compactification.*[1] Just like the Stone–Čech compactification we will see that a space can only have

[1]Hans Freudenthal (1905–1990) was a Jewish, German-born Dutch mathematician at the University of Amsterdam. He was suspended from his duties by the Nazis during the war. After the war he was reinstated to his former position.

one Freudenthal compactification. We begin by reminding ourselves that,

> *a space is said to be zero-dimensional if and only if every point has an open neighborhood base of clopen sets.*

While,

> *a space is totally disconnected if and only if every connected component is a singleton set.*

Both of these are related to either connectedness or disconnectedness properties. By Theorem 20.24 "a locally compact Hausdorff space is zero-dimensional if and only if it is totally disconnected".

Definition 26.1 Given a locally compact Hausdorff compactification of the space, S, the *Freudenthal compactification*, ϕS, of S is the maximal compactification whose outgrowth, $\phi S \backslash S$, is totally disconnected. Equivalently ϕS is the maximal compactification whose outgrowth, $\phi S \backslash S$, is zero-dimensional. By "maximal" we mean that, if $\mathscr{Z} = \{\alpha_i S : i \in I\}$ is the family of all compactifications of S with zero-dimensional remainder, $\phi S \in \mathscr{Z}$ and $\alpha_i S \preceq \phi S$, for all i.

There is another way to visualize the Freudenthal compactification, ϕS, of a space S. In the following theorem, we show that ϕS is the unique compactification of S obtained by collapsing the connected components of $\beta S \backslash S$ to points. This theorem guarantees that a maximal compactification with zero-dimensional outgrowth exists in the partially ordered set of all compactifications. Furthermore, there is only one such compactification (up to equivalence).

Theorem 26.2 *Let S be a locally compact Hausdorff space and let αS be a compactification of S. Let $\pi_{\beta \to \alpha} : \beta S \to \alpha S$ be the corresponding continuous map which fixes the points of S. The expression, ϕS, denotes the Freudenthal compactification of S.*

(a) *If $\alpha S \backslash S$ is zero-dimensional then, for any $p \in \alpha S \backslash S$, $\pi_{\beta \to \alpha}^{\leftarrow}(p)$ is a union of connected components of $\beta S \backslash S$.*

(b) *Suppose that, for each $p \in \alpha S \backslash S$, $\pi^{\leftarrow}_{\beta\to\alpha}(p)$ is a connected component of $\beta S \backslash S$. Then αS is equivalent to the Freudenthal compactification, ϕS, of S.*

Proof. Let S be a locally compact Hausdorff space and αS be a compactification of S.

(a) We are given that the outgrowth, $\alpha S \backslash S$, is zero-dimensional. We are required to show that, for each point $p \in \alpha S \backslash S$, $\pi^{\leftarrow}_{\beta\to\alpha}(p)$ is a union of connected components of $\beta S \backslash S$.

Let $p \in \alpha S \backslash S$ and C be a connected component in $\beta S \backslash S$ which intersects $\pi^{\leftarrow}_{\beta\to\alpha}(p)$. It suffices to show that $C \subseteq \pi^{\leftarrow}_{\beta\to\alpha}(p)$.

Suppose $C \nsubseteq \pi^{\leftarrow}_{\beta\to\alpha}(p)$. Then there exists $t \in C \backslash \pi^{\leftarrow}_{\beta\to\alpha}(p)$. Then there exists a point, q, in $\alpha S \backslash S$ distinct from p such that $t \in \pi^{\leftarrow}_{\beta\to\alpha}(q)$. Since $\alpha S \backslash S$ is zero-dimensional, there is some clopen neighborhood U of q in $\alpha S \backslash S$ which does not contain p. Then

$$t \in \pi^{\leftarrow}_{\beta\to\alpha}(q) \subseteq \pi^{\leftarrow}_{\beta\to\alpha}[U]$$

where $\pi^{\leftarrow}_{\beta\to\alpha}[U] \cap \pi^{\leftarrow}_{\beta\to\alpha}(p) = \varnothing$. Since $\pi_{\beta\to\alpha}$ is continuous, $\pi^{\leftarrow}_{\beta\to\alpha}[U]$ is a clopen set in $\beta S \backslash S$ which only intersects a portion of C. This contradicts the fact that C is a connected component of $\beta S \backslash S$. So, $C \backslash \pi^{\leftarrow}_{\beta\to\alpha}(p)$ must be empty. This means that $C \subseteq \pi^{\leftarrow}_{\beta\to\alpha}(p)$. Since every point of $\pi^{\leftarrow}_{\beta\to\alpha}(p)$ belongs to a component of $\beta S \backslash S$ which is entirely contained in $\pi^{\leftarrow}_{\beta\to\alpha}(p)$, $\pi^{\leftarrow}_{\beta\to\alpha}(p)$ must be a union of connected components. We are done for part (a).

(b) We are given that, for each $p \in \alpha S \backslash S$, $\pi^{\leftarrow}_{\beta\to\alpha}(p)$ is a connected component of $\beta S \backslash S$. We are required to show that αS and ϕS are equivalent compactifications.

A function which collapses the connected components of a compact Hausdorff set, $\beta S \backslash S$, to points in $\alpha S \backslash S$ produces a totally disconnected set (equivalently a zero-dimensional space) (see the example on page 454). Then $\alpha S \backslash S$ is zero-dimensional.

We have shown in part (a) that $\pi_{\beta\to\alpha}$ pulls back points to unions of components. Then, for each $p \in \alpha S \backslash S$, $\pi^{\leftarrow}_{\beta\to\alpha}(p)$ is the "union" of a single component. Then $\alpha S \backslash S$, cannot be partitioned any further without slicing into a connected component. Then αS must be the maximal compactification with zero-dimensional outgrowth. So, $\alpha S \equiv \phi S$. As required.

26.2 Perfect Compactification: Definition

We have seen above that, if αS has zero-dimensional outgrowth, $\alpha S\backslash S$, then

> *for all* $p \in \alpha S\backslash S$, $\pi_{\beta\to\alpha}{}^{\leftarrow}(p)$ *is the union of connected components of* $\beta S\backslash S$.

From this fact we concluded that there is only one maximal compactification, ϕS, with zero-dimensional outgrowth, namely the compactification, αS, for which the fibers of $\pi_{\beta\to\alpha}|_{\beta S\backslash S}$ is a single connected component of $\beta S\backslash S$. It is called the Freudenthal compactification.

Those compactifications, αS, with zero-dimensional outgrowth must then be "less than or equal to" the Freudenthal compactification, ϕS.

There may be compactifications, αS, whose outgrowth, $\alpha S \setminus S$, satisfies the property,

$$\textit{for all } p \in \alpha S\backslash S,\ \pi_{\beta\to\alpha}{}^{\leftarrow}(p) \textit{ is connected}$$

but is not zero-dimensional. This could occur only if $\pi_{\beta\to\alpha}{}^{\leftarrow}(p)$ is a connected subset of at most one component.

Such compactifications are referred to as "perfect" compactifications. We formally define this.

Definition 26.3 Let γS be a Hausdorff compactification of S and $\pi_{\beta\to\gamma} : \beta S \to \gamma S$ be the continuous map which projects βS onto γS. The compactification, γS, is said to be a *perfect compactification* provided $\pi_{\beta\to\gamma}{}^{\leftarrow}(p)$ is connected for every $p \in \gamma S\backslash S$.

The Freudenthal compactification, ϕS, is an example of a perfect compactification. In the family of all perfect compactifications of S, it is in fact the smallest one. So perfect compactifications form a family of compactifications which are gathered "above" ϕS, while the ones with zero-dimensional outgrowth form a family which are gathered "below" ϕS.

If for a locally compact Hausdorff space, S, $\beta S\backslash S$ is zero-dimensional then βS is both the Freudenthal compactification and the only perfect compactification of S.

The main result in this chapter is an algebraic characterization of a perfect compactification.

26.3 Algebraic Subrings of $C^*(S)$

Our algebraic characterization will involve properties of a particular type of subring of $C^*(S)$, called an "algebraic subring". Setting up the stage for the proof of the characterization first requires a proper understanding the concept of an algebraic subring and some of its properties. Characterizing the topological property of "perfect compactification", αS, with an algebraic property of a subring of $C_\alpha(S)$, requires a bit of mathematical prowess. We begin with the following definition. Three lemmas will follow.

Definition 26.4 Let S be a locally compact Hausdorff space and $C^*(S)$ be the ring of all bounded continuous real-valued functions on S.

(a) We will say that $\mathscr{A}$ is an *algebraic subring* of $C^*(S)$ if $\mathscr{A}$ is a subring which contains all of the constant functions and all functions, f, such that $f^2 = f \cdot f \in \mathscr{A}$.
(b) Given a subring, $\mathscr{A}$, of $C^*(S)$, a subset T of S on which every function in $\mathscr{A}$ is constant is called a *stationary set of the subring,* $\mathscr{A}$. If T is a stationary set and, for any $p \in S\backslash T$, there is a function $f \in \mathscr{A}$ which is not constant on $T \cup \{p\}$ then we say that T is a *maximal stationary set of* $\mathscr{A}$. Equivalently, if T is a stationary set of $\mathscr{A}$ containing y and $T = \{x \in S : f(x) = f(y), \text{ for all } f \in \mathscr{A}\}$ then T is a *maximal stationary set of* $\mathscr{A}$.

By definition, if T is a maximal stationary set of the subring, $\mathscr{A}$, of $C(S)$ then every function in $\mathscr{A}$ is constant on T. Then, for $f \in \mathscr{A}$, $f[T]$ is some singleton set, say $\{t_f\}$, and so $T \subseteq Z(f - t_f)$. Then $T \subseteq \cap\{Z(f - t_f) : f \in \mathscr{A}\}$. Since T is maximal with respect to this property, then

$$T = \cap\{Z(f - t_f) : f \in \mathscr{A}\}$$

a closed subset of S. So, a maximal stationary set, T, of $\mathscr{A}$ is an intersection of zero sets generated by the functions in $\mathscr{A}$.

The following lemma offers yet another way of describing the maximal stationary set of subring.

Lemma 26.5 *Given a space, S, let $\mathscr{G}$ be a subring of $C(S)$. For any point $p \in e_{\mathscr{G}}[S]$, $e_{\mathscr{G}}{}^{\leftarrow}(p)$ is a maximal stationary set of $\mathscr{G}$.*

Proof. Let $\mathscr{G}$ is a subring of $C(S)$ and let p be a point in $e_{\mathscr{G}}[S] \subseteq \prod_{f\in\mathscr{G}}\mathbb{R}_f$. Then, for a given $x \in S$, p is of the form

$$\begin{aligned} p &= e_{\mathscr{G}}(x) \\ &= \langle f(x)\rangle_{f\in\mathscr{G}} \\ &= \langle p_f\rangle_{f\in\mathscr{G}} \\ &\in \textstyle\prod_{f\in\mathscr{G}}\mathbb{R}_f \end{aligned}$$

That is, $p = \langle p_f\rangle_{f\in\mathscr{G}}$ where the element p_f is the image of the given x under $f : S \to \mathbb{R}$.

If $p = \langle p_f\rangle_{f\in\mathscr{G}}$ is in the range, $e_{\mathscr{G}}[S]$, of S then

$$e_{\mathscr{G}}{}^{\leftarrow}(p) = \{x \in S : e_{\mathscr{G}}(x) = p\}$$

contains at least one point, say, a.

Let

$$[a] = \{x \in S : e_{\mathscr{G}}(x) = e_{\mathscr{G}}(a) = p\}$$

Then, $e_{\mathscr{G}}[[a]] = \{p\}$, so $e_{\mathscr{G}}$ is constant on $[a]$. In fact, for each $f \in \mathscr{G}$,

$$f(x) = f(a) = p_f$$

for all $x \in [a] \subseteq S$. So, f is constant on $[a]$ for each $f \in \mathscr{G}$. Then $[a]$ is a stationary set of $\mathscr{G}$. If $x \notin [a]$ then $e_{\mathscr{G}}(x) \neq e_{\mathscr{G}}(a)$. So, $e_{\mathscr{G}}{}^{\leftarrow}(p) = [a]$ is a maximal stationary set of $\mathscr{G}$. As required.

Lemma 26.6 *Let T be a connected subset of a topological space, S, and $\mathscr{G}$ be the subset of $C^*(S)$ which contains only those functions which are constant on T. Then $\mathscr{G}$ is an algebraic subring which is closed with respect to the uniform norm topology.*

Proof. We are given that $\mathscr{G}$ contains precisely all those functions which are constant on the connected subspace T.

Clearly, $\mathscr{G}$ is closed under multiplication and addition and so $\mathscr{G}$ is a subring of $C^*(S)$.

Claim: That $\mathscr{G}$ is algebraic. Let g be a function on S such that $g^2[T] = k$. It suffices to show that $g \in \mathscr{G}$. For every $x \in T$, $g(x)$ is $\sqrt{k}$ or $-\sqrt{k}$. By hypothesis, T is connected and so the range $g[T]$ cannot contain two elements. So, g must be constant on T and so belongs to $\mathscr{G}$. We deduce that $\mathscr{G}$ is algebraic, as claimed.

Claim: That $\mathscr{G}$ is closed with respect to the uniform norm. Suppose h is a limit point of $\mathscr{G}$. Then there is a sequence $\{t_i : i \in \mathbb{N}\}$ in $\mathscr{G}$ which converges to h with respect to the uniform norm topology. Since each t_i is constant on T, h is constant on T and so belongs to $\mathscr{G}$. So, $\mathscr{G}$ contains it's limit points and so is closed with respect to the uniform norm.

In the next lemma, we establish a relationship between a topological property of a maximal stationary set of a subring, $\mathscr{A}$, and an algebraic property of this subring, for the case where S is compact.

Lemma 26.7 *Let S be a compact Hausdorff space and $\mathscr{A}$ be an algebraic subring of $C^*(S)$. If T is a maximal stationary set of $\mathscr{A}$ then T is a connected subset of S.*

Proof. We are given that S is compact and Hausdorff and T is a maximal stationary set of the algebraic subring, $\mathscr{A}$, of $C^*(S)$.

Suppose $x \in S$ and

$$e_{\mathscr{A}}(x) = t = \langle\!\langle t_f \rangle\!\rangle_{f \in \mathscr{A}}$$

Let $T = e_{\mathscr{A}}^{\leftarrow}(\langle\!\langle t_f \rangle\!\rangle_{f \in \mathscr{A}})$. Then $f(x) \in f[T] = \{t_f\}$, for each $f \in \mathscr{A}$. This means that

$$T = \cap\{f^{\leftarrow}\{t_f\} : f \in \mathscr{A}\}$$

is the maximal stationary set of $\mathscr{A}$.

We are required to show that T is connected.

Suppose T is *not* connected. That is, suppose

$$T = F_1 \cup F_2$$

where F_1 and F_2 are disjoint non-empty closed subsets of T. Since S is compact, both F_1 and F_2 are compact. We will show that this leads to a contradiction.

There exist disjoint open subsets, U_1 and U_2, of S, such that

$$F_1 \subseteq U_1$$
$$F_2 \subseteq U_2$$

So,

$$\begin{aligned} T &= \cap\{f^{\leftarrow}(t_f) : f \in \mathscr{A}\} \\ &= \cap\{f^{\leftarrow}(t_f, t_f + \varepsilon) : f \in \mathscr{A}, \varepsilon \in \mathbb{R}^+\} \qquad (\dagger) \\ &= \cap\{f^{\leftarrow}[t_f, t_f + \varepsilon] : f \in \mathscr{A}, \varepsilon \in \mathbb{R}^+\} \\ &\subseteq U_1 \cup U_2 \end{aligned}$$

Step 1: We claim that there exists at least one function, $h : S \to \mathbb{R}$, in $\mathscr{A}$ and one ε such that

$$e_{\mathscr{A}}{}^{\leftarrow}(t) = T \subseteq h^{\leftarrow}[t_h,\ t_h + \varepsilon] \subseteq U_1 \cup U_2$$

Proof of claim in step 1: Suppose no such function $h : S \to \mathbb{R}$ exists.

That is, for all f and ε, there is an $x \in S \setminus (U_1 \cup U_2)$ such that $x \in f(x) \in [t_f,\ t_f + \varepsilon]$.

Then, for all f, ε,

$$f^{\leftarrow}[t_f,\ t_f + \varepsilon] \cap [S \setminus (U_1 \cup U_2)] \neq \varnothing \text{ and } T \subseteq f^{\leftarrow}[t_f,\ t_f + \varepsilon]$$

This means that the collection of closed subsets of S

$$\mathscr{C} = \{f^{\leftarrow}[t_f,\ t_f + \varepsilon] \ : f \in \mathscr{A}, \varepsilon \in \mathbb{R}^+\} \cup \{S \setminus (U_1 \cup U_2)\}$$

satisfies the finite intersection property.

Since S is compact, there exists

$$q \in \cap\{f^{\leftarrow}[t_f,\ t_f + \varepsilon] \cap [S \setminus (U_1 \cup U_2)] : f \in \mathscr{A}, \varepsilon \in \mathbb{R}^+\}$$

This contradicts $T = \cap\{f^{\leftarrow}[t_f + \varepsilon] : f \in \mathscr{A}, \varepsilon \in \mathbb{R}^+\}$.

Then there must exist at least one continuous function, $h : S \to \mathbb{R}$, in $\mathscr{A}$ and one ε such that

$$T = F_1 \cup F_2 \subseteq h^{\leftarrow}[t_h, \; t_h + \varepsilon] \subseteq U_1 \cup U_2 \; (*)$$

as claimed.

Step 2: Suppose $h \in \mathscr{A}$ and $\varepsilon > 0$ such that $h^{\leftarrow}[t_h, \; t_h + \varepsilon] \subseteq U_1 \cup U_2$. Let

$$W_1 = h^{\leftarrow}[t_h, \; t_h + \varepsilon] \cap U_1 \quad \text{(Which contains } F_1)$$
$$W_2 = h^{\leftarrow}[t_h, \; t_h + \varepsilon] \cap U_2 \quad \text{(Which contains } F_2)$$

Since U_1 and U_2 are disjoint, then $W_1 \cap U_1$ and $W_2 \cap U_2$ are disjoint neighborhoods of F_1 and F_2, respectively. So,

$$T = F_1 \cup F_2 \subseteq W_1 \cup W_2$$

See that

$$\begin{aligned} h^{\leftarrow}(t_h + \varepsilon) \subseteq h^{\leftarrow}[t_h, t_h + \varepsilon] & \\ = h^{\leftarrow}[t_h, t_h + \varepsilon] \cap (U_1 \cup U_2) & \qquad \text{(by (*))} \\ = [h^{\leftarrow}[t_h, t_h + \varepsilon] \cap U_1] \; \cup \; [h^{\leftarrow}[t_h, t_h + \varepsilon] \cap U_2] & \\ = W_1 \cup W_2 & \end{aligned}$$

so there exists $y \in W_1 \cup W_2$ such that

$$h(y) = t_h + \varepsilon \; (**)$$

Step 3: For $h \in \mathscr{A}$ and $\varepsilon > 0$ given above, we define a real-valued function $r : S \to \mathbb{R}$ on S, as follows

$$r(x) = (t_h + \varepsilon) - h(x) \quad \text{if } x \in W_1$$
$$r(x) = h(x) - (t_h + \varepsilon) \quad \text{if } x \in S \backslash W_1$$

Claim: We claim that r is a continuous real-valued function on S.

— By (∗∗), there exists $y \in W_1 \cup W_2$ such that $r(y) = 0$.
— Moreover, see that $t_h + \varepsilon$ is maximal in $[t_h,\ t_h + \varepsilon]$, therefore,

$$r[W_1] \subseteq [0, \infty)$$
$$r[S \backslash W_1] \subseteq (-\infty, 0]$$

where $r(y) = (t_h + \varepsilon) - (t_h + \varepsilon) = 0$.
— See that the function r is well-defined and continuous on each of the pieces, W_1 and $S \backslash W_1$.
— Also note that $r(y) = 0$ is a boundary point of each of the subsets $r[W_1]$ and $r[S \backslash W_1]$.
— So, r is continuous on S. As claimed.

Claim A: We now claim that the function r (defined above) belongs to the subring $\mathscr{A}$.

Proof of claim A: Given that $h \in \mathscr{A}$ and $\mathscr{A}$ is a subring, then $(t_h + \varepsilon) - h(x) \in \mathscr{A}$. Note that $r^2 = ((t_h + \varepsilon) - h(x))^2$. Since $\mathscr{A}$ is a subring, then $r^2 \in \mathscr{A}$. Since $\mathscr{A}$ is algebraic and $r^2 \in \mathscr{A}$, then $r \in \mathscr{A}$. As claimed.

Claim B: We claim that r is not constant on T.

Proof of claim B: Recall from (†) that

$$T = F_1 \cup F_2 = \cap\{f^{\leftarrow}(t_f, t_f + \varepsilon) : f \in \mathscr{A}, \varepsilon \in \mathbb{R}^+\} \subseteq h^{\leftarrow}(t_h, t_h + \varepsilon)$$

See that $r(x) = 0$ if and only if $h(x) = t_h + \varepsilon$; that is, if $x \in h^{\leftarrow}(t_h + \varepsilon)$.

Since $F_1 \cup F_2 \subseteq h^{\leftarrow}(t_h, t_h + \varepsilon)$, r is not zero on $F_1 \cup F_2$. We know that

$$r[F_1] \subseteq r[W_1] \subseteq [0, \infty)$$
$$r[F_2] \subseteq r[W_2] \subseteq r[S \backslash W_1] \subseteq [0, \infty)$$

then $r > 0$ on F_1 and $r < 0$ on F_2. So r is not constant on T, as claimed.

Since T is maximal stationary set of $\mathscr{A}$ and $r \in \mathscr{A}$, then r must be constant on T. So, we have a contradiction.

We conclude that T must be connected. We are done.

26.4 An Algebraic Characterization of "Perfect Compactification"

Having established that (for a compact space S),

> *all maximal stationary sets of an algebraic subring are connected,*

we are ready to roll out the proof of the following characterization of perfect compactifications.

Theorem 26.8 *Let γS be a compactification of S. Then γS is a perfect compactification if and only if $C_\gamma(S)$ is an algebraic subring of $C^*(S)$.*

Proof. ($\Leftarrow$) We are given that $\mathscr{G} = C_\gamma(S)$ where $C_\gamma(S)$ is an algebraic subring of $C^*(S)$.

We are required to show that γS is a perfect compactification. To do this it suffices to show that, for any $p \in \gamma S$, $\pi_{\beta\to\gamma}{}^{\leftarrow}(p)$ is connected.

Given that $C_\gamma(S)$ is algebraic, it is easily verified, that $C_\gamma(S)^\beta$ must also be an algebraic subring of $C(\beta S)$. By Lemma 21.12, $\pi_{\beta\to\gamma} = e_{\mathscr{G}}{}^{\gamma\leftarrow} \circ e_{\mathscr{G}}{}^{\beta}$. So, for $p \in \gamma S \backslash S$,

$$\begin{aligned}\pi_{\beta\to\gamma}{}^{\leftarrow}(p) &= [e_{\mathscr{G}}{}^{\gamma\leftarrow} \circ e_{\mathscr{G}}{}^{\beta}]^{\leftarrow}(p)\\ &= e_{\mathscr{G}}{}^{\beta\leftarrow}(e_{\mathscr{G}}{}^{\gamma}(p))\\ &= \text{a maximal stationary set of } \mathscr{G}^\beta \text{ in } \beta S\end{aligned}$$

Since $\mathscr{G}^\beta$ is an algebraic subring of $C(\beta S)$, then by Lemma 26.7, the maximal stationary set, $\pi_{\beta\to\gamma}{}^{\leftarrow}(p)$, of $\mathscr{G}^\beta$ is connected. So, for any $p \in \gamma S$, $\pi_{\beta\to\gamma}{}^{\leftarrow}(p)$ is connected. By definition, γS is a perfect compactification. As required for the direction, $\Leftarrow$.

($\Rightarrow$) Suppose γS is a perfect compactification. By definition of perfect compactification, $\pi_{\beta\to\gamma}{}^{\leftarrow}(p)$ is connected for every $p \in \gamma S \backslash S$.

We are required to show that $C_\gamma(S)$ is an algebraic subring of $C^*(S)$.

Let $\mathscr{G} = C_\gamma(S)$. To show that $\mathscr{G}$ is an algebraic subring of $C^*(S)$ it suffices to show that $\mathscr{G}^\beta$ is an algebraic subring of $C(\beta S)$.

For $p \in \gamma S$, let $\mathscr{F}_p$ denote the set of all functions in $C(\beta S)$ which are constant on the connected set, $\pi_{\beta\to\gamma}{}^{\leftarrow}(p)$. By Lemma 26.6, $\mathscr{F}_p$ is an algebraic subring of $C(\beta S)$.

Since $\pi_{\beta\to\gamma} = e_{\mathscr{G}}{}^{\gamma\leftarrow} \circ e_{\mathscr{G}}{}^{\beta}$ (by Lemma 21.12), then $\pi_{\beta\to\gamma}{}^{\leftarrow}(p) = e_{\mathscr{G}}^{\beta\leftarrow}(e_{\mathscr{G}}{}^{\gamma}(p))$. So,

$$e_{\mathscr{G}}^{\beta}[\pi_{\beta\to\gamma}{}^{\leftarrow}(p)] = e_{\mathscr{G}}{}^{\gamma}(p)$$

It follows that $\mathscr{G}^{\beta} \subseteq \mathscr{F}_p$. It is easily verified that the intersection of a family of algebraic subrings is an algebraic subring. So, $\cap\{\mathscr{F}_p : p \in \gamma S\}$ is an algebraic subring of $C(\beta S)$. So,

$$\mathscr{G}^{\beta} \subseteq \cap\{\mathscr{F}_p : p \in \gamma S\}$$

If we can show that $\cap\{\mathscr{F}_p : p \in \gamma S\} \subseteq \mathscr{G}^{\beta}$, then we are done.

Let $f \in \cap\{\mathscr{F}_p : p \in \gamma S\}$. If $g \in C(\gamma S)$ is such that $g(\pi_{\beta\to\gamma}(x)) = f(x)$ for all $x \in \beta S$, then $g|_S = f|_S$; so $f|_S \in C_{\gamma}(S) = \mathscr{G}$. Hence, $\cap\{\mathscr{F}_p : p \in \gamma S\} \subseteq \mathscr{G}^{\beta}$.

We conclude that $\mathscr{G}^{\beta} = C_{\gamma}(S)^{\beta} = \cap\{\mathscr{F}_p : p \in \gamma S\}$, an algebraic subring of $C(\beta S)$.

So, $C_{\gamma}(S)$ is an algebraic subring of $C^*(S)$. We are done with the direction $\Rightarrow$.

Example 1. Recall that the Freudenthal compactification is the maximal compactification in the family of all compactifications whose outgrowth is zero-dimensional (equivalently, totally disconnected). The Freudenthal compactification can also be seen as the minimal "perfect compactification" of a space. In the following proposition, we provide an example of a space whose Freudenthal compactification is βS. In such a case, S has only one perfect compactification, βS.

The proposition involves the space, $\mathbb{N}$. Suppose A is a subset of $\mathbb{N}$. We confirm that $\mathrm{cl}_{\beta\mathbb{N}}A$ and $\mathrm{cl}_{\beta\mathbb{N}}\mathbb{N}\backslash A$ are clopen in $\beta\mathbb{N}$. To see this simply note that A is a zero-set. Then there is a continuous function, $f : \mathbb{N} \to \{0,1\}$, such that $A = Z(f-1)$. Then f extends continuously to $f^{\beta} : \mathrm{cl}_{\beta\mathbb{N}}A \to \{0,1\}$ where

$$f^{\beta}[\mathrm{cl}_{\beta\mathbb{N}}A] = \mathrm{cl}_{\mathbb{R}}f[A] = \{1\}$$

Then

$$\text{cl}_{\beta\mathbb{N}} A = Z(f^\beta - 1) = \beta\mathbb{N} \backslash Z(f^\beta - 0)$$

is clopen in $\beta\mathbb{N}$.

Proposition 26.9 *The Stone–Čech compactification, $\beta\mathbb{N}$, of $\mathbb{N}$ is both the Freudenthal compactification and the minimal perfect compactification.*

Proof. To show that $\beta\mathbb{N}$ is the Freudenthal compactification of $\mathbb{N}$, it suffices to show that $\beta\mathbb{N}\backslash\mathbb{N}$ is zero-dimensional, equivalently, totally disconnected. We have shown that subspaces of totally disconnected spaces are themselves totally disconnected. So, it will suffice to show that $\beta\mathbb{N}$ is totally disconnected.

Let C be a connected component of $\beta\mathbb{N}$. Suppose p and q are points in C. We claim that $p = q$. If $p \neq q$ then they are limits of two distinct z-ultrafilters $\mathscr{Z}_p$ and $\mathscr{Z}_q$ in $Z[\mathbb{N}]$. Then $\mathscr{Z}_p$ contains a zero-set $Z(f - 1)$ which does not belong to $\mathscr{Z}_q$. Then p belongs to the clopen zero-set $Z(f^\beta - 1)$ which does not contain q. Since C is connected, then p must equal q. We conclude that $\beta\mathbb{N}$ is totally disconnected. Then $\beta\mathbb{N}\backslash\mathbb{N}$ is zero-dimensional.

It is the maximal compactification with zero-dimensional outgrowth and so is Freudenthal. It also is the smallest compactification which is perfect.

Since the Freudenthal compactification is the smallest of perfect compactifications, no compactification $\alpha\mathbb{N} \prec \beta\mathbb{N}$ is perfect. So, by the above characterization, no proper subring of $C^*(\mathbb{N})$ is algebraic.

Concept Review

1. Define the Freudenthal compactification of a space.
2. How can βS be used to construct the Freudenthal compactification ϕS?
3. What does it mean to say that a compact space is a perfect compactification of S?

4. State what we mean when we that $\mathscr{A}$ is an algebraic subring of $C^*(S)$.
5. If $\mathscr{A}$ is an algebraic subring of $C^*(S)$ what is a stationary set of $\mathscr{A}$? What does it mean to say that a subset, T, is a maximal stationary set of $\mathscr{A}$?
6. State an algebraic characterization (involving $C^*(S)$) of a perfect compactification.

Chapter 27

Spaces Whose Elements Are Sequences

Abstract
In this chapter, we investigate a family of topological spaces whose elements are sequences. These spaces of sequences are vector spaces equipped with a topology which is determined by norms. The norms we will study are ℓ_p-norms and the ℓ_∞-norm. We will verify when the ℓ_p-spaces are separable, second countable, Lindelöf and sequentially compact. The ℓ_∞-spaces of sequences will be shown to be complete. Along the way we will prove two important inequalities: Hölder's inequality and Minkowski's inequality for the ℓ_p-spaces.

27.1 Definitions: ℓ_p-Spaces

To begin we remind ourselves that a sequence of points can be viewed as a function; one whose domain is $\mathbb{N}$. If S is a normed topological space, the functions studied here will be S-valued. We will represent the set of all S-valued sequences as

$$F(\mathbb{N}, S)$$

These are more commonly expressed as

$$F(\mathbb{N}, S) = \{\{s_0, s_1, s_2, s_3 \ldots\} : s_i \in S\} = \{\langle s_i \rangle_{i \in \mathbb{N}} : s_i \in S\}$$

the family of all sequences whose elements belong to the space S. The topology defined on $F(\mathbb{N}, S)$ depends on the topology of S.

Given a normed vector space $(S, +, \cdot, \| \;\|)$ and the set $F(\mathbb{N}, S)$, an "ℓ_p-space" and the "ℓ_p-norm" are defined as follows.

Definition 27.1 Let S be a normed vector space $(S, +, \cdot, \| \;\|)$ and $F(\mathbb{N}, S)$ denote the set of all sequences, $f = \{s_0, s_1, s_2, s_3, \ldots\} = \langle s_i \rangle_{i \in \mathbb{N}}$, whose elements, s_i, belong to S. Let p be any real number greater than or equal to 1. The ℓ_p-norm of $f = \langle s_i \rangle_{i \in \mathbb{N}}$ is defined as

$$\|f\|_p = \|\langle s_i \rangle_{i \in \mathbb{N}}\|_p = \left[\sum_{i=0}^{\infty} \|s_i\|^p \right]^{1/p}$$

Let $(\mathscr{S}_b, \| \;\|_p)$ denote the family of those sequences, $\langle s_i \rangle_{i \in \mathbb{N}}$, in $F(\mathbb{N}, S)$ such that

$$\|\langle s_i \rangle_{i \in \mathbb{N}}\|_p < \infty$$

The expression, "ℓ_p-space", refers to the (bounded) normed space, $(\mathscr{S}_b, \| \;\|_p)$.

The use of the lower-case cursive form, ℓ, in the definition, is standard. When we refer to "ℓ_p-norm" we are speaking of an infinite family of norms on $F(\mathbb{N}, S)$. These are distinguished from each other by the chosen value of p in $[1, \infty)$. They are viewed as belonging to the same family of norms because the same formula to used to compute them.

The space S may be the space, $\mathbb{R}$, or the complex numbers, $\mathbb{C}$, equipped with the usual norm. We should be careful not to confuse the norm, $\| \;\|$, on S with the chosen ℓ_p-norm, $\| \;\|_p$, on $F(\mathbb{N}, S)$.

The definition of the ℓ_p-norm can be modified so as to include the set of n-tuples, $F(\{1, 2, , 3 \ldots, n\}, S) = S^n$ where

$$\|\langle s_i \rangle_{i=1\ldots n}\|_p = \left[\sum_{i=1}^{n} \|s_i\|^p \right]^{1/p}$$

The only significant difference is the symbol, n, which replaces, "∞". Everything else flows identically. The case where, $p = 2$, turns out to be the Euclidean norm on S^n.

When applied to $\mathbb{R}$, the ℓ_p-norm is a generalization of the absolute value since

$$\|x\|_2 = [x^2]^{1/2} = |x|$$

On $\mathbb{R}^2$,

$$\|(a,b)\|_2 = \sqrt{a^2+b^2}$$

The ℓ_p-norms on $\mathbb{R}^n$ flow naturally to ℓ_p-norms on $\mathbb{R}^\infty$ (viewed as infinite sequences of real numbers).

27.1.1 *The ℓ_p-norm is a valid norm for all $p \geq 1$*

Having defined the ℓ_p-norm on the family of S-valued sequences, $F(\mathbb{N}, S)$, in this way we have yet to verify that it satisfies the three norm axioms N1, N2 and N3 for all values of p. Proving that any ℓ_p-norm satisfies the first two norm axioms is straightforward. Let $f = \langle s_i \rangle_{i\in\mathbb{N}}$.

N1: $\|f\|_p < \infty$ (by definition) and $\|f\|_p = [\sum_{i=0}^{\infty} \|s_i\|^p\,]^{1/p} = 0 \Leftrightarrow$ $s_i = 0$ for all $i \in \mathbb{N}$.

N2:

$$\begin{aligned}
\|\alpha f\|_p &= \left[\sum_{i=0}^{\infty} \|\alpha s_i\|^p\right]^{1/p} \\
&= \left[\sum_{i=0}^{\infty} |\alpha|^p \|s_i\|^p\right]^{1/p} \\
&= \left[\lim_{n\to\infty} \sum_{i=0}^{n} |\alpha|^p \|s_i\|^p\right]^{1/p} \\
&= \left[\lim_{n\to\infty} |\alpha|^p \sum_{i=0}^{n} \|s_i\|^p\right]^{1/p} \\
&= [|\alpha|^p]^{1/p} \left[\lim_{n\to\infty} \sum_{i=0}^{n} \|s_i\|^p\right]^{1/p} \\
&= |\alpha| \left[\sum_{i=0}^{\infty} \|s_i\|^p\right]^{1/p} = |\alpha| \|f\|_p
\end{aligned}$$

N3: The triangle inequality for ℓ_p-norms, also called the Minkowski inequality, is more difficult to prove. It is in fact quite tricky. It's proof is the main subject of the next subsection on inequalities.

27.2 Two Fundamental Inequalities: Hölder's and Minkowski's Inequalities

To following inequalities are important. The proof of Minkowski's inequality (the triangle inequality for $\| \|_p$) must be preceded by "Hölder's inequality".

The following lemma helps set up the proof of Hölder's inequality.

Lemma 27.2 *Lemma for Hölder's Inequality For any* $t \in (0,1)$ *and any* $A > 0$, $B > 0$

$$A^t B^{1-t} \leq tA + (1-t)B$$

Proof. We are given that $A > 0$, $B > 0$. For values of $t \in (0,1)$, we define the functions $f(t)$ and $g(t)$ as

$$f(t) = tA + (1-t)B = (A-B)t + B$$
$$g(t) = A^t B^{1-t}$$

To prove the lemma it suffices to show that $g(t) \leq f(t)$ on $(0,1)$.

If we let $a = \ln A$ and $b = \ln B$, then

$$A = e^a$$
$$B = e^b$$

The functions $f(t)$ and $g(t)$ then take on the form (Figure 27.1),

$$f(t) = te^a + (1-t)e^b = (e^a - e^b)t + e^b$$
$$g(t) = (e^a)^t (e^b)^{(1-t)} = e^{ta+(1-t)b} = e^{(a-b)t+b}$$

where $f(t)$ represents a straight line joining the points $(0, e^b)$ and $(1, e^a)$ and $g(t)$ represents an exponential function joining the same points $(0, e^b)$ and $(1, e^a)$.

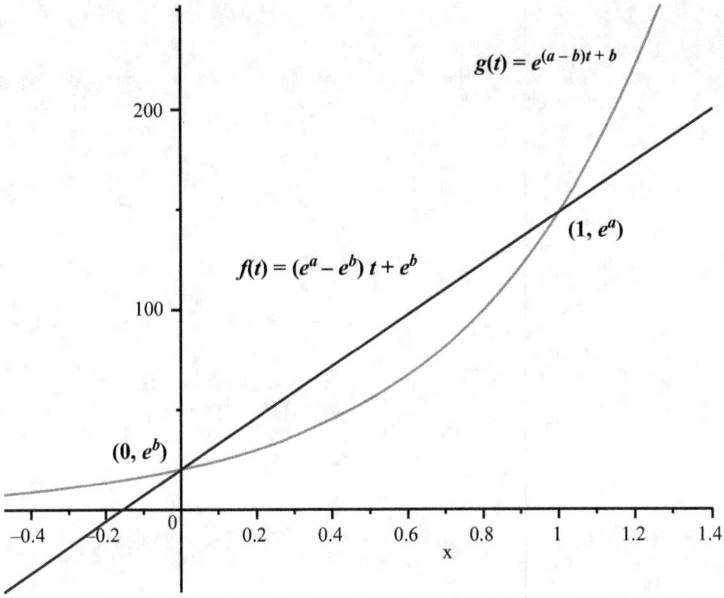

Figure 27.1. The graph of $f(t)$ and $g(t)$ when $a > b$. See Figure 27.2 for case $a < b$.

Regardless of whether $a > b$ or $b > a$ the function $g(t)$ is concave upwards. In order for $g(t)$ to intersect $f(t)$ at the points $(0, e^b)$ and $(1, e^a)$ it must be that $g(t) \leq f(t)$ on the interval, $(0, 1)$. Equality holds only if $a = b$.

So, $A^t B^{1-t} \leq tA + (1-t)B$ on $(0, 1)$, as required.

Theorem 27.3 *Hölder's Inequality Let S be a normed vector space $(S, +, \cdot, \| \ \|)$ and $F(\mathbb{N}, S)$ denote the set of all sequences, $f = \langle s_i \rangle_{i \in \mathbb{N}}$, whose elements, s_i, belong to S. Let p be any real number greater than or equal to 1. The ℓ_p-norm of $f = \langle s_i \rangle_{i \in \mathbb{N}}$ is defined as*

$$\|f\|_p = \|\langle s_i \rangle_{i \in \mathbb{N}}\|_p = \left[\sum_{i=0}^{\infty} \|s_i\|^p\right]^{1/p}$$

Let p be any real number > 1 and $f = \langle s_i \rangle_{i \in \mathbb{N}}$, $g = \langle t_i \rangle_{i \in \mathbb{N}} \in \ell_p$. Let $q = \frac{p}{p-1}$. Then,

$$\sum_{i=0}^{\infty} \|s_i\| \|t_i\| \leq \|f\|_p \|g\|_q$$

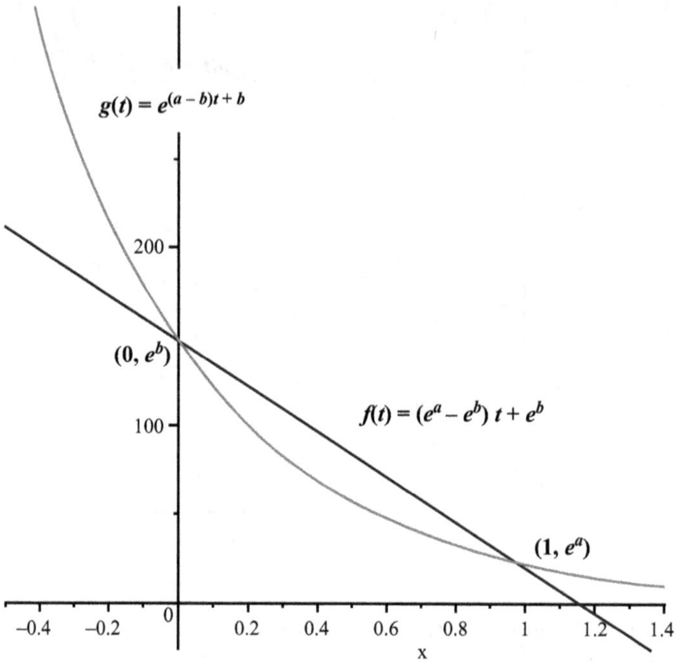

Figure 27.2. The graph of $f(t)$ and $g(t)$ when $a < b$.

Proof. Note that $q = \frac{p}{p-1}$ implies $\frac{1}{p} + \frac{1}{q} = 1$. If either f or g is 0 then both sides are zero and so the theorem holds true. Suppose then that neither f nor g is the zero sequence.

Suppose $A = \|s_i\|^p / \|f\|_p^p$ and $B = \|t_i\|^q / \|g\|_q^q$.

Let $t = 1/p$. Then $q = \frac{p}{p-1}$ implies $1 - t = 1/q$. Then $A^t B^{1-t} \leq tA + (1-t)B$ (this follows from the lemma).

$$A^t B^{1-t} \leq tA + (1-t)B \Rightarrow A^{1/p} B^{1/q} \leq \frac{A}{p} + \frac{B}{q}$$

$$\Rightarrow \frac{\|s_i\|}{\|f\|_p} \frac{\|t_i\|}{\|g\|_q} \leq \frac{\|s_i\|^p}{p\|f\|_p^p} + \frac{\|t_i\|^q}{q\|g\|_q^q}$$

implies

$$\sum_{i=0}^{\infty} \frac{\|s_i\|}{\|f\|_p} \frac{\|t_i\|}{\|g\|_q} \leq \frac{1}{p} \sum_{i=0}^{\infty} \frac{\|s_i\|^p}{\|f\|_p^p} + \frac{1}{q} \sum_{i=0}^{\infty} \frac{\|t_i\|^q}{\|g\|_q^q}$$

$$
\begin{aligned}
&= \frac{1}{p}\frac{\sum_{i=0}^{\infty}\|s_i\|^p}{\|f\|_p^p} + \frac{1}{q}\frac{\sum_{i=0}^{\infty}\|t_i\|^q}{\|g\|_q^q} \\
&= \frac{1}{p}\frac{\|f\|_p^p}{\|f\|_p^p} + \frac{1}{q}\frac{\|g\|_q^q}{\|g\|_q^q} \\
&= \frac{1}{p} + \frac{1}{q} \\
&= 1 \quad \text{(By hypothesis)}
\end{aligned}
$$

From which we obtain

$$\sum_{i=0}^{\infty} \|s_i\|\|t_i\| \leq \|f\|_p\|g\|_q$$

This proves Hölder's inequality.

Hölder's inequality can now be used as a tool to prove Minkowski's inequality, $\|f+g\|_p \leq \|f\|_p + \|g\|_p$ (which is the triangle inequality for the $\| \ \|_p$ norm).

Theorem 27.4 (The Minkowski Inequality). *Let p be any real number greater or equal to 1 and $(S, \| \ \|)$ be a normed vector space. Let $f = \langle s_i \rangle_{i\in\mathbb{N}}$, $g =< \langle t_i \rangle_{i\in\mathbb{N}} \in \ell_p$ where the entries belong to S. Then,*

$$\|f+g\|_p \leq \|f\|_p + \|g\|_p$$

Proof. We are given that $p \in [1, \infty)$ and $(S, \| \ \|)$ be a normed vector space and

$$f = \langle s_i \rangle_{i\in\mathbb{N}}, \ g = \langle t_i \rangle_{i\in\mathbb{N}} \in \ell_p$$

Case $p = 1$: If $p = 1$, then, since $\sum_{i=0}^{n}\|s_i + t_i\| \leq \sum_{i=0}^{n}\|s_i\| + \sum_{i=0}^{n}\|t_i\|$, for all n, $\|f+g\|_1 \leq \|f\|_1 + \|g\|_1$. Then the statement holds true for this particular case.

Case $p > 1$: Suppose now that $p > 1$. Let $q = \frac{p}{p-1}$.

Claim: We claim that if $f = \langle s_i \rangle_{i\in\mathbb{N}}$,

$$\left\| \, \langle \|s_i\|^{p-1} \rangle_{i\in\mathbb{N}} \, \right\|_q = \|f\|_p^{p-1} \quad (*)$$

Proof of claim: Suppose $f = \langle s_i \rangle_{i\in\mathbb{N}}$ and we are given $q = p/(p-1)$.

$$\begin{aligned}
\left\| \langle \|s_i\|^{p-1} \rangle_{i\in\mathbb{N}} \right\|_q &= \left[\sum_{i=0}^{\infty} \left| \|s_i\|^{p-1} \right|^q \right]^{1/q} \\
&= \left[\sum_{i=0}^{\infty} \|s_i\|^{q(p-1)} \right]^{1/q} \\
&= \left[\sum_{i=0}^{\infty} \|s_i\|^p \right]^{1/q} \\
&= \left[\left[\sum_{i=0}^{\infty} \|s_i\|^p \right]^{1/p} \right]^{p/q} \\
&= \|f\|_p^{p/q} \\
&= \|f\|_p^{p-1} \quad (\text{Since } q = p/(p-1))
\end{aligned}$$

So, $\left\| \langle \|s_i\|^{p-1} \rangle_{i\in\mathbb{N}} \, \right\|_q = \|f\|_p^{p-1}$, as claimed.

The inequality is now proven in the following chain of inequalities.

$$\begin{aligned}
\|f+g\|_p^p &= \left(\left[\sum_{i=0}^{\infty} \|s_i + t_i\|^p \right]^{1/p} \right)^p \\
&= \sum_{i=0}^{\infty} \|s_i + t_i\|^{p-1} \|s_i + t_i\| \\
&\leq \sum_{i=0}^{\infty} \|s_i + t_i\|^{p-1} (\|s_i\| + \|t_i\|) \text{ (Triangle inequality for } \| \ \|.) \\
&= \left[\sum_{i=0}^{\infty} \|s_i + t_i\|^{p-1} \|s_i\| \right] + \left[\sum_{i=0}^{\infty} \|s_i + t_i\|^{p-1} \|t_i\| \right]
\end{aligned}$$

$$\leq \left(\left\| \langle \|s_i + t_i\|^{p-1} \rangle_{i\in\mathbb{N}} \right\|_q \cdot \|f\|_p \right)$$
$$+ \left(\left\| \langle \|s_i + t_i\|^{p-1} \rangle_{i\in\mathbb{N}} \right\|_q \cdot \|g\|_p \right) \quad \text{(By Hölder's twice)}$$
$$= \left(\left\| \langle \|s_i + t_i\|^{p-1} \rangle_{i\in\mathbb{N}} \right\|_q \right) \cdot [\|f\|_p + \|g\|_p]$$
$$= \| f + g\|_p^{p-1} \cdot [\|f\|_p + \|g\|_p] \quad \text{(By claim *)}$$

Then

$$\|f + g\|_p = \left(\frac{1}{\| f + g \|_p^{p-1}} \right) \|f + g\|_p^p \leq \|f\|_p + \|g\|_p$$

from which we obtain the desired triangle inequality.

27.3 A Few Simple Topological Properties of ℓ_p-Spaces

We now investigate topological properties possessed by ℓ_p-spaces. If we are given an ℓ_p-space, $\mathscr{S} = F(\mathbb{N}, S)$, the norm of a sequence, $f = \langle s_i \rangle_{i\in\mathbb{N}}$, in $\mathscr{S}$ is defined as

$$\|f\|_p = \left[\sum_{i=0}^{\infty} \|s_i\|^p \right]^{1/p}$$

where the computation of $\|f\|_p$ involves the norm of its entries and so the topology on $\mathscr{S}$ will depend on the topology of S.

Since normed spaces are metrizable spaces, then $(\mathscr{S}, \| \ \|_p)$ can be viewed as a metric space. Then $\mathscr{S}$ is normal (by Theorem 9.19) and first countable (by Theorem 5.9).

The following theorem illustrates why $\mathscr{S}$ is separable whenever S is separable. If equipped with an ℓ_p-norm, $(\mathscr{S}, \| \ \|_p)$ is an ℓ_p-space whose elements are sequences with entries belonging to S.

Theorem 27.5 *Suppose $(S, \| \ \|)$ is a normed vector space and $\mathscr{S}$ represents the ℓ_p-space, $(F(\mathbb{N}, S), \| \ \|_p)$ of all infinite sequences with entries from S. If S is separable and $p > 1$, then $\mathscr{S}$ is a separable topological space (with respect to the ℓ_p-norm).*

Proof. Since S is separable, S contains a dense countable subset, $D = \{d_i : i \in \mathbb{N}\}$, of S. Let

$$\begin{aligned}\mathscr{D} = \{\langle x_i\rangle_{i\in\mathbb{N}} &: x_i \in D \text{ for finitely many i's and } x_i \\ &= 0 \text{ otherwise}\} \subseteq \mathscr{S}\end{aligned}$$

We claim that $\mathscr{D}$ is dense in $\mathscr{S}$. Let $f = \langle s_i\rangle_{i\in\mathbb{N}} \in \mathscr{S}$. It suffices to show that, for any $\varepsilon > 0$, the ε-ball centered at f,

$$B_\varepsilon(f) = \{y \in \mathscr{S} : \|y - f\|_p < \varepsilon\}$$

contains a sequence in $\mathscr{D}$.

Since $\|f\|_p = [\sum_{i=0}^{\infty} \|s_i\|^p]^{1/p}$ is a number, then

$$\lim_{n\to\infty} \sum_{i=n}^{\infty} \|s_i\|^p = 0$$

Let $\varepsilon > 0$. We can then choose n such that $\sum_{i=n+1}^{\infty} \|s_i\|^p < \varepsilon^p/2$.

Since D is dense in S, every open ball, $B_\varepsilon(s_i)$, with center s_i meets D, and so, for a given n, for each $i = 0$ to n, we can find $d_i \in D$ such that $\|s_i - d_i\| < \varepsilon^p/2n$.

For $i > n$, set $d_i = 0$. Let $d = \{d_i : i \in \mathbb{N}\} = \{d_0, d_1, d_2, \ldots d_n, 0, 0, 0, \ldots\}$. Then $d \in \mathscr{D}$.

Then

$$\begin{aligned}\|f - d\|_p^p &= \sum_{i=0}^{\infty} \|s_i - d_i\|^p \\ &= \sum_{i=0}^{n} \|s_i - d_i\|^p + \sum_{i=n+1}^{\infty} \|s_i - 0\|^p \\ &< \sum_{i=0}^{n} \frac{\varepsilon^p}{2n} + \frac{\varepsilon^p}{2} \\ &= \frac{n\varepsilon^p}{2n} + \frac{\varepsilon^p}{2} = \varepsilon^p\end{aligned}$$

So, $\|f-d\|_p < \varepsilon$. Since $d \in \mathscr{D}$, then $B_\varepsilon(f) \cap \mathscr{D} \neq \varnothing$. We conclude that $\mathscr{D}$ is dense in $\mathscr{S}$, so the ℓ_p-space, $\mathscr{S}$, is separable. This completes the proof.

If S is separable, since ℓ_p-spaces are both metrizable and separable, then (by Theorem 16.3) they are both second countable and Lindelöf. Also since $\mathscr{S}$ is separable metrizable, it is realcompact (see the example on page 577).

Example 1. Show that an ℓ_p-space is not, generally, sequentially compact.

Solution: Let $\mathscr{S} = F(\mathbb{N}, S)$ be an ℓ_p-space and choose a $\kappa \in S$ such that $\|\kappa\| = 1$. Let

$$U = \{f \in \mathscr{S} : \text{ all entries are 0, except one term is } \kappa\}$$

We will construct a sequence in $F(\mathbb{N}, S)$ with no converging subsequence.

For each n let $f_n = \langle s_{(n,j)} \rangle_{j \in \mathbb{N}}$ be the sequence in U whose nth term is κ. Then

$$U = \{f_n : n \in \mathbb{N}\}$$

If $n \neq k$,

$$\begin{aligned} \|f_n - f_k\|_p &= \left[\sum_{j=0}^{\infty} \|s_{(n,j)} - s_{(k,j)}\|^p \right]^{1/p} \\ &= (1^p + 1^p)^{1/p} \\ &= 2^{1/p} \end{aligned}$$

Let $\varepsilon = (2^{1/p})/2$. If $g \in U$ is the limit of a subsequence of U, then $\|f_n - g\|_p = 2^{1/p}$ must be $< \varepsilon$ for infinitely many terms, f_n. This can obviously not be. Then there is no converging subsequence for $\{f_n : n \in \mathbb{N}\}$. So $\mathscr{S}$ cannot be sequentially compact.

By Theorem 17.2, in metric spaces, sequentially compact, compact and countably compact are equivalent topological properties. So, $\mathscr{S}$ is not compact.

27.4 The Sup-Norm on $F(\mathbb{N}, S)$

We know that for each real value of $p \in [1, \infty)$, $\|f\|_p$ is a valid norm on $F(\mathbb{N}, S)$.

We now wonder if there is a natural way we can extend the value of p so as to include $p = \infty$, In a sense, we are wondering whether the limit, $\lim_{p\to\infty} \|f\|_p$ is a number. If so, is this number a norm of f which we could represent as "$\|f\|_\infty$". How would we define it?

To investigate this, let $(S, \| \ \|)$ be a normed vector space, let $f = < s_i >_{i\in\mathbb{N}} \in F(\mathbb{N}, S)$ and $p \geq 1$, let

$$k(n) = \sup\{\|s_i\| : i = 0, 1, 2, \ldots, n\}$$

Then $\{k(n) : n \in \mathbb{N}\}$ is an increasing sequence of numbers.

Let

$$\beta = \sup\{\|s_i\| : i \in \mathbb{N}\}$$

To ensure that β is a number we will assume that each sequence, $\langle s_i \rangle_{i\in\mathbb{N}}$, is bounded with respect to $\| \ \|$. Now, $k(n) \leq \beta$ for all values of n. If $\sup\{k(n) : n \in \mathbb{N}\} < \beta$ then there exists t such that $\sup\{k(n) : n \in \mathbb{N}\} < t < \beta$. Then $k(n) < t$ for all $n \in \mathbb{N}$. Then $\|s_i\| < t < \beta$ for all $i \in \mathbb{N}$ contradicting the definition of β.

So,

$$\lim_{n\to\infty} k(n) = \beta$$

We claim that, if f is bounded in $F(\mathbb{N}, S)$, $\lim_{p\to\infty} \|f\|_p = \beta$.

Proof of claim: Let $\varepsilon > 0$. Let $f = < s_i >_{i\in\mathbb{N}}$ be a bounded element of $F(\mathbb{N}, S)$. Since $\lim_{n\to\infty} k(n) = \beta$, then there exists an $M \in \mathbb{N}$ such that

$$n > M \qquad \Rightarrow \qquad |k(n) - \beta| < \varepsilon/3$$

Let

$$g(n,p) = \left[\sum_{i=0}^{n} \|s_i\|^p\right]^{1/p}$$

See that, for any choice of n,

$$\begin{aligned}
\lim_{p\to\infty} g(n,p) &= \lim_{p\to\infty}\left[\sum_{i=0}^{n}\|s_i\|^p\right]^{1/p} \\
&= \lim_{p\to\infty}\left[\sum_{i=0}^{n-m}\frac{\|s_i\|^p}{k(n)^p}+m\right]^{1/p} k(n) \quad (\text{Where } m\geq 1) \\
&= (1)k(n) \\
&= k(n)
\end{aligned}$$

Then there exists $\kappa > 0$ such that $p > \kappa$ implies, for any choice of n

$$|\, g(n,p) - k(n)| < \varepsilon/3$$

Then, if $n > M$ and $p > \kappa$,

$$\begin{aligned}
|g(n,p)-\beta| &= |g(n,p)-k(n)+k(n)-\beta| \\
&\leq |g(n,p)-k(n)|+|k(n)-\beta| \\
&< \varepsilon/3-\varepsilon/3 \\
&< \varepsilon
\end{aligned}$$

If $\alpha = \max\{M,\kappa\}$, when $n, \kappa > \alpha$, $|g(n,p)-\beta| < \varepsilon$.
Then

$$\lim_{p\to\infty}\lim_{n\to\infty} g(n,p) = \lim_{p\to\infty}\beta = \beta$$

This implies,

$$\begin{aligned}
\lim_{p\to\infty}\|f\|_p &= \lim_{p\to\infty}\left[\sum_{i=0}^{\infty}\|s_i\|^p\right]^{1/p} \\
&= \lim_{p\to\infty}\left[\lim_{n\to\infty}\sum_{i=0}^{n}\|s_i\|^p\right]^{1/p}
\end{aligned}$$

$$= \lim_{p\to\infty} \lim_{n\to\infty} \left(\left[\sum_{i=0}^{n} \|s_i\|^p \right]^{1/p} \right)$$
$$= \lim_{p\to\infty} \lim_{n\to\infty} g(n,p)$$
$$= \beta$$

So, $\lim_{p\to\infty} \|f\|_p = \beta$. This establishes the claim.

If $f = \langle s_i \rangle_{i\in\mathbb{N}} \in F(\mathbb{N}, S)$, it will be convenient to represent $\beta = \sup\{\|s_i\| : i \in \mathbb{N}\}$ as

$$\lim_{p\to\infty} \|f\|_p = \|f\|_\infty$$

We formally define the *sup-norm* (also called the ∞*-norm*), $\|f\|_\infty$.

Definition 27.6 Let S be a normed vector space. We say a sequence $\langle s_i \rangle_{i\in\mathbb{N}}$ in $F(\mathbb{N}, S)$ is *bounded* if there is a k such that $\|s_i\| \leq k$ for all $i \in \mathbb{N}$. Let $\mathscr{S}_b$ be the set of all bounded sequences with elements from S. If $f = \langle s_i \rangle_{i\in\mathbb{N}}$, we define the *sup-norm*, $\|f\|_\infty$, of f as

$$\|f\|_\infty = \sup\{\|s_i\| : i \in \mathbb{N}\}$$

The sup-norm on $\mathscr{S}_b$ is also referred as the ℓ_∞*-norm*. If $\mathscr{S}_b$ is equipped with the norm $\|\ \|_\infty$ we refer to $\mathscr{S}_b$ as being an ℓ_∞*-space*.

Verifying that $\|\ \|_\infty$ is a valid norm is straightforward. Verifying this is left as an exercise for the reader.

27.5 Topologizing $F(\mathbb{N}, S)$ by Defining Convergence

Once chosen, a norm on $F(\mathbb{N}, S)$ precisely determines how sequences of sequences in $F(\mathbb{N}, S)$ will converge. The norm will topologize $F(\mathbb{N}, S)$. But defining norms on $F(\mathbb{N}, S)$ is not the only way to topologize $F(\mathbb{N}, S)$.

We can also view $F(\mathbb{N}, S)$ as an infinite product, $S^{\mathbb{N}}$ (see Definition 12.6 on page 305). We normally assign on infinite products either the "product topology" or the "box topology". The box topology on $S^{\mathbb{N}}$

is equivalent to the topology obtained by assigning the ℓ_∞-norm on $F(\mathbb{N}, S) = S^{\mathbb{N}}$. That is, if $\lim_{n\to\infty} \|g - f_n\|_\infty = 0$ then the sequence, $\{f_n : n \in \mathbb{N}\}$, will

converge uniformly to the sequence g.

On the other hand, assigning the product topology on $F(\mathbb{N}, S) = S^{\mathbb{N}}$ produces a topological space independent of any norm defined on $F(\mathbb{N}, S)$. It is the topology with which the closed subsets are defined by "pointwise convergence". This means that, if $f_n = \langle s_{n_i}\rangle_{i\in\mathbb{N}}$, and for each i,

$$\lim_{n\to\infty} s_{n_i} = t_i$$

then the sequence $g = \langle t_i\rangle_{i\in\mathbb{N}}$ represents the

pointwise limit of the sequence $\{f_n : n \in \mathbb{N}\}$

Unless explicitly specified otherwise, the default topology on $F(\mathbb{N}, S)$ is the one determined by uniform convergence. Note that, if a sequence in $F(\mathbb{N}, S)$ converges uniformly, then it also converges pointwise. The converse does not hold true.

27.6 Completeness of $(F(\mathbb{N}, S), \|\ \|_\infty)$

Recall that a normed vector space, $(S, \|\ \|)$, is said to be *complete* if every Cauchy sequence in $(S, \|\ \|)$ converges to a point which is contained in S.

In the following example, we verify that, whenever $(S, \|\ \|)$ is complete, the normed vector space $(F(\mathbb{N}, S)_b, \|\ \|_\infty)$ is complete.[1]

Example 2. Verify that, if $(S, \|\ \|)$ is a complete normed space, $\mathscr{S}_b = (F(\mathbb{N}, S)_b, \|\ \|_\infty)$ is a complete normed vector space.

Solution: Let $C = \{f_n : n \in \mathbb{N}\}$ be a Cauchy sequence of sequences in the ℓ_∞-space, $\mathscr{S}_b$. Since C is Cauchy, it is bounded, so there is a real number, K, such that $\|f_n\|_\infty < K$, for all $n \in \mathbb{N}$ (as shown in the above example).

[1]The subscript b indicates "bounded" to ensure that the sup-norm is never ∞.

For each n, let $f_n = \langle s_{n_i} \rangle_{i \in \mathbb{N}}$. We are required to show that $\{f_n : n \in \mathbb{N}\}$ converges to some sequence, h, in the ℓ_∞-space, $\mathscr{S}_b$.

Let $\varepsilon > 0$. Since $\{f_n : n \in \mathbb{N}\}$ is Cauchy, there exists $N \in \mathbb{N}$ such that, $m, n > N \Rightarrow \|f_n - f_m\|_\infty < \varepsilon$.

Then, when $m, n > N$, and each i,

$$\begin{aligned} \| s_{n_i} - s_{m_i} \| &\leq \sup_{i \in \mathbb{N}} \{ \| s_{n_i} - s_{m_i} \| \} \\ &= \| \langle s_{n_i} - s_{m_i} \rangle_{i \in \mathbb{N}} \|_\infty \\ &= \| \langle s_{n_i} \rangle_{i \in \mathbb{N}} - \langle s_{m_i} \rangle_{i \in \mathbb{N}} \|_\infty \\ &= \| f_n - f_m \|_\infty \\ &< \varepsilon \end{aligned}$$

Then, for each i, when $m, n > N$, $\| s_{n_i} - s_{m_i} \| < \varepsilon$. Then for each i, $\{s_{n_i} : n \in \mathbb{N}\}$ is a Cauchy sequence in S. Since S is a complete space, for each i, $\{s_{n_i} : n \in \mathbb{N}\}$ converges to some $t_i \in S$ (with respect to $\| \ \|$). That is, for each i, there exists N_i, such that,

$$n > N_i \qquad \Rightarrow \qquad \|t_i - s_{n_i}\| < \varepsilon$$

Let $h = \{t_i : i \in \mathbb{N}\}$.

We view h as a candidate for the limit of the Cauchy sequence, $\{f_n : n \in \mathbb{N}\}$. To prove this we must first show that $\|h\|_\infty \in \mathbb{R}$, and then show that $\|h - f_n\|_\infty < \varepsilon$ for all $n > N$ for some N.

For our chosen ε, for $i \in \mathbb{N}$, and any $n > N_i$,

$$\begin{aligned} \|t_i\| &= \|t_i - s_{n_i} + s_{n_i}\| \\ &\leq \|t_i - s_{n_i}\| + \|s_{n_i}\| \\ &< \varepsilon + \|f_n\|_\infty \quad (\text{Since } n > N_i.) \\ &\leq \varepsilon + K \end{aligned}$$

Then $\|h\|_\infty \leq \sup\{\|t_0\|, \|t_1\|, \ldots \|t_{N_i}\|, \varepsilon + K\} \in \mathbb{R}$. So, $\|h\|_\infty$ is a number.

We now claim that $\{f_n : n \in \mathbb{N}\}$ converges to h.

Suppose $m, n > N$. See that, independently of the value of i,

$$\begin{aligned}
\|t_i - s_{m_i}\| &= \|(\lim_{n\to\infty} s_{n_i}) - s_{m_i}\| \\
&= \lim_{n\to\infty} \|s_{n_i} - s_{m_i}\| \\
&\le \sup_{i\in\mathbb{N}}\{\|s_{n_i} - s_{m_i}\|\} \quad \text{(since norms are continuous)} \\
&= \|f_n - f_m\|_\infty \\
&< \varepsilon
\end{aligned}$$

Then $\sup_{i\in\mathbb{N}}\{\|t_i - s_{m_i}\|\} \le \varepsilon$. So, $\|h - f_m\|_\infty \le \varepsilon$. Therefore, $\{f_n : n \in \mathbb{N}\}$ converges to h.

27.7 Topologizing Other Families of Functions

For two normed topological spaces S and T, let $F(S,T)$ denote the family of all functions mapping S into T. Whenever the functions on S are continuous and real-valued the expression, $C(S,\mathbb{R})$, specifies this fact. It is standard to express this set more concisely as $C(S)$. We can topologize $C(S)$ in a way that is analogous to the method used to topologize the family of all sequences, $F(\mathbb{N},S)$. We do this by defining a similar norm on $C[a,b]$ called, an L_p-norm. The symbol, ℓ_p-norm, is reserved specifically for the set, $F(\mathbb{N},S)$, or the family of all n-tuples, S^n, of elements from S.

Definition 27.7 Let $C[a,b]$ denote the set of all real-valued continuous functions on $[a,b]$ and $p \in [1,\infty)$. If $f \in C[a,b]$ we define,

$$\|f\|_p = \left[\int_a^b |f(x)|^p\right]^{1/p}$$

We refer to this norm on $C[a,b]$ as an L_p-norm and we refer to $(C[a,b], \| \|_p)$ as an "L_p-space".

We must again verify that $\| \|_p$ satisfies the three norm axioms. Verifying that it satisfies the N1 and N2 axioms is straightforward.

Just as for ℓ_p-norms the triangle inequality, $\|f+g\|_p \leq \|f\|_p + \|g\|_p$, for L_p-norms is referred to as the Minkowski inequality. It is proven by first obtaining an equivalent version of Hölders inequality.

Theorem 27.8 *Hölder's inequality. Let p be any real number strictly greater than 1 and $f, g \in (C[a,b], \| \ \|_p)$. Let $q = \frac{p}{p-1}$. Then,*

$$\int_a^b |f(x)| \cdot |g(x)|\, dx \leq \|f\|_p \|g\|_q$$

Proof. The proof mimics the proof of Hölder's inequality for sequences replacing the summation sign, $\sum_{i=0}^{\infty}$ with the integration sign.

Theorem 27.9 *Minkowski's inequality Let p be any real number strictly greater than 1 and $f,g \in (C[a,b],\| \ \|_p)$. Then, $\|f+g\|_p \leq \|f\|_p + \|g\|_p$. That is,*

$$\left[\int_a^b |f(x)+g(x)|^p\, dx\right]^{1/p} \leq \left[\int_a^b |f(x)|^p\, dx\right]^{1/p} + \left[\int_a^b |g(x)|^p\, dx\right]^{1/p}$$

Proof. The proof mimics the proof of Minkowski's inequality for sequences replacing the summation sign, $\sum_{i=0}^{\infty}$ with the integration sign.

Definition 27.10 Let $C(K,\mathbb{R})$ be the set of all (bounded) continuous real-valued function on the compact set K. If $f \in C(K,\mathbb{R})$, we define the *sup-norm*, $\|f\|_\infty$, of f as

$$\|f\|_\infty = \sup\{|f(x)| : x \in K\}$$

Example 3. Consider the expression

$$\alpha(x) = x^{x^{x^{x^{\cdot^{\cdot^{\cdot}}}}}}$$

How should we interpret this expression? Are there values of x, for which this expression has meaning?

Proposed solution: Let $C = \{f_n : n \in \mathbb{N}\backslash\{0\}\}$ be a sequence of functions recursively defined as follows. $f_1(x) = x$ and for $n > 1$

$$f_{n+1}(x) = x^{f_n(x)}$$

Then

$$f_1(x) = x$$

$$f_2(x) = x^x$$

$$f_3(x) = x^{x^x}$$

$$f_4(x) = x^{x^{x^x}}$$

$$\vdots$$

forms an infinite family of continuous functions on the interval $(0, \infty)$.

We now reformulate the question: Are there values of x such that the sequence $\{f_n(x) : n \in \mathbb{N}\backslash\{0\}\}$ converges pointwise to $\alpha(x)$?

There is at least one such value for x. Observe that $f_n(1) = 1$ for all n; hence, $\lim_{n\to\infty} f_n(1) = 1$. Are there others?

Case 1. We analyze the behavior of the functions, $\{f_n(x) : n > 0\}$, for values of $x \geq 1$.

Fact 1: For each n, $f_n(x)$ is increasing on the part of its domain where $x \geq 1$. This can be verified by seeing that $f_2(x)$ has a positive derivative whenever $x \geq 1$. Then by mathematical induction, we can conclude that, for all n, $f_n(x)$ has a positive derivative, whenever $x \geq 1$.

Fact 2: For each $a \geq 1$, $M_a = \{f_n(a) : n \in \mathbb{N}\backslash\{0\}\}$ is an increasing sequence of numbers. This is verified by applying mathematical induction on M_a.

We first determine the interval $[1, a)$ such that

$$\lim_{n\to\infty} f_n(x)$$

is a number for each x in $[1, a)$. At such points

$$\begin{aligned} y &= \lim_{n\to\infty} f_{n+1}(x) \\ &= \lim_{n\to\infty} x^{f_n(x)} \\ &= x^{\lim_{n\to\infty} f_n(x)} \\ &= x^y \end{aligned}$$

So the expression, $y = x^y$, represents a curve which is the pointwise limit of the sequence $\{f_n(x) : n > 0\}$ for certain values of x greater than or equal to 1.

For what values of $x \geq 1$ can we accept $y = x^y$ as a representation of the expression $\alpha(x)$? We verify if the curve, $y = x^y$, has an unbounded derivative with respect to x. Observe that $y = x^y$ if and only if $x = e^{\frac{\ln y}{y}}$.

Differentiating both sides with respect to x we obtain

$$\frac{dy}{dx} = \frac{y^2}{(1 - \ln y)e^{\frac{\ln y}{y}}}$$

We see that the curve of $y = x^y$ has a defined derivative whenever y belongs to $[1, e)$ but gets steeper as it approaches e from the left.[2] Solving for x in the expression, $e = x^e$, we obtain $x = e^{1/e}$.[3]

So, $y = x^y$ can suitably represent $\alpha(x)$ for values of x in $[1, e^{1/e})$.

Case 2. We analyze the behavior of the functions, $\{f_n(x) : n > 0\}$, for values of x in $(0, 1]$.

The functions $\{f_n(x) : n > 1\}$ are not defined at $x = 0$ but converge pointwise to the curve $y = x^y$ for x in $(0, 1]$. The derivative of the curve, $y = x^y$, with respect to x shows that it is smooth and is positive with respect to x on $(0, 1]$. So, $y = x^y$ suitably represents $\alpha(x)$ on $(0, 1]$.

---

[2] Also see that the the slope of the tangent line is negative for $y > e$.

[3] Note that e does not bound all values of $f_n(x)$, just the values of $f_n(x)$ where $\{f_n(x) : n > 0\}$ converges.

The limit expression,

$$\lim_{n\to\infty} \{f_n(a) : n > 0\}$$

has no (unique) numerical value for $a \in [e^{1/e}, \infty)$.

Concept Review

1. Let $F(\mathbb{N}, \mathbb{R})$ denote the set of all sequences of real numbers. Define the ℓ_p-norm and ℓ_p-space.
2. State the Minkowski inequality for the ℓ_p-space, $F(\mathbb{N}, \mathbb{R})$.
3. State Hölder's inequality for ℓ_p-spaces.
4. Define the sup-norm, $\|f\|_\infty$, on $\mathscr{S}_b$ the set of all bounded sequences in $F(\mathbb{N}, \mathbb{R})$.
5. Describe a condition on S which guarantees that $(F(\mathbb{N}, S)_b, \| \; \|_\infty)$ is complete.

Chapter 28

Completing Incomplete Metric Spaces

Abstract
Some metric spaces are not complete. The space $\mathbb{Q}$ is the example which immediately comes to mind.Every irrational number is the limit of a Cauchy sequence in $\mathbb{Q}$, so $\mathbb{R}$ is the smallest non-compact complete metric space which densely contains $\mathbb{Q}$. We investigate how we can complete a metric space by isometrically and densely embedding it in a complete metric space.

28.1 Introduction

Recall that a metric space, S, is said to be complete if every Cauchy sequence in S converges to a point in S. Not all metric spaces are complete. Simply witness the subspace of rational numbers. But we can densely embed the rationals in the complete metric space $\mathbb{R}$. We will show that, likewise, any incomplete metric space, S can be densely embedded in some complete metric space.

28.2 A Space Whose Elements are Cauchy Sequences of S

To achieve this, we will first gather together all Cauchy sequences of the metric space (S, ρ) to form a set,

$$\mathscr{S} = \{\alpha \in \mathscr{P}(S) : \alpha = \{\alpha_n : n \in \mathbb{N}\} \text{ is a Cauchy sequence in } S\}$$

An equivalence relation on $\mathscr{S}$: If $\alpha = \{\alpha_n : n \in \mathbb{N}\}$ and $\gamma = \{\gamma_n : n \in \mathbb{N}\}$ are *both* Cauchy sequences in S, we will say "α is equivalent to γ" and write $\alpha \sim \gamma$, if and only if $\lim_{n\to\infty} \rho(\alpha_n, \gamma_n) = 0$. The relation $\sim$ is easily seen to be an equivalence relation on $\mathscr{S}$ where $[\alpha] = \{\gamma \in \mathscr{S} : \gamma \sim \alpha\}$ forms an equivalence class and the family of subsets

$$[\mathscr{S}] = \{[\alpha] : \alpha \in \mathscr{S}\}$$

partitions $\mathscr{S}$. For any $\eta \in [\alpha]$, $[\eta] = [\alpha]$.

Observations. We are given that $\alpha = \{\alpha_n\}$ and $\gamma = \{\gamma_n\}$ are *both* Cauchy sequences in (S, ρ).

(1) We claim that if $\gamma \in [\alpha]$ and $\varepsilon > 0$ then there exist M such that $n, m > M$ implies $\rho(\alpha_n, \gamma_m) < \varepsilon$. Let $\varepsilon > 0$ and $\gamma \in [\alpha] \in [\mathscr{S}]$. Since $\gamma \sim \alpha$, then (by definition of $\sim$) $\lim_{n\to\infty} \rho(\gamma_n, \alpha_n) = 0$. There exists, $M > 0$ such that $n, m > M$ implies, $\rho(\alpha_n, \gamma_n) < \varepsilon/2$ and $\rho(\gamma_n, \gamma_m) < \varepsilon/2$ (since γ is Cauchy). Then, if $n, m > M$,

$$\begin{aligned}\rho(\alpha_n, \gamma_m) &\leq \rho(\alpha_n, \gamma_n) + \rho(\gamma_n, \gamma_m)\\ &< \varepsilon/2 + \varepsilon/2\\ &< \varepsilon\end{aligned}$$

So, if $\gamma \in [\alpha]$, there exists M such that

$$\rho(\alpha_n, \gamma_m) < \varepsilon \text{ when } n, m > M \quad (**)$$

as claimed.

(2) Also, suppose $\lim_{n\to\infty} \alpha_n = x \in S$, and $\gamma \in [\alpha]$. Then, if $\varepsilon > 0$, there exists, M, such that if $n, m > M$,

$$\begin{aligned}\rho(\gamma_n, x) &\leq \rho(\gamma_n, \alpha_m) + \rho(\alpha_m, x))\\ &< \varepsilon/2 + \varepsilon/2 \quad \text{(By (**))}\\ &< \varepsilon\end{aligned}$$

So, "if α converges to x in S, all Cauchy sequences in $[\alpha]$ converge to x".

28.3 Defining a Suitable Metric on the Set $[\mathscr{S}]$

We now define a metric δ on $[\mathscr{S}] = \{[\alpha] : \alpha \in \mathscr{S}\}$ as follows: For $[\alpha], [\beta] \in [\mathscr{S}]$,

$$\delta([\alpha], [\beta]) = \inf \{ \lim_{n\to\mathbb{N}} \rho(\eta_n, \gamma_n) : \eta \in [\alpha], \gamma \in [\beta]\}$$

(a) We claim that δ is well-defined.

Proof of claim: We prove this in two steps.

(1) We begin by showing that $\delta([\alpha], [\beta])$, has a numerical value.

To do this we show that, for the Cauchy sequences α, η in $[\alpha]$ and β, γ in $[\beta]$, the sequence $\{\rho(\eta_n, \gamma_n) : n \in \mathbb{N}\}$ is Cauchy in $\mathbb{R}$. If so, then $\lim_{n\to\mathbb{N}} \rho(\eta_n, \gamma_n)$ is a positive number (since $\mathbb{R}$ is complete); hence, $\inf\{\lim_{n\to\mathbb{N}} \rho(\eta_n, \gamma_n) : \eta \in [\alpha], \gamma \in [\beta]\}$ is a number and so $\delta([\alpha], [\beta]) \in \mathbb{R}$.

Let $\varepsilon > 0$. Then there exists N such that $k, j > N$ implies $\rho(\eta_j, \eta_k) < \varepsilon/2$ and $\rho(\gamma_k, \gamma_j) < \varepsilon/2$.

$$\begin{aligned}
\rho(\eta_j, \gamma_j) &\leq \rho(\eta_j, \eta_k) + \rho(\eta_k, \gamma_j) \\
&\leq \rho(\eta_j, \eta_k) + [\rho(\eta_k, \gamma_k) + \rho(\gamma_k, \gamma_j)] \\
\rho(\eta_j, \gamma_j) - \rho(\eta_k, \gamma_k) &\leq \rho(\eta_j, \eta_k) + \rho(\gamma_k, \gamma_j) \\
&\quad \times \text{(Subtract } \rho(\eta_k, \gamma_k) \text{ on both sides)} \\
&< \varepsilon/2 + \varepsilon/2 = \varepsilon \text{ (By (**))} \\
\rho(\eta_k, \gamma_k) &\leq \rho(\eta_k, \eta_j) + \rho(\eta_j, \gamma_k) \\
&\leq \rho(\eta_k, \eta_j) + [\rho(\eta_j, \gamma_j) + \rho(\gamma_j, \gamma_k)] \\
\rho(\eta_k, \gamma_k) - \rho(\eta_j, \gamma_j) &\leq \rho(\eta_k, \eta_j) + \rho(\gamma_j, \gamma_k) \\
&\quad \times \text{(Subtract } \rho(\eta_j, \gamma_j) \text{ on both sides)} \\
&< \varepsilon/2 + \varepsilon/2 = \varepsilon \text{ (By (**))}
\end{aligned}$$

Then $k, j > N$ implies,

$$|\rho(\eta_j, \gamma_j) - \rho(\eta_k, \gamma_k)| < \varepsilon$$

So, $\{\rho(\eta_j,\gamma_j) : j \in \mathbb{N}\}$ is Cauchy and so is bounded in $\mathbb{R}$, say by K. So, $\lim_{j\to\infty}\rho(\eta_j,\gamma_j) \leq K$ and since $\mathbb{R}$ is complete $\lim_{j\to\infty}\rho(\eta_j,\gamma_j)$ is a real number greater than zero. Then $\delta([\alpha],[\beta]) \in \mathbb{R}$.

(2) Next, we verify that, for the Cauchy sequences η, γ in $[\alpha]$ and $[\beta]$, respectively, $\delta([\eta],[\gamma]) = \delta([\alpha],[\beta])$. So, the value of $\delta([\alpha],[\beta])$ does not depend on the choice of representatives η and γ.

For all j, n,

$$\rho(\eta_{(j)_n},\gamma_{(j)_n}) \leq \rho(\eta_{(j)_n},\alpha_{(j)_n}) + \rho(\alpha_{(j)_n},\beta_{(j)_n}) + \rho(\beta_{(j)_n},\gamma_{(j)_n})$$

$$\Rightarrow$$

$$\lim_{n\to\infty}\rho(\eta_{(j)_n},\gamma_{(j)_n}) \leq \lim_{n\to\infty}\rho(\eta_{(j)_n},\alpha_{(j)_n}) + \lim_{n\to\infty}\rho(\alpha_{(j)_n},\beta_{(j)_n}) + \lim_{n\to\infty}\rho(\beta_{(j)_n},\gamma_{(j)_n})$$

$$\Rightarrow$$

$$\lim_{n\to\infty}\rho(\eta_{(j)_n},\gamma_{(j)_n}) \leq 0 + \lim_{n\to\infty}\rho(\alpha_{(j)_n},\beta_{(j)_n}) + 0$$

$$\Rightarrow$$

$$\delta([\eta],[\gamma]) \leq \delta([\alpha],[\beta])$$

Similarly

$$\delta([\alpha],[\beta]) \leq \delta([\eta],[\gamma])$$

So, $\delta([\alpha],[\beta]) = \delta([\eta],[\gamma])$.
So, $\delta([\alpha],[\beta])$ is well-defined, as claimed.

(3) Next we verify that $\delta : [\mathscr{S}] \times [\mathscr{S}] \to \mathbb{R}^+$, is a valid metric.

Indiscernibility of identicals: Clearly, if $[\alpha] = [\beta]$ then $\delta([\alpha],[\beta]) = 0$.

Identity of indiscernibles: Suppose $\delta([\alpha], [\beta]) = 0$. Then $\lim_{n\to\infty}\rho(\eta_n,\gamma_n) = 0$ for some $\eta \in [\alpha],\ \gamma \in [\beta]$. This implies $\eta \in [\gamma]$, so $[\alpha] = [\beta]$.

Symmetry of δ follows immediately from the symmetry of ρ.

Triangle inequality: Let $[\alpha]$, $[\beta]$, and $[\varphi]$ be three elements of $[\mathscr{S}]$.

Let $\mu = \{\mu_n : n \in \mathbb{N}\} \in [\varphi]$,

$$\delta(\{\mu\}, [\alpha]) = \inf\{\lim_{n\to\infty} \rho(\mu_n, \eta_n) : \eta \in \alpha,\ \mu \in \{\mu\}\}$$

$$\delta(\{\mu\}, [\beta]) = \inf\{\lim_{n\to\infty} \rho(\mu_n, \gamma_n) : \gamma \in \beta,\ \mu \in \{\mu\}\}$$

For $i \in \mathbb{N}\backslash\{0\}$ there exist $\eta_{(i)} \in [\alpha]$ and $\gamma_{(i)} \in [\beta]$ such that

$$\lim_{n\to\infty} \rho(\mu_n, {\eta_{(i)}}_n) < \delta(\{\mu\}, [\alpha]) + 1/i$$

$$\lim_{n\to\infty} \rho(\mu_n, {\gamma_{(i)}}_n) < \delta(\{\mu\}, [\beta]) + 1/i$$

Then, for all $i \in \mathbb{N}$, and the chosen $\mu \in [\varphi]$,

$$\begin{aligned}
\delta([\alpha], [\beta]) &= \inf\{\lim_{n\to\mathbb{N}} \rho(\eta_n, \gamma_n) \ :\ \eta \in [\alpha],\ \gamma \in [\beta]\} \\
&\le \lim_{n\to\infty} \rho({\eta_{(i)}}_n, {\gamma_{(i)}}_n) \\
&\le \lim_{n\to\infty} \rho({\eta_{(i)}}_n, \mu_n) + \lim_{n\to\infty} \rho(\mu_n, {\gamma_{(i)}}_n) \text{ (Triangle inequality)} \\
&< \delta([\alpha], \{\mu\}) + 1/i + \delta(\{\mu\}, [\beta]) + 1/i
\end{aligned}$$

This implies

$$\delta([\alpha], [\beta]) \le \delta([\alpha], \{\mu\}) + \delta(\{\mu\}, [\beta])$$

This inequality holds true independently of the choice of $\mu \in [\varphi]$. So,

$$\delta([\alpha], [\beta]) \le \delta([\alpha], [\varphi]) + \delta([\varphi], [\beta])$$

This establishes the triangle inequality property.

Then $\delta : [\mathscr{S}] \times [\mathscr{S}] \to \mathbb{R}^+$, is a valid metric, as claimed. So, $([\mathscr{S}], \delta)$ is a metric space.

28.4 The Space $([\mathscr{S}], \delta)$ Contains an Isometric Copy of (S, ρ)

If α is a sequence which converges to $x \in S$ then α is Cauchy and so $[\alpha] \in [\mathscr{S}]$. More significantly, the constant sequence $\mathbf{x} = \{x, x, x, \ldots\}$ belongs to $[\alpha]$ and so $[\alpha] = [\mathbf{x}]$. If $\mathbf{y} = \{y, y, y, \ldots\}$ and

$\mathbf{x} = \{x, x, x, \ldots\}$ are distinct sequences, then $[\mathbf{y}] \neq [\mathbf{x}]$. It follows that the function $f : (S, \rho) \to ([\mathscr{S}], \delta)$ defined as

$$f(x) = [\mathbf{x}]$$

maps S one-to-one into $[\mathscr{S}]$.

We claim the function f is also isometric. It suffices to show that $\delta(\, f(x), f(y)\,) = \rho(x, y)$. Suppose $x, y \in S$ and $\eta \in [\mathbf{x}]$, $\gamma \in [\mathbf{y}]$. Then

$$\rho(\eta_n, \gamma_n) \leq \rho(\eta_n, x) + \rho(x, y) + \rho(y, \gamma_n)$$

implies

$$\begin{aligned}\lim_{n\in\infty} \rho(\eta_n, \gamma_n) &\leq \lim_{n\in\infty} \rho(\eta_n, x) + \lim_{n\in\infty} \rho(x, y) + \lim_{n\in\infty} \rho(y, \gamma_n)\\ &= 0 + \rho(x, y) + 0\\ &= \rho(x, y)\end{aligned}$$

Then

$$\begin{aligned}\delta([\mathbf{x}], [\mathbf{y}]) &= \inf \{\rho(x, y) : \eta \in [\mathbf{x}], \gamma \in [\mathbf{y}]\}\\ &= \rho(x, y)\end{aligned}$$

So,

$$\delta(f(x), f(y)) = \rho(x, y)$$

So, the function f is isometric, as claimed.

28.5 The Space $([\mathscr{S}], \delta)$ Is Complete

If $[\mathscr{S}]$ is complete and contains a dense isometric copy of S (as we will soon show) then it can be viewed as a completion of S.

Proposition 28.1 *The metric space, $([\mathscr{S}], \delta)$, is complete.*

Proof. We are given a metric space (S, ρ). We defined $\mathscr{S}$ as being the set whose elements, $\alpha = \{\alpha_n : n \in \mathbb{N}\}$, are the Cauchy sequences in S. We are given the set $[\mathscr{S}] = \{[\alpha] : \alpha \in \mathscr{S}\}$ where $\beta \in [\alpha]$ if and only if $\beta = \{\beta_n : n \in \mathbb{N}\} \in \mathscr{S}$ and $\lim_{n\to\infty} \rho(\alpha_n, \beta_n) = 0$. If $[\alpha], [\beta] \in [\mathscr{S}]$,

$$\delta([\alpha], [\beta]) = \inf\{ \lim_{n\to\mathbb{N}} \rho(\eta_n, \gamma_n) : \eta \in [\alpha],\ \gamma \in [\beta]\}$$

is a metric.

Let

$$[\mathscr{U}] = \{[\alpha_{(j)}] : j \in \mathbb{N}\}]$$

be a Cauchy sequence in $[\mathscr{S}]$ with $\alpha_{(j)} = \{{\alpha_{(j)}}_n : n \in \mathbb{N}\}$ as representative of the class, $[\alpha_{(j)}]$. We are required to show that $[\mathscr{U}]$ converges to some $[\zeta]$ in $[\mathscr{S}]$.

28.5.1 Construction of a Cauchy sequence ζ: Let $\varepsilon > 0$.

Since $[\mathscr{U}]$ is Cauchy in $[\mathscr{S}]$, there exists M such that $j, k > M \Rightarrow \delta([\alpha_{(j)}], [\alpha_{(k)}]) < \varepsilon/3$. That is, for any $j, k > M$,

$$\inf\{ \lim_{n\to\infty} \rho({\eta_{(j)}}_n, {\eta_{(k)}}_n) : \eta_{(j)} \in [\alpha_{(j)}],\ \eta_{(k)} \in [\alpha_{(k)}]\} \ < \varepsilon/3$$

Then there exists at least one pair $j, k > M$ such that,

$$\lim_{n\to\infty} \rho({\alpha_{(j)}}_n, {\alpha_{(k)}}_n) < \varepsilon/3.$$ [1]

So, for this j, k, there exists N_1 such that $n > N_1$ implies,

$$\rho({\alpha_{(j)}}_n, {\alpha_{(k)}}_n) < \varepsilon/3 \quad (*)$$

Consider the triangle inequality expression for ρ,

$$\rho({\alpha_{(k)}}_k, {\alpha_{(j)}}_j) \leq \rho({\alpha_{(k)}}_k, {\alpha_{(k)}}_n) + \rho({\alpha_{(k)}}_n, {\alpha_{(j)}}_n) + \rho({\alpha_{(j)}}_n, {\alpha_{(j)}}_j)$$

By $(*)$ there is $j, k > M$ and N_1, such that

$$n > N_1 \qquad \Rightarrow \qquad \rho({\alpha_{(j)}}_n, {\alpha_{(k)}}_n) < \varepsilon/3$$

---

[1]For, if $\lim_{n\to\infty} \rho({\alpha_{(j)}}_n, {\alpha_{(k)}}_n) \geq \varepsilon/3$, for all $j, k > M$ $\delta([\alpha_{(j)}], [\alpha_{(k)}]) = \inf\{\lim_{n\to\infty} \rho({\alpha_{(j)}}_n, {\alpha_{(k)}}_n) : \eta_{(j)} \in [\alpha_{(j)}],\ \eta_{(k)} \in [\alpha_{(k)}]\} \geq \varepsilon/3$.

Since $a_{(k)} = \{\alpha_{(k)_n} : n \in \mathbb{N}\}$ is Cauchy, there exists N_2 such that,

$$n, k > N_2 \qquad \Rightarrow \qquad \rho(\alpha_{(k)_k}, \alpha_{(k)_n}) < \varepsilon/3$$

Since $a_{(j)} = \{\alpha_{(j)_n} : n \in \mathbb{N}\}$ is Cauchy in S, there exists N_3 such that,

$$j, n > N_3 \qquad \Rightarrow \qquad \rho(\alpha_{(j)_j}, \alpha_{(j)_n}) < \varepsilon/3$$

Then for, $k, j, n > N = \max\{M, N_1, N_2, N_3\}$

$$\begin{aligned}\rho(\alpha_{(k)_k}, \alpha_{(j)_j}) &\leq \rho(\alpha_{(k)_k}, \alpha_{(k)_n}) + \rho(\alpha_{(k)_n}, \alpha_{(j)_n}) + \rho(\alpha_{(j)_n}, \alpha_{(j)_j}) \\ &< \varepsilon/3 + \varepsilon/3 + \varepsilon/3 \\ &= \varepsilon\end{aligned}$$

Then $\cup\{a_{(j)} : j \in \mathbb{N}\}$ contains a sequence $\zeta = \{a_{(j)_j} : j \in \mathbb{N}\}$ in S which is a Cauchy sequence and so belongs to $\mathscr{S}$. Then $[\zeta] \in [\mathscr{S}]$.

Claim: We claim that the Cauchy sequence, $[\mathscr{U}] = \{[\alpha_{(j)}] : j \in \mathbb{N}\}$, in $[\mathscr{S}]$ converges to $[\zeta]$.

Proof of claim: It suffices to show that $\lim_{j\to\infty} \delta([\alpha_{(j)}], [\zeta]) = 0$.

Let $\varepsilon > 0$.

Since $\zeta = \{\alpha_{(0)_0}, \alpha_{(1)_1}, \alpha_{(2)_2}, \ldots\}$ is a Cauchy sequence in S, there exists M_1 such that $j, k > M_1$ implies $\rho(\alpha_{(j)_j}, \alpha_{(k)_k}) < \varepsilon/2$.

Since $a_{(j)} = \{\alpha_{(j)_n} : n \in \mathbb{N}\}$ is Cauchy in S, there exists M_2 such that $j, k > M_2$ implies $\rho(\alpha_{(j)_k}, \alpha_{(j)_j}) < \varepsilon/2$.

By triangle inequality,

$$\rho(\alpha_{(j)_k}, \alpha_{(k)_k}) \leq \rho(\alpha_{(j)_k}, \alpha_{(j)_j}) + \rho(\alpha_{(j)_j}, \alpha_{(k)_k})$$

If $j, k > M = \max\{M_1, M_2\}$ then

$$\begin{aligned}\rho(\alpha_{(j)_k}, \alpha_{(k)_k}) &\leq \rho(\alpha_{(j)_k}, \alpha_{(j)_j}) + \rho(\alpha_{(j)_j}, \alpha_{(k)_k}) \\ &< \varepsilon/2 + \varepsilon/2 \\ &= \varepsilon\end{aligned}$$

So

$$j, k > M \qquad \Rightarrow \qquad \rho(\alpha_{(j)_k}, \alpha_{(k)_k}) < \varepsilon$$

Since $\inf\{\lim_{k\to\infty} \rho(\eta_{(j)_k}, \eta_{(k)_k}) : \eta_{(j)} \in [\alpha_{(j)}], \eta_{(k)} \in [\zeta]\} < \varepsilon$, then $\lim_{j\to\infty} \delta([\alpha_{(j)}], [\zeta]) = 0$.

We conclude that $[\mathscr{U}]$ converges to $[\zeta]$, as claimed.

With the next result we complete the proof of the statement:

Every metric space has a completion.

Proposition 28.2 *The metric space (S, ρ) is dense in $([\mathscr{S}], \delta)$.*

Proof. We are given a metric space (S, ρ) and $\mathscr{S}$, the set of all Cauchy sequences in S. If $\alpha \in \mathscr{S}$, $[\alpha] = \{\eta \in \mathscr{S} : \lim_{n\to\infty} \rho(\eta_n, \alpha_n) = 0\}$. Let $[\mathscr{S}] = \{[\alpha] : \alpha \text{ is Cauchy in } S\}$. For $\mathbf{x} = \{x, x, x, \ldots\}$ in S let

$$[S] = \{[\mathbf{x}] : x \in S\}$$

in $[\mathscr{S}]$. The function $f(x) = [\mathbf{x}]$, embeds an isometric copy, $f[S]$, of S into $[\mathscr{S}]$.

For $\alpha = \{\alpha_n : n \in \mathbb{N}\}$, suppose $[\alpha] \in [\mathscr{S}] \setminus [S]$. To show that $[S]$ is dense in $[\mathscr{S}]$, it will suffice to show that there is a sequence, $\{[\mathbf{a}_{(n)}] : n \in \mathbb{N}\}$, in $[S]$ which converges to $[\alpha]$ with respect to the metric δ.

We define $\{[\mathbf{a}_{(n)}] : n \in \mathbb{N}\}$ in terms of $\alpha = \{\alpha_n : n \in \mathbb{N}\}$ as

$$\begin{aligned}
\mathbf{a}_{(0)} &= \{\alpha_0, \alpha_0, \alpha_0, \ldots .\} \\
\mathbf{a}_{(1)} &= \{\alpha_1, \alpha_1, \alpha_1, \ldots .\} \\
\mathbf{a}_{(2)} &= \{\alpha_2, \alpha_2, \alpha_2, \ldots .\} \\
&\vdots \\
\mathbf{a}_{(k)} &= \{\alpha_k, \alpha_k, \alpha_k, \ldots .\} \\
&\vdots
\end{aligned}$$

We claim that $\{[\mathbf{a}_{(n)}] : n \in \mathbb{N}\}$ converges to $[\alpha]$.

Proof of claim: Let $\varepsilon > 0$. It suffices to show there is a N_0 such that $k > N_0$ implies $\delta([\alpha], [\mathbf{a}_k]) < \varepsilon$.

Since α is Cauchy in S, there exists N_0, such that

$$m, n > N_0 \Rightarrow \rho(\alpha_n, \alpha_m) < \varepsilon$$

Suppose $k > N_0$ and $n > N_0$. Then

$$\begin{aligned}\delta([\alpha], [\mathbf{a}_k]) &= \inf \{ \lim_{n\to\mathbb{N}} \rho(\eta_n, \gamma_n) : \eta \in [\alpha],\ \gamma \in [\mathbf{a}_k]\} \\ &= \inf \{ \lim_{n\to\mathbb{N}} \rho(\alpha_n, \alpha_k) : \eta \in [\alpha],\ \gamma \in [\mathbf{a}_k]\} \\ &= \rho(\alpha_n, \alpha_k) \\ &< \varepsilon\end{aligned}$$

Then $\{[\mathbf{a}_{(n)}] : n \in \mathbb{N}\}$ converges to $[\alpha]$.

So, (S, ρ) is dense in $([\mathscr{S}], \delta)$.

Chapter 29

The Uniform Space and the Uniform Topology

Abstract
This section serves as a brief introduction to uniform space. Given a non-empty set S we define a family of subsets, $\mathscr{U}$, of $\mathscr{P}(S \times S)$ which satisfies a set of axioms. The family, $\mathscr{U}$, is called a uniformity on S and the pair $(S, \mathscr{U})$, is called a uniform space. We then show that any metric space, (S, ρ), generates a uniform space, $(S, \mathscr{U}_\rho)$, which inherits some of its metric space properties. Even though $(S, \mathscr{U})$ is not a topological space, the uniformity $\mathscr{U}$, generates a topology $\tau_{\mathscr{U}}$, on S. A topological space, (S, τ), which can be obtained from a uniformity, $\mathscr{U}$, is called a uniformizable topological space.

29.1 Introduction

A topological space generalizes many concepts and properties associated to metric spaces. But, some properties closely associated to the structure of a metric space do not transfer easily to a less structured topological space. For example, notions associated to "Cauchy sequences" or "uniform continuity" are not meaningful in the context of a topological space even though such tools can be useful in describing some topological properties. To see this, witness how the function $g(x) = 1/x$ on $\mathbb{R}\backslash\{0\}$ maps $\mathbb{R}\backslash\{0\}$ homeomorphically onto itself. The sequence $T = \{1, 1/2, 1/3, \ldots\}$ is a Cauchy sequence in the domain of g. But its image, $g[T] = \{1, 2, 3, \ldots\}$, is not Cauchy in $\mathbb{R}\backslash\{0\}$. So being "Cauchy" is not a topological property and so cannot be adequately expressed in this mathematical context.

In this section, we discuss a space which, like topological space, generalizes metric spaces, but in slightly different way.

We begin by reviewing a few basic notions associated with relations.

Basic definitions and notation: Recall that for a given non-empty set S, a subset U of $S \times S$ is referred to as a (binary) *relation* on S. If U is a relation on S,

$$\text{dom } U = \{x \in S : \text{ there is a } y \in S \text{ such that } (x, y) \in U\}$$
$$\text{im } U = \{y \in S : (x, y) \in U \text{ for some } x \in S\}$$

If U is a relation on S, then its inverse, U^{-1}, is defined as

$$U^{-1} = \{(x, y) : (y, x) \in U\}$$

Composition of relations: If U and V are relations on S

$$U{\circ}V = \{(a, b) : (a, c) \in V \text{ and } (c, b) \in U \text{ for some } c \in \text{im } V\}$$

The relation

$$\Delta(S) = \{(x, x) : x \in S\}$$

is referred to as the *diagonal* of $S \times S$.

Symmetric relations: The relation, U, is said to be *symmetric* if and only if $U = U^{-1}$. Alternatively, U is symmetric whenever...

$$(a, b) \in U \text{ if and only if } (b, a) \in U$$

Definition 29.1 Let S be a non-empty set. A *uniform space* is a pair $(S, \mathscr{U})$ where $\mathscr{U}$ is a non-empty subset of $\mathscr{P}(S \times S)$ whose elements satisfy the following five axioms:

U1. $\Delta(S) \subseteq U$ for all $U \in \mathscr{U}$.
U2. If $U, V \in \mathscr{U}$, then $U \cap V \in \mathscr{U}$. That is, $\mathscr{U}$ is closed under finite intersections.
U3. If $U \in \mathscr{U}$ then $V{\circ}V \subseteq U$ for some $V \in \mathscr{U}$.

U4. If $U \in \mathscr{U}$ then $E^{-1} \subseteq U$ for some $E \in \mathscr{U}$.

U5. If $U \in \mathscr{U}$ and $U \subseteq V \in \mathscr{P}(S \times S)$, then $V \in \mathscr{U}$. That is, $\mathscr{U}$ is closed under supersets.

A subset, $\mathscr{U}$, of $\mathscr{P}(S \times S)$ which satisfies these five axioms is called a *uniformity* for S.

It may be helpful to view each element of a uniformity, $\mathscr{U}$, of S, as a region in $S \times S$ surrounding the diagonal, $\Delta(S)$. A pair of points a, b in S may be considered as being "near" each other if and only if the pair (a, b) in $S \times S$ is "near" the diagonal, $\Delta(S)$.

In spite of its nomenclature, a uniform space, $(S, \mathscr{U})$, is *not* a topological space. We will however soon define a topology, $\tau_{\mathscr{U}}$, on S which is derived from the structure, $\mathscr{U}$. Remember that, even though a topology and a uniformity are both families of subsets, a topology is a subset of $\mathscr{P}(S)$ while a uniformity is a subset of $\mathscr{P}(S \times S)$.

Like a topology on S, a uniformity, $\mathscr{U}$, for S is not unique.

Also note that, since $\mathscr{U}$ is non-empty and is closed both under finite intersections and supersets (U2 and U5), it is a filter of sets on $S \times S$.

Example 1. The largest uniformity on S, is the family

$$\mathscr{U}_d = \{U \in \mathscr{P}(S \times S) : \Delta(S) \subseteq U\}$$

It is called the *discrete uniformity.*

Example 2. The smallest uniformity on S is the singleton set,

$$\mathscr{U}_i = \{S \times S\}$$

It is called the *trivial uniformity* or the *indiscrete uniformity.*

Proposition 29.2 *Let S be a non-empty set and $\mathscr{U}$ be a uniformity on S. If $U \in \mathscr{U}$, then U^{-1} also belongs to $\mathscr{U}$ on S.*

Proof. We are given a set S and an element, U, of a uniformity, $\mathscr{U}$. By property U4, $\mathscr{U}$ contains an element V such that $V^{-1} \subseteq U$. Then $V = (V^{-1})^{-1} \subseteq U^{-1}$. Since $V \in \mathscr{U}$ and $V \subseteq U^{-1}$, then, by U5, $U^{-1} \in \mathscr{U}$, as required.

Note that,

$$U \in \mathscr{U} \Rightarrow [U \cap U^{-1} \in \mathscr{U} \text{ and is symmetric}] \quad (\dagger)$$

To see this, note that if $(a, b) \in U \cap U^{-1}$ then $(a, b) \in U$, so $(b, a) \in U^{-1}$; if $(b, a) \notin U$, then $(a, b) \notin U^{-1}$, contradiction. So $(b, a) \in U$. Then $U \cap U^{-1}$ is symmetric.

29.2 A Base for a Uniformity

To facilitate our investigation of uniform spaces, $(S, \mathscr{U})$, we first briefly review some of the mechanisms used to generate various topologies on a set S. There are surprisingly few axioms that govern the family, τ, which defines a topological space. They simply state that τ contains both $\varnothing$ and S, is closed under finite intersections and closed under arbitrary unions. We also know that any given topology, τ, on S, contains a distinguished subfamily, $\mathscr{B}$, referred to as a *base* for τ, where,

> *for every $x \in S$ and $x \in U \in \tau$, there is a $B \in \mathscr{B}$ such that $x \in B \subseteq U$.*

It allows us to recover all open sets simply by taking arbitrary unions of elements of $\mathscr{B}$.

We also saw that, given any (untopologized) set S and $\mathscr{B} \subseteq \mathscr{P}(S)$, the property,

> *if $\mathscr{B}$ covers S and for $A, B \in \mathscr{B}$, if $x \in A \cap B$, there is $C \in \mathscr{B}$ such that $x \in C \subseteq A \cap B$.*

characterizes a base for a topology on S. By this we mean that, if $\mathscr{B}$ satisfies this property, it generates a topology, τ, on S by collecting the union of all subsets of $\mathscr{B}$. More specifically,

$$\tau = \{U \in \mathscr{P}(S) : U = \cup\{C : C \in \mathscr{C}\} \text{ for some subset } \mathscr{C} \text{ of } \mathscr{B}\}$$

The definition we provide for the base of a uniformity will, in some ways, be similar to the one given for a base for a topology.

Definition 29.3 Let S be a non-empty set and $\mathscr{U}$ be a uniformity on S.

A subfamily $\mathscr{B}$ of $\mathscr{U}$ is a *base* for $\mathscr{U}$, if,

for every $U \in \mathscr{U}$, *there is a* $B \in \mathscr{B}$ *such that* $B \subseteq U$

A subfamily, $\mathscr{S}$, of $\mathscr{U}$ is a *subbase* for $\mathscr{U}$, if the family of finite intersections of the members of $\mathscr{S}$ forms a base for $\mathscr{U}$.

The characterization we will obtain for the uniformity base, $\mathscr{B}$, is a slightly weakened version of axioms U1 to U4. The uniformity, $\mathscr{U}$, is then generated from $\mathscr{B}$ by applying the axiom U5 to $\mathscr{B}$.

Theorem 29.4 *Let S be a non-empty set and $\mathscr{B}$ be a non-empty subset of $\mathscr{P}(S \times S)$.*

The family, $\mathscr{B}$, is a base for some uniformity on S if and only if, $\mathscr{B}$ satisfies U1, U2, U3 and U4 where U2* is a slightly weaker version of U2:*

U2 : "If $U, V \in \mathscr{B}$, then $W \subseteq U \cap V \in \mathscr{U}$ for some $W \in \mathscr{B}$"*

Proof. We are given a set S.

($\Leftarrow$) Suppose $\mathscr{B}$ is a subset of $\mathscr{P}(S \times S)$ which satisfies properties, U1, U2*, U3, and U4. We are required to show that $\mathscr{B}$ is a base for some uniformity on S.

Let

$$\mathscr{U}_{\mathscr{B}} = \{V \in \mathscr{P}(S \times S) : V \text{ contains some } B \in \mathscr{B}\}$$

— By definition of $\mathscr{U}_{\mathscr{B}}$ every V in $\mathscr{U}_{\mathscr{B}}$ contains an element of $\mathscr{B}$.
— Furthermore, if $B \in \mathscr{B}$, $B \subseteq B$ so $\mathscr{B}$ is a subfamily of $\mathscr{U}_{\mathscr{B}}$.

It now suffices to show that the family, $\mathscr{U}_{\mathscr{B}}$ generated by $\mathscr{B}$, is a uniformity on S. We do this by verifying that it satisfies the properties U1 to U5.

U1. Since $\Delta(S) \subseteq B$ for all $B \in \mathscr{B}$ and $B \subseteq V$ for some V in $\mathscr{U}_{\mathscr{B}}$, then every element of $\mathscr{U}_{\mathscr{B}}$ contains $\Delta(S)$. So, $\mathscr{U}_{\mathscr{B}}$ satisfies U1.

U2. Let $U, V \in \mathscr{U}_{\mathscr{B}}$. Then $B \subseteq U$, $W \subseteq V$ for some $B, W \in \mathscr{B}$. There exists $Z \in \mathscr{B}$ such that $Z \subseteq B \cap W$ (by U2*). Then $Z \subseteq U \cap V$. By definition of $\mathscr{U}_{\mathscr{B}}$, $U \cap V \in \mathscr{U}_{\mathscr{B}}$. So, $\mathscr{U}_{\mathscr{B}}$ satisfies U2.

U3. Let $U \in \mathscr{U}_{\mathscr{B}}$. Then $B \subseteq U$ for some $B \in \mathscr{B}$. Since $\mathscr{B}$ satisfies U3, $W \circ W \subseteq B$ for some $W \in \mathscr{B}$ where $W \in \mathscr{U}_{\mathscr{B}}$. So, $W \circ W \subseteq U$. We conclude that $\mathscr{U}_{\mathscr{B}}$ satisfies U3.

U4. Let $U \in \mathscr{U}_{\mathscr{B}}$. Then $B \subseteq U$ for some $B \in \mathscr{B}$. Since $\mathscr{B}$ satisfies U4, $W^{-1} \subseteq B$ for some $W \in \mathscr{B}$ where $W \in \mathscr{U}_{\mathscr{B}}$. Then $W^{-1} \subseteq U$. So, $\mathscr{U}_{\mathscr{B}}$ satisfies U4.

U5. Let $U \in \mathscr{U}_{\mathscr{B}}$ and $V \in \mathscr{P}(S \times S)$ such that $U \subseteq V$. Then $B \subseteq U \subseteq V$ for some $B \in \mathscr{B}$. Then $B \subseteq V$. By definition of $\mathscr{U}_{\mathscr{B}}$, $V \in \mathscr{U}_{\mathscr{B}}$. So, $\mathscr{U}_{\mathscr{B}}$ satisfies U5.

We conclude that if $\mathscr{B}$ is a subset of $\mathscr{P}(S \times S)$ which satisfies properties, U1, U2*, U3 and U4, then it is a base for the uniformity, $\mathscr{U}_{\mathscr{B}}$.

($\Rightarrow$) Suppose now that $\mathscr{B}$ is a base for the uniformity $\mathscr{U}$ on S. Then $\mathscr{B}$ is a subfamily of $\mathscr{U}$ where for every $V \in \mathscr{U}$ there is a $B \in \mathscr{B}$ such that $B \subseteq V$. It suffices to show that $\mathscr{B}$ satisfies U1, U2*, U3, U4.

That $\mathscr{B}$ satisfies U1 and U2* is immediate.

Suppose $B \in \mathscr{B}$. Since $B \in \mathscr{U}$, there is a $V \in \mathscr{U}$ such that $V \circ V \subseteq B$ for some $V \in \mathscr{U}$. Since $\mathscr{B}$ is a base of $\mathscr{U}$, there is a $W \in \mathscr{B}$ such that $W \subseteq V$. Then $W \circ W \subseteq V \circ V \subseteq B$. So, $\mathscr{B}$ satisfies U3.

Suppose $B \in \mathscr{B}$. Since $B \in \mathscr{U}$, there is a $V \in \mathscr{U}$ such that $V^{-1} \subseteq B$ for some $V \in \mathscr{U}$. Since $\mathscr{B}$ is a base of $\mathscr{U}$, there is a $W \in \mathscr{B}$ such that $W \subseteq V$. Then $W^{-1} \subseteq V^{-1} \subseteq B$. So, $\mathscr{B}$ satisfies U4.

We are done.

The above result guarantees that...

> *if* $\mathscr{B} \subseteq \mathscr{P}(S \times S)$ *satisfies* U1, U2*, U3, *and* U4 *then the collection,* $\mathscr{U}_{\mathscr{B}}$, *of all supersets of elements of* $\mathscr{B}$ *is a uniformity on* S.

The following result shows that, for any uniformity $\mathscr{U}$ on S, the family of all symmetric relations in $\mathscr{U}$ forms a base for $\mathscr{U}$.

Theorem 29.5 *Let S be a non-empty set and $\mathscr{U}$ be a uniformity on S with base $\mathscr{B}$. Suppose*

$$\mathscr{B}_{sym} = \{V \in \mathscr{U} : V \text{ is symmetric}\}$$

Then $\mathscr{B}_{sym}$ forms a base for the uniformity $\mathscr{U}$.

Proof. Let $\mathscr{U}$ be a uniformity on S with base $\mathscr{B}$ and $U \in \mathscr{B} \subseteq \mathscr{U}$. By Proposition 29.2, $U^{-1} \in \mathscr{U}$. Let $B = U \cap U^{-1}$. By U2, $B \in \mathscr{U}$. If $(a, b) \in U \cap U^{-1}$ then $(a, b) \in U$ and so $(b, a) \in U^{-1}$. Since $(a, b) \in U^{-1}$, then $(b, a) \in U$. Then $(b, a) \in U \cap U^{-1}$. So, $B = U \cap U^{-1}$ is symmetric. Since $B \subseteq U$, $\mathscr{B}_{\text{sym}}$ forms a base for the uniformity $\mathscr{U}$, as required.

Example 3. Consider the set of real numbers $\mathbb{R}$. For $\kappa > 0$, let

$$B_\kappa = \{(a, b) \in \mathbb{R} \times \mathbb{R} : |a - b| < \kappa\}$$

Verify that the collection,

$$\mathscr{B} = \{B_\kappa : \kappa > 0\}$$

forms a base for some uniformity on $\mathbb{R}$.

Solution: We verify that $\mathscr{B}$ satisfies the four base properties for a uniformity.

U1. Since $(x, x) \in B_\kappa$ for all κ, then $\mathscr{B}$ satisfies U1.

U2*. For κ_1 and κ_2 larger than 0, let $\kappa_3 = \min\{\kappa_1, \kappa_2\}$. If $(x, y) \in B_{\kappa_3}$ then (x, y) are strictly within a distance of κ_3 from each other. So, $(x, y) \in B_{\kappa_1} \cap B_{\kappa_2}$. Then $B_{\kappa_3} \subseteq B_{\kappa_1} \cap B_{\kappa_2}$. It follows that $\mathscr{B}$ satisfies U2*.

U3. Let $B_\kappa \in \mathscr{B}$. We claim there exists λ such that $B_\lambda \circ B_\lambda \subseteq B_\kappa$. Let $\lambda = \kappa/4$. Recall that

$$U \circ V = \{(u, v) : (u, y) \in V \text{ and } (y, v) \in U \quad \text{for some } y \in \text{im } V.\}$$

Let $(u, v) \in B_{\kappa/4} \circ B_{\kappa/4}$. We claim that $|u - v| < \kappa$.

$$\begin{aligned}(u, v) \in B_{\kappa/4} \circ B_{\kappa/4} &\Rightarrow \exists\, z \in \text{ im} \\ &\quad \times B_{\kappa/4} \text{ such that } (u, z) \in B_{\kappa/4} \\ &\quad \times \text{ and } (z, v) \in B_{\kappa/4} \\ &\Rightarrow |u - z| < \kappa/4 \text{ and } |z - v| \in \kappa/4 \\ |u - v| &\leq |u - z| + |z - v| \\ &< \kappa/4 + \kappa/4 \\ &= \kappa/2 < \kappa\end{aligned}$$

Then $(u, v) \in B_\kappa$. So $B_{\kappa/4} \circ B_{\kappa/4} \subseteq B_\kappa$. We conclude that $\mathscr{B}$ satisfies U3.

U4. Let $B_\kappa \in \mathscr{B}$. Since $B_\kappa = \{(x, y) : |x - y| < \kappa\} = \{(x, y) : |y - x| < \kappa\} = B_\kappa^{-1}$, then $\mathscr{B}$ satisfies U4.
Then

$$\mathscr{B} = \{B_\kappa : \kappa > 0\}$$

is a base which generates a uniformity on $\mathbb{R}$. Then the uniformity on $\mathbb{R}$ is

$$\mathscr{U} = \{V \in \mathscr{P}(\mathbb{R} \times \mathbb{R}) : B_\kappa \subseteq V \text{ for some } \kappa > 0\}$$

Definition 29.6 The uniformity on $\mathbb{R}$ generated by the base, $\mathscr{B} = \{B_\kappa : \kappa > 0\}$, illustrated in the above example is called the *usual* or the *standard* uniformity on $\mathbb{R}$.

29.3 A Uniform Space, $(S, \mathscr{U})$, Generated by a Metric, ρ, on S

We see that a uniform space, $(S, \mathscr{U})$, has slightly more structure than a topological space (S, τ). In the following example, we verify that a metric, ρ, on a set S will always generate a uniformity on S.

Example 4. Let (S, ρ) be a metric space. For each $\kappa > 0$, let

$$B_\kappa = \{(x, y) \in S \times S : \rho(x, y) < \kappa\}$$

be a subfamily of $\mathscr{P}(S \times S)$ generated by the metric ρ. Verify that the family, $\mathscr{B}_\rho = \{B_\kappa : \kappa > 0\}$, is a base which generates a uniformity, $\mathscr{U}_\rho$, on S.

Solution: It will suffice to show that $\mathscr{B}_\rho$ satisfies properties U1, U2*, U3, and U4. If so, then, by Theorem 29.4, $\mathscr{B}_\rho$ generates the uniformity,

$$\mathscr{U}_\rho = \{V \in \mathscr{P}(S \times S) : V \text{ contains some } B_\kappa \in \mathscr{B}_\rho\}$$

Since $\rho(x, x) = 0 < \kappa$ for all $x \in S$, then $\Delta(S) \subseteq B_\kappa$ for each κ. This is property U1.

For B_{κ_1}, $B_{\kappa_2} \in \mathscr{B}_\rho$, let $\kappa = \min\{\kappa_1, \kappa_2\}/2$. Then $(x, y) \in B_\kappa$ implies $\rho(x, y) < \kappa$. Since $\kappa < \kappa_1$ and $\kappa < \kappa_2$, then $(x, y) \in B_{\kappa_1} \cap B_{\kappa_2}$. So, $B_\kappa \subseteq B_{\kappa_1} \cap B_{\kappa_2}$. Then there exists κ such that $B_\kappa \subseteq B_{\kappa_1} \cap B_{\kappa_2}$. This is property U2*.

Let $B_\kappa \in \mathscr{B}_\rho$. We claim there exists λ such that $B_\lambda \circ B_\lambda \subseteq B_\kappa$. Let $\lambda = \kappa/4$. Recall that

$$U \circ V = \{(u, v) : (u, y) \in V, (y, v) \in U, \quad \text{for some } y \in \text{im } V.\}$$

Let $(u, v) \in B_{\kappa/4} \circ B_{\kappa/4}$. We claim that $\rho(u, v) < \kappa$.

$$\begin{aligned} (u, v) \in B_{\kappa/4} \circ B_{\kappa/4} &\Rightarrow \text{ there is } z \in \text{ im } B_{\kappa/4} \\ &\quad \times \text{ such that } (u, z) \in B_{\kappa/4}, (z, v) \in B_{\kappa/4} \\ &\Rightarrow \rho(u, z) < \kappa/4 \text{ and } \rho(z, v) \in \kappa/4 \\ \rho(u, v) &\leq \rho(u, z) + \rho(z, v) \\ &< \kappa/4 + \kappa/4 \\ &= \kappa/2 < \kappa \end{aligned}$$

Then $(u, v) \in B_\kappa$. So, $B_{\kappa/4} \circ B_{\kappa/4} \subseteq B_\kappa$. We conclude that $\mathscr{B}$ satisfies U3.

Let $B_\kappa \in \mathscr{B}_\rho$. Recall $B_\kappa = \{(x, y) : \rho(x, y) < \kappa\} = \{(x, y) : \rho(y, x) < \kappa\} = B_\kappa^{-1}$. Then $\mathscr{B}_\rho$ satisfies U4.

Then $\mathscr{B}_\rho$ is a base which generates a uniformity on (S, ρ).

Definition 29.7 Given a metric space, (S, ρ), the uniformity, $\mathscr{U}_\rho$, generated by ρ is called the *metric uniformity*.

This example shows that every metric, ρ, on a space S, will generate a uniformity, $\mathscr{U}_\rho$. In Example 3, the uniformity described on $\mathbb{R}$ is simply the metric uniformity generated by the metric $\rho(a, b) = |a - b|$ on $\mathbb{R}$.

Example 5. Show that the metric 2ρ generates the same uniformity on S as the metric space ρ.

Solution: It will suffice to show that $\mathscr{U}_\rho = \mathscr{U}_{2\rho}$.

Suppose $V \in \mathscr{U}_\rho$. Then $V \in \mathscr{P}(S \times S)$ contains some $B_\lambda \in \mathscr{B}_\rho$. There exist $B_\mu \in \mathscr{B}_{2\rho}$ such that $B_\mu \subseteq B_\lambda \subseteq V$. So, $V \in \mathscr{U}_{2\rho}$. Then $\mathscr{U}_\rho \subseteq \mathscr{U}_{2\rho}$.

Suppose $V \in \mathscr{U}_{2\rho}$. Then $V \in \mathscr{P}(S \times S)$ contains some $B_\lambda \in \mathscr{B}_{2\rho}$. There exist $B_\mu \in \mathscr{B}_\rho$ such that $B_\mu \subseteq B_\lambda \subseteq V$. So, $V \in \mathscr{U}_\rho$. Then $\mathscr{U}_{2\rho} \subseteq \mathscr{U}_\rho$.

Then ρ and 2ρ generate the same uniformity on S.

Since a single uniformity does not always distinguish distinct metric spaces, uniform spaces have less structure than metric spaces.

29.4 The Uniform Topology, $\tau_{\mathscr{U}}$, Generated by a Uniformity, $\mathscr{U}$, on a Set

It is important to keep in mind that a uniformity, $\mathscr{U}$, on a set is not, by itself, a topology on S. It is not even a subset of $\mathscr{P}(S)$. We will show, however, that we can use the elements of a uniformity, $\mathscr{U}$, on S to define, in a natural way, a corresponding topology, $\tau_{\mathscr{U}}$ on S.

Recall that, if B is an element of a uniformity $\mathscr{U}$ on S, then $B : S \to S$ is a subset of $S \times S$ which contains $\Delta(S)$. So, for each $B \in \mathscr{U}$, the image of $\{x\}$ under B is a subset of S which contains the point x. Let $B(x)$ denote the image of $\{x\}$. The collection

$$\{B(x) : B \in \mathscr{U}\}$$

is a family of subsets of S each containing x. Since, for each $B \in \mathscr{U}$, the image of B at x is $B(x) = \{y \in S : (x, y) \in B\}$, $B(x)$ can also be viewed as

$$B(x) = \pi_2 [(\{x\} \times S) \cap B]$$ [1]

Theorem 29.8 *Let S be a non-empty set, $\mathscr{U}$ be a uniformity on S, and $\mathscr{B}$ be a base for $\mathscr{U}$. For $x \in S$, and B in $\mathscr{U}$, let $B(x)$ denote the image of $\{x\}$ (in S) under the relation, B. Let*

$$\tau_{\mathscr{U}} = \{U \in \mathscr{P}(S) : \textit{for each } x \in U \textit{ there exists } B \in \mathscr{U} \textit{ such that } B(x) \subseteq U\}$$

Then $\tau_{\mathscr{U}}$ is a topology on S with base $\mathscr{B}_\tau = \{B(x) : x \in S$ and $B \in \mathscr{B}\}$.

---

[1] Where $\pi_2(a, b) = b$.

Proof. We are given a set S and a uniformity $\mathscr{U}$ on S and the family of subsets,
$\tau_{\mathscr{U}} = \{U \in \mathscr{P}(S) :$ for every $x \in U$ there exists $B \in \mathscr{U}$ such that $B(x) \subseteq U\}$.

O1. Clearly $\varnothing \in \tau_{\mathscr{U}}$. Let $x \in S$. Since $S \times S \in \mathscr{U}$, $x \in (S \times S)(x) = S \subseteq S$. So $S \in \tau_{\mathscr{U}}$.

O2. Let $U, V \in \tau_{\mathscr{U}}$. We claim $U \cap V \in \tau_{\mathscr{U}}$. Suppose $x \in U \cap V$. It suffices to show there is a $W \in \mathscr{U}$ such that $W(x) \subseteq U \cap V$.

By definition of $\tau_{\mathscr{U}}$ there exists B_1, B_2 in $\mathscr{U}$ such that $x \in B_1(x) \subseteq U$ and $x \in B_2(x) \subseteq V$. By U2, $B_1 \cap B_2 \in \mathscr{U}$.

$$\begin{aligned} z \in (B_1 \cap B_2)(x) \quad &\Rightarrow \quad (x, z) \in B_1 \cap B_2 \\ &\Rightarrow \quad z \in B_1(x) \text{ and } z \in B_2(x) \\ &\Rightarrow \quad z \in B_1(x) \cap B_2(x) \\ &\Rightarrow \quad z \in U \cap V \end{aligned}$$

Then $(B_1 \cap B_2)(x) \subseteq U \cap V$. So $U \cap V \in \tau_{\mathscr{U}}$.

O3. Let $\{U_i : i \in A\}$ be a family of elements of $\tau_{\mathscr{U}}$. If $x \in \cup\{U_i : i \in A\}$ then $x \in U_j$ for some $j \in A$. Then there exists $B \in \mathscr{U}$ such that $x \in B(x) \subseteq U_j \subseteq \cup\{U_i : i \in A\}$. So, $\tau_{\mathscr{U}}$ is closed under arbitrary unions.

So, $\tau_{\mathscr{U}}$ is a topology. By definition of $\tau_{\mathscr{U}}$, for each $U \in \tau_{\mathscr{U}}$ and $x \in U$ there is a $V \in \mathscr{U}$ such that $x \in V(x) \subseteq U$. Then there is a $B \in \mathscr{B}$ such that $B \subseteq V$ and so $x \in B(x) \subseteq V(x) \subseteq U$. We conclude that

$$\mathscr{B}_\tau = \{B(x) : x \in S \text{ and } B \in \mathscr{B}\}$$

forms a base for $\tau_{\mathscr{U}}$. (Note that, in the proof, we can choose as base $\mathscr{B}$ for the uniformity, $\mathscr{U}$, the family, $\mathscr{B} = \mathscr{B}_{\text{sym}}$, if necessary.)

Definition 29.9 Let S be a non-empty set and $\mathscr{U}$ be a uniformity on S. The topology on S, $\tau_{\mathscr{U}}$, generated by the uniformity, $\mathscr{U}$, is called a *uniform topology*.

Let (S, τ) be a topological space. If

$$\tau = \tau_{\mathscr{U}}$$

for some uniformity, $\mathscr{U}$, on S, then (S, τ) is said to be *uniformizable* topological space.

In the literature, the words "uniform space" is often used to refer to both $(S, \mathscr{U})$ and $(S, \tau_{\mathscr{U}})$. It is usually clear from the context to which space the author is referring to.

Theorem 29.8 guarantees that given a uniformity on a set S we can generate a corresponding topology on S:

$$S \longrightarrow (S, \mathscr{U}) \longrightarrow (S, \tau_{\mathscr{U}})$$

Finally, note that, for a given set S, we can speak of a base $\mathscr{B}$ for the uniformity $\mathscr{U}$ and a base $\mathscr{B}$ for the topology $\tau_{\mathscr{U}}$. Even though we tend to use the same symbol, $\mathscr{B}$, for both bases, the reader should determine, by the context, which one it refers to.

For convenience we juxtapose both concepts.

$\mathscr{B}$ is a base for $\mathscr{U}$: For every $U \in \mathscr{U}$, there is a $B \in \mathscr{B}$ such that $B \subseteq U$.

$\mathscr{B}_\tau$ is a base for $\tau_{\mathscr{U}}$: $\mathscr{B}_\tau = \{B(x) : x \in S \text{ and } B \in \mathscr{B}$, a base for $\mathscr{U}\}$.

By Definition 29.9, a topological space, (S, τ), derived from some uniformity, $\mathscr{U}$, on S (that is, where $\tau = \tau_{\mathscr{U}}$) is said to be "uniformizable". Are there any topological spaces that are not uniformizable? That is, is there a topology, τ, on S for which there is no uniformity, $\mathscr{U}$, such that $\tau = \tau_{\mathscr{U}}$? The answer is yes. We will show in Theorem 29.17 that a topological space derived from a uniformity is a regular space (i.e., $T_1 + T_3$).[2] So, a non-regular space is not uniformizable.

Example 6. Consider the usual uniformity, $\mathscr{U}$, on $\mathbb{R}$ which is generated by the base, $\mathscr{B} = \{B_\kappa : \kappa > 0\}$, given in example three (where

---

[2]In fact, a uniformizable space is always completely regular. We will not prove this in this book. For a proof the reader is referred to Theorem 38.2 of Willard where it is shown that, for (S, τ), "uniformizable $\Leftrightarrow$ completely regular".

$B_\kappa = \{(a,b) \in \mathbb{R} \times \mathbb{R} : |a-b| < \kappa\}$). By Theorem, 29.8, the uniformity, $\mathscr{U}$, generates a topology $\tau_{\mathscr{U}}$ on $\mathbb{R}$. Verify that $\tau_{\mathscr{U}}$ is the usual topology, τ, on $\mathbb{R}$. That is,

$$\mathbb{R} \longrightarrow (\mathbb{R}, \mathscr{U}) \longrightarrow (\mathbb{R}, \tau_{\mathscr{U}}) = (\mathbb{R}, \tau)$$

Solution: Let $\mathscr{U}$ be the usual uniformity on $\mathbb{R}$ and τ denote the usual topology on $\mathbb{R}$. Then $\mathscr{U}$ is generated by the base, $\mathscr{B} = \{B_\kappa : \kappa > 0\}$, where B_κ represents all pairs (a,b) such that $|a-b| < \kappa$.

Claim 1: We claim that $\tau_{\mathscr{U}} \subseteq \tau$.

Proof of claim: Let U be an open neighborhood of $x \in \mathbb{R}$ with respect to $\tau_{\mathscr{U}}$. By definition of $\tau_{\mathscr{U}}$, there exists $B_\kappa \in \mathscr{B}$ such that $x \in (B_\kappa)(x) \subseteq U$.

See that $(B_\kappa)(x) = (x - \kappa/2, x + \kappa/2)$.

So, U is a union of basic open intervals. So, $U \in \tau$. We conclude that $\tau_{\mathscr{U}} \subseteq \tau$, as claimed.

Claim 2: We claim that $\tau \subseteq \tau_{\mathscr{U}}$.

Proof of claim: Let A be a non-empty open subset in $\mathbb{R}$ and $x \in A$. Then there exists ε such that $(x-\varepsilon, x+\varepsilon) \subseteq A$. Consider $B_{\varepsilon/2}$, a base element of $\mathscr{U}$. Then $(B_{\varepsilon/2})(x) \in \tau_{\mathscr{U}}$. By Claim 1, $\tau_{\mathscr{U}} \subseteq \tau$; then $(B_{\varepsilon/2})(x) \in \tau$.

See that,

$$\begin{aligned} x &\in (x - \varepsilon/4, x + \varepsilon/4) \\ &= \pi_2\left[(\{x\} \times S) \cap B_{\varepsilon/2}\right] \\ &= (B_{\varepsilon/2})(x) \\ &\subseteq (x-\varepsilon, x+\varepsilon) \\ &\subseteq A \end{aligned}$$

Then the family, $\mathscr{B} = \{B(x) : x \in \mathbb{R}, B \in \mathscr{U}\}$ forms a base for τ. By Theorem 29.8, $\mathscr{B}$ is also a base for $\tau_{\mathscr{U}}$. So, $A \in \tau_{\mathscr{U}}$. We conclude that $\tau \subseteq \tau_{\mathscr{U}}$, as claimed.

So, $\tau_{\mathscr{U}} = \tau$.

We now examine, in a more general context, how the open subsets of the topology, $\tau_{\mathscr{U}}$, relate to the uniformity, $\mathscr{U}$, which generates it.

Theorem 29.10 *Let S be a non-empty set and $\mathscr{U}$ be a uniformity on S. Let $\tau_{\mathscr{U}}$ be the topology on S generated by $\mathscr{U}$. If $T \subseteq S$, then*

$$int_{\tau_{\mathscr{U}}} T = \{x \in T : x \in B(x) \subseteq T \text{ for some } B \in \mathscr{U}\}$$

Proof. We are given a set S and a uniformity $\mathscr{U}$ on S.

For each $B \in \mathscr{U}$, $B(x) = \pi_2[(\{x\} \times S) \cap B]$ and $\tau_{\mathscr{U}}$ is the topology generated by the base $\mathscr{B} = \{B(x) : x \in S \text{ and } B \in \mathscr{U}\}$.

Let $K = \{x \in T : x \in B(x) \subseteq T \text{ for some } B \in \mathscr{U}\}$. By definition $K \subseteq T$. We are required to show that $K = \text{int}_{\tau_{\mathscr{U}}} T$.

We first show that K is open in S (hence, $K \subseteq \text{int}_{\tau_{\mathscr{U}}} T$).

Let $x \in K$. Then there exists $U \in \mathscr{U}$ such that $x \in U(x) \subseteq T$. There exists $V \in \mathscr{U}$ such that $V \circ V \subseteq U$. We claim $V(x) \subseteq T$; if so $x \in V(x) \subset K$ and so K is open.

Let $u \in V(x)$. Then we have $V(u) \subseteq (V \circ V)(x)$. We then obtain the chain of inclusions,

$$\begin{aligned} u &\in V(u) \\ &\subseteq (V \circ V)(x) \\ &\subseteq U(x) \\ &\subseteq T \end{aligned}$$

So, $u \in T$ implies $V(x) \subseteq T$, as claimed.

Then $x \in V(x) \subset K$, which implies K is open.

Then $K \subseteq \text{int}_{\tau_{\mathscr{U}}} T$.

We now show $\text{int}_{\tau_{\mathscr{U}}} T \subseteq K$. Let $x \in \text{int}_{\tau_{\mathscr{U}}} T$. Since $\mathscr{B}$ is a base for $\tau_{\mathscr{U}}$ and $\text{int}_{\tau_{\mathscr{U}}} T$ is open, there exists $B \in \mathscr{U}$ such that $x \in B(x) \subseteq \text{int}_{\tau_{\mathscr{U}}} T \subseteq T$. So, $x \in K$.

Then $\text{int}_{\tau_{\mathscr{U}}} T = K$.

The following results exhibits another way of expressing the closure of a set in $(S, \tau_{\mathscr{U}})$.

In the statement, we use the following notation.

Notation: Let $\mathscr{U}$ be a uniformity on S. If $T \subseteq S$ then

$$U[T] = \cup_{x \in T}\{U(x)\}$$

Theorem 29.11 *Let S be a non-empty set and $\mathscr{U}$ be a uniformity on S. Let $\tau_{\mathscr{U}}$ be the topology on S generated by $\mathscr{U}$. Let $T \subseteq S$. Then the closure, $cl_{\tau_{\mathscr{U}}} T$, of T with respect to the uniform topology $\tau_{\mathscr{U}}$ can be expressed as, then*

$$cl_{\tau_{\mathscr{U}}} T = \bigcap_{U \in \mathscr{U}} \{U[T]\}$$

Proof. We are given a set S and a uniformity $\mathscr{U}$ on S from which we obtain the space $(S, \tau_{\mathscr{U}})$ equipped with the uniform topology. Let $T \subseteq S$.

We know that $\mathscr{B} = \{B(x) : x \in S \text{ and } B \in \mathscr{B}_{\text{sym}}\}$ forms a base of $\tau_{\mathscr{U}}$.

We are required to show that $x \in \mathrm{cl}_{\tau_{\mathscr{U}}} T$ if and only if $x \in \bigcap_{U \in \mathscr{U}} \{U[T]\}$.

To say that "$x \in \mathrm{cl}_{\tau_{\mathscr{U}}} T$" is equivalent to saying that "$B(x)$ intersects T for all $B \in \mathscr{B}$".

To say that "$B(x)$ intersects T for all $B \in \mathscr{B}$" is equivalent to saying that "x belongs to $B^{-1}[T]$ for all $B \in \mathscr{B}$". But since B is symmetric, we can replace "$B^{-1}[T]$" with "$B[T]$".

So we can rewrite the second statement as:

"The set, $B(x)$ intersects T for all $B \in \mathscr{B}$ if and only if, $x \in \bigcap_{U \in \mathscr{U}} \{U[T]\}$".

The desired conclusion follows.

Example 7. Suppose $(S, \mathscr{U})$ is the uniform space equipped with the discrete uniformity. Show that the corresponding topological space is equipped with the discrete topology, $\tau_{\mathscr{U}} = \mathscr{P}(S)$.

Solution: By definition, $\mathscr{U} = \{U \in \mathscr{P}(S \times S) : \Delta(S) \subseteq U\}$, is the discrete uniformity. By definition, a base for $\mathscr{U}$ is a subfamily, $\mathscr{B}$, such that for every $U \in \mathscr{U}$, there is a $B \in \mathscr{B}$ such that $B \subseteq U$.

We choose,

$$\mathscr{B} = \{\Delta(S)\}$$

By definition, for the given $\mathscr{B}$,

$$\mathscr{B}_\tau = \{B(x) : x \in S \text{ and } B \in \mathscr{B} \text{ a base for } \mathscr{U}\}$$

forms a base for $(S, \tau_{\mathscr{U}})$.

For each $x \in S$,

$$\begin{aligned} B(x) &= \pi_2[(\{x\} \times S) \cap B] \\ &= \pi_2[(\{x\} \times S) \cap \Delta(S)] \\ &= \{x\} \end{aligned}$$

Then $\{\{x\} : x \in S\}$ forms a base for the uniform topology. This base generates $\tau_{\mathscr{U}} = \mathscr{P}(S)$.

Example 8. Suppose $(S, \mathscr{U})$ is the uniform space equipped with the single element uniformity, $\mathscr{U} = \{S \times S\}$. Show that the corresponding topological space is equipped with the trivial topology, $\tau_{\mathscr{U}} = \{\varnothing, S\}$.

Solution: We are given that $\mathscr{U} = \{S \times S\}$, so $\mathscr{B} = \{S \times S\}$ is a base for $\mathscr{U}$. Now, $\mathscr{B}_\tau = \{B(x) : x \in S \text{ and } B \in \mathscr{B} \text{ a base for } \mathscr{U}\}$. For each $x \in S$, $B(x) = \pi_2[(\{x\} \times S) \cap (S \times S)] = S$. So, $\tau_{\mathscr{U}} = \{\varnothing, S\}$.

Example 9. For each $\kappa \in \mathbb{R}$, let

$$B_\kappa = \Delta(\mathbb{R}) \cup [\kappa, \infty) \times [\kappa, \infty)$$

Suppose $(\mathbb{R}, \mathscr{U})$ is the uniform space where $\mathscr{U}$ has as base,

$$\mathscr{B} = \{B_\kappa : \kappa \in \mathbb{R}\}$$

Show that the corresponding topological space, $(\mathbb{R}, \tau_{\mathscr{U}})$, is the discrete space.

Solution: We are given the family of subsets $\mathscr{B} = \{B_\kappa : \kappa \in \mathbb{R}\}$ is a base for $\mathscr{U}$. It is easily verified that $\mathscr{B}$ satisfies the properties required for it to serve as a base for some uniformity, $\mathscr{U}$, on S.

Now $\tau_{\mathscr{U}} = \{U \in \mathscr{P}(S) :$ for each $x \in U$ there exists $B \in \mathscr{U}$ such that $B(x) \subseteq U\}$.

If $x < \kappa$ then $B(x) = \{x\}$; so $\{x\}$ is an open base element of $\mathbb{R}$. So, $(\mathbb{R}, \tau_{\mathscr{U}})$ is a discrete topological space.

The following result will be a useful tool in proofs to come.

Proposition 29.12 *For any $M \in \mathscr{U}$, there exists $W \in \mathscr{B}$ such that $W_\circ W_\circ W \subseteq M$.*

Proof. We are given a set S, a uniformity $\mathscr{U}$ on S. For any $M \in \mathscr{U}$, by U3, there is a $B \in \mathscr{B}$ such that $B_\circ B \subseteq M$. There is a $D \in \mathscr{B}$ such that $D_\circ D \subseteq B$. Since

$$D_\circ D \subseteq B$$
$$(D \cap B)_\circ(D \cap B) \subseteq D_\circ D \subseteq B$$
$$(D \cap B)_\circ(D \cap B)_\circ(D \cap B) \subseteq (D \cap B)_\circ B \subseteq B_\circ B \subseteq M$$

there exists $W = D \cap B \in \mathscr{B}$ such that $W_\circ W_\circ W \subseteq M$.

This proves the proposition.

29.5 On the Product Topology Derived from $(S, \tau_{\mathscr{U}}) \times (S, \tau_{\mathscr{U}})$

Since a uniformity, $\mathscr{U}$, is subfamily of $\mathscr{P}(S \times S)$, every element, M, in $\mathscr{U}$ can be viewed as a subset of the Cartesian product, $(S, \tau_{\mathscr{U}}) \times (S, \tau_{\mathscr{U}})$, equipped with the product topology. We will see that every element, M, in $\mathscr{U}$ on S is a neighborhood of the diagonal, $\Delta(S)$ (with respect to this product topology). That is, $\Delta(S) \subseteq \text{int}_{S\times S}M$. To confirm this fact, it will suffice to show that, for all $M \in \mathscr{U}$, $\text{int}_{S\times S}M \in \mathscr{U}$. For, if this is so, it will follow from U1 that $\Delta(S) \subseteq \text{int}_{S\times S}M$, for all $M \in \mathscr{U}$.

Theorem 29.13 *Let S be a non-empty set and $\mathscr{U}$ be a uniformity on S. If $M \in \mathscr{U}$ then $\text{int}_{S\times S}M \in \mathscr{U}$.*

Proof. We are given a set S, a uniformity $\mathscr{U}$ on S and the uniform topology,

$$\tau_{\mathscr{U}} = \{U \in \mathscr{P}(S) : \text{ for each } x \in U \text{ there exists } B \in \mathscr{U} \text{ such that } B(x) \subseteq U\}$$

on S. Recall (from Theorem 29.8) that $\mathscr{B} = \{B(x) : x \in S \text{ and } B \in \mathscr{B}_{\text{sym}}\}$ forms a base for the topology, $\tau_{\mathscr{U}}$, on S (where $\mathscr{B}_{\text{sym}}$ denotes all symmetric relations in $\mathscr{U}$).

An open base element of $S \times S$ (equipped with the product topology) is of the form $U(x) \times V(y)$ for some $U, V \in \mathscr{B} \subseteq \mathscr{U}$. We will begin by establishing the following three facts.

Fact 1. For any $M \in \mathscr{U}$,

$$\text{int}_{S\times S}M = \{(x,y) : (x,y) \in W(x) \times W(y) \subseteq M \text{ for some } W \in \mathscr{B}\}$$

Proof. By U2, $U \cap V \in \mathscr{U}$ Then $(U \cap V)(x) \subseteq U(x)$ and $(U \cap V)(y) \subseteq V(y)$.

Then $(U \cap V)(x) \times (U \cap V)(y) \subseteq U(x) \times V(y)$.

Let $M \in \mathscr{U}$. We are required to show that $\text{int}_{S\times S}M \in \mathscr{U}$. The set, $\text{int}_{S\times S}M$, can be expressed as,

$$\text{int}_{S\times S}M = \{(x,y) : (x,y) \in U(x) \times V(y) \subseteq M \text{ for } U, V \in \mathscr{B}\}$$

Since $(U \cap V)(x) \subseteq U(x)$ and $(U \cap V)(y) \subseteq V(y)$, then $(U \cap V)(x) \times (U \cap V)(y) \subseteq U(x) \times V(y)$.

By U2, $U \cap V \in \mathscr{U}$. Also, if $U, V \in \mathscr{B}$ then $U \cap V \in \mathscr{B}$. If we let $W = U \cap V$, the set, $\text{int}_{S\times S}M$, can then be expressed as,

$$\text{int}_{S\times S}M = \{(x,y) : (x,y) \in W(x) \times W(y) \subseteq M \text{ for some } W \in \mathscr{B}\}$$

This proves Fact 1.

Fact 2. For any $M \in \mathscr{U}$, there exists $W \in \mathscr{B}$ such that $W \circ W \circ W \subseteq M$.

This fact is proven in Proposition 29.12.

Fact 3. For any $U, W \in \mathscr{U}$ where W is known to be symmetric,

$$W \circ U \circ W = \bigcup_{(x,y)\in U} \{W(x) \times W(y)\}$$

Proof. Consider the composition $W{\circ}U{\circ}W$ where W is symmetric in $\mathscr{U}$. Recall that

$$W{\circ}U{\circ}W = \{(u,v) : (u,x) \in W \text{ and } (y,v) \in W \\ \text{for some } (x,y) \in U\} \quad (*)$$

See that, if $(u,v) \in W$, then $u \in W(x)$ and $v \in W(y)$ for some $(x,y) \in U$. So, $(u,v) \in W(x) \times W(y)$. Then, for some $(x,y) \in U$, $W \subseteq W(x) \times W(y)$.

Conversely, if, for some $(x,y) \in U$, $(a,b) \in W(x) \times W(y)$, then $a \in W(x)$ and $b \in W(y)$. So, $(a,x) \in W$ and $(b,y) \in W$. Since W is symmetric, $(y,b) \in W$. Since $(x,y) \in U$, then $(a,b) \in W$, so $W(x) \times W(y) \subseteq W$.

We conclude that, whenever W is symmetric, there is $(x,y) \in U$ such that

$$W = W(x) \times W(y)$$

Restating (*) we get,

$$W{\circ}U{\circ}W = \{(u,v) \in S{\times}S : (u,v) \in W(x){\times}W(y) \text{ for some } (x,y) \in U\}$$

Equivalently,

$$W{\circ}U{\circ}W = \bigcup_{(x,y)\in U} \{W(x) \times W(y)\}$$

This proves Fact 3.

We now combine these few facts, to proceed as follows. Let $M \in \mathscr{U}$.

By Fact 1,

$$\text{int}_{S\times S}M = \{(x,y) : (x,y) \in W(x) \times W(y) \subseteq M \text{ for some } W \in \mathscr{B}\}$$

If we prove that $W \subseteq \text{int}_{S\times S}M$, then by applying U5, we will be done.

By Fact 2, there exists $W \in \mathscr{B}$ such that $W{\circ}W{\circ}W \subseteq M$.

By Fact 3,

$$W{\circ}W{\circ}W = \bigcup_{(x,y)\in W} \{W(x) \times W(y)\} \subseteq M$$

Then...

$$(x, y) \in W \Rightarrow (x, y) \in W(x) \times W(y) \subseteq \text{int}_{S\times S}M$$

Then every point of W belongs to $\text{int}_{S\times S}M$. So, $W \subseteq \text{int}_{S\times S}M$. Since $W \in \mathscr{B}$, by U5, $\text{int}_{S\times S}M \in \mathscr{U}$. We are done.

We know that $\mathscr{B} = \{B \in \mathscr{U} : B \text{ is symmetric }\}$ forms a base for $\mathscr{U}$. This means that, for every $U \in \mathscr{U}$, there is a $B \in \mathscr{B}$ such that $B \subseteq U$. By the above theorem, not only is $\text{int}_{S\times S}B \subseteq B$, but $\text{int}_{S\times S}B \in \mathscr{U}$. This proves the following corollary.

Corollary 29.14 *Let S be a non-empty set and $\mathscr{U}$ be a uniformity on S. The the family of all $(S \times S)$-open symmetric elements of $\mathscr{U}$ forms a base for $\mathscr{U}$.*

We now describe an alternate way for representing the closure (with respect to the product topology on $S \times S$) of the elements of $\mathscr{U}$.

Theorem 29.15 *Let S be a non-empty set and $\mathscr{U}$ be a uniformity on S. Let $\mathscr{B}_{sym}$ denote the base of all symmetric elements of $\mathscr{U}$. If $M \in \mathscr{U}$ then*

$$cl_{S\times S}M = \bigcap_{B\in\mathscr{B}_{sym}} \{B \circ M \circ B\}$$

Proof. We are given a set S, a uniformity $\mathscr{U}$ on S and the base, $\mathscr{B}_{\text{sym}}$, of all symmetric elements of $\mathscr{U}$.

Let $M \in \mathscr{U}$. Then by U5, $\text{cl}_{S\times S}M \in \mathscr{U}$. We keep in mind that M is a subset of $(S, \tau_{\mathscr{U}}) \times (S, \tau_{\mathscr{U}})$, equipped with the product topology. We are required to prove that $\text{cl}_{S\times S}M = \bigcap_{B\in\mathscr{B}_{\text{sym}}}\{B \circ M \circ B\}$. It suffices to show that,

$$B(x) \times B(y) \text{ intersects } M \text{ for all } B \in \mathscr{B}_{\text{sym}} \text{ if and only if}$$

$$(x, y) \in \bigcap_{B\in\mathscr{B}_{\text{sym}}} B \circ M \circ B$$

Let $(x, y) \in S \times S$. Let $B(x) \times B(y)$ be a basic open neighborhood of (x, y) (where $B \in \mathscr{B}_{\text{sym}}$).

To say that "$B(x) \times B(y)$ intersects M" is equivalent to saying that "there is some $(u, v) \in M$ such that $(x, y) \in B(u) \times B(v)$".

To say that "there is some $(u, v) \in M$ such that $(x, y) \in B(u) \times B(v)$" is equivalent to saying that "$(x, y) \in \bigcup_{(u,v)\in M} B(u) \times B(v)$".

By Fact 3 proven in the previous theorem, we can express the right-hand side, "$(x, y) \in \bigcup_{(u,v)\in M} B(u) \times B(v)$" as,

$$(x, y) \in B{\circ}M{\circ}B$$

Combining the two statements we obtain,

$$B(x) \times B(y) \text{ intersects } M \text{ for all } B \in \mathscr{B}_{\text{sym}} \Leftrightarrow (x, y) \in \bigcap_{B\in\mathscr{B}_{\text{sym}}} B{\circ}M{\circ}B$$

We conclude that, if $M \in \mathscr{U}$,

$$\text{cl}_{S\times S}M = \bigcap_{B\in\mathscr{B}_{\text{sym}}} \{B{\circ}M{\circ}B\}$$

29.6 Separation Properties Inherent to $(S, \tau_{\mathscr{U}})$

The uniform space, $(S, \tau_{\mathscr{U}})$, inherits from the uniformity $\mathscr{U}$ on S some separation properties. We can show immediately that the uniformity $\mathscr{U}$ on S guarantees that $(S, \tau_{\mathscr{U}})$ is T_3.

Theorem 29.16 *Let S be a non-empty set and $\mathscr{U}$ be a uniformity on S. Let $\tau_{\mathscr{U}}$ be the topology on S generated by $\mathscr{U}$. The space, $(S, \tau_{\mathscr{U}})$, is T_3.*[3]

Proof. We are given a set S, a uniformity $\mathscr{U}$ on S and the derived topological space, $(S, \tau_{\mathscr{U}})$, equipped with the uniform topology.

[3] Recall that a space is T_3 if, given a closed set F and a point $x \notin F$, there exists disjoint open neighborhoods U, V such that $F \subseteq U$ and $x \in V$.

Fact: The family of all $(S \times S)$-closed symmetric sets in $\mathscr{U}$ forms a base for $\mathscr{U}$.

Proof of fact: Suppose $U \in \mathscr{U}$. By Proposition 29.12, there exists $V \in \mathscr{B}_{\text{sym}}$ such that $V \circ V \circ V \subseteq U$.

By Theorem 29.15,

$$\text{cl}_{S\times S}V = \bigcap_{B\in\mathscr{B}_{\text{sym}}} \{B \circ V \circ B\}$$

which is, by U5, an element of $\mathscr{U}$.

Then

$$\text{cl}_{S\times S}V \subseteq V \circ V \circ V \subseteq U$$

So we can declare that there exists a $(S\times S)$-closed $W = \text{cl}_{S\times S}V \in \mathscr{U}$ such that $W \subseteq V$. Then $W \cap W^{-1}$ is a closed and symmetric element of $\mathscr{U}$ contained in U.

We conclude that the family of all $(S \times S)$-closed symmetric sets in $\mathscr{U}$ forms a base for $\mathscr{U}$. This establishes the stated fact.

In what follows we will represent this base by, $\mathscr{B}_{\text{cl-sym}}$

Let F be closed in S (with respect to $\tau_{\mathscr{U}}$) and $x \in S\backslash F$. Then $S\backslash F$ is open in S.

To show that $(S, \tau_{\mathscr{U}})$ is T_3 it suffices to show that x has a closed neighborhood which misses F. Since $S\backslash F$ is open, there exists an open neighborhood V of x such that $x \in V \subseteq S\backslash F$.

A base $\mathscr{B}_\tau$ for $\tau_{\mathscr{U}}$ is of the form

$$\mathscr{B}_\tau = \{B(x) : x \in S \text{ and } B \in \mathscr{B}_{\text{cl-sym}}, \text{ a base for } \mathscr{U}\}$$

We have seen that there is closed symmetric $W \in \mathscr{B}_{\text{cl-sym}}$ such that $x \in W(x) \subseteq S\backslash F$. Since $W(x) = \pi_2[(\{x\} \times S) \cap W]$, and π_2 is open and one-to-one on the closed subset $\{x\} \times S) \cap W$ of $\{x\} \times S)$, then $W(x)$ is a closed neighborhood of x. Since $W(x)$ misses F, we conclude that $(S, \tau_{\mathscr{U}})$ is T_3.

Corollary 29.17 *Let S be a non-empty set and $\mathscr{U}$ be a uniformity on S. Let $\tau_{\mathscr{U}}$ be the topology on S generated by $\mathscr{U}$. The space, $(S,\tau_{\mathscr{U}})$,* *is regular* ($T_1 + T_3$) *if and only if*

$$\bigcap_{U \in \mathscr{U}} \{U\} = \Delta(S)$$

Proof. We are given a set S, a uniformity $\mathscr{U}$ on S and the derived topological space, $(S, \tau_{\mathscr{U}})$, equipped with the uniform topology.

($\Rightarrow$) Suppose S is regular. Then S is T_3 and T_1. This implies that the points in $(S, \tau_{\mathscr{U}})$ are closed. By Theorem 29.11,

$$\{x\} = \text{cl}_{\tau_{\mathscr{U}}} \{x\} = \bigcap_{U \in \mathscr{U}} \{U(x)\}$$

Suppose $(a, b) \in \bigcap_{U \in \mathscr{U}} \{U\}$. Then $b \in \bigcap_{U \in \mathscr{U}} \{U(a)\}$. So b must equal a. We conclude that $\bigcap_{U \in \mathscr{U}} \{U\} = \Delta(S)$.

($\Leftarrow$) Suppose that $\bigcap_{U \in \mathscr{U}} \{U\} = \Delta(S)$. Then for each $x \in S$, $\{x\} = \bigcap_{U \in \mathscr{U}} \{U(x)\} = \text{cl}_{\tau_{\mathscr{U}}} \{x\}$. So, $\{x\}$ is closed and hence, $(S, \tau_{\mathscr{U}})$ is T_1. We have shown that $(S, \tau_{\mathscr{U}})$ is T_3, so, if $\bigcap_{U \in \mathscr{U}} \{U\} = \Delta(S)$, the space is regular ($T_3 + T_1$).

29.7 Uniform Continuity on a Uniform Space $(S, \mathscr{U})$

The uniformity $\mathscr{U}$ defined on a set S will allow us to introduce the notion of "uniform continuity" on the resulting uniform space, $(S, \mathscr{U})$.

Definition 29.18 Let $(S, \mathscr{U})$ and $(T, \mathscr{V})$ be two sets equipped with the uniformities, $\mathscr{U}$ and $\mathscr{V}$, respectively. Let $f : S \to T$ be a function mapping S into T. Let $g_f : S \times S \to T \times T$ be the function defined as

$$g_f(x, y) = (f(x), f(y))$$

We will say that a function $f : (S, \mathscr{U}) \to (T, \mathscr{V})$ is *uniformly continuous relative to* $\mathscr{U}$ *and* $\mathscr{V}$ if and only if for any $V \in \mathscr{V}$ the set

$$g_f^{\leftarrow}[V] = \{(x, y) \in S \times S : (f(x), f(y)) \in V\}$$

is an element of the uniformity, $\mathscr{U}$.

Equivalently, $f : S \to T$ is uniformly continuous if and only if for any $V \in \mathscr{V}$ there is an element $U \in \mathscr{U}$ such that $g_f[U] \subseteq V$.

Example 10. Recall that the *usual uniformity* on $\mathbb{R}$ is the uniformity with base $\mathscr{B} = \{B_\varepsilon : \varepsilon > 0\}$, where $B_\varepsilon = \{(a, b) \in \mathbb{R} \times \mathbb{R} : |a - b| < \varepsilon\}$. Let $(S, \mathscr{U})$ be a uniform space and $f : S \to \mathbb{R}$ be a function mapping S into $\mathbb{R}$, where $(\mathbb{R}, \mathscr{V})$ is equipped with the usual uniformity, $\mathscr{V}$. By definition, $f : S \to \mathbb{R}$ is uniformly continuous on S if and only if for any $\varepsilon > 0$, there exists $U \in \mathscr{U}$ such that $g_f[U] \subseteq B_\kappa$. That is, for any pair, $(x, y) \in U$, $|f(x) - f(y)| < \varepsilon$. Note that the choice of U is independent of the location of (x, y) in $S \times S$.

In the following example, a little bit closer to home, we examine what uniform continuity means for a function $f : \mathbb{R} \to \mathbb{R}$.

Example 11. Let $(\mathbb{R}, \mathscr{U})$ be the uniform space for $\mathbb{R}$ equipped with the usual uniformity with the base $\mathscr{B} = \{B_\varepsilon : \varepsilon > 0\}$. Let $f : \mathbb{R} \to \mathbb{R}$ be a function mapping $\mathbb{R}$ into $\mathbb{R}$. By definition, f is uniformly continuous on $\mathbb{R}$ if and only if for any $B_\varepsilon \in \mathscr{U}$, there is a $B_\delta \in \mathscr{U}$ such that $g_f[B_\delta] \subseteq B_\varepsilon$. Equivalently, for any $\varepsilon > 0$ there exists a $\delta > 0$ such that $|x - y| < \delta$ implies $|f(x) - f(y)| < \varepsilon$.

Note that the composition of two uniformly continuous functions is easily verified to be uniformly continuous.

We will immediately formally define the following three concepts.

Definition 29.19 If, for a given one-to-one function, $f : (S, \mathscr{U}) \to (T, \mathscr{V})$, both f and $f^{\leftarrow}$ are known to be uniformly continuous then we say that the function $f : S \to T$ is a *uniform isomorphism* from between S and T.

If $f : S \to T$ is a uniform isomorphism then we will say that $(S, \mathscr{U})$ and $(T, \mathscr{V})$ are *uniformly isomorphic.*

A property on $(S, \mathscr{U})$ which is shared by all uniform spaces which are uniformly isomorphic to $(S, \mathscr{U})$ is said to be a *uniform invariant.*

29.8 Uniform Continuity Implies Continuity

Continuity is always defined with respect to some topology. If we are given a function $f : S \to T$ which is known to be uniformly continuous with respect to the uniformities $\mathscr{U}$ and $\mathscr{V}$, respectively, what can say about the same function $f : S \to T$ mapping S to T with respect to the corresponding topologies $\tau_{\mathscr{U}}$ and $\tau_{\mathscr{V}}$. We answer this question with the following theorem.

Theorem 29.20 *Let S and T be sets with corresponding uniformities, $\mathscr{U}$ and $\mathscr{V}$, respectively. If $f : S \to T$ is a uniformly continuous function with respect to the uniformities $\mathscr{U}$ and $\mathscr{V}$ then this same function is a continuous function with respect to the uniform topologies, $\tau_{\mathscr{U}}$ and $\tau_{\mathscr{V}}$.*

Proof. We are given sets S and T with respective uniformities $\mathscr{U}$ and $\mathscr{V}$. Suppose $f : S \to T$ is a uniformly continuous function.

To prove continuity of f it suffices to show f is continuous at each point $x \in S$.

Let $x \in S$ and let V be a base element of $\mathscr{V}$ such that $f(x) \in V(f(x))$, a neighborhood of $f(x)$ in T. It suffices to produce a neighborhood W, of x such that $f[W] \subseteq V(f(x))$. (See Definition 6.2 and 6.3.) By the definition of "$f : S \to T$ is uniformly continuous", $g_f^{\leftarrow}[V]$ is an element of $\mathscr{U}$ (with $\Delta(S)$ in its $(S \times S)$-interior).

See that,

$$\begin{aligned} f^{\leftarrow}[V(f(x))] &= \{y \in S : f(y) \in V(f(x))\} \\ &= \{y \in S : (f(x), f(y)) \in V\} \\ &= \{y \in S : g_f(x, y) \in V\} \\ &= \left[g_f^{\leftarrow}[V]\right](x) \quad \text{(Where } g_f^{\leftarrow}[V] \in \mathscr{U}) \end{aligned}$$

Since $\left[g_f^{\leftarrow}[V]\right](x)$ is a neighborhood of x, then $f^{\leftarrow}[V(f(x))]$ is a neighborhood of x.

Since $f[f^{\leftarrow}[V(f(x))]] = V(f(x))$, f is continuous at x. Since x was arbitrarily chosen in S, f is continuous on all of S.

It follows from the theorem that a function $f : S \to T$ which is a uniform isomorphism can also be seen as a homeomorphism mapping S onto T.

Uniform continuity on a uniform space generalizes the notion of uniform continuity from one metric space to another. In the case of a metric space, it is more often discussed without explicitly introducing the concept of uniformities on a set. To see this we present the following example.

Example 12. Suppose (M_1, ρ_1) and (M_2, ρ_2) are two metric spaces. Suppose $f : M_1 \to M_2$ is a function satisfying the condition,

> *for every* $\varepsilon > 0$, *there is a* $\delta > 0$ *such that* $f[B_\delta(x)] \subseteq B_\varepsilon(f(x))$, *for all* $x \in S$.

In this case, the condition imposed on the existence of δ is stronger than the one required in the definition of continuity at x (where the value of δ depends on both the value of ε and the point, x). It easily seen here that if f satisfies the given condition then f is continuous at each point.

We compare this to the same function $f : M_1 \to M_2$ where the sets M_1 and M_2 are equipped with the metric uniformities, $\mathscr{U}$ and $\mathscr{V}$, respectively. A metric uniformity has as base, $\mathscr{B} = \{B_\kappa : \kappa > 0\}$ where $B_\kappa = \{(x, y) : \rho(x, y) < \kappa\}$. By using the language and notation for uniformities, we can say that f is uniformly continuous if,

> for any $\varepsilon > 0$, there is $\delta > 0$ such that, if $(x, y) \in B_\delta \in \mathscr{U}$, then $(f(x), f(y)) \in B_\varepsilon \in \mathscr{V}$. Equivalently, $f[B_\delta(x)] \subseteq B_\varepsilon(f(x))$, for all $x \in S$.

Chapter 30

The Stone–Weierstrass Theorem

Abstract
In this section, we revisit the relationship between algebraic structures of $C^*(S)$ and the family of Hausdorff compactifications of S. We provide a detailed proof of the Stone–Weierstrass Theorem. One consequence of this important theorem is allowing us to view the family of all compactifications as a lattice of topological spaces.

30.1 Introduction

In this section, for a topological space S, the symbol $C_b(S)$ represents all real-valued bounded functions on S. We will view $(C_b(S), +, \cdot)$ as a *ring of bounded functions*. We equip $C_b(S)$ with the sup-norm,

$$\|f\|_\infty = \sup\{|f(x)| : x \in S\}$$

It is an easy exercise to show that $\| \ \|_\infty$ is a valid norm on $C_b(S)$. So, $C_b(S)$ can be viewed as a metric space, $(C_b(S), \rho)$, where $\rho(f, g) = \sup\{|f(x) - g(x)| : x \in S\}$. In Theorem 1.11, we proved that $(C_b(S), \| \ \|_\infty)$ is a complete space. That is, all Cauchy sequences in $C_b(S)$ converge to an element of $C_b(S)$.

Suppose that $(A, \leq)$ is a partially ordered set (meaning the binary operation, $\leq$, is reflexive, antisymmetric and transitive). A partially ordered set, $(A, \leq)$, is referred to as being a

lattice.

if, for every $a, b \in A$, $a \vee b = \sup\{a, b\} \in A$ and $a \wedge b = \inf\{a, b\} \in A$ (with respect to $\leq$). The symbol, $\vee$, is read as "join" while the symbol, $\wedge$, is read as "meet". The set A is said to be a *complete lattice* if, for any non-empty $B \subseteq A$, $\vee B = \sup B$ exists in A and $\wedge B = \inf B$ exists in A. The lattice derived from $(A, \leq)$ is denoted as

$$(A, \leq, \vee, \wedge)$$

The set $(B, \leq, \vee_*, \wedge_*)$ is said to be a *sublattice* of $(A, \leq, \vee, \wedge)$ if $(B, \leq_*) \subseteq (A, \leq)$ and $(B, \leq, \vee_*, \wedge_*)$ is itself a lattice with respect to the binary relations, $\vee_*$ and $\wedge_*$, it inherits from $(A, \leq, \vee, \wedge)$. That is, if $a, b \in B$, $a \leq_* b$ if and only if $a \leq b$ and $a \vee_* b = a \vee b$ and $a \wedge_* b = a \wedge b$.

Example 1. For a topological space S, the set $C(S)$ denotes the set of all continuous real-valued functions on S. For $f, g \in C(S)$, we will say $f \leq g$ if $f(x) \leq g(x)$ for all $x \in S$. Then $(C(S), \leq)$ is a partial ordering. If

$$h(x) = \{\max\{f(x), g(x)\} : x \in S\} = (f \vee g)(x)$$
$$k(x) = \{\min\{f(x), g(x)\} : x \in S\} = (f \wedge g)(x)$$

$(C(S), \leq, \vee, \wedge)$ can easily be verified to be a lattice. To show this use the identities,

$$f \vee g = \frac{f + g + |f - g|}{2} \quad (*)$$
$$f \wedge g = \frac{f + g - |f - g|}{2} \quad (**)$$

For further reference, we include the following identity which is deduced by these.

$$f \wedge g = f + g - (f \vee g) \quad (***)$$

Analogous definitions apply to $C^*(S)$ (the set of all continuous bounded functions) to obtain the lattice $(C^*(S), \leq, \vee, \wedge)$.[1]

The subset $\mathscr{F}$ of $C^*(S)$ is said to be a *subring* of the ring, $(C^*(S), +, \cdot)$, if it closed under $+$ and $\cdot$. When we say that the subring, $\mathscr{F}$, of the ring, $C^*(S)$, is complete, we mean that it is complete

[1]Note that neither $(C(S), \leq, \vee, \wedge)$ nor $(C^*(S), \leq, \vee, \wedge)$ need be a complete lattice unless extra conditions are attached to S.

with respect to the sup-norm, $\| \ \|_\infty$. We would like to prove the Stone–Weierstrass Theorem. But first we will need in the following two lemmas, in which we associate the metric space notion of "completeness of $\mathscr{F}$ with respect to $\| \ \|_\infty$", to the algebraic notion of "$\mathscr{F}$, a sublattice of $C^*(S)$".

Lemma 30.1 *The Stone–Weierstrass lemma one. Let S be a topological space. Let $\mathscr{F}$ be a complete subring of $(C^*(S), +, \cdot)$ which contains all constant functions. Then, if $f \in \mathscr{F}$, $|f| \in \mathscr{F}$.*

Proof. We are given that $\mathscr{F}$ is a complete subring of $(C^*(S), +, \cdot)$ which contains all constant functions. Let $f \in \mathscr{F}$. We are required to show that $|f|$ belongs $\mathscr{F}$. Since f is bounded, there exists $k \in \mathbb{R}$ such that $|f(x)|_\infty < k$ on S. Then $\|f/k\|_\infty < 1$. If we show that $|f/k| \in \mathscr{F}$ then so does $|f|$. So let us suppose, without loss of generality, that $\|f\|_\infty < 1$. Then $|f(x)| < 1$ on S

Consider the binomial expansion of the expression $h(x) = (1-x)^{1/2}$,

$$(1-x)^{1/2} = \sum_{n=0}^{\infty} C(1/2, n)x^n$$

Convergence of this series to $(1-x)^{1/2}$ on $[0,1]$ is known to be uniform (see page 305). The power series $\sum_{n=0}^{\infty} C(1/2, n)x^n$ can be seen as a sequence of polynomials which converges uniformly to $(1-x)^{1/2}$ on $[0,1]$. This means that there is a polynomial, $p(x)$, such that

$$\left|(1-x)^{1/2} - p(x)\right| < \varepsilon \quad \text{on } [0,1]$$

Let $x = 1 - u^2$. Substituting this term in the above expression we obtain,

$$\left|\sqrt{u^2} - p(1-u^2)\right| < \varepsilon, \quad \text{for } u \text{ in } [-1,1]$$

from which we obtain $\big|\,|u| - p(1-u^2)\big| < \varepsilon$, for u in $[-1,1]$. We substitute $f(x)$ into u, to obtain

$$\big|\,|f(x)| - p(1-f(x)^2)\big| < \varepsilon, \quad \text{for } f(x) \text{ in } [-1,1]$$

and so,

$$\| |f| - p_\circ(1 - f^2)\|_\infty \leq \varepsilon$$

Let $h = p_\circ(1 - f^2) \in \mathscr{F}$. Now, since p is a polynomial and $\mathscr{F}$ is a subring which contains all constant functions, then $h \in \mathscr{F}$. So every ε-ball, $B_\varepsilon(|f|)$ (with respect to $\| \|_\infty$), contains an element, h, of $\mathscr{F}$. So $|f|$ is a limit point of $\mathscr{F}$. Since $\mathscr{F}$ is complete, the function, $|f|$, must belong to $\mathscr{F}$. As required.

Lemma 30.2 *The Stone–Weierstrass lemma two. Let S be a topological space. Let $\mathscr{F}$ be a complete subring of $(C^*(S), +, \cdot)$ which contains all constant functions. Then $(\mathscr{F}, \vee, \wedge)$ is a sublattice of $(C^*(S), \vee, \wedge)$.*

Proof. We are given that $\mathscr{F}$ is a complete subring of $(C^*(S), +, \cdot)$ which contains all constant functions. Let $f, g \in \mathscr{F}$. To show that $\mathscr{F}$ is a sublattice we are required to show that $f \vee g$ and $f \wedge g$ both belong $\mathscr{F}$

Consider the two identities (*) and (***) (given above)

$$f \vee g = \frac{f + g + |f - g|}{2}$$

$$f \wedge g = f + g - (f \vee g)$$

In the first lemma, we showed that if $f \in \mathscr{F}$ then $|f| \in \mathscr{F}$. Since $\mathscr{F}$ is a subring, then both $f \vee g$ and $f \wedge g$ both belong $\mathscr{F}$. So, $\mathscr{F}$ is a sublattice of $(C^*(S), \leq, \vee, \wedge)$. As required.

In the following theorem, we refer to a subset $\mathscr{F}$ of $C^*(S)$ which separates points of S. This means that, "for every pair of distinct points p and q in S, there is a function $h \in \mathscr{F}$ such that $h(p) \neq h(q)$".

While studying the Stone–Weierstrass theorem below, the reader may have a vague impression of having previously seen a similar statement (presented in 21.11). Compare the statements 21.11 to 30.3, the conditions and the conclusions that follow from these conditions. See the slight differences in how we approach the statements.

Most of the technical work for the proof of the following theorem is done in the preceding two lemmas.[2]

Theorem 30.3 *The Stone–Weierstrass theorem. Let S be a compact topological space. Let $\mathscr{F}$ be a complete subring of $C^*(S)$ which contains the constant functions. If $\mathscr{F}$ separates the points of S then $\mathscr{F} = C^*(S)$.*

Proof. Suppose S is a compact space and $f \in C^*(S)$. Also suppose that $\mathscr{F}$ is a complete subring of $C^*(S)$ which separates points of S and which contains all constant functions, $\{r \in C^*(S) : r(x) = r\,\forall\, x, r \in \mathbb{R}\}$. To show that $\mathscr{F} = C^*(S)$ it will suffice to show that, for any $\varepsilon > 0$, $B_\varepsilon(f) \cap \mathscr{F} \neq \varnothing$ with respect to $\|\ \|_\infty$. (For this would imply f is a limit point of $\mathscr{F}$ and since $\mathscr{F}$ is complete, $f \in \mathscr{F}$.)

If f is constant then, since $\mathscr{F}$ contains all constant functions, then $f \in \mathscr{F}$. So, we will assume, without loss of generality, that f is not constant; then $\inf\{f(x) : x \in S\} < \sup\{f(x) : x \in S\}$.

Step 1. Choose $p, q \in S$. By hypothesis, there exists $h_{pq} \in \mathscr{F}$ such that $h_{pq}(p) \neq h_{pq}(q)$. We define the function g_{pq} as,

$$g_{pq}(x) = \left(\frac{h_{pq}(x) - h_{pq}(q)}{h_{pq}(p) - h_{pq}(q)}\right) f(p) - \left(\frac{h_{pq}(x) - h_{pq}(p)}{h_{pq}(p) - h_{pq}(q)}\right) f(q)$$

Verify that, for such choices of p and q, $g_{pq}(p) = f(p)$ and $g_{pq}(q) = f(q)$. Since $\mathscr{F}$ is a subring which contains the constant functions and $h_{pq} \in \mathscr{F}$, then $g_{pq} \in \mathscr{F}$.

Step 2. Let $\varepsilon > 0$. Let

$$U_{pq} = \{x \in S : (g_{pq} - f)(x) < \varepsilon\}$$
$$V_{pq} = \{x \in S : (g_{pq} - f)(x) > -\varepsilon\}$$

---

[2]The Theorem 21.11 states: Suppose (S, τ) is a completely regular space and $\mathscr{F}$ is a *subalgebra* of $C^*(S)$ which separates points and closed sets in S. Suppose $\mathscr{F}$ generates the compactification $\alpha S = e^{\beta}_{\mathscr{F}}[\beta S] = \text{cl}_{\mathcal{T}} e_{\mathscr{F}}[S]$. If $\mathscr{F}$ satisfies the properties, (1) $\mathscr{F}$ contains $C_\omega(S)$ and (2) every $f \in C_\alpha(S)$ is equivalent to some function $g \in \mathscr{F}$, then $\mathscr{F} = C_\alpha(S)$.

For what follows we will fix $q \in S$. Note that, since $g_{pq}(q) = f(q)$,

$$q \in V_{pq} \text{ (independently of the value of } p)$$

Similarly, see that $g_{pq}(p) = f(p)$, so $p \in U_{pq}$. Since $U_{pq} = (g_{pq} - f)^{\leftarrow}[(-\infty, \varepsilon)]$ is open for any p, then $\{U_{pq} : p \in S\}$ is an open cover of S. Since S is compact, there is a subset, $\{p_1, p_2, p_3, \ldots, p_n\}$, of S such that $\{U_{p_1q}, U_{p_2q}, \ldots, U_{p_nq}\}$ covers all of S. Let $V_q = \cap\{V_{p_1q}, V_{p_2q}, \ldots, V_{p_nq}\}$. Then $q \in V_q$.

We define, for our fixed q, the function $g_q : S \to \mathbb{R}$ as,

$$g_q(x) = \wedge\{g_{p_1q}(x), g_{p_2q}(x), \ldots, g_{p_nq}(x)\}$$

By Lemma 30.2, since $\mathscr{F}$ is a complete subring which contains the constant functions, $\mathscr{F}$ is a sublattice. So, $g_q \in \mathscr{F}$.

Then for any $x \in S$, $x \in U_{p_kq}$ for some $k \in \{1, 2, 3, \ldots, n\}$. For this k and x we have,

$$g_q(x) \leq g_{p_kq}(x) < f(x) + \varepsilon \tag{1}$$

This expression is independent of the value of x in S, so we can write more generally that, *independent of the chosen value* $q \in S$,

$$g_q < f + \varepsilon, \quad \text{for all } q \in S. \tag{2}$$

Let $y \in V_q$ where $V_q = \cap\{V_{p_1q}, V_{p_2q}, \ldots, V_{p_nq}\}$ is a non-empty open subset of S. Then there is some $p_tq \in \{p_1q, p_2q, \ldots, p_nq\}$ such that $g_q(y) = g_{p_tq}(y)$ and so,

$$g_q(y) > f(y) - \varepsilon \tag{3}$$

Step 3. Since $q \in V_q$ for each $q \in S$, then $\{V_q : q \in S\}$ forms an open cover of S, so there exist $\{q_1, q_1, \ldots q_m\}$ such that $\{V_{q_1}, V_{q_2}, \ldots V_{q_m}\}$ covers S.

Let $x \in S$. Then $x \in V_{q_b}$ for some $q_b \in \{q_1, q_2, \ldots, q_m\}$.

Let

$$g = \vee\{g_{q_1}, g_{q_2} \cdots g_{q_m}\}$$

Then $g \in \mathscr{F}$ (by the lemma) and there is some $q_a \in \{g_{q1}, g_{q2} \cdots g_{qm}\}$ such that

$$g(x) = g_{q_a}(x) > f(x) - \varepsilon \tag{4}$$

Since this expression is independent of the value of q and x, we can conclude that

$$f - \varepsilon < g \tag{5}$$

Step 4. From inequality (2) we have $g_q < f + \varepsilon$ (on S) *for all values of* q. Then

$$g = \vee\{g_{q1}, g_{q2} \cdots g_{qm}\} < f + \varepsilon \tag{6}$$

Then, combining inequalities (5) and (6),

$$f - \varepsilon < g < f + \varepsilon$$

This is equivalent to stating "for arbitrary $\varepsilon > 0$, $g \in B_\varepsilon(f)$ with respect to $\| \ \|_\infty$".

Since, by hypothesis, $\mathscr{F}$ is complete, it is closed; then $f \in \mathscr{F}$. So, $C^*(S) \subseteq \mathscr{F}$; it follows that $\mathscr{F} = C^*(S)$, as required.

30.2 A Consequence of the *Stone–Weierstrass Theorem*

Let $\mathscr{C} = \{\alpha_i S : i \in I\}$ represent the family of all compactifications of a locally finite completely regular space. We will partially order the family, $\mathscr{C}$, by defining "$\preceq$" in the following way:

> *If $\alpha_i S$ and $\alpha_j S$ belong to $\mathscr{C}$, we will write*
>
> $$\alpha_i S \preceq \alpha_j S$$
>
> *if and only if there is a continuous function $f : \alpha_j S \to \alpha_i S$, mapping $\alpha_j S \backslash S$ onto $\alpha_i S \backslash S$ which fixes the points of S.*

See that with $\preceq$ thus defined, $(\mathscr{C}, \preceq)$ forms a partially ordered set of compact topological spaces.

Notation: Suppose two compactifications αS and γS are such that $\alpha S \preceq \gamma S$, then by definition, there is a continuous function $f : \gamma S \to \alpha S$, mapping $\gamma S \backslash S$ onto $\alpha S \backslash S$ which fixes the points of S. We will represent this continuous function f as,

$$\pi_{\gamma \to \alpha} : \gamma S \to \alpha S$$

The subscript of the function $\pi_{\gamma \to \alpha}$ explicitly expresses which compactification is larger than (or equal to) the other. Note that a pair of compactifications need not necessarily be comparable in the sense that one need not necessarily by "less than" the other.

Notation: For any compactification αS let $C_\alpha(S) = \{f|_S : f \in C(\alpha S)\}$. Clearly, $C_\alpha(S) \subseteq C^*(S)$. If S is locally compact and completely regular then it has a one-point compactification, $\omega S = S \cup \{\omega\}$ (see 21.15). We will adopt the following notation:

$$C_\omega(S) = \{f|_S : f \in C(\omega S)\}$$

Note that, since ωS is compact and Hausdorff then it is normal (see 14.3). Then it is completely regular and so $C(\omega S)$ separates points and closed sets of ωS. Clearly, it also contains all constant functions. It is also a complete ring, with respect to $\| \; \|_\infty$ (see 1.11). Then it is easily verified that $C_\omega(S)$ also separates points and closed sets of S (remember that since S is locally compact it is open in ωS).

Also note that $f \in C_\omega(S) \subseteq C^*(S)$ if and only if f is constant on the compliment of some compact subset of S (since, if it extends to ωS, it must be constant on the outgrowth, $\{\omega\}$).

Compare the following statement to the one in Theorem 21.11.[3]

Theorem 30.4 *Let S be locally compact and completely regular. If $C^*(S)$ contains a subring, $\mathscr{F}$, which is complete and contains $C_\omega(S)$ then $\mathscr{F} = C_\alpha(S)$ for some compactification, αS, of S.*

[3]Theorem 21.11 states: Suppose (S, τ) is a completely regular space and $\mathscr{F}$ is a *subalgebra* of $C^*(S)$ which separates points and closed sets in S. Suppose $\mathscr{F}$ generates the compactification $\alpha S = e^{\beta}_{\mathscr{F}}[\beta S] = \mathrm{cl}_{T} e_{\mathscr{F}}[S]$. If $\mathscr{F}$ satisfies the properties, (1) $\mathscr{F}$ contains $C_\omega(S)$ and (2) every $f \in C_\alpha(S)$ is equivalent to some function $g \in \mathscr{F}$, then $\mathscr{F} = C_\alpha(S)$.

Proof. We are given that S is locally compact completely regular, that $\mathscr{F}$ is complete subring of $C^*(S)$ and contains $C_\omega(S)$. Since ωS is compact and $C(\omega S)$ separates points and closed subsets of ωS, then $C_\omega(S)$ separates points and closed sets of S. Then $\mathscr{F}$ separates points and closed sets in S. Every function f in $\mathscr{F}$ extends to a function $f^\beta \in C(\beta S)$. Then $\mathscr{F}^\beta = \{f^\beta : f \in \mathscr{F}\} \subseteq C(\beta S)$.

By the embedding theorem I (Theorem 7.17) the evaluation map

$$e_{\mathscr{F}} : S \to \prod_{f\in\mathscr{F}} f[S]$$

embeds S into $T = \prod_{f\in\mathscr{F}} \mathbb{R}$. Then $e_{\mathscr{F}^\beta}[\beta S] = \mathrm{cl}_T e_{\mathscr{F}}[S]$ can be viewed as a compact space which densely contains a homeomorphic copy of S. We can then represent $\mathrm{cl}_T e_{\mathscr{F}}[S]$ as a compactification, $\alpha S = \mathrm{cl}_T e_{\mathscr{F}}[S]$. By the Stone–Weierstrass theorem, $\mathscr{F}^\alpha = C^*(\alpha S)$. Then $\mathscr{F} = C_\alpha(S)$.

By the theorem above, we can associate to each complete subring, $\mathscr{F}$, of $C^*(S)$, which contains $C_\omega(S)$, a compactification αS of S such that $\mathscr{F} = C_\alpha(S)$.

Conversely, for each compactification γS we can associate a unique complete subring, $C_\alpha(S)$, of $C^*(S)$, which contains $C_\omega(S)$. (If S is locally compact, the ring $C_\alpha(S)$ contains $C_\omega(S)$ so every function, f in ωS extends to all compactifications of S).

For compactifications αS and γS, $\alpha S \preceq \gamma S$ if and only if $C_\alpha(S) \subseteq C_\gamma(S)$.

We define $C_\alpha(S) \vee C_\gamma(S) =$ the smallest complete subring, $C_\delta(S)$, of $C^*(S)$ which contains both $C_\alpha(S)$ and $C_\gamma(S)$. We can equivalently say,

$$C_\alpha(S) \vee C_\gamma(S) = \langle C_\alpha(S), C_\gamma(S)\rangle$$

to represent the subring generated by $C_\alpha(S)$ and $C_\gamma(S)$.[4]

[4]It is the intersection of all subrings of $C^*(S)$ which contains both $C_\alpha(S)$ and $C_\gamma(S)$.

We define $C_\alpha(S) \wedge C_\gamma(S) = C_\alpha(S) \cap C_\gamma(S)$, the largest subring, $C_\delta(S)$, of $C^*(S)$ which is contained in both $C_\alpha(S)$ and $C_\gamma(S)$.

We can then view $\mathscr{C}$ as a lattice, $(\mathscr{C}, \vee, \wedge)$ where

$$\alpha S \vee \gamma S = \delta S \text{ if and only if } C_\alpha(S) \vee C_\gamma(S) = C_\delta(S)$$

$$\alpha S \wedge \gamma S = \delta S \text{ if and only if } C_\alpha(S) \cap C_\gamma(S) = C_\delta(S)$$

Chapter 31

Metrizability

Abstract
In this section, we prove two important metrizability theorems. The Urysohn metrization theorem is proved for the special case where the space is known to be separable. This is of particular interest since it exposes the strategy used to prove the more general Nagata–Smirnov metrization theorem.

31.1 Introduction

We have previously discussed topological spaces, (S, τ), on which we have defined a metric producing a metric topology which is equivalent to τ. Such spaces are said to be *metrizable*. We know that metric spaces are first countable (5.9), normal (9.19) and paracompact (19.11). So metrizable spaces must at least be first countable, normal and paracompact. Determining those specific properties on a topological space which characterize metrizability is the question we will investigate here.

31.2 The Urysohn Metrization Theorem

We begin by discussing metrizability of a certain class of topological spaces. The following results help us characterize those separable spaces which are metrizable. We begin by proving two lemmas and a theorem.

Lemma 31.1 *Let $\{(S_n, \rho_n) : n \in \mathbb{N}\}$ be a countable family of metric spaces. Suppose that, for each $n \in \mathbb{N}$, $\sup\{\rho_n(x, y) : x, y \in S_n\} \leq 1$. Let $S = \prod_{n\in\mathbb{N}} S_n$. Let $\rho : S \times S \to \mathbb{R}$ be defined as*

$$\rho(\langle x_n \rangle_{n\in\mathbb{N}}, \langle y_n \rangle_{n\in\mathbb{N}}) = \sum_{n\in\mathbb{N}} \frac{\rho_n(x_n, y_n)}{2^n}$$

Then $\rho : S \times S \to \mathbb{R}$ is a metric on S. Furthermore, the topology on $S = \prod_{n\in\mathbb{N}} S_n$ derived by the metric ρ is the product topology.

Proof. This statement has been proven in an example which appears on page 162.

Lemma 31.2 *Let S be a T_1 topological space. Suppose $\{S_n : i \in \mathbb{N}\}$ is a countably infinite indexed family where, for each n, $S_n = [0, 1]$ is equipped with the usual topology. Let $\mathscr{F} = \{f_n : n \in \mathbb{N}\}$ be a set of* continuous functions $f_n : S \to S_n = [0, 1]$.

(a) *If $\mathscr{F}$* separates points and closed sets *of its domain, S, then the evaluation map,*

$$e(x) = \langle f_n(x) \rangle_{n\in\mathbb{N}} \in \prod_{n\in\mathbb{N}} f_n[S] \subseteq \prod_{n\in\mathbb{N}} [0, 1]$$

with respect to $\mathscr{F}$, is both continuous and one-to-one on S.

(b) *Furthermore, the function, $e : S \to \prod_{n\in\mathbb{N}} S_n$, maps S homeomorphically onto*

$$e[S] \subseteq \prod_{n\in\mathbb{N}} f_n[S] \subseteq \prod_{n\in\mathbb{N}} S_n$$

Hence, this evaluation map embeds a homeomorphic copy of S into the (metrizable[1]) product space, $\prod_{n\in\mathbb{N}} [0, 1]$.

Proof. This lemma is simply restating the Embedding Theorem I, 7.17, for the case where the codomain of each f_n is $[0, 1]$.

---

[1]By Lemma 31.1.

The two above lemmas allow us to streamline the proof of *Urysohn's metrization theorem.*

Theorem 31.3 *The Urysohn metrization theorem. Let S be a regular $(T_1 + T_3)$ second countable topological space. Then S is embedded in $\prod_{n\in\mathbb{N}}[0,1] = [0,1]^{\mathbb{N}}$ (with the product topology). So second countable regular topological spaces are metrizable spaces.*

Proof. We are given that S is a regular and second countable topological space. Then, by Theorem 16.2,[2] S is Lindelöf and, by Theorem 16.6,[3] S is normal.

If we can show that there exists a function which embeds S into the metrizable product space, $\prod_{n\in\mathbb{N}}[0,1]$, then S can be viewed as a subspace of a metric space and so must be metrizable.

Since S is second countable there is a countable base, $\mathscr{B} = \{B_i : i \in \mathbb{N}\}$ of open sets in S. We define the countably infinite set

$$\mathscr{A} = \{(U,V) : U, V \in \mathscr{B}, \text{cl}_S U \subseteq V\}$$

in $\mathscr{B} \times \mathscr{B}$.

Since S is normal, for each pair (U,V), there is a continuous function $f_{(U,V)} : S \to [0,1]$ such that

$$f_{(U,V)}[\text{cl}_S U] = \{0\}$$
$$f_{(U,V)}[S \backslash V] = \{1\}$$

Let $\mathscr{F} = \{f_{(U,V)} : (U,V) \in \mathscr{A}\}$.

Claim: We claim that $\mathscr{F}$ separates points and closed sets.

Proof of claim: Let F be a closed subset of S and $x \in S\backslash F$. We can choose $V \in \mathscr{B}$ such that $x \in V \subseteq S\backslash F$ and choose $U \in \mathscr{B}$ such that $x \in \text{cl}_S U \subset V$. Then $(U,V) \in \mathscr{A}$. Let $f_{(U,V)}$ be the member of $\mathscr{F}$ which corresponds to this choice of (U,V) where $x \in Z(f)$ and $F \subseteq S\backslash V \subseteq Z(f-1)$. So, $\mathscr{F}$ separates the point x and the closed subset F of S. This establishes the claimed.

---

[2]Which states that second countable spaces are Lindelöf.

[3]Which states that regular Lindelöf spaces are normal.

Then, by the Embedding Theorem I, 7.17, the evaluation map, $e_{\mathscr{F}}$, embeds a homeomorphic copy of S in the metrizable space, $\prod_{n\in\mathbb{N}}[0,1]$.

It follows that S is metrizable.

Corollary 31.4 *Let S be a T_1-space. The following statements are equivalent.*

(1) *The space S is regular and second countable.*
(2) *The space S is homeomorphically embedded in $[0,1]^{\mathbb{N}}$.*
(3) *The space S is both metrizable and separable.*

Proof. We are given that S be a T_1-space.

(1) $\Rightarrow$ (2) That a regular and second countable space is embedded in $[0,1]^{\mathbb{N}}$ is proven in the theorem immediately above.

(2) $\Rightarrow$ (3) We are given that S is embedded in $[0,1]^{\mathbb{N}}$. In Lemma 31.1, we proved that $[0,1]^{\mathbb{N}}$ is metrizable. Since $[0,1]$ is second countable (example on page 103), so is $[0,1]^{\mathbb{N}}$ (7.9). Since S is embedded in $[0,1]^{\mathbb{N}}$, S is second countable. By Theorem 5.10, S is separable. So, S is both metrizable and separable.

(3) $\Rightarrow$ (1) Suppose S is metrizable and separable. Metrizable spaces are normal (9.19) and so are regular. A metric space which is separable is second countable (5.11). So, S is regular and second countable.

So, all second countable regular spaces are separable metrizable spaces.

31.3 The Nagata–Smirnov Metrization Theorem

The statement in the following theorem involves a collection of subsets referred to as being a "locally finite collection".

Definition 31.5 A *locally finite collection*, $\mathscr{C}$, of subsets of a topological space, S, is one for which every point in S has at least one neighborhood which meets only finitely many elements of $\mathscr{C}$.

The collection of sets, $\mathscr{D}$ is said to be *σ-locally finite* if and only if $\mathscr{D}$ is the countable union of locally finite families of locally finite sets.

The collection of sets, $\mathscr{A}$ is said to be a *discrete family of sets* is one for which every point in S has at least one neighborhood which meets at most one elements of $\mathscr{A}$.

In Lemma 6.16, we show that it satisfies the interesting property:

$$\textit{If } \mathscr{C} \textit{ is locally finite, } cl_S[\cup\mathscr{C}] = \cup\{cl_S C : C \in \mathscr{C}\} \qquad (*)$$

That is, if p is an accumulation point of $\cup\mathscr{C}$ it is an accumulation point of some set in $\mathscr{C}$. The statement, (*), is invoked in the proof of the statements below.

In the following statements we prove two interesting properties of those regular spaces which have a σ-locally finite base.

— In the first lemma, we show that, if a regular space S has a σ-locally finite base, then every non-empty open subset of S is the countable union of open subsets which is also a countable union of the closures of these same subsets.

— In the second lemma, we show that, if a regular space S has a σ-locally finite base, then it is a normal space.

Lemma 31.6 *Suppose the space S is regular and has a σ-locally finite base. That is, S has a base, $\mathscr{B} = \cup_{n\in\mathbb{N}}\mathscr{B}_n$, where each $\mathscr{B}_n$ is locally finite. Let W be an open subset of S. Then*

$$W = \cup\{U_{nW} : n \in \mathbb{N}\} = \cup\{cl_S U_{nW} : n \in \mathbb{N}\}$$

where each U_{nW} is open in S.

Proof. Let $\mathscr{B} = \cup_{n\in\mathbb{N}}\mathscr{B}_n$ be a σ-locally finite base of the regular space S and W be some non-empty open subset in S. Let

$$\mathscr{U}_W = \{B \in \mathscr{B} : B \subseteq \mathrm{cl}_S B \subseteq W\}$$

Each B in $\mathscr{U}_W$ belongs to $\mathscr{B}_n$, for some n.

For each m, we define,

$$\begin{aligned}\mathscr{U}_{mW} &= \{B \in \mathscr{U}_W : B \in \mathscr{B}_m\} \\ &= \mathscr{U}_W \cap \mathscr{B}_m\end{aligned}$$

For each $m \in \mathbb{N}$, we define the subset, U_{mW}, as

$$U_{mW} = \cup\{B : B \in \mathscr{U}_{mW} = \mathscr{U}_W \cap \mathscr{B}_m\}$$

We first claim that $W \subseteq \cup\{\text{cl}_S U_{nW} : n \in \mathbb{N}\}$.

Proof of claim: For any $q \in W$, regularity of S guarantees that there is a $B \in \mathscr{B}$ such that $q \in B \subseteq \text{cl}_S B \subseteq W$. So $B \in \mathscr{U}_W$. Then $q \in B \in \mathscr{U}_W \cap \mathscr{B}_m$ for some m.

So there is an m such that $q \in U_{mW}$. Therefore $W \subseteq \cup \{U_{nW} : n \in \mathbb{N}\}$. Since $U_{nW} \subseteq \text{cl}_S U_{nW}$ for each n, $W \subseteq \cup\{\text{cl}_S U_{nW} : n \in \mathbb{N}\}$, as claimed.

It now suffices to show $\cup\{\text{cl}_S U_{nW} : n \in \mathbb{N}\} \subseteq W$. This is the part of proof where the locally finite property is involved.

For each $n \in \mathbb{N}$

$$\begin{aligned}\text{cl}_S U_{nW} &= \text{cl}_S \left[\cup\{B : B \in \mathscr{U}_{nW}\}\right] \\ &= \cup\{\text{cl}_S B : B \in \mathscr{U}_{nW}\} \\ &\qquad \text{(Each } \mathscr{U}_{nW} \text{ is locally finite and (*) above)} \\ &\subseteq W \quad \text{(Since } \text{cl}_S B_x \subseteq W\text{)}\end{aligned}$$

So, $\cup\{\text{cl}_S U_{nW} : n \in \mathbb{N}\} \subseteq W$.

We conclude that if S is regular, has a σ-locally finite base, $\mathscr{B} = \cup\{\mathscr{B}_n : n \in \mathbb{N}\}$, and W is open in S then

$$W = \cup\{U_{nW} : n \in \mathbb{N}\} = \cup\{\text{cl}_S U_{nW} : n \in \mathbb{N}\} \quad (**)$$

where each U_{nW} is the union of open sets in $\mathscr{B}$. This completes the proof of the lemma.

Note: An easy (and interesting) consequence of the result (**) is the following fact:

If S is regular with a σ-locally finite base then every closed subset of S is a G_δ.

To see this, consider a closed subset, F, of S.

Then, by $(**)$, $S\backslash F = \cup\{\mathrm{cl}_S U_{n_{S\backslash F}} : n \in \mathbb{N}\}$ a countable union of closed sets. Then $F = \cap\{S\backslash \mathrm{cl}_S U_{n_{S\backslash F}} : n \in \mathbb{N}\}$, a countable intersection of open sets, a G_δ.

Lemma 31.7 *A regular space S which has a σ-locally finite base is a normal space.*

Proof. We are given that S is regular and has a σ-locally finite base, $\mathscr{B}$. That is,

$$\mathscr{B} = \cup_{n\in\mathbb{N}\backslash\{0\}} \mathscr{B}_n$$

where each $\mathscr{B}_n$ is locally finite.

Suppose A and K are disjoint non-empty closed subsets. To show that S is normal it suffices to show that A and K are respectively contained in disjoint open subsets of S.

Since S is regular and has a σ-locally finite base, by Lemma 31.6

$$S\backslash K = \cup\{U_{n_{S\backslash K}} : n \in \mathbb{N}\} = \cup\{\mathrm{cl}_S U_{n_{S\backslash K}} : n \in \mathbb{N}\} \quad (\dagger)$$

$$S\backslash A = \cup\{V_{n_{S\backslash A}} : n \in \mathbb{N}\} = \cup\{\mathrm{cl}_S V_{n_{S\backslash A}} : n \in \mathbb{N}\} \quad (\dagger\dagger)$$

Where $U_{n_{S\backslash K}} = \cup\{B : B \in \mathscr{U}_{S\backslash K} \cap \mathscr{B}_n\}$ and $V_{n_{S\backslash A}} = \cup\{B : B \in \mathscr{U}_{S\backslash A} \cap \mathscr{B}_n\}$.

Let

$$U = \cup\{U_{n_{S\backslash K}} : n \in \mathbb{N}\}$$

$$V = \cup\{V_{n_{S\backslash A}} : n \in \mathbb{N}\}$$

Then

$$K \subseteq V$$

$$A \subseteq U$$

Now, U and V are open neighborhoods of A and K, but they may not be disjoint.

For each n, we replace $U_{n_{S\backslash K}}$ and $V_{n_{S\backslash A}}$ with $U^*_{n_{S\backslash K}}$ and $V^*_{n_{S\backslash A}}$ defined as follows:

$$U^*_{n_{S\backslash K}} = U_{n_{S\backslash K}} \backslash \cup\{\text{cl}_S V_{i_{S\backslash A}} : i \leq n\}$$
$$V^*_{n_{S\backslash A}} = V_{n_{S\backslash A}} \backslash \cup\{\text{cl}_S U_{i_{S\backslash K}} : i \leq n\}$$

If

$$\begin{aligned} x \in A \quad &\Rightarrow \quad x \in S\backslash K \\ &\Rightarrow \quad x \in U_{m_{S\backslash K}}, \text{ for some } m \quad (\text{By } (\dagger)) \\ x \in A \quad &\Rightarrow \quad x \notin S\backslash A \\ &\Rightarrow \quad x \notin \text{cl}_S V_{n_{S\backslash A}}, \text{ for all } n \quad (\text{By } (\dagger\dagger)) \\ &\text{so} \\ x \in A \quad &\Rightarrow \quad x \in U^*_{n_{S\backslash K}} \end{aligned}$$

Similarly, $x \in K \Rightarrow x \in V^*_{n_{S\backslash A}}$.

Then, for each n,

$$A \subseteq U^*_{n_{S\backslash K}}$$
$$K \subseteq V^*_{n_{S\backslash A}}$$

Let

$$U^* = \cup\{U^*_{n_{S\backslash K}} : n \in \mathbb{N}\}$$
$$V^* = \cup\{V^*_{n_{S\backslash A}} : n \in \mathbb{N}\}$$

Clearly, U^* and V^* are both open and neighborhoods of A and K, respectively.

We claim that $U^* \cap V^* = \varnothing$. If so, S is normal, as required.

Proof of claim: Suppose $x \in U^* \cap V^*$. Then $x \in U^*_{j_{S\backslash K}} \cap V^*_{k_{S\backslash A}}$ for some j and k.

That is,

$$x \in \left[U_{j_{S\backslash K}} \backslash \cup\{\text{cl}_S V_{i_{S\backslash A}} : i \leq j\} \cap V_{k_{S\backslash A}} \backslash \cup\{\text{cl}_S U_{i_{S\backslash K}} : i \leq k\}\right] \quad (*)$$

Suppose without loss of generality that $j \leq k$.

From $(*)$, $x \in U_{j_{S\setminus K}} \cap S\setminus\cup\{\mathrm{cl}_S U_{i_{S\setminus K}} : i \leq k\}$.[4]

So, $x \in U_{j_{S\setminus K}}$ but $x \notin \mathrm{cl}_S U_{j_{S\setminus K}}$, a contradiction.

Then $U^* \cap V^* = \varnothing$ as claimed. So, S is normal.

Theorem 31.8 *A regular space S with a σ-locally finite topology is metrizable.*

Proof. We are given that S is regular and has a σ-locally finite base, $\mathscr{B}$. That is,

$$\mathscr{B} = \cup_{n\in\mathbb{N}\setminus\{0\}}\mathscr{B}_n$$

where each $\mathscr{B}_n$ is locally finite. Then, by the previous lemma, S is normal. To prove metrizability of S it will suffice to produce a continuous function which embeds S into a metric space.

If $U \in \mathscr{B}$, the expression "m_U" declares that the base element U belongs to $\mathscr{B}_m$. For example, the label 2_U refers to the base element U which belongs to $\mathscr{B}_2$.

Normality of S guarantees that, if $m_B \in \mathscr{B}$, there exists continuous "separating" function denoted as "f_{m_B}" mapping S into $[0,1]$ such that, for a point $u \in B$

$$f_{m_B}[S\setminus B] = \{0\} \quad \text{and} \quad f_{m_B}(u) = q_m \in (0, 1/m) \qquad (\dagger)$$

More generally, we can construct a family of continuous functions

$$\mathscr{F} = \{f_{n_B} : n \in \mathbb{N}\setminus\{0\}, B \in \mathscr{B}\}$$

where

$$f_{n_B}[B] \subseteq (0, 1/n)$$
$$f_{n_B}(x) = 0, \quad \text{if } x \notin B$$

We claim that $\mathscr{F}$ separates points and closed subsets of S.

To see this, let K be a closed subset of S and $z \in S\setminus K$. Since $\mathscr{B}$ is an open base for S, then there exits $B_m \in \mathscr{B}$ such that $z \in B_m \subseteq S\setminus K$. Then there is $f_{m_B} \in \mathscr{F}$ such that $f_{m_B}(z) = q_z \in (0, 1/m)$ and

---

[4] $A\setminus B \cap C\setminus D = A \cap C \cap S\setminus B \cap S\setminus D$.

$f_{m_B}[S \backslash B_m] = \{0\}$. So, $\mathscr{F}$ separates points and closed subsets of S, as claimed.

See that $e_{\mathscr{F}}$ is a one-to-one mapping: Let $J = \{n_B : n \in \mathbb{N} \backslash \{0\}, B \in \mathscr{B}_n\}$. If $j \in J$, let $[0, j]_j$ denote the range of f_j.

Let

$$T = \prod_{j \in J} [0,1]_j = [0,1]^J$$

Since $\mathscr{F}$ separates points and closed sets, then the evaluation map, $e_{\mathscr{F}} : S \to T$, defined as

$$e_{\mathscr{F}}(x) = \langle\!\langle f_j(x) \rangle\!\rangle_{j \in J}$$

is a one-to-one function mapping S into $T = \prod_{j \in J}[0,1]_j$. (Argue as in 7.17).

To prove continuity of $e_{\mathscr{F}}$, we will use the topology on S which is generated by $\mathscr{B}$. Rather than use the product topology for T we will choose the *uniform topology* induced by the uniform metric (see Example 3 on page 151):

$$\rho(<x_j>_{j \in J}, <a_j>_{j \in J}) = \sup_{j \in J} \{|x_j - a_j|\}$$

where, in this particular case,

$$\begin{aligned} \rho(e_{\mathscr{F}}(x), e_{\mathscr{F}}(a)) &= \rho(<f_j(x)>_{j \in J}, <f_j(a)>_{j \in J}) \\ &= \sup_{j \in J} \{|f_j(x) - f_j(a)|\} \end{aligned}$$

Then T can be viewed as a metric space with metric, ρ.

Claim: We claim that the function $e_{\mathscr{F}} : S \to \prod_{j \in J}[0,1]_j$, where $\prod_{j \in J}[0,1]_j$ is equipped with the uniform metric, is a continuous function.

Proof of claim: Let $\varepsilon > 0$ and $a \in S$.

To show that $e_{\mathscr{F}}$ is continuous at a, it suffices to find an open neighborhood $V(a)$ in S such that $x \in V(a)$ implies

$$\rho(e_{\mathscr{F}}(x), e_{\mathscr{F}}(a)) = \sup_{n_B \in J} \{|f_{n_B}(x) - f_{n_B}(a)|\} < \varepsilon$$

Let $N \in \mathbb{N}$, such that $1/N \leq \varepsilon/2$.

Since $\mathscr{B}_N$ is locally finite, there exists an S-open neighborhood, $U_N(a)$, of a such that $U_N(a)$ intersects at most finitely many base elements, say $\mathscr{E}_N = \{B_i : i = 1 \text{ to } m\}$ in $\mathscr{B}_N$.

Then,

$$B \notin \mathscr{E} \quad \Rightarrow \quad f_{N_B}[U_N(a)] = \{0\} \quad (\text{Since } B \cap U_N(a) = \varnothing)$$

$$B \in \mathscr{E} \quad \Rightarrow \quad f_{N_B}[U_N(a) \cap B] \subseteq (0, 1/N) \quad (\text{By } \dagger)$$

Since $f_{N_B} : S \to [0,1]$ is continuous, there exists an open neighborhood, $V_N(a) \subseteq U_N(a)$, such that,

$$x \in V_N(a) \text{ implies } |f_{N_B}(x) - f_{N_B}(a)| \leq \min\{1/N, \varepsilon/2\}$$

is satisfied.

(a) For each of the values of i, $i = 1, 2, \ldots, N-1$ and $a \in S$ we repeat the procedure to prove the existence of $\{V_1(a), V_2(a), \ldots, V_{N-1}(a)\}$. For $V_i(a) \subseteq U_i(a)$,

$$x \in V_i(a) \text{ implies } |f_{i_B}(x) - f_{i_B}(a)| \leq \min\{1/i, \varepsilon/2\} \leq \varepsilon/2 < \varepsilon$$

Let

$$V(a) = V_1(a) \cap V_2(a) \cap \cdots \cap V_N(a)$$

Then, for $i \leq N$,

$$x \in V(a) \text{ implies } |f_{i_B}(x) - f_{i_B}(a)| \leq \min\{1/i, \varepsilon/2\} < \varepsilon \quad (\dagger\dagger)$$

Then,

$$x \in V(a) \quad \Rightarrow \quad \sup_{\{i_B \,:\, i \leq N\}} |f_{i_B}(x) - f_{i_B}(a)| < \varepsilon \quad (**)$$

(B) For values of i such that $i > N$ let $V_i(a) = V(a)$

$$\begin{aligned} x \in V(a) \Rightarrow\ & |f_{i_B}(x) - f_{i_B}(a)| \\ & \leq \min\{1/i, \varepsilon/2\} < \min\{1/N, \varepsilon/2\} \leq \varepsilon/2 < \varepsilon \quad (\dagger\,\dagger\,\dagger) \end{aligned}$$

Then,

$$x \in V(a) \quad \Rightarrow \quad \sup_{\{i_B \,:\, i \geq N\}} |f_{i_B}(x) - f_{i_B}(a)| < \varepsilon \quad (***)$$

Combining (∗∗) and (∗∗∗) we conclude that for $a \in S$,

$$x \in V(a) \Rightarrow \rho(e_{\mathscr{F}}(x), e_{\mathscr{F}}(a)) = \sup_{n_B \in J} |f_{n_B}(x) - f_{n_B}(a)| < \varepsilon$$

So, $e_{\mathscr{F}}$ is continuous at $a \in S$. Since a is arbitrary, $e_{\mathscr{F}} : S \to T$ is continuous on S, as claimed.

We now show that

$$e_{\mathscr{F}}(x) = \langle f_j(x) \rangle_{j \in J} \in T = \prod_{j \in J} [0,1]_j$$

is an open function.

Let U be a non-empty open subset of S with the point $u \in U$. Then $F = S \setminus U$ is closed in S. Since $\mathscr{F}$ separates points and closed sets, there exists $k \in J$ such that $f_k(u) \notin \text{cl}_{[0,1]} f_k[F]$. Then $f_k(u) \in [0,1] \setminus \left[\text{cl}_{[0,1]} f_k[F]\right]$.

We now show that $e_{\mathscr{F}}[U]$ is open in $\prod_{j \in J} [0,1]_j$.

Note that,

$$\begin{aligned}(\pi_k \circ e_{\mathscr{F}})(u) &= \pi_k(e_{\mathscr{F}}(u)) \\ &= \pi_k(\langle f_j(u) \rangle_{j \in J}) \\ &= f_k(u) \\ &\in [0,1] \setminus \left[\text{cl}_{[0,1]} f_k[F]\right]\end{aligned}$$

Since $e_{\mathscr{F}}(u) \in \pi_k^{\leftarrow}\left[[0,1] \setminus \left[\text{cl}_{[0,1]} f_k[F]\right]\right]$ and since π_k is continuous, then $\pi_k^{\leftarrow}\left[[0,1] \setminus \left[\text{cl}_{[0,1]} f_k[F]\right]\right]$ is an open neighborhood of $e_{\mathscr{F}}(u)$ in $\prod_{j \in J} [0,1]$. It now suffices to show that

$$\pi_k^{\leftarrow}\left[[0,1] \setminus \left[\text{cl}_{[0,1]} f_k[F]\right]\right] \subseteq e_{\mathscr{F}}[U]$$

Suppose $e_{\mathscr{F}}(a) \in \pi_k^{\leftarrow}\left[[0,1] \setminus \left[\text{cl}_{[0,1]} f_k[F]\right]\right]$.

$$\begin{aligned}& e_{\mathscr{F}}(a) \in \pi_k^{\leftarrow}\left[[0,1] \setminus \left[\text{cl}_{[0,1]} f_k[F]\right]\right] \\ \Rightarrow \quad & (\pi_k \circ e_{\mathscr{F}})(a) \in \pi_k\left[\pi_k^{\leftarrow}\left[[0,1] \setminus \text{cl}_{[0,1]} f_k[F]\right]\right] \\ \Rightarrow \quad & f_k(a) \in \left[[0,1] \setminus \text{cl}_{[0,1]} f_k[F]\right] \\ \Rightarrow \quad & f_k(a) \notin \text{cl}_{[0,1]} f_k[F] \\ \Rightarrow \quad & a \in S \setminus F = S \setminus (S \setminus U) = U \\ \Rightarrow \quad & e_{\mathscr{F}}(a) \in e_{\mathscr{F}}[U]\end{aligned}$$

So, $\pi_k^{\leftarrow}\left[[0,1]\backslash\left[\text{cl}_{[0,1]}f_k[F]\right]\right]$ is an open neighborhood of $e_{\mathscr{F}}(u)$ which is entirely contained in $e[U]$. We conclude $e_{\mathscr{F}}[U]$ is open in $\prod_{j\in J}[0,1]$.

So, $e_{\mathscr{F}}$ embeds S into the metric space, T.
So, S is metrizable.

In the following theorem, we invoke a result (19.8) previously proven in the chapter on paracompactness. It states:

> *If S is metrizable space and $\mathscr{U}$ is an open cover of S then there is an open cover, $\mathscr{E}$, which both refines $\mathscr{U}$ and is σ-locally finite.*

Recall that a collection of sets, $\mathscr{E}$, *refines* a collection of sets, $\mathscr{U}$, if every element of $\mathscr{E}$ is contained in some element of $\mathscr{U}$.

Theorem 31.9 *The Nagata–Smirnov metrization theorem.*[5] *The space S is metrizable if and only if S is regular and has a σ-locally compact base of open sets.*

Proof. Let S be a topological space.

($\Leftarrow$) The proof that regular spaces which have a σ-locally compact base of open sets are metrizable is presented in Theorem 31.8.

($\Rightarrow$) Suppose S is metrizable. Then there exists a metric, ρ, such that (S,ρ) is equivalent to S. Since metric spaces are regular, so is S. We are simply required to find a σ-locally finite basis for S.

To do this, we will construct an open cover of S which will allow us to invoke Lemma 19.8 (cited above) to arrive at the required result. For $n\in\mathbb{N}$, let $\mathscr{U}_n=\{B_{1/n}(x): x\in S\}$; it is, clearly, an open cover of S. Then, by Lemma 19.8, there is a σ-locally finite cover, $\mathscr{E}_n$, of S which refines $\mathscr{U}_n$.

[5] Jun-iti Nagata (1925–2007) was a Japanese mathematician specializing in topology. Yuri Mikhailovich Smirnov (b. 1921, Kaluga, d. 2007, Moscow) was a Soviet and Russian mathematician, specializing in topology. This theorem was proved independently by Nagata in 1950 and Smirnov in 1951.

Let $\mathscr{E} = \cup\{\mathscr{E}_n : n = 1, 2, 3, \ldots\}$ be a σ-locally finite cover of S which refines $\mathscr{U} = \cup\{\mathscr{U}_n : n = 1, 2, 3, \ldots\}$. We claim that $\mathscr{E}$ is a base for S. Suppose $a \in S$ and $B_\varepsilon(a)$ is an open ε-ball centered at a with respect to ρ. Then there exists $k \in \mathbb{N}$, such that $1/k < \varepsilon/2$. Now, $\mathscr{B}_k$ covers S so there is $B \in \mathscr{B}_k$ such that $a \in B$. The diameter of B is less than $(1/k)/2 < \varepsilon/2$. So, $B \subseteq B_\varepsilon(a)$. Then $\mathscr{E}$ is a base for S as claimed. Then $\mathscr{E}$ is a σ-locally finite base for S, as required.

Chapter 32

The Stone Space

Abstract
In this section, we discuss those partially ordered sets called lattices and Boolean algebras along with some of their properties. Given a Boolean algebra, B, we construct the set, $\mathscr{S}(B)$, of all ultrafilters on B and topologize it. We call the resulting topological space, $(\mathscr{S}(B), \tau)$, the Stone space. We show that the Stone space, $(\mathscr{S}(B), \tau)$, is a compact, zero-dimensional Hausdorff space. We then show that any Boolean algebra, B, is isomorphic to $\mathscr{B}(\mathscr{S}(B))$, the set of all clopen subsets of $(\mathscr{S}(B), \tau)$.

32.1 The Lattice of a Partially Ordered Set

The main objective of this chapter is to derive a topological representation of a Boolean algebra. Since we would like to make this chapter as self-contained as possible, we will begin by working our way from the topic of partially ordered sets, to lattices, complete lattices, lattice filters and ultrafilters and finally to Boolean algebras. Along the way we will gather useful tools which will allow us to prove the main results of this chapter.

Given a partially ordered set $(P, \leq)$ there may be pairs of elements a, b in P whose least upper bound or greatest lower bound does not belong to P. This section will involve partially ordered sets in which every pair of elements in P have lower and upper bounds contained in P (with respect to $\leq$). We refer to this type of set as a *lattice* of the partially ordered set. We discussed the notion of a lattice of a partially ordered set before, on page 671. For convenience, and, hopefully, to render this section a bit more "self-contained" we review the basics of these notions here.

Definition 32.1 A partially ordered set $(P, \leq)$ is called a *lattice* if, for all pairs a, b in P, $a \vee b = \text{lub}\{a, b\}$ and $a \wedge b = \text{glb}\{a, b\}$ (with respect to the partial order, $\leq$) both exist in P. The lattice derived from $(P, \leq)$ is denoted as

$$(P, \leq, \wedge, \vee)$$

The binary relation "$\wedge$" is often expressed as "meet" and "$\vee$" as "join".

Definition 32.2 If B is a non-empty subset of the lattice $(P, \leq, \wedge, \vee)$, then

$$\vee B = \text{lub}\{B\}$$

$$\wedge B = \text{glb}\{B\}$$

both with respect to $\leq$.
Note that $\vee B$ and $\wedge B$ may or may not be an element of P. A lattice $(P, \leq, \wedge, \vee)$ is said to be a

complete lattice

if for any non-empty subset B of P, both $\vee B$ and $\wedge B$ exist as elements of P.

Example 1. Let S be a set and $(\mathscr{P}(S), \subseteq)$ be a partially ordered set ordered by inclusion. Let A and B be any pair of subsets of S (read, $A, B \in \mathscr{P}(S)$). We define $A \vee B = A \cup B$ and $A \wedge B = A \cap B$. For $\mathscr{U} \subseteq \mathscr{P}(S)$, $\vee\mathscr{U} = \cup\{A \in \mathscr{P}(S) : A \in \mathscr{U}\}$ (read, "$\vee\mathscr{U}$ is the union of all subsets, A, of S such that $A \in \mathscr{U}$") is an element of $\mathscr{P}(S)$. Also define $\wedge\mathscr{U} = \cap\{A \in \mathscr{P}(S) : A \in \mathscr{U}\}$ (read, "$\wedge\mathscr{U}$ is the intersection of all subsets, A, of S such that $A \in \mathscr{U}$") is an element of $\mathscr{P}(S)$. In this case, with $\vee$ and $\wedge$ defined as $\cup$ and $\cap$, respectively,

$$(\mathscr{P}(S), \subseteq, \cup, \cap)$$

is a complete lattice.

Example 2. We are given a topological space (S, τ).

(a) Then

$$(\tau, \subseteq, \cap, \cup)$$

is a lattice of sets since it is closed under finite intersections and unions. But, since infinite $\cap\{U : U \in \mathscr{F} \subseteq \tau\}$ need not be open, it is not a complete lattice.

(b) If φ denotes the set of all closed subsets of S then

$$(\varphi, \subseteq, \cap, \cup)$$

is a lattice, but not a complete lattice (since infinite collections of open subsets need not have an intersection which is an element of τ).

(c) If $Z[S]$ denotes the set of all zero-sets in S then

$$(Z[S], \subseteq, \cap, \cup)$$

is a lattice, but not a complete lattice.[1]

Example 3. Let (S, τ) be a topological space. We partially order the elements of τ with "$\subseteq$". If A and B are both members of τ, we define $\vee$ and $\wedge$ as,

$$A \vee B = A \cup B$$
$$A \wedge B = \text{int}_S(A \cap B)$$

Also, if $\psi \subseteq \tau$,

$$\vee\psi = \cup\{A \in \mathscr{P}(S) : A \in \psi\} \in \tau$$
$$\wedge\psi = \text{int}_S\left(\cap\{A \in \mathscr{P}(S) : A \in \psi\}\right) \in \tau$$

---

[1]For example, suppose $\mathbb{R}$ is equipped with the discrete topology and $\mathbb{R} \cup \{\omega\}$ is its one-point compactification. To verify that arbitrary intersections of zero-sets in $Z[\mathbb{R} \cup \{\omega\}]$ need not be a zero-set in $Z[\mathbb{R} \cup \{\omega\}]$, suppose $F = \{0, \omega\}$. For any open neighborhood U of F, $\mathbb{R} \backslash U$ must be compact and so finite. So, F cannot be a G_δ and so could not be a zero-set. For each $x \in \mathbb{R} \backslash \{0\}$, let $U_x = (\mathbb{R} \cup \{\omega\}) \backslash \{x\}$. Then each U_x is a zero-set in $\mathbb{R} \cup \{\omega\}$. Then $\cap\{U_x : x \notin F\} = F$. Then F is a the intersection of zero-sets but not itself a zero-set.

In this case, if $A, B \in \tau$, $A \vee B = A \cup B \in \tau$ and $A \wedge B = \text{int}_S(A \cap B) = A \cap B \in \tau$. So, $(\tau, \subseteq, \vee, \wedge)$ is a lattice. Then

$$(\tau, \subseteq, \vee, \wedge)$$

(with $\wedge$ and $\vee$ as defined above) is a complete lattice.

Example 4. Let S be a topological space. Let

$$\mathscr{B}(S) = \{A \subseteq S : A \text{ is clopen in } S\}^2$$

partially ordered by inclusion, $\subseteq$. If A and B are clopen subsets of S we define $A \vee B = A \cup B$ and $A \wedge B = A \cap B$. Then $(\mathscr{B}(S), \subseteq, \cup, \cap)$ is a lattice in $(\mathscr{P}(S), \subseteq)$. But it need not necessarily be complete (since arbitrary unions of clopen sets as well as arbitrary intersections of clopen sets need not be clopen).

Example 5. Let $(\mathscr{B}, \leq)$ be a partially ordered set and $(\mathscr{B}, \leq, \vee, \wedge)$ be an associated lattice. Suppose that $\mathscr{B}$ has a least element, say L. That is, $L \leq B$ for all $B \in \mathscr{B}$. Suppose that, $\vee\mathscr{M} \in \mathscr{B}$ for all non-empty subsets $\mathscr{M}$ of $\mathscr{B}$.

Show that $\wedge\mathscr{M}$ must then also belong to $\mathscr{B}$ for all non-empty subsets $\mathscr{M}$ of $\mathscr{B}$.

Solution: We are given that $(\mathscr{B}, \leq)$ is a partially ordered set which has a least element L. We are also given that $\vee\mathscr{M} \in \mathscr{B}$ for all non-empty $\mathscr{M} \subseteq \mathscr{B}$.

Let $\mathscr{M} \subseteq \mathscr{B}$ where $\mathscr{M} \neq \varnothing$. We are required to show that $\wedge\mathscr{M} \in \mathscr{B}$.

Let

$$\mathscr{N} = \{N \in \mathscr{B} : N \leq M \text{ for all } M \in \mathscr{M}\}$$

Since $L \in \mathscr{B}$ and $L \leq M$ for all $M \in \mathscr{M}$, then $L \in \mathscr{N}$. So, $\mathscr{N} \neq \varnothing$. Since $\mathscr{N} \neq \varnothing$, then, by hypothesis, $\vee\mathscr{N}$ exists in $\mathscr{B}$.

For $N \in \mathscr{N}$, $N \leq M$ for all $M \in \mathscr{M}$; so $M \geq N$ for all $M \in \mathscr{M}$. This means N is a lower bound of $\mathscr{M}$. This holds true for each $N \in \mathscr{N}$. Then the lub$\mathscr{N} = \vee\mathscr{N}$ is a lower bound of $\mathscr{M}$. Then

$$\vee\mathscr{N} \leq \text{glb}\mathscr{M} = \wedge\mathscr{M}$$

[2]A set is clopen if it is simultaneously open and closed in S.

We claim that $\vee\mathcal{N} = \text{glb}\mathcal{M}$. For suppose A is a lower bound of $\mathcal{M}$. That is, suppose $A \leq M$ for all $M \in \mathcal{M}$. Then $A \in \mathcal{N}$. This implies $A \leq \vee\mathcal{N}$. So, $\vee\mathcal{N}$ is a lower bound of $\mathcal{M}$ which greater than or equal to the lower bound A. Then $\vee\mathcal{N}$ is the greatest lower bound of $\mathcal{M}$. So, $\vee\mathcal{N} = \wedge\mathcal{M}$, as claimed.

By hypothesis, $\vee\mathcal{N} \in \mathcal{B}$ so $\wedge\mathcal{M} \in \mathcal{B}$, as required.

32.2 Regular Open Sets Revisited

We revisit the notion of regular open sets in order to provide a more intricate example of a complete lattice. For convenience, we restate the definition of "regular open set" previously discussed on page 81.

Definition 32.3 Let S be a topological space. An *open* subset, B, of S is said to be *regular open in S* if

$$B = \text{int}_S(\text{cl}_S(B))$$

The set of all regular open subsets of S will be denoted as $\mathcal{Ro}(S)$.

See that $(\mathcal{Ro}(S), \subseteq)$ is a partially ordered subset of $(\tau, \subseteq)$.

For example, $B = (-5, 0) \cup (0, 5)$ is open in $\mathbb{R}$ but *not* regular open in $\mathbb{R}$ (since $\text{int}_S(\text{cl}_S B) = (-5, 5) \neq B$).

Observation. If B is an open subset of S it is always the case that

$$B \subseteq \text{int}_S\text{cl}_S B \subseteq \text{cl}_S B \quad (*)$$

since $B \subseteq \text{cl}_S B$ implies $B = \text{int}_S B \subseteq \text{int}_S\text{cl}_S B$.

Example 6. Show that for *any* subset B of S, $\text{int}_S\text{cl}_S B \in \mathcal{Ro}(S)$.

Solution: Note that, for any set B in S, since $\text{int}_S\text{cl}_S B$ is open in S, $\text{int}_S\text{cl}_S B \subseteq \text{int}_S\text{cl}_S(\text{int}_S\text{cl}_S B)$ (since $\text{int}_S\text{cl}_S B$ is open in S and by

(∗)). Also

$$\begin{aligned}\text{int}_S\text{cl}_S(\text{int}_S\text{cl}_S B) &\subseteq \text{int}_S\text{cl}_S(\text{cl}_S B)\\ &= \text{int}_S\text{cl}_S B\end{aligned}$$

implies

$$\text{int}_S\text{cl}_S B = \text{int}_S\text{cl}_S(\text{int}_S\text{cl}_S B)$$

so, for any B,

$$\text{int}_S\text{cl}_S B \in \mathscr{R}o(S)$$

The set $\mathscr{R}o(S)$ is then the set of all open subsets B of S which can be expressed in the form $B = \text{int}_S\text{cl}_S B$.

When the elements of a set are subsets of a certain type, partially ordered by "$\subseteq$", it does not automatically follow that $\vee$ and $\wedge$ represent $\cup$ and $\cap$, respectively. Consider the following statement which verifies that $(\mathscr{R}o(S), \subseteq, \vee, \wedge)$ forms a complete lattice in τ.

Theorem 32.4 *Let (S, τ) be a topological space and $(\mathscr{R}o(S), \subseteq)$ represent the family of all regular open sets in S partially ordered by inclusion, $\subseteq$. Suppose that, for $A, B \in \mathscr{R}o(S)$ and $\mathscr{M} \subseteq \mathscr{R}o(S)$,*

$$\begin{aligned}A \wedge B &= A \cap B\\ \vee\mathscr{M} &= \mathit{int}_S\,\mathit{cl}_S(\cup\mathscr{M})\end{aligned}$$

Then $(\mathscr{R}o(S), \subseteq, \vee, \wedge)$ is a complete lattice in $(\tau, \subseteq)$.

Proof. Given S be a topological space.

Suppose A and B belong to $\mathscr{R}o(S)$.

Step 1: We claim that $A \wedge B = A \cap B \in \mathscr{R}o(S)$.

By definition of $\mathscr{R}o(S)$, $A = \text{int}_S\text{cl}_S A$ and $B = \text{int}_S\text{cl}_S B$. Since A and B are open, then so is $A \wedge B = A \cap B$. Then

$A \wedge B \subseteq \text{int}_S\text{cl}_S(A \wedge B)$. Also,

$$\begin{aligned}\text{int}_S\text{cl}_S(A \wedge B) &= \text{int}_S\text{cl}_S(A \cap B)\\ &\subseteq \text{int}_S(\text{cl}_S A \cap \text{cl}_S B) \quad \text{By Theorem 4.6}\\ &= \text{int}_S\text{cl}_S A \cap \text{int}_S\text{cl}_S B\\ &= A \cap B\\ &= A \wedge B\end{aligned}$$

Therefore, $\text{int}_S\text{cl}_S(A \wedge B) \subseteq A \wedge B$. We conclude that

$$A \wedge B = \text{int}_S\text{cl}_S(A \wedge B) \quad (*)$$

So, $A \wedge B$ is a regular open set.

Step 2: Let $\mathscr{M} \subseteq \mathscr{R}o(S)$ where $\mathscr{M}$ is non-empty. That is $M = \text{int}_S\text{cl}_S M$ for all $M \in \mathscr{M}$. We claim that $\vee\mathscr{M} = \text{int}_S\text{cl}_S(\vee\mathscr{M})$. If so then $\vee\mathscr{M}$ is regular open.

Proof of claim: Since $\vee\mathscr{M} = \text{int}_S\text{cl}_S(\cup\mathscr{M})$, $\vee\mathscr{M}$ is open in S. So,

$$\vee\mathscr{M} \subseteq \text{int}_S\text{cl}_S(\vee\mathscr{M})$$

Also,

$$\begin{aligned}\text{int}_S\text{cl}_S(\vee\mathscr{M}) &= \text{int}_S\text{cl}_S[\text{int}_S\text{cl}_S(\cup\mathscr{M}]\\ &\subseteq \text{int}_S\text{cl}_S[\text{cl}_S(\cup\mathscr{M})]\\ &= \text{int}_S\text{cl}_S(\cup\mathscr{M})\\ &= \vee\mathscr{M}\end{aligned}$$

So, $\text{int}_S\text{cl}_S(\vee\mathscr{M}) \subseteq \vee\mathscr{M}$.

We conclude that $\vee\mathscr{M} = \text{int}_S\text{cl}_S(\vee\mathscr{M})$, as claimed.

We now claim that $\wedge\mathscr{M} = \text{int}_S\text{cl}_S(\wedge\mathscr{M})$.

First note that $\mathscr{R}o(S)$ has a *smallest* element: Since $\text{int}_S\text{cl}_S\varnothing = \text{int}_S\varnothing = \varnothing$ and $\varnothing \subseteq U$ for all $U \in \mathscr{R}o(S)$, then $\varnothing$ is the smallest regular open set with respect to $\subseteq$.

We have shown in Example 4 above that for any lattice, $(\mathscr{B}, \leq, \vee, \wedge)$, which has a least element and $\vee\mathscr{M} \in \mathscr{B}$ (for all non-empty $\mathscr{M} \subseteq \mathscr{B}$) then $\wedge\mathscr{M} \in \mathscr{B}$ for all non-empty $\mathscr{M} \subseteq \mathscr{B}$. The lattice $(\mathscr{R}o(S), \subseteq, \vee, \wedge)$ satisfies precisely these conditions. So, $\wedge\mathscr{M} = \text{int}_S\text{cl}_S(\wedge\mathscr{M})$, as claimed.

Then $\wedge\mathscr{M} \in \mathscr{R}o(S)$ for all non-empty $\mathscr{M} \subseteq \mathscr{R}o(S)$.
We conclude that $(\mathscr{R}o(S), \subseteq, \vee, \wedge)$ is a complete lattice.

32.2.1 *The set $\mathscr{R}o(S)$ is an open base for* some *topology on S*

Given a topological space, (S, τ), by definition, $\varnothing$ and S belong to $\mathscr{R}o(S)$. On page 81, we showed that

$$\mathscr{R}o(S) \text{ is closed under finite intersections}$$

Then, if $x \in S$ and $x \in A \cap B$ where $\{A, B\} \subseteq \mathscr{R}o(S)$, since $A \cap B \in \mathscr{R}o(S)$, then $\mathscr{R}o(S)$ satisfies the "base property". Then, by Theorem 5.4,

$$\mathscr{R}o(S) \text{ is an open base for } \textit{some} \text{ topology on } S$$

32.2.2 *Some notation involved in the space generated by regular open sets*

We will denote the topology generated by $\mathscr{R}o(S)$ by

$$\tau_s$$

So, given any topological space (S, τ) we can construct by using $\mathscr{R}o(S)$, the topological space

$$(S, \tau_s)$$

Note that τ_s is strictly weaker than τ (since there are open sets which are not regular open). So, (S, τ) may differ from (S, τ_s).

32.2.3 *If (S, τ) is Hausdorff then so is (S, τ_s)*

We claim that if (S, τ) is Hausdorff then so is (S, τ_s). If (S, τ) is Hausdorff, τ separates points of S. Then, for $x, y \in S$, there exists disjoint $U, V \in \tau$ containing x and y, respectively. Then x, y belong to $\text{int}_S\text{cl}_S(U)$ and $\text{int}_S\text{cl}_S(V)$, respectively (since $U \subseteq \text{int}_S\text{cl}_S(U)$ and $V \subseteq \text{int}_S\text{cl}_S(V)$).

Suppose $A = \text{int}_S\text{cl}_S U \cap \text{int}_S\text{cl}_S V$. See that $\text{cl}_S U \cap V = \varnothing$ so $\text{int}_S\text{cl}_S U \cap V = \varnothing$. Since $\text{int}_S\text{cl}_S U \cap V = \varnothing$ and $\text{int}_S\text{cl}_S V \subseteq \text{cl}_S V$, $A \subseteq \text{cl}_S V \backslash V$. Then the open subset A must be empty. So, if U and V separates x and y then so do the open base elements $\text{int}_S\text{cl}_S(U)$ and $\text{int}_S\text{cl}_S(V)$ of τ_s. So, (S, τ_s) is Hausdorff, as claimed.

So,

$$\text{If } (S, \tau) \text{ is Hausdorff then } (S, \tau_s) \text{ is Hausdorff}$$

32.2.4 *Sublattices*

Note that every element of $(\mathscr{B}(S), \subseteq, \cap, \cup)$ (the clopen subsets of S) is an element of $(\mathscr{R}o(S), \subseteq, \vee, \wedge)$. We say that $(\mathscr{B}(S), \subseteq, \cap, \cup)$ is a

sublattice

of $\mathscr{R}o(S)$.

The lattice $(\mathscr{R}o(S), \subseteq, \vee, \wedge)$ cannot be viewed as a "sublattice" of $(\mathscr{P}(S), \subseteq, \cap, \cup)$ since $A \vee B = \text{int}_S\text{cl}_S(A \cup B) \neq A \cup B$.

32.3 Filters and Ultrafilters in a Lattice of Sets

Suppose S is a set and the family, $\mathscr{P}(S)$, of all subsets of S is partially ordered by inclusion "$\subseteq$". In what follows the set, L, denotes a particular subset of $\mathscr{P}(S)$ which is partially ordered by "$\subseteq$" and which contains both $\varnothing$ and S.

The definition of "lattice filter" is analogous to the one of "filter of sets".

Definition 32.5 Let $(L \subseteq, \vee, \wedge)$ be a lattice of sets and $\mathscr{F} \subseteq L$. The subfamily $\mathscr{F}$ of L is called an *L-filter base* if it satisfies the two conditions,

(1) $\mathscr{F}$ is non-empty,
(2) $\mathscr{F}$ is such that, for non-empty $A, B \in \mathscr{F}$ there exists $D \neq \varnothing$ in $\mathscr{F}$ such that $D \subseteq A \wedge B \in \mathscr{F}$.

If an L-filter base $\mathscr{F}$ also satisfies the condition,

$$\text{If } A \in \mathscr{F} \text{ and } A \subseteq C \text{ for some } C \in L \text{ then } C \in \mathscr{F}$$

then we say that $\mathscr{F}$ is an *L-filter*. Given an L-filter base, $\mathscr{F}$, it can always generate an L-filter which we will denote as, $\mathscr{F}^*$. The lattice L is itself an L-filter. A filter, $\mathscr{F}$, is said to be a *proper L-filter* if $\mathscr{F} \neq L$. Note that $\mathscr{F}$ is a proper L-filter if and only if $\varnothing \notin \mathscr{F}$.[3]

Definition 32.6 Let $(L, \subseteq, \vee, \wedge)$ be a lattice of sets in $\mathscr{P}(S)$. An *L-ultrafilter* is a proper filter $\mathscr{F} \subseteq L$ which is not properly contained in any other proper filter in L. If the filter $\mathscr{F}$ is such that $\wedge\{F : F \in \mathscr{F}\} \neq \varnothing$ then we say that the filter $\mathscr{F}$ is a *fixed L-ultrafilter*. Those L-ultrafilters which are not fixed are said to be *free L-ultrafilters.*

Example 7. Given a topological space (S, τ) the family of all zero-sets, $(Z[S], \subseteq, \cap, \cup)$ forms a lattice of sets contained in $\mathscr{P}(S)$. It contains both $\varnothing$ and S.

If $\mathscr{F}$ is a non-empty subset of $Z[S]$ such that, for Z_1 and Z_2 in $\mathscr{F}$, $Z_1 \cap Z_2 = Z_3 \in \mathscr{F}$, and whenever $Z \in \mathscr{F}$ and $Z \subseteq Z^*$ for some $Z^* \in Z[S]$ then $Z^* \in \mathscr{F}$, the collection $\mathscr{F}$ is a

z-filter.

The concept of z-filters has been discussed extensively in 14.10. If we apply the notation suggested in the definition we could also say "$Z[S]$-filter".

Also, instead of saying "$Z[S]$-ultrafilter" we can write z-ultrafilter.

We present the following associated facts about L-ultrafilters.

Theorem 32.7 *Let S be a topological space and $(L, \subseteq, \vee, \wedge)$ be a lattice in $(\mathscr{P}(S), \subseteq)$. Suppose $\mathscr{F}$ is a proper L-filter.*

(a) *The L-filter, $\mathscr{F}$, can be extended to an L-ultrafilter.*
(b) *The L-filter, $\mathscr{F}$, is an L-ultrafilter if and only if, for every $A \in \mathscr{P}(S)$, either $A \in \mathscr{F}$ or $S \backslash A \in \mathscr{F}$.*

---

[3]Since if $\varnothing \in \mathscr{F}$, $\varnothing \subseteq F$ for all $F \in L$, so $\mathscr{F} = L$.

Proof.

(a) We are given that $L \subseteq \mathscr{P}(S)$ and $\mathscr{F}$ is a proper L-filter. Let

$$\mathscr{H} = \{\mathscr{M} : \mathscr{M} \text{ is a proper } L\text{-filter such that } \mathscr{F} \subseteq \mathscr{M}\}$$

We partially order $\mathscr{H}$ with $\subseteq$. Let $\mathscr{C}$ be a chain in $(\mathscr{H}, \subseteq)$. Then $\cup\{C : C \in \mathscr{C}\}$ is an upper bound of $\mathscr{C}$ with respect to $\subseteq$. So every chain in $\mathscr{H}$ has an upper bound. By Zorn's lemma, $(\mathscr{H}, \subseteq)$ has a maximal element. That is, $\mathscr{H}$ contains a filter, $\mathscr{F}^*$, which is not properly contained in any other filter. Since $\mathscr{F}^* \in \mathscr{H}$, $\mathscr{F} \subseteq \mathscr{F}^*$. Then $\mathscr{F}$ can be extended to an L-ultrafilter, as required.

(b) ($\Rightarrow$) We are given that $\mathscr{F}$ is an L-ultrafilter in $L \subseteq \mathscr{P}(S)$ and that $A \subseteq S$. We are required to show that either $A \in \mathscr{F}$ or $S\backslash A \in \mathscr{F}$.

Suppose neither A nor $S \backslash A$ belongs to $\mathscr{F}$. Let

$$\mathscr{H}_A = \{B \in L : A \cap F \subseteq B \text{ for some } F \in \mathscr{F}\}$$

We claim that $\mathscr{H}_A$ is a proper L-filter base which contains $\mathscr{F}$ as a proper subset.

Proof of claim: See that, if $U \in \mathscr{F}$, $A \cap U \subseteq U$, an element of $\mathscr{F}$. Then $U \in \mathscr{H}_A$, so $\mathscr{F} \subseteq \mathscr{H}_A$.

Also, since $A \cap F \subseteq A$ for all $F \in \mathscr{F}$, $A \in \mathscr{H}_A$. Then, $A \in \mathscr{H}_A \backslash \mathscr{F}$. Then $\mathscr{F}$ is a *proper* subset of $\mathscr{H}_A$.

Suppose $B, D \in \mathscr{H}_A$. We now show that $B \wedge D \in \mathscr{H}_A$.

Then $B, D \in L$, $A \cap F_1 \subseteq B$ and $A \cap F_2 \subseteq D$ for some $F_1, F_2 \in \mathscr{F}$.

Then $A \cap (F_1 \wedge F_2) \subseteq A \cap F_1 \subseteq B$ and $A \cap (F_1 \wedge F_2) \subseteq A \cap F_2 \subseteq D$. Then $A \cap (F_1 \wedge F_2) \subseteq B \cap D \subseteq \text{glb}\{B, D\} = B \wedge D$. Since $B \wedge D \in L$, then $B \wedge D \in \mathscr{H}_A$.

Also, see that $\varnothing \notin \mathscr{H}_A$. For, if it was, $A \cap F = \varnothing$ for some F and so $F \subseteq S \backslash A$ which would imply $S \backslash A \in \mathscr{F}$, a contradiction. So, $\mathscr{H}_A$ is proper.

Hence, since $\mathscr{H}_A$ is proper, is non-empty and is closed under finite applications of $\wedge$, it is an L-filter base, as claimed.

Then $\mathscr{H}_A$ will generate an L-filter $\mathscr{H}_A^*$. Then $\mathscr{F} \subset \mathscr{H}_A^*$ which contradicts the fact that $\mathscr{F}$ is an L-ultrafilter.

Then the L-ultrafilter $\mathscr{F}$ must intersect either A or $S \backslash A$.

($\Leftarrow$) Conversely, for $L \subseteq \mathscr{P}(S)$, suppose that, for every $A \subseteq S$, either $A \in \mathscr{F}$ or $S \backslash A \in \mathscr{F}$. We are required to show that $\mathscr{F}$ is an ultrafilter. Suppose not. That is, suppose that there exists a proper filter, $\mathscr{H}$, such that $\mathscr{F} \subset \mathscr{H}$. Then there exists some non-empty A such $A \in \mathscr{H} \backslash \mathscr{F}$. Then $S \backslash A$ cannot belong to $\mathscr{F}$, for, if it did, then $A \cap (S \backslash A) = \varnothing$ must belong to $\mathscr{H}$, contradicting the fact that $\mathscr{H}$ is a proper filter. Hence, $\mathscr{F}$ must be an L-ultrafilter.

A word of caution: For the characterization of an L-ultrafilter above, things will work out a bit differently in the case $L = \mathscr{R}o(S)$, as the following theorem shows.

Theorem 32.8 *Let S be a topological space. We have shown that $(\mathscr{R}o(S), \subseteq, \vee, \cap)$ is a complete lattice in $(\tau, \subseteq)$. An $\mathscr{R}o(S)$-filter, $\mathscr{F}$, is an $\mathscr{R}o(S)$-ultrafilter if and only if, for any $A \in \mathscr{R}o(S)$, either A or $S \backslash cl_S(A)$ belongs to $\mathscr{F}$.*

Proof. ($\Rightarrow$) Suppose $\mathscr{F}$ is an $\mathscr{R}o(S)$-ultrafilter and $A \in \mathscr{R}o(S)$. Let $F \in \mathscr{F}$.

Case 1: If $F \cap A = \varnothing$ then (since F is open) $F \subseteq S \backslash \mathrm{cl}_S A$. Since

$$\begin{aligned} S \backslash \mathrm{cl}_S A &= \mathrm{int}_S(S \backslash A) \\ &= \mathrm{int}_S(S \backslash \mathrm{int}_S \mathrm{cl}_S A) \\ &= \mathrm{int}_S \mathrm{cl}_S(S \backslash \mathrm{cl}_S A) \end{aligned}$$

$F \subseteq S \backslash \mathrm{cl}_S A \in \mathscr{R}o(S)$.

Case 2: Suppose $F \cap A \neq \varnothing$ for all $F \in \mathscr{F}$. Suppose $A \notin \mathscr{F}$. Let

$$\mathscr{H} = \{B \in \mathscr{R}o(S) : A \cap F \subseteq B\}$$

Then $\mathscr{F} \subseteq \mathscr{H}$ and $A \in \mathscr{H} \backslash \mathscr{F}$. As shown in the theorem above, $\mathscr{H}$ is a filter base which generates a filter $\mathscr{H}^*$ in $(\mathscr{R}o(S), \subseteq)$. Since $\mathscr{F} \subset \mathscr{H}^*$, this contradicts the fact that $\mathscr{F}$ is an $\mathscr{R}o(S)$-ultrafilter. So, A must belong to $\mathscr{F}$.

($\Leftarrow$) Suppose that for any $A \in \mathscr{R}o(S)$, either A or $S \backslash \mathrm{cl}_S(A)$ belongs to $\mathscr{F}$. See that the filter $\mathscr{F}$ extends to an $\mathscr{R}o(S)$-ultrafilter $\mathscr{F}^*$. Suppose $A \in \mathscr{F}^*$. If $A \notin \mathscr{F}$ then $S \backslash \mathrm{cl}_S A \in \mathscr{F} \subseteq \mathscr{F}^*$ implying that $A \cap (S \backslash \mathrm{cl}_S A) = \varnothing \in \mathscr{F}^*$, a contradiction. So $A \in \mathscr{F}$. Then $\mathscr{F}^* \subseteq \mathscr{F}$ which implies that $\mathscr{F}$ is an $\mathscr{R}o(S)$-ultrafilter.

It can similarly be shown that a τ-filter, $\mathscr{F}$, is a τ-ultrafilter if and only if, for any $A \in \tau$, either A or $S\backslash \mathrm{cl}_S(A)$ belongs to $\mathscr{F}$.

32.4 Boolean Algebras

The definition which follows is a generalization of the notion of a "lattice", $(P \leq, \vee, \wedge)$, for a partially ordered set discussed earlier in this section. We can also approach the notion of a lattice independent of a predetermined partial ordering on a given set and then determine a partial ordering in terms of its properties.

Definition 32.9 Let B be any non-empty set. Suppose there is defined on B two binary operations $\wedge$ and $\vee$ such that $a, b \in B$ implies $a \vee b \in B$ and $a \wedge b \in B$. Suppose $\wedge$ and $\vee$ also satisfy the properties:

(a) $a = a \vee a$ and $a = a \wedge a$ (Reflexivity)

(b) $a \wedge b = b \wedge a$ and $a \vee b = b \vee a$ (Commutativity)

(c) $a \wedge (b \wedge c) = (a \wedge b) \wedge c$; $a \vee (b \vee c) = (a \vee b) \vee c$ (Associativity)

A set $(B, \vee, \wedge)$ along with two such binary operations is also referred to as a lattice.

A lattice $(B, \vee, \wedge)$ is said to be a *distributive lattice* if, in addition to the three properties stated above, the following property is satisfied:

(d) $\forall x, y, z \in B, x \vee (y \wedge z) = (x \vee y) \wedge (x \vee z)$ and $x \wedge (y \vee z) = (x \wedge y) \vee (x \wedge z)$

The lattice $(B, \vee, \wedge)$ is said to be a

complemented lattice

if it has an element we refer to as "maximal element", denoted by 1, and an element we refer to as "minimal element", denoted by 0, such that, for every $x \in B$ there exists a *unique* x' such that

$$x \vee x' = 1 \quad \text{and} \quad x \wedge x' = 0^4$$

A complemented distributive lattice is referred to as being a *Boolean algebra*. A Boolean algebra is denoted as

$$(B, \vee, \wedge, 0, 1, ')$$

Exercise: A. Suppose $(B, \vee, \wedge)$ is a set equipped with binary operations $\vee$ and $\wedge$. Suppose that $\vee$ and $\wedge$ are reflexive, commutative, associative and distributive binary operations. Show that

$$a \wedge b = a \qquad \Leftrightarrow \qquad a \vee b = b$$

Example 8. Suppose we are given a Boolean algebra, $(B, \vee, \wedge)$, as defined above. That is, the binary operations $\vee$ and $\wedge$ are reflexive, commutative, associative and distributive. Suppose we define the binary operation, $\leq$, on B as:

$$a \leq b \qquad \Leftrightarrow \qquad a \wedge b = a \qquad \Leftrightarrow \qquad a \vee b = b$$

Show that $\leq$ thus defined is a partial ordering of $(B, \vee, \wedge)$.

Solution: Reflexivity: Since $a \wedge a = a$ and $a \vee a = a$, then $a \leq a$.

Antisymmetry: Suppose $a \leq b$ and $b \leq a$. Then $a \vee b = b$ and $b \vee a = a$. By property in part (b) of the definition, $a \vee b = b \vee a$. So, $a = b$.

Transitivity: Suppose $a \leq b$ and $b \leq c$.
Then

$$b \wedge c = b, \;\; b \vee c = c \quad \text{and} \quad a \wedge b = a, \, a \vee b = b$$

So,

$$a \wedge c = (a \wedge b) \wedge c = a \wedge (b \wedge c) = a \wedge b = a$$

$$a \vee c = a \vee (b \vee c) = (a \vee b) \vee c = b \vee c = c$$

Since $a \wedge c = a$ and $a \vee c = c$, then $a \leq c$.

---

[4]Note that in this context and with the information given it is not meaningful to say $x' = S \backslash x$.

Hence, $\leq$ partially orders B. In such a case, we can write

$$(B, \leq, \vee, \wedge)$$

as a partially ordered lattice.

Exercise: B. Suppose $(B, \leq)$ is a lattice where

$$a \leq b \quad \Leftrightarrow \quad a \wedge b = a \quad \Leftrightarrow \quad a \vee b = b$$

where $a \vee b = \sup\{a, b\}$ and $a \wedge b = \inf\{a, b\}$. Show that $\vee$ and $\wedge$ are reflexive, commutative, associative and distributive binary operations. So, B is a Boolean algebra.

32.4.1 *A few properties related to complements*

(a) Note that in $(B, \leq, \vee, \wedge, 0, 1, ')$, for any $y \in B$,

$$0 \leq y \quad \Leftrightarrow \quad 0 \vee y = y \quad \Leftrightarrow \quad 0 \wedge y = 0$$

$$y \leq 1 \quad \Leftrightarrow \quad y \vee 1 = 1 \quad Lra \quad y \wedge 1 = y$$

(b) The expression x'' is to be interpreted as $(x')'$. Note that, since $x' \wedge x = 0 = x' \wedge x''$ and $x' \vee x = 1 = x'' \vee x'$ and the complement of x' and of x is unique, then

$$x'' = x$$

(c) Also, by definition of 0 and 1, $0 \wedge 1 = 0$ and $0 \vee 1 = 1$. This implies

$$0' = 1 \quad \text{and} \quad 1' = 0$$

(d) Consider $z = x \wedge y$ and $w = x' \vee y'$. Then

$$\begin{aligned} z \vee w &= (x \wedge y) \vee (x' \vee y') \\ &= [(x \wedge y) \vee x'] \vee [(x \wedge y) \vee y'] \\ &= (\,[x \vee x'] \wedge [y \vee x']\,) \vee (\,[x \vee y'] \wedge [y \vee y']\,) \\ &= (\,1 \wedge [y \vee x']\,) \vee (\,[x \vee y'] \wedge 1\,) \\ &= [y \vee x'] \vee [x \vee y'] \\ &= [y \vee y'] \vee [x \vee x'] \\ &= 1 \end{aligned}$$

Similarly, $z \wedge w = 0$.

So,

$$(x \wedge y)' = x' \vee y'$$

(e) Suppose $x \leq y$. Then $x \wedge y = x$. So, $(x \wedge y)' = x' \vee y' = x'$. It follows that $y' \leq x'$.

Exercise: C. Show that if $(B, \leq, \vee, \wedge, 0, 1)$ is a distributive lattice with maximum and minimum elements 1 and 0 respectively then, for every element $x \in B$, its complement, x', is unique.

Example 9. The lattices $(\mathscr{P}(S), \subseteq, \cup, \cap, S, \varnothing, ')$ and $(\mathscr{B}(S), \subseteq, \cup, \cap, S, \varnothing, ')$ where

$$\mathscr{B}(S) = \{A \subseteq S : A \text{ is clopen in } S\}$$

are the simplest examples of Boolean algebras.

In both cases $(\mathscr{P}(S), \subseteq, \cup, \cap, S, \varnothing, ')$ and $(\mathscr{B}(S), \subseteq, \cup, \cap, S, \varnothing, ')$ we can say

$$A' = S \backslash A$$

32.4.2 *The set* $(\mathscr{R}o(S), \subseteq, \vee, \cap)$ *is a Boolean algebra*

Recall that, for a given topological space, S, $(\mathscr{R}o(S), \subseteq, \vee, \cap)$ is a complete lattice. But is it also a Boolean algebra? It is. But this is not at all obvious. We verify this.

To show that $(\mathscr{R}o(S), \subseteq, \vee, \cap)$ is a Boolean algebra we must show that it is (1) a complemented lattice and (2) a distributive lattice.

See that $(\mathscr{R}o(S), \subseteq, \vee, \cap)$ contains $0 = \varnothing$ and $1 = S$.

We claim that the set $\mathscr{R}o(S)$ is a complemented lattice: Let $A \in \mathscr{R}o(S)$. Then $A = \text{int}_S \text{cl}_S A$. See that

$$\begin{aligned} A \wedge (S \backslash \text{cl}_S A) &= A \cap (S \backslash \text{cl}_S A) \\ &= \varnothing \\ A \vee (S \backslash \text{cl}_S A) &= \text{int}_S \text{cl}_S (A \cup (S \backslash \text{cl}_S A)) \\ &= \text{int}_S S \\ &= S \end{aligned}$$

So, $A' = S \backslash \text{cl}_S A$. (It is erroneous to interpret A' as meaning $S \backslash A$.)

The element A' belongs to $\mathscr{Ro}(S)$ since

$$\begin{aligned} S\backslash \mathrm{cl}_S A &= S\backslash \mathrm{cl}_S \mathrm{int}_S \mathrm{cl}_S A \\ &= \mathrm{int}_S \mathrm{cl}_S (S\backslash \mathrm{cl}_S A) \end{aligned}$$

So, $\mathscr{Ro}(S)$ is closed under complements, as claimed.

Finally, we show that $(\mathscr{Ro}(S), \subseteq, \vee, \cap)$ is a distributive lattice. Let $A, B, C \in \mathscr{Ro}(S)$.

$$\begin{aligned} A \cap (B \vee C) &= \mathrm{int}_S \mathrm{cl}_S A \cap \mathrm{int}_S \mathrm{cl}_S (B \cup C) \\ &= \mathrm{int}_S \mathrm{cl}_S (A \cap (B \cup C)) \\ &= \mathrm{int}_S \mathrm{cl}_S (\,(A \cap B) \cup (A \cap C)\,) \\ &= (A \cap B) \vee (A \cap C) \end{aligned}$$

Proceed similarly to show that $A \vee (B \cap C) = (A \vee B) \cap (A \vee C)$.

Then $(\mathscr{Ro}(S), \subseteq, \vee, \cap)$ is a Boolean algebra.

32.5 Boolean Filters and Ultrafilters

We are given that $(B, \leq, \vee, \wedge, 0, 1, ')$ is a Boolean algebra (a complemented distributive lattice where $a \leq b \;\Leftrightarrow\; b = a \vee b \;\Leftrightarrow\; a = a \wedge b$). Like lattices $(L, \subseteq, \vee, \wedge)$, given a Boolean algebra $(B, \leq, \vee, \wedge, 0, 1, ')$, the set B has special subsets we call B-filters and B-ultrafilters. The following concepts and properties are slight generalizations of ones we have already seen, so they will seem familiar to readers.

Definition 32.10 Suppose that, for a given set, B, $(B, \leq, \vee, \wedge, 0, 1, ')$ is a Boolean algebra. Then $\mathscr{F} \subseteq B$ is a

B-filter base

if $\mathscr{F}$ is non-empty and is such that, for any $x, y \in \mathscr{F}$, there is a $z \in \mathscr{F}$ where $z \neq 0$ and $z \leq x \wedge y$.

If $\mathscr{F}$ is a B-filter base which also satisfies the property,

$$[\, x \in \mathscr{F} \text{ and } x \leq y\,] \qquad \Rightarrow \qquad y \in \mathscr{F}$$

$\mathscr{F}$ is called a *B-filter*.

We say that $\mathscr{F}$ is a B-*ultrafilter* if $\mathscr{F}$ is a proper filter and $\mathscr{F}$ is not properly contained in any other proper B-ultrafilter.

Note that a B-filter or a B-ultrafilter is a subset of the Boolean algebra, B, and so can be seen as a particular element of $\mathscr{P}(B)$.

Keep in mind the fact that B-filters and B-ultrafilters are subsets of B (not subsets of $\mathscr{P}(B)$).

We will emphasize three important B-ultrafilter characterizations. Of particular interest: For an ultrafilter $\mathscr{F}$, if x is an element of B which meets every element of $\mathscr{F}$ then x must belong to $\mathscr{F}$. Also if $x \in B$, either x or its complement belongs to $\mathscr{F}$.

Theorem 32.11 *Let $\mathscr{F}$ be a proper B-filter in the Boolean algebra, $(B, \leq, \vee, \wedge, 0, 1, ')$. Then the following statements are equivalent.*

(1) *The set $\mathscr{F}$ is a (proper) B-ultrafilter.*
(2) *If $a \in B$ and $a \wedge y \neq 0$ for all $y \in \mathscr{F}$ then $a \in \mathscr{F}$.*
(3) *Whenever $x \vee y \in \mathscr{F}$, either $x \in \mathscr{F}$ or $y \in \mathscr{F}$.*
(4) *For any $x \in B$, either $x \in \mathscr{F}$ or $x' \in \mathscr{F}$.*

Proof. We are given that $\mathscr{F}$ is a B-filter in the Boolean algebra,$(B, \leq, \vee, \wedge, 0, ')$.

($1 \Rightarrow 2$) Suppose the set $\mathscr{F}$ is a B-ultrafilter and that $a \in B$ is such that $a \wedge y \neq 0$ for each $y \in \mathscr{F}$. We are required to show that $a \in \mathscr{F}$.

Let

$$U = \{z \in B : z \geq a \wedge y \neq 0 \text{ for some } y \in \mathscr{F}\}$$

We claim that $U = \mathscr{F}$.

Since $z \in \mathscr{F} \Rightarrow a \wedge z \leq z \Rightarrow z \in U$

$$\mathscr{F} \subseteq U$$

See that, since $a \geq a \wedge y$ for all $y \in \mathscr{F}$, then $a \in U$.

It now suffices to show that U is a B-filter. (If so, since $\mathscr{F}$ is an ultrafilter which is contained in U, $\mathscr{F} = U$.)

We claim that U_a is a proper B-filter.

See that $0 \notin U$ (since if $0 \in U \Rightarrow 0 \geq a \wedge y \neq 0$, for $y \in \mathscr{F}$, contradicting $0 \leq z$ for all $z \in B$).

Suppose $u, v \in U$. There must exist $b, c \in \mathscr{F}$ such that $u \geq a \wedge b$ and $b \geq a \wedge c$. Then

$$\begin{aligned} u \wedge v &\geq (a \wedge b) \wedge (a \wedge c) \\ &= a \wedge (b \wedge c) \end{aligned}$$

implies $u \wedge v \in U$ (since $b \wedge c \in \mathscr{F}$).

Also, $u \in U$ and $v \geq u$, $v \geq u \geq a \wedge b$ for some $b \in \mathscr{F}$. So $v \in U$.

We conclude that U is a proper B-filter which contains the ultrafilter, $\mathscr{F}$, as claimed.

So,

$$\mathscr{F} = U$$

Since $a \in U$, $a \in \mathscr{F}$, as required.

(2 ⇒ 3) Suppose that, if $a \in B$ is such that $a \wedge y \neq 0$ for each $y \in \mathscr{F}$, then $a \in \mathscr{F}$. We are required to show that, whenever $x \vee y \in \mathscr{F}$, either $x \in \mathscr{F}$ or $y \in \mathscr{F}$.

Suppose $x \vee y \in \mathscr{F}$. Suppose that neither x nor y belongs to $\mathscr{F}$. Then by hypothesis, x and y each miss at least one element of $\mathscr{F}$. That is, there exists $a, b \in \mathscr{F}$ such that $a \wedge x = 0$ and $b \wedge y = 0$. Since both $x \vee y$ and $a \vee b$ belong to $\mathscr{F}$, then so does $(x \vee y) \wedge (a \vee b)$.

But

$$\begin{aligned} (x \vee y) \wedge (a \vee b) &= (x \wedge a \wedge b) \vee (y \wedge a \wedge b) \\ &\leq (x \wedge a) \vee (y \wedge b) \\ &= 0 \vee 0 \\ &= 0 \end{aligned}$$

So, $0 \in \mathscr{F}$, contradicting that $\mathscr{F}$ is proper. So, either x or $y \in\in \mathscr{F}$.

(3 ⇒ 4) Suppose that whenever $x \vee y \in \mathscr{F}$, either $x \in \mathscr{F}$ or $y \in \mathscr{F}$.

Since $x \vee y' = 1 \in \mathscr{F}$, then either x or x' belongs to $\mathscr{F}$.

(4 ⇒ 1) Suppose $\mathscr{F}$ is a B-filter such that for $x \in B$, either x or x' belongs to $\mathscr{F}$. We are required to show that $\mathscr{F}$ is a B-ultrafilter. Suppose $\mathscr{U}$ is a B-ultrafilter which contains $\mathscr{F}$. It suffices to show that $\mathscr{F} = \mathscr{U}$.

If $x \in \mathscr{U}$ then, by hypothesis, either x or x' belongs to $\mathscr{F}$. If $x' \in \mathscr{F}$ then $x' \in \mathscr{U}$. So $x \wedge x' = 0 \in \mathscr{U}$. This is a contradiction (since by hypothesis $\mathscr{U}$ is a proper filter). Then $x \in \mathscr{F}$. So, $\mathscr{F} = \mathscr{U}$. Then $\mathscr{F}$ is a B-ultrafilter.

Definition 32.12 Suppose we are given two Boolean algebras, $(B_1, \leq_1, \vee_1, \wedge_1, 0, 1,')$ and $(B_2, \leq_2, \vee_2, \wedge_2, 0, 1,')$ and a function f which maps elements of B_1 to elements of B_2. We say that $f : B_1 \to B_2$ is a *Boolean algebra homomorphism* if, for any $x, y \in B$,

(1) $f(x \vee_1 y) = f(x) \vee_2 f(y)$
(2) $f(x \wedge_1 y) = f(x) \wedge_2 f(y)$
(3) $f(x') = f(x)'$

The function $f : B_1 \to B_2$ is a *Boolean isomorphism* if f is a bijection and both f and $f^{\leftarrow}$ are Boolean homomorphisms.

Note. See that if $f : (B_1, \leq_1, \vee_1, \wedge_1, 0, 1,') \to (B_2, \leq_2, \vee_2, \wedge_2, 0, 1,')$ is a Boolean homomorphism then

$$\begin{aligned} f(1) &= f(1 \vee 1') \\ &= f(1) \vee f(1') \\ &= f(1) \vee f(1)' \\ &= 1 \end{aligned}$$

Similarly, $f(0) = f(0 \wedge 0') = f(0) \wedge f(0)' = 0$.

32.6 The Set $\mathscr{S}(B)$ of All B-Ultrafilters on B

Given a Boolean algebra, $(B, \leq, \vee, \wedge, 0, 1)$, a B-ultrafilter, $\mathscr{U}$, is a subset of B and so is an element of $\mathscr{P}(B)$. So the B-ultrafilter, $\mathscr{U}$, can be seen as an element of the Boolean algebra,

$$(\mathscr{P}(B), \subseteq, \cup, \cap, B, \varnothing)$$

Not every element of $\mathscr{P}(B)$ is a B-ultrafilter. The subset of $\mathscr{P}(B)$ which is made of B-ultrafilters is denoted as

$$(\mathscr{S}(B),\ \subseteq,\ \cup,\ \cap,\ B,\ \varnothing)$$

where B is the largest B-ultrafilter and $\varnothing$ is the smallest (with respect to the partial order $\subseteq$).

A function $\varphi : B \to \mathscr{P}(\mathscr{S}(B))$:
For each element, x, of a Boolean algebra $(B, \leq, \vee, \wedge, 0, 1, ')$, we define

$$\varphi_x = \{\mathscr{U} \in \mathscr{S}(B) : x \in \mathscr{U}\} \in \mathscr{P}(\mathscr{S}(B))$$

It represents the set of all B-ultrafilters on B which contain the element x. Then for each $x \in B$,

$$\varphi_x \in \mathscr{P}(\mathscr{S}(B))$$

We define the function,

$$\varphi : (B,\ \leq,\ \vee,\ \wedge,\ 0,\ 1) \ \to\ (\mathscr{P}(\mathscr{S}(B)),\ \subseteq,\ \cap,\ \cup,\ \mathscr{S}(B),\ \varnothing)$$

as follows:

$$\varphi(x) = \varphi_x$$

The largest set of all B-ultrafilters is $\mathscr{S}(B)$.

See that $\varphi : B \to \mathscr{P}(\mathscr{S}(B))$ is a well-defined function. For suppose $\varphi_x \neq \varphi_y$. Without loss of generality, suppose $\mathscr{U} \in \varphi_x$ and $\mathscr{U} \notin \varphi_y$. Then $x \in \mathscr{U}$ and $y \notin \mathscr{U}$. So, $x \neq y$.

Theorem 32.13 *The function*

$$\varphi : (B,\ \leq,\ \vee,\ \wedge,\ 0,\ 1) \ \to\ (\mathscr{P}(\mathscr{S}(B)),\ \subseteq,\ \cap,\ \cup,\ \mathscr{S}(B),\ \varnothing)$$

defined as $\varphi(x) = \varphi_x$ *satisfies the following four properties.*

(a) *If* $x, y \in B$, $\varphi_{x\wedge y} = \varphi_x \cap \varphi_y$.
(b) *If* $x, y \in B$, $\varphi_{x\vee y} = \varphi_x \cup \varphi_y$.
(c) $\varphi_0 = \varnothing$ *and* $\varphi_1 = \mathscr{S}(B)$.
(d) $\varphi_{x'} = \mathscr{S}(B)\backslash\varphi_x$. *We can write* $\varphi_{x'} = \varphi'_x$.

Then the function $\varphi : B \to \mathscr{P}(\mathscr{S}(B))$ *is a Boolean homomorphism.*

Proof.

Proof of property (a): We are given that $\varphi_{x\wedge y}$ is the family of all ultrafilters which contain $x \wedge y$. We claim that every ultrafilter in $\varphi_{x\wedge y}$ contains x. If so then $\varphi_{x\wedge y} \subseteq \varphi_x$. Let $\mathscr{V}$ be an ultrafilter in $\varphi_{x\wedge y}$. We will show that $\mathscr{V} \in \varphi_x$. If $z \in \mathscr{V}$ then $z \wedge (x \wedge y) \neq 0$. Then, for any element z in $\mathscr{V}$, $z \wedge x \neq 0$. Then every element of $\mathscr{V}$ meets x. So, $\mathscr{V} \in \varphi_x$. We conclude that $\varphi_{x\wedge y} \subseteq \varphi_x$, as claimed.

By applying similar reasoning, every ultrafilter in $\varphi_{x\wedge y}$ contains y. So,

$$\varphi_{x\wedge y} \subseteq \varphi_x \cap \varphi_y$$

Let $\mathscr{V} \in \varphi_x \cap \varphi_y$. Then $\mathscr{V}$ is an ultrafilter which contains x and contains y. We claim that $\mathscr{V} \in \varphi_{x\wedge y}$.

Since $\mathscr{V}$ is an ultrafilter which contains both x and y, then it must contain $x \wedge y$. Let $z \in \mathscr{V}$. Then $z \wedge x \neq 0$ and $z \wedge y \neq 0$. So $\mathscr{V} \in \varphi_{x\wedge y}$.

Therefore $\varphi_x \cap \varphi_y \subseteq \varphi_{x\wedge y}$, as claimed.

We conclude

$$\varphi_{x\wedge y} = \varphi_x \cap \varphi_y$$

Proof of property (b): Let $\mathscr{V} \in \varphi_{x\vee y}$. Then $x \vee y \in \mathscr{V}$. Then x or y belongs to $\mathscr{V}$ (Theorem 32.11). Suppose without loss of generality that $x \in \mathscr{V}$. Then $\mathscr{V} \in \varphi_x$. We conclude that $\mathscr{V} \in \varphi_x \cup \varphi_y$. So, $\varphi_{x\vee y} \subseteq \varphi_x \cup \varphi_y$.

Suppose $\mathscr{V} \in \varphi_x \cup \varphi_y$. Suppose without loss of generality that $\mathscr{V} \in \varphi_x$. Let $z \in \mathscr{V}$. Then $z \wedge x \neq 0$. So, $z \wedge (x \vee y) = (z \wedge x) \vee (z \wedge y)$ cannot be 0. Then $\mathscr{V} \in \varphi_{x\vee y}$. We conclude that $\varphi_x \cup \varphi_y \subseteq \varphi_{x\vee y}$.

Finally $\varphi_x \cup \varphi_y = \varphi_{x\vee y}$.

Proof of property (c): Since B-ultrafilters are proper filters, no ultrafilter can contain the 0-element of B; then

$$\varphi(0) = \varphi_0 = \varnothing$$

Also, 1 belongs to all B-ultrafilters hence, $\varphi(1) = \varphi_1 = \mathscr{S}(B)$.

Proof of property (d): See that, $\varnothing = \varphi_0 = \varphi_{x\wedge x'} = \varphi_x \cap \varphi_{x'}$.

Also, $\mathscr{S}(B) = \varphi_1 = \varphi_{x\vee x'} = \varphi_x \cup \varphi_{x'}$. It follows that

$$\varphi_{x'} = \mathscr{S}(B)\backslash\varphi_x = \varphi_x'$$

That $\varphi : B \to \mathscr{P}(\mathscr{S}(B))$ is a Boolean homomorphism follows from properties (a), (b), and (d).

The proof of the theorem is complete.

32.7 Topologizing $\mathscr{S}(B)$ to Obtain the Stone Space of B

We will now use the set,

$$\varphi[B] = \{\varphi_x : x \in B\} \subseteq \mathscr{P}(\mathscr{S}(B))$$

to topologize the set, $\mathscr{S}(B)$, of all B-ultrafilters.

Theorem 32.14 *Let $(B, \leq, \vee, \wedge, 0, 1, ')$ be a Boolean algebra. Then the set*

$$\mathscr{M} = \{\varphi_x : x \in B\}$$

is an open base for some topology, τ, on the set $\mathscr{S}(B)$.

Proof. Let $\mathscr{M} = \{\varphi_x : x \in B\} \subseteq \mathscr{P}(\mathscr{S}(B))$ where $\varphi_x = \{\mathscr{U} \in \mathscr{S}(B) : x \in \mathscr{U}\}$. To show that $\mathscr{M}$ is an open base topology on $\mathscr{S}(B)$ it suffices to show that $\mathscr{M}$ satisfies the "base property". (See Theorem 5.4.)

We claim that $\cup\mathscr{M} = \mathscr{S}(B)$. The set $\cup\mathscr{M}$ is a union of sets whose elements are B-ultrafilters. So, $\cup\mathscr{M} \subseteq \mathscr{S}(B)$. If $\mathscr{U} \in \mathscr{S}(B)$ and $x \in \mathscr{U}$ then $\mathscr{U} \in \varphi_x \in \cup\mathscr{M}$. So, $\mathscr{S}(B) \subseteq \cup\mathscr{M}$. Then $\cup\mathscr{M} = \mathscr{S}(B)$, as claimed.

Since each ultrafilter in $\mathscr{S}(B)$ belongs to some φ_x, the elements of $\mathscr{M}$ cover $\mathscr{S}(B)$.

Suppose φ_x and φ_y belong to $\mathscr{M}$. Since B is a Boolean algebra, then $x \wedge y \in B$. So, $\varphi_{x\wedge y} \in \mathscr{M}$. Also see that for any B-ultrafilter, $\mathscr{U}$,

$$x \wedge y \in \mathscr{U} \qquad \Leftrightarrow \qquad x, y \in \mathscr{U}$$

Then $\mathscr{U} \in \varphi(x \wedge y)$ if and only if $\mathscr{U}$ belongs to both φ_x and φ_y. Then

$$\varphi(x \wedge y) = \varphi(x) \cap \varphi(y)$$

This implies that, for any $x, y \in B$ other than 0 and 1, there exists a B-ultrafilter, $\mathscr{U}$, such that

$$\mathscr{U} \in \varphi_{x \wedge y} \subseteq \varphi_x \cap \varphi_y$$

We conclude that $\mathscr{M}$ satisfies the "base property" with respect to the set $\mathscr{S}(B)$. Then, by 5.4, $\mathscr{M}$ is an open base for a topology, say τ, on $\mathscr{S}(B)$, as required.

The previous theorem states that, given a Boolean algebra $(B, \leq, \vee, \wedge, 0, 1, ')$, we can topologize, $\mathscr{S}(B)$, the set of all B-ultrafilters, by using the elements of $\mathscr{M}$ as an open base for $\mathscr{S}(B)$. That is, for each $x \in B$,

...the set, φ_x, is an open base element for a topology on $\mathscr{S}(B)$

Notation: The topology on $\mathscr{S}(B)$ generated by $\mathscr{M}$ will be denoted as

$$\tau_{\mathscr{M}}$$

to form the topological space

$$(\mathscr{S}(B), \tau_{\mathscr{M}})$$

It is also important to note that if $\varphi_x \in \mathscr{M}$ there is in $\mathscr{M}$ an element $\varphi_{x'} = \varphi'_x = \mathscr{S}(B) \backslash \varphi_x$.

Since both φ_x and $\varphi'_x = \varphi_{x'}$ are open base elements and $\varphi_{x'}$ is the $\mathscr{S}(B)$-complement of φ_x, then φ_x and φ'_x are clopen in $(\mathscr{S}(B), \tau_{\mathscr{M}})$.

So,

... *all elements φ_x of the base $\mathscr{M}$ are clopen sets in $(\mathscr{S}(B), \tau_{\mathscr{M}})$*

Definition 32.15 Let $(B, \leq, \vee, \wedge, 0, 1, ')$ be a Boolean algebra and let $\mathscr{S}(B) = \{\mathscr{U} : \mathscr{U}$ is a B-ultrafilter $\}$. When equipped with the topology, $\tau_{\mathscr{M}}$, the set

$$(\mathscr{S}(B), \tau_{\mathscr{M}})$$

is referred to as the

Stone space of B.[5]

Recall that, if (S, τ) is some topological space then

$$(\mathscr{B}(S), \subseteq, \cup, \cap, \varnothing, S)$$

represents the Boolean algebra of all clopen sets in S.

We summarize the main steps in the construction of a Stone space of a Boolean algebra.

(1) We are given a Boolean algebra

$$(B, \leq, \vee, \wedge, 0, 1, ')$$

(2) From $(B, \leq, \vee, \wedge, 0, 1, ')$ we form those subsets, $\mathscr{U}$, of B we call B-ultrafilters. So each B-ultrafilter, $\mathscr{U}$ is an element of $\mathscr{P}(B)$.
(3) We can construct the set, $\mathscr{S}(B)$, of all B-ultrafilters.
(4) For each $x \in B$, we construct the set, φ_x, of all ultrafilters in $\mathscr{S}(B)$ which contain the element x. So, $\varphi_x \in \mathscr{P}(\mathscr{S}(B))$.
(5) We construct the family, $\mathscr{M} = \{\varphi_x : x \in B\}$, a subset of $\mathscr{P}(\mathscr{S}(B))$ which satisfies the "base property" and so forms a base for a topology, $\tau_{\mathscr{M}}$ on $\mathscr{S}(B)$.
(6) The topological space $(\mathscr{S}(B), \tau_{\mathscr{M}})$ generated by $\mathscr{M}$ is called the Stone space.
(7) Every element φ_x in $\mathscr{M}$ is shown to be a clopen subset of $(\mathscr{S}(B), \tau_{\mathscr{M}})$. So $\mathscr{M} \subseteq \mathscr{B}(\mathscr{S}(B))$.

32.8 Stone's Representation Theorem

The *Stone representation theorem* states that any Boolean algebra can be represented as a Boolean algebra of all clopen subsets of some compact zero-dimensional Hausdorff space.

---

[5]Named after the American mathematician, Marshall Harvey Stone (April 8, 1903 to January 9, 1989). He significantly contributed to real analysis, functional analysis, topology and the study of Boolean algebras. In 1936, he published a long paper that included Stone's representation theorem for Boolean algebras.

Theorem 32.16 *Let* $(B, \leq, \vee, \wedge, 0, 1, ')$ *be a Boolean algebra and*

$$\mathscr{S}(B) = \{\mathscr{U} : \mathscr{U} \text{ is a } B\text{-ultrafilter on } B\}$$

For each $x \in B$, *let*

$$\varphi_x = \{\mathscr{U} \in \mathscr{S}(B) : x \in \mathscr{U}\}$$

The function, $\varphi : B \to \mathscr{P}(\mathscr{S}(B))$, *is defined as* $\varphi(x) = \varphi_x$ *(the family of all B-ultrafilters which contain x). Let* $\tau_{\mathscr{M}}$ *denote the topology on* $\mathscr{S}(B)$ *which is generated by*

$$\mathscr{M} = \{\varphi_x : x \in B\}$$

where each φ_x *has been shown to be clopen in* $\mathscr{S}(B)$.

The Stone space of B space

$$(\mathscr{S}(B), \tau_{\mathscr{M}})$$

is a compact zero-dimensional Hausdorff topological space.[6]

Proof. Let $(B, \leq, \vee, \wedge, 0, 1, ')$ be a Boolean algebra.

We are given that $(\mathscr{S}(B), \tau_{\mathscr{M}})$ is the topological space generated by the open base $\mathscr{M} = \{\varphi_x : x \in B\}$.

By definition, the topological space is zero-dimensional (since $\mathscr{M}$ is an open base of clopen sets). So, we need only show that $\mathscr{S}(B)$ is Hausdorff and compact.

We claim that $(\mathscr{S}(B), \tau_{\mathscr{M}})$ is Hausdorff. Let $\mathscr{U}_1$ and $\mathscr{U}_2$ be distinct B-ultrafilters in $\mathscr{S}(B)$. Then there exists some $x \in \mathscr{U}_2$ which does not belong to $\mathscr{U}_1$. Then x' must belong to $\mathscr{U}_1$. It follows that $\mathscr{U}_2 \in \varphi_x$ and $\mathscr{U}_1 \in \varphi_{x'}$. Since $x \wedge x' = 0$ implies $\varphi_x \cap \varphi_{x'} = \varnothing$, it follows that $\mathscr{S}(B)$ is a Hausdorff topological space, as claimed.

We now claim that $(\mathscr{S}(B), \tau_{\mathscr{M}})$ is a compact space.

Proof of claim: Since $\mathscr{M} = \{\varphi_x : x \in B\}$ is a base of clopen sets in $\mathscr{S}(B)$, $\mathscr{M}$ is a base for closed sets in $\mathscr{S}(B)$ and so every closed set in $\mathscr{S}(B)$ is the intersection of elements in $\mathscr{M}$.

[6]Recall that a space is zero-dimensional if it has a base of clopen sets.

If F is a closed subset of $\mathscr{S}(B)$ let

$$T_F = \{x \in B : F \subseteq \varphi_x\}$$

That is, $F = \cap\{\varphi_x : x \in T_F\}$.

Suppose $\{F_i : i \in I\}$ is a family of closed subsets of $\mathscr{S}(B)$ which satisfies the finite intersection property. To show compactness it suffices to show that $\cap\{F_i : i \in I\} \neq \varnothing$.

Then $\{F_i : i \in I\}$ is a $\mathscr{S}(B)$-filter base of closed subsets of $\mathscr{S}(B)$.

See that, for each i, $F_i = \cap\{\varphi_x : x \in T_{F_i}\}$ and $T_{F_i} = \{x \in B : F_i \subseteq \varphi_x\}$.

Let

$$T = \cup\{T_{F_i} : i \in I\}$$

a non-empty subset of B.

So,

$$\begin{aligned}\cap\{F_i : i \in I\} &= \cap_{i\in I}[\cap\{\varphi_x : x \in T_{F_i}]\}\\ &= \cap\{\varphi_x : x \in \cup_{i\in I} T_{F_i}\}\\ &= \cap\{\varphi_x : x \in T\}\end{aligned}$$

We claim that T is B-filter in B.

Proof of claim: To see this, suppose $a, b \in T$. Then $a \in T_{F_j}$ and $b \in T_{F_k}$ for some $j, k \in I$.

This means $F_j \subseteq \varphi_a$ and $F_k \subseteq \varphi_b$. It follows that $F_j \cap F_k \subseteq \varphi_a \cap \varphi_b = \varphi_{a\wedge b}$. Since $\{F_i : i \in I\}$ satisfies the finite intersection property, then $\varphi_{a\wedge b} \neq \varnothing$. This means that that $0 < a \wedge b \neq 0$. By definition, T is a B-filter base.

Suppose $c \in T$ and $c \leq d$ for some d in B. Then $F_j \subseteq \varphi_c \subseteq \varphi_d$ for some j. So, $d \in T_{F_j} \subseteq T$. This implies that T is a B-filter, as claimed.

Let ψ be the unique (proper) B-ultrafilter which contains the B-filter, T.

Since $T \subseteq \psi$, then

$$\begin{aligned}\cap\{\varphi_x : x \in \psi\} &\subseteq \cap\{\varphi_x : x \in T\}\\ &= \cap\{\varphi_x : x \in \cup_{i\in I} T_{F_i}\}\\ &= \cap\{F_i : i \in I\}\end{aligned}$$

Suppose a belongs to the B-ultrafilter ψ. Recall that φ_a is the set of all B-filters that contain a. Then $\psi \in \varphi_a$. So,

$$\psi \in \cap\{\varphi_x : x \in \psi\}$$

We then see that $\cap\{\varphi_x : x \in \psi\}$ is non-empty so $\cap\{F_i : i \in I\}$ is non-empty.

We have shown that, if $\{F_i : i \in I\}$ satisfies the finite intersection property, then $\cap\{F_i : i \in I\}$ is non-empty. We can then conclude that $(\mathscr{S}(B), \tau_{\mathscr{M}})$ is compact, as required.

Theorem 32.17 *Let $(B, \leq, \vee, \wedge, 0, 1, ')$ be a Boolean algebra and*

$$\mathscr{S}(B) = \{\mathscr{U} : \mathscr{U} \textit{ is a } B\textit{-ultrafilter on } B\}$$

Suppose $\mathscr{B}(\mathscr{S}(B))$ is the set of all *clopen subsets of $(\mathscr{S}(B), \tau_{\mathscr{M}})$. Then*

$$\mathscr{M} = \mathscr{B}(\mathscr{S}(B))$$

Proof. Let $(B, \leq, \vee, \wedge, 0, 1, ')$ be a Boolean algebra.

We are required to show that $\mathscr{M} = \mathscr{B}(\mathscr{S}(B))$. We already know that $\mathscr{M} \subseteq \mathscr{B}(\mathscr{S}(B))$. So, it will suffice to show that $\mathscr{B}(\mathscr{S}(B)) \subseteq \mathscr{M}$. let U be a clopen subset of $\mathscr{S}(B)$. Then

$$U = \cup\{\varphi_y : y \in V\}$$

for some $V \subseteq B$. We claim that $U \in \mathscr{M}$.

We have shown that $\mathscr{S}(B)$ is a compact Hausdorff topological space, hence, U must be compact. Then $\{\varphi_y : y \in V\}$ is an open cover the compact set U. So there exists a finite subcover $\{\varphi_y : y = y_1, y_2, \ldots, y_k\}$ such that $U = \cup\{\varphi_y : y = y_1, y_2, \ldots, y_k\}$. Since

$$U = \cup\{\varphi_y : y = y_1, y_2, \ldots, y_k\} = \varphi_{y_1 \vee y_2 \vee \cdots \vee y_k} \in \mathscr{M}$$

(see properties of φ above). Then $\mathscr{B}(\mathscr{S}(B)) \subseteq \mathscr{M}$. So, $\mathscr{B}(\mathscr{S}(B)) = \mathscr{M}$, as required.

Definition 32.18 A Boolean algebra $(B, \leq, \vee, \wedge, 0, 1, ')$ is said to have a

topological representation

if there exists a a topological space, S, and a Boolean isomorphism $f : B \to \mathscr{B}(S)$ which maps B onto the Boolean algebra, $(\mathscr{B}(S), \subseteq, \cup, \cap, \varnothing, S)$, of all clopen subsets of some topological space, S.

The following theorem relates a Boolean algebra, B, to the topological space, $(\mathscr{S}(B), \tau_{\mathscr{M}})$, which is generated from B. We know that the space, $\mathscr{S}(B)$, has a base of clopen sets (with respect to the topology, $\tau_{\mathscr{M}}$). From $\mathscr{S}(B)$ we gather together all clopen subsets of $\mathscr{S}(B)$ to form the Boolean algebra,

$$(\mathscr{B}(\mathscr{S}(B)), \subseteq, \cup, \cap, 0, 1)$$

of all clopen subsets of $\mathscr{S}(B)$. We show that B can be mapped isomorphically onto $\mathscr{B}(\mathscr{S}(B))$. By definition, $\mathscr{B}(\mathscr{S}(B))$ will then serve as a "topological representation" of B.

Theorem 32.19 *The Stone representation theorem. Let $(B, \leq, \vee, \wedge, 0, 1, ')$ be a Boolean algebra and*

$$\mathscr{S}(B) = \{\mathscr{U} : \mathscr{U} \text{ is a } B\text{-ultrafilter on } B\}$$

For each $x \in B$, let

$$\varphi_x = \{\mathscr{U} \in \mathscr{S}(B) : x \in \mathscr{U}\}$$

The function, $\varphi : B \to \mathscr{P}(\mathscr{S}(B))$, is defined as $\varphi(x) = \varphi_x$ (the family of all B-ultrafilters which contain x). Let $\tau_{\mathscr{M}}$ denote the topology on $\mathscr{S}(B)$ which is generated by

$$\mathscr{M} = \{\varphi_x : x \in B\}$$

where each φ_x has been shown to be clopen in $\mathscr{S}(B)$.

The function $\varphi : B \to \mathscr{B}(\mathscr{S}(B))$ is a Boolean isomorphism mapping mapping B onto $\mathscr{B}(\mathscr{S}(B))$.

Proof. Let $(B, \leq, \vee, \wedge, 0, 1, ')$ be a Boolean algebra.

We are required to show that the function $\varphi : B \to \mathscr{B}(\mathscr{S}(B))$ is a Boolean isomorphism onto $\mathscr{B}(\mathscr{S}(B))$. We have shown in the list of properties of φ, that

$$\varphi_{x\wedge y} = \varphi_x \cap \varphi_y$$
$$\varphi_{x\vee y} = \varphi_x \cup \varphi_y$$
$$\varphi_{x'} = \mathscr{S}(B)\backslash\varphi_x$$

Since we have shown that $\mathscr{M} = \mathscr{B}(\mathscr{S}(B))$, $\varphi : B \to \mathscr{B}(\mathscr{S}(B))$ is a homomorphism mapping B onto $\mathscr{B}(\mathscr{S}(B))$.

We now show that *the function* $\varphi : B \to \mathscr{B}(\mathscr{S}(B))$ *is one-to-one.*

Suppose x and y are distinct points in B. It suffices to show that $\varphi_x \neq \varphi_y$.

Since $x \neq y$, then one of the two statements $x \leq y$, $y \leq x$ must be false. We will assume, without loss of generality, that $x \not\leq y$.

Claim: We claim that, since $x \not\leq y$, then $x \wedge y' \neq 0$.

Proof of claim: Suppose that $x \not\leq y$ and $x \wedge y' = 0$. See that

$$\begin{aligned}
x' = x' \vee 0 \quad &\Rightarrow \quad x' = x' \vee (x \wedge y') \\
&\Rightarrow \quad x' = (x' \vee x) \wedge (x' \vee y') \quad (B \text{ is distributive}) \\
&\Rightarrow \quad x' = 1 \wedge (x' \vee y') \\
&\Rightarrow \quad x' = (x' \vee y') \\
&\Rightarrow \quad y' \leq x' \\
&\Rightarrow \quad x \leq y
\end{aligned}$$

We have a contradiction. The source of the contradiction is our supposition that $x \wedge y' = 0$. So, $x \wedge y' \neq 0$, as claimed.

Consider the collection $\mathscr{F} = \{u \in B : x \wedge y' \leq u\}$. Clearly, $x \wedge y'$, x, $y' \in \mathscr{F}$.

Now, $\mathscr{F}$ is easily seen to be a B-filter. So there exists $\mathscr{U} \in \mathscr{S}(B)$ such that $\mathscr{F} \subseteq \mathscr{U}$. Since $x, y' \in \mathscr{U}$, $\mathscr{U} \in \varphi_x \cap \varphi_{y'} = \varphi_{x\wedge y'}$. Then $\mathscr{U}$ belongs to φ_x but cannot belong to φ_y.

So, $\varphi_x \neq \varphi_y$. So $\varphi : B \to \mathscr{B}(\mathscr{S}(B))$ is one-to-one.

So, $\varphi : B \to \mathscr{B}(\mathscr{S}(B))$ is an isomorphism onto $\mathscr{B}(\mathscr{S}(B))$, as required.

Chapter 33

Baire Spaces

Abstract
In this section, we present a brief introduction to a particularly important class of topological spaces called "Baire spaces". The properties possessed by members of this class turned out to be relevant in many applied fields, in particular in analysis. We will define such spaces and describe a few characterizations and properties. This is followed by a few examples.

33.1 Definitions

On a first reading, the definition of the class of Baire spaces may appear odd or of little interest and may lead the reader to wonder whether it is worth investing mental energy in studying such properties. This first impression may change once we start investigating characterizations and seeking examples of Baire spaces and non-Baire spaces.

A non-empty set, A, where

$$\text{int}_S \text{cl}_S A = \varnothing$$

is said to be *nowhere dense.* Since every point in the closure of a nowhere dense set is a boundary point of the set, such sets are often perceived as being "thin". Note that arbitrary unions of "thin" sets may not be "thin", if we gather enough of these together.

For example, consider a countable set $S = \{x_1, x_2, x_3, \ldots\}$ equipped with the cofinite topology. Let $\{F_1, F_2, F_3, \ldots\}$ be a collection of subsets where $F_i = \{x_i\}$. Then each F_i is closed in S (since if $x \in S \backslash F_i$, $S \backslash F_i$ is an open neighborhood of x which misses F_i). If $x \in F_i$ any basic open set containing x cannot be entirely contained in F_i. So, $\text{int}_S \text{cl}_S F_i = \text{int}_S F_i = \varnothing$. Then each F_i is nowhere dense in S (…is "thin"). But

$$\cup\{F_i : i = 1, 2, 3, \ldots\} = \cup\{S\backslash\{x_i\} : i = 1, 2, 3, \ldots\} = S$$

So, $\cup\{F_i : i = 1, 2, 3, \ldots\}$ is not nowhere dense ("not thin").

Definition 33.1 A topological space, S, is a *Baire space* (or simply said to be Baire) if and only if, the union of any countable family, $\{F_i : i \in \mathbb{N}\}$, of closed subsets with empty interior has empty interior.

A subset U of S is said to be *first category in S* if U can be expressed as a union, $\cup\{A_i : i \in \mathbb{N}\}$ of a countable family of nowhere dense subsets (that is, $\text{cl}_S A_i$ has empty interior).[1]

A subset U of S which is not of the first category in S is said to be of the *second category in S*.[2]

When viewed in this way, a first category set may be perceived as being the union of countably many "thin" sets.

A second category set, U, is one which, when expressed as a countable union, $\cup\{A_i : i \in \mathbb{N}\}$, of sets, at least one A_i has a closure with non-empty interior.

The following examples will help internalize the meaning of the above definitions.

Example 1. Suppose S is a Baire space and U is any non-empty open subset of S. Show that U cannot be a first category set in S.

---

[1]The expression *A is first category* also appears in texts in the form of the following equivalent definition: "The set, A, *is first category in S* if and only if $S\backslash\text{cl}_S A$ is dense in S". The set A if first category in S if and only if it is "meager" in S.

[2]A set A is *non-meager in S* if and only if it is a set of second category. A set A is said to be *residual* if it is the complement of a meager set.

Solution: Suppose S is Baire and U is a first category open subset. That is, suppose U can be expressed as the union, $\cup\{A_i : i \in \mathbb{N}\}$, of subsets where $\text{cl}_S A_i$ has empty interior. Then, since S is Baire and $\text{cl}_S A_i = \varnothing$ for each i, $\cup\{\text{cl}_S A_i : i \in \mathbb{N}\}$ must have empty interior. So,

$$\begin{aligned}\text{int}_S U &= U \\ &= \cup\{A_i : i \in \mathbb{N}\} \\ &\subseteq \cup\{\text{cl}_S A_i : i \in \mathbb{N}\}\end{aligned}$$

Since $\text{int}_S U \subseteq \cup\{\text{cl}_S A_i : i \in \mathbb{N}\}$, $\cup\{\text{cl}_S A_i : i \in \mathbb{N}\}$ has interior, we have a contradiction. So every non-empty open subset of a Baire space is second category in S.

33.2 Characterizations

The definition of a Baire space often appears in texts in different forms. The following proposition shows that the alternate definitions are equivalent.

Theorem 33.2 *Let S be a topological space. The following statements are equivalent.*

(1) *The space S is a Baire space.*
(2) *The intersection of all members of a countable family, $\{U_i : i \in \mathbb{N}\}$, of dense and open subsets of S, is a dense subset of S.*
(3) *Every non-empty open subset of S is second category.*

Proof. We are given that S is a topological space.

$(1 \Rightarrow 2)$ Suppose S is a Baire space as defined above. Let $\{U_i : i \in \mathbb{N}\}$ be a countable family of open dense subsets of S. We claim $\cap\{U_i : i \in \mathbb{N}\}$ is dense in S.

For each i, $S \backslash U_i$ is a closed subset of S with empty interior (since U_i is dense in S). Since S is a Baire space,

$$E = \cup\{S \backslash U_i : i \in \mathbb{N}\}$$

also has empty interior. So, $S\backslash E$ is dense in S. Then,

$$\begin{aligned} S\backslash E &= S\backslash \cup\{S\backslash U_i : i \in \mathbb{N}\} \\ &= \cap\{S\backslash [S\backslash U_i] : i \in \mathbb{N}\} \\ &= \cap\{U_i : i \in \mathbb{N}\} \end{aligned}$$

So, $\cap\{U_i : i \in \mathbb{N}\}$ is dense in S. As required.

($2 \Leftarrow 1$) Suppose the intersection of all members of a countable family, $\{U_i : i \in \mathbb{N}\}$, of dense and open subsets of S, is a dense subset of S. Let $\{F_i : i \in \mathbb{N}\}$ be a family of closed subsets of S such that F_i has empty interior in S. Then $S\backslash F_i$ is a dense open subset of S. We claim that $\cup\{F_i : i \in \mathbb{N}\}$ has empty interior. To show this it suffices to show that $S\backslash[\cup\{F_i : i \in \mathbb{N}\}]$ is dense in S.

See that

$$S\backslash[\cup\{F_i : i \in \mathbb{N}\}] = \cap\{S\backslash F_i : i \in \mathbb{N}\}$$

an intersection of countably many dense open subsets of S. By hypothesis, $\cap\{S\backslash F_i : i \in \mathbb{N}\}$ is dense in S. So, $\text{int}_S[\cup\{F_i : i \in \mathbb{N}\}]$ is empty in S.

We can conclude that S is a Baire space.

($1 \Rightarrow 3$) The proof is presented in the example following the definition, above.

($3 \Rightarrow 1$) Suppose every non-empty open subset of S is second category. This means that, if U is an open subset of S such that $U = \cup\{A_i : i \in \mathbb{N}\}$, then $\text{cl}_S A_i$ has non-empty interior for at least one i.

Let $\{F_i : i \in \mathbb{N}\}$ be a family of closed sets in S each with empty interior. To show that S is Baire, it suffices to show that $\cup\{F_i : i \in \mathbb{N}\}$ has empty interior.

Let us suppose that $U = \text{int}_S[\cup\{F_i : i \in \mathbb{N}\}]$ is non-empty. Since $U \subseteq \cup\{F_i : i \in \mathbb{N}\}$, $U = \cup\{V_i \cap F_i : i \in \mathbb{N}\}$ for some V_i's. Since every open subset of S is hypothesized to be second category, $\text{int}_S\text{cl}_S(V_j \cap F_j)$ is non-empty for some j. Since $\text{cl}_S(V_j \cap F_j) \subseteq F_j$, then $\text{int}_S\text{cl}_S(V_j \cap F_j) \subseteq \text{int}_S F_j$. This contradicts our hypothesis stating that each F_i has empty interior. The source of our contradiction is our supposition that $\text{int}_S[\cup\{F_i : i \in \mathbb{N}\}]$ is non-empty. Then $\text{int}_S[\cup\{F_i : i \in \mathbb{N}\}]$ must me empty so S is a Baire space.

33.3 Baire Category Theorems

At the first reading, a few questions which may come to mind are "How common are Baire spaces?", "Are they hard to recognize?" Baire spaces are, in fact, quite common as we shall see. The following results show, by describing some of their most basic characteristics, that Baire spaces are quite common. We will then provide a few examples.

Lemma 33.3 *Suppose S is a regular space. Let $\mathscr{F} = \{F_i : i \in \mathbb{N}\}$ be a countable collection of closed subsets, each with empty interior. Let U be any open subset of S. Then S contains a countably infinite nested collection, $\mathscr{U} = \{U_i : i \in \mathbb{N}\}$, of non-empty open subsets and an infinite sequence, $\{u_i : i \in \mathbb{N}\}$ such that, for each $i \in \mathbb{N}$* (Figure 33.1),

$$u_i \in U_i \subseteq cl_S U_i \subseteq S \backslash F_i \quad \text{and} \quad cl_S U_i \subseteq U_{i-1} \backslash F_i \subseteq U \backslash F_i$$

Proof. We are given that S is a regular topological space and $\mathscr{F} = \{F_i : i \in \mathbb{N}\}$, is a countably infinite collection of closed subsets each with empty interior. Let U be *any* non-empty open subset of S.

Since F_0 has empty interior, $U \not\subseteq F_0$, so there exists $u_0 \in U \backslash F_0$. Since $\{u_0\} \cap F_0 = \varnothing$ and S is regular, there exists some open set U_0,

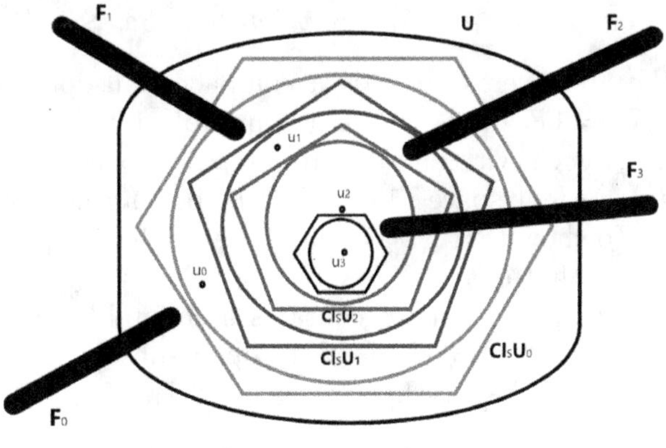

Figure 33.1. For Lemma 33.3.

such that

$$u_0 \in U_0 \subseteq \mathrm{cl}_S U_0 \subseteq S \backslash F_0 \quad \text{and} \quad \mathrm{cl}_S U_0 \subseteq U \backslash F_0$$

Since F_1 has empty interior, $U_0 \not\subseteq F_1$, so there exists $u_1 \in U_0 \backslash F_1$.

Since $\{u_1\} \cap F_1 = \varnothing$ and S is regular, there exists some open set, U_1 such that

$$u_1 \in U_1 \subseteq \mathrm{cl}_S U_1 \subseteq S \backslash F_1 \quad \text{and} \quad \mathrm{cl}_S U_1 \subseteq U_0 \backslash F_1 \subseteq U \backslash F_1$$

We inductively construct a nested set, $\mathscr{U} = \{U_i : i \in \mathbb{N}\}$, of open subsets, by using the recipe just described. Suppose U_{i-1} is an open subset such that

$$u_{i-1} \in U_{i-1} \subseteq \mathrm{cl}_S U_{i-1} \subseteq S \backslash F_{i-1} \text{ and } \mathrm{cl}_S U_{i-1} \subseteq U_{i-2} \backslash F_{i-1} \subseteq U \backslash F_{i-1}$$

Since F_i has empty interior, $U_{i-1} \not\subseteq F_i$, so there exists $u_i \in U_i \backslash F_i$

Since $\{u_i\} \cap F_i = \varnothing$ and S is regular, there exists some open set, U_i such that

$$u_i \in U_i \subseteq \mathrm{cl}_S U_i \subseteq S \backslash F_i \quad \text{and} \quad \mathrm{cl}_S U_i \subseteq U_{i-1} \backslash F_i \subseteq U \backslash F_i \qquad (*)$$

Then $\mathscr{U}$ is a collection of open subsets of S which satisfies, for all $n \in \mathbb{N}$, the property described in (*). This is the property required for the lemma.

Theorem 33.4 *Suppose S is a compact Hausdorff space. Let $\mathscr{F} = \{F_i : i \in \mathbb{N}\}$ be a countable collection of closed subsets, each with empty interior. Then $\cup\mathscr{F}$ has empty interior. So, S is Baire.*

Proof. We are given that S is a compact Hausdorff topological space and $\mathscr{F} = \{F_i : i \in \mathbb{N}\}$, is a countably infinite collection of closed subsets, each with empty interior. Since S is compact Hausdorff, S is a normal space (14.3). Let U be *any* non-empty open subset of S. To show that $\cup\mathscr{F} = \cup\{F_i : i \in \mathbb{N}\}$ has empty interior it suffices to show that $U \not\subseteq \cup\mathscr{F}$.

By the lemma, the space S contains a countably infinite collection, $\mathscr{U} = \{U_i : i \in \mathbb{N}\}$, of non-empty open subsets and an infinite sequence, $\{u_i : i \in \mathbb{N}\}$ such that, for each $i \in \mathbb{N}$,

$$u_i \in U_i \subseteq \mathrm{cl}_S U_i \subseteq S \backslash F_i \quad \text{and} \quad \mathrm{cl}_S U_i \subseteq U_{i-1} \backslash F_i \subseteq U \backslash F_i$$

Then the collection $\mathscr{U}^* = \{\mathrm{cl}_S U_i : i \in \mathbb{N}\}$, is a nested collection of closed subsets such that $\mathrm{cl}_S U_i \subseteq U_{i-1} \subseteq \mathrm{cl}_S U_{i-1}$ for all i. Since $\mathscr{U}^*$ satisfies the FIP and S is compact, then $\cap\{\mathrm{cl}_S U_i : i \in \mathbb{N}\} \neq \varnothing$. So, there exists $z \in \cap\{\mathrm{cl}_S U_i : i \in \mathbb{N}\} \neq \varnothing$.

Since $z \in \mathrm{cl}_S U_i \subseteq U \backslash F_i$ for all $i \in \mathbb{N}$, then

$$z \in \cap\{U \backslash F_i : i \in \mathbb{N}\} = U \backslash \cup\{F_i : i \in \mathbb{N}\}$$

Then $U \not\subseteq \cup\mathscr{F}$. Since U is arbitrarily chosen open subset of S, then we conclude that $\cup\mathscr{F}$ has empty interior. So, S is Baire.

Theorem 33.5 *The Baire category theorem I. Complete metric spaces are Baire.*

Proof. We are given that S is a complete metric space, equipped with the metric ρ. Then S is regular (example on page 225). Let $\mathscr{F} = \{F_i : i \in \mathbb{N}\}$ be a countably infinite collection of closed subsets, each with empty interior. Since S is a metric space, S is a normal space (9.19). Let U be *any* non-empty open subset of S. To show that $\cup\mathscr{F} = \cup\{F_i : i \in \mathbb{N}\}$ has empty interior it suffices to show that $U \not\subseteq \cup\mathscr{F}$.

By Lemma 33.3, the space S contains a countably infinite collection, $\mathscr{U} = \{U_i : i \in \mathbb{N}\}$, of non-empty open subsets and an infinite sequence, $\{u_i : i \in \mathbb{N}\}$ such that, for each $i \in \mathbb{N}$,

$$u_i \in U_i \subseteq \mathrm{cl}_S U_i \subseteq S \backslash F_i \quad \text{and} \quad \mathrm{cl}_S U_i \subseteq U_{i-1} \backslash F_i \subseteq U \backslash F_i \qquad (*)$$

The set $\mathscr{U}$ can be constructed so that

$$\mathrm{diam}(U_0) = \sup\{\rho(x, y) : x, y \in U_1\} \leq 1$$

and for each $i > 1$,

$$\mathrm{diam}(U_i) = \sup\{\rho(x, y) : x, y \in U_i\} < 1/i$$

while still maintaining the property described in (*).

Then the collection $\mathscr{U}^* = \{\mathrm{cl}_S U_i : i \in \mathbb{N}\}$, is a nested collection of closed subsets such that $u_i \in \mathrm{cl}_S U_i \subset U_{i-1} \subseteq \mathrm{cl}_S U_{i-1}$, for all i.

Let $\varepsilon > 0$. We can choose N such that $i > N$ implies $\mathrm{diam}(\mathrm{cl}_S U_i) \leq 1/i < \varepsilon$. Since the collection $\mathscr{U}^*$ is nested, for

$i, j > N$, $\rho(u_i, u_j) < \varepsilon$. Then the sequence $\{u_i : i \in \mathbb{N}\}$ is Cauchy in S. Since S is complete, the sequence $\{u_i : i \in \mathbb{N}\}$ converges to some point z in $\cap\mathscr{U}^*$. So, $\cap\mathscr{U}^*$ is non-empty. For the reasons stated at the end of the previous theorem, $z \in U\backslash\cup\mathscr{F}$. So, $U \not\subseteq \cup\mathscr{F}$. So $\cup\mathscr{F}$ has empty interior. We conclude that S is a Baire space.

Theorem 33.6 *The Baire Category theorem II. Locally compact Hausdorff spaces are Baire spaces.*

Proof. We are given that S is a locally compact Hausdorff space.

Then S is completely regular (by 18.8) and S is an open subset of βS (by 21.16). Since βS is compact and Hausdorff then βS is Baire (by 33.5).

Let $\mathscr{F} = \{F_i : i \in \mathbb{N}\}$ be a countably infinite collection of closed subsets, each with empty interior. We consider the collection $\mathscr{F}^* = \{\text{cl}_{\beta S} F_i : i \in \mathbb{N}\}$. If U is an open subset of βS such that $U \subseteq \text{cl}_{\beta S} F_i$ then $U \cap S$ is an open subset of S contained in F_i contradicting the hypothesis that F_i has empty interior in S. So, $\{\text{cl}_{\beta S} F_i : i \in \mathbb{N}\}$ is a collection of closed subsets of βS each of which has empty interior.

Since βS is compact Hausdorff, it is Baire. So, $\cup\mathscr{F}^*$ has empty interior. Suppose V is an open subset (in S) of $\cup\mathscr{F}$. Since S is open in βS, V is open (in βS) contrary to the fact that $\cup\mathscr{F}^*$ has empty interior. So, $\cup\mathscr{F}$ has empty interior. We conclude that S is a Baire space.

As an example of an application of the Baire Category Theorem we consider the following proposition.

Recall that an point, p, in a space S is called an *isolated point* if $\{p\}$ is open in S.

Proposition 33.7 *Suppose S is a complete metric space which contains no isolated points. Then S is an uncountably large set.*

Proof. Let S be a complete metric space which contains no isolated points. Suppose S is countable. By Theorem 33.5, S is Baire.

Suppose $S = \{p_0, p_1, p_2, p_3, \ldots\}$ where each p_i is a point in S and $\{p_i\}$ is closed in S. Note that $\text{int}_S\{p_i\} = \varnothing$, for each i. Since S is Baire, then $S = \cup\{\{p_i\} : i \in \mathbb{N}\}$ has empty interior, contradiction. So, S is an uncountable set.

In an example presented on page 53, we showed that the set, $\mathbb{Q}$, of all rationals is an F_σ (the countable union of closed sets). We verify, that $\mathbb{Q}$ is *not* a G_δ (the countable intersection of open sets).

Proposition 33.8 *The set of all rational numbers cannot be a G_δ.*

Proof. Since $\mathbb{R}$ is a complete metric space, $\mathbb{R}$ is Baire. By Theorem 33.2, the countable intersection of dense open subsets of $\mathbb{R}$ is dense in $\mathbb{R}$. Since $\mathbb{Q}$ is countable, we can also express it as $\mathbb{Q} = \{x_i : i \in \mathbb{N}\}$.

Suppose $\mathbb{Q}$ is a G_δ. That is, suppose $\mathbb{Q}$ can be expressed as

$$\mathbb{Q} = \cap\{U_i : i \in \mathbb{N}\}$$

where each U_i is open in $\mathbb{R}$. If $x \notin U_i$ every open neighborhood U of x meets $\mathbb{Q}$ and so must intersect U_i. This implies that each U_i is dense in $\mathbb{R}$.

For each i, consider the subset $V_i = U_i \cap \mathbb{R}\backslash\{x_i\}$. The set V_i is open and dense in $\mathbb{R}$. See that $\cap\{V_i : i \in \mathbb{N}\} = \varnothing$. This contradicts the fact that $\mathbb{R}$ is Baire.

Then $\mathbb{Q}$ cannot be a G_δ.

Proposition 33.9 *The open subset of a Baire space is a Baire space.*

Proof. We are given that S is a Baire space and U is open subspace of S.

We are required to show that U is Baire.

Let $\mathscr{F} = \{F_i : i \in \mathbb{N}\}$ be a countably infinite collection of closed subsets of U, each with empty interior (in U). It suffices to show that $\cup\mathscr{F}$ has empty interior (in U).

We consider the collection

$$\mathscr{F}^* = \{\text{cl}_S F_i : i \in \mathbb{N}\}$$

We claim that each $\text{cl}_S F_i$ has empty interior in S. If V is an open subset of U such that $V \subseteq \text{cl}_S F_i$ then $V \cap U$ is a non-empty open subset of U contained in F_i contradicting the hypothesis that F_i has empty interior in U. So, $\{\text{cl}_S F_i : i \in \mathbb{N}\}$ is a collection of closed subsets each of which has empty interior in S, as claimed.

Since S is a Baire space, $\cup\mathscr{F}^*$ has empty interior. To show that has $\cup\mathscr{F}$ empty interior in U we will suppose V is an open subset (in U) of $\cup\mathscr{F}$. Since U is open in S, $V \cap U$ is open (in S) contrary to the fact that $\cup\mathscr{F}^*$ has empty interior. So, $\cup\mathscr{F}$ has empty interior. We conclude that U is a Baire space.

Example 2. Show that, if T is a Baire space and T is dense in S, then S is a Baire space.

Solution: Suppose that T is a dense Baire subspace of a space S. We must show that S is Baire.

Suppose $\mathscr{F} = \{F_i : i \in \mathbb{N}\}$, is a countably infinite collection of closed subsets of S, each with empty interior (in S). We must show that $\cup\mathscr{F}$ has empty interior.

We consider the collection,

$$\mathscr{F}^* = \{F_i \cap T : i \in \mathbb{N}\}$$

of closed subsets of T.

We claim each $F_i \cap T$ has empty interior in T. Suppose that V^* is an open subset of T such that $V^* \subseteq F_i \cap T$. Then there exists an open subset V in S such that $V^* = V \cap T$. Then $V^* \subseteq \text{int}_S \text{cl}_S V^* \subseteq F_i$.

Since V is non-empty and T is dense in S, $\text{int}_S \text{cl}_S V^* \subseteq \text{int}_S F_i$ contradicting our hypothesis. Then each $F_i \cap T$ has empty interior in T, as claimed.

Since T is a Baire space, then $\cup\mathscr{F}^*$ has empty interior.

If W is an open subset in S such that $W \subseteq \cup\mathscr{F}$ then $W \cap T \subseteq \cup\mathscr{F}^*$. Since T is a Baire space, $W \cap T$ must be empty. Since T is dense in S, $\text{int}_S \cup \mathscr{F}$ is empty. So, S is a Baire space.

Example 3. The space, $\mathbb{Q}$, is not a Baire space.

Solution: Since $\mathbb{Q}$ is a countable set, it can be enumerated $\mathbb{Q} = \{x_i : i \in \mathbb{N}\}$. So, $\mathbb{Q} = \cup\{\{\mathbf{x}_i\} : i \in \mathbb{N}\}$, the countable union of sets with empty interior in $\mathbb{Q}$ (with respect to the subspace topology). Since

$$\text{int}_{\mathbb{Q}}[\cup\{\{\mathbf{x}_i\} : i \in \mathbb{N}\}] = \mathbb{Q} \neq \varnothing$$

$\mathbb{Q}$ is not a Baire space.

Example 4. Let f be a function mapping $\mathbb{R}$ into $\mathbb{R}$ which is continuous on some non-empty subset of $\mathbb{R}$. Let

$$Q_f = \{x \in \mathbb{R} : f \text{ is continuous at } x\}$$

be the "largest subset of $\mathbb{R}$ on which f is continuous". Show that Q_f is a G_δ.

Solution: We are given that $f : \mathbb{R} \to \mathbb{R}$ is continuous on some non-empty subset of $\mathbb{R}$. For each $i \in \mathbb{N}\backslash\{0\}$, let

$$U_i = \cup\{f^{\leftarrow}[B_{1/i}(z)] : f^{\leftarrow}[B_{1/i}(z)] \text{ is open, for all } z \in \mathbb{R}\}$$

Then, for each $i \in \mathbb{N}\backslash\{0\}$, U_i is an open subset of $\mathbb{R}$.

Let

$$U = \cap\{U_i : i \in \mathbb{N}\backslash\{0\}\}$$

Then U is a G_δ set in $\mathbb{R}$.

We claim that $U = Q_f$. We first verify that $U \subseteq Q_f$. To do this we verify that f is continuous on U. The function f is continuous on U if f pulls back basic open sets in $\mathbb{R}$ to open subsets of U (6.4).

For all z and all $i \in \mathbb{N}\backslash\{0\}$, $f^{\leftarrow}[B_{1/i}(z)]$ is open in $U = \cap\{U_i : i \in \mathbb{N}\backslash\{0\}\} \subseteq \mathbb{R}$. So, f is continuous on U. Then $U \subseteq Q_f$.

Conversely, if $x \in Q_f$ and $f(x) = z$, $f^{\leftarrow}[B_{i/i}(z)]$ is open in Q_f so $x \in f^{\leftarrow}[B_{i/i}(z)] \subseteq U$. Then $Q_f \subseteq U$.

Then

$$Q_f = U = \cap\{U_i : i \in \mathbb{N}\backslash\{0\}\}$$

Since U is a G_δ, the largest subset of $\mathbb{R}$ on which f is continuous is a G_δ.

Example 5. Is there a function $f : \mathbb{R} \to \mathbb{R}$ which is continuous precisely on $\mathbb{Q}$ (and discontinuous on $\mathbb{R}\backslash\mathbb{Q}$)?

Solution: In the previous example, we showed that, given a function $f : \mathbb{R} \to \mathbb{R}$, the largest subset of $\mathbb{R}$ on which f is continuous is a G_δ set. So, if f is continuous on $\mathbb{Q}$ (and discontinuous elsewhere) the set $\mathbb{Q}$ is a G_δ. We showed in Proposition 33.8 that $\mathbb{Q}$ is not a G_δ. Contradiction! We must then conclude that there can be no real-valued function on $\mathbb{R}$ which is continuous only on $\mathbb{Q}$ (and discontinuous elsewhere).

Chapter 34

The Class of F-Spaces

Abstract
In this section, we introduce the class of topological spaces called, F-spaces. In order to better follow the contents of this chapter the reader is encouraged to first review the contents of Chapters 21, 23, and 24.

34.1 Definition

The notion of those topological spaces, S, called F-spaces is introduced in the text *Rings of all Continuous Functions* by Gillman and Jerison.[1] We begin by presenting the topological definition.

Definition 34.1 A completely regular topological space S is called an *F-space* if every cozero-set in S is C^*-embedded in S.

---

[1]In that text an F-space, S, is introduced in the form of a particular algebraic property of the ring, $C(S)$, of all continuous real-valued functions:

> *An F-space is a topological space S for which every finitely generated ideal in $C(S)$ is a principal ideal.*

We will discuss those spaces called F-spaces, strictly from a topological point of view.

34.2 A Few Topological Properties and Characterizations of F-Spaces

Recall that the "cozero-set of f" in S, denoted by $\mathrm{Cz}(f)$, is the set $S\backslash Z(f) = \{x \in S : f(x) \neq 0\}$ (also denoted as $\mathrm{coz}(f)$). Suppose $f \in C^*(S)$. See that the subsets $f^{\leftarrow}[(-\infty, 0)]$ and $f^{\leftarrow}[(0, \infty)]$ (the two subsets of S which are mapped by f to negative and positive numbers, respectively) have the following representations

$$f^{\leftarrow}[(-\infty, 0)] = \mathrm{Cz}(f \wedge 0) = \{x \in S : f(x) < 0\}$$
$$f^{\leftarrow}[(0, \infty)] = \mathrm{Cz}(f \vee 0) = \{x \in S : f(x) > 0\}$$

Then, if $f \in C^*(S)$, the domain of f can be partitioned into three sets,

$$S = \mathrm{Cz}(f \wedge 0) \;\cup\; Z(f) \;\cup\; \mathrm{Cz}(f \vee 0)$$

Even if the two subsets of S, $\mathrm{Cz}(f \wedge 0)$ and $\mathrm{Cz}(f \vee 0)$, are disjoint, they need not be completely separated. If they are for all f, then certain interesting topological properties on S will hold true.

The following properties on S, will be useful when discussing its compactifications.

Lemma 34.2 *Suppose S is a completely regular space.*

(a) *If, for every $g \in C^*(S)$, $Cz(g \wedge 0)$ and $Cz(g \vee 0)$ are completely separated sets in S, then S is an F-space.*
(b) *If, for every $g \in C^*(S)$, $Cz(g \wedge 0)$ and $Cz(g \vee 0)$ are completely separated sets in S, then βS is an F-space.*
(c) *If disjoint cozero-sets in S are completely separated then S is an F-space.*
(d) *If disjoint cozero-sets in S are completely separated then βS is an F-space.*

Proof. Suppose S is a completely regular space.

(a) We are given that, for every $g \in C^*(S)$, $\mathrm{Cz}(g \wedge 0)$ and $\mathrm{Cz}(g \vee 0)$ are completely separated sets.
Let $H = \mathrm{Cz}(h)$ be a cozero-set in S.

Let A and B be completely separated subsets of H. To show that H is C^*-embedded in S it suffices to show that A and B are completed separated (in S) and then invoke the Urysohn extension theorem.

There exists some function,

$$k : H \to [-1, 1]$$

in $C^*(H)$ such that

$$k[A] = \{1\} \quad \text{and} \quad k[B] = \{-1\}$$

Then

$$\begin{aligned} A \subseteq k^{\leftarrow}[\{1\}] \quad &\subseteq k^{\leftarrow}[(0, \infty)] \quad = \mathrm{Cz}(k \vee 0) \\ B \subseteq k^{\leftarrow}[\{-1\}] &\subseteq k^{\leftarrow}[(-\infty, 0)] = \mathrm{Cz}(k \wedge 0) \end{aligned}$$

With this in mind, we construct a separating function over S for A and B. We define $f : S \to \mathbb{R}$ as:

$$f(x) = \begin{cases} 0 & \text{for } x \text{ in } S \backslash \mathrm{Cz}(h) \\ k(x) \cdot |h(x)| & \text{for } x \text{ in } \mathrm{Cz}(h) \end{cases}$$

Claim 1: We claim that f is continuous on S.

Proof of claim: See that S is the union of the two S-closed subsets $Z(k) \cup Z(h)$ and $\mathrm{cl}_S\mathrm{Cz}(\mathrm{k})$. The function f is constant on $Z(k) \cup Z(h)$ and so $f|_{Z(k)\cup Z(h)}$ is continuous.
For every sequence (net) A in $\mathrm{Cz}(k)$ converging to a limit point a in $\mathrm{cl}_S\mathrm{Cz}(k) \backslash \mathrm{Cz}(k)$, $f[A]$ converges to $f(a) = 0$. So, $f|_{cl_S\mathrm{Cz}(k)}$ is continuous. Then f is continuous on S (Pasting lemma), as claimed.

Claim 2: We claim that f completely separates A and B in S.

Proof of claim: For the given continuous function $f : S \to \mathbb{R}$

$$f[S \backslash \mathrm{Cz}(h)] = \{0\}$$

and

$$\begin{aligned} f[A] &\subseteq f[k^{\leftarrow}[(0, \infty)]] \\ &= k[k^{\leftarrow}[(0, \infty)]] \;\cdot\; |h|\,[k^{\leftarrow}[(0, \infty)] \\ &\subseteq \;(0, \infty) \end{aligned}$$

$$\begin{aligned} f[B] &\subseteq f[k^{\leftarrow}[(-\infty,0)]] \\ &= k[k^{\leftarrow}[(-\infty,0)]] \cdot |h|\,[k^{\leftarrow}[(-\infty,0)] \\ &\subseteq (-\infty,0) \end{aligned}$$

implies

$$\begin{aligned} A &\subseteq f^{\leftarrow}[(0,\infty)] = \mathrm{Cz}(f \vee 0) \\ B &\subseteq f^{\leftarrow}[(-\infty,0)] = \mathrm{Cz}(f \wedge 0) \end{aligned}$$

By hypothesis, $\mathrm{Cz}(f \wedge 0)$ and $\mathrm{Cz}(f \vee 0)$ are completely separated in S and so A and B must also be completed separated in S, as claimed.

By Urysohn's extension lemma, H is C^*-embedded in S. So, S is an F-space, as required.

(b) Suppose that, for every $g \in C^*(S)$, $\mathrm{Cz}(g \wedge 0)$ and $\mathrm{Cz}(g \vee 0)$ are completely separated sets in S. We are required to show that βS is an F-space.

Let $h \in C(\beta S)$. Then $\mathrm{Cz}(h \wedge 0)$ and $\mathrm{Cz}(h \vee 0)$ are disjoint cozero-sets in βS.

Claim: We claim that $\mathrm{Cz}(h \wedge 0)$ and $\mathrm{Cz}(h \vee 0)$ are completely separated in βS. If so then, by part (a), βS is an F-space.

Proof of claim: By hypothesis, $\mathrm{Cz}(h|_S \wedge 0)$ and $\mathrm{Cz}(h|_S \vee 0)$ are completely separated sets in S. Then there is some $t \in C^*(S)$ such that

$$\begin{aligned} \mathrm{Cz}(h|_S \wedge 0) &\subseteq Z(t) \\ \mathrm{Cz}(h|_S \vee 0) &\subseteq Z(t-1) \end{aligned}$$

See that $\mathrm{Cz}(h \wedge 0) \subseteq Z(t^{\beta})$: For, if $y \in \mathrm{Cz}(h \wedge 0) \setminus Z(t^{\beta})$, then $\mathrm{Cz}(h \wedge 0) \setminus Z(t^{\beta})$ is an open neighborhood of y contained in $\beta S \setminus S$. Similarly $\mathrm{Cz}(h \vee 0) \subseteq Z(t^{\beta}-1)$. So, $\mathrm{Cz}(h \wedge 0)$ and $\mathrm{Cz}(h \vee 0)$ are completely separated in βS, as claimed.

By part (a), βS is an F-space.

(c) Suppose disjoint cozero-sets in S are completely separated. Then, for every $g \in C^*(S)$, $\mathrm{Cz}(g \wedge 0)$ and $\mathrm{Cz}(g \vee 0)$ are completely separated sets in S.

Then, by part (a), S is an F-space.

(d) Suppose disjoint cozero-sets in S are completely separated. Then, for every $g \in C^*(S)$, $\text{Cz}(g \wedge 0)$ and $\text{Cz}(g \vee 0)$ are completely separated sets in S. Then, by part (b), βS is an F-space.

Theorem 34.3 *A completely regular space S is an F-space if and only if disjoint cozero-sets of S are completely separated in S.*

Proof. Suppose S is a completely regular space.

($\Leftarrow$) We are given that disjoint cozero-sets in S are completely separated. By part c) of Lemma 34.2 the space S is an F-space.

($\Rightarrow$) Suppose S is an F-space.

Let $U = \text{Cz}(h)$ and $V = \text{Cz}(t)$ be disjoint cozero-sets in S.

We will show that the sets U and V are completely separated in S.

Given that,

$$\begin{aligned} U \cup V &= S\backslash Z(h) \cup S\backslash Z(t) \\ &= S\backslash [Z(h) \cap Z(t)] \\ &= S\backslash Z(h^2 + t^2) \\ &= \text{Cz}(h^2 + t^2) \end{aligned}$$

then $U \cup V$ is also a cozero-set in S.

Let $f : U \cup V \to \{0, 1\}$ be a function such that $f[U] = \{0\}$ and $f[V] = \{1\}$. Since U and V are disjoint, f is continuous on $U \cup V$. Since S is an F-space and $U \cup V$ is a cozero-set, f extends to f^* on S where $f^*|_{U\cup V} = f$. Then

$$\begin{aligned} \text{Cz}(h) &\subseteq Z(f) \subseteq Z(f^*) \\ \text{Cz}(t) &\subseteq Z(f-1) \subseteq Z(f^*-1) \end{aligned}$$

So the disjoint cozero-sets, $U = \text{Cz}(h)$ and $V = \text{Cz}(t)$, are completely separated by f^* in S, as required.

Theorem 34.4 *Let S be a completely regular F-space. Then every C^*-embedded subset of S is itself an F-space.*

Proof. Suppose S is an F-space and H is a C^*-embedded subspace of S.

Let $f \in C^*(H)$. Then, by hypothesis, there exists $f^* \in C^*(S)$ such that $f^*|_H = f$.

The space S and the subspace H can be expressed as the disjoint unions,

$$\begin{aligned} H &= \mathrm{Cz}(f \vee 0) \ \cup Z(f) \ \cup \mathrm{Cz}(f \wedge 0) \\ S &= \mathrm{Cz}(f^* \vee 0) \cup Z(f^*) \cup \mathrm{Cz}(f^* \wedge 0) \end{aligned}$$

Since S is an F-space, then, by Theorem 34.3, $\mathrm{Cz}(f^* \vee 0)$ and $\mathrm{Cz}(f^* \wedge 0)$ are completely separated in S. That is, there exists $t \in C^*(S)$ such that

$$\mathrm{Cz}(f^* \vee 0) \subseteq Z(t)$$
$$\mathrm{Cz}(f^* \wedge 0) \subseteq Z(t-1)$$

Then for $t|_H \in C^*(H)$,

$$\mathrm{Cz}(f \vee 0) \subseteq Z(t|_H)$$
$$\mathrm{Cz}(f \wedge 0) \subseteq Z(t|_H - 1)$$

So, $\mathrm{Cz}(f \vee 0)$ and $\mathrm{Cz}(f \wedge 0)$ are completely separated in H.

Then, by Theorem 34.2(a), H is an F-space, as required.

Corollary 34.5 *Let S be a completely regular space. Then, if βS is an F-space then S is an F-space.*

Proof. By Theorem 34.4, if βS is an F-space then, since S is C^*-embedded in βS, S is an F-space.

Theorem 34.6 *Let S be a completely regular F-space. Then βS is also an F-space.*

Proof. Let S be an F-space. The set, S, is dense and C^*-embedded in βS.

We are required to show that βS is an F-space.

Let

$$H = \text{Cz}(h) = S \backslash Z(h)$$

be a cozero-set in βS for some $h \in C^*(\beta S)$.

Then $\text{Cz}(h|_S) = \text{Cz}(h) \cap S$ is a non-empty cozero-subset of S.

To show that βS is an F-space it suffices to show that H is C^*-embedded in βS.

Let $g : H \to \mathbb{R}$ be a bounded continuous real-valued function on H.

Since S is an F-space, $\text{Cz}(h|_S)$ is C^*-embedded in S.

Then, for $g : H \to \mathbb{R}$,

$$g|_{\text{Cz}(h)} : \text{Cz}(h) \to \mathbb{R}$$

extends to a function

$$[g|_{\text{Cz}(h)}]^* : S \to \mathbb{R}$$

Since S is C^*-embedded in βS, then $[g|_{\text{Cz}(h)}]^*$ extends to

$$[g|_{\text{Cz}(h)}]^{*\beta} : \beta S \to \mathbb{R}$$

where $[g|_{\text{Cz}(h)}]^{*\beta}$ agrees with g on H.

We conclude that H is C^*-embedded in βS; so βS is an F-space, as required.

We summarize the above results in the following corollary.

Corollary 34.7 *Let S be a completely regular space. The following are equivalent.*

(a) *The space S is an F-space.*
(b) *Disjoint cozero-sets of S are completely separated in S.*
(c) *For every $f \in C^*(S)$, $Cz(f \vee 0)$ and $Cz(f \wedge 0)$ are completely separated sets.*
(d) *The compactification βS is an F-space.*

Proof. We are given that S is a completely regular space.

The proof of (a) $\Leftrightarrow$ (b) is given in part (a) Theorem 34.3.

The proof of (b) $\Rightarrow$ (c) is obvious.

The proof of (c) $\Rightarrow$ (d) is given in part (b) of Theorem 34.2.

The proof of (d) $\Leftrightarrow$ (a) is given in Theorem 34.5 and 34.6.

In the following proof we will invoke the four previously established results listed below.

— Theorem 23.2, states that if N is C-embedded in S then N is C^*-embedded in S.

— Theorem 23.3 states that a subspace U of S which can be mapped homeomorphically onto a closed subset of $\mathbb{R}$ is C-embedded in S.

— Corollary 23.6 states that, if N is a C-embedded copy of $\mathbb{N}$ in a locally compact set S and $Z(f^\beta)$ is a zero-set in βS, such that $(\mathrm{cl}_{\beta S}N)\backslash N \subseteq Z(f^\beta) \cap \beta S\backslash S$, then

$$(\mathrm{cl}_{\beta S}N)\backslash N \subseteq \mathrm{int}_{\beta S\backslash S}(\,Z(f^\beta) \cap \beta S\backslash S\,)$$

— Theorem 24.3, states that a space S is realcompact if and only if, for any $p \in \beta S \setminus S$, there is a function, $h \in C^*(S)$, such that $p \in Z(h^\beta) \subseteq \beta S\backslash S$.

— Corollary 34.7 states that a space S is an F-space if and only if for every $f \in C^*(S)$, the two corresponding subsets,

$$\mathrm{Cz}(h \vee 0) = f^{\leftarrow}[(0,\infty)] \quad \text{and} \quad \mathrm{Cz}(h \wedge 0) = f^{\leftarrow}[(-\infty,0)]$$

are completely separated in S.

Theorem 34.8 *Let S be a completely regular locally compact and realcompact space. Then $\beta S\backslash S$ is an F-space.*

Proof. We are given that the completely regular space S is locally compact and realcompact.

Since S is locally compact, S is open in βS, so $\beta S\backslash S$ is a compact set.

Let $h \in C(\beta S\backslash S)$.

To show that $\beta S\backslash S$ is an F-space, it will suffice to show that the two subsets, $\mathrm{Cz}(h \vee 0) = h^{\leftarrow}[(0,\infty)]$ and $\mathrm{Cz}(h \wedge 0) = h^{\leftarrow}[(-\infty,0)]$ of $\beta S \setminus S$, are completely separated in $\beta S \setminus S$. We can then invoke

Corollary 34.7, part (c). To show that they are completely separated it will suffice to show that that their closures, $\mathrm{cl}_{\beta S \setminus S}\mathrm{Cz}(h \vee 0)$ and $\mathrm{cl}_{\beta S \setminus S}\mathrm{Cz}(h \wedge 0)$, have empty intersection.

Since the compact subsets of a completely regular space are C^*-embedded (see example on page 492), then $\beta S \setminus S$ is a C^*-embedded subset of βS. Then $h : \beta S \setminus S \to \mathbb{R}$ extends to

$$h^\beta : \beta S \to \mathbb{R}$$

where

$$h = h^\beta|_{\beta S \setminus S}$$
$$h^\beta|_S \in C^*(S)$$

See that,

$$\beta S \setminus S = \mathrm{Cz}(h \vee 0) \ \cup \ Z(h) \ \cup \ \mathrm{Cz}(h \wedge 0)$$

a union of three pairwise disjoint sets.

Suppose p is any point in $Z(h)$. Our plan is to show that $p \in \mathrm{int}_{\beta S \setminus S} Z(h)$; if so then $Z(h) = \mathrm{int}_{\beta S \setminus S} Z(h)$ and $\mathrm{cl}_{\beta S \setminus S}\mathrm{Cz}(h \vee 0) = \mathrm{Cz}(h \vee 0)$. By applying a similar reasoning it will follow that $\mathrm{cl}_{\beta S \setminus S}\mathrm{Cz}(h \wedge 0) = \mathrm{Cz}(h \wedge 0)$. Then, $\mathrm{Cz}(h \wedge 0)$ and $\mathrm{Cz}(h \vee 0)$ would be completely separated.

Step 1: Since S is realcompact, then, by Theorem 24.3, there is a function t in $C^*(S)$ such that $p \in Z(t^\beta)$ where $Z(t^\beta)$ is entirely contained in $\beta S \setminus S$. Let $g \in C^*(S)$ be such that $Z(g^\beta) = Z(h^\beta) \cap Z(t^\beta)$.[2] Then $p \in Z(g^\beta) \subseteq \beta S \setminus S$.

Step 2: We define the function $f : S \to \mathbb{R}$ as

$$f = 1/g$$

Since g does not vanish on S, the function f is well-defined and continuous on S.

Since g^β is continuous, if A is any net (sequence) in S converging to a point x in $Z(g^\beta)$, then $g[A]$ converges to $g^\beta(x) = 0$. Then f is an unbounded function on S.

[2]Note that if $g^\beta = |h^\beta| + |t^\beta|$ then $Z(g^\beta) = Z(h^\beta) \cap Z(t^\beta)$.

See that the function $f : S \to \omega\mathbb{R}$[3] extends continuously to $f^{\beta(\omega)}\colon \beta S \to \omega\mathbb{R}$, where

$$f^{\beta(\omega)}[Z(g^\beta)] = \{\infty\} \quad \text{and} \quad f^{\beta(\omega)\leftarrow}\{\infty\} = Z(g^\beta)$$

Step 3: For each $i \in \mathbb{N}$, choose $m_i \in [2i-1, 2i+1] \cap f[S]$, if non-empty and not equal to m_{i-1}. Since f is unbounded, we can choose infinitely many such m_i's to construct the subset

$$M = \{m_i : i \in \mathbb{N}\}$$

in $f[S] \subseteq \mathbb{R}$. See that...

— M is an unbounded subset of $\mathbb{R}$,
— Each $\{m_i\}$ is a clopen subset of M
— The set M is a closed subset of $\mathbb{R}$ (since M contains all its limit points).

So M is a closed unbounded copy of $\mathbb{N}$ in $f[S]$.

Claim 1: We claim that, S contains a closed C^*-embedded copy, N, of $\mathbb{N}$.

Proof of claim: For each $i \in \mathbb{N}$, choose $n_i \in f^{\leftarrow}(m_i)$ in S. Let $N = \{n_i : i \in \mathbb{N}\}$. Then the function, $f|_N : N \to \mathbb{R}$, maps N homeomorphically onto the copy, M, of $\mathbb{N}$ in $f[S]$. Since $f[N] = M$ contains all its limit points in $\mathbb{R}$, then so does N. We then have that N is a closed copy of $\mathbb{N}$ in S.

Since f maps N onto a closed subset of $\mathbb{R}$, we then invoke Theorem 23.3 to conclude that N is C-embedded copy of $\mathbb{N}$ in S. By Theorem 23.2, N is C^*-embedded in S. This establishes the Claim 1.

Step 4: See that $\mathrm{cl}_{\beta S}N = \beta N$ with $\beta N \backslash N \subseteq \beta S \backslash S$.[4]

Claim 2: We claim that $\beta N \backslash N = Z(g^\beta)$.

Proof of claim: If $x \in \beta N \backslash N$, x is an accumulation point of the set $N \subseteq S$. Since f is unbounded on N then $g^\beta(x) = 0$. So $\beta N \backslash N \subseteq Z(g^\beta)$.

If $x \in Z(g^\beta)$, then $g^\beta(x) = 0$ is a limit point of M. Therefore $f^{\beta(\omega)}(x) = \infty$. Then $x \in \mathrm{cl}_{\beta S}N \backslash N$. So $Z(g^\beta) \subseteq \beta N \backslash N$.

---

[3]The set $\omega\mathbb{R}$ is the one-point compactification, $\mathbb{R} \cup \{\infty\}$, of $\mathbb{R}$.
[4]As shown in the example on page 512.

We conclude that $\beta N\backslash N = Z(g^\beta)$ as claimed.

Step 5: We now invoke Theorem 23.6 to deduce that

$$Z(g^\beta) = \beta N\backslash N \subseteq \mathrm{int}_{\beta S\backslash S} Z(g^\beta)$$

Since $p \in Z(g^\beta) \subseteq Z(h)$, then $p \in \mathrm{int}_{\beta S\backslash S} Z(h)$.

We can then deduce that $Z(g^\beta) = \mathrm{int}_{\beta S\backslash S} Z(h)$ and so $\mathrm{cl}_{\beta S\backslash S}$ $\mathrm{Cz}(h \vee 0) = \mathrm{Cz}(h \vee 0)$. By applying a similar reasoning , $\mathrm{cl}_{\beta S\backslash S}\mathrm{Cz}(h \wedge 0) = \mathrm{Cz}(h \wedge 0)$. Then $\mathrm{Cz}(h \vee 0)$ and $\mathrm{Cz}(h \wedge 0)$ are completely separated in $\beta S\backslash S$.

By 34.7, $\beta S\backslash S$ is an F-space as required.

Theorem 34.9 *Let S be a completely regular F-space and $\{x_i : i \in \mathbb{N}\}$ be a sequence in S which converges to $y \in S$. Then X is constant on some tail-end of X.*

Proof. We are given that S is a completely regular F-space and $X = \{x_i : i \in \mathbb{N}\}$ is a sequence in S converging to y. Suppose X is not constant on any tail-end of X. Then there exists a subsequence $V = \{x_{k(i)} : i \in \mathbb{N}\}$ such that $x_{k(i)} \neq x_{k(u)}$ if $u < i\}$. Letting $c_i = x_{k(i)}$, $V = \{c_i : i \in \mathbb{N}\}$ is a subsequence of distinct elements converging to y.

Let

$$A = \{c_{j(i)} : i \in \mathbb{N}\}$$
$$B = \{c_{t(i)} : i \in \mathbb{N}\}$$

be two subsequences of X where $j(i) = 2i$ and $t(i) = 2i + 1$. Then A and B are two infinitely countable disjoint subsets of S where $\mathrm{cl}_S A = A \cup \{y\}$ and $\mathrm{cl}_S B = B \cup \{y\}$. Also $A \cap \mathrm{cl}_S B = \varnothing = B \cap \mathrm{cl}_S A$, but $\mathrm{cl}_S A \cap \mathrm{cl}_S B = \{y\}$.

In an example on page 259 we have shown that there exists disjoint cozero-sets U and V which contain A and B, respectively. By Corollary 34.7, part (b), since S is an F-space, U and V are completely separated. Therefore there exists $f \in C^*(S)$ such that $U \subseteq f^{\leftarrow}(0)$ and $V \subseteq f^{\leftarrow}(1)$. Then $\mathrm{cl}_S U \subseteq f^{\leftarrow}(0)$ and $\mathrm{cl}_S V \subseteq f^{\leftarrow}(1)$. This means that $\mathrm{cl}_S U \cap \mathrm{cl}_S V = \varnothing$. This cannot be since $\mathrm{cl}_S A \subseteq \mathrm{cl}_S U$ and $\mathrm{cl}_S B \subseteq V$ and $\mathrm{cl}_S A \cap \mathrm{cl}_S B = \{y\}$. Then X must be constant on some tail-end of X.

Appendix

A.1 Axioms and Set Theory Statements

I Axioms and Classes: 1 — Classes, Sets and Axioms

Axiom A1 (Axiom of extent): For the classes x, A and B, $[A = B] \Leftrightarrow [x \in A \Leftrightarrow x \in B]$.

Axiom A2 (Axiom of class construction): Let $P(x)$ designate a statement about x which can be expressed entirely in terms of the symbols $\in, \vee, \wedge, \neg, \Rightarrow, \forall$, brackets and variables x, y, $z, \ldots$, A, $B, \ldots$ Then there exists a class C which consists of all the elements x which satisfy $P(x)$.

Axiom A3 (Axiom of pair): If A and B are sets, then the doubleton $\{A, B\}$ is a set.

Axiom A4 (Axiom of subsets): If S is a set and ϕ is a formula describing a particular property, then the class of all sets in S which satisfy this property ϕ is a set. More succinctly, every subclass of a set of sets is a set. (Also called the Axiom of comprehension, Axiom of separation or Axiom of specification).

Axiom A5 (Axiom of power set): If A is a set then the power set $P(A)$ is a set.

Axiom A6 (Axiom of union): If $\mathscr{A}$ is a set of sets then $\bigcup_{C \in \mathscr{A}} C$ is a set.

Axiom A7 (Axiom of replacement): Let A be a set. Let $\phi(x, y)$ be a formula which associates to each element x of A a set y in such a way that, whenever both $\phi(x, y)$ and $\phi(x, z)$ hold true, $y = z$. Then there exists a set B which contains all sets y such that $\phi(x, y)$ holds true for some $x \in A$.

Axiom A8 (Axiom of infinity): There exists a non-empty set A that satisfies the condition: "$X \in A$" $\Rightarrow$ "$X \cup \{X\} \in A$". (A set satisfying this condition is called a *successor set* or an *inductive set.*)

Axiom A9 (Axiom of regularity): Every non-empty set A contains an element x whose intersection with A is empty.

Axiom of choice: For every set $\mathscr{A}$ of non-empty sets there is a rule f which associates to every set A in $\mathscr{A}$ an element $a \in A$.

I Axioms and Classes: 2 — Constructing Classes and Sets

Theorem 2.1 For any class C, $C = C$. If it is not true that $A = B$ we will write $A \neq B$.

Definition 2.2 If A and B are classes (sets) we define $A \subseteq B$ to mean that every element of A is an element of B. That is, $A \subseteq B$ iff $x \in A \Rightarrow x \in B$ If $A \subseteq B$ we will say that A is a *subclass* (*subset*) of B. If $A \subseteq B$ and $A \neq B$ we will say that A is a *proper subclass* (*proper subset*) of B and write $A \subset B$ when we explicitly want to say $A \neq B$.

Theorem 2.3 If C, D, and E are classes (sets) then:

(a) $C = C$.
(b) $C = D \Rightarrow D = C$.
(c) $C = D$ and $D = E \Rightarrow C = D$.
(d) $C \subseteq D$ and $D \subseteq C \Rightarrow C = D$.
(e) $C \subseteq D$ and $D \subseteq E \Rightarrow C \subseteq E$.

Theorem 2.4 There exists a class which is not an *element.*

Definition 2.5 The Axiom 2 states that $C = \{x : x \neq x\}$ is a class. It contains no elements. We will call the class with no elements the *empty class* and denote it by $\varnothing$.

Theorem 2.6 For any class C, $\varnothing \subseteq C$.

Theorem 2.7 Let S be a set. Then:

(a) $\varnothing \subseteq S$ and so $\varnothing$ is a set.
(b) The set S is an element. Hence, all sets are elements.

Definition 2.8 If A is a set then we define the *power set of* A as being the class $\mathscr{P}(A)$ of all subsets of A. It can be described as follows: $\mathscr{P}(A) = \{X : X \subseteq A\}$.

II Class Operations: 3 — Operations on Classes and Sets

Definition 3.1 Let A and B be classes (sets). We define the *union* $A \cup B$ of the class A and the class B as

$$A \cup B = \{x : (x \in A) \vee (x \in B)\}$$

That is, the element $x \in A \cup B$ iff $x \in A$ or $x \in B$. If $\mathscr{A}$ is a non-empty class of classes then we define the *union of all classes in* $\mathscr{A}$ as

$$\bigcup_{C \in \mathscr{A}} C = \{x : x \in C \text{ for some } C \in \mathscr{A}\}$$

That is, the element $x \in \bigcup_{C \in \mathscr{A}} C$ iff there exists $C \in \mathscr{A}$ such that $x \in C$.

Definition 3.2 Let A and B be classes (sets). We define the *intersection* $A \cap B$ of the class A and the class B as

$$A \cap B = \{x : (x \in A) \wedge (x \in B)\}$$

That is, the element $x \in A \cap B$ iff $x \in A$ and $x \in B$. If $\mathscr{A}$ is a non-empty class of classes then we define the *intersection of all classes in* $\mathscr{A}$ as

$$\bigcap_{C \in \mathscr{A}} C = \{x : x \in C \text{ for all } C \in \mathscr{A}\}$$

That is, the element $x \in \bigcap_{C \in \mathscr{A}} C$ iff $x \in C$ for every class C in $\mathscr{A}$.

Definition 3.3 We will say that two classes (sets) C and D are *disjoint* if the two classes have no elements in common. That is, the classes C and D are disjoint if and only if $C \cap D = \varnothing$.

Definition 3.4 The *complement*, C', of a class (set) C is the class of all elements which are not in C. That is, if C is a class, then

$$C' = \{x : x \notin C\}$$

Hence, $x \in C'$ iff $x \notin C$. Given two classes (sets) C and D, the *difference* $C - D$, of C and D, is the class

$$C - D = C \cap D'$$

The *symmetric difference*, $C \triangle D$, is the class

$$C \triangle D = (C - D) \cup (D - C)$$

Theorem 3.5 Let C and D be classes (sets). Then,

(a) $C \subseteq C \cup D$
(b) $C \cap D \subseteq C$

Theorem 3.6 Let C and D be classes (sets). Then,

(a) $C \cup (C \cap D) = C$
(b) $C \cap (C \cup D) = C$

Theorem 3.7 Let C be a class (a set). Then $(C')' = C$.

Theorem 3.8 *DeMorgan's laws.* Let C and D be classes (sets). Then,

(a) $(C \cup D)' = C' \cap D'$
(b) $(C \cap D)' = C' \cup D'$

Theorem 3.9 Let C, D, and E be classes (sets). Then,

(a) $C \cup D = D \cup C$ and $C \cap D = D \cap C$ (*Commutative laws*)
(b) $C \cup C = C$ and $C \cap C = C$ (*Idempotent laws*)
(c) $C \cup (D \cup E) = (C \cup D) \cup E$ and $C \cap (D \cap E) = (C \cap D) \cap E$ (*Associative laws*)
(d) $C \cup (D \cap E) = (C \cup D) \cap (C \cup E)$ and $C \cap (D \cup E) = (C \cap D) \cup (C \cap E)$ (*Distribution*)

Theorem 3.10 Let A be a class (a set) and $\mathscr{U}$ denote the class of all elements.

(a) $\mathscr{U} \cup A = \mathscr{U}$
(b) $A \cap \mathscr{U} = A$
(c) $\mathscr{U}' = \varnothing$
(d) $\varnothing' = \mathscr{U}$
(e) $A \cup A' = \mathscr{U}$

Theorem 3.11 Let $\mathscr{A}$ be a non-empty class (set).

(a) $\left(\bigcup_{C\in\mathscr{A}} C\right)' = \bigcap_{C\in\mathscr{A}} C'$
(b) $\left(\bigcap_{C\in\mathscr{A}} C\right)' = \bigcup_{C\in\mathscr{A}} C'$

Theorem 3.12 Let D be a class and $\mathscr{A}$ be a non-empty class (set) of classes.

(a) $D \cap \left(\bigcup_{C\in\mathscr{A}} C\right) = \bigcup_{C\in\mathscr{A}}(D \cap C)$
(b) $D \cup \left(\bigcap_{C\in\mathscr{A}} C\right) = \bigcap_{C\in\mathscr{A}}(D \cup C)$

Theorem 3.13 Let $\{B_{(i,j)} : i = 1, 2, 3, \ldots,\ j = 1, 2, 3, \ldots\}$ be a set of sets Then $\cup_{i=1}^{\infty}(\cap_{j=1}^{\infty} B_{(i,j)}) = \cap_{j=1}^{\infty}(\cup_{i=1}^{\infty} B_{(i,j)})$.

II Class Operations: 4 — Cartesian Products

Definition 4.1 Let c and d be elements. We define the *ordered pair* (c, d) as $(c, d) = \{\{c\}, \{c, d\}\}$.

Theorem 4.2 Let $a, b, c,$ and d be classes (which are elements). Then $(a, b) = (c, d)$ iff $a = c$ and $b = d$.

Alternate definition 4.3 If c and d are classes define (c, d) as follows: $(c, d) = \{\{c, \varnothing\}, \{d, \{\varnothing\}\}\}$.

Definition 4.4 Let C and D be two classes (sets). We define the *Cartesian product*, $C \times D$, as follows: $C \times D = \{(c, d) : c \in C \text{ and } d \in D\}$.

Lemma 4.5 Let C and D be two classes (sets). Then the *Cartesian product*, $C \times D$, of C and D satisfies the property: $C \times D \subseteq \mathscr{P}(\mathscr{P}(C \cup D))$.

Corollary 4.6 If C and D are classes, then the *Cartesian product*, $C \times D$, is a class. If C and D are sets, then $C \times D$ is a set.

Theorem 4.7 Let C, D, E, and F be a classes. Then

(a) $C \times (D \cap E) = (C \times D) \cap (C \times E)$
(b) $C \times (D \cup E) = (C \times D) \cup (C \times E)$
(c) $(C \cap E) \times D = (C \times D) \cap (E \times D)$
(d) $(C \cup E) \times D = (C \times D) \cup (E \times D)$
(e) $(C \cup D) \times (E \cup F) = (C \times E) \cup (D \times E) \cup (C \times F) \cup (D \times F)$
(f) $(C \cap D) \times (E \cap F) = (C \times E) \cap (D \times E) \cap (C \times F) \cap (D \times F)$

Theorem 4.8 If $C \subseteq D$ and $E \subseteq F$, then $C \times E \subseteq D \times F$.

Theorem 4.9 Given three classes (sets) S, U and V there is a one-to-one correspondence between the two classes (sets) $S \times (U \times V)$ and $(S \times U) \times V$.

Theorem 4.10 For classes c, d, e and f, if $(c, d) = \{\{c, \varnothing\}, \{d, \{\varnothing\}\}\}$ and $(e, f) = \{\{e, \varnothing\}, \{f, \{\varnothing\}\}\}$ then $(c, d) = (e, f)$ iff $c = e$ and $d = f$.

III Relations: 5 — Relations on a Class or Set

Definition 5.1

(a) We will call any subset R of ordered pairs in $\mathscr{U} \times \mathscr{U}$ a *binary relation.*
(b) We will say that R is *binary relation on a class C* if R is a subclass (subset) of $C \times C$. In such cases we will simply say that R is a *relation in* C or *on* C.
(c) If A and B are classes (sets) and R is a subclass (subset) of $A \times B$ then R can be viewed as a relation on $A \cup B$.

Definitions 5.2 Let C be a class (a set). (a) The relation $\in_C = \{(x, y) : x \in C, y \in C, x \in y\}$ is called the *membership relation on* C. (b) The relation

$$Id_C = \{(x, y) : x \in C, y \in C, x = y\}$$

is called the *identity relation on* C.

Definitions 5.3 Let R be a relation on a class (set) C. The *domain of* R is the class, $\operatorname{dom} R = \{x : x \in C \text{ and } (x, y) \in R \text{ for some } y \in C\}$. The *image of* R is the class, $\operatorname{im} R = \{y : y \in C \text{ and } (x, y) \in R \text{ for some } x \in C\}$. The word *range* *of* R is often used instead of "the image of R". If $R \subseteq A \times B$ is viewed as a relation on $A \cup B$, then $\operatorname{dom} R \subseteq A$ and $\operatorname{im} R \subseteq B$.

Definition 5.4 Let C be a class (a set) and let R be a relation defined in C. The inverse, R^{-1}, of the relation R is defined as follows:

$$R^{-1} = \{(x, y) : (y, x) \in R\}$$

Definition 5.5 Let C be a class (a set) and let R and T be two relations in C. We define the relation $T \circ R$ as follows : $T \circ R = \{(x, y)$: there exists some $z \in \operatorname{im} R$ such that $(x, z) \in R$ and $(z, y) \in T\}$

III Relations: 6 — Equivalence Relations and Order Relations

Definition 6.1 Let S be a class and R be a relation on S.

(a) We say that R is a *reflexive* relation on S if, for every $x \in S$, $(x, x) \in R$.
(b) We say that R is a *symmetric* relation on S if, whenever $(x, y) \in R$ then $(y, x) \in R$.
(c) We say that R is an *anti-symmetric* relation on S if, whenever $(x, y) \in R$ and $(y, x) \in R$ then $x = y$.
(d) We say that R is an *asymmetric* relation on S if, whenever $(x, y) \in R$ then $(y, x) \notin R$.
(e) We say that R is a *transitive* relation on S if, whenever $(x, y) \in R$ and $(y, z) \in R$ then $(x, z) \in R$.
(f) We say that R is an *irreflexive* relation on S if, for every $x \in S$, $(x, x) \notin R$.
(g) We say that R satisfies the property of *comparability* on S if, for every $x, y \in S$ where $x \neq y$, either $(x, y) \in R$ or $(y, x) \in R$.

Definition 6.2 Let S be a class and R be a relation on S. We say that R is an *equivalence relation* on S if R is simultaneously (1) reflexive, (2) symmetric, and (3) transitive on S.

Definition 6.3 Let S be a class.

(a) *Non-strict order relation.* The relation R is a non-strict order relation on S if it is simultaneously reflexive (aRa holds true for any a in S), antisymmetric (if aRb and bRa then $a = b$)

and transitive (aRb and bRc implies aRc) on S. A non-strict order relation, R, on S is said to be a *non-strict linear order relation* if, for every pair of elements a and b in S, either $(a, b) \in R$, $(b, a) \in R$ or $a = b$. That is, every pair of elements are comparable under R.[1] A non-strict ordering, R, on S which is not linear is said to be a *non-strict partial ordering relation* on S.[2]

(b) *Strict order relation.* The relation R is a strict order relation if it is simultaneously irreflexive ($(a, a) \notin R$), asymmetric ($(a, b) \in R \Rightarrow (b, a) \notin R$) and transitive on S. If every pair of distinct elements, a and b, in S are comparable under a strict order relation, R, then R is a *strict linear ordering* on S. Those strict orderings which are not linear are called *strict non-linear orderings* or, more commonly, *strict partial ordering relation.*

A non-strict partial order R on S always induces a strict partial order R^* by defining $aR^*b \Rightarrow [aRb \text{ and } a \neq b]$. Similarly, a strict partial order R on S always induces a non-strict partial order $R^\dagger$ by defining $aR^\dagger b \Rightarrow [aRb \text{ or } a = b]$.

Definition 6.4 Let S be a class and R be a partial ordering or a strict ordering relation on S. If R is a partial ordering relation $(a, b) \in R$ is represented as $a \leq b$ and if R is a strict ordering relation $(a, b) \in R$ is represented as $a < b$. If $a \leq b$ and $a \neq b$, we will simply write $a < b$.

(a) A subset of S which is linearly ordered by R is called a *chain* in S. If R linearly orders S then S is a linearly ordered subset of itself and so is a chain.

(b) An element a of S is called a *maximal* element of S if there does not exist an element b in S such that $a < b$. An element a of S is called a *minimal* element of S if there does not exist an element b in S such that $b < a$.

[1]A class on which is defined a linear ordering R is also said to be *fully ordered* or *totally ordered* by R. In certain branches of mathematics "linearly ordered set" is abbreviated as *l.o.set* or simply called *loset.*

[2]In certain branches of mathematics "partially ordered set" is abbreviated as *p.o.set* or simply called a *poset.*

(c) An element m in S is called the *minimum element of* S if $m \leq a$ ($m < a$) for all $a \in S$. An element M in S is called the *maximum element of* S if $a \leq M$ ($a < M$) for all $a \in S$.

III Relations: 7 — The Partition of a Set Induced by an Equivalence Relation

Notation 7.1 Let R be an equivalence relation on a set S and let $x \in S$. Then the set S_x is defined as follows: $S_x = \{y : (x, y) \in R\}$. That is, S_x is the set of all elements y in S such that y is related to x under R.

Theorem 7.2 Let R be an equivalence relation on a set S. Let x and y be two elements in S which are not related under R. Then any element z in S which is related to x cannot be related to y.

Theorem 7.3 Let R be an equivalence relation on a set S. Let x and y be two elements in S which are not related under R. Then $S_x \cap S_y = \varnothing$.

Theorem 7.4 Let R be an equivalence relation on a set S. Let x and y be two elements in S which are related under R. Then $S_x = S_y$.

Theorem 7.5 Let R be an equivalence relation on a set S. For every $x \in S$ there exists some $y \in S$ such that $x \in S_y$.

Theorem 7.6 Let R be an equivalence relation on a set S. Then $\bigcup_{x \in S} S_x = S$.

III Relations: 8 — On Partitions and Quotient Sets of a Set

Definition 8.1 Let S be a set. We say that a set $\mathscr{C} \subseteq \mathscr{P}(S)$ forms a *partition of* S if $\mathscr{C}$ satisfies the 3 properties:

(1) $\bigcup_{A \in \mathscr{C}} A = S$.
(2) If A and $B \in \mathscr{C}$ and $A \neq B$ then $A \cap B = \varnothing$.
(3) $A \neq \varnothing$ for all $A \in \mathscr{C}$.

Definition 8.2 Let S be a set on which an equivalence relation R is defined.

(a) Each element S_x of $\mathscr{S}_R = \{S_x : x \in S\}$ is called an *equivalence class of x under R* or an *equivalence class induced by the relation R*.
(b) The set $\mathscr{S}_R = \{S_x : x \in S\}$ of all equivalence classes induced by the relation R is called the *quotient set of S induced by R*. The set $\mathscr{S}_R$ is more commonly represented by the symbol S/R. So $S/R = \{S_x : x \in S\}$. From here on we will use the more common notation, S/R.

Theorem 8.3 Let S be a set and $\mathscr{C}$ be a partition of S. Let $R_{\mathscr{C}}$ be the relation such that $(x, y) \in R_{\mathscr{C}}$ iff $\{x, y\} \subseteq S$ for some S in $\mathscr{C}$. Then $R_{\mathscr{C}}$ is an equivalence relation on S.

IV Functions: 9 — Functions: A Set-Theoretic Definition

Definition 9.1 A function from A to B is a triple $\langle f, A, B\rangle$ satisfying the following properties:

(1) A and B are classes and $f \subseteq A \times B$.
(2) For every $a \in A$ there exists $b \in B$ such that $(a, b) \in f$.
(3) If $(a, b) \in f$ and $(a, c) \in f$ then $a = c$.

Definition 9.2 If $f : A \to B$ is a function and $D \subseteq A$ then we say that the function $f : D \to C$ is *a restriction of f to D*. In this case we will use the symbol $f|_D$ to represent the restriction of f to D. Note that, if $D \subseteq A$, then we can write $f|_D \subseteq f$ since $f|_D = \{(x, y) : x \in D \text{ and } (x, y) \in f\} \subseteq f$.

Theorem 9.3 Let $f : A \to B$ be a function and suppose $A = C \cup D$. Then $f = f|_c \cup f|_D$.

Theorem 9.4 Two functions $f : A \to B$ and $g : A \to B$ are equal if and only if $f(x) = g(x)$ for all $x \in A$.

Definitions 9.5 Let $f : A \to B$ be a function.

(a) We say that "f maps A *onto* B" if $\operatorname{im} f = B$. We often use the expression "$f : A \to B$ is *surjective*" instead of the words *onto B*.
(b) We say that "f maps A *one-to-one into* B" if whenever $f(x) = f(y)$ then $x = y$. We often use the expression "$f : A \to B$ is *injective*" instead of the words *one-to-one into B*.

(c) If the function $f : A \to B$ is both one-to-one and onto B then we can simply say that f is "one-to-one and onto". Another way of conveying this is to say that f is *bijective*, or f is a *bijection*. So "injective + surjective ⇒ bijective".

(d) Two classes (or sets) A and B for which there exists some bijective function $f : A \to B$ are said to be in *one-to-one correspondence*.

IV Functions: 10 — Compositions of Function

Definition 10.1 Suppose $f : A \to B$ and $g : B \to C$ are two functions such that the image of the function f is contained in the domain of the function g. Let $h = \{(x, z) : y = f(x) \text{ and } z = g(y) = g(f(x))\ \}$. Thus $(x, z) \in h$ if and only if $(x, z) = (x, g(f(x))$. We will call h the *composition of* g *and* f, and denote it as $g \circ f$.

Theorem 10.2 Let $f : A \to B$ and $g : B \to C$ be two functions such that the image of the function f is contained in the domain of the function g. Then the composition of g and f, $(g \circ f) : A \to C$, is a function.

Theorem 10.3 Let $f : A \to B$, $g : B \to C$ and $h : C \to D$ be three functions. Then $h \circ (g \circ f) = (h \circ g) \circ f$.

Theorem 10.4 Let $f : A \to B$. Then $I_B \circ f = f$ and $f \circ I_A = f$.

Definition 10.5 Let $f : A \to B$. If $g : B \to A$ is a function satisfying $g \circ f = I_A$ then we will call g an "*inverse of* f" and denote it as f^{-1}.

Theorem 10.6 Let $f : A \to B$ be a one-to-one onto function.

(a) An inverse function $f^{-1} : B \to A$ of f exists.
(b) The function f^{-1} is one-to-one and onto.
(c) The function $f^{-1} : B \to A$ satisfies the property $f \circ f^{-1} = I_B$.
(d) The inverse function, f^{-1}, of f is unique.

Definition 10.7 A function $f : A \to B$ which is one-to-one and onto is called an *invertible function*.

Theorem 10.8 Let $f : A \to B$ and $g : B \to C$ be two onto-to-one and onto functions.

(a) The function $g \circ f$ is also one-to-one and onto.
(b) The inverse, $g \circ f$, is $(g \circ f)^{-1} = f^{-1} \circ g^{-1}$

IV Functions: 11 — Images and Inverse Images of Sets

Definition 11.1 Let A and B be sets and suppose $f : A \to B$ is a function acting on A. If S is a subset of $A = \operatorname{dom} f$ we define the expression $f[S]$ as follows: $f[S] = \{y : y = f(x) \text{ for some } x \in S\}$. We will say that $f[S]$ is the *image of the set S under f.*

Definitions 11.2 Let $f : A \to B$ be a function where A and B are sets. We define $f^{\leftarrow} : \mathscr{P}(B) \to \mathscr{P}(A)$ as $f^{\leftarrow}(X) = Y$ iff $Y = \{y : y \in A, f(y) \in X\}$. In particular, $f^{\leftarrow}(\{x\}) = Y$ iff $Y = \{y : y \in A, f(y) = x\}$. We will refer to it as the *set-valued inverse function $f^{\leftarrow}$.*

Theorem 11.3 Let $f : A \to B$ be a function. Then $f^{\leftarrow} : \mathscr{P}(\operatorname{im} f) \to \mathscr{P}(A)$ is a one-to-one function on its domain $\mathscr{P}(\operatorname{im} f)$.

Theorem 11.4 Let $f : A \to B$ be a function mapping the set A to the set B. Let $\mathscr{A}$ be a set of subsets of A and $\mathscr{B}$ be a set of subsets of B. Let $D \subseteq A$ and $E \subseteq B$. Then:

(a) $f\left[\bigcup_{S \in \mathscr{A}} S\right] = \bigcup_{S \in \mathscr{A}} f[S]$
(b) $f\left[\bigcap_{S \in \mathscr{A}} S\right] \subseteq \bigcap_{S \in \mathscr{A}} f[S]$ with equality only if f is one-to-one
(c) $f[A - D] \subseteq B - f[D]$ with equality only if f is one-to-one and onto B
(d) $f^{\leftarrow}\left(\bigcup_{S \in \mathscr{B}} S\right) = \bigcup_{S \in \mathscr{B}} f^{\leftarrow}(S)$
(e) $f^{\leftarrow}\left(\bigcap_{S \in \mathscr{B}} S\right) = \bigcap_{S \in \mathscr{B}} f^{\leftarrow}(S)$
(f) $f^{\leftarrow}(B - E) = A - f^{\leftarrow}(E)$

IV Functions: 12 — Equivalence Relations Defined by Functions

Definition 12.1 Let $f : A \to B$ be a function which maps a set A into a set B. We define an equivalence relation R_f on A as follows: Two elements a and b are related under R_f if and only if $\{a, b\} \subseteq f^{\leftarrow}(x)$ for some x in $\operatorname{im} f$. The quotient set of A induced by R_f is then $A/R_f = \mathscr{A}_{R_f} = \{f^{\leftarrow}(\{x\}) : x \in f[A]\}$ We will refer to R_f as the *equivalence relation determined by*

f and A/R_f (or $\mathscr{A}_{R_f}$) as the *quotient set of* A *determined by* f.

Theorem 12.2 Let $f : S \to T$ be a function where S and T are sets. There exists an onto function $g_f : S \to S/R_f$ and a one-to-one function $h_f : S/R_f \to T$ such that $h_f \circ g_f = f$. The function, $h_f \circ g_f = f$, is called the *canonical decomposition of* f.

V From Sets to Numbers: 13 — The Natural Numbers

Definition 13.1 For any set x, we define the *successor* x^+, of x as $x^+ = x \cup \{x\}$.

Definition 13.2 If x is a set then $x^+ = x \cup \{x\}$. A set A is called an *inductive set* if it satisfies the following two properties:

(a) $\varnothing \in A$.
(b) $x \in A \Rightarrow x^+ \in A$.

Definition 13.3 We define the *set of all natural numbers*, $\mathbb{N}$, as the intersection of all inductive sets. That is $\mathbb{N} = \{x : x \in I$ for any inductive set $I\}$.

Theorem 13.4 Let A be a subset of $\mathbb{N}$. If A satisfies the two properties:

(a) $0 \in A$
(b) $m \in A \Rightarrow m^+ \in A$

then $A = \mathbb{N}$.

Corollary 13.5 (The Principle of mathematical induction.) Let P denote a particular set property. Suppose $P(n)$ means "the property P is satisfied depending on the value of the natural number n". Let

$$A = \{n \in \mathbb{N} : P(n) \text{ holds true}\}$$

If A satisfies the two properties:

(a) $0 \in A$. That is $P(0)$ holds true,
(b) $(n \in A) \Rightarrow (n^+ \in A)$. That is, $P(n)$ holds true $\Rightarrow P(n^+)$ holds true.

then $A = \mathbb{N}$. That is, $P(n)$ holds true for all natural numbers n.

Definition 13.6 A set S which satisfies the property "$x \in S \Rightarrow x \subset S$" is called a *transitive set.*

Theorem 13.7 The non-empty set S is a transitive set if and only if the property "$x \in y$ and $y \in S$" $\Rightarrow$ "$x \in S$".

Theorem 13.8 The set $\mathbb{N}$ of natural numbers is a transitive set.

Theorem 13.9

(a) For natural numbers n, m, $m \in n \Rightarrow m \subseteq n$. Hence, every natural number is a transitive set.
(b) For any natural number n, $n \neq n^+$.
(c) For any natural number n, $n \notin n$.
(d) For any distinct natural numbers n, m, $m \subset n \Rightarrow m \in n$.

Theorem 13.10 Let m and n be distinct natural numbers.

(a) If $m \subset n$ then $m^+ \subseteq n$.
(b) *All natural numbers are comparable.* Either $m \subset n$ or $n \subset m$. Equivalently, $m \in n$ or $n \in m$. Hence, both "$\subset$" and "$\in$" linearly order $\mathbb{N}$.
(c) There is no natural number m such that $n \subset m \subset n^+$.

Theorem 13.11 Every natural number has an immediate predecessor. If k and n are natural numbers such that $k^+ = n$ then k is called an *immediate predecessor* of n. For any non-zero natural number n, $k = \bigcup_{m \subset n} m$ is an immediate predecessor of n.

Theorem 13.12 *Unique immediate predecessors.* Any non-zero natural number has a *unique* immediate predecessor.

Theorem 13.13 (The Principle of mathematical induction: second version.) Let P denote a particular property. Suppose $P(n)$ means "the property P is satisfied depending on the value of the natural number n". Let

$$A = \{n \in \mathbb{N} : P(n) \text{ holds true}\}$$

Suppose that, for any natural number n,

$$P(k) \text{ is true for all } k < n \Rightarrow P(n) \text{ is true}$$

Then $P(n)$ holds true for all natural numbers n.

V From Sets to Numbers: 14 — The Natural Numbers as a Well-Ordered Set

Notation 14.1 We define the relation "$\in_=$" on $\mathbb{N}$ as follows:

$$m \in_= n \quad \text{if and only if } m = n \text{ or } m \in n$$

If $m \in_= n$ and we want to state explicitly that $m \neq n$ we write $m \in n$.

Theorem 14.2 Let $(S, \leq)$ be a linearly ordered set. Suppose $T \subseteq S$. The element q is *a least element of* T *with respect to* "$\leq$" if $q \in T$ and $q \leq m$ for all $m \in T$. If S is equipped with a strict linear ordering "$<$" a *least element of* T *with respect to* $<$ is an element $q \in T$ such that $q < m$ for all $m \in T$ where $m \neq q$. The set $(S, \leq)$ is said to be *well-ordered* with respect to "$\leq$" if every non-empty subset T of S contains its least element with respect to $\leq$. Similarly, the set $(S, <)$ is said to be *well-ordered* with respect to "$<$" if every non-empty subset T of S contains its least element with respect to $<$.

Theorem 14.3 The natural numbers $\mathbb{N}$ is a strict $\in$-well-ordered set.

Corollary 14.4 Every natural numbers n is a $\in$-well-ordered set.

Theorem 14.5 Any bounded non-empty subset of $(\mathbb{N}, \in)$ has a maximal element.

Definition 14.6 Consider the set $\{1, 2\}^{\mathbb{N}}$ of all functions mapping natural numbers to 1 or 2. We define the *lexicographic order* "$<$" on $\{1, 2\}^{\mathbb{N}}$ as follows: For any two elements $f = \{a_0, a_1, a_2, a_3, \ldots\}$ and $g = \{b_0, b_1, b_2, b_3, \ldots\}$ in $\{1, 2\}^{\mathbb{N}}$, $f = g$ if and only if $a_i = b_i$ for all $i \in \mathbb{N}$ and $f < g$ if and only if for the first two unequal corresponding terms a_i and b_i, $a_i \in b_i$. A lexicographic ordering can similarly be defined on $S^{\mathbb{N}}$ where S is any subset of $\mathbb{N}$.

V From Sets to Numbers: 15 — Arithmetic of the Natural Numbers

Definition 15.1 Let m be a fixed natural number. Addition of a natural number n with m is defined as the function

$r_m : \mathbb{N} \to \mathbb{N}$ satisfying the two conditions

$$r_m(0) = m$$
$$r_m(n^+) = [r_m(n)]^+$$

The expression $m+n$ as simply another way of writing $r_m(n)$. Thus,

$$r_m(0) = m \quad \Leftrightarrow \quad m + 0 = m \qquad (3)$$
$$r_m(n^+) = [r_m(n)]^+ \quad \Leftrightarrow \quad m + n^+ = (m+n)^+ \qquad (4)$$

Theorem 15.2 Let m be a fixed natural number and let $r_m : \mathbb{N} \to \mathbb{N}$ be a function satisfying the two properties

$$\begin{cases} r_m(0) &= m \\ r_m(n^+) &= [r_m(n)]^+ \end{cases}$$

Then r_m is a well-defined function on $\mathbb{N}$.

Definition 15.3 For any natural number m, multiplication with the natural number m is defined as the function $s_m : \mathbb{N} \to \mathbb{N}$ satisfying the two conditions

$$s_m(0) = 0$$
$$s_m(n^+) = s_m(n) + m$$

We define the expression mn and $m \times n$ as alternate ways of writing $s_m(n)$. Thus,

$$s_m(0) = 0 \Leftrightarrow m0 = m \times 0 = 0 \qquad (5)$$
$$s_m(n^+) = s_m(n) + m \Leftrightarrow mn^+ = mn + m = m \times n + m \qquad (6)$$

Theorem 15.4 Let m be a fixed natural number and let $s_m : \mathbb{N} \to \mathbb{N}$ be a function satisfying the two properties

$$\begin{cases} s_m(0) &= 0 \\ s_m(n^+) &= s_m(n) + m \end{cases}$$

Then s_m is a well-defined function on $\mathbb{N}$.

Theorem 15.5 For any two natural numbers m and n, $m \in_= n$ if and only if there exists a *unique* natural number k such that $n = m + k$.

Definition 15.6 For any two natural numbers m and n such that $m \leq n$, the *unique* natural number k satisfying $n = m + k$ is called the *difference between n and m* and is denoted by $n - m$. The operation "$-$" is called *subtraction*.

V From Sets to Numbers: 16 — The Integers $\mathbb{Z}$ and the Rationals $\mathbb{Q}$

Theorem 16.1 Let $Z = \mathbb{N} \times \mathbb{N}$. Let R_z be a relation on Z that is defined as follows: $(a, b)R_z(c, d)$ if and only if $a + d = b + c$. Then R_z is an equivalence relation on Z.

Corollary 16.2 Let $Z = \mathbb{N} \times \mathbb{N}$ be equipped with the equivalence relation R_z defined as:

$$(a, b)R_z(c, d) \qquad \Leftrightarrow \qquad a + d = b + c$$

For each $n \in \mathbb{N}$ let $[(0, n)]$ and $[(n, 0)]$ denote the R_z-equivalence classes containing the elements $(0, n)$ and $(n, 0)$ respectively. Then the quotient set induced by R_z can be expressed as

$$Z/R_z = \{[(0, n)] : n \in \mathbb{N}\} \cup \{[(n, 0)] : n \in \mathbb{N}\}$$

Definitions 16.3 The set of *integers*, $\mathbb{Z}$, is defined as:

$$\begin{aligned}\mathbb{Z} = Z/R_z &= \{[(a, b)] : a, b \in \mathbb{N}\} \\ &= \{[(0, n)] : n \in \mathbb{N}\} \cup \{[(n, 0)] : n \in \mathbb{N}\}\end{aligned}$$

(a) *Negative integers*: The set of *negative integers* is defined as being the set

$$\mathbb{Z}^- = \{[(0, n)] : n \in \mathbb{N}\}$$

Positive integers: The set of *positive integers* is defined as being the set

$$\mathbb{Z}^+ = \{[(n, 0)] : n \in \mathbb{N}\}$$

The elements of the form $[(0,n)]$ can be represented by $-n = [(0,n)]$ while the elements of the form $[(n,0)]$ can be represented as $n = [(n,0)]$.

(b) *Order relation on* $\mathbb{Z}$: We define a relation $\leq_z$ on $\mathbb{Z}$ as follows: $[(a,b)] \leq_z [(c,d)]$ if and only if $a+d \leq b+c$. It is a routine exercise to show that $\leq_z$ is a linear ordering of $\mathbb{Z}$.

(c) *Addition on* $\mathbb{Z}$: We must sometimes distinguish between addition of natural numbers and addition of integers. Where there is a risk of confusion we will use the following notation: "$+_n$" means addition of natural numbers while "$+_z$" means addition of integers.

Addition $+_z$ on $\mathbb{Z}$ is defined as:

$$[(a,b)] +_z [(c,d)] = [(a +_n c,\ b +_n d)]$$

(d) *Opposites of integers*: The *opposite* $-[(a,b)]$ of $[(a,b)]$ is defined as

$$-[(a,b)] = [(b,a)]^3$$

(e) *Subtraction on integers*: Subtraction "$-_z$" on $\mathbb{Z}$ is defined as:

$$[(a,b)] -_z [(c,d)] = [(a,b)] + (-[(c,d)])^4$$

(f) *Multiplication of integers*: Multiplication $\times_z$ on $\mathbb{Z}$ is defined as

$$[(a,b)] \times_z [(c,d)] = [(ac+bd, ad+bc)].^5$$

In particular, $[(0,n)] \times_z [(m,0)] = [(0+0, 0+nm)] = [(0,nm)] = -[(nm,0)]$ and $[(n,0)] \times_z [(m,0)] = [(nm,0)]$.

[3]Note that $-[(n,0)] = [(0,n)] = -n$.

[4]When there is no risk of confusion with subtraction of other types of numbers we will simply use "$-$".

[5]Note that the "center dot" can be used instead of the "$\times_z$" symbol. When there is no risk of confusion with multiplication of other types of numbers we will simply use "$\times$".

(g) *Absolute value of an integer*: The absolute value, $|n|$, of an integer n is defined as

$$|n| = \begin{cases} n & \text{if } 0 \leq_z n \\ -n & \text{if } n <_z 0 \end{cases}$$

(h) *Equality of two integers*: If (a, b) and (c, d) are ordered pairs which are equivalent under the relation R_z, then the R_z-equivalence classes $[(a, b)]$ and $[(c, d)]$ are equal sets. To emphasize that they are equal sets under the relation R_z we can write

$$[(a, b)] =_z [(c, d)]$$

(i) *Distribution properties*: If $[(a, b)]$, $[(c, d)]$ and $[(e, f)]$ are integers then

$$\begin{aligned} &[(a, b)] \times_z ([(c, d)] +_z [(e, f)]) \\ &=_z [(a, b)] \times_z [(c, d)] +_z [(a, b)] \times_z [(e, f)] \end{aligned}$$

and

$$\begin{aligned} &([(c, d)] +_z [(e, f)]) \times_z [(a, b)] \\ &=_z [(c, d)] \times_z [(a, b)] +_z [(e, f)] \times_z [(a, b)] \end{aligned}$$

Theorem 16.4 Let $Q = \mathbb{Z} \times \mathbb{Z}^*$ where $\mathbb{Z}^* = \mathbb{Z} - \{0\}$. Let R_q be a relation on Q defined as follows: $(a, b)R_q(c, d)$ if and only if $a \times_z d = b \times_z c$. Then R_q is an equivalence relation on Q.

Definitions 16.5 The set of *rational numbers*, $\mathbb{Q}$, is defined as:

$$\mathbb{Q} = Q/R_q = \{[(a, b)] : a \in \mathbb{Z}, b \in \mathbb{Z}^*\}^6$$

The expression $[(a, b)]$ is normally written in the form $\frac{a}{b}$.

(a) We define a relation $\leq_q$ on $\mathbb{Q}$ as follows: If b and d are both positive, $[(a, b)] \leq_q [(c, d)]$ if and only if $a \times_z d \leq_z b \times_z c$.

[6] Recall that a and b is shorthand for expressions of the form $[(0, a)]$ or $-[(0, b)]$.

(b) Addition $+_q$ on $\mathbb{Q}$ is defined as:

$$[(a,b)] +_q [(c,d)] = [(ad +_z bc,\ b \times_z d)]$$

(c) Subtraction $-_q$ on $\mathbb{Q}$ is defined as:

$$[(a,b)] -_q [(c,d)] = [(a,b)] +_q [(-c,d)]$$

(d) Multiplication $\times_q$ on $\mathbb{Q}$ is defined as

$$[(a,b)] \times_q [(c,d)] = [(a \times_z c, b \times_z d)]$$

(e) *Equality of two rational numbers*: If (a,b) and (c,d) are ordered pairs of integers $(b, d \neq 0)$ which are equivalent under the relation R_q, then the R_q-equivalence classes $[(a,b)]$ and $[(c,d)]$ are equal sets. To emphasize that they are equal sets under the relation R_q we can write

$$[(a,b)] =_q [(c,d)]$$

(f) *Opposites of rational numbers.* If (a,b) is an ordered pair of integers $(b \neq 0)$ and $[(a,b)]$ is its R_q-equivalence class then the opposite of the rational number $[(a,b)]$ is defined as $[(-a,b)]$ and is denoted as $-[(a,b)] =_q [(-a,b)]$.

Theorem 16.6 Suppose a and b are positive integers where $b \neq 0$ and $[(a,b)]$ is an R_q equivalence class. Then
(a) $[(-a,-b)] = \frac{-a}{-b} = \frac{a}{b} = [(a,b)]$.
(b) $-[(a,b)] =_q [(-a,b)] = \frac{-a}{b} = \frac{a}{-b} = [(a,-b)]$

V From Sets to Numbers: 17 — Dedekind Cuts: "Real Numbers Are Us!"

Definition 17.1 For any real number r, let ${}_{\mathbb{R}}S_r$ denote the interval $(-\infty, r)$ in $\mathbb{R}$. It is the subset of all real numbers strictly smaller than r. The subset ${}_{\mathbb{R}}S_r$ is called an *initial segment in* $\mathbb{R}$. For any real number r, the subset ${}_{\mathbb{Q}}S_r = (-\infty, r) \cap \mathbb{Q} = {}_{\mathbb{R}}S_r \cap \mathbb{Q}$ is called an *initial segment in* $\mathbb{Q}$. For each of ${}_{\mathbb{R}}S_r$ and ${}_{\mathbb{Q}}S_r$ the real number r is the least upper bound of $(-\infty, r)$ and $(-\infty, r) \cap \mathbb{Q}$ respectively. Note that r may be a non-rational number even for ${}_{\mathbb{Q}}S_r$ a proper subset of $\mathbb{Q}$.

Definition 17.2 The elements ${}_{\mathbb{Q}}S_r = (-\infty, r) \cap \mathbb{Q}$ of the set $\mathscr{D} = \{{}_{\mathbb{Q}}S_r : r \in \mathbb{R}\}$ are called *Dedekind cuts.*

Definition 17.3 The set of all Dedekind cuts $\mathscr{D}$, linearly ordered by inclusion with addition $+$ and multiplication $\times$ as described above is called the *real numbers.* Those Dedekind cuts which do not have a least upper bound in $\mathbb{Q}$ are called *irrational numbers.*

Lemma 17.4 The union of a set of Dedekind cuts is either $\mathbb{Q}$ or a Dedekind cut.

Theorem 17.5 Every non-empty subset of $\mathbb{R}$ which has an upper bound has a least upper bound.

VI Infinite Sets: 18 — Infinite Sets versus Finite Sets

Definition 18.1 A set S is said to be an *infinite set* if S has a *proper* subset X such that a function $f : S \to X$ maps S one-to-one onto X. If a set S is not infinite then we will say that it is a *finite set.*

Theorem 18.2 Basic properties of infinite and finite sets.

(a) The empty set is a finite set.
(b) Any singleton set is a finite set.
(c) Any set which has a subset which is infinite must itself be infinite.
(d) Any subset of a finite set must be finite.

Theorem 18.3 Let $f : X \to Y$ be a one-to-one function mapping X onto Y. The set Y is infinite if and only if the set X is infinite.

Corollary 18.4 The one-to-one image of a finite set is finite.

Lemma 18.5 Let S be an infinite set and $x \in S$. Then $S - \{x\}$ is an infinite set.

Theorem 18.6 Every natural number n is a finite set.

Corollary 18.7 [AC] A set S is finite if and only if S is empty or in one-to-one correspondence with some natural number n.

Theorem 18.8 *The recursively constructed function theorem.* Let S be a set, $k : \mathscr{P}(S) \to S$ be a well-defined function on $\mathscr{P}(S)$ and $f \subseteq \mathbb{N} \times S$ be a relation. We write $f(n) = a$ if and only if $(n, a) \in f$. Let $m \in S$. Suppose the relation f satisfies the two properties

$$\begin{cases} f(0) = m & \Rightarrow (0, m) = (0, f(0)) \in f \\ (n, f(n)) \in f & \Rightarrow (n+1, k(S - \{f(0), f(1), \ldots, f(n)\}) \\ & = (n+1, f(n+1)) \in f \end{cases}$$

Then f is a well-defined function on $\mathbb{N}$.

Theorem 18.9 [AC] A set S is an infinite set if and only if it contains a one-to-one image of the $\mathbb{N}$.

Theorem 18.10 If the set S is a finite set and $f : S \to X$ is a function, then $f[S]$ is finite.

Theorem 18.11 If a set S contains n elements then $\mathscr{P}(S)$ contains 2^n elements. Hence, if a set S is a finite set, then the set $\mathscr{P}(S)$ is finite.

VI Infinite Sets: 19 — Countable and Uncountable Sets

Definition 19.1 Two sets A and B are said to be *equipotent* sets if there exists a one-to-one function $f : A \to B$ mapping one onto the other. If A and B are *equipotent* we will say that "A is equipotent to B" or "A is equipotent with B".

Definition 19.2 *Countable sets* are those sets that are either finite or equipotent to $\mathbb{N}$. All infinite sets which are not countable are called *uncountable sets.*

Theorem 19.3 A subset of a countable set is countable.

Theorem 19.4 Any finite product, $\mathbb{N} \times \mathbb{N} \times \cdots \times \mathbb{N}$, of $\mathbb{N}$ is a countable set.

Lemma 19.5 Suppose f maps an infinite countable set A onto a set $B = f[A]$. Then B is countable.

Theorem 19.6 Let $\{A_i : i \in S \subseteq \mathbb{N}\}$ be a countable set of non-empty countable sets A_i. Then $\bigcup_{i\in S} A_i$ is countable.

Theorem 19.7 The set of all real numbers $\mathbb{R}$ is uncountable.

VI Infinite Sets: 20 — Properties of the Equipotence Relation

Theorem 20.1 Let $\mathscr{S}$ be a class of sets. The equipotence relation R_e on $\mathscr{S}$ is an equivalence relation on $\mathscr{S}$.

Theorem 20.2 Suppose A, B, C, and D are sets such that $A \sim_e B$ and $C \sim_e D$ where $A \cap C = \varnothing = B \cap D$. Then $(A \cup C) \sim_e (B \cup D)$.

Theorem 20.3 Suppose A, B, C and D are sets such that $A \sim_e B$ and $C \sim_e D$. Then $A \times C \sim_e B \times D$.

Corollary 20.4 Suppose A and B are infinite sets.

(1) If $\{A, B\} \subset [\mathbb{N}]_e$ then $A \times B \in [\mathbb{N}]_e$.
(2) If $\{A, B\} \subset [\mathbb{R}]_e$ then $A \times B \in [\mathbb{R}]_e$.

Theorem 20.5 If $\{A_i : i = 0, 1, 2, \ldots, n\}$ is a set of n non-empty countable sets then $\prod_{i=0}^{n} A_i$ is countable for all n.

Theorem 20.6 Suppose S is and infinite set and T is a countable set such that $S \cap T = \varnothing$. Then $S \sim_e S \cup T$.

Theorem 20.7 If the sets A and B are equipotent, then so are their associated power sets $\mathscr{P}(A)$ and $\mathscr{P}(B)$.

Theorem 20.8 Any non-empty set S is embedded in its power set $\mathscr{P}(S)$. But *no* subset of S is equipotent with $\mathscr{P}(S)$.

Definition 20.9 We will say that the non-empty set A is *properly embedded* in the set B if A is equipotent to a proper subset of B but *B is not equipotent to A or any of its subsets.* To represent the relationship "A is properly embedded in B" we will write

$$A \hookrightarrow_e B$$

If A and B are sets such that A is equipotent to a subset of B (where B may, or may not, be embedded in A), we will say that A is *embedded* in B. To represent the relationship "A is embedded in B" we will write

$$A \hookrightarrow_{e\sim} B$$

Definition 20.10 Let $\mathscr{S} = \{S : S \text{ is a set}\}$ and $\mathscr{E} = \{[S]_e : S \in \mathscr{S}\}$. Let $[A]_e$ and $[B]_e$ be elements of $\mathscr{E}$. We write

$$[A]_e <_e [B]_e$$

if and only if $A \hookrightarrow_e B$. We write

$$[A]_e \leq_e [B]_e$$

if and only if $A \hookrightarrow_{e\sim} B$.

Proposition 20.11 Let S be any set. Suppose $\mathscr{P}^0(S) = S$, $\mathscr{P}^1(S) = \mathscr{P}(S)$ and $\mathscr{P}^n(S) = \mathscr{P}(\mathscr{P}^{n-1}(S))$ for all $n \geq 1$. The set

$$\{[S]_e, [\mathscr{P}(S)]_e, [\mathscr{P}^2(S)]_e, [\mathscr{P}^3(S)]_e, \ldots, [\mathscr{P}^n(S)]_e \ldots\}$$

forms an infinite $<_e$-ordered chain of distinct classes in $\mathscr{E}$.

Theorem 20.12 For every any non-empty set S, $2^S = \{\chi_T : T \in \mathscr{P}(S)\} \sim_e \mathscr{P}(S)$

Theorem 20.13 The set $\mathbb{R}$ is embedded in $\mathscr{P}(\mathbb{N})$.

Theorem 20.14 The set $\mathscr{P}(\mathbb{N})$ is embedded in $\mathbb{R}$.

VI Infinite Sets: 21 — The Shröder–Bernstein Theorem

Theorem 21.1 (The Schröder–Bernstein theorem) If S and T are infinite subsets where S is embedded in T and T is embedded in S then S and T are equipotent.

Lemma 21.2 Let T be a proper subset of the set S and $f : S \to T$ be a one-to-one function mapping S into T. Then there exists a one-to-one function $h : S \to T$ mapping S **onto** T.

Theorem 21.3 The set $\mathbb{R}$ of all real numbers is equipotent to $\mathscr{P}(\mathbb{N})$.

Theorem 21.4 The sets $\mathbb{N}^{\mathbb{N}}$ and $\mathbb{R}$ are equipotent.

Definition 21.5 If A and B are two sets, then the symbol B^A refers to the set of all functions mapping A into B.

VII Cardinal Numbers: 22 — An Introduction to Cardinal Numbers

Postulate 22.1 There exists a class of sets $\mathscr{C}$ which satisfies the following properties:

(1) Every natural number n is an element of $\mathscr{C}$.
(2) Any set $S \in \mathscr{S}$ is equipotent with precisely one element in $\mathscr{C}$.

The sets in $\mathscr{C}$ are called *cardinal numbers*. When we say that a set S *has cardinality* κ we mean that $\kappa \in \mathscr{C}$ and that $S \sim_e \kappa$. If the set S has cardinality κ, we will write $|S| = \kappa$.

Definition 22.2 If S and T are sets and $\kappa = |S|$ and $\lambda = |T|$ then we define *addition*"+", *multiplication* "×" and *exponentiation* of two cardinal numbers as follows:

(a) If $S \cap T = \varnothing$,

$$\kappa + \lambda = |S \cup T|$$

(b)
$$\kappa \times \lambda = |S \times T|$$

(c)
$$\kappa^\lambda = |S^T|$$

where S^T represents the set of all functions mapping T into S (as previously defined). That is, $|S|^{|T|} = |S^T|$. For convenience we define $0^\lambda = 0$ and $\kappa^0 = 1$.

Theorem 22.3 The class $\mathscr{C}$ of all cardinal numbers is a proper class.

VII Cardinal Numbers: 23 — Arithmetic of Cardinal Numbers

Theorem 23.1 Addition on $\mathscr{C}$ is well-defined. That is, if S_1, S_2, T_1 and T_2 are sets such that $\kappa = |S_1| = |S_2|$ and $\lambda = |T_1| = |T_2|$, then $|S_1 \cup T_1| = \kappa + \lambda = |S_2 \cup T_2|$.

Theorem 23.2 Let κ, λ, ϕ, and ψ be any four cardinal numbers. Then

(a) $\kappa + \lambda = \lambda + \kappa$ (Commutativity of addition)
(b) $(\kappa + \lambda) + \phi = \kappa + (\lambda + \phi)$ (Associativity of addition)

(c) $\kappa \leq \kappa + \lambda$
(d) $\kappa \leq \lambda$ and $\phi \leq \psi \Rightarrow \kappa + \phi \leq \lambda + \psi$.

Theorem 23.3 Multiplication on $\mathscr{C}$ is well-defined. That is, if S_1, S_2, T_1 and T_2 are sets such that $\kappa = |S_1| = |S_2|$ and $\lambda = |T_1| = |T_2|$, then $|S_1 \times T_1| = \kappa \times \lambda = |S_2 \times T_2|$.

Theorem 23.4 Let κ, λ, ϕ, and ψ be any three cardinal numbers. Then

(a) $\kappa \times \lambda = \lambda \times \kappa$. (Commutativity of multiplication)
(b) $(\kappa \times \lambda) \times \phi = \kappa \times (\lambda \times \phi)$. (Associativity of multiplication)
(c) $\kappa \times (\lambda + \phi) = (\kappa \times \lambda) + (\kappa \times \phi)$. (Left-hand distributivity)
(d) $\lambda > 0 \Rightarrow \kappa \leq (\kappa \times \lambda)$.
(e) $\kappa \leq \lambda$ and $\phi \leq \psi \Rightarrow \kappa \times \phi \leq \lambda \times \psi$.
(f) $\kappa + \kappa = 2 \times \kappa$.
(g) $\kappa + \kappa \leq \kappa \times \kappa$ when $\kappa \geq 2$.

VII Cardinal Numbers: 24 — Exponentiation of Cardinal Numbers

Theorem 24.1 Exponentiation on $\mathscr{C}$ is well-defined. That is, if S, S^*, T and T^* are sets such that $|S| = |S^*|$ and $|T| = |T^*|$, then $|S^T| = |S^{*T^*}|$.

Theorem 24.2 Let κ, λ, and ϕ be any three cardinal numbers. Then

(a) $\kappa^{\lambda+\phi} = \kappa^\lambda \times \kappa^\phi$
(b) $(\kappa^\lambda)^\phi = \kappa^{\lambda\times\phi}$
(c) $(\kappa \times \lambda)^\phi = \kappa^\phi \times \lambda^\phi$

Theorem 24.3 Let κ, λ, and α be infinite cardinal numbers. Then

(a) $\kappa \leq \kappa^\lambda$
(b) $\alpha \leq \kappa \Rightarrow \alpha^\lambda \leq \kappa^\lambda$
(c) $\alpha \leq \lambda \Rightarrow \kappa^\alpha \leq \kappa^\lambda$

VII Cardinal Numbers: 25 — Sets of Cardinality c

Theorem 25.1 Let $\mathbb{C}$ denote the set of all complex numbers and $\mathbb{J}$ denote the set of all irrational numbers. Let n denote the cardinality of a non-empty finite set.

(a) The cardinality of $\mathbb{R}^n$ is c
(b) The cardinality of $\mathbb{C}$ is c
(c) The cardinality of $\mathbb{J}$ is c

Theorem 25.2

(a) Let $\mathscr{S}_{\mathbb{R}}$ denote the set of all countably infinite sequences of real numbers. The cardinality of $\mathscr{S}_{\mathbb{R}}$ is c.
(b) Let $\mathscr{S}_{\mathbb{N}}$ denote the set of all countably infinite sequences of natural numbers. The cardinality of $\mathscr{S}_{\mathbb{N}}$ is c.
(c) Let $\mathbb{N}^{\mathbb{N}}_{(1-1)}$ denote the set of all one-to-one functions mapping $\mathbb{N}$ to $\mathbb{N}$. The cardinality of $\mathbb{N}^{\mathbb{N}}_{(1-1)}$ is c.
(d) Let $\mathbb{R}^{\mathbb{N}}_{(1-1)}$ denote the set of all one-to-one functions mapping $\mathbb{N}$ to $\mathbb{R}$. Then the cardinality of $\mathbb{R}^{\mathbb{N}}_{(1-1)}$ is c.

Proposition 25.3 The Cantor set has cardinality c.

VIII Ordinal Numbers: 26 — Well-Ordered Sets

Theorem 26.1 Let $f : T \to S$ be a one-to-one function mapping T onto S. If T is a well-ordered then T induces a well-ordering on S. In particular, every countable set can be well-ordered.

Definition 26.2 Given a *well-ordered* set $(S, \leq)$, a proper subset U satisfying the property

$$[u \in U \text{ and } x \leq u] \qquad \Rightarrow \qquad [x \in U]$$

is called an *initial segment* of S. In this definition the strict order relation $<$ can be used instead of $\leq$ without altering the meaning of "initial segment".

Theorem 26.3 If $(S, \leq)$ is a well-ordered set then every initial segment in S is of the form $S_a = \{x \in S : x < a\}$ for some $a \in S$.

Definition 26.4 Let $f : (S, \leq_S) \to (T, \leq_T)$ be a function mapping a well-ordered class, $(S, \leq_S)$, onto a well-ordered class, $(T, \leq_T)$. Note that the symbols $\leq_S$ and $\leq_T$ will allow us to distinguish between the order relations applied to each set S and T.

(a) We will say that the function f is *increasing* on $(S, \leq_S)$ if

$$(x \leq_S y) \qquad \Rightarrow \qquad (f(x) \leq_T f(y))$$

(b) We will say that the function f is *strictly increasing* on $(S, \leq_S)$ if

$$(x <_S y) \quad \Rightarrow \quad (f(x) <_T f(y))$$

A strictly increasing function must be one-to-one.

(c) If $f : (S, \leq_S) \to (T, \leq_T)$ is strictly increasing then f is said to be an *order isomorphism* mapping S into T.

If there exists an *onto* order isomorphism between the two well-ordered classes, $(S, \leq_S)$ and $(T, \leq_T)$, we will say that the classes are *order isomorphic*, or that a function maps S *order isomorphically* onto T.

If there exists an *onto* order isomorphism between the two well-ordered classes $(S, \leq_S)$ and $(T, \leq_T)$ we will say that the classes are *order isomorphic* or that a function maps S *order isomorphically* onto T.

Theorem 26.5 Let $(S, \leq_S)$ and $(T, \leq_T)$ be two well-ordered sets.

(a) The inverse of an order isomorphism is an order isomorphism.

(b) If $f : (S, \leq_S) \to (S, \leq_S)$ is a *strictly increasing* function mapping S into itself then $f(x) \geq x$ for all $x \in S$.

(c) The set S cannot be order isomorphic to an initial segment of itself.

(d) If $f : (S, \leq_S) \to (S, \leq_S)$ is an order isomorphism then f is the identity function.

(e) If $f : (S, \leq_S) \to (T, \leq_T)$ and $g : (S, \leq_S) \to (T, \leq_T)$ are order isomorphisms mapping S onto T then $f = g$.

(f) Suppose $f : (S, \leq_S) \to (T, \leq_T)$ is an order isomorphism mapping S onto an initial segment of T. Then S and T cannot be order isomorphic.

Notation 26.6 Let S and T be two well-ordered sets. Then the expression

$$S \sim_{\text{wo}} T$$

means "S and T are order isomorphic". The expression

$$S <_{\text{wo}} T$$

means "$S \sim_{\text{wo}} T_a$" where T_a is some *initial segment* of T. The expression

$$S \leq_{\text{wo}} T$$

means "$S \sim_{\text{wo}} T$ or $S <_{\text{wo}} T$".

Theorem 26.7 Let $(S, \leq_S)$ and $(T, \leq_T)$ be two well-ordered sets. Then either $S \leq_{\text{wo}} T$ or $T \leq_{\text{wo}} S$.

Proposition 26.8 For every natural number n, the lexicographically ordered set $S = \{1, 2, \ldots, n\} \times \mathbb{N}$ is well-ordered.

VIII Ordinal Numbers: 27 — Ordinal Numbers: Definition and Properties

Definition 27.1 Let S be a set. If S satisfies the two properties,

(1) S a transitive set,
(2) S is strictly $\in$-well-ordered

then S is called an *ordinal number.*

Notation 27.2 When viewed as an ordinal number, $\mathbb{N}$ will be represented by the lower-case Greek letter ω.

Theorem 27.3 If α is an ordinal number then so is its successor $\alpha^+ = \alpha \cup \{\alpha\}$.

Definition 27.4 Suppose the set S is $<$-ordered. We say that an element y in S is an *immediate successor* of the element x if $x < y$ and there does not exist any element z in S such that $x < z < y$. We say that x is an *immediate predecessor* of y if y is an immediate successor of x.

Theorem 27.5 Let α be an ordinal number greater than zero.

(a) Every element x of the ordinal α is an initial segment of α.
(b) The ordinal α is an initial segment of some ordinal.
(c) Every initial segment x in α is an ordinal number.
(d) Every element of the ordinal α is an ordinal number.

Proposition 27.6 Any infinite ordinal not equal to ω contains ω.

Proposition 27.7 Let α and β be distinct ordinal numbers. If $\alpha \subset \beta$, then $\alpha \in \beta$.

Lemma 27.8 If the ordinals α and β are order isomorphic, then $\alpha = \beta$.

Theorem 27.9 The relation "$\in$" linearly orders the class of all ordinals.

Definition 27.10 An ordinal α which does not contain a maximal element is called a *limit ordinal.*

Proposition 27.11 If U is a non-empty set of ordinals which contains a maximal element β with respect to $\in$, then the union, $\cup\{\alpha : \alpha \in U\}$, is equal to the maximal ordinal, β, of U.

Theorem 27.12 If U is a set of ordinals which does *not* contain a maximal element with respect to "$\in$", then $\gamma = \cup\{\alpha : \alpha \in U\}$ is a limit ordinal which is not contained in U. Furthermore, γ is the $\in$-least ordinal which contains all elements of U.

Corollary 27.13 Let U be a non-empty set of ordinals which contains no maximal element. If U satisfies the "initial segment property", then U is the limit ordinal $\cup\{\alpha : \alpha \in U\}$.

Definition 27.14 Let T be a non-empty subset of an ordered set $(S, <)$. If u is an upper bound of the set T and, for any other upper bound v of T, $u \leq v$, then we say that u is the *least upper bound* of T. We also abbreviate the expression by writing $u = \text{lub}\, T$ or $u = \text{lub}(T)$.

Theorem 27.15 Let γ be a non-zero ordinal number. The following are equivalent:

(1) The ordinal γ is a limit ordinal
(2) The ordinal γ is such that $\gamma = \cup\{\alpha : \alpha \in \gamma\}$
(3) The ordinal γ is such that $\text{lub}(\gamma) = \gamma$

VIII Ordinal Numbers: 28 — Properties of the Class of Ordinal Numbers

Theorem 28.1 The class, $\mathscr{O}$, of ordinal numbers is a strict $\in$-linearly ordered class.

Theorem 28.2 The class $\mathscr{O}$ of all ordinal numbers is $\in$-well-ordered.

Theorem 28.3 A set S is an initial segment of $\mathscr{O}$ if and only if S is an ordinal number.

Theorem 28.4 The class $\mathscr{O}$ of all ordinal numbers is not a set.

Theorem 28.5 Principle of transfinite induction. Let $\{x_\alpha : \alpha \in \mathscr{O}\}$ be a class whose elements are indexed by the ordinals. Let P denote a particular element property. Suppose $P(\alpha)$ means "the element x_α satisfies the property P". Suppose that, for any $\beta \in \mathscr{O}$,

$$\text{“}P(\alpha) \text{ is true } \forall\ \alpha \in \beta\text{” implies “}P(\beta) \text{ is true”}$$

Then $P(\alpha)$ holds true for all ordinals $\alpha \in \mathscr{O}$.

Corollary 28.6 Transfinite induction. Version 2. Let $\{x_\alpha : \alpha \in \mathscr{O}\}$ be a class whose elements are indexed by the ordinals. Let P denote a particular element property. Suppose $P(\alpha)$ means "the element x_α satisfies the property P". Suppose that:

(1) $P(0)$ holds true,
(2) $P(\alpha)$ holds true implies $P(\alpha + 1)$ holds true,
(3) If β is a limit ordinal, "$P(\alpha)$ is true for all $\alpha \in \beta$ implies $P(\beta)$ is true".

Then $P(\alpha)$ holds true for all ordinals α.

Theorem 28.7 Let S be a $<$-well-ordered set. Then S is order isomorphic to some ordinal number $\alpha \in \mathscr{O}$. Furthermore this order isomorphism is unique.

Definition 28.8 Let S be a $<$-well-ordered set. If α is the unique ordinal which is order isomorphic to S then we will say that S is of *order type* α, or of *ordinality* α. of S is α. If S is of ordinality α, we will write ${}^{\text{ord}}S = \alpha$.

Lemma 28.9 Hartogs' lemma. Let S be any set. Then there exists an ordinal α which is not equipotent with S or any of its subsets.

Theorem 28.10 There exists an uncountable ordinal.

Corollary 28.11 The class $\omega_1 = \{\alpha \in \mathscr{O} : \alpha \text{ is a countable ordinal}\}$ is the least uncountable ordinal.

Definition 28.12 Let S be any set. Let

$$U = \{\alpha \in \mathscr{O} : \ \alpha \text{ not equipotent to any subset of } S\}$$

By Hartogs' lemma the class U is non-empty. Since $\mathscr{O}$ is $\in$-well-ordered, U contains a unique least ordinal $h(S)$. We will call the ordinal $h(S)$ the *Hartogs number of* S. Then h can be viewed as a class function which associates each set S in the class of all sets to a unique ordinal number α in the class of all ordinals $\mathscr{O}$.

Theorem 28.13 There exists a strictly increasing class $\{\omega_\alpha : \alpha \in \mathscr{O}\}$ of pairwise non-equipotent infinite ordinals all of which are uncountable except for $\omega_0 = \omega$.

Theorem 28.14 Let $\{\omega_\alpha : \alpha \in \mathscr{O}\}$ be the class of ordinals as defined in the previous theorem.

(a) Every element of $\{\omega_\alpha : \alpha \in \mathscr{O}\}$ is a limit ordinal.
(b) For every ordinal α, either $\alpha \in \omega_\alpha$ or $\alpha = \omega_\alpha$.

Theorem 28.15 The Transfinite recursion theorem. Let W be a well-ordered class and $f : W \to W$ be a class function mapping W into W. Let $u \in W$. Then there exists a unique class function $g : \mathscr{O} \to W$ which satisfies the following properties:

(a) $g(0) = u$
(b) $g(\alpha^+) = f(g(\alpha)), \ \forall \alpha \in \mathscr{O}$
(c) $g(\beta) = \text{lub}\{g(\alpha) : \alpha \in \beta\}, \ \forall$ limit ordinals β

VIII Ordinal Numbers: 29 — Initial Ordinals: "Cardinal Numbers Are Us!"

Definition 29.1 We say that β is an *initial ordinal* if it is the least of all ordinals equipotent with itself. That is, β is an *initial ordinal* if $\alpha \in \beta \Rightarrow \alpha \not\sim_e \beta$.

Lemma 29.2 The class of all initial ordinals is a subclass of $\mathscr{I}$.

Theorem 29.3 The class $\mathscr{I}$ is precisely the class of all initial ordinals.

Theorem 29.4 [AC] The Well-ordering theorem. Every set can be well-ordered.

Theorem 29.5 The class of all initial ordinals $\mathscr{I} = \omega_0 \cup \{\omega_\alpha : \alpha \in \mathscr{O}\}$ satisfies the following properties:

(1) Every set S is equipotent to exactly one element in $\mathscr{I}$.
(2) Two sets S and T are equipotent if and only if they are equipotent to the same element of $\mathscr{I}$.
(3) The class $\mathscr{I}$ is $\in$-linearly ordered.

Definition 29.6 Cardinal numbers. An ordinal is called a *cardinal number* if and only if this ordinal is an initial ordinal. The class, $\mathscr{I}$, is also referred to as the class, $\mathscr{C}$, of all cardinal numbers.

Lemma 29.7 For any infinite cardinal number κ, define a relation $<_*$ on $\kappa \times \kappa$ as follows. For pairs (α, β) and (γ, ψ) of *ordinal* pairs in $\kappa \times \kappa$,

$$(\alpha, \beta) <_* (\gamma, \psi) \begin{cases} \alpha \cup \beta \in \gamma \cup \psi \\ \text{or} \\ \beta \in \psi \text{ when } \alpha \cup \beta = \gamma \cup \psi \\ \text{or} \\ \alpha \in \gamma \text{ when } \alpha \cup \beta = \gamma \cup \psi \text{ and } \beta = \psi \end{cases}$$

Then $<_*$ well-orders $\kappa \times \kappa$.

Theorem 29.8 [AC] For any ordinal α, $\aleph_\alpha \times \aleph_\alpha = \aleph_\alpha$.

Corollary 29.9 Let κ be an infinite cardinal and $\{A_\alpha : \alpha \in \beta\}$ be a set of non-empty sets indexed by the elements of the ordinal $\beta \in_= \kappa$ where $|A_\alpha| \in_= \kappa$ for all $\alpha \in \beta$. Then $|\cup \{A_\alpha : \alpha \in \beta\}| \in_= \kappa$.

Corollary 29.10 For any infinite cardinal number κ, $\kappa^\kappa = 2^\kappa$.

Definition 29.11

(a) We say that an infinite cardinal number, $\aleph_\gamma$, is a *successor cardinal* if the index, γ, has an immediate predecessor (i.e., $\gamma = \beta + 1$, for some β). The expression, $\aleph_{\alpha+} = \aleph_{\alpha+1}$, denotes a successor cardinal. We say that an infinite cardinal number, $\aleph_\gamma$, is a *limit cardinal* if γ is a limit ordinal (i.e., $\gamma = \text{lub}\{\alpha : \alpha \in \gamma\}$).
(b) We say that a limit cardinal $\aleph_\gamma$ is a *strong limit cardinal* if $\aleph_\gamma$ is uncountable and $\{2^{\aleph_\alpha} : \alpha \in \gamma\} \subseteq \aleph_\gamma$.

Theorem 29.12 [GCH] Every uncountable limit cardinal is a strong limit cardinal.

Theorem 29.13 [AC] There exists a strong limit cardinal number.

Definition 29.14

(a) We say that an infinite cardinal $\aleph_\gamma$ is a *singular cardinal number* if $\aleph_\gamma$ is the least upper bound of a strictly increasing sequence of ordinals, $\{\alpha_\kappa : \kappa \in \beta\}$, indexed by the elements of some ordinal β in $\aleph_\gamma$.

(b) An infinite cardinal $\aleph_\gamma$ is said to be a *regular cardinal number* if it is not a singular cardinal number. That is, there does not exist an ordinal, β, in $\aleph_\gamma$ such that $\aleph_\gamma = \text{lub}\{\alpha_\kappa : \kappa \in \beta\}$.

Theorem 29.15 Every infinite successor cardinal, $\aleph_{\alpha^+} = \aleph_{\alpha+1}$, is a regular cardinal.

Theorem 29.16 Let γ be an infinite cardinal number. If γ is a singular cardinal then the cardinal number $\aleph_\gamma$ is a singular cardinal.

Definition 29.17 A regular cardinal number which a limit cardinal is called an *inaccessible cardinal.* A regular cardinal number which is a strong limit cardinal is called a *strongly inaccessible cardinal.*

Definition 29.18 Let $\aleph_\gamma$ be an infinite cardinal. We say the *cofinality of* $\aleph_\gamma$ is β and write $\text{cf}(\aleph_\gamma) = \beta$ if β is the least ordinal in $\aleph_\gamma$ which indexes an increasing set of ordinals $\{\theta_\alpha : \alpha < \beta\}$ such that $\aleph_\gamma = \text{lub}\{\theta_\alpha : \alpha < \beta\}$. If no such β exists in $\aleph_\gamma$ then we say the cofinality $\text{cf}(\aleph_\gamma)$ of is $\aleph_\gamma$ and write $\text{cf}(\aleph_\gamma) = \aleph_\gamma$.

Theorem 29.19 The cofinality $\text{cf}(\aleph_\gamma)$ of an infinite cardinal number $\aleph_\gamma$ is a cardinal number. Hence, $\text{cf}(\aleph_\gamma)$ is the smallest cardinality of all sets which are cofinal in $\aleph_\gamma$.

Theorem 29.20 The cofinality $\text{cf}(\varphi)$ of an infinite cardinal number φ is a regular cardinal.

Theorem 29.21 If κ is an infinite cardinal and $\text{cf}(\kappa) \leq \lambda$, then $\kappa < \kappa^\lambda$.

IX More on Axioms: Choice, Regularity, and Martin's Axiom: 30 — Axiom of Choice

Theorem 30.1 Suppose $\mathscr{S}$ is a finite set of non-empty sets whose union is the set M. Then there exists a function $f : \mathscr{S} \to M$ which maps each set to one of its elements.

Theorem 30.2 Let AC* denote the statement:

For any set $\mathscr{S} = \{S_\alpha : \alpha \in \gamma\}$ of non-empty sets, $\Pi_{\alpha\in\gamma} S_\alpha$ is non-empty.

The Axiom of choice holds true if and only if AC* holds true.
The Axiom of choice holds true if and only if AC* holds true.

Theorem 30.3 The statement "Every set is well-orderable" holds true if and only if the Axiom of choice holds true.

Theorem 30.4 [AC] Any infinite set can be expressed as the union of a pairwise disjoint set of infinite countable sets.

Theorem 30.5 [AC] Let $(X, <)$ be a partially ordered set. If every chain of X has an upper bound then X has a maximal element.

Theorem 30.6 Suppose that those partially ordered sets $(X, <)$ in which every chain has an upper bound must have a maximal element. Then given any subset $\mathscr{S} \subseteq \mathscr{P}(S) - \varnothing$ there exists a choice function $f : \mathscr{S} \to S$ which maps each set in $\mathscr{S}$ to one of its elements.

Theorem 30.7 [ZL] Every vector space has a basis.

IX More on Axioms: Choice, Regularity, and Martin's Axiom: 31 — Axiom of Regularity and Cumulative Hierarchy

Theorem 31.1 The Axiom of regularity holds true if and only if every non-empty set S contains a *minimal* element with respect to the membership relation "$\in$".

Theorem 31.2 [Axiom of regularity] No set is an element of itself.

Definition 31.3 We say that a class is *well-founded* if it does not contain an infinite descending chain of sets. That is, there does not exist an infinite sequence $\{x_n : n \in \omega\}$ such that $\cdots \in x_4 \in x_3 \in x_2 \in x_1 \in x_0$.

Theorem 31.4 [AC] The *Axiom of regularity* and the statement "*Every set is well-founded*" are equivalent statements.

Definition 31.5 Let x be a set. The transitive closure of x is a set t_x satisfying the following three properties:

(1) The set t_x is a transitive set
(2) $x \subseteq t_x$
(3) t_x is the $\subseteq$-least transitive set satisfying properties 1 and 2

Theorem 31.6 Let x be a set. Then there exists a smallest transitive set t_x which contains all elements of x. That is, if s is a transitive set such that $x \subseteq s$, then $x \subseteq t_x \subseteq s$.

Definition 31.7 Define the class function $f : \mathscr{S} \to \mathscr{S}$ as $f(S) = \mathscr{P}(S)$. The elements of the class $\{V_\alpha : \alpha \in \mathscr{O}\}$ belong to the image of the class function $g(\alpha) = V_\alpha$ recursively defined as follows:

$$
\begin{aligned}
g(0) = V_0 &= \varnothing &&= 0 \\
g(1) = V_1 &= f(V_0) &&= \mathscr{P}(0) = \{\varnothing\} = 1 \\
g(2) = V_2 &= f(V_1) &&= \mathscr{P}(1) = \{\{\varnothing\}, \varnothing\} = 2^1 = 2 \\
g(3) = V_3 &= f(V_2) &&= \mathscr{P}(2) = \{\{\{\{\varnothing\}, \varnothing\}\}, \\
& && \quad \{\{\varnothing\}\}, \{\varnothing\}, \varnothing\} \\
g(4) = V_4 &= f(V_3) &&= \mathscr{P}(4) \text{ (16 elements)} \\
\vdots \quad & \quad \vdots \\
g(\alpha^+) = V_{\alpha^+} &= f(V_\alpha) &&= \mathscr{P}(V_\alpha) \\
\vdots \quad & \quad \vdots \\
\text{If } \lambda = \text{limit ordinal}, \quad g(\lambda) = V_\lambda &= \cup_{\alpha \in \lambda} V_\alpha \\
\vdots \quad & \quad \vdots
\end{aligned}
$$

The class $\{V_\alpha : \alpha \in \mathscr{O}\}$ is called the *Cumulative hierarchy of sets*. We define $V = \cup_{\alpha \in \mathscr{O}} V_\alpha$.

Lemma 31.8 (a) Each set V_α is a transitive set.
(b) If $\alpha \in \beta$ then $V_\alpha \in V_\beta$. Hence, $V_\alpha \subset V_\beta$.

Lemma 31.9 For any non-empty set B, $B \subset V$ implies $B \in V$.

Theorem 31.10 For every set x, $x \in V = \cup_{\alpha \in \mathscr{O}} V_\alpha$.

Theorem 31.11 Let $V = \cup_{\alpha \in \mathscr{O}} V_\alpha$ be the class of sets constructed as described above. If V contains all sets then every set has a $\in$-minimal element.

Theorem 31.12 Let $V = \cup_{\alpha \in \mathscr{O}} V_\alpha$ be the class of sets constructed as described above.

(a) The rank of the empty set $\varnothing$ is zero.
(b) If $U \in V_\beta$ then $\text{rank}(U) < \beta$; hence, $U \notin V_{\text{rank}(U)}$ for all sets U. Conversely, $\text{rank}(U) < \beta \Rightarrow U \in V_\beta$.
(c) If U and V are sets such that $U \in V$ then $\text{rank}(U) < \text{rank}(V)$.
(d) If γ is an ordinal then $\text{rank}(\gamma) = \gamma$.

Proposition 31.13 Let β be a limit ordinal. If a and b are two elements of V_β. Then $\{a, b\}$ is an element of V_β. That is, V_β satisfies the property described by the Axiom of pair.

Proposition 31.14 Let β be any ordinal. If U is an element of V_β then $\cup\{x : x \in U\} \in V_\beta$. That is, V_β satisfies the property described by the Axiom of union.

Proposition 31.15 Let β be a limit ordinal. If U is an element of V_β then V_β also contains a set $Y = \mathscr{P}(U)$ such that $S \subseteq U$ implies $S \in Y$. That is, V_β satisfies the property described by the Axiom of power set.

Proposition 31.16 Let α be any ordinal. For any two elements x and y of V_α, if for all $z \in V_\alpha (z \in x \Leftrightarrow z \in y)$, then x and y are the same set.

Proposition 31.17 Let α be an ordinal number such that $\alpha > \omega_0$. Then $\omega_0 \in V_\alpha$.

Proposition 31.18 Let α be an ordinal number. Then the Axiom of subsets holds true in V_α.

Proposition 31.19 The set V_{ω_0} satisfies the property described by the Axiom of replacement.

Proposition 31.20 Let γ be a limit ordinal. Then the set V_γ satisfies the property described by the Axiom of choice.

Proposition 31.21 Let α be an ordinal. Then the set V_α satisfies the property described by the Axiom of construction.

IX More on Axioms: Choice, Regularity, and Martin's Axiom: 32 — Martin's Axiom

Definition 32.1 Let $(P, \leq)$ be a partially ordered set. If P contains no uncountable strong antichain then $(P, \leq)$ is said to satisfy the *countable chain condition.* In this case, we say that $(P, \leq)$ satisfies the *ccc* or that $(P, \leq)$ is a *ccc* partial order.

Definition 32.2 Let $(P, \leq)$ be a partially ordered set. Let D be a subset of P such that for every element p in P there exists an element d in D such that $d \leq p$. A subset D satisfying this property is said to be *dense in the partial ordering* $(P, \leq)$.

Definition 32.3 Let F be a subset of a partially ordered set $(P, \leq)$. If F is non-empty and satisfies the two properties, (1) If x and y belong to F there exists z in F which is less than or equal to both x and y (i.e., F is a *filter base* or *downward directed*), (2) if x belongs to F and x is less than or equal to an element y of P, then y belongs to F (i.e., F is upward closed). A filter in $(P, \leq)$ is a *proper filter* if it is not all of P. If $x \in P$ then the set of all elements above x is called a *principal filter with principal element* x. Such a filter is the smallest filter which contains x.

Theorem 32.4 MA(κ): Let κ be an infinite cardinal and $(P, \leq)$ be a non-empty partially ordered set satisfying the *countable chain condition.* Let $\mathscr{D} = \{D \in \mathscr{P}(P) : D \text{ is dense in } P\}$ such that $|\mathscr{D}| \leq \kappa$. Then there is a proper filter $F \subseteq P$ such that, $F \cap D \neq \varnothing$ for every set $D \in \mathscr{D}$. The statement MA($\aleph_0$) holds true in *ZFC*.

Theorem 32.5 The statement MA($2^{\aleph_0}$) fails in *ZFC*.

Definition 32.6 *Martin's axiom*, MA, is defined as being MA(κ) where κ satisfies $\aleph_0 \leq \kappa < 2^{\aleph_0}$

Theorem 32.7 [MA] Suppose κ is an infinite cardinal such that $\kappa < 2^{\aleph_0}$. If X is a Hausdorff compact space with *ccc* and $\{U_\alpha : \alpha \leq \kappa\}$ is a family of open dense subsets of X then $\cap\{U_\alpha : \alpha\} \neq \varnothing$.

X Ordinal Numbers Arithmetic: 33 — Addition

Definition 33.1 Let $(S, \leq_S)$ and $(T, \leq_T)$ be two *disjoint* well-ordered sets. We define the relation "$\leq_{S\cup T}$" on $S \cup T$ as follows:

(a) $u \leq_{S\cup T} v$ if $\{u, v\} \subseteq S$ and $u \leq_S v$.
(b) $u \leq_{S\cup T} v$ if $\{u, v\} \subseteq T$ and $u \leq_T v$.
(c) $u \leq_{S\cup T} v$ if $u \in S$, $v \in T$.

Theorem 33.2 Let $(S, \leq_S)$ and $(T, \leq_T)$ be two *disjoint* well-ordered sets. Then the relation $\leq_{S\cup T}$ well-orders the set $S \cup T$.

Definition 33.3 Let α and β be two ordinal numbers. Let $(S, \leq_S)$ and $(T, \leq_T)$ be two *disjoint* well-ordered sets of order type α and β respectively.[7] We define $\alpha + \beta$ as follows:

$$\alpha + \beta = {}^{\text{ord}}(S \cup T,\ \leq_{S\cup T})$$

Theorem 33.4 Let $(S, \leq_S)$, $(T, \leq_T)$ and $(U, \leq_U)$, $(V, \leq_V)$ be two pairs of *disjoint* well-ordered sets such that

$$^{\text{ord}}S = \alpha = {}^{\text{ord}}U$$
$$^{\text{ord}}T = \beta = {}^{\text{ord}}V$$

Then ${}^{\text{ord}}(S \cup T, \leq_{S\cup T}) = \alpha + \beta = {}^{\text{ord}}(U \cup V, \leq_{U\cup V})$. Hence, addition of ordinal numbers is well-defined.

Theorem 33.5 Let α, β, and γ be three ordinal numbers. Then:

(a) $(\alpha + \beta) + \gamma = \alpha + (\beta + \gamma)$ (Addition is associative)
(b) For any ordinal $\gamma > 0$, $\alpha < \alpha + \gamma$
(c) For any ordinal γ, $\gamma \leq \alpha + \gamma$

[7]Addition can also be defined inductively as follows: For all α and β, (a) $\beta + 0 = \beta$, (b) $\beta + (\alpha + 1) = (\beta + \alpha) + 1$, (c) $\beta + \alpha = \text{lub}\{\beta + \gamma : \gamma < \alpha\}$ whenever α is a limit ordinal.

(d) $\alpha < \beta \Rightarrow \alpha + \gamma \leq \beta + \gamma$
(e) $\alpha < \beta \Rightarrow \gamma + \alpha < \gamma + \beta$
(f) $\alpha + \beta = \alpha + \gamma \Rightarrow \beta = \gamma$ (Left term cancellation is acceptable)
(g) $\alpha + 0 = \alpha$

Theorem 33.6 Let β be a limit ordinal. Then, for any ordinal, α,

$$\alpha + \beta = \sup\{\alpha + \gamma : \gamma < \beta\}$$

X Ordinal Numbers Arithmetic: 34 — Multiplication

Definition 34.1 Let $(S, \leq_S)$ and $(T, \leq_T)$ be two well-ordered sets. We define the *lexicographic ordering* on the Cartesian product $S \times T$ as follows:

$$(s_1, t_1) \leq_{S\times T} (s_2, t_2) \text{ provided } \begin{cases} s_1 <_S s_2 \\ \quad\text{or} \\ s_1 = s_2 \quad \text{and} \quad t_1 \leq_T t_2 \end{cases}$$

Theorem 34.2 Let $(S, \leq_S)$ and $(T, \leq_T)$ be two well-ordered sets. The lexicographic ordering of the Cartesian product $S \times T$ is a well-ordering.

Theorem 34.3 If the well-ordered sets S_1 and S_2 are order isomorphic and the well-ordered sets T_1 and T_2 are order isomorphic then the lexicographically ordered Cartesian products $S_1 \times T_1$ and $S_2 \times T_2$ are order isomorphic.

Definition 34.4 Let α and β be two ordinals with set representatives A and B respectively. We define the multiplication $\alpha \times \beta$ as:

$$\alpha \times \beta = {}^{\text{ord}}(B \times A)$$

The product $\alpha \times \beta$ is equivalently written as $\alpha\beta$, (respecting the order). Note the order of the terms in the Cartesian product $B \times A$ is different from the order $\alpha \times \beta$ of their respective ordinalities.

Theorem 34.5 Let α, β, and γ be three ordinal numbers. Then:

(a) $(\gamma\beta)\alpha = \gamma(\beta\alpha)$ (Multiplication is associative)
(b) For any $\gamma > 0$, $\alpha < \beta \Rightarrow \gamma\alpha < \gamma\beta$

(c) $\gamma(\alpha + \beta) = \gamma\alpha + \gamma\beta$ (Left-hand distribution is acceptable)
(d) For any $\gamma > 0$, $\gamma\alpha = \gamma\beta \Rightarrow \alpha = \beta$ (Left-hand cancellation is acceptable)
(e) $\gamma 0 = 0$
(f) For any limit ordinal $\beta \neq 0$, $\alpha\beta = \sup\{\alpha\gamma : \gamma < \beta\}$.

Definition 34.6 Let γ be any non-zero ordinal. We define the γ-based exponentiation function $g_\gamma : \mathscr{O} \to \mathscr{O}$ as follows:

(1) $g_\gamma(0) = 1$
(2) $g_\gamma(\alpha^+) = g_\gamma(\alpha)\gamma$
(3) $g_\gamma(\alpha) = \text{lub}\{g_\gamma(\beta) : \beta < \alpha\}$ whenever α is a limit ordinal.

Whenever $\gamma \neq 0$ we represent $g_\gamma(\alpha)$ as γ^α. Then $\gamma^{\alpha+1} = \gamma^\alpha\gamma$. If $\gamma = 0$ we define $\gamma^\alpha = 0^\alpha = 0$.

Theorem 34.7 Let α, β, and γ be three ordinal numbers. Then, assuming $\gamma > 1$,

$$\alpha < \beta \quad \Leftrightarrow \quad \gamma^\alpha < \gamma^\beta$$

Theorem 34.8 Let α, β, and γ be three ordinal numbers where $\alpha \neq 0$.

(a) $\gamma^\beta\gamma^\alpha = \gamma^{\beta+\alpha}$
(b) $(\gamma^\beta)^\alpha = \gamma^{\beta\alpha}$

A.2 Boolean Algebras and Martin's Axiom

Definition A.1 A partially ordered set $(P, \leq)$ is called a *lattice* if $a \vee b = \max\{a, b\}$ and $a \wedge b = \min\{a, b\}$ both exist in P for all pairs a, b in P.

Definition A.2 If B is a subset of a partially ordered set, $(P, \leq)$, $\vee B$ denotes the least upper bound of B and $\wedge B$ denotes the greatest lower bound of B (both with respect to $\leq$). Note that $\vee B$ and $\wedge B$ may or may not be an element of B. A lattice $(P, \leq)$ is said to be a *complete lattice* if for any non-empty subset B of P, both $\vee B$ and $\wedge B$ exist and belong to P.

Definition A.3 Let X be a topological space. A subset B is said to be *regular open in* X if $B = \text{int}_X(\text{cl}_X(B))$. The set of all regular open subsets of X will be denoted as $\mathscr{R}o(X)$.

Theorem A.4 Let X be a topological space. Then $(\mathscr{Ro}(X), \subseteq, \vee, \cap)$ is a complete lattice in $(\tau(X), \subseteq)$.

Definition A.5 Let $(L, \leq, \vee, \wedge)$ be a lattice. An *L-ultrafilter* is a proper filter $\mathscr{F}$ in L which is not properly contained in any other proper filter in L. If the filter $\mathscr{F}$ is such that $\cap\{F : F \in \mathscr{F}\} \neq \varnothing$ then we say that the filter $\mathscr{F}$ is a *fixed ultrafilter.* Ultrafilters which are not fixed are said to be *free ultrafilters.*

Theorem A.6 Let X be a topological space.

(a) Suppose $\mathscr{F}$ is a proper L-filter where $(L, \subseteq, \vee, \wedge)$ is a lattice in $(\mathscr{P}(X), \subseteq)$. Then $\mathscr{F}$ can be extended to an L-ultrafilter.

(b) Suppose $\mathscr{F}$ is an L-filter in $(L, \subseteq, \vee, \wedge)$ a lattice in $(\mathscr{P}(X), \subseteq)$. Then $\mathscr{F}$ is an L-ultrafilter if and only if for every $A \subseteq X$, either $A \in \mathscr{F}$ or $X - A \in \mathscr{F}$.

Theorem A.7 Let X be a topological space. Then $(\mathscr{Ro}(X), \subseteq, \vee, \cap)$ is a complete lattice in $(\tau(X), \subseteq)$. An $\mathscr{Ro}(X)$-filter, $\mathscr{F}$, is an $\mathscr{Ro}(X)$-ultrafilter if and only if, for any $A \in \mathscr{Ro}(X)$, either A or $X - \mathrm{cl}_X(A)$ belongs to $\mathscr{F}$.

Definition A.8 A lattice $(L, \vee, \wedge)$ is said to be a *distributive lattice* if, for any x, y, and z in L, $x \vee (y \wedge z) = (x \vee y) \wedge (x \vee z)$ and $x \wedge (y \vee z) = (x \wedge y) \vee (x \wedge z)$.

— The lattice $(L, \vee, \wedge)$ is said to be a *complemented lattice* it has a maximum element, denoted by 1, and a minimum element, denoted by 0, and for every $x \in L$ there exists a unique x' such that $x \vee x' = 1$ and $x \wedge x' = 0$.

— A complemented distributive lattice is referred to as being a *Boolean algebra.* A Boolean algebra is denoted as $(B, \leq, \vee, \wedge, 0, 1, ')$ when we explicitly want to express what the maximum and minimum elements are.

Definition A.9 Suppose we are given two lattices $(B_1, \leq_1, \vee_1, \wedge_1, 0, 1, ')$ and $(B_2, \leq_2, \vee_2, \wedge_2, 0, 1, ')$ and a function f which maps elements of B_1 to elements of B_2. We say that $f : B_1 \to B_2$ is a *Boolean homomorphism* if, for any $x, y \in B$,

(1) $f(x \vee_1 y) = f(x) \vee_2 f(y)$,
(2) $f(x \wedge_1 y) = f(x) \wedge_2 f(y)$,
(3) $f(x') = f(x)'$.

The function $f : B_1 \to B_2$ is a *Boolean isomorphism* if f is a bijection and both f and $f^{\leftarrow}$ are Boolean homomorphisms.

Definition A.10 Let $(B, \leq, \vee, \wedge, 0, 1, ')$ be a Boolean algebra. Let $\mathscr{S}(B) = \{\mathscr{U} : \mathscr{U}$ is a B-ultrafilter$\}$. We define the function $f_B : B \to \mathscr{P}(\mathscr{S}(B))$ as follows: $f_B(x) = \{\mathscr{F} \in \mathscr{S}(B) : x \in \mathscr{U}\}$.

Theorem A.11 Let $(B, \leq, \vee, \wedge, 0, 1, ')$ be a Boolean algebra. Then the set $\{f_B(x) : x \in B\}$ is a base for the open sets of some topology, $\tau(\mathscr{S}(B))$, on the set $\mathscr{S}(B)$ of all B-ultrafilters.

Theorem A.12 Let $(B, \leq, \vee, \wedge, 0, 1, ')$ be a Boolean algebra.

(1) The function $f_B : B \to \mathscr{P}(\mathscr{S}(B))$ is a Boolean homomorphism mapping B into $\mathscr{P}(\mathscr{S}(B))$.
(2) For every $x \in B$, $f_B(x)$ is clopen in $\mathscr{S}(B)$. Hence, $f_B[B] \subseteq \mathscr{B}(\mathscr{S}(B))$ (the set of all clopen sets in $\mathscr{S}(B)$).
(3) The function $f_B : B \to \mathscr{S}(B)$ is a Boolean isomorphism mapping B into $\mathscr{B}(\mathscr{S}(B))$ (the set of all clopen sets in $\mathscr{S}(B)$).
(4) The topological space $(\mathscr{S}(B), \tau(\mathscr{S}(B)))$ where $\tau(\mathscr{S}(B))$ is the topology generated by the open base $\{f_B(x) : x \in B\}$ is a compact zero-dimensional Hausdorff topological space.
(5) The Boolean isomorphism $f_B : B \to \mathscr{S}(B)$ maps B onto $\mathscr{B}(\mathscr{S}(B))$ (the set of all clopen sets in $\mathscr{S}(B)$).

Theorem A.13 Let κ be an infinite cardinal number such that $\kappa < 2^{\aleph_0}$. Then the following are equivalent:

(1) (Martin's axiom MA) If $(P, \leq)$ is a partially ordered set satisfying *ccc* and $\mathscr{D} = \{D_\alpha : \alpha \leq \kappa\}$ is a family of dense subsets of P, then there exists a filter $\mathscr{F}$ on P such that $\mathscr{F} \cap D_\alpha \neq \varnothing$ for each $\alpha \leq \kappa$.
(2) If X is a compact Hausdorff topological space satisfying *ccc* and $\mathscr{D} = \{D_\alpha : \alpha \leq \kappa\}$ is a family of dense open subsets of X, then $\cap\{D_\alpha : \alpha \leq \kappa\} \neq \varnothing$.
(3) If $(B, \leq, \vee, \wedge, ')$ is a Boolean algebra with the *ccc* property and $\mathscr{D} = \{D_\alpha : \alpha \leq \kappa\}$ is a family of dense subsets of B, then there exists a filter $\mathscr{F}$ on B such that $\mathscr{F} \cap D_\alpha \neq \varnothing$ for each $\alpha \leq \kappa$.

(4) If $(P, \leq)$ is a partially ordered set satisfying *ccc* and $|P| \leq \kappa$ and $\mathscr{D} = \{D_\alpha : \alpha \leq \kappa\}$ is a family of dense subsets of P, then there exist a filter $\mathscr{F}$ on P such that $\mathscr{F} \cap D_\alpha \neq \varnothing$ for each $\alpha \leq \kappa$.

Theorem A.14 Let κ be a cardinal such that $\aleph_0 \leq \kappa < 2^{\aleph_0}$. Let X be a Hausdorff topological space satisfying *ccc* such that $\{x \in X : x \text{ has a compact neighborhood}\}$ is dense in X. Suppose that $\mathscr{D} = \{D_\alpha : \alpha \leq \kappa\}$ is a family of dense open subsets of X. Then $\cap\{D_\alpha : \alpha \leq \kappa\}$ is dense in X if and only if Martin's axiom holds true.

Bibliography

André, R., Axioms and set theory, www.math.uwaterloo.ca/~randre/1aase t_theory_140613.pdf, 2015.

André, R., When does the family of singular compactifications form a complete lattice? *Rocky Mountain J. Math.*, 27(4) (Fall 1997), 979–1007.

André, R., Compactifying a convergence space with functions, *Int. J. Math. Math. Sci.*, 19(4) (1996), 657–666, https://core.ac.uk/download/pdf/208409171.pdf.

Bell, M.G. and Ginsburg, J., First countable Lindelöf extensions of uncountable discrete spaces, *Can. Math. Bull.*, 23 (1980), 397–399.

Caterino, A., Faulkner, G.D., and Vipera, M.C., Two applications of singular sets to the theory of compactifications, Rendiconti Dell'Institue di Mat. dell' Univ. di Thieste (1989).

Chandler, R.E., *Hausdorff Compactifications*, Marcel Dekker, New York, 1976.

Chandler, R.E., Cain G.L., and Faulkner, G.D., Singular sets and remainders, *Trans. Amer. Math Soc.*, 268 (1981), 161–171.

Chandler, R.E. and Faulkner, G.D., Singular compactifications: The order structure, *Proc. Am. Math. Soc.*, 100(2) (1987), 377–382.

Comfort, W.W. and Negropontis, S., *The Theory of Ultrafilters*, Springer-Verlag, New York, 1974.

Comfort, W.W., Retractions and other continuous maps, from βX onto $\beta X \setminus X$, *Trans. Amer. Math. Soc.*, 114 (1965), 843–847.

Devlin, K.J., *Fundamentals of Contemporary Set Theory*, Springer-Verlag, New York, 1974.

Diamond, B., Algebraic subrings and perfect compactifications, *Topol. Appl.*, 39 (1991), 217–228.

Dugundji, J., *Topology*, Allyn & Bacon, Boston, Massachusetts, 1967.

Enderton, H.B., *Elements of Set Theory*, Academic Press, Inc., California, 1977.

Engelking, R., *General Topology*, Polish Scientific Publishers, Warsaw, 1977.

Gillman, L. and Jerison, M., *Rings of Continuous Functions*, Van Nostrand, Princeton, New Jersey, 1960.

Ginsburg, J. and Saks, V., Some applications of ultrafilters in topology, *Pac. J. Math.*, 57 (1975), 403–417.

Glicksberg, I., Stone–Čech compactifications of products, *Trans. Am. Math. Soc.*, 90 (1959), 369–382.

Hall, D.W. and Spencer, G.L., *Elementary Topology*, John Wiley & Sons, Inc., New York, 1955.

Henricksen, M. and Isbell, J.R., Some properties of compactifications, *Duke Math. J.*, 25 (1958), 83–106.

Hrbacek, K. and Jech, T., *Introduction to Set Theory*, Marcel Dekker, Inc., New York and Basel, 1984.

Isbell, J.R., Zero-dimensional spaces, *Tohoku Math. J.*, 7 (1955), 1–8.

Kelley, J.L., *General Topology*, Springer-Verlag, New York, 1991.

Monk, J.D., *Introduction to Set Theory*, McGraw-Hill, New York, 1969.

Munkres, J.R., *Topology*, Prentice Hall Inc., 2000.

Ponomarev, V.I., On paracompact spaces and their continuous mappings, *Dokl. Aked. Nauk SSSR*, 143 (1962), 46–49 — *Soviet Math. Dokl.*, 3 (1962), 347–350.

Porter, J.R. and Woods, R.G., *Extensions and Absolutes of Hausdorff Spaces*, Springer-Verlag, New York, 1988.

Rayburn, M.C., On Hausdorff compactifications, *Pac. J. Math.*, 44 (1973), 707–714.

Rudin, M.E., The box product of countably many compact metric spaces, *General Topol Appl.*, 2 (1972), 293–298.

Roitman, J., *Introduction to Modern Set Theory*, John Wiley & Sons, New York, 1990.

Rudin, M.E., A new proof that metric spaces are paracompact, *Proc. Am. Math. Soc.*, 20 (1969), 603.

Simmons, G.F., *Introduction to Topology and Modern Analysis*, Robert E. Krieger Publishing Company, Florida, 1983.

Steen, L.A. and Seebach, J.A. Jr., *Counterexamples in Topology*, Springer-Verlag, New York, 1986.

Stephenson, R. M. Jr., Pseudocompact spaces, *Trans. Am. Math. Soc.*, 134 (1968), 437–448.

Thomas, J., A regular space not completely regular, *Am. Math. Month.*, 76 (1969), 181–182.

Thrivikraman, T., On compactifications of Tychonoff spaces, *Yokohama Math. J.*, 20 (1972), 99–106.

Walker, R.C., *The Stone–Čech Compactification*, Springer-Verlag, New York, 1974.

Willard, S., *General Topology*, Dover Publication Inc., New York, 2004.

Weese, M. and Winfried, J., *Discovering modern set theory. I. The basics, volume 8 of graduate studies in mathematics*, (1996), xvii–210. American Mathematical Society, Providence.

Index

G

H

T

U

W

Z

www.ingramcontent.com/pod-product-compliance
Lightning Source LLC
LaVergne TN
LVHW021055220926
848484LV00021B/919